ANNUAL REVIEW
OF ENERGY

EDITORIAL COMMITTEE (1976)

P. L. AUER
J. DARMSTADTER
J. P. HOLDREN
J. M. HOLLANDER
M. K. SIMMONS

Responsible for the organization of Volume 1
(Ad Hoc Editorial Committee, 1974)

J. P. HOLDREN
J. M. HOLLANDER
W. K. H. PANOFSKY
S. H. SCHURR
G. T. SEABORG
M. K. SIMMONS

Assistant Editor	T. SNYDER
Indexing Coordinator	M. A. GLASS
Subject Indexer	S. WILLY

ANNUAL REVIEW OF ENERGY

JACK M. HOLLANDER, *Editor*
Lawrence Berkeley Laboratory

MELVIN K. SIMMONS, *Associate Editor*
Lawrence Berkeley Laboratory

VOLUME 1, 1976

Volume 1 of the *Annual Review of Energy* was supported by grant OEP74-11470 A02 from the National Science Foundation.

ANNUAL REVIEWS INC.   4139 EL CAMINO WAY   PALO ALTO, CALIFORNIA 94306

ANNUAL REVIEWS INC.
Palo Alto, California, USA

COPYRIGHT © 1976 BY ANNUAL REVIEWS INC.
ALL RIGHTS RESERVED

International Standard Book Number: 0-8243-2301-7

Annual Reviews Inc. and the editors of its publications assume no responsibility for the statements expressed by the contributors to this Review.

REPRINTS

The conspicuous number aligned in the margin with the title of each article in this volume is a key for use in ordering reprints. Available reprints are priced at the uniform rate of $1 each postpaid; the minimum acceptable reprint order is 10 reprints and/or $10.00, prepaid. A quantity discount is available.

FILMSET BY TYPESETTING SERVICES LTD., GLASGOW, SCOTLAND
PRINTED AND BOUND IN THE UNITED STATES OF AMERICA

ANNUAL REVIEWS INC. is a nonprofit corporation established to promote the advancement of the sciences. Beginning in 1932 with the *Annual Review of Biochemistry*, the Company has pursued as its principal function the publication of high quality, reasonably priced Annual Review volumes. The volumes are organized by Editors and Editorial Committees who invite qualified authors to contribute critical articles reviewing significant developments within each major discipline.

Annual Reviews Inc. is administered by a Board of Directors whose members serve without compensation.

BOARD OF DIRECTORS
1976

Dr. J. Murray Luck
Founder Emeritus, Annual Reviews Inc.
Department of Chemistry
Stanford University

Dr. Esmond E. Snell
President, Annual Reviews Inc.
Department of Biochemistry
University of California, Berkeley

Dr. Joshua Lederberg
Vice President, Annual Reviews Inc.
Department of Genetics
Stanford University Medical School

Dr. William O. Baker
President
Bell Telephone Laboratories

Dr. James E. Howell
Graduate School of Business
Stanford University

Dr. Wolfgang K. H. Panofsky
Director
Stanford Linear Accelerator Center

Dr. John Pappenheimer
Department of Physiology
Harvard Medical School

Dr. Colin S. Pittendrigh
Department of Biological Sciences
Stanford University

Dr. Alvin M. Weinberg
Director, Institute for Energy Analysis
Oak Ridge Associated Universities

Dr. Harriet Zuckerman
Department of Sociology
Columbia University

Annual Reviews are published in the following sciences: Anthropology, Astronomy and Astrophysics, Biochemistry, Biophysics and Bioengineering, Earth and Planetary Sciences, Ecology and Systematics, Energy, Entomology, Fluid Mechanics, Genetics, Materials Science, Medicine, Microbiology, Nuclear Science, Pharmacology and Toxicology, Physical Chemistry, Physiology, Phytopathology, Plant Physiology, Psychology, and Sociology. In addition, two special volumes have been published by Annual Reviews Inc.: *History of Entomology* (1973) and *The Excitement and Fascination of Science* (1965).

PREFACE

Among the diverse manifestations of man's cultural development few are more profound than his use of energy. He discovered early that he could extract the energy stored in organic materials, and he used this energy to provide warmth, to forge his tools, to do some of the work of producing food for himself and his family. In time he learned how to make use of energy on a larger scale, first by means of water and wind power and then by the combustion of fossil fuels. Creating societies based on the high productivity that energy-intensive industry made possible, he wrought great social revolutions. He discovered that energy and mineral resources were cheap to recover and apparently limitless in abundance, so their use grew rapidly and a very high level of material affluence was reached. But in its use of energy society took little note of the environment that sustained it. Eventually man perceived that the growing accumulation of waste was degrading his milieu and threatening his well-being. Thus there came about an *environmental crisis*. Shortly thereafter he perceived also that the earth's supplies of fuels are not limitless, nor are they distributed equitably, nor are they readily replaceable with substitute fuels or novel technologies. As the supply of available fuels diminished, the economic value of energy increased and there arose energy-related tensions that degraded the political environment. Thus there came about an *energy crisis*.

The energy crisis of the early 1970s was not a crisis in the usual sense. It was rather the beginning of an era when man first fully realized the magnitude of the energy-resource-environment problem, when he realized that this problem, which took many years to develop, will also take many years to solve; yet it demands his attention without delay. It was the moment when he began to grapple seriously with the problem, to try to find short-term expedients to ameliorate it and long-term programs to solve it.

As a companion to its sibling volumes in many other topics of scientific inquiry, *Annual Review of Energy* is a natural product of the new Era of Energy. We dedicate this series to a continuing review and discussion of the significant issues related to energy: the technologies of energy generation and end-use; regional and global energy systems; environmental and societal impacts of energy systems; the economics and politics of energy; and scientific and research frontiers in energy. The philosophy of *Annual Review of Energy* will follow the tradition of Annual Reviews in its aim to present selective and scholarly reviews and analyses of the literature in important fields of research. Additionally, however, it will be novel in its commentary on the state of new, interdisciplinary areas for which a robust literature has yet to develop.

Annual Review of Energy will not attempt to cover the entire spectrum of energy topics each year, but will develop a cyclical inclusivity that will allow thorough review of selected subject areas each year. This first volume is the overture to the series and, as such, has been designed with a broad scope, broader than subsequent volumes will have as the series reaches equilibrium. Volume 1 constitutes an overview of the energy system of a particular geopolitical region, that of the United States. In place of extensive technical detail, the reviews in Volume 1 emphasize issues, technical and other, that are playing a central role in the development of the energy system of the United States—issues that affect and are affected by US energy policy. In the same manner, and with the same philosophy, Volume 2 will review salient aspects and issues of the global energy system.

Energy is one of the most pervasive commodities of the contemporary world. In the United States, about 10% of the gross national product is involved in the production and distribution of energy. The flow of energy in the United States takes place within a complex system of interacting components that involve both the public and private sectors. The cycle of this system consists essentially of extracting from natural sources the requisite fuels, transporting and processing them, converting some of the energy stored in them to a variety of useful forms, distributing this energy to the end-use sectors of the economy, and disposing of or re-using the waste products of the cycle. Because domestic sources of fuels are not adequate at present, some energy enters the system by importation, either in the form of raw fuels or as finished products.

A convenient way to visualize the major components of the energy system is the network diagram shown in Figure 2 of the review by Hoffman and Wood. This diagram identifies the principal pathways of energy flow from source to consumer. Another type of representation, designed to show the relative magnitudes of energy flows from resources to end-uses, is the diagram given by Cook, reproduced as Figure 6 of the review by Schipper. Many features of the energy system that are basic to the issues reviewed in this volume can be extracted from these drawings.

This volume introduces the *Annual Review of Energy* and provides a comprehensive review of the US energy system. It opens with a historical discussion of the growth of energy use by man, a dissertation on how we became ensnared in an "energy trap." This review ranges in its description over the full spectrum of energy and resource supplies and uses, past and future, and provides an introduction to the reviews of the specific topics that follow. It concludes by pointing the way to what the author defines as "solfus."

The chapters that follow not only describe the components of the US energy system, but also seek to bring out the spectrum of issues that are involved in determining its future. Of these, domestic energy supply is of paramount importance, and accordingly we have included twelve reviews of components of the delivery system, either in use or under development. Some, like coal and nuclear energy, can have increasing impact in the next decade or two; others, like solar and fusion energy, clearly belong to the longer-term future. Geothermal energy, oil shale, synthetic fuels, and waste materials have potential impact in an intermediate time period. The technologies of energy storage, including the use of hydrogen as a secondary fuel, are of great importance in determining an optimum mix of energy sources in the system; these are reviewed in this volume, as is also the topic of advanced energy conversion, which is of central importance to the efficiency of the supply system. It was our intention to include a review of the domestic oil and gas technologies, but unfortunately the review paper on this topic was incomplete at the time of our publication deadline. The subject is, however, covered in the introductory review of "energy in our future."

Many technical, economic, and institutional factors are involved in determining the demand for energy. To understand them, and to be able to forecast demand into the future, are among the principal challenges of energy analysis. The cost of energy is a prime factor influencing demand, as are the related aspects of governmental regulation of energy supply and price. Many of the new energy technologies are land-intensive, and the question of governmental policy with regard to utilization of federal lands and resources is of importance to their development. Also vital is an assessment of the capital market for the energy industry, because of the high capital-intensiveness of new energy facilities. All of these issues are reviewed in this volume.

The demand for energy is also a function of life-style and of the efficiency with which energy is used, factors that are of course not independent of energy cost. Because of rising energy costs and uncertainties of supply, increasing attention is being paid to the need for our energy system to become more efficient, to provide greater utility or well-being from each unit of energy consumed. Many avenues can lead to moderation of the growth of energy demand, and these are called generically "energy conservation." In this volume, three chapters are devoted to energy conservation: one emphasizes the technologies of conservation; the second, the economics of conservation; and the third, social and institutional factors related to conservation.

That the use of energy has been beneficial to the human condition is without question; that it also imposes hazards to that condition is becoming increasingly understood. Probably the ultimate limit on our use of energy will not be imposed by exhaustion of fuel supply, but rather by undesirable side effects of energy use that are not, and cannot be, under our complete control. Although societies will attempt to maximize the ratio of benefit to risk as they plan their energy futures, there are widely disparate views on how to assess risk and how to interpret the data ("facts") that provide input to that assessment. Because of the importance of the subject of the impacts of energy use, we have included five related reviews in this volume. The first discusses the general question of environmental effects of energy systems, and the second focuses on their effects on human health. A third review assesses the economic costs of environmental pollution. In a fourth review the theory of risk/benefit analysis is discussed. Because of the special societal interest, the growing literature, and the wide range of responsible opinion on nuclear safety and risk, a separate review of this topic is included in this volume, and there will be continuing coverage in future volumes.

The recent energy crisis, which brought to public attention the growing dependence of the United States on foreign energy supplies, evoked a characteristic American response: a national goal of energy "independence" or "self-sufficiency."

The imperative of this goal fuels the country's entire energy program. It is important to understand the events and decisions that shaped the US energy system and those that led to the energy crisis of 1973. Besides, it is interesting history. These are described in some detail in the review of "energy self-sufficiency." This chapter also deals with the emergence of national energy policies and programs. The decisions made by the federal government as to priorities for research and development determine to a considerable degree both the rate at which various energy technologies will be developed and also the end-use efficiencies of energy use; thus they will have a profound effect on the character of the US energy system in the future.

Although this volume focuses on the domestic energy system in the United States, that system does not stand alone but is an integral part of the larger and more complex global energy system. Therefore it is fitting that Volume 1 end with a bridge to Volume 2, in which that larger system will be assessed. The closing review provides this link by describing generically the many issues and options of the international energy system that will be covered in detail in Volume 2 of *Annual Review of Energy*.

THE EDITOR

CONTENTS

OVERVIEW
    Energy in our Future, *Harrison Brown*   1

ENERGY SUPPLY AND DISTRIBUTION: RESOURCES AND TECHNOLOGIES
    Coal: Energy Keystone, *Richard A. Schmidt and George R. Hill*   37
    Production of High-Btu Gas from Coal, *H. R. Linden, W. W. Bodle, B. S. Lee, and K. C. Vyas*   65
    Clean Liquids and Gaseous Fuels from Coal for Electric Power, *S. B. Alpert and R. M. Lundberg*   87
    The Nuclear Fuel Cycle, *E. Zebroski and M. Levenson*   101
    Solar Energy, *Frederick H. Morse and Melvin K. Simmons*   131
    Geothermal Energy, *Paul Kruger*   159
    Oil Shale: The Prospects and Problems of an Emerging Energy Industry, *Stephen Rattien and David Eaton*   183
    Nuclear Fusion, *R. F. Post*   213
    Waste Materials, *Clarence G. Golueke and P. H. McGauhey*   257
    Hydrogen Energy, *D. P. Gregory and J. B. Pangborn*   279
    Energy Storage, *Fritz R. Kalhammer and Thomas R. Schneider*   311
    Advanced Energy Conversion, *Edward V. Somers, Daniel Berg, and Arnold P. Fickett*   345

ENERGY AND THE ECONOMY
    The Energy Industry and the Capital Market, *William E. Pelley, Richard W. Constable, and Herbert W. Krupp*   369
    The Price of Energy, *Diana E. Sander*   391
    Energy System Modeling and Forecasting, *Kenneth C. Hoffman and David O. Wood*   423

ENERGY CONSERVATION
    Raising the Productivity of Energy Utilization, *Lee Schipper*   455
    Potential for Energy Conservation in Industry, *Charles A. Berg*   519
    Social and Institutional Factors in Energy Conservation, *Paul P. Craig, Joel Darmstadter, and Stephen Rattien*   535

IMPACTS OF ENERGY ON ENVIRONMENT, HEALTH, AND SAFETY
    Social and Environmental Costs of Energy Systems, *Robert J. Budnitz and John P. Holdren*   553
    Health Effects of Energy Production and Conversion, *C. L. Comar and L. A. Sagan*   581

Economic Costs of Energy-Related Environmental Pollution, *Lester B. Lave and Lester P. Silverman* ... 601
Philosophical Basis for Risk Analysis, *Chauncey Starr, Richard Rudman, and Chris Whipple* ... 629
Safety of Nuclear Power, *J. M. Hendrie* ... 663

ENERGY POLICY AND POLITICS
Energy Self-Sufficiency, *Peter L. Auer* ... 685
Energy Regulation: A Quagmire for Energy Policy, *William O. Doub* ... 715
Federal Land and Resource Utilization Policy, *John A. Carver, Jr.* ... 727

INTERNATIONAL ASPECTS OF ENERGY
International Energy Issues and Options, *Mason Willrich* ... 743

INDEXES
Author Index ... 773
Subject Index ... 780

Copyright 1976. All rights reserved

ENERGY IN OUR FUTURE

×11001

Harrison Brown
Division of Humanities and Social Sciences, California Institute of Technology, Pasadena, California 91125

THE GROWTH OF HUMAN ENERGY DEMAND

Proto-human creatures emerged upon the earth some two million years ago. For the greater part of their existence they lived much as did the other animals about them; they gathered edible plants and hunted other animals. Their primary need for energy was in the form of food to nourish their bodies and amounted to about 3000 Calories[1] daily. The controlled use of fire, which greatly extended the variety of foods that could be eaten and the range of human habitation, increased per capita energy consumption to about 8000 Calories per day, corresponding to the heat that would be released by burning a little over 1 kilogram of coal per day or somewhat over 400 kilograms of coal each year. In view of the fact that the earth in its natural state could hardly support more than about ten million food-gatherers (1), the maximum consumption of energy by humans in preagricultural times probably amounted to no more than the equivalent of about four million tons of coal annually.

The invention of agriculture and animal domestication made it possible for several hundred persons to be supported by an area of land that formerly could support only one. But although this development resulted in a tremendous upsurge in human population, it initially had little effect upon per capita energy demands. As beasts of burden partially replaced humans in tasks requiring the concentration of energy such as plowing, perhaps an additional 4000 Calories were added to the average individual's energy consumption. This brought the total energy requirement to about 12,000 Calories, or the equivalent of burning about 600 kilograms of coal per person each year. By the beginning of the Christian era, the population of human beings had grown to approximately 250 million persons (2), corresponding to a worldwide energy demand equivalent to burning about 150 million tons of coal annually.

The demands for energy continued to grow in substantial measure as a result of the emergence of new technologies, which made possible the manufacture of objects of metal or pottery. The water mill and the wind mill appeared and their use gradually spread. Eventually the appearance of the breast-strap or postilion harness in the Middle Ages greatly increased the efficiency of beasts of burden with

[1] The food Calorie or large calorie equals 1000 standard calories.

the result that animals quickly replaced humans as sources of power for the concentration of energy. By that time total energy consumption was probably equivalent to burning about 1300 kilograms of coal per person per year. As the population of human beings during the Middle Ages was about 400 million persons (2), total human energy demand amounted to the equivalent of burning about 500 million tons of coal annually.

Coal was probably utilized in Europe as early as the twelfth century AD. But with the linking of coal to iron by the Darby family in the eighteenth century and with the design of engines that could be moved by steam, energy demands increased extremely rapidly. By 1785 the steam engine had been harnessed to power looms and spinning machines. It was only a matter of time before the steam engine was applied to both land and sea transport and eventually to the farm. By the late nineteenth century coal was providing the greater part of the 75,000 Calories per day required by the average person who lived in industrial society (3). This corresponds to burning nearly four tons of coal per person each year.

The Industrial Revolution led to numerous technological competitions in which new technologies emerged and displaced old ones. The steam-powered iron ship, for example, appeared in the late eighteenth century and competed with the wind-powered wooden clipper ship. As the population of the former increased, that of the latter decreased (see Figure 1). By 1890 the tonnages of the two types of ships were equal. Not long afterward the sailing ship as a practical merchant vessel was virtually extinct.

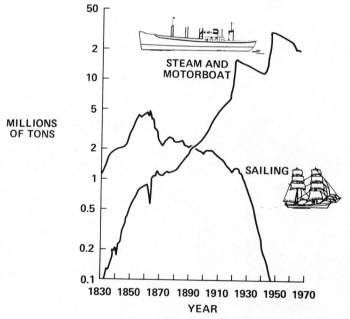

Figure 1 Tonnage of US merchant vessels.

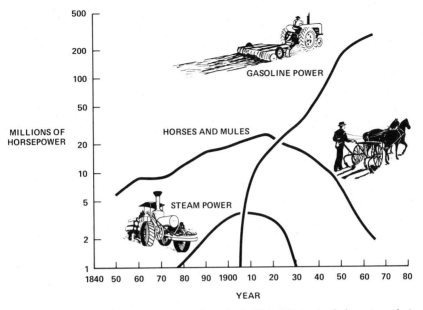

Figure 2 Horsepower of prime movers on farms in the United States (excludes automotive).

This evolutionary competitive process in which a new technological species drove an older species to extinction brought many benefits to humankind. Perhaps most important, travel times between continents were lessened considerably. Transport costs (including inventory costs) were likewise lessened. Much heavier loads could be accommodated. But all of this was at the expense of greatly increased consumption of energy per unit of material transported. Monetary costs went down; energy costs went up.

A second example of technological competition was the replacement of horses and mules by steam and gasoline power as prime movers on the farm. Prior to the Civil War in the United States, horses and mules were the primary means for concentrating energy. In the nation as a whole, which was primarily agrarian, there was approximately one horse or mule for every four persons; this situation prevailed for many decades. But steam power was introduced to the farm shortly after the Civil War and by the turn of the century was clearly providing major competition to horses. Had the internal combustion engine not been introduced to the farm in 1906, the steam engine undoubtedly would have driven the horse to extinction as a working farm animal. Instead, by 1930 the internal combustion engine had driven the steam engine to extinction on the farm and by 1970 the farm work animal was also virtually extinct[2] (see Figure 2).

[2] In 1972 farm work animals accounted for less than 0.5% of the available horsepower of prime movers on farms in the United States.

This major technological transition led to a considerable increase in farm production per man-hour worked and to decreased costs in monetary terms. But the energy costs, both to power the new equipment and to manufacture it, were greatly increased.

As technological innovations followed each other with breathtaking rapidity, and as society adjusted itself to the innovations, energy expenditures rose rapidly. Often innovations did not compete directly with existing technologies but made it possible to do things that had not been possible previously. Not infrequently, innovations were introduced as luxuries—the telephone and the automobile, for example. But as society evolved around these innovations, they were transformed from luxuries to necessities. Some, like the telephone, required small amounts of direct energy; others, like the automobile, required great amounts of energy—but together such innovations led to an exponential growth of energy consumption.

By 1970 the average person in the United States was consuming about 250,000 Calories per day, about 80 times as much as had been required by individual

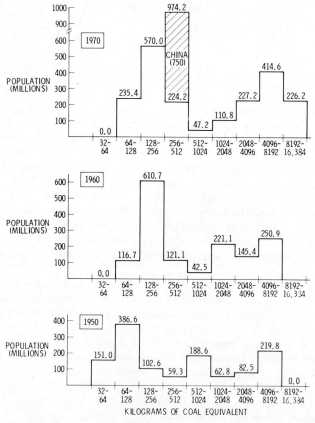

Figure 3 Population distribution of per capita consumption of energy.

primitive humans. This corresponds to the annual consumption of approximately 11 tons of coal per person—about six times as much as the average for the population of the world as a whole. World energy consumption had grown to the equivalent of about 7000 million tons of coal each year or approximately 1800 times the total human need of preagricultural times.

THE FISSIONING OF HUMAN SOCIETY

Industrialization, which was based on steam power and associated with increased wealth (as measured in monetary terms) and increased consumption of metals and energy, emerged in the United Kingdom, spread westward and eastward to North America and Europe, then to Japan and the USSR. These areas, together with Australia and New Zealand, became very rich in comparison with the rest of the world.

Prior to World War II the world was only slowly separating into rich countries and poor countries. Consumption levels among the world's people appear to have been a continuum with most persons being very poor, a few being very rich, and the rest, numbering more than the rich but fewer than the poor, being somewhere in between. Since World War II, however, a striking pattern has evolved amounting to no less than a fissioning of human society into two quite separate and distinct cultures—the culture of the rich and the culture of the poor, with very few people living between these two extremes.

The evolution of this process is dramatically illustrated by the changing pattern of per capita energy consumption between 1950 and 1970 (4). Figure 3 shows the numbers of people living at various levels of energy consumption in 1950, 1960, and 1970, expressed in kilograms of coal equivalent per capita. The levels of energy consumption are shown on a logarithmic scale, in which each step increases by a factor of two. The figure clearly shows a general spread of levels of per capita energy consumption in 1950, the emergence of a bimodal distribution by 1960, and the evolution of a very clear bimodal distriution by 1970. No reasonably authoritative data are available for the People's Republic of China prior to 1970.

A nearly identical pattern of evolution can be seen when we examine per capita steel consumption or per capita gross national product (in constant dollars). The two peaks provide a convenient division of nations into the "rich" and "poor" categories, as shown in Table 1.

Table 1 Classification of nations

	Per Capita Energy Consumption in 1970 (kg)	Per Capita Steel Consumption in 1970 (kg)	Per Capita GNP in 1973 (US Dollars)	Population in 1973 (Millions)
Rich nations	2048–16,384	256–1024	1280–5120	977 (25.4%)
Intermediate nations	1024– 2048	64– 256	640–1280	248 (6.4%)
Poor nations	64– 1024	2– 64	40– 640	2629 (68.2%)

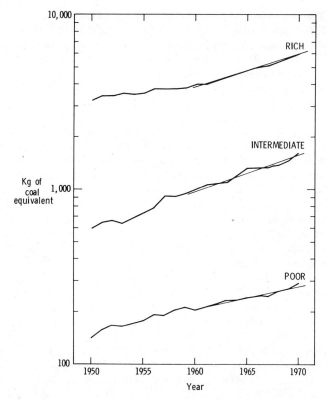

Figure 4 Per capita consumption of energy by economic grouping.

The change of the average per capita energy consumption with time (excluding China) is shown separately for the rich and the poor countries in Figure 4. In large measure as a result of rapid population growth in the poor countries, the curves showing the changes in per capita energy consumption have been diverging at an

Table 2 Average population and consumption levels for energy and steel in 1970

	Classification of Nations		
	Rich	Intermediate	Poor
Population (millions)	954	234	2440
Energy (10^6 metric tons)	5680	384	717
Steel (10^6 metric tons)	507	38	51
Per capita energy (kg/person)	6010	1610	293
Per capita steel (kg/person)	537	158	21
Per capita GNP (US dollars 1973)	2720	846	169

appreciable rate. In 1960 the average person in a rich country consumed 18.5 times as much energy as a person in a poor country. By 1970 this ratio had grown to 20.5. The averages of the quantities discussed above for the year 1970 are shown in Table 2.

Figure 5 shows the changes in per capita energy consumption on a regional basis. It is significant that the USSR (the primary contributor to the line labeled Eastern Europe) has already overtaken Western Europe, but neither is converging very rapidly with the rising per capita energy consumption in North America. Japan is approaching the per capita energy consumption of Europe. Largely because of rapid population growth, per capita energy consumption in Asia and Latin America is appreciably less than in the industrialized regions. A graphic comparison of the populations and levels of consumption of fossil fuels of the rich and poor countries is shown in Figure 6.

If the current rates of growth of energy and population were to continue for 50 years beyond 1970 (which is no further in the future than 1920 was in the past), per capita consumption of energy in coal equivalents would rise to 54 tons in the rich countries and to 1.4 tons in the poor. By that time the population of the world

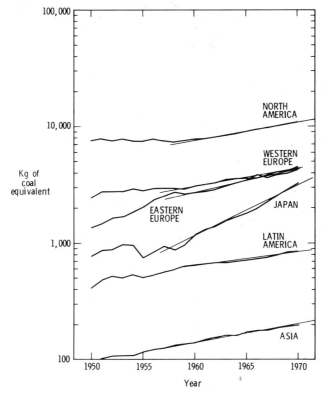

Figure 5 Regional per capita consumption of energy.

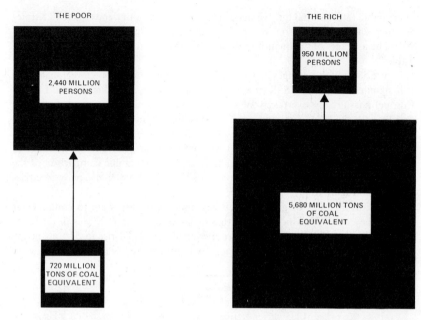

Figure 6 Energy consumption and population in the rich and poor nations.

would be 10.5 billion, about 1.4 billion of whom would be rich and 8.5 billion of whom would be poor. Consumption of energy in the poor countries alone would rise to a level considerably higher than total consumption in the world today, while total world consumption of energy would approach the equivalent of 100 billion tons of coal annually. The ultimate consequences of the actual occurrence of such developments are awesome to contemplate.

THE EVOLUTION OF THE "ENERGY TRAP"

In addition to the food necessary to maintain one's body, an individual in an industrial society requires energy for a variety of purposes. In the household, office, and store, hot water, cooking facilities, light, space heating, and air conditioning are necessary. Industries require power for their machines, heat for chemical reactions, coke for reducing iron ore, oil for petrochemicals. Transportation requires power. Agriculture requires power and energy in the form of fertilizers and other chemicals. Depending upon the geographic location of the user, the resources available, and the state of technology, one or more basic sources of energy could be used: wood, coal, petroleum, natural gas, hydropower, nuclear power, earth heat, wind, and direct solar radiation. The decision is determined in large measure by economics, but also by convenience and to some extent by security considerations.

For countless millennia, solar radiation, operating through the mechanism of photosynthesis, was humanity's primary energy source. This continued to be true for some time after the onset of the Industrial Revolution. Charcoal (made from wood) remained the preferred, but increasingly expensive, agent for reducing iron ore to the metal. Households were prodigious consumers of fuel wood for cooking and heating. In 1850 wood supplied 90% of the fuel used in the United States, with coal accounting for the balance (5). Nevertheless coal quickly became the primary industrial fuel, useful as it was for steel manufacture and the generation of steam power.

By 1885 the railroads had become the most important users of coal in the United States, closely followed by coke ovens. By 1910 fuel wood accounted for little more than 10% of energy consumption in the United States, with coal accounting for nearly 80%. Oil, natural gas, and hydropower accounted for the balance.

Oil first came into substantial use in the United States about 1860. At that time its cost (measured in 1970 dollars) was about $20 per barrel (6). Its initial use, in the form of kerosene, was in lamps to provide light and later in stoves to furnish heat. But a series of technological developments rapidly changed the pattern of use. For example, the development of the incandescent electric bulb decreased the demand for kerosene. Early in this century fuel oil came into widespread use to power steamships. This was followed in turn by the rapid upsurge in use of the internal combustion engine after World War I and by the use of oil for residential and commercial heating.

By 1948 the price of US oil had fallen from the original $20 per barrel to about $3.50 per barrel (all in 1970 dollars) (6). By that time coal was being purchased by the electric utilities at an average price of about $12 per ton (7). At these prices oil was about one sixth more expensive than coal if one were to judge the two fuels solely on the basis of their respective energy contents. In spite of this differential, however, petroleum use increased extremely rapidly. In part this was due to the emergence of the internal combustion engine and, associated with it, the immense popularity of the automobile. But in addition, oil actually displaced coal in a number of markets. The railroads turned to diesel power; households turned to heating oil; power companies turned to fuel oil.

The problem of competition between fuels is an extremely complex one involving many factors such as transportation costs, ease of handling, efficiency of use, and cleanliness, in addition to original cost. Oil is a liquid and can be transported by pipeline and tanker considerably less expensively than coal can be transported by train or slurry pipeline. Oil lends itself to automatic handling more readily than does coal. Usually petroleum can be burned more cleanly than can coal with respect to emissions of particulate matter and oxides of sulfur.

The availability of natural gas in large quantities also contributed substantially to the changing energy picture. Gas is associated with petroleum and should be a premium fuel in the sense that it provides heat with high efficiency and with unequalled convenience and cleanliness. For many years, due to a lack of transmission facilities, the price of natural gas could not reflect this premium value. But as the vast network of gas pipelines evolved in the United States, the price came closer to reflecting the real value.

In 1948 the price of natural gas at the well was about one eighth the cost of coal per unit of energy content and practically free when compared with petroleum (8). But the cost of transporting and distributing gas is substantial, with the result that for many years the main cost of gas to the customer was the transportation cost. Between 1950 and 1970 natural gas accounted for more than half of the growth in total energy supply in the United States (9), and the price of the fuel at the well increased rapidly. As demand for gas has increased, frequent difficulties have been encountered stemming from a progressively more acute shortage of fuel at the well.

In the latter part of the nineteenth century the total energy demand in the United States increased at a modest rate of 2–3% per year. The rate of growth increased after 1900, then decreased following World War I and became negative during the Great Depression. Since World War II growth has been considerably more rapid and in the decade 1960–1970 averaged nearly 4.8% annually. As total needs grew, competition between fuels led to the emergence of oil as the primary fuel in the United States. In 1950 oil and coal were of about equal importance. During the following quarter century oil made great inroads so that by 1973 coal accounted for only 18%. Oil provided 46% of the nation's energy, natural gas was the second most important fuel, providing 31%, and hydropower 4% (10).

For several decades the United States exported considerable quantities of petroleum products. However the gap between imports and exports gradually narrowed, and after World War II the United States entered a dramatic new era in which it imported more petroleum than it exported. By 1960 the United States was importing 18% of its petroleum requirements. By 1970 this had risen to 23% and by 1973, to 35%. Since 1970 virtually all of the growth in US energy consumption has come from petroleum imports (11).

A major reason for this development was price. Vast reservoirs of oil that could be extracted at low cost were found in the Middle East and Venezuela. Indeed, by the early 1970s the incremental cost of Middle Eastern oil at dockside in the Persian Gulf was 10¢ per barrel, including a 20% return on all investments. Needless to say, the oil companies that were operating in the Middle East did not sell the crude oil for 10¢ per barrel. For a lengthy period the Persian Gulf price was $1.25 per barrel, giving a profit or "rent" of $1.15 after deducting the 10¢ cost (12). Even so, this was a bargain, for transportation charges were kept low in part by improved tanker technology. Under these circumstances imports rose rapidly—a development of considerable concern to domestic producers.

In the early years of the development of Middle Eastern oil, the companies holding concessions in the region kept most of the "rent" or extra profit. But gradually over the years the governments of the Middle Eastern countries increased their shares. Following the creation of the Organization of Petroleum Exporting Countries (OPEC) in 1960, these countries managed to keep most of the profit and put themselves in a position of being able to dictate what the price would be.

Although the growing dependence of the United States on imported oil was a matter of concern to many students of the situation as early as 1950, the sequence of events in Western Europe and in Japan was even more serious. Although both areas have sources of coal, neither is endowed with appreciable petroleum deposits. The availability of inexpensive Middle Eastern oil allowed Western Europe and

Japan to become dependent on imported petroleum for their survival. As recently as 1956 both Japan and Western Europe obtained only one fifth of their energy from imported petroleum (13). The remainder of Japan's needs were satisfied by coal, largely produced domestically. In Europe too, domestic coal provided the balance of the needed energy. However, by 1970 petroleum filled about 70% of Japan's energy requirements and coal provided the remainder. All of the petroleum was imported, as was one half of the coal. In the same year imported petroleum filled about one half of Western Europe's energy needs. By 1973 Western Europe was importing about 15 million barrels of petroleum each day, while Japan and the United States were each importing about 6 million barrels daily (14).

Thus the stage was set for collective action by OPEC, which by 1973 consisted of Saudi Arabia, Iran, Venezuela, Nigeria, Libya, Kuwait, Iraq, United Arab Emirates, Algeria, Indonesia, Qatar, Ecuador, and Gabon (an associate member). Prior to the Arab-Israeli war of October 1973 the price of OPEC oil was about $2.00 per barrel. By the end of 1973 the price had been raised to $3.44 and to more than $10.00 by the end of 1974. In addition, the Arab oil-producing nations cut off oil supplies to the United States and other countries during the Arab-Israeli war; the embargo lasted until March 1974.

The collective OPEC action coupled with the embargo caused a number of serious dislocations throughout the greater part of the world. The industrialized nations had to pay greatly increased sums for fuel to keep their societies functioning, and this in turn led to substantial balance of payments problems. Vastly increased sums of money flowed from rich countries to OPEC countries and in particular to Arab countries (who have the most oil). In 1974 Saudi Arabia alone earned nearly $30 billion, corresponding to about $3500 for every inhabitant. About $80 billion flowed to the Arab states plus Iran, which collectively have a population of over 70 million persons. Although more fortunate than some, all of these countries suffer from the symptoms typical of other underdeveloped countries such as high rates of illiteracy and high birthrates and death rates.

Such large sums of money flowing to so few developing countries immediately raised numerous questions. What fraction of their new-found wealth could they spend effectively for development? How could they invest those funds which they could not spend quickly? How large would their military expenditures become? Could the vast sums of surplus money be recycled in such a way as to ease the balance of payments problems of the rich petroleum-importing countries? Above all, could the rich petroleum-importing countries find a way out of the trap in which they had permitted themselves to become ensnared?

For the moment the trap that has snared the importing countries is extremely strong. It isn't that other petroleum deposits won't be found—they probably will be. It isn't that other forms of energy can't be effectively put to use—they certainly can be. It isn't that per capita energy consumption can't be decreased without lowering living comfort—this certainly can be done. The difficulty is time, which is a commodity that cannot be purchased. Time is needed to intensify exploration. Time is needed to undertake the necessary research and development. Time is needed to build new plants, to relocate industries, to change ways of life.

The rich importing countries have little choice but to push in these new directions as rapidly as they can. But for the next decade at least, they are destined to remain in the trap that they unthinkingly helped build themselves. From that trap they can look forward to further price increases and intensified political pressures—and backward to actions they wish they had taken 10, 20, or 30 years ago.

THE "FOOD TRAP" AND ITS RELATIONSHIP TO THE "ENERGY TRAP"

Just as the major industrialized nations of the world were for the most part self-sufficient with respect to energy prior to World War II, they were also able to grow sufficient basic foodstuffs to satisfy most of their needs at the levels of individual consumption existing at the time. But since then the situation has changed dramatically, in large measure because of an explosion in agricultural productivity in North America.

A part of the explosion stemmed from the fact that one liter of gasoline burned in a one-horsepower engine yields the energy equivalent of seven man-days of hard physical labor. In the United States that is currently the equivalent of buying labor at a cost of 2.5¢ per man-day.

Another part of the explosion stemmed from the intensive application of science and technology to the improvement of crop yields per hectare. Still another part involved the considerable improvement in efficiency of transformation of cereals such as corn into animal products such as eggs, milk, and meat.

High and increasing levels of agricultural production per hectare have resulted from the genetic selection of plant varieties best suited for particular environments, from irrigation, and from liberal applications of fertilizers and pesticides. As one result of these and other developments, the yield of corn in the United States was more than doubled between the early 1950s and the 1970s (15).

But these tremendous increases in crop yields have depended in part upon very large inputs of energy from fossil fuels. A recent analysis of the energy inputs to corn production indicates that while in 1945 we utilized about 0.25 Calories of fossil fuel energy for very edible Calorie of corn grown, by 1970 this ratio had grown to 0.35 fossil fuel Calories per corn Calorie. In 1970 gasoline together with the energy required to make nitrogenous fertilizers accounted for 60% of the fossil fuel energy input to corn production (16).

In relation to other US energy expenditures, the energy inputs to the agricultural sector are not large. Recent studies (17–20) indicate that somewhere between 3% and 4% of the total US energy budget is used directly and indirectly on farms. Of this, over 50% is consumed directly in the form of fuel for tractors and other machinery and electricity. The balance is consumed indirectly: about 20% for producing the fertilizers, another 20% for manufacturing the farm machinery, and about 10% for miscellaneous inputs such as irrigation.

The greater part of the energy input to agriculture is for the cultivation of crops. When we consider that a large proportion of the crops are fed to animals for the production of meat, milk, and eggs, we appreciate that animal products are energetically quite expensive. Indeed, it would appear that about three fourths of the energy input to food grain for domestic consumption in the United States goes into the production of animal products (20).

Thus, the tremendous inputs of science and technology (including large inputs of energy from fossil fuels), together with the unusually large ratio of fertile arable land to people in North America, led to an unprecedented capacity for the production of surplus food. Americans were able to feed themselves, on the average, a very rich bill of fare and still export agricultural products to others at prices that most rich countries were willing and able to pay. Indeed, US agricultural production became so large, relative to demand, that requests for cash purchases overseas could easily be met and substantial shipments of foodstuffs could be made to poor countries under Public Law 480, which permits payment for the food to be made in local currency. In addition, in order to avoid excessive production surpluses, the US government for many years paid its farmers considerable sums of money to keep substantial areas of good farmland idle.

Concurrent with these developments, the people of Europe and Japan were becoming wealthier. Generally, the rich eat more animal products than do the poor; unfortunately animals are not very efficient converters of plant calories into meat calories. About 16 Calories of feed are required to produce 1 Calorie of beef. Chicken and hogs are more efficient than cattle; they require, respectively, 5 and 8 Calories of feed per Calorie of meat produced (21). Thus as people have consumed more meat, their apparent per capita consumption of cereals has increased. The per capita consumption of cereals in the United States now exceeds 800 kilograms per person, most of which is fed to animals. This is about 4.5 times larger than per capita consumption of cereals in India, where most of the grains are eaten directly by people.

Figure 7 shows the changes in per capita consumption of cereals in various parts of the world. There is some indication that the level of consumption has stopped increasing in the United States. But in other parts of the rich regions of the world, notably Europe, the USSR, and Japan, per capita demands for cereals are increasing rapidly and may eventually approach demand in the United States.

Gradually the demands for cereals that were created by growing affluence in Western Europe and Japan outstripped production, and these areas became substantial importers, primarily from North America. About 85% of the North American production is concentrated in the United States. The exports of Australia, the only other important net exporter, are only about 6% of North America's. The external demands placed on North American cereal production have fluctuated from year to year and have depended in part on the harvests in the importing countries and in part on the financial terms that exporters and importers are willing to agree upon. Most major regions of the world are now net importers of cereals, in part as a result of growing affluence in the rich countries and growing population in the poor ones.

World grain reserves are usually measured in terms of the quantities actually in storage at the beginning of a new harvest. During the past 15 years these reserves have varied from a high of 155 million tons to a low of about 90 million tons in 1974 or about 7% of annual world production. In recent years a convergence of trends has resulted in pressures to draw down reserves and to place idled cropland back into production.

In addition to the twin pressures of growing affluence and population, plant diseases and adverse weather conditions have taken their toll in crop yields in various parts of the world. As a result of these pressures the United States started to bring its idle acres back into production in the late 1960s. By 1974 no government payments were made for keeping cropland idle. As reserves have decreased and demands have increased, food prices have soared. The food price increases were accelerated in 1973–1974 by the rapid increases in the price of crude oil.

Increased prices have seriously limited the ability of many poor countries to buy needed food. At the same time, food aid to the poor countries has been drastically curtailed. Between 1972 and 1974 wheat shipments under Public Law 480 were decreased by a factor of 4.6 and corn shipments were curtailed by a factor of 2.7.

In certain of the poor countries such as India, which are dependent on petroleum imports, the sudden increase of petroleum prices effectively disrupted the so-called

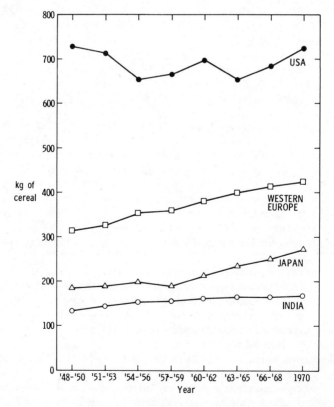

Figure 7 Per capita consumption of cereals.

Green Revolution, which basically involves the use of plant varieties that can make use of large applications of chemical fertilizers and water. India must import petroleum to manufacture fertilizers and to run its water pumps. In view of the fact that its nitrogen-fixation capacity is inadequate, India must also import fertilizers, primarily from Japan. Because food production is inadequate, India must also import cereals. All of this has placed enormous strains on India's balance of payments. And even if India were able to afford to purchase all of the fertilizer needed, the country would have difficulty obtaining it at any price. In May 1974, for example, Japan announced its decision to decrease exports of fertilizer to India and China by 15%.

Here it is important to emphasize again that the Green Revolution is highly dependent on energy inputs, primarily in the form of fertilizers, other agricultural chemicals, and irrigation. If all of the land now being cultivated in India were devoted to the new high-yielding varieties of wheat and rice, and labor-intensive methods of cultivation were retained at the same time, the total national energy budget would be increased by about 25% for fertilizers alone. Considering the large amount of irrigation that would be necessary, a doubling of the national energy budget would not be an unreasonable requirement. But where would this energy be obtained, if not from imported petroleum? And how would India be expected to pay for that energy?

Thus, food is like petroleum in that many nations, both rich and poor, have permitted themselves to become dependent on imports. As with petroleum, exports are now controlled by a relatively few countries. In both cases the importing countries are extremely vulnerable to the decisions of the exporters. And of all nations, the poor ones are the most vulnerable to major perturbations in the system.

THE LIFE EXPECTANCY OF PETROLEUM AND NATURAL GAS IN THE UNITED STATES

Since most major energy-consuming nations now find themselves dependent on petroleum, it is important that the life expectancy of that resource be carefully evaluated. Obviously the time that remains before petroleum is exhausted will depend on how much there is under the ground (or sea) and how rapidly we extract and consume it.

With respect to the ultimate production of crude oil in the world, the petroleum exploration and extraction experience in the United States is vast and directly applicable to the world problem. Although crude oil production in the United States accounts today for little more than 15% of world production, as recently as 1950 US production was more than one half of the world total. About one third of all of the oil ever extracted from the earth has been produced in the conterminous United States. It seems likely that about one half of the oil that can potentially be recovered from conterminous US deposits (excluding shale oil) has already been extracted. By contrast, only about one eighth of the petroleum likely to be extracted eventually from deposits in the rest of the world has been recovered thus far.

It is generally believed that petroleum and natural gas are derived from organic debris that was buried in sediments under oceans and seas during the geologic past. Those sediments were subsequently transformed into sedimentary rocks, and petroleum and natural gas are now found only in or adjacent to basins filled with such rocks. Sedimentary rocks are porous and the pores are generally filled with water, except when the water has been displaced by petroleum or natural gas. Under the influence of heat, pressure, and gravity, the oil and gas tend to accumulate in limited spatial regions that to some extent can be located by geological and geophysical procedures.

If one knows statistically how much petroleum is extractable from a sedimentary basin of a certain type, then by analogy one can estimate the oil content per unit volume of similar sedimentary basins elsewhere. The geographical location and extent of the sedimentary basins in the land areas of the world are now reasonably well known. Given that information, combined with the available extraction experience, one can make a rough estimate of the petroleum deposits in the world that are yet to be discovered.

In the United States, outside of Alaska, reservoirs of petroleum have been found associated with a number of sedimentary basins, but two areas stand out as being of greatest importance. The first embraces large parts of Texas and Louisiana and extends into the Gulf of Mexico. The second, in California, embraces the Los Angeles and San Joaquin Basins and extends into the continental shelf. These areas account for well over 80% of the oil thus far extracted from the 48 contiguous states and for nearly 90% of US proved reserves outside of Alaska. In 1969–1970, discovery of a third major petroleum reservoir in Alaska increased US proved crude oil reserves by about 30%.

Table 3 shows the situation with respect to US petroleum production and reserves as of 1974 (22, 23). With respect to the history that led to the present situation, it is important that we recognize several critical developments.

First, within the United States proper the annual rate of petroleum production increased exponentially from 1875 to the Great Depression at a rate of 8.4% per year. After the Depression and World War II the growth of production continued to be exponential but at a reduced rate of increase. As can be seen from Figure 8, the rate of production passed through a peak in 1970 and has been decreasing ever

Table 3 US production and reserves of crude oil (billions of barrels)

Location	Cumulative Production[a]	Proved Reserves[a]	1974 Production
Texas and vicinity	72.53	18.70	2.36
California	16.78	3.49	0.32
Alaska	0.63	10.11	0.07
Other states	19.43	3.00	0.46
Total	109.37	35.30	3.21

[a] As of January 1974.

since. The difference between demand and production has been met by imports, which in 1974 filled about 30% of the US petroleum need.

Second, within the United States, excluding Alaska, the proved reserves of petroleum, like the rate of production, increased exponentially for many years. As can be seen from Figure 9, the rate of increase lessened, becoming zero during the period of 1959–1967. Since 1967 the proved reserves have decreased steadily.

Third, within the United States, excluding Alaska, the ratio of proved reserves to production was about 12 years for some time after World War II. Since 1963 the ratio has declined steadily, reaching about 8 years in 1974. This useful index represents the number of years that petroleum would last if production were to remain the same and if no new discoveries were to be made. Generally it is both impractical and undesirable to extract oil from a field at an annual rate exceeding 10% of the remaining reserves. Extraction at a greater rate usually prejudices ultimate recovery.

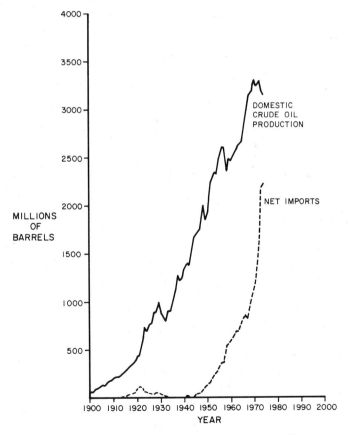

Figure 8 US production and net imports of crude oil.

As Hubbert pointed out many years ago (24), when an exhaustible resource is extracted from the ground in the face of increasing demand, the shape of the production curve should look something like the recorded production curve of crude oil in the United States shown in Figure 8. In the full cycle the production begins at zero and eventually returns to zero when the resource is exhausted. The technology of production requires that the early phase of production be an exponential rate of increase and the declining phase an exponential rate of decrease. At some point in between, the rate of increase of production starts to lessen considerably, eventually reaching zero as the production curve passes through a maximum.

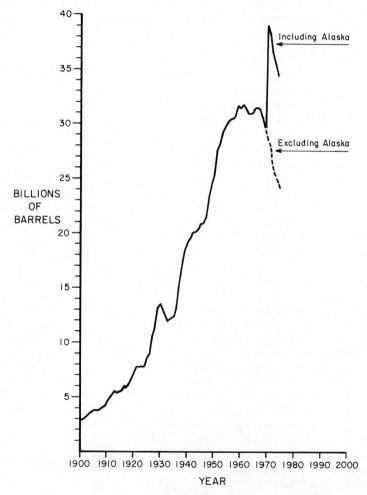

Figure 9 US proved reserves of crude oil (estimates as of 12/31/yr).

In a later discussion of the mathematical relationships between production, discovery, and reserves, Hubbert (25) suggests that the peak rates in both petroleum discovery and production occur at about the halfway points of cumulative discoveries and production. His analysis suggests that the peak of discovery rate in the United States, excluding Alaska, was reached about 1957 when cumulative discoveries amounted to 85.3 billion barrels. The amount of crude oil that might ultimately be extracted from the ground, assuming existing extraction efficiencies, would then be twice this or about 170 billion barrels. Similarly, when the production rate of crude oil in the United States, exclusive of Alaska, peaked in 1970, cumulative production was 98.8 billion barrels. Twice this is 198 billion barrels. Hubbert emphasizes that such figures are only approximate, stressing that it is quite possible for the peak rates of discovery or of production to occur somewhat earlier or later than the halfway point. He suggests that a more accurate approximation can be obtained by analyzing the curve of oil discovered per unit length of exploratory drilling. The data show a long-term decline in the rate of discovery with cumulative drilling and suggest 165 billion barrels as an upper limit for the amount of crude oil ultimately to be extracted.

In any event, it makes little difference whether such estimates are high or low. The trends make it seem reasonably clear that we in the United States are approaching the end of our conventional oil resources. If we assume that the upper limit of crude oil to be extracted ultimately from the conterminous United States, including the continental shelves, will be 200 billion barrels, then, since about 134 billion barrels have already been discovered as of January 1974, fewer than 66 billion barrels are yet to be found. Ten percent of the total available oil had been extracted by 1937, about 33 years before the production peak. If we assume that the production curve will be symmetrical, 90% of the extractable petroleum originally in place will have been extracted by the year 2003.

Thus, with very little question, if present per capita levels of energy consumption in the United States are to be maintained, increasing proportions of the energy must be derived from sources other than domestic petroleum reserves. Alaskan oil, which is supposed to start flowing in 1977, will help to some extent. But the quantities that might ultimately be extractable from the North Slope would satisfy present US petroleum needs for only about a decade at most.

Here it must be pointed out that the figures for reserves and quantities of crude oil that might be ultimately recoverable are based on our continuing to apply the extraction technologies now in general use. It has been estimated (26) that although a firm figure for the average recovery of oil in the United States is not available, about 30% appears reasonable. This means that about 200 billion barrels of petroleum lie unrecoverable in existing worked fields using current technologies. A drastic improvement in extraction efficiency, perhaps made possible by greatly increased petroleum prices, would make available a supply of crude oil that is larger than any other single source within the United States. It is generally recognized, however, that this will be extremely difficult to develop and is unlikely to have a major effect upon crude oil supplies within the next decade when problems of petroleum supply are likely to become particularly acute.

Associated with petroleum are substantial quantities of natural gas (principally methane) and condensable hydrocarbons known as natural gas liquids, which can be extracted as liquids during the production of natural gas. The problem of estimating the ultimate quantities of natural gas and natural gas liquids that can be extracted is essentially the same as that for crude oil. In the case of natural gas, proved reserves peaked in 1967. As was true of petroleum, reserves jumped in 1970 because of the Alaskan discoveries, but have dropped steadily since that time. Production of natural gas rose steadily until 1971, then leveled. Production dropped in 1974 and the downward trend appears to be continuing in 1975. Since natural gas liquids are produced along with natural gas, the production and reserve curves are similar, although the ratio of gas to liquid appears to be decreasing with time.

A panel of the Committee on Mineral Resources and the Environment of the National Academy of Sciences has analyzed the numerous estimates of potentially extractable hydrocarbons in the United States, including Alaska and the continental shelves (26). The panel concludes that the hydrocarbon resource base of the United States approximates 113 billion barrels of crude oil and natural gas liquids combined and 15 trillion cubic meters of gas. Table 4 was constructed by combining these figures with the energy conversion factor between petroleum and natural gas and by making use of reasonable estimates for the amount of natural gas liquids in natural gas. Although the estimate for the ultimate extractable quantity of crude oil is somewhat greater than that estimated by using the techniques of Hubbert, it is nevertheless well within reasonable bounds. Something like the equivalent of 500 billion barrels of petroleum appears to be ultimately extractable. Of this, somewhat over 40% has already been removed.

An appreciation of the significance of Table 4 is essential for an understanding of the difficult energy situation now confronting the United States. We are clearly pushing against the upper limit of our domestic extractable hydrocarbon resources. The upward jump of reserves brought about by the Alaskan discoveries can be only temporary; reserves are destined to continue their downward path. Production of hydrocarbons will also continue downward after a brief upsurge following the completion of the Alaskan oil pipeline. There will be another smaller jump upward upon completion of several gas lines about 1979. Since our energy demands are

Table 4 Readily extractable hydrocarbons in the United States[a] (billions of barrels of petroleum liquids or liquid equivalents of natural gas)

	Cumulative Production[b]	Proved Reserves[b]	Undiscovered Resources	Yet to Be Extracted	Total	1974 Production
Crude oil	109	35	98	133	242	3.2
Natural gas liquids	13	7	15	22	35	0.6
Natural gas	82	45	95	140	222	4.0
Total	204	87	208	295	499	7.8

[a] Includes Alaska and the continental shelves.
[b] As of January 1974.

not likely to decrease during the next few years, although the rate of increase may, we must compensate for the decreased domestic production either by utilizing greater quantities of other energy resources in the United States or by importing greater quantities of crude oil and other hydrocarbons from other countries.

There are certainly a number of domestic energy resources that would in principle be used to compensate for these impending shortages. I have already mentioned the substantial quantities of oil that have been left behind in "exhausted" reservoirs. US resources of petroleum in oil shale are extremely large—possibly enough to last 7000 years at present rates of consumption of crude oil and 3000 years at present rates of total consumption of hydrocarbons. Our proved coal reserves are enormous and could satisfy present US energy needs for about a millennium. There are almost infinite quantities of nuclear energy that could be tapped by use of existing technology. Added to this, solar energy can be used for a multiplicity of purposes including space heating and cooling, heating of water, and even the generation of power.

Where, then, lies the problem?

The problem lies largely in the "energy trap" in which we have become ensnared. The energy trap has

1. caused us to take extremely low energy costs for granted and to accept them as the norm to be expected for a very long time in the future;
2. deterred the development of a broadly based industrial expertise in the use of alternative energy sources;
3. caused us to give low priority to research programs aimed at developing alternative energy supplies;
4. prevented us from investing adequate amounts of capital in known technological processes for utilizing alternative energy sources safely on a substantial scale.

There are of course numerous subsidiary problems, many of them of an environmental nature. There are many real and potential environmental hazards associated with the increased use of coal, oil shale, and uranium for power. But in principle, at least, most of these problems are soluble if we are willing to pay enough for our energy. Clearly a large part of the environmental component of our energy problem can be directly related to our definition of acceptable energy prices.

In any event, it is highly unlikely that energy sources other than hydrocarbons will be in a dominant position in the US energy picture in the next two decades. The internal combustion engine is not likely to disappear quickly. The capital now invested in equipment for utilizing hydrocarbon liquids and gases is enormous and is likely to be used to the term of its normal life expectancy. Of course many actions can be taken to hasten the changeover from petroleum and natural gas to coal, uranium, and the sun. But here there are two key words that are to some extent related to each other: time and price.

Any major changes in primary energy sources will require substantial alterations in governmental policies including energy pricing policies, very large capital investments, and the resolution of many legal and political problems as well as the solution of a number of technical difficulties. It seems unlikely that these changes, urgent as they are, will be brought about very quickly, even if the changes were in some way guaranteed to be profitable. But today the potential developers of a coal gasification plant, a nuclear electric plant, or a scheme for extracting oil from shale on a large scale have no reasonable assurances of profitability, particularly when the environmental variables are taken into consideration.

Eventually, of course, there is no doubt that either the shift must be made or we perish. The questions are: When and under what circumstances will the changes take place?

In the short term—the next 10 to 20 years—there seems to be little doubt that the United States, like Western Europe and Japan, is destined to be a major importer of petroleum and petroleum products. It now appears that imports in 1975 will amount to one third of total hydrocarbon demand. As domestic production of petroleum and natural gas falls during the next two years, imports must increase still further. The opening of the trans-Alaska oil pipeline in 1977 and of gas pipelines from Alaska, perhaps in 1979, will relieve the shortage somewhat. But shortly thereafter imports will rise again beyond 33% of the present total requirement for hydrocarbons, to perhaps 65% by 1985 (27). And imports seem destined to continue to rise until such time as we shift clearly and on a truly large scale from hydrocarbons to other energy sources. Beyond about 1995 we will have no choice, for world availability of hydrocarbons will probably be on the decline and competition for the remaining resources will be severe.

THE GLOBAL LIFE EXPECTANCY OF CRUDE OIL AND NATURAL GAS

Hubbert (25) analyzed several estimates of ultimate recovery of world crude oil by geographical area and concluded that somewhere between 1350 and 2100 billion barrels can be extracted in the long run. In 1972 H. R. Warman (28, 29), Chief Geologist of the British Petroleum Company, reviewed the situation and concluded that the "independent work by my colleagues in British Petroleum and myself suggest a figure around 1800 billion barrels to be a reasonable maximum." Although Warman's figure has been criticized (30), it has also been strongly defended (31). In a report issued in 1975 by the National Academy of Science's Commission on Natural Resources (26) it was estimated that "1130 billion barrels of crude oil yet to be discovered is deemed reasonable." This implies that a total of some 2000 billion barrels of crude oil can ultimately be extracted. For the purposes of this discussion the differences between these estimates have little significance. Let us use here the estimates of crude oil given by the Academy panel.

World resources of natural gas (including natural gas liquids) can be roughly estimated on the basis of the US experience with respect to the ratio of extractable gas to crude oil. These figures suggest that the ultimate amount of natural gas that will be extracted is equivalent in energy content to about 2300 billion barrels of crude oil. By similar reasoning, the ultimate extractable volume of natural gas liquids would be about 440 billion barrels.

Here it must be noted that in past decades a large proportion of the natural

Table 5 Readily extractable hydrocarbons in the world (billions of barrels of petroleum liquids or liquid equivalents of natural gas)

	Cumulative Production[a]	Proved Reserves[a]	Undiscovered Resources	Yet to Be Extracted	Total	1974 Production
Crude oil	300	560	1140	1700	2000	20.4
Natural gas liquids	120[b]	70	250	320	440	0.9
Natural gas	620[b]	380	1300	1680	2300	8.5[c]
Total	1040	1010	2690	3700	4740	30.0

[a] As of January 1974.
[b] Including flared and otherwise unrecovered gases and liquids.
[c] Marketed.

gas and associated liquids released from the earth has been lost because of lack of adequate storage and distribution facilities. A substantial fraction of the gas has been "flared" into the atmosphere. With time, however, increasing proportions of the available gas have been marketed. In the United States today only 1% of the produced gas is flared. In the rest of the world, however, more than one fifth of the liberated gas is wasted in this manner.

These estimates, summarized in Table 5, indicate that almost 5000 billion barrels of crude oil might eventually be extracted from the earth as a combination of crude oil, natural gas, and natural gas liquids. About one fifth of the total has already been extracted. Another one fifth is in the form of proved reserves. The remaining three fifths are yet to be discovered.

World production and proved reserves of crude oil are shown in Figures 10 and 11. The combination of the two curves plus the estimated quantities of crude oil and natural gas that might ultimately be extracted suggest that petroleum as well as total hydrocarbon production will probably peak by the year 2000 and possibly as early as 1990.

As the world approaches the peak of production of natural hydrocarbons there will be numerous adjustments in patterns of energy consumption and in national economies. Some nations will be affected very little; others will undergo revolutionary transformations. The nations most affected will be today's major importers and exporters. The least affected nations will be those that are at present self-sufficient with respect to hydrocarbon fuels and that do not need to export them either for their economic survival or for their development. Among the industrialized countries, the Soviet Union is in the most favorable position. The petroleum that can ultimately be extracted in the USSR appears to be greater than that in the United States, including Alaska, and a smaller proportion of the total has already been extracted.

Among the developing countries that are not substantial exporters and that seem destined to be self-sufficient for many years in the future, China, Mexico, and Colombia are outstanding. The developing nations that are major exporters include the 13 members of OPEC; all are attempting to use their oil revenues to expedite development. As their reserves become depleted they must decide how much of the remaining petroleum they intend to reserve for their own purposes.

Although the United States must clearly import increasing proportions of its hydrocarbons, and although its position does not appear to be as favorable as that of the USSR, it is in a very good position when compared with other industrialized parts of the world, particularly Western Europe and Japan. The discoveries in the North Sea will help Europe to some extent, but there will be no respite from the need for imports, given present patterns of consumption. The entire industrialized world must look forward to increasing levels of hydrocarbon imports until such time as it is able to shift over to other more abundant forms of energy.

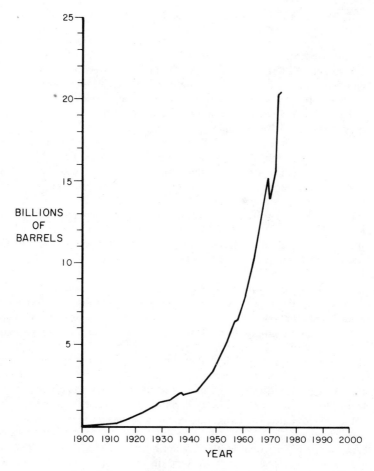

Figure 10 World production of crude oil.

THE AVAILABILITY OF ALTERNATIVE ENERGY RESOURCES

The alternative energy resources available are vast. Indeed, they are so very large that for as long a time as technological capabilities are retained there need never be a shortage of energy, even if energy consumption worldwide increases by another order of magnitude or two and even if we speak in terms of a life expectancy for human civilization of millions of years. The main difficulty is that all other forms of available energy are less convenient than oil and natural gas. Virtually all are more expensive than conventional hydrocarbons. Some are more difficult to extract. Some present environmental hazards. Most require extremely large installations and capital investments. Indeed, we have been greatly spoiled by the easy availability of petroleum.

With respect to fossil fuels there are substantial reserves of oil shale, tar sands, and coal already identified and even greater reserves probably existing but not yet discovered. Table 6 shows the estimated quantities of these fuels and compares

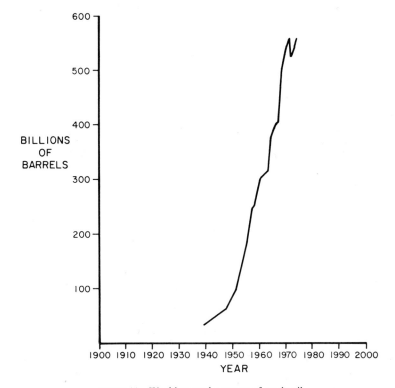

Figure 11 World proved reserves of crude oil.

Table 6 Estimated resources of shale oil, tar sand oil, and coal compared with resources of conventional hydrocarbons

	United States		World	
	Barrels of Oil (Billions)	Equivalent Metric Tons Coal (Billions)	Barrels of Oil (Billions)	Equivalent Metric Tons Coal (Billions)
Conventional hydrocarbons	500	69	5000	690
Identified reserves of shale oil (32)	2000	280	3000	415
Undiscovered reserves of shale oil (32)	25,500	3530	340,000	47,000
Incomplete reserves of tar sand oil (26)	30	4	750	140
Minable coal and lignite (25, 32)		1500		7600
Total		5380		55,800

them with the estimated quantities of conventional hydrocarbons that will eventually be extracted.

The hydrocarbon in tar sands is a mixture of viscous liquids that cannot be extracted by means of wells. Techniques have been developed, however, for mining and extracting the material on a large scale, and these are now being applied. As the oil from these sands belongs to the same chemical family as crude oil, it can be processed by existing oil refineries.

The principal hydrocarbon in oil shales is kerogen, which is a solid. In most processes the kerogen is converted to either gaseous or liquid fuel and then concentrated. Several conversion processes have been developed (33), and in view of the rising prices of conventional oil and gas, some of the processes appear to approach competitiveness. Yields from the shales that are considered to be reserves or potential reserves range from 10 to 100 US gallons of hydrocarbon per ton. Colorado oil shale containing 30 gallons of hydrocarbon per ton is now being seriously viewed as competitive with conventional petroleum.

One of the principal drawbacks to the use of coal is its sulfur content, which pollutes the atmosphere when the coal is burned. In recent years considerable effort has been directed toward developing processes that will convert coal to various clean fuels (33). Methods now exist for converting coal to combustible gas, to synthetic hydrocarbon liquids, or to methyl alcohol. Of course coal can also be burned directly to generate electricity, but unless the fuel is relatively free of sulfur, special provision must be made to remove the sulfur dioxide formed during combustion. In addition, the particulate matter formed by the ash must be removed to prevent pollution of the atmosphere.

A further difficulty with the utilization of oil shales and coal is that of carrying out the mining operations on an adequate scale. Strip or surface mining is far less costly than underground mining, but in the process the topsoil and other parts of the natural environment are destroyed. The task of minimizing the destruction and returning the mined land to a productive condition is both difficult and costly. In addition, huge quantities of crushed rock and other residues must be disposed of in some way.

On the positive side, however, if these problems are all solved in such a way that the products are not prohibitively expensive, the proved and suspected reserves of oil shale and coal would satisfy human needs for a very long time in the future. But the quantities available would by no means last forever. At present rates of energy consumption all fossil fuels combined would last several thousand years. But both population and per capita demands for energy are increasing and seem destined to continue increasing in the years ahead. Under these circumstances the production of fossil fuels would probably peak in another 200 to 300 years if other sources of energy were not utilized on a large scale in the meantime.

As is the case with petroleum, the most favorable known deposits of oil shale and coal are not equitably distributed around the world. The best of the known oil shale deposits are in the United States and Brazil. On the other hand there is good reason to believe that oil shale deposits, particularly those of lower grade, should be widely distributed among the continents. Coal, on the other hand, is quite unevenly distributed as can be seen in Figure 12. The USSR and to a lesser extent the United States loom as the giants of the coal world, accounting together for about three fourths of the world's deposits.

One possible danger associated with the expanded use of hydrocarbons and coal lies in the fact that the carbon dioxide in the atmosphere equilibrates very slowly with the bicarbonate of the deep ocean. The deep sea turns over only every several

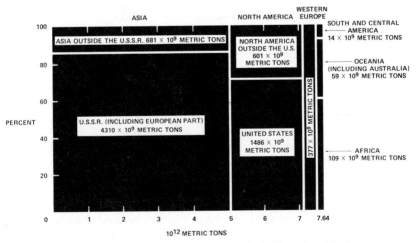

Figure 12 Estimates of world resources of minable coal and lignite.

hundred years. Thus a large part (about half) of the carbon dioxide remains in the atmosphere. Apparently as a result of the combustion of fossil fuels, the carbon dioxide concentration in the atmosphere has increased from an average of about 290 ppm to 322 ppm in 1970. The average concentration is expected to reach 375 ppm by the year 2000 (34). An eventual doubling of carbon dioxide concentration would by no means be out of the question. Theoretical studies indicate that a doubling of the concentration could effect an increase of the temperature near the earth's surface by about 2°C. Such a change could trigger other mechanisms, possibly leading to irreversible climatic effects.

This single aspect of greatly increased consumption of fossil fuels should be monitored very closely on a worldwide basis. Any clear physical or theoretical indication of emerging adverse effects may make it advisable to lessen substantially the global rate of fossil fuels consumption.

Thus at some time man's consumption of fossil fuels will start decreasing. It is too early to say whether this change will come about because of decreasing availability of fossil fuels in the ground, because of prohibitively high costs of mining and conversion, or because of adverse environmental effects. But even before we reach that time, it seems probable that we will be using nuclear or solar power, or perhaps both, on a very large scale.

Hydropower is by no means yet being fully utilized. If we assume the present observed average rate of flow, which is available 95% of the year, about 10,000 billion kilowatt-hours could be generated each year (35). About 13% of this is already being utilized. With present technology, about 4 billion tons of coal would be required each year to generate that power—only a little over half the present world energy demand.

In some parts of the world tides can be harnessed, but the potential does not seem to amount to more than a few percent of the potential of hydropower (25). The power that can be potentially generated by winds appears to be even smaller than the generating capacity of the tides (35).

Geothermal resources could prove important in many parts of the world including the United States. There are wide variations in the resource estimates even in the United States, where the geology of thermal areas is reasonably well known. The highest of the US estimates suggests that an electricity-generating capacity of 400 billion watts based on geothermal resources could be developed in the western United States by 1995 (36). This would correspond to the rate of electrical generation from the expenditure of about 1.4 billion tons of coal annually. This is close to, but slightly smaller than, the present annual US energy budget.

Although the estimate cited for the United States is probably optimistic, it seems clear nevertheless that geothermal energy might be a very important supplement to other energy resources, at least over a period of 50 to 100 years. This is particularly true in light of the relatively nonpolluting qualities of the resource.

When we consider the entire heat content of the earth's crust, as distinct from thermally active locations, a different picture emerges, for tremendous quantities of energy are available in principle—enough to last for over a million years at present rates of consumption. But the technical difficulties are enormous. At this stage no real conclusion can be drawn as to whether or not geothermal energy can be exploited on a truly large scale. If it can be so exploited it must be considered along with fission, fusion, and solar energy as one of humanity's primary long-term energy options.

Electricity is now being produced in many parts of the world from nuclear fuels, and nuclear reactors of the present type, which feed primarily upon uranium-235, will undoubtedly grow in number. The total amount of energy likely to be produced by such reactors is, in the long run, resource-limited. As only about 1% of the total energy available in the uranium is utilized, the quantities of uranium needed are large. Quantities of uranium that can be obtained for $30 per kilogram or less are no more than one million tons. Perhaps an additional five million tons could be obtained at costs under $100 per kilogram (37). It is likely that for as long as nuclear technologies are employed that make use of such a small fraction of the total energy available, the spread of nuclear power will be basically limited by the cost of uranium.

On the other hand, it now seems likely that in the near future breeder reactors will be available that will feed on plutonium derived from the most common isotope of uranium (uranium-238), releasing as much as 60% of the available energy. From a technological point of view, such reactors might be in operation in a number of localities by the early 1990s. It is also possible that breeder reactors using thorium (which is more abundant than uranium) will be available around the turn of the century.

The deployment of fast breeder reactors will have a profound effect upon the uranium supply problem, for even very expensive uranium can be viewed as economical. When uranium costs as high as $1000 per pound can be seriously considered, vast reserves of uranium, as well as thorium, can be identified that would not have been seriously considered under current conditions. For example, to take an extreme case, it has been demonstrated (38) that about 30% of the uranium and thorium in an average granite is situated in such a way that it can be washed out of the pulverized rock by using dilute acid. Considering the energy cost of quarrying, pulverizing, and treating a granite in this way, the energy "profit" from a ton of average granite is equivalent to about 15 tons of coal. But there are huge outcrops of granites in various parts of the world that contain considerably higher than average concentrations of uranium and thorium. The energy potentially available from such rocks in the United States is thousands of times larger than that of all oil shales and coal combined.

Thus the fast breeder reactor gives us access to a supply of energy that can last for thousands and perhaps millions of years. However, there are numerous problems that must be solved. There are problems of reactor safety—plutonium dispersed in the atmosphere can be lethal. There are problems of waste disposal—huge quantities of radioactive by-products will be generated. Last, but by no means least, there are problems of preventing plutonium from falling into hands of unscrupulous persons. Not much plutonium is needed to make a bomb of substantial explosive force.

Beyond the fast breeder lies fusion, perhaps first attained by reacting deuterium with tritium. It is not as yet technologically feasible to achieve this on a large scale, but there are signs that this may eventually be achieved. However, contrary to popular opinion, this does not open up new fuel reserves that are much greater than those offered by fission. In fusion involving tritium, the supply of lithium (from which tritium is made) is the limiting factor.

The potential supply of solar power and heat is practically unlimited. Its effective utilization suffers from the fact that it is of relatively low intensity, variable in its availability, and not available in any one location for the entire day. In spite of these difficulties, however, the prospects for its use on a large scale seem reasonably hopeful. Already it is in use to some extent in some regions of the world for heating water and for space heating, and it seems likely that such uses will spread fairly rapidly. Numerous investigations are now in progress on the possibilities of generating electrical power by concentrating the energy with mirrors, by using thermocouples or photoelectric cells, and by using biological processes.

With nearly 25% of the energy consumed in the United States used to heat and cool buildings, this particular aspect of solar energy is receiving high priority. The University of Delaware's solar house has been in operation for two years. The best systems tested there have provided at least 70% of the heating and cooling needs of the building (39).

In the area of power, studies are being made of arrays of reflectors that concentrate heat onto a receiver located atop a central tower. Here a fluid is heated to high temperature, which in turn drives a turbo generator. With regard to the photovoltaic cell, major efforts are being made to lower installation costs. The US Energy Research and Development Administration hopes eventually to reduce the capital cost of an array to about $500 per kilowatt of installed capacity (39).

Thus, although great difficulties can be anticipated, it now appears that humanity need never suffer from a shortage of available energy. We will be faced with considerably higher costs than we have experienced during the Golden Age of petroleum and natural gas. We will be confronted by environmental problems that are far more difficult to solve than those presented by oil and gas. But to compensate for these difficulties we have available far more energy than we can use even in the foreseeable long-range future when presumably our total energy needs will be far greater than they are today.

ENERGY: THE KEY TO NATURAL RESOURCE ABUNDANCE

Basically the problem of providing adequate resources for the perpetuation and expansion of civilization is the problem of providing adequate quantities of energy of the right type at the right place at the right time. This is true no matter whether the resource is food, minerals, structural materials, clean air, or energy itself.

It is often said that "it takes money to make money." Similarly it takes energy to make energy. When our energy comes from petroleum, oil fields must be found, wells must be drilled, and the oil must be pumped to the surface and transported to the refinery where it is modified to·suit the needs of the users. All of this requires energy, not only to drill, pump, and move the material, but to make the equipment, pipes, rails, trains, and cracking plants. Vast quantities of energy are required to mine and transport coal, to lay pipelines for gas, to construct and move oil tankers, to enrich isotopes and build nuclear reactors.

Today about a quarter of our energy consumption in the United States is accounted for in the mining, extraction, conversion, and transportation of fuels. As we move away from oil and natural gas toward oil shales, coal, and other more sophisticated energy sources, this proportion will naturally become larger. It is estimated that by the year 2000 more than one third of our energy consumption will be for the production of energy.

The production of more food, whether in the United States or in India, necessitates the expenditure of more energy for the manufacture and distribution of fertilizers and pesticides and for the distribution of water for irrigation. Water itself is becoming energetically more expensive as we move it greater distances and find it increasingly necessary to remove pollutants. In some parts of the world we now obtain our water from the sea and as this practice spreads, energy costs will increase still further.

With respect to minerals we find it necessary to dig ever deeper into the earth and to process ores of lower and lower grade. At one time in the United States we had very large deposits of high-grade iron ore. As time passed our better grades were consumed, so we moved to lower-grade ores such as taconites that require beneficiation and agglomeration to make them usable. Naturally this requires increased expenditures of energy. An alternative is to import high-grade iron ore from other countries. This, too, results in higher energy costs, in this case for transportation.

When copper was first smelted, very high grade deposits of ore could be found near the surface of the ground. As time passed and the easily accessible high-grade ore disappeared, it became necessary to process ores of lower grade and to dig more deeply into the ground. That process has continued to this day. As recently as 1900 the average grade of copper ore mined in the United States was 4.0%. By 1910 this had dropped below 2.0% and during World War II it dropped below 1.0%. In 1970 the average grade was 0.60%, and the vast open-pit copper mines were the largest artificial holes ever to have been dug (32). Clearly the energy costs involved with moving and processing these vast quantities of materials are substantial.

Is there any limit to how low the grade of ore can be and still provide the metals and other substances necessary for the continued functioning of our society? *Here the ultimate answer is that there is no limit, provided the necessary quantities of energy are available for undertaking the physical and chemical operations.* In practice, of course, there are limits to grade, below which we need not concern ourselves. Those limits are represented by the most extensive and yet the lowest-grade ore deposits on the surface of the earth: seawater and ordinary sedimentary and igneous rocks. In principle, a high level of civilization could be nourished for many millions of years by those substances. From them society could obtain all that it needs, including the energy itself. But the energy costs per capita of such a system would obviously be much higher than those of today.

As grades of ore decrease, as other resources such as wood, water, and fertilizers become more costly and scarce, and as problems of pollution of air, water, and land become more acute, there will be increasing pressures to recycle the materials that flow through our agro-industrial network. Already we are recycling substantial quantities of iron, copper, and some other metals. In some cities paper is recycled or burned for energy. But in our society as a whole our recycling effectiveness is very low.

In the years ahead we will probably approach asymptotically the condition of nearly complete recycling, where in effect we recognize that except for energy we live in a closed system. All organic wastes will be reclaimed for their water and for their chemical potential, whether for fertilizers, other useful chemical compounds, or energy. The metals from solid wastes will be recovered. Potentially dangerous pollutants will be isolated before they can be injected into the atmosphere, lakes, or rivers.

All of these steps will increase substantially our energy expenditures. At the same time those increased expenditures can insure for all humanity a cleaner environment and a steady flow of the raw materials necessary for the smooth functioning of our complex technological society.

ENERGY PRICING IN PERSPECTIVE

How much should we be willing to pay for our supplies of energy? Are the economics of energy really functioning for long-term human benefit when such economics make it so attractive for us to consume the greater part of the existing petroleum and natural gas in the world before we learn how to use coal and oil shale effectively? Are the economics working for our benefit when they dissuade us from investing adequate quantities of capital in the technologies of utilizing coal and oil shale? Are they working for our benefit when they encourage us to burn useful, complicated hydrocarbon molecules to heat water and buildings and to generate power, when such molecules could be used more effectively for the manufacture of a vast array of complex and extremely useful petrochemicals?

The combination of the low cost and convenience of oil led us into the energy trap. Yet why should crude oil cost so little? Should its cost be based simply upon its ease of extraction? In the days when the incremental extraction costs in the Middle East were 10¢ per barrel and the oil companies were charging $1.25 per barrel for Middle Eastern oil, the companies in effect declared that there need be little relationship between price and cost. Under these circumstances why, then, was the world so shocked in 1973–1974 when the OPEC countries collectively and arbitrarily increased the price of crude oil by a factor of four?

In retrospect the world was shocked because it was badly hurt economically by the action rather than because of any fundamental wrongdoing on OPEC's part—with the single qualification that the action was perhaps taken too suddenly. The principle of arbitrariness had already been established.

Not long after the OPEC action, the then Secretary-General of the organization attended a private conference in New York. After some of the American participants had made a number of critical remarks about OPEC's action, the Secretary-General asked an extremely important question. "How," he asked, "does one place a fair value on a resource which is under the ground, which is being extracted, and which will one day be gone, never to reappear?" Few persons in the room were prepared to give an answer, but most grasped the point that there was nothing magical, or for that matter even rational, about the pre-1973 price of crude oil.

Clearly, from the point of view of the consumer, the fairness of a price will be related in some way to the cost of doing business in the absence of the resource. If there were no crude oil available, hydrocarbon fuels would be obtained by conversion of oil shale or coal to synthetic fuels. If the price of crude oil were raised to a level greater than the cost of converting oil shale or coal, and if the price were predictably maintained, with little question conversion plants would be built in profusion in many parts of the world.

Of course, when the price of crude oil was raised so rapidly in 1973–1974 the consumer nations could not have taken any meaningful action for the simple reason that options other than that of accepting the increased prices (or perhaps engaging in military action) were simply not viable. A great deal of time is required to develop the necessary technology, to accumulate the needed (and substantial) capital, and to build the plants that are essential to achieve at least some semblance of energy independence.

Obviously an economic dislocation, such as that which accompanied the sudden fourfold rise in energy costs, is uncomfortable. The only way that we in the United States, as well as other major importing nations, can protect ourselves against similar dislocations in the future is to develop options that are truly viable. This would mean using, on a substantial scale, energy sources other than crude oil, natural gas, or other import-dependent technologies.

If such options are to be developed and put into operation on a meaningful scale, the entire price structure of our various energy resources must be reexamined and drastically altered. Here the term *price* is used to include all costs to the consumer including taxes, less government subsidies.

Ideally, the revised price structure and associated guarantees would make oil shale and coal competitive with crude oil and natural gas (both domestic and imported) for the production of liquid fuels and petrochemicals. Nuclear power (fission) would be made competitive with coal for the generation of electricity. Solar energy would be made competitive with fossil fuels for space heating and cooling. In this way all of our major near-term energy possibilities would be developed and utilized on a substantial scale.

Of course, under such circumstances our energy costs would be greater than they are today but not dramatically so. Estimated costs of producing synthetic fuels from coal and oil shale are in a range less than $12 per barrel (33), and even making due allowances for underestimation, the cost is not far from the current price of imported crude oil. The real problem is to establish the price guarantees that will enable the producers of synthetic crude to invest the very large amounts of capital necessary and to protect their investment should the price of crude oil drop drastically. Once this is done and facilities are established on a large scale, further major increases in the prices of crude oil charged by the exporters would simply not be possible.

Here we must keep in mind the fact that even if energy costs (in constant dollars) were to go considerably beyond the present price of imported crude oil, the results need not be catastrophic. In 1971 fuel costs represented something like 2.5% of the US gross national product, which is not very large, particularly when we consider the amount of wealth that is generated by the energy.

THE ROAD TO SOLFUS

In the history of society's utilization of energy there has been a sequence of replacement of one energy resource by another, brought about in part by technological change. Wood was displaced by coal, and coal was in large measure displaced by crude oil, then by natural gas. Each of these transitions has involved the replacement of one dominant energy source by another, which then became dominant, only to be displaced by yet another fuel. Energy from nuclear fission is usually cited as the most likely candidate for replacing oil and natural gas, with fission in turn eventually giving way to some virtually infinite and relatively safe energy source such as solar radiation or nuclear fusion. Häfele has dubbed this ultimate energy source "solfus" (39).

There are many reasons, however, for doubting that the future will be as simple as the past. In the first place, it is by no means clear what future energy demands are likely to be. In the second place, although nature has endowed us with vast energy resources, none are devoid of serious technological and sociological problems, all of which must be resolved if the resources are to be effectively utilized.

With respect to total basic energy demand, one can take the conservative view that the world population will stop growing when it reaches ten billion persons and that per capita demand for energy will stop growing when it reaches the average of that of the rich nations today—the equivalent of about six metric tons of coal per person per year. This would give a total energy demand of about 60 billion tons of coal annually—nearly ten times the present world consumption. We could also take the ultraconservative view that all nations can learn to use energy more efficiently and that living patterns can be adjusted so that ultimate per capita energy needs will be reduced, perhaps by as much as a factor of two. This would give a total world demand of about 30 billion tons of coal annually. For many reasons it seems doubtful that world demands could be stabilized at much less than this in the absence of a catastrophe, which would of course result in a world energy demand close to zero.

On the other hand, we have seen that the energy costs of maintaining the status quo in the present rich countries will undoubtedly rise because of the increased energy that must be used to extract necessary raw materials, to recycle used materials, and to combat pollution. These requirements could easily triple the current per capita needs in the rich countries and give rise to an ultimate world energy demand equivalent to 180 billion tons of coal annually.

From a radical point of view, the reclaiming of seawater and arid lands could well result in a doubling of the 10 billion persons suggested by present demographic trends. To go further, there is little reason for us to suppose that in the long run an expenditure of the equivalent of 20 tons of coal per person each year (twice the present US level) would really satisfy human wants. An ultraradical futurist might well say (and for plausible reasons) that human beings could not possibly achieve their full potential unless they were each allocated the equivalent of 100 tons of coal each year.

Thus, ruling out catastrophe, we see that estimates of eventual world annual levels of energy consumption ranging from the present equivalent of 7 billion to 2000 billion tons of coal are all plausible depending upon how conservative or radical our outlook might be.

Although these estimates do not tell us precisely when solfus will take over as our major fuel, they do tell us that when we view the energy situation in historical perspective, the road to solfus, like the road to catastrophe, is not a very long one and, like the road to hell, is paved with good intentions. Nevertheless it is essential that we recognize the critical importance of our traveling that road carefully.

With respect to our future energy options, it seems essential that we keep all of them open by developing and gaining experience from each of them. The difficulties presented by coal and shale are considerable. Fission may prove to be a dead end in more ways than one. Geothermal power may prove impracticable on a truly large scale. It may turn out that we can't afford the capital costs of nuclear power. And fusion may turn out to be impossible to achieve in a practical way.

Further, the needs of nations will differ considerably, at least in the short run. Europe and Japan may need nuclear power on a large scale while the United States is powered by coal and shale and the USSR is still powered by oil. And the diverse energy sources themselves are more useful for certain applications than for others. As time passes, crude oil, natural gas, oil shale, and coal should be increasingly reserved for chemical use as distinct from the production of heat and power. Nuclear energy is particularly suited for power production and solar energy for heating and cooling. We must attempt to achieve the maximum flexibility with the mix of energy sources available to us.

Thus, it appears that success with respect to the foreseeable problems of energy lies in diversification rather than specialization. In the short run we must recognize that there is no single solution—even if we admit that in the long run solfus is the answer.

Literature Cited

1. Matras, J. 1973. *Populations and Societies*, p. 17. Englewood Cliffs, NJ: Prentice-Hall
2. See Ref. 1, p. 21
3. *Sci. Am.* Sept. 1971, Vol. 225. (The entire issue is devoted to energy.)
4. Brown, H. 1975. Population growth and affluence: the fissioning of human society. *Q. J. Econ.* 89:236–46
5. Landsberg, H. H., Schurr, S. H. 1968. *Energy in the United States: Sources, Uses and Policy Issues*, p. 30. New York: Random House
6. Fisher, J. C. 1974. *Energy Crises in Perspective*, p. 44. New York: Wiley
7. See Ref. 6, p. 60
8. Schurr, S. H., Netschert, B. C. 1960. *Energy in the American Economy, 1850–1975*, pp. 545–49. Baltimore: Johns Hopkins Univ. Press
9. Ford Foundation. 1974. *Exploring Energy Choices: A Preliminary Report*

of the Ford Foundation's Energy Policy Project, p. 13. New York: Ford Foundation
10. See Ref. 9, p. 124
11. See Ref. 9, p. 1
12. See Ref. 6, p. 66
13. Organization for Economic Cooperation and Development. 1972. *Statistics of Energy 1956–1970*. Paris: OECD
14. *International Economic Report of the President, Transmitted to Congress March 1975*. Washington DC: GPO
15. National Academy of Sciences. 1975. *Agricultural Production Efficiency*, p. 6. Washington DC: Nat. Acad. Sci.
16. Pimentel, D. et al. 1973. Food production and the energy crisis. *Science* 182:443–49
17. Hirst, E. 1974. Food-related energy requirements. *Science* 184:134–38
18. Steinhart, J. S., Steinhart, C. E. 1974. Energy use in the U.S. food system. *Science* 184:307–16
19. Heichel, G. H. 1974. Energy needs and food yields. *Technol. Rev.* 76:18–25
20. Heichel, G. H., Frink, C. R. 1975. Anticipating the energy needs of American agriculture. *J. Soil Water Conserv.* 30:48–53
21. See Ref. 15, p. 116
22. *Oil Gas J.* Jan. 27, 1975, p. 108
23. *Oil Gas J.* April 7, 1975, pp. 45–46
24. Hubbert, M. K. 1949. Energy from fossil fuels. *Science* 109:103–9
25. Hubbert, M. K. 1969. In *Resources and Man*, pp. 157–239. Nat. Acad. Sci. San Francisco: Freeman
26. National Academy of Sciences, Commission on Natural Resources. 1975. *Mineral Resources and the Environment*. Washington DC: Nat. Acad. Sci.
27. National Petroleum Council. 1972. *U.S. Energy Outlook, Summary Report*, p. 7. Washington DC: Nat. Pet. Counc.
28. Warman, H. R. 1972. The future of oil. *Geogr. J.* 138 (3):287–97
29. Warman, H. R. 1973. How much oil is left. *Environment and Change* 2:164–73
30. Odell, P. R. 1973. The future of oil: a rejoinder. *Geogr. J.* 139 (3):436–54
31. Simpson, R. D. H. 1973. Further remarks on the future of oil. *Geogr. J.* 139 (3):455–59
32. US Geological Survey. 1973. *United States Mineral Resources*. Prof. Pap. 820. Washington DC: GPO
33. Vyas, K. C., Bodle, W. W. March 24, 1975. Coal and oil-shale conversion looks better. *Oil Gas J.*, pp. 45–54
34. Study of Man's Impact on Climate (SMIC). 1974. *Inadvertent Climate Modification*, pp. 233–40. Cambridge, Mass: SMIC
35. Parker, A. 1975. World energy resources. *Energy Policy* 3 (1):58–66
36. Berman, E. R. 1975. *Geothermal Energy*, pp. 35–39. Park Ridge, NJ: Noyes Data Corp.
37. InterTechnology Corporation. Nov. 1971. *The U.S. Energy Problem*, Vol. II: Appendices, Part A, p. E-53. Warrenton, Va: InterTechnology Corp.
38. Brown, H., Silver, L. T. 1955. *The Possibilities of Obtaining Long-Range Supplies of Uranium, Thorium and Other Substances from Igneous Rocks*. Presented at UN Int. Conf. Peaceful Uses Atomic Energy, New York, Pap. No. P/850
39. Häfele, W. Nov. 1974. *Future Energy Resources*. Res. Rep. RR-74-20. Laxenburg, Austria: Int. Inst. Appl. Syst. Anal.

Copyright 1976. All rights reserved

COAL: ENERGY KEYSTONE ✕11002

Richard A. Schmidt and George R. Hill
Electric Power Research Institute, Palo Alto, California 94303

INTRODUCTION

Coal provided the energy resource base for the historical industrialization of the American economy and is being called on once more to satisfy many of the nation's future needs. Coal is truly a keystone, which Webster's dictionary defines as "something on which associated things depend for support."[1] Still, the current, actively renewed interest in coal as an essential factor in the American economy comes after decades of indifference or neglect. Although the expectation for rejuvenation of the coal industry and a return to emphasis on coal for meeting energy requirements can probably be realized, there are many vital factors influencing coal extraction and preparation for utilization. These factors must be assessed and understood if coal is again to fulfill a central role in the nation's energy system. Toward such an understanding, this chapter seeks to provide a brief review of essential information on coal resources and reserves, coal production technology, problems facing the electric utilities in coal production and utilization, and an assessment of future conditions in coal development.

Coal and Electric Utilities

Over the past three decades, the electric utility industry has become the largest consumer of coal (Figure 1). In the early 1940s, coal was used in relatively equal amounts by retail markets, manufacturing, coking, railroads, and electric utilities. This diversity in coal use changed substantially during the late 1940s, 1950s, and 1960s. During these times, the railroad market for coal virtually disappeared and was replaced by oil; the retail market shrank because coal was largely replaced by gas; and less coal was used in manufacturing and coking because of technological improvements leading to greater efficiency in processing.

Following a period of adjustment within the coal industry, the lost markets were replaced by the rapid growth of coal use by electric utilities. In a period of slightly more than 30 years, this use of coal increased from about 50 million tons in 1940 to about 400 million tons in 1974. As a result, the electric utility industry consumes now roughly two thirds of the present total national production of coal.

[1] This is entirely appropriate, as *keystone* is a term long associated with coal, stemming from the early days of the nation's development when most coal was produced in Pennsylvania, the Keystone State.

Figure 1 also suggests estimated future requirements in coal production. The electric utility industry's coal use is projected to be 600 million tons in 1980, 900 million tons in 1990, and 1200 million tons in 2000. Note that these projections do not include coal to supply power generation units reconverted from oil to coal or new coal plants for gasification or liquefaction. The present uncertain status of

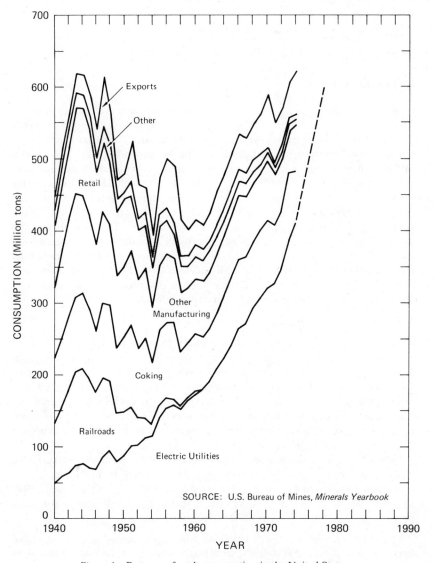

Figure 1 Patterns of coal consumption in the United States.

reconversion and the rapidly escalating capital costs for conversion plants compound the uncertainty in estimating future coal production for these uses.

Coal production adequate to satisfy the estimated requirements depends on several essential factors of technology and practice that influence or control the extent and pace of developments. The principal factors may be summarized as (a) quantity and quality of reserves, (b) extraction operations, (c) treatment and preparation facilities, (d) transportation arteries and equipment, (e) labor to operate each stage of the production sequence, and (f) capital for establishment of new installations or modernization of existing operations. Depending on conditions present at a given locale or in a given circumstance, any or all of the above factors could prove to be a constraint to timely and economical development of coal deposits. The effect of such constraints could be to limit the rate at which coal production capacity may be adapted to meet the projected requirements. Accordingly, this review covers the principal aspects of the above factors to provide a basis for assessing possible future conditions regarding coal production for use in the electric utility industry.

COAL PRODUCTION TECHNOLOGY

Coal production has been a part of the American scene for so long that it has been taken for granted. Also, it may be mistakenly assumed that coal production is a backward industry that persists in the use of outmoded technology in disregard of the prospective benefits from aerospace-type approaches applied to a conceptually simple activity. The extent to which such views might be held is an indication of a serious lack of knowledge about coal extraction.

Far from being a set of cut-and-dried operations, coal extraction as an enterprise is replete with uncertainties in fundamental aspects such as deposit characteristics, mining conditions, labor relations, market economics, equipment performance, transportation availability, financial security, and governmental regulations. This array of natural uncertainties and institutional problems is such that coal production is described in the profession as "winning" (1). The term reflects the struggle of the coal producers against the imperfectly understood processes of nature that created the deposits they labor to exploit.

Coal Resources and Reserves: Quantity

The location, magnitude, and quality of coal deposits must be known with some precision if development is to be considered seriously. Extensive geological field work and painstaking study is required to delineate the deposits and to estimate the amount of resources that are present. Mining engineers then use these data to estimate what portion of the resources can be recovered and can thereby constitute a coal reserve. This important distinction between resources and reserves is often overlooked, with the result that the amount of usable coal could be misstated.

Apart from the obvious problem in terminology, the most serious problem in determining both coal resources and recoverable reserves is that much of the basic technical field work in the public domain was performed many years ago. Many estimates are founded on work done in the early part of this century without the advantage of modern techniques and instrumentation and thus are often lacking the degree of refinement and specificity necessary in assessment of potentials for the large-scale, costly developments of the present. An intensive effort to remedy this is to be carried out by the present programs of the US Geological Survey and the US Bureau of Mines. The Geological Survey is primarily responsible for determination of the nation's resources, while the Bureau of Mines provides estimates of recoverable reserves. The data on coal resources and reserves resulting from their work represent the only authoritative account of the nation's coal deposits, and these data are used to describe the situation apparent at present.

The total remaining coal resources of the United States to a depth of 3000 feet, determined by mapping and exploration, are about 1580 billion tons.[2] The total comprises, in rough terms, the following: 686 billion tons bituminous coal, 424 billion tons subbituminous coal, and 450 billion tons lignite. Clearly this is a substantial amount. However, as is shown in this section, not all of the coal resources can be recovered, thereby reducing the quantity of usable reserves. Additionally, as shown in the next section, the different heating values and quality of coals further reduce the magnitude of the available coal reserves.

To be considered targets for development, coal seams must be of relatively high quality (ideally, high heating value and low sulfur content), occurring in relatively thick, uniform, and extensive beds that are close enough to the earth's surface to permit application of economical mining technology. As part of the planning for Project Independence, a new evaluation of the nation's coal deposits was made by the US Bureau of Mines and the US Geological Survey. A new term was coined: the *reserve base*, defined as "the quantity of in-place coals calculated under specified depth and thickness criteria." This new term is intended to be intermediate between resources (the total stock of coal in the ground) and reserves (the quantity of coal that actually can be recovered). Note that the reserve base represents 100% of the coal in place that meets the specified criteria; no allowance for incomplete recovery has been made. Actually, the amount of coal that can be recovered from a given deposit varies from about 25% to 90% of the coal in place. Therefore, for a reasonable estimate of the amount of recoverable coal, the reserve-base data must be reduced by an appropriate amount to allow for incomplete recovery. The reserve-base data represent the most recent estimate of the quantity of US coal deposits that may be developed in the future. Tables 1 and 2 present reserve-base data together with the author's estimate of recoverable reserves for both underground and surface mining.

Recoverable reserves for underground mining were estimated for this review by applying a recovery factor to the reserve base. Although the recovery of coal reserves in a given deep mine may be about 50%, there are substantial areas — between mines, under cities, in pockets too small for economical recovery, too

---

[2] This estimate includes coals to a depth of about 3000 feet. Additional coals are known to occur down to about 6000 feet, but these are not considered further in this review because prospects for their extraction are regarded as speculative.

badly faulted, etc—where mining is either impossible or otherwise precluded. The effect of these and other conditions is to lower the amount of recoverable reserves. Because of the influence of the above factors, it is assumed in this analysis that coal recovery by underground mining from the reserve base will be, at a maximum, only about 25% of the total coal in place. Clearly, this recoverability factor is uncertain; it could go as low as about 20% or as high as about 35% of the total coal in place. It is considered unlikely, however, according to the US Bureau of Mines, that recoverability from the reserve base would be as high as 50%. Table 1 shows that the estimated amount of coal recoverable through underground mining is nearly 75 billion tons, with about two thirds of the total comprised of eastern and midwestern bituminous coal, about one third subbituminous coal, and minor amounts of anthracite. This estimate, recognized as conservative, could be increased slightly by applying a different recoverability factor. In view of the several uncertainties in estimating coal reserves, it is believed prudent to be conservative at present so that developments of finite resources of coal can be carefully planned and evaluated in advance.

For surface minable reserves, the estimates of the US Bureau of Mines (2) have been adopted (Table 2). These data show that the estimated recoverable coal through surface mining is nearly 45 billion tons. More than half of the total is subbituminous coal, nearly one-third bituminous coal, and the remainder lignite.

The total estimate of recoverable coal resources indicated by the above data is 120 billion tons, or about 8% of the total US coal resources. Clearly, the quantity of recoverable reserves is still substantial. However, as is discussed next, factors determining coal quality limit the potential use of recoverable coal reserves.

Coal Resources and Reserves: Quality

The reserve estimates given in Table 1 are based on the simple addition of coal tonnages, without regard to heating value. To an electric utility, the heating value of coal (expressed as the number of Btu's per pound or per ton) is the important parameter, not merely tonnage alone. In generating a given amount of heat for production of electricity, a utility would have to use significantly larger amounts of low-Btu coal than of high-Btu coal. Unfortunately, during consumption of the additional tonnage of low-Btu coal to make up the Btu difference, the sulfur content of the additional tonnage is emitted, and consequently the effective sulfur content of the coals is increased. Employing a standardized heating value for coals used by electric utilities, Rieber (3) found that "conventional estimates of both known resources and recoverable reserves of low sulfur coal are grossly overstated." In particular, Rieber's analysis shows a small increase in the estimates of bituminous coal resources/reserves and a large reduction in the estimates of subbituminous coal and lignite. A significant portion of US coal resources and reserves, which are normally considered to be low in sulfur content, are reclassified to higher sulfur categories. For example, known recoverable reserves in the lowest sulfur category (≤ 0.7 wt. % sulfur) are reduced from a conventional estimate of 68.2 billion tons to 16.4 billion tons on a consistent Btu-sulfur-adjusted basis (Table 3). Should coal demand be such as to require a doubling over present

Table 1 Estimated deep minable coal resources and reserves (million short tons)[a,b]

Region/State	Anthracite Reserve Base[c]	Deep Minable Reserves[d]	Bituminous Reserve Base	Deep Minable Reserves	Subbituminous Reserve Base	Deep Minable Reserves	Lignite Reserve Base	Deep Minable Reserves
Appalachian								
Alabama			1,798	450				
Georgia			1	*[e]				
Kentucky, eastern			9,467	2,367				
Maryland			902	226				
North Carolina			31	8				
Ohio			17,423	4,356				
Pennsylvania	7,030	1,758	22,789	5,697				
Tennessee			667	167				
Virginia	138	35	2,833	708				
West Virginia			34,378	8,595				
Total	7,168	1,793	90,289	22,574	0	0	0	0
Midwest								
Arkansas	96	24	306	77				
Illinois			53,442	13,361				
Indiana			8,949	2,237				
Iowa			2,885	721				
Kentucky, western			8,720	2,180				
Michigan			118	30				
Missouri			6,074	1,519				
Oklahoma			860	215				
Total	96	24	81,354	20,340	0	0	0	0

West						
Alaska	28				4,246	1,062
Colorado		7	9,227	2,307	4,745	1,186
Montana			1,384	346	63,781	15,945
New Mexico	2	*e	1,527	382	607	152
Oregon					1	*e
Utah			3,780	945		
Washington			251	63	1,185	296
Wyoming			4,524	1,131	23,030	5,758
Total	30	7	20,693	5,174	97,595	24,399
Grand Total	7,294	1,824	192,336	48,088	97,595	24,399

[a] The maximum depth for all ranks of coal except lignite is 1000 feet. Lignite beds that can be mined by underground methods are not included. Seams greater than the following thickness are included: bituminous coal and anthracite, 28 inches or more; subbituminous coal, 60 inches or more.
[b] These data include 100% of the coal in place; no recoverability factor has been included.
[c] Source: US Bureau of Mines (1a).
[d] Source: Electric Power Research Institute, data from the US Bureau of Mines and the US Geological Survey.
[e] * = negligible.

Table 2 Estimated surface minable coal resources and reserves (million short tons)[a,b,c]

Region/State	Anthracite Reserve Base[d]	Anthracite Strippable Reserves[e]	Bituminous Reserve Base	Bituminous Strippable Reserves	Subbituminous Reserve Base	Subbituminous Strippable Reserves	Lignite Reserve Base	Lignite Strippable Reserves
Appalachian								
Alabama			157	134			1,027	300[g]
Kentucky, eastern			3,450	781				
Maryland			146	21				
North Carolina			*f	*f				
Ohio			3,654	1,033				
Pennsylvania	90		1,091	752				
Tennessee			320	74				
Virginia			679	256				
West Virginia			5,212	2,118				
Total	90	0	14,709	5,169	0	0	1,027	300
Midwest								
Arkansas			231	149			32	25
Illinois			12,223	3,247				
Indiana			1,674	1,096				
Kansas			1,388	375				
Kentucky, western			3,904	977				
Michigan			1	1				
Missouri			3,414	1,160				
Oklahoma			434	111				
Texas							3,272	1,309
Total	0	0	23,269	7,116	0	0	3,304	1,334

COAL: ENERGY KEYSTONE 45

West								
Alaska			1,201		5,902	3,926	296	100[g]
Arizona				480	350	275[g]		
Colorado			870	500				
Montana			250	*[f]	35,431	3,400	7,131	3,497
New Mexico					2,008	1,500[g]	16,003	2,079
North Dakota			*[f]	*[f]	*[f]			
Oregon								*[f]
South Dakota			262	150	500	135	428	160
Utah							8	*[f]
Washington					23,674	13,971		
Wyoming								
Total	0	0	2,583	1,130	67,865	23,207	23,866	5,836
Grand Total	90	0	40,561	13,415	67,865	23,207	28,197	7,470

[a] The maximum depth for all ranks of coal except lignite is 1000 feet. Only the lignite beds that can be mined by surface methods are included; depths are less than about 120 feet. Seams greater than the following thickness are included: bituminous coal and anthracite, 28 inches or more; subbituminous coal and lignite, 60 inches or more.
[b] These data include 100% of the coal in place; no recoverability factor has been included.
[c] Source: US Bureau of Mines (2).
[d] Source: US Bureau of Mines (1a).
[e] Source: Electric Power Research Institute, data from the US Bureau of Mines and the US Geological Survey.
[f] * = negligible.
[g] Estimated amount.

Table 3 Summary comparison of estimated coal resources and reserves ($\leq 0.7\%$ sulfur) January 1, 1965 (million short tons)[a]

Coal Province	Conventional Estimates (Tonnage Only)		Standardized Estimates (Normalized for Btu/Sulfur)	
	Resources	Reserves	Resources	Reserves
Appalachian				
Bituminous	37,320	4,105	44,784	4,926
Anthracite	12,550	1,630	14,056	1,826
Interior				
Bituminous	445	70	472	74
Rockies				
Bituminous	45,215	4,585	48,581	4,925
Subbituminous	181,670	19,470	46,329	4,632
Lignite	344,620	37,905	0	0
West Coast				
Bituminous	900	80	855	76
Subbituminous	3,780	340	0	0
Total	626,500	68,185	155,077	16,459

[a] Source: Rieber (3).

levels by 1985, the cumulative production would be on the order of 20 billion tons. With conventional reserve estimates close to 50 billion tons of coal would remain, even after such expanded production. However, under the revised, standardized reserve estimates, known recoverable reserves of lowest-sulfur coal would be completely exhausted before 1985, and there would even be a deficit!

The reduction in conventional reserve estimates is concentrated in the western states (Table 3), where nearly 85% of these reserves are shifted to higher sulfur categories through the standardization procedures. The reserves most affected are the coals of lower heating value (subbituminous and lignite), which are present in the largest tonnages.

These data are of exceptional significance for long-range policy planning. There seems to be no practical alternative to use of bituminous coals of high heating value and high sulfur content, simply because of the scarcity of low-sulfur coal of equivalent heating value. Revision of air quality standards would be the most immediate way to achieve continued use of certain high-sulfur coals and expanded use of others, with or without flue gas desulfurization devices. Alternatively, high-sulfur coals could be gasified or liquefied to produce "clean fuels"; it remains to be seen whether such substitute fuels would be restricted to markets other than electric utilities because of the inherent properties of the converted fuels and shortages of natural fluid hydrocarbons. For example, the requirements of the petrochemical industry may necessitate dedication of gaseous fuels (both natural and from coal conversion) to that use rather than to meet electric utility needs.

In view of the higher heating value of eastern/midwestern reserves, it is reasonable to conclude that the main coal supply for most electric utility needs will come from these sources rather than from those of the West. While the western resources may be developed to a heretofore unprecedented degree, the energy from these sources probably will serve indigenous western needs. The low heating value of western coals, high cost of transportation, and lack of available water in the West could act to constrain the pace and degree of development of those resources.

The different coal characteristics and geological conditions occurring in eastern and western coalfields exert major influences upon the manner of their development. In the ideal case, mining methods should be tailored to the characteristics of the deposits to obtain optimum efficiency of operation and resource recovery. In large measure, however, the evolution of mining technology has been to emphasize the development of eastern coal deposits and then to attempt to extend such technology to the coalfields of the West. It is not unreasonable to suggest that there may be a limit to which eastern-oriented technology (surface or deep mining) can be modified for application to the different conditions of the West. Whether such a limit exists, and, if so, what consequences for resource recovery or costs (or both) may obtain, should be determined at an early date. One result of such an analytical process may be to identify needs for research and development to better work the western coal deposits.

EXTRACTION AND PREPARATION

Historically, coal extraction technology evolved from a sequence of trial-and-error activites. Successful practices were often refined, mechanized, and institutionalized by reserve owners, mine operators, suppliers of equipment, and regulatory agencies.[3] These practices resulted in productivity achievable at modern American coal mines that is unrivaled by other nations. Unfortunately, the fact that US coal mine productivity is reported by federal agencies in terms of averages tends to mask the true accomplishments of the coal production industry in advancing mining technology. The actual situation at progressive modern coal mines is revealed by individual and state reports that provide the requisite data for analysis. For example, a recent survey by Straton (5) showed that the productivity at many large deep mines in the East and Midwest was substantially greater than the national average (even after allowance for decreases in productivity attributed to the provisions of the Federal Coal Mine Health and Safety Act of 1969).

The principal technical factors influencing future coal mine productivity and reserve recovery that pertain to deep mining, surface mining, and coal preparation are described below, together with constraints on improvements in coal production.

DEEP MINING   Coal is recovered through deep mining using conventional, continuous, and longwall mining methods. The conventional mining method employs a set of specialized equipment that performs specific tasks in an established sequence. Although capable of production of significant tonnages of relatively clean

---

[3] See, for example, Hedley et al (4).

coal, the sequential nature of conventional mining results in a nominal productivity less than that achievable by other methods.

The continuous mining method differs from the conventional method in that a single machine is employed mechanically to remove coal from the working face; separate steps of drilling, undercutting, blasting, and loading are not required. However, the continuous mining method usually results in more impurities in the coal because the mining machine is not selective of roof or floor material; this "refuse" must be removed from the coal through processing at the surface. Continuous miners are capable of high productivity, but result in significant amounts of dust and fine coal particles that represent a health hazard to unprotected workers. Efforts to control such dust have resulted in reduced mining rates in deep mines using this method.

The longwall mining method, as the name suggests, works a long panel of coal, often from 200 to 500 feet across and up to 1500 feet long. Coal is removed from this panel by a shearer or plow that is moved across the coal face. The longwall method is capable of the highest productivity known in deep mining, but it is limited to coal reserves where physical conditions are favorable. Also, longwall mining may generate large volumes of dust, especially when working in friable coal seams.

In spite of the recent developments in deep mining technology, a number of problems remain. Many mines still do not employ the technology capable of the highest productivity or resource recovery, largely because of capital and labor constraints, and because of the institutional lag time from the introduction of new equipment to its adoption for actual operation. Even for many mines that have the latest mining equipment, the operations are such that full productive capacity is rarely realized. For example, a continuous mining machine commonly cannot operate at its capacity because of a lack of continuous coal haulage equipment and the need to wait for roof bolting to be accomplished to protect workers against falls.

These technical problems require attention if the nation's deep minable reserves are to be developed fully. The problems discussed refer mainly to the bituminous coal reserves that are being worked by deep mining methods. The coal seams are relatively thin (5–7 feet thick). The deep subbituminous coal and lignite reserves of the plains and mountain states are much thicker (seams 10 feet thick and thicker), and it appears that an entirely new deep mining technology must be developed for effective recovery of these reserves.

SURFACE MINING   Surface coal mining methods have advanced greatly in recent years, primarily as a result of the introduction of giant excavation and haulage equipment. This large-scale equipment has extended the range of surface mining and allowed the recovery of seams once thought to be unminable. The increased capacity of surface mining equipment has occurred in area strip mining, contour strip mining, and auger mining. Most surface mining production is from area strip mining, which is practiced on relatively flat terrain and in which the largest equipment is employed. Contour strip mining is carried out in mountainous terrain, as is auger mining; these methods account for relatively less tonnage.

Although surface mining methods are inherently less hazardous to workers than deep mining, they cause disturbances to the surface environment that are matters of intense public concern. These disturbances include erosion, acid drainage, and scarred landscapes, which have frequently defied attempts at rehabilitation of mined lands for some other purpose. Public attitudes toward the effects of past surface mining practices may lead to prohibition of surface coal mining in certain areas in spite of encouraging progress in some recent reclamation efforts. The consequence of possible prohibition of surface mining under certain conditions would be to reduce the remaining recoverable coal reserves. Detailed studies would be required to define with precision the magnitude of reserve reduction by prospective restrictions upon mining (by either deep or surface methods).

PREPARATION/REFUSE DISPOSAL At one time, practically all bituminous coal was placed on the market as run-of-mine (6). Now, more than half the bituminous coal is processed and cleaned or washed to improve its quality, by removing waste material and other impurities introduced in mining, and to reduce the sulfur and ash content. "It must be borne in mind that washing reduces the tonnage that goes on the market." (7) As for most minerals, coal production is traditionally reported by amount of salable material shipped from the mines. At present, therefore, production figures commonly reflect the smaller amount of cleaned coal rather than the total material mined.

Coal preparation practices are directly related to mining methods. Continuous miners are highly efficient machines capable of excavating large volumes of material in a relatively short time. However, they do not discriminate between coal, slate, or roof and bottom materials, and even slight variances in coal seam thickness and character (or operator action) can lead to removal of significant amounts of waste materials. These waste materials are brought to the surface with coal, but must be removed in the preparation or cleaning process. Thus, although a continuous miner can mine at a rapid rate, its output normally requires additional treatment before a salable product is reached. The advantage of the continuous miner is that such cleaning and treatment can be done at the surface, and on a high volume of material. Large amounts of refuse must also be disposed of at the surface. Both the mining equipment and the cleaning facilities are complex and costly, and review of the *Keystone Coal Industry Manual* (8) suggests that such methods are limited, for practical purposes, to larger companies that are in a better financial position to undertake the investment required for such high-capacity operations.

To clean the coals, a variety of mechanical processes are employed that rely upon the difference in density to separate coal from other rocks and minerals. Cleaning methods are not completely efficient, and roughly 15–20% of raw coal may be lost. This lost coal and waste material is termed "refuse" and is discarded near coal cleaning plants, which presents both physical hazards to local inhabitants and adverse environmental impacts to the area. Refuse is comprised of rocks, minerals, and coal. Its heating value varies, but commonly is about 3000–5000 Btu/lb. If a way could be found to use this material, this would be the fastest, most direct way to increase coal production and also to mitigate negative social and physical effects.

The 30-year period since 1940 shows a steady, nearly tenfold increase in refuse

(Figure 2). The amount of uncleaned coal was substantially reduced in the mid-to-late 1950s and early 1960s (in other words, most coal was cleaned). This was a time of relatively low production, and it is likely that cleaning was employed to present the highest quality product to remaining consumers. The amount of uncleaned coal increased again in the late 1960s and continues to increase in the 1970s. This may be because of the fact that cleaning causes coal losses and diminishes the total available to be sold. In a period of heavy demand for coal, certain consumers such as electric utilities might have been willing to accept uncleaned coal to ensure their supply. Also, in time of increasing coal prices, they may have found that the added cost for cleaning coal was more than they could afford to pay.

AN INITIAL ASSESSMENT OF THE FUTURE OF COAL MINING   This review of the state of the art of mining technology would be incomplete without an assessment of its possible future. The present underground production is nearly evenly divided between conventional and continuous mining, with a relatively small amount contributed by longwall methods. Roughly 20 years passed from the introduction of continuous mining until it achieved about half of the total production of deep-mined coal. If this pattern is repeated by longwall technology, then it will be early in the 1990s before longwall deep mining accounts for a comparable part of coal production. While the pace of introduction of longwall technology could be accelerated through government action[4] or incentives, it does not appear likely that longwall mining will account for substantial production in this country in the immediate future.

The lag in introduction of new technology means that, for all practical purposes, the nation must rely upon existing methods of deep coal mining for the remainder of the present century. To be sure, these existing methods—conventional, continuous, and longwall—can be improved upon, refined, and even automated to some degree to provide increased production efficiency as well as improved health and safety. Still, even these improvements would be just that, and the basic mining methods would remain essentially as known at present.

In terms of R&D on coal mining, this assessment suggests that it will be concerned with incremental improvements in present systems to realize benefits in productivity and health/safety factors. This will be especially true for R&D sponsored by the coal producers and, similarly, that undertaken by agencies of the federal government. Such an emphasis is logical in terms of the economic, political, and social interests of the research-sponsoring organizations. Yet, substantial areas for innovation in deep mining R&D will remain to be pursued by other organizations; in particular, work on novel methods to work thin eastern deposits, as well as work on perfecting a technology for recovery of thick western seams. Overall, this is a somewhat pessimistic conclusion, but it is consistent with patterns of recent experience and with the inherent uncertainties characteristic of coal extraction by underground mining methods.

[4] It was noted by Mr. John Corcoran of Consolidation Coal Company that revision of existing regulations against subsidence and for leaving barrier pillars around oil and gas wells would be required prior to the advent of a real increase in longwall mining (9).

Figure 2 shows that the proportion of surface mining to the total national production increased steadily over the last 30 years, until it now accounts for roughly half of the total. During this time, the capacities of surface mining equipment increased steadily, culminating in giant overburden excavators and coal haulage units. This equipment was developed through the time-honored coal mining approach of trial, error, and refinement. As a result, a strong technological base has been established.

The capacities of existing surface mining equipment thus cover a wide range and appear capable of conforming to most prospective mining situations. We suggest that the giant equipment developed in the 1960s represents a practical upper limit to capacity that will probably be duplicated rarely, if at all, in the future. The reason for this has been stated concisely as follows (10):

> While (giant) equipment is physically capable of moving deeper and deeper overburden, the performance increase is not always proportional to the investment cost. For example, a 100 cubic yard shovel operating at a depth of 92 feet can handle somewhat

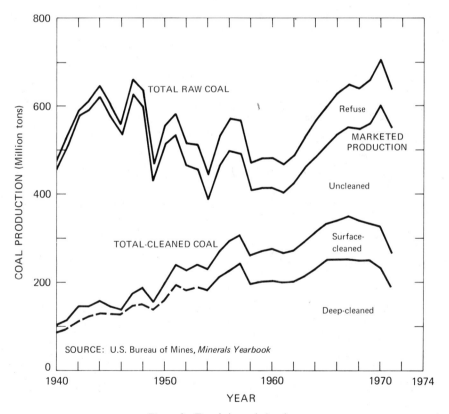

Figure 2 Trends in coal cleaning.

more than twice the volume of a 45 cubic yard shovel operating in 52 feet of overburden. But operating under these conditions for the same thickness of coal, the net increase in coal availability would be less than 20 percent.

An additional, nontrivial concern is that a giant machine represents most productive capacity at a given property; failure of the machine could curtail or, at best, seriously impair production.

It seems likely that future surface mining for the remainder of the century will employ the array of equipment types in existence at present. Surely, improvement in equipment can be expected to increase efficiency and operation, but it seems reasonable to conclude that the basic equipment types and classes have been perfected through the developments of recent years, and that the resulting technology will be employed in prospective future operations.

In terms of R&D on surface mining methods, there appears to be little real need for much effort in developing giant excavating equipment. Rather, effort appears to be required in perfecting the manner in which such equipment is employed, to minimize adverse environmental impact and optimize coal recovery.

TRANSPORTATION AND DISTRIBUTION

Few electric utility plants are located adjacent to coal mines, and coal transportation facilities are an essential part of the overall system of coal production and consumption. Actually, a mutual dependency exists, because coal represents a principal source of revenue for the transportation operators; for example, coal comprised about one quarter of the revenue freight handled by Class I railroads in 1973 and nearly one quarter of the total handled by inland waterways.

Several modes of transportation are employed in delivery of coal from mines to electric utility plants (Table 4). As the table shows, roughly half of the coal for use by electric utilities is transported by rail, with another quarter of the total transported by barge, and the remainder by truck or other means. There are several useful recent reviews of the basic technology of coal transportation and distribution,[5] and this paper does not attempt to duplicate those works. Instead, a brief discussion of some of the key problem areas in the principal transportation modes is presented, with emphasis on the impacts these problem areas have on the electric utility industry.

Rail

The railroad industry of the United States has always been an essential link between coal producers and consumers. At present, however, the industry faces serious problems that may constrain its ability to ensure delivery of coal supplies for electric utility use, among which are the following:

1. *Roadbeds* on many vital rail systems are in such poor condition that, if usable at all, trains must travel under "slow rules," increasing transport time and costs, and leading to inefficiencies in equipment distribution. This latter point is critical with respect to coal shipment, because many mines lack storage facilities for their coal

[5] See, for example, (11, 12).

Table 4 Transportation of coal for electric utility use, 1972 (million short tons)[a]

Method	Tonnage
Rail	195.0
River barge	76.5
Great Lakes barge	17.6
Tidewater barge	2.4
Truck	48.4
Tramway, conveyor, private railroad	42.3
Total	382.2

[a] Source: US Bureau of Mines (7).

production. Unless such mines have rail cars available for loading, it may be impractical or impossible for them to operate at all.

Roadbed conditions, and costs for improving them, have been cited as justification for abandonment of certain track especially in the northeastern part of the nation. This presents a potential problem of access to either mining areas or consumers. Careful assessment of this rationale should be made both from the larger viewpoint of access and from the standpoint of increased efficiency of rail operations.

2. *Hopper cars* are in short supply; in 1974 there were only about 354,000 hopper cars owned by leading coal carriers, down 40,000 from the total in 1970. Only about 334,000 of these cars were listed as serviceable (13). Clearly, with the smaller number of cars, any delays in transit or in turnaround could adversely impact operations.

In order to meet future requirements, the National Academy of Engineering estimated that 150,000 new hopper cars of 100-ton capacity would be required. However, the number of new and rebuilt hopper cars installed by US railroads in 1973 was only about 2600, the lowest amount in years.

3. *Locomotives* appear not to be as serious a problem as hopper cars, for the numbers of locomotives have remained relatively constant. However, many locomotives in service today are relatively old and, to some extent, underpowered. There will be a continuing need to retire older units and replace them with new locomotives.

4. *Institutional factors* such as work rules and other regulatory practices exert a strong influence on the rail transport of coal for electric utility use. In a practical sense, these nontechnical factors exert effective control over the manner of coal movement. Although it will be important to search for technological improvements in rail transportation, it is no exaggeration to conclude that without attention to these institutional factors, benefits from advanced technology may go unrealized.

Barge

Waterborne shipment of coal is the least expensive way of transportation when origin and destination points permit. However, barge transportation is not without its problem areas:

1. *Cargo handling* or transfer from one mode of transportation to another represents an important bottleneck that delays shipments and markedly increases costs. Improved technology for efficient cargo handling and transfer, installed at a few localities, needs to be employed more widely to eliminate this problem.

2. *Capacity* of main waterways could pose a serious constraint to increased use of barge transportation. This may require widening and deepening of existing channels and even the construction of new facilities for navigation and water control. Clearly, such efforts would be both time consuming and costly, although they may be necessary to satisfy future transportation requirements.

Other Transportation Methods

Other transportation methods employed to move coal from mine to consumer are mainly short-haul methods, including trucks, trains, conveyors, etc. These methods will probably continue to play a relatively small but locally crucial role in future coal transport.

Increasing attention is being given to the possible application of slurry pipelines for coal transport; one such pipeline is in operation from Black Mesa in Arizona. In the last four years, there has been a fourfold increase in the capacity of commercial slurry pipeline systems.

The principal constraint upon development of slurry pipelines is the availability of sufficient water to meet the pipeline's needs. There is serious concern in several western states that slurry pipelines would deplete available water supplies to the detriment of established agriculture, communities, or industry. Although the concept of the slurry pipeline has many attractive features, particularly in the West where it would supplement existing rail systems for coal transportation, this fundamental problem—availability of necessary water—could serve to constrain its application in the West. In the East, where there is abundant water, there already exist well-developed rail and barge systems serving electric utility needs, and it remains to be seen whether a slurry pipeline system could be economically developed in this area in the near future.

It is reasonable to conclude that, despite the array of problems facing the coal transportation sector, at least for the immediate future (and probably for the remainder of the present century), the mix of transportation modes will be essentially as it is at present. Rail will predominate, followed by barge and truck, with minor volumes handled by other modes such as slurry pipelines. There seems to be no alternative to rejuvenation of the nation's railroads and waterways (both technically and institutionally) to ensure reliable delivery of essential coal supplies to meet electric utility requirements.

EXTERNAL EFFECTS OF COAL PRODUCTION

Coal development and distribution operations are responsible for external effects upon the human, social, and physical environments, and these effects are of intense public concern. The external effects of coal production may be categorized broadly as worker-related effects and environment-related effects. Worker-related effects include those factors, conditions, or events that influence the health, safety, or well-being of workers as they pursue their occupation; these effects are principally associated with underground mining. Environmental effects are those changes to the physical environment that are caused by coal production activities; they are mainly attributed to surface mining. Clearly, any industrial activity can produce conditions in which risks are presented to workers or to environmental quality. The problem in coal production is to recognize the nature of risks and to formulate positive approaches to deal with them.

Deep mining results in impacts on the physical environment that are matters of public concern. These include (among others) acid mine drainage, subsidence, mine fires, and disposal of wastes from coal preparation or treatment. Important as these impacts may be, they pale in comparison to the human and social effects of deep mining. All mining operations are risky because they expose the workers to hazards stemming from the imperfectly understood processes of nature that formed the deposits. However, deep mining for coal is the most hazardous occupation in the nation. In 1970, large deep mines in the United States had about 0.5 fatalities per million tons of production. Small deep mines were four times more hazardous, having nearly two fatalities per million tons. Deep mining was similarly hazardous in terms of nonfatal injuries, with about 75 nonfatal injuries per million tons of production. Although the industry's safety record has increased significantly in the last two years, it still is the most hazardous occupation known.

Efforts to improve coal mine safety are underway in response to the provisions of the Federal Coal Mine Health and Safety Act of 1969. Strong health and safety standards are being enforced by both federal and state agencies. Safety training for all miners and supervisors is being improved and expanded. It was shown by Theodore Barry and Associates (14) that experience with a particular mining task is a critical factor influencing accidents (the less experience with a task, the greater the risk of injury). Efforts are underway to correct this situation by providing more intensive safety training and instruction.

Surface mining is inherently less hazardous than deep mining to the workers employed in operations. The hazards are not comparable in intensity to those encountered in deep mining, such as roof falls, methane release, and coal dust explosions. Although surface mining does require ordinary prudence to avoid injuries and even fatalities, the record of the recent past indicates that the numbers and rates of injuries or fatalities are but a fraction of those related to deep mining. Manageable as the industrial safety aspects of surface mining may be, the impacts of surface mining operations upon the physical environment have often been severe. As a result, the environmental effects of surface coal mining have become matters of intense public concern, both in areas of the East where such operations have been in progress for many years and in areas of the West where new operations are planned for the future.

Techniques employed for rehabilitation of surface mined lands include (a) planning, (b) spoil handling and treatment, (c) revegetation, and (d) management and monitoring. Surface mining, by any criterion, is a drastic environmental change effected by man's technologies. An area disturbed by mining seldom returns to the same ecological equilibrium that existed prior to the disturbance. Rehabilitation techniques include monitoring of mining techniques, spoil reshaping, control of erosion, prevention of damage to the hydrologic system, revegetation, and careful management. The rehabilitation of a specific site will depend on the physical characteristics of the site and the post-mining land use objectives.

Predevelopment planning seeks to identify the likely effects of mining operations in advance, so that these effects may be dealt with systematically. A careful assessment of proposed exploration and mining proposals should include geologic, topographic, hydrologic, physical, and chemical characteristics of the area as well as an analysis of the wildlife and the archeological, historical, and cultural values and a listing of any special, critical, or unique features.

It is becoming clear that, in the future, any decision to surface mine for coal in a particular area must include a strong commitment to rehabilitate the land concurrently.

FACTORS CONSTRAINING COAL DEVELOPMENT

The realization that there are limitations on the rate at which the coal industry can expand has been expressed in several previous studies.[6]

In a 1971 report, the US Bureau of Mines pointed out that "... in addition to coal's quality characteristics that may restrict its use in certain markets, other limiting factors such as manpower and transportation availability, and adequate mine production and preparation and cleaning plant capacity are directly related to coal supply and availability" (17).

The principal factors constraining coal extraction may then be categorized as (a) quantity and quality of reserves; (b) mining, production, and transportation; and (c) labor and capital availability. Each of these factors is described briefly below.

Reserves

Coal reserves required to support large-capacity, modern mines are substantial. To be minable with efficient, mechanized equipment through either deep or surface mining methods, the reserves must be in a contiguous block and of sufficient magnitude to support the mining operations (and/or the plant(s) that consume the coal) for at least 20 years. For example, with a 5-foot-thick seam in the eastern coalfields, and a requirement for about 15,000 tons a day to supply a plant, the total reserve over a 20-year lifetime would be about 180 million tons; in terms of area, this would cover roughly 45 square miles.

It would be no simple task to assemble such large blocks of reserves if one had to start from scratch. This is especially true in the East where there are a large number of small reserve blocks owned by different persons. Still, even in the East there are sizable, contiguous areas of coal reserves held or controlled by coal mining companies as a result of acquisitions or leases dating back to the early part of this century or before. These reserves could possibly be targets for development to supply coal to existing or projected coal-processing units in the future. An important

[6] See, for example, (15–18).

question clouding that potential use of eastern coal reserves is the extent to which the reserves may be committed to special-purpose uses for which there is no effective substitute at present.

Large blocks of coal-bearing lands in the West are owned by the federal government. The public lands are typically interspersed with lands held by private organizations in a checkerboard pattern according to alternate sections. Although the problem for the coal developer who seeks to work western coal deposits is simplified in that he has only a few coal-owing organizations to deal with, the fact that the largest acreages are publically owned means that major questions of public policy need to be decided before rights to develop the reserves can be conveyed and operations begun. Although these are obvious targets for future development, it remains to be seen whether the questions regarding access to public coal lands in the West will be resolved in a fashion and at a time to permit orderly development to proceed.

Market conditions also exert an important influence on the manner of coal reserve development, especially for less readily mined deposits.[7]

> The vigor in which the less accessible reserves are recovered by the mining industry depends largely on the condition of the coal market at the time of mining. Hence, during a buyer's market, the commercially-oriented mining industry is compelled to mine the easier and less costly reserves. Conversely, during a seller's market, the need to rapidly expand production results in more difficult mining and higher cost coal as few obstacles are encountered in finding markets. Hence, a seller's market tends to enhance the recovery of reserves while a buyer's market does not (17).

It is also vital to examine coal quality. However, Leonard (19) noted that "in close parallel with the reserve problem is the need to develop better methods for the characterization of coal seams and associated lithotypes, based on drill core data, once an area is selected for mining." In this context, further data are required on physical as well as chemical characteristics. The emphasis on sulfur content should not overshadow the importance of other constituents that could influence coal's efficiency or cost of use. As noted by the Bureau of Mines (17), "To a great degree, increased knowledge of the inherent coal characteristics allows boiler designers to capitalize on those characteristics and to design units specifically to take advantage of heretofore undesirable coals."

Although the Bureau of Mines attempted to relate broad coal properties to types of combustion systems (17), the amounts and locations of coal reserves having these properties remain to be established. As a result, it is possible that future engineering efforts could succeed in developing an efficient process or apparatus that could not operate to its optimum design conditions because of a lack of coal having the requisite properties. Clearly, the availability of coal reserves with specific properties and characteristics needs to be determined with considerable precision in order to make most effective use of coal. The present state of knowledge about the quality of coal reserves thus represents a constraint on expansion of coal extraction in the future.

[7] See, for example, (16, 17).

Production

After coal reserves have been discovered, delineated, and evaluated, "...the next step is to select a mining method that is physically, economically, and environmentally adaptable to recovering (the coal) from the deposit....the spatial characteristics of the (coal) and surrounding rock limit the methods that can be employed to mine it" (20). With regard to coal mining methods, it is cautioned that (21)

> A popular thesis suggests that the proper approach to selecting the method of mining should be rooted in the inverse solution, whereby the limitations of various methods in use serve as criteria to eliminate most of these methods from consideration in the specific case.... Although such a negative approach unquestionably is valid in discarding obviously inapplicable mining systems from consideration, it falls short in defining the best system, which in practice quite often turns out to be a variation of a textbook standard, or a combination of two or more such standards.

Mining methods must be developed to work particular deposits. This represents a constraint to development of new technologies, because the special conditions of individual sites will be the ultimate determinant of their successful application.

Labor and Capital Availability

Prospects for expansion in coal extraction face serious constraints in the areas of manpower and capital availability.

LABOR   No matter what technology is employed in coal production, the availability of men with sufficient ability to be trained as miners and with a willingness to pursue mining occupations will be a determining factor in the level of production realized. Moreover, the total labor picture must include the workers in supporting industries who provide goods and services used by miners in their jobs. Thus, even the large numbers of new miners estimated to be required (18) are probably understated because the labor requirements of supporting industries are not included. Important as the need for new miners and new personnel may be, "Even more crucial is the shortage of professionally-trained people" (16). Mining engineers and geologists are particularly in short supply, and positive actions will be required to alleviate potentially adverse effects on production from shortages in technical personnel.

Technology can be employed in dealing with labor problems, at least up to a certain point. Although mechanized mining equipment is highly complex, requiring very skilled personnel for its maintenance and repair, it is nevertheless true that its operation can be performed by relatively unskilled labor. "Based on actual experience, another main advantage (of longwall mining) is that men with little or no mining experience can be trained much more quickly than is possible with the other types of coal mining. If the current shortage of skilled miners continues, this can be vitally important" (22). This suggests that the skill of miners required for operations is inversely proportional to the degree of mechanization represented by mining technology. Longwall mining is highly mechanized, a fact that permits the use of relatively unskilled labor (or labor that can be trained to perform relatively simple tasks). Conventional mining, at the other extreme, is less highly mechanized and demands highly skilled and experienced workers to perform the production operations. Thus, the manpower problem facing the coal industry is critically related to the technologies employed in operations. As such, the problem may be tractable in part, through R&D and technological improvements; still, even R&D can be a constraint as discussed in the following section.

CAPITAL   Expansion of coal production capacity will require large capital investments. This situation "... has resulted in an extremely serious problem with respect to securing adequate financing, whether for the expansion of existing facilities or the opening of new mines.... It is important that the coal industry recognize that it is going to be competing in the financial markets with every other borrower...." (23). This point was also made by E. B. Leisenring, Jr., of Westmoreland Coal Company,[8] who pointed out that assumption of risks by utilities would make bankers more willing to put up equity loans for mine construction. However, J. L. Williams, Jr., of the Tennessee Valley Authority, maintained that while the coal purchaser should share some risks with the producer, coal companies are making unreasonable demands (23):[9]

> Escalation provisions, designed to cover every possible facet of increased cost, are being insisted upon.... In some instances, *suppliers want clauses designed to guarantee that return on investment will never go below 20%*. [Italics added.] Some suppliers want to renegotiate prices periodically, in order to keep the contract in line with current market prices, regardless of whether the cost of producing the coal has changed.

It is interesting that coal producers would seek to establish a floor of 20% return on investment, as indicated above; this suggests that their present rate of return is, in fact, greater than 20%. Clearly, this conclusion is at variance with the conventional wisdom about the relative profitability of coal-producing companies. Nevertheless, it is consistent with a recent analysis of profits in mining industries (24), which found that the return on shareholders' investments for a limited sample of coal-producing companies was about 30%, or three times that of the typical producer of other mineral commodities.

Research and Development

Research and development activities are addressed to solving technical or economic problems and are not commonly regarded as a constraint. Yet, Schweitzer (25) found that, with respect to coal production, "... industry representatives felt that near-term problems might be obscured by the great emphasis being placed on research and development...." Therefore, R&D could itself be a constraint to growth in coal extraction, mainly because "the prospects for new technologies defy conclusive analysis; research is risky because we cannot be sure what it will produce" (26). As noted above, coal extraction as an enterprise is replete with uncertainties.

[8] Cited by Phillips (23).
[9] See footnote 8.

With all these risk-producing uncertainties, one could logically ask why coal developers should add a further risk through support of R&D when the outcome of that work would be in itself uncertain.

In an attempt to minimize their risk, coal producers have relied upon evolutionary growth in mining technology, mainly through trial and error. New developments by equipment manufacturers that were proved successful through tests and performance demonstrations under operating conditions have been gradually adopted by the coal producers through further trials and errors. Engineering work has resulted in advancements in the state of mining technology, but it is not accurate to describe such work as R&D. A possible explanation for the coal producer's limited involvement in R&D is that such activities represent a compounding of uncertainties and an introduction of a completely avoidable risk.

Under normal circumstances of moderate coal demand, such risks would remain beyond the considerations of coal producers who would be, logically, preoccupied with the array of problems and natural uncertainties associated with coal production. However, the situation is completely changed under extraordinary circumstances where society in general and coal consumers in particular are preoccupied with the achievement and maintenance of a secure domestic supply of coal to meet present or projected energy requirements. Under such conditions of strong societal demand, producers, consumers, and the public may be willing to support R&D leading to greater assurance of supply or moderation of costs. The guidance of coal producers would be sought for development of R&D programs to make them most relevant to actual coal production problems.[10] The coal producers, who are quite knowledgeable about the technical problems they face, are principal sources of information about the R&D needs of their industry, and their expertise is essential to formulation of a national program.

Adoption of new technologies by the coal industry must be carried out with great care and over an extended period of time to avoid severe economic and structural dislocations. Large coal mines have been increasingly developed in recent years and comprise more than half of total production today. Every indication is that large mines will represent an even larger share of total production in the future. These mines are established only under long-term contracts that cover supplies to particular consumers for 20–30 years. Thus, coal producers have invested in existing coal-extraction technologies for their existing and new mines, and they have made firm plans to employ these technologies at least for the duration of their consumer's requirements, in many cases throughout the remainder of the present century.

In short, the coal industry's structure and practice appears to limit its ability to implement major technological change, unless the new technology is demonstrated and available at the inception of a given operation. Although it is possible to estimate schedules for performance of R&D,[11] it is probably hazardous to plan

[10] See, for example, (27).

[11] See, for example, (28). It is probably unrealistic to plan, as did the US Environmental Protection Agency, for "major decisions on total coal production increases" in mid-1980. Those decisions have probably already been made. The only question to be decided is the timing of such developments and the implications of timing on possible technologies.

investments that anticipate successful results of such work at any particular time. As a result, the coal industry is forced to rely on existing technology (or incremental optimization changes to that technology) in planning for future operations. The consequence of this situation is to limit the flexibility of the coal producers to benefit from some R&D efforts. The extent and manner of this restriction should be assessed in comparison to other industries; it would appear that the coal in industry is somewhat unique with respect to its adoption of new technologies, and this requires further investigation.

Coal consumers, moreover, are faced with an equally difficult situation. They bear the risk that the R&D expenditures undertaken on their behalf could prove imperfectly successful, and that despite expenditure of substantial sums, the condition of their coal supply may change but little, either because of failure in the R&D itself or in the limited application of the R&D results by the coal-producing organizations and supporting industries. It is imperative, therefore, that coal-consuming industries have a thorough understanding of the problems in coal production that may be addressed through R&D, as well as the prospects for realization of real benefits in comparison with anticipated costs. This is true for R&D funded by public as well as private entities.

An Electric Utility Perspective on Coal Production and Related R&D

In the past, as noted by Christensen (29), "For the electric plant...there are definite advantages in relying upon the large (coal) company, providing, of course, that the coal is of satisfactory rank and can be had at low transport costs." Nevertheless, Christensen points out that the utilities "...may find ways of using shipments from many mines if the incentives are sufficient."

The fact that these remarks were made in a time of abundant coal supply is significant. In a buyer's market, the steady demand of the electric utilities would "...make them especially attractive customers to coal companies, since it contributes to the possibility of continuing operation of mines and, in turn, to the lowering of mining costs. For this reason, (the utilities) possess a strategic advantage in bargaining in coal markets" (29).

The situation changes in a seller's market, which exists at present. Continuous coal usage by the electric utilities, to justify special high-volume coal-burning equipment, could be a problem when coal for such facilities is in short supply at reasonable costs. In this case, the utility, which requires reliable supplies of coal to continue operations, is at a strategic disadvantage to the coal suppliers. In some recent circumstances, this has led to considerable independence of coal producers, and relative indifference of certain coal producers to the plight of certain utility customers seeking supply even at inflated prices.

SUMMARY

A brief review of critical factors influencing coal development for electric utility use was presented.

The quantity and quality of US coal reserves available for development was assessed. It was concluded that only about 8% of the total resources determined by mapping and exploration are recoverable. Further, the heating value and sulfur content of reserves further limit developments, particularly for the lower-heating-value coals of the western states.

Geological factors have influenced the development of both deep and surface mining technologies. Development methods and practices are characterized by trial and error, which leads to mining technologies that are capable of high productivity and resource recovery.

At least for the remainder of this century, coal development methods will be limited to the basic technologies that exist at present, although incremental improvements may be achieved through advanced R&D. Transportation and distribution of coal will continue to be largely by rail, with barge shipments employed where possible and where intermodal transport problems can be overcome.

External effects of coal production will continue to receive considerable attention. Mining operations, in particular, pose risk to man or his environment. These risks are presently concentrated, respectively, in deep mining and surface mining. The thrust of future advanced R&D on these effects will probably be toward sufficient controls over external effects of principal concern at present.

A complex array of interrelated problems faces the electric utilities with regard to coal production. These problems are in realization of fuel supplies of adequate quality and quantity at reasonable prices. A number of factors constrain progress toward resolving these problems; among these constraints are quality and quantity of reserves, development technology, availability of labor and capital, and research and development. Unless properly conceived and conducted, R&D could prove to be a constraint upon progress, because the outcome of R&D efforts is uncertain and represents another risk in an already uncertain enterprise. Attention to the interrelations among components of the coal production system is required to alleviate the impact of this additional constraint. This will ensure effective use of coal resources as the nation's energy keystone in meeting future energy requirements.

Literature Cited

1. US Bureau of Mines. 1968. *Dictionary of Mining, Mineral, and Related Terms*, p. 1240. Washington DC: US Bur. Mines
1a. US Bureau of Mines. 1974. *Demonstrated Coal Reserve Base of the United States on January 1, 1974*. Mineral Industry Survey, June, 1974
2. US Bureau of Mines. 1971. *Strippable Reserves of Bituminous Coal and Lignite in the United States*. Inform. Circ. 8531
3. Rieber, M. 1973. *Low Sulfur Coal: A Revision of Reserve and Supply Estimates*. Univ. Ill., Center Adv. Comput. CAC Doc. No. 88
4. Hedley, D. G. F., Cochrane, T. S., Grant, F. 1971. Effectiveness of bolting in coal and salt strata, p. 30. In *Proc. Conf. Underground Mining Environ.*, Univ. Mo. (Rolla) Act. 27–29
5. Straton, J. W. 1972. *Productivity and Cost Changes 1969–1971 Resulting from PL 91-173*. Presented at Annual Meeting, Am. Inst. Mining, Metall., and Petrol. Eng., San Francisco
6. Moore, E. S. 1940. *Coal*. New York: Wiley
7. US Bureau of Mines. 1972. *Minerals Yearbook*. Washington DC: US Bur. Mines
8. *Keystone Coal Industry Manual*. 1974. New York: McGraw-Hill. 859 pp.
9. Corcoran, J. 1973. In *Mining Congr. J.*, pp. 69–70
10. Risser, H. E. 1968. *Coal Strip Mining — Is It Reaching a Peak?* Presented at

Fall Meeting, Soc. Mining Engineers, Minneapolis, Minnesota
11. *Coal Age.* 1974. Special issue on Transportation for Coal. Vol. 79, No. 7
12. National Coal Association. 1974. *Coal Traffic Annual, 1974.* 66 pp.
13. National Coal Association. 1974. *Coal Traffic Annual, 1974,* pp. 25–27
14. Theodore Barry and Associates. 1971. *Industrial Engineering Study of Hazards Associated with Underground Coal Mine Production.* Prepared for US Bur. Mines, open file report
15. Shell Oil Company. 1972. *The National Energy Outlook.* New York
16. National Petroleum Council, Coal Task Group. 1973. *Coal Availability.* 287 pp.
17. US Bureau of Mines. 1971. *Restrictions on the Uses of Coal,* Nat. Tech. Info. Serv. Pub. No. 202 168. 57 pp.
18. National Academy of Engineering. 1974. *US Energy Prospects—An Engineering Viewpoint.* Washington DC: Nat. Acad. Sci. 141 pp.
19. Leonard, J. W., Mitchell, D. R. 1968. *Coal Preparation.* New York: Am. Inst. Mining, Metall., and Petrol. Eng.
20. Morrison, R. G. K., Russell, P. L. 1973. Selecting a mining method—rock mechanics, other factors, pp. 9-1 to 9-22. In *SME Mining Engineering Handbook,* ed. A. B. Cummins, I. A. Given, Vol. 1, Sec. 9. New York: Am. Inst. Mining, Metall., Petrol. Eng.
21. Robinson, N. II. 1973. Underground mining systems and equipment, pp. 12-31 to 12-45. In *SME Handbook,* Vol. 1, Sec. 12
22. Robertson, N. 1974. *Financing Energy Needs.* Presented at 1973 Mining Conv./Environ. Show of Am. Mining Congr., Denver, Colo.
23. Phillips, J. G. June 29, 1974. Coal industry problems hamper production goals. *Nat. J. Rep.,* pp. 951–61
24. Just, E. Dec. 1973. Metal mining profits are inadequate. *Mining Congr. J.,* pp. 29–31
25. Schweitzer, S. A. 1974. *The Limits to Kentucky Coal Output: A Short-Term Analysis.* Univ. Kentucky Inst. Mining and Minerals Res., TR 81–74 IMMR2
26. Gordon, R. L. 1973. *Coal's Role in the Age of Environmental Concern.* Presented at MIT Energy Conf., Cambridge, Mass.
27. Gouse, S. W., Rubin, E. S. 1974. *A Program of Research, Development and Demonstration for Enhancing Coal Utilization to Meet National Energy Needs.* Rep. of CMU/NSF-RANN Workshop Adv. Coal Technol., Carnegie-Mellon Univ., Pittsburgh, Pa.
28. US Environmental Protection Agency. 1974. *R&D Energy Task Force Program, Standard Timelines, Manuscript*
29. Christensen, C. L. 1962. *Economic Redevelopment in Bituminous Coal.* Cambridge, Mass: Harvard Univ. Press

Copyright 1976. All rights reserved

PRODUCTION OF HIGH-BTU GAS FROM COAL

✕11003

H. R. Linden, W. W. Bodle, B. S. Lee, and K. C. Vyas
Institute of Gas Technology, Chicago, Illinois 60616

INTRODUCTION

Natural gas in recent years has supplied about one third of the total energy requirements of the United States. Until 1968, yearly additions to natural gas reserves exceeded net production, but, in that year, production exceeded additions by a substantial margin. Since then, the situation has deteriorated to the point where the Federal Power Commission now estimates that, because of inadequate supplies, contract curtailments by interstate pipelines will reach 2.4 trillion cubic feet (cf) for the 12-month period ending August 31, 1975. This corresponds to about 11% of total natural gas consumption.

One course for improving the situation is to supplement the natural gas supply by producing high-Btu gas from the extensive coal supplies available in the United States. In a recent study, the American Gas Association reported that it has identified 176 potential sites with sufficient uncommitted coal reserves and water to supply synthetic gas plants of 250 million cf/day capacity. These sites, 20% of which are east of the Mississippi River, contain a total of 42 billion tons of recoverable coal, enough to make 542 trillion cf of high-Btu gas.

For these reasons, recent activities in coal gasification have been concerned to a large extent with production of high-heating-value gas of pipeline quality. Such gas contains 90% or more of methane, is practically free of carbon monoxide and sulfur compounds, and has a heating value of over 900 Btu/standard cubic foot (scf).

Although no commercial-size coal gasification plants to produce high-Btu gas are in operation, a number of planned projects have been announced. A summary of the salient features of such plants and their status as of May 1, 1975, is shown in Table 1. At present, intensive research and development activity is under way to improve the technology, as summarized in Table 2.

FUNDAMENTAL REACTIONS FOR COAL GASIFICATION

The conversion of coal to pipeline gas is explained most readily in terms of the composition of the raw material and the product. A typical bituminous coal con-

Table 1 Announced commercial and demonstration coal gasification plants (as of May 1, 1975)

Controlling Company(s)[a]	Site	Process	Coal Feed, tons/day	Plant Output, million CF/day	Status
A. HIGH-Btu Gas PROJECTS					
El Paso Natural Gas Co.	Four Corners Area, New Mexico	Lurgi Gasification	28,250	288	El Paso Natural Gas Co. plans to construct and operate the Burnham Coal Gasification Complex at the Navajo Indian Reservation. During the week of April 20, 1975, the FPC granted the request by El Paso to defer decision on the proposal until water and coal supply contracts are resolved.
WESCO, Texas Eastern Transmission Corp., and Pacific Lighting Corp. (Utah International Corp.)	Four Corners Area, New Mexico	Lurgi Gasification With Methanation	102,500	1000 (4 plants)	The firms plan to construct and operate four plants on the Navajo Indian Reservation near Farmington, N.M. Utah International will supply the coal and water. First plant estimated project cost is $853 million. The FPC has allowed a $1.38/1000 CF gas price for the 6-month start-up period.
Panhandle Eastern Pipe Line Co. (Peabody Coal Co.)	Eastern Wyoming	Lurgi Gasification With Methanation	25,000	270	Plant operation is anticipated in the 1978-1980 period, assuming timely receipt of all required governmental authorizations.
Natural Gas Pipeline Co. of America	Dunn County, North Dakota	Lurgi Gasification With Methanation	108,500 (4 mines)	1000 (4 plants)	The company plans to build at least 4 and possibly 8 gasification plants. Coal will be mined from 110,000 acres of leases in North Dakota. The first plant is scheduled to go on-line in 1982.
American Natural Gas Co. (North American Coal Corp.)	Beulah-Hazen Area, North Dakota	Lurgi Gasification With Methanation	--	275	Initial gas production is scheduled for late 1981 based on FPC authorization in first quarter of 1976. Recent plant cost estimated at $778 million; mine cost, $126 million; gas cost, $2.50/1000 CF. Interest during construction will be provided by a surcharge on Michigan-Wisconsin gas sales.
Northern Natural Gas Co. Cities Service Gas Co. (Peabody Coal Co.)	Powder River Basin, Montana	Lurgi Gasification With Methanation	--	1000 (4 plants)	Northern Natural and Cities Service plan to construct four 250 million CF/day coal gasification plants. Peabody Coal has agreed to supply about 500 million tons of coal, and the gas companies are negotiating for another like amount. Through 1975, $10 to $11 million will be spent for preliminary development. Construction of the first plant could start in 1976-1977 with operation in 1979-1980.
Texas Gas Transmission Corp. (Consolidation Coal Co.) Commonwealth of Kentucky	Western Kentucky	--	--	80	Texas Gas Transmission and the State of Kentucky have signed an agreement to build an 80 million CF/day gasification plant expandable to 250 million CF/day. A Federal Government commitment to fund a portion of capital and operating costs is being sought and is an essential condition for construction.

Company	Location	Process	Capacity	Comments
Colorado Interstate Gas Corp. (Westmoreland Coal Co.)	Southeast Montana	--	25,000	250	Colorado Interstate has an option on 300 million tons of coal and 10,000 acre-feet per year of water to be supplied by Westmoreland for development of a coal gasification project.
The Columbia Gas System, Inc.	Illinois	--	--	300	Columbia Gas has agreed to exchange a 50% interest in 43,400 acres of its 300,000 acres of West Virginia coal lands for a 50% interest in 35,000 acres of Illinois coal lands held by Exxon's Carter Oil Co. The Illinois coal will be held by Columbia for coal gasification pending development of an economically and technically sound coal gasification process.
Consolidated Natural Gas Co.	Southwest Pennsylvania	--	--	--	The company has purchased about 450 million tons of recoverable coal reserves in southwest Pennsylvania for eventual use as feedstock for coal gasification.
Pennsylvania Gas and Water Co.	Pennsylvania	HYGAS® or similar one	5,000	80	The company has proposed to the Office of Coal Research a plan for financing and operating a demonstration plant.
Southern Natural Gas Co.	Illinois	--	--	250	Southern Natural Gas Co. has acquired an option to purchase coal reserves in the Illinois Basin from Consolidation Coal Co. The option will not be exercised unless the FPC allows inclusion of the cost of the coal reserves in Southern's rate base.
Texas Eastern Transmission Corp. (Peabody Coal Co.)	Southern Illinois	--	--	250	Texas Eastern has obtained tentative dedication of a large reserve of Peabody's southern Illinois coal while a feasibility study of a gasification plant is being made.
Panhandle Eastern Pipe Line Co. (Peabody Coal Co.)	Southern Illinois	Lurgi Gasification With Methanation	--	--	Feasibility studies under way.
Exxon Corp. (Carter Oil)	Northern Wyoming	--	--	--	Feasibility study started in late 1973. State and Federal coal leases committed.
El Paso Natural Gas Co.	Southwestern North Dakota	--	--	--	Four plants are projected with first to be in operation by 1981. Reserves of 2 billion tons of coal are under lease.
Island Creek Coal Co.	Kentucky	--	--	--	Island Creek Coal Co. is negotiating with an undisclosed utility.
Texaco, Inc.	Wyoming	--	--	--	2 billion tons of coal reserves obtained from Reynolds Metals. Plant may be for either gas or liquids.
B. COMBINATION GAS AND LIQUID PROJECT					
COALCON Co. (Chemico and Union Carbide)	--	Union Carbide	2,400	22.4 (plus 3900 bbl/day liquid fuel)	Plant cost will be $237.2 million. ERDA will fund an initial $21 million for plant design. The balance will be shared equally by the Federal Government and industry. Plant operation is scheduled for 1979.

[a] Mining partners in parentheses.

Source: *Gas Supply Review.*

Table 2 Summary of high-Btu coal gasification R & D projects

Name of Process	Owner(s) or Contractor and Site	Description of Process	Coal Type	Consumption, tons/day	Plant Output, million CF/day	Status and Funding
Lurgi Pressure Gasification (Lurgi Gesellschaft fur Warme und Chemotechnik m.b.H.)	El Paso Natural Gas Co. (Four Corners Area, N.M.)	A bed of crushed coal is introduced to the gasifier vessel through lock hoppers and travels downward as a moving bed. The process operates at pressures up to 450 psi. Steam and oxygen are introduced through a revolving steam-cooled rate that also removes ash at the bottom of the gasifier. The hydrogen-rich gas passes up through the coal bed, producing some methane by hydrogenation of coal. The product gas is purified and methanated to produce 972 Btu/SCF gas.	Subbituminous	800	35 (Raw synthesis gas)	A single Lurgi gasifier will be installed to test improvements in the process. Goals are operation at 20% above design capacity, and 30% above design pressure, gasifier mechanical improvements, coal fines gasification, and improved pollution control. Capital investment will be $19 million; first-year operating cost will be $5 million. Construction will require 13 months after necessary approvals have been obtained. The FPC has granted intermediate approval for the project.
Lurgi Pressure Gasification	Conoco Methanation Co. and Scottish Gas Board (Westfield, Scotland)	Methanation of purified low-Btu product gas from a Lurgi pressure gasifier to demonstrate the commercial feasibility of the combination of Lurgi gasification and methanation to produce a 950+ Btu/SCF gas.	--	--	2.5	Conoco Methanation Co. designed and constructed facilities to carry out a 1-yr test at a total cost of $6 million. Fourteen U.S. companies sponsored the test, which was successfully completed in September 1974. The construction contractor was Woodall-Duckham Ltd.
Lurgi Pressure Gasification	South African Coal, Oil, and Gas Corp. and Lurgi (Sasolburg, South Africa)	Methanation of the product gas from a Lurgi gasifier to produce 950+ Btu/SCF gas.	--	--	--	An experimental program has been completed to demonstrate commercial feasibility of methanation. A slipstream from the SASOL plant will be used as feed to the methanator.
COGAS	COGAS Development Co. (FMC Corp., Panhandle Eastern Pipe Line Co., Tenneco Inc., Consolidated Natural Gas Service Co., Republic Steel Corp.) (Princeton, N.J.)	Coal is charged to a series of fluidized-bed reactors with increasing stage temperatures to pyrolyze the coal and drive off volatile fractions in each stage. After separation of the oils and gas produced in the initial processing, the residual char reacts with steam (heat supplied indirectly by combustion of char with air) to produce a synthesis gas that is purified and methanated to a 950+ Btu/SCF gas. The operating pressure is 50 psig. The oil by-product is hydrotreated to make a synthetic crude oil.	Subbituminous and bituminous	--	--	A 100-ton of coal/day pilot plant built at Leatherhead, England, uses a process in which char is gasified and ash is removed as slag. The $7 million Princeton, N.J. facility, originally a $4.5 million COED pilot plant completed in 1970, removed residue as fly ash and has been shut down; the Leatherhead process is indicated as superior. COGAS Development Co. is evaluating process alternatives and economics for an 800-1000 ton/day coal gasification/liquefaction facility.

Process	Developer	Description	Coal type	Size	Notes
HYGAS®	Institute of Gas Technology (Chicago, Illinois)	Ground, dried coal is pre-treated with air, slurried with by-product oil, and fed to a two-stage fluidized-bed hydrogasifier operating at 1000-1500 psia; hydrogen-rich gas for the reaction can be furnished by processes using electric energy or oxygen, or by the steam-iron process. Gas from the reactor is purified and methanated to produce 950+ Btu/SCF gas.	All U.S. coal types	75 (pilot plant)	1.5	Pilot plant in operation. Preliminary demonstration, plant design complete. Original cost of pilot plant was approximately $9.5 million. Project funded under joint ERDA/A.G.A. program. A run of 27 days completed in March 1974, using lignite and externally supplied hydrogen to produce high-Btu gas. Steam-oxygen gasification section to produce hydrogen has been added, and a fully integrated operation successfully performed in April 1975.
Steam-Iron Process	Institute of Gas Technology (Chicago, Illinois)	A fluidized-bed process in which producer gas, made from steam, air, and programmer char is used to reduce a circulating stream of iron oxide. The reduced iron oxide is subsequently oxidized with steam to produce a hydrogen-rich gas for the hydrogasifier.	Hydrogasified char	36 (pilot plant)	2.0 (44% hydrogen)	Pilot plant under construction for completion near end of 1975. Project is funded at $18.2 million under the joint ERDA/A.G.A. program.
CO_2 Acceptor	Conoco Coal Development Co. (Pilot plant constructed and operated by Stearns-Roger Corp.) (Rapid City, S.D.)	Coal is charged to a gasifier vessel at 150 psi. Dolomite (the "Acceptor") is added to the gasifier where it reacts with carbon dioxide. "Acceptor" regeneration and ash removal are carried out in an air-blown regenerator vessel where spent char is combusted. The product gas can be purified and methanated to produce a 950+ Btu/SCF gas.	Lignite and subbituminous	30 (pilot plant)	Up to 2 (375 Btu/CF)	Original cost about $9.3 million. Ten-day run completed in September 1974. Methanation stage recently added now in initial start-up.
BI-GAS	Bituminous Coal Research, Inc. (Homer City, Pa.) (Pilot plant to be managed by Phillips Petroleum Co. and operated by Stearns-Roger Corp.)	Coal is introduced into a reactor where it is contacted and partially gasified with hydrogen-rich gas produced in the lower section of the reactor by the slagging gasification of recycle char with oxygen and steam. The pressure of operation is 1000 psia. The product gas can be purified and methanated to produce 950+ Btu/SCF gas.	All U.S. coal types	120 (pilot plant)	2, 3	Construction by Stearns-Roger Corp. scheduled for completion in fall 1975. Plant cost will be about $35 million. Sponsored under joint ERDA/A.G.A. program.

Table 2 (continued)

Name of Process	Owner(s) or Contractor and Site	Description of Process	Coal Type	Coal Consumption, tons/day	Plant Output, million CF/day	Status and Funding
Kellogg Molten Salt Process	M. W. Kellogg Company	In the latest version of this process, oxygen, steam, and coal are injected into a high-purity alumina reaction vessel where molten sodium carbonate catalyzes the coal for complete gasification. The product gas can be purified and methanated to produce 950+ Btu/SCF gas.	All U.S. coal types	--	--	OCR funded a bench-scale program from 1964 to 1967. Total expenditures were $1.7 million. Major difficulties were experienced with materials of construction. OCR ceased sponsorship because of this problem, budgetary restrictions, and assignment of higher priorities to other coal gasification processes. M. W. Kellogg has carried out additional bench-scale development work since 1967, but support has not yet been obtained for construction of a large-scale pilot plant.
Simplified Coal Gasification Process	Garrett Research and Development Co. (Sponsored by Colorado Interstate Gas Co.)	Coal is introduced into a single reactor at atmospheric pressure where gas is produced by rapid devolatilization of coal. Large amounts of excess char are produced.	All U.S. coal types	--	--	Early stage of development. Tested in small pilot plant in La Verne, Calif.
None	Gulf General Atomic and Stone & Webster	Coal is dissolved and the solution hydrogasified to produce gas that is purified and methanated to 950+ Btu/SCF gas. Heat supply from nuclear reactor.	All U.S. coal types	--	--	Feasibility study completed by State of Oklahoma under $300,000 funding by ERDA. Support being sought for 2-yr R&D program to cost $650,000 for first phase.
ATGAS	Applied Technology Corp.	Coal is injected into a molten-iron bath where steam and oxygen react with the carbon to produce hydrogen and carbon monoxide. The gases are then methanated to produce 950+ Btu/SCF gas.	All U.S. coal types	--	--	Emphasis now being placed on low-Btu gas production.
Synthane	ERDA (Pilot plant to be operated by Lummus)	Coal is introduced into a single reactor after pretreatment with oxygen at pressure. Hydrogen-rich gas for the reaction is produced by use of oxygen and steam in the reactor. The process operates at 500-1000 psia. The product gas can be purified and methanated to produce 950+ Btu/SCF gas.	All U.S. coal types	72 (Pilot plant)	1.2	Construction by Rust Engineering Co. is essentially complete and operation is starting. Estimated cost is $13 million.
Hydrane	ERDA (Pittsburgh, Pa.)	Raw coal reacts directly with hydrogen in a free-fall zone, and the intermediate char formed reacts further with hydrogen in a fluidized-bed. Residual char gasified with steam and oxygen to produce syngas for hydrogen. Research goals are a reactor product stream with 70% methane content.	All U.S. coal types	--	--	A 10 lb/hr integrated pilot plant is in operation. Scale-up to a 24 ton/day pilot plant to be built at Morgantown, W. Va., is planned to cost $25 million.

Process	Company	Coal Type	Scale	Description
	Exxon Corp. (Baytown, Texas)		--	Exxon has spent $20 million on coal gasification and liquefaction since 1966 and is committing $10 million for additional R&D. First phase of gasification experiments finished in 1974. Next phase, construction of $75 to $80 million pilot plant charging 500 tons/day of coal postponed indefinitely.
Agglomerating Ash Gasification Process	Battelle-Columbus Laboratories (West Jefferson, Ohio) (Process developed by Union Carbide)			Company states that process does not use oxygen. No other details are available.
		Bituminous	25 (Pilot plant)	Coal or char is combusted with air in one fluidized bed and is steam-gasified in a separate fluidized bed. The heat for the gasification reaction is provided by circulation of hot ash agglomerates from the burner to the gasifier. The gasification section will operate at 100 psi. The synthesis gas can be purified and methanated to produce 950+ Btu/SCF gas.
Liquid Phase Methanation	Chem Systems Inc. (Hackensack, N.J.)		0.8 (Synthesis gas)	The synthesis gas flows through the reactor with an inert liquid that picks up heat in the fluidized catalyst bed. The vaporized inert fluid is condensed and recycled. Methanation takes place in a single pass. The catalyst life is estimated longer than 1 year.
			--	Pilot plant under construction by Chemico is scheduled for completion in fall 1975. Funded in joint ERDA/A.G.A. program.
			--	Chem Systems was awarded a $1.9 million contract under the joint ERDA/A.G.A. program. During the past 2-1/2 years the liquid phase methanation process has been under development. A skid-mounted transportable pilot plant is planned for operation by June 1975.
Lurgi Gasification	American Natural Gas Co. and Natural Gas Pipeline Co., at South Africa Coal, Oil, and Gas Corp. with Lurgi (Sasolburg, South Africa)	North Dakota lignite	--	Lurgi pressure gasification plus methanation of the product gas from a Lurgi gasifier to produce 950+ Btu/SCF gas using the Lurgi hot-gas recycle process.
			--	An experimental program to demonstrate commercial feasibility of methanation. A slipstream from the SASOL plant was used as feed to the methanator. 12,000 tons of North Dakota lignite was transported to South Africa for testing.
Catalytic Methanation	Catalyst & Chemicals Inc. with El Paso Natural Gas, COGAS Development Co., and WESCO (Louisville, Ky.)		--	The adiabatic fixed-bed process utilizes 3 beds and recycles product gas as inert diluent.
			--	A 3-yr pilot plant program recently completed. 4000-hr life tests were run on two different catalysts. Results indicate catalyst life of over 2 years.
Lurgi Slagging Gasifier	Conoco Methanation Co. and British Gas Corp. (Westfield, Scotland)		--	Modified Lurgi gasifier operates at temperatures to cause slagging of coal ash. This allows operation at lower steam rates, higher throughput, and higher thermal efficiency.
			--	Conoco will coordinate project, British Gas Corp. will be project operator, and Lurgi will provide technical assistance. 3-yr test period. The process was tested on a pilot-plant scale during the 1962-64 period by BGC. The project cost estimated at $10 million.
GE Fixed-Bed Gasifier	General Electric Co., Schenectady, N.Y.		--	Unit is a fixed-bed reactor using extrusion for injection of coal to reactor. Uses inert diluting agents such as silicon carbide and coal ash to reduce swelling and caking of coal.
			--	Work to date has been done in a feasibility unit operating at atmospheric pressure and consuming about 500 lb of coal per day. Successful runs, using low-grade coals from Illinois and Missouri have been made. Feasibility studies have been carried out at atmospheric pressure for production of SNG from caking coals. A demonstration gasifier operating at 300 psig is planned within 1 year.
Modified Lurgi Gasification	American Gas Association by British Gas Corporation (Westfield, Scotland)		--	Lurgi gasifier is modified to include rabble-arm stirrers that make it applicable to use with caking coals.
			--	Tests concluded in April 1974, successfully gasified U.S. caking coals (Illinois and Pittsburgh). Caking coals required substantially more steam and oxygen than noncaking. Fines in all coal types reduced gasifier capacity.

tains 75% carbon, 5% hydrogen, and approximately 20% of undesirable constituents, which are mostly ash and sulfur that must be removed or discarded in the course of processing. In contrast, natural gas or methane contains 75% carbon, 25% hydrogen, and only negligible quantities of undesirable constituents. So to convert coal to gas, one must either add a lot of hydrogen or reject a lot of carbon. The more efficient way is to add as much hydrogen as possible and to reject as little carbon as possible.

The overall chemistry of coal gasification is very simple:

$$\text{coal} + \text{water} \rightarrow \text{methane} + \text{carbon dioxide},$$
$$CH_{0.8} + 0.8H_2O \rightarrow 0.6CH_4 + 0.4CO_2.$$

In this case, 40% of the carbon is rejected as CO_2. This is just the amount required to remove the oxygen from water—the source of the additional hydrogen required to add to the remaining coal.

In practice, it is not feasible to convert coal directly to methane by reaction with water. Instead, the conversion is carried out in steps. In one approach, in which methane is produced from synthesis gas, coal is first reacted with steam and oxygen at a relatively high temperature (1900–2500°F) to produce hydrogen and carbon oxides. The reaction of carbon with steam, producing carbon monoxide and hydrogen (synthesis gas), is highly endothermic:

$$C + H_2O \rightarrow CO + H_2. \qquad 1.$$

Reaction 1 does not occur unless the required heat of reaction is supplied to the system. Usually, this heat is supplied by burning some of the coal:

$$C + O_2 \rightarrow CO_2. \qquad 2.$$

Whatever carbon supplies heat in this way is not available for methane production or for removing oxygen from water to supply hydrogen. The percentage of carbon in the coal converted to methane is reduced accordingly. Therefore, in the first stage of the steam-oxygen gasification route to methane production, some of the coal reacts with steam to produce carbon monoxide and hydrogen, and some of the coal reacts directly with oxygen to supply the heat of reaction. Note that the carbon

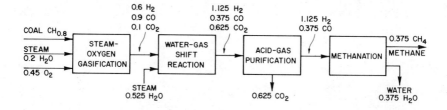

Figure 1 Production of pipeline gas by synthesis gas methanation.

must be burned with pure oxygen; if air were used for the combustion, the diluent nitrogen would appear in the product gas.

In the second stage of this version of the process, the water-gas shift reaction is carried out:

$$CO + H_2O \rightarrow CO_2 + H_2. \qquad 3.$$

Reaction 3 is controlled so that the product gas contains hydrogen and carbon monoxide in the ratio of about 3:1, which is the proper proportion for the final stage of the process.

Following the water-gas shift reaction, both the sulfur pollutants and the carbon dioxide manufactured during Reactions 2 and 3 are removed by acid-gas cleaning. After both the water-gas shift reaction and acid-gas cleanup, the gas contains only carbon monoxide and hydrogen, which are in the proper ratio for catalytic methanation:

$$CO + 3H_2 \rightarrow H_2O + CH_4. \qquad 4.$$

Water is removed from the gas by dehydration, and the high-Btu, methane-rich product is directly substitutable for natural gas.

This sequence of reactions is illustrated schematically in Figure 1. The primary inefficiencies with the system are that the methanation reaction (Reaction 4) is highly exothermic and that all the product methane is produced by catalytic methanation. Significant heat is released from the process at this point, but this heat is released at such low temperatures that it is of little value for the rest of the process. It cannot, for example, be recovered for use in support of Reaction 1 because Reaction 1 occurs at a much higher temperature. Although some of this heat can be used to raise steam, much of the heat is discarded, constituting a process inefficiency.

In contrast to the system just described, most modern gasification technologies

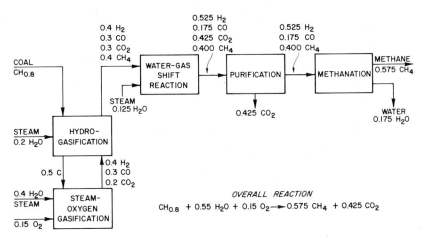

Figure 2   Production of pipeline gas by hydrogasification plus methanation.

employ the concept known as hydrogasification. In this system, the incoming coal is initially reacted in a reactor with a hydrogen-rich gas to form substantial amounts of methane directly:

$$2CH_{0.8} + 1.2H_2 \rightarrow CH_4 + C. \qquad 5.$$

The hydrogen-rich gas for hydrogasification is manufactured from steam utilizing the char leaving the hydrogasifier reactor. For example, the hydrogen may be produced in a steam-oxygen gasifier, as illustrated in Figure 2.

The block flow diagrams of Figures 1 and 2 have the same overall reaction for the formation of methane from coal. The difference in the two simplified flow sheets is the location of the methane-producing steps. The key to the increased efficiency of the modern coal-to-high-Btu gas processes (Figure 2) is hydrogasification (Reaction 5), in which appreciable quantities of methane are formed directly in the primary gasifier. The heat released by methane formation is at a high enough temperature level to be used in the steam-carbon reaction to produce hydrogen. Consequently, less oxygen is used to produce heat for the steam-carbon reaction, and less heat is lost in the low-temperature methanation step. These factors lead to a higher overall process efficiency (65–70% in contrast to 50–55% by synthesis-gas methanation).

GENERAL ASPECTS OF GASIFICATION SYSTEMS

The rates and degrees of conversion for the various reactions involved in gasification are functions of temperature, pressure, gas composition, and the nature of the coal or char being gasified. The rate of any reaction is higher at higher temperature, but the equilibrium of the reaction may be favored by either high or low temperatures. The effect of pressure on the rate of reaction depends on the specific reaction. Some reaction rates are favored by higher pressure such as the carbon-hydrogen reaction to produce methane. For the production of methane, higher pressure (1000 psia) and the lower temperature (1400–1700°F) are the most favorable conditions. Low pressure and high temperature, on the other hand, favor production of CO and H_2.

Aside from the operating conditions, most gasification systems can be classified in four categories according to the method of contacting gaseous and solid streams: 1. suspension- or entrained-bed reactor (for example, Koppers-Totzek, Texaco, Babcock and Wilcox, and BI-GAS), 2. fixed- or moving-bed reactor (for example, Lurgi and Wellman-Galusha), 3. fluidized-bed reactor (for example, Winkler, HYGAS®, and Synthane), and 4. molten bath reactor (Kellogg and ATGAS).

In the conventional suspension-bed reactor, solids and gaseous streams are contacted cocurrently, and high temperatures are required to achieve the complete reaction of coal and gases in relatively short periods of time. To achieve high temperature in the reactor requires relatively large amounts of oxygen. Ash is removed from the gasifier as slag. Operations of this type are not generally sensitive to the type of coal (noncaking or caking).

In the moving-bed gasifier, coal fed to the top moves downward slowly countercurrently to a stream of oxygen and steam that are fed from the bottom. Temperature at the bottom of the reactor is higher than at the top. Devolatilization of coal takes place at low temperature, and, consequently, relatively large amounts of heavy liquid hydrocarbons are produced in the gasifier. Ash is removed from the bottom of the gasifier. In this type of reactor, it is necessary to control the bed porosity to enable the gases to flow uniformly through the coal bed. The coal feed must, therefore, be of fairly uniform size and contain a minimum amount of fines. Residence time of coal in the reactor is much higher than in the suspension-bed type.

The fluidized-bed gasifier allows intimate mixing and contacting of gas and solids and provides a relatively long residence time compared with a suspension-bed gasifier. The gasifying medium (oxygen, steam, hydrogen) is fed through a bottom distributor plate and acts as the fluidizing medium. Dry ash is removed continuously from the fluid bed.

An essential element in the gasification of coal is the supply of heat. The direct method of generating heat and synthesis gas is the partial oxidation of char with steam and oxygen. Several other methods for indirectly providing the heat needed to produce synthesis gas from coal char are being developed; these include electrical resistance heating in a fluid bed, the use of molten salt or metal as a heat-transfer medium, and the use of a circulating stream of dolomite. Another method of producing a hot hydrogen-enriched gas stream is the cyclical reduction and oxidation of iron ore. The decomposition of steam in the oxidation step produces a hot hydrogen and unreacted steam mixture for the hydrogasification step.

The type of coal being gasified is important to the operation. The suspension-type gasifier can handle any type of coal, but if the coal fed to a moving-bed or fluidized-bed gasifier is of the caking variety, special measures must be undertaken so that the coal does not agglomerate during the gasification operation. In the case of the fluidized-bed operation, the coal is ground to a size small enough that it can be pretreated by mild oxidation at about 750°F to destroy its caking tendency. This is not possible in a moving-bed operation because of the larger size of lump required. Mildly caking coal has been gasified successfully in a moving bed, with revolving arms that stir the bed to prevent agglomeration.

The chemical composition, volatile content, and moisture content of coal will affect its processing during gasification. The wide variation in the composition of typical coals is shown in Table 3. Eastern bituminous coals are generally of the caking variety while western subbituminous and lignite are noncaking. The free-swelling index shown is a measure of the caking tendency.

Gasification systems to produce high-Btu gas are operated, preferably, at elevated pressure. To introduce coal feedstock into the gasifier, lock hoppers have been successfully utilized in commercial plants up to pressures of 450 psi. Another method of feeding coal at elevated pressure is as a pumped slurry in oil or water. Successful operation of such systems has been demonstrated on a pilot-plant scale.

RAW GAS UPGRADING

In all of the processes for production of high-Btu gas from coal, the raw gas from the gasifier contains not only carbon dioxide as an undesirable constituent but also,

Table 3 Typical coal properties

Constituents [analysis as received (wt %)]	Type of Coal				
	Montana Subbituminous	Illinois No. 6 Coal	Navajo Coal	West Kentucky No. 11 Seam Coal	Pittsburgh Seam Coal	Ohio High-Sulfur Coal
Carbon	53.15	64.89	49.19	68.12	73.8	69.5
Hydrogen	3.62	4.49	3.60	4.67	5.2	4.7
Nitrogen	0.66	1.26	0.85	1.37	1.5	1.2
Sulfur	0.51	3.93	0.69	3.48	1.6	3.8
Oxygen	14.48	8.14	10.15	7.27	8.0	6.5
Ash + other	5.58	10.79	19.27	6.69	7.4	10.2
Moisture	22.00	6.50	16.25	8.40	2.5	4.1
Total	100.00	100.00	100.00	100.00	100.0	100.0
Heating value (Btu/lb)	8,815	12,603	8,664	12,330	12,700	12,630
Free-swelling index		2.5–4.5		2.5–4.5	7.5–8.0	2.0–5.0

in most cases, significant amounts of hydrogen sulfide, ammonia, and light oils. Depending on the gasification conditions, certain amounts of fly ash or coal dust, tars, heavy oils, phenols, and organic sulfur compounds (COS and CS_2) are also present.

If the raw gas exits the gasifier at an elevated temperature, it is advantageous to utilize the heat in the gas. This may be accomplished in a waste-heat boiler. Tars and heavy oils are removed by cooling. A direct water wash, often in a venturi scrubber, is employed to remove ammonia, light oils, phenols, and dust.

Sulfur (mostly H_2S) may be removed either before or after the water-gas shift operation; if after, a sulfur-resistant shift catalyst is used. In any case, all sulfur must be removed before methanation since the nickel catalyst utilized for methanation has a very low tolerance for sulfur. Many processes may be employed for the removal of the acid gases. These may be either physical absorption (for example, methanol) or chemical reaction (for example, amine solutions). The H_2S removed can then be converted to elemental sulfur. Because the H_2S concentration in the gas is low compared with the concentration of CO_2, it is often desirable to carry out the acid-gas removal in two steps. H_2S is first removed to yield a stream containing about 25% H_2S and 75% CO_2, which can be efficiently converted to elemental sulfur. A second step then removes the remaining CO_2.

Following the acid-gas removal, shift, and methanation, the high-Btu gas is dried (generally by glycol wash) and, if necessary, compressed to pipeline pressure. The finished gas generally contains less than 0.1% CO, practically no H_2S, and about 5% H_2.

SPECIFIC GASIFICATION PROCESSES

Lurgi

Of the currently available, fully commercial, coal gasification techniques, Lurgi is the only one being considered applicable to production of synthetic pipeline gas. Other commercial processes, such as Koppers-Totzek, Wellman-Galusha, and Winkler, are undesirable because they operate at pressures only slightly above atmospheric pressure and have high oxygen requirements and low conversion efficiencies. Thirteen commercial Lurgi plants—utilizing a total of 63 fixed-bed, high-pressure, nonslagging, steam-oxygen gasifiers—have been built. Such units are capable of producing up to 20 million cf/day of 400–450 Btu/cf heating-value gas at pressures up to 450 psig. This medium-heating-value gas can be upgraded by catalytic methanation to pipeline gas. In a one-year testing program completed in September 1974 at Westfield, Scotland, the feasibility of the methanation step using gas from a Lurgi generator was demonstrated. This unit produced an average of 2.1 million scf/day at 979 Btu/scf during the test.

The Lurgi gasifier receives coal through a lock hopper system. The coal flows downward as a fixed bed; consequently, the coal size must be $>\frac{1}{4}$ inch to provide sufficient opening for good gas distribution. Even then, the permissible gas flow rate is low. Production is only approximately 10 million cf/day of high-Btu gas equivalent per gasifier compared with 80 million for newer processes. Lurgi is the only fixed-bed reactor system being considered for coal gasification.

Heat is supplied by burning residual carbon at the bottom of the reactor with oxygen and steam. This produces a mixture of hydrogen and carbon monoxide as well as heat. As the hydrogen-rich gas flows upward through the coal bed, methane is formed and the coal is devolatilized.

The crude gas leaving the gasifier at temperatures between 700°F and 1100°F (depending on type of coal) contains tar, oil, naphtha, phenols, and ammonia, plus coal and ash dust. Quenching with oil removes tar, oil, and dust. Shifting, acid-gas removal, methanation, and drying produce a clean, high-Btu gas.

Caking bituminous coals, prevalent in the United States east of the Mississippi River, cannot normally be used in the Lurgi gasifier without special steps. One method is to introduce rabble arms that stir the bed and prevent agglomeration. The stirring arms, which require water cooling, are attached to a power-driven centrally located vertical shaft. During 1973 and 1974, a full-size Lurgi gasifier in Westfield, Scotland, was modified for operation on American caking coals. Four American coals (Illinois Nos. 5 and 6, Pittsburgh No. 8, and Rosebud, Wyoming) were shipped to Scotland in amounts of about 4000 tons each for test purposes. The first three of these coals are caking; the Rosebud coal is noncaking. For each coal, tests were conducted using both coarse-graded (2–7% below $\frac{1}{8}$ inch) and fine-graded or simulated run-of-mine (15–20% below $\frac{1}{8}$ inch) materials. It was found that the Rosebud coal behaved similarly to the Scottish coal that is also noncaking. However, the caking coals required 50–70% more steam and 45–55% more oxygen than the Rosebud per unit volume of gas produced. No difficulties were encountered in gasifying the Rosebud coarse-graded, but channeling in the gasifier bed and dust carry-over were excessive in the first test with fine-graded. This difficulty was eliminated in a later test made with the stirrers removed. Both Illinois coarse-graded coals were gasified without difficulty at feed rates of 70–75% of the noncaking type. The maximum feed rate in these cases was limited by the steam quantity available. With Illinois simulated run-of-mine, indications of channeling suggest that, because of the high content of fines, 67–68% of normal would be a maximum feasible feed rate. Initial difficulties, caused by agglomeration with Pittsburgh coarse-graded, were overcome by discontinuing recycle tar feed to the gasifier and reducing the stirrer speed to a minimum; coal feed in this case was at 54% of normal (limited by steam quantity available). With Pittsburgh simulated run-of-mine, at a feed rate of 57% of normal, indications of channeling were even more pronounced than in the Illinois case. Added steam and oxygen requirements for the caking-type coals are estimated to decrease the overall conversion efficiency to about 80% of that obtainable when gasifying noncaking coals.

Another development program is currently under way at Westfield to operate the commercial-size Lurgi gasifier in such a way that ash will be removed as slag. If the unit can be operated with the ash in a molten state, the temperature in the combustion zone can be increased several hundred degrees, thereby allowing higher combustion rates and increased heat-transfer and reaction rates in the other two zones in the gasifier. Small-scale tests have indicated that a major advantage of slagging operation in the Lurgi fixed-bed gasification is the reduction in the amount of excess steam required. Steam fed to the unit is reduced by as much as 80%, which greatly reduces the gas flow through the beds. It is understood that gas production per unit of cross-sectional area may be increased by more than twofold. On the other hand, the high slagging temperature requires 10% more oxygen and shifts the gas composition toward $CO + H_2$ (relatively less CH_4).

New Processes Under Development

Four new processes to produce high-Btu gas from coal, intended to improve on Lurgi as the basic gasification step, are in a relatively advanced stage of development and incorporate an initial hydrogasification step. They are: HYGAS® (Institute of Gas Technology), CO_2 Acceptor (Conoco Coal Development Corporation), Synthane (ERDA), and BI-GAS (Bituminous Coal Research, Inc.).

All of these new processes have the following common features:

1. continuous fluid-bed or entrained-bed gasification at elevated pressure equivalent to or higher than Lurgi;
2. more effective contacting of raw or lightly pretreated coal with hot, hydrogen-rich gas at pressure to form more methane directly by hydrogasification of the coal than in Lurgi;
3. the use of less oxygen, or none at all, in the production of the hydrogen-rich (synthesis) gas from the residual char remaining after the primary gasification step; and
4. the need for a final catalytic methanation step to achieve a production of pipeline quality. The percentage of the total methane formed in this step, however, varies substantially from process to process.

Table 4 shows compositions of raw gas obtained from these gasification processes and also for the Lurgi and Kellogg molten salt gasifiers. Methane produced in the primary gasifier is shown as a percentage of the total methane produced as the ultimate high-Btu gas. This percentage affects, to a large degree, the overall conversion efficiency obtainable in the process. No comparable data are available from the COGAS process pilot plant. This process is being developed on a proprietary basis by a group of companies and is also in an advanced stage of development. (See Table 2.)

HYGAS®

This process is under development. Construction of the HYGAS® pilot plant began in Chicago in August 1969. The first test was conducted in October 1971. A run of 27 days was completed in March 1974, using lignite and externally supplied hydrogen. All parts of the plant were operated simultaneously to produce high-Btu gas. Following the addition of a steam-oxygen gasification section to the plant, the entire plant has again been operated (in April 1975) on a fully integrated and self-sustained basis to produce pipeline-quality gas from lignite.

In this process, coal is first crushed to $<\frac{1}{8}$ inch. For caking bituminous coal, the crushed coal is pretreated. Pretreatment with air at atmospheric pressure and 700–800°F in a fluidized-bed reactor destroys the agglomerating tendencies, thus eliminating the possibility that agglomerated coal will plug process equipment. For noncaking lignite and subbituminous coals, pretreatment is not needed.

The prepared coal then is slurried with a light oil (a by-product of the process),

Table 4 Typical raw gas compositions

	Process							
	HYGAS®			BI-GAS	CO$_2$ Acceptor	Lurgi	Synthane	Molten Salt
	Electro-thermal	Oxygen Gasifier	Steam-Iron					
Gas composition (mol %)								
CO	21.3	18.0	7.4	22.9	14.1	11.8	10.5	26.0
CO$_2$	14.4	18.5	7.1	7.3	5.5	16.3	18.2	10.3
H$_2$	24.2	22.8	22.5	12.7	44.6	22.7	17.5	34.8
H$_2$O	17.1	24.4	32.9	48.0	17.1	41.7	37.1	22.6
CH$_4$	19.9	14.1	26.2	8.1	17.3	6.5	15.4	5.8
C$_2$H$_6$	0.8	0.5	1.0	—	0.37	0.4	0.5	—
H$_2$S + COS	1.3	0.9	1.5	0.7	0.03	0.2	0.3	0.2
N$_2$ + Ar	—	—	—	0.3	0.2	0.2	0.5	0.3
Other	1.0	0.8	1.4	—	0.8	0.2	—	—
Total	100.0	100.0	100.0	100.0	100.00	100.0	100.0	100.0
Higher heating value (dry basis) (Btu/scf)	437	374	565	378	400	322	405	329
Methane made in the gasifier (%)	66	60	80	48	55	46	70	28
Ultimate methane per 100 moles of raw gas	32.7	25.2	35.4	17.0	32.6	15.8	23.3	21.0

and the slurry is pumped to 1000–1500 psi and injected into a fluidized drying bed at the top of the vertical multistage HYGAS® unit, where the oil is driven off and recovered for reuse. This mode of transferring solids into a high-pressure environment may prove superior to the use of lock hoppers at these high pressures.

The oil-free coal now passes downward through two hydrogasification stages, which provide countercurrent treatment. (Coal passes down; gases produced pass upward and are drawn off.) This arrangement permits time and temperature control for each step to optimize processing conditions.

In the first hydrogasification stage, dried coal is flash-heated to the reaction temperature (1200–1300°F) by dilute phase contact with hot reaction gas and recycled hot char. Volatile matter in the coal as well as "active" carbon are converted to methane in a few seconds. "Active" carbon, which is present during the first moments of gasification, gasifies at a rate greater than 10 times that of the carbon in the less reactive char.

The solids pass down to the second HYGAS® stage and enter a dense-phase fluidized-bed reactor at temperatures of 1700–1800°F, where formation of methane from partially depleted coal char continues simultaneously with a steam-carbon reaction that produces hydrogen and carbon monoxide.

Hot gases produced in the lower hydrogasification stage rise, passing through the first-stage hydrogasification reactor and into the fluidized drying bed, where much of their heat is used to dry the feed coal. After leaving the hydrogasifier, the raw gas is shifted to the proper $H_2:CO$ ratio in preparation for methanation; the oils, carbon dioxide, unreacted steam, sulfur compounds, and other impurities are removed; and the purified gas is catalytically methanated. Sulfur is recovered in elemental form.

The residual coal char is used to produce the hydrogen-rich gas required in the process by one of two different techniques under development at the Institute of Gas Technology (IGT). These processes are

1. OXYGEN GASIFIER   Heat for the reaction of steam with char to form H_2 and CO is supplied by combustion of a portion of the char with oxygen in a fluid bed.
2. STEAM-IRON PROCESS   Hydrogen is produced in an oxidizer by reaction of steam with reduced iron oxide. The oxidized solid is then reduced by a producer gas obtained by combustion of char with air and circulated back to the oxidizer.

Another method for providing heat for the reaction of steam with char to form H_2 and CO is to pass an electric current through a fluid bed of char. An electrothermal gasifier for this purpose was developed to a full pilot-plant scale earlier in the program and was successfully tested in a batch operation. Work on this method of hydrogen-rich gas production was discontinued because the economics do not appear, at this time, to be as favorable as the oxygen or steam-iron systems.

Both the electrothermal and the steam-iron systems eliminate the need for an oxygen supply to the gasifier.

CO_2 Acceptor

Another process in pilot-plant operation is the Conoco Coal Development CO_2 Acceptor process. It is similar to the electrothermal and steam-iron versions of the

HYGAS® process in that it eliminates the need for oxygen in the supply of heat through a novel technique. A pilot plant with a nominal capacity of 1 ton/hr of North Dakota lignite was completed early in 1971, and operation is now under way.

In this process, the heat source for the endothermic steam-coal reaction is hot calcined dolomite (the acceptor). The acceptor also releases additional heat when the calcium oxide forms calcium carbonate by combining with the carbon dioxide evolved in the steam-carbon and carbon monoxide shift reactions taking place in the gasifier. The spent acceptor is continuously regenerated by reconverting the calcium carbonate to calcium oxide and carbon dioxide, utilizing residual carbon from the gasification step to supply the necessary heat by combustion with air, thus eliminating the need for oxygen. Because of chemical reaction limitations, the CO_2 Acceptor process is designed primarily for lignite.

Crushed and dried coal is fed to a fluidized-bed gasifier at about 1500°F and 150 psi where dolomite is used to remove CO_2 and H_2S. Spent dolomite and spent char are conveyed to the regenerator, where the char is burned with air at about 1900°F to calcine the spent dolomite. Cyclones separate char, ash, and dolomite fines from the hot regenerator-effluent gas that is used for power recovery and steam production.

Raw gas from the gasifier passes through a heat-recovery section, water quench, and acid-gas removal. The purified gas has an $H_2:CO$ ratio of about 3.2, so that no shift is required before methanation. Dehydration of the methanated gas produces pipeline gas.

Synthane

A gasification pilot plant having a capacity of 72 tons/day is nearing completion at Bruceton, Pennsylvania. Production of high-Btu gas is designed to be 1.2 million scf/day. A key feature of the Synthane process is pretreatment of caking coal in a fluid bed at 800°F with steam and oxygen at full gasification pressure. Pretreatment gases can, therefore, be combined with the main gasifier product.

In this process, crushed and dried coal is fed to the fluidized-bed pretreater (if necessary) through lock hoppers. About 12% of the total steam and oxygen required in the process is fed to the pretreater.

The gasifier operates at about 1800°F and 500–1000 psi with steam and oxygen fed to the bottom of a dense-phase fluidized bed. Char, containing roughly 30% of the carbon from the original coal, is removed and sent to the power plant.

Dust particles are removed from the raw product gas in cyclones. Tar is removed by water wash. Cleaned gas goes to the shift converter. The overall $H_2:CO$ ratio is raised from 1.7 to 3.0. Gas from the shift converter is treated to remove CO_2 and H_2S. The purified gas is methanated and dehydrated to produce pipeline-quality gas.

BI-GAS

A 120 ton/day coal gasification pilot plant to produce 2.4 million scf/day is under construction at Homer City, Pennsylvania. Completion is scheduled for late 1975.

In this process, coal is crushed, dried, and pulverized (70% through 200 mesh). Coal and steam are fed to the upper reactor by upward-directed concentric injecting nozzles that are spaced around the shell so that streams do not strike the refractory but impinge in the gas space. Here the coal is rapidly heated to 1700°F by hot gas from the lower reactor. Devolatilization of coal occurs, producing methane and char. The residence time of the coal is a few seconds.

Entrained char, separated from raw gas in a cyclone, is then fed into the lower section of the gasifier, where preheated oxygen (1200°F) and steam are reacted with the char at a temperature of about 2700°F, causing slagging of the ash. Molten slag from the gasifier drains into quench water.

Raw gas from the upper reactor passes through a waste-heat recovery system. The bulk of entrained char is separated in a cyclone, and the remainder in a wet scrubber or sand filter. After shift, the gas is purified to remove CO_2 and H_2S, methanated, and dried to produce pipeline-quality gas.

Since the first stage of this process employs an entrained- rather than a fixed- or fluidized-bed system, it is expected that all types of coal (caking as well as noncaking) will be amenable without prior treatment.

COGAS

A pilot plant located in Princeton, New Jersey, with a capacity of $2\frac{1}{2}$ tons of char feed/day, was started up in May 1974. In Leatherhead, England, a pilot plant with a capacity of 50 tons of char feed/day was started up in April 1974.

This process converts coal to gas and oil products. The first step consists of a multistage fluidized-bed operation in which the coal is pyrolyzed to produce pyrolysis oil and gas plus a substantial quantity of low-volatile, reactive char. Heat for the pyrolysis is supplied by hot synthesis gas (from char gasification) and by recycled hot char. Char from pyrolysis is fed to a gasifier, where it is reacted with steam to produce synthesis gas.

Heat for gasification is supplied in one of two ways. One system (at the Princeton plant) uses a circulating inert heat carrier such as a ceramic or a pelletized coal ash. This material is heated by mixing with char fines that are then combusted with air. The solids are then transported back to the gasifier, where they shower through the bed of fluidized char being gasified. In this system, ash is removed as fly ash.

The second method of supplying gasification heat, which is installed at the Leatherhead plant, utilizes circulating char heated by combustion gases from burning of char fines with air. In this system, ash is removed as slag.

After removal of oil and light hydrocarbons, pyrolysis gas, which contains about 35% methane, is combined with synthesis gas from the steam gasifier. The combined gas is compressed from about 50 psig to an intermediate pressure, shifted, purified of sulfur compounds and CO_2, methanated to form high-Btu gas, and finally compressed to pipeline pressure.

Union Carbide–Battelle

A 25 ton/day pilot plant is under construction at West Jefferson, Ohio. A basic feature of this process is that it provides heat for the reaction of carbon and steam by circulation of hot coal-ash agglomerates. These are produced by the combustion of char with air.

Crushed (smaller than 35 mesh) dried coal is injected near the bottom of a fluidized-bed zone in the gasifier. Hot ash agglomerates (2000°F) from the combustor enter the gasifier near the top. The lighter coal particles migrate upward. Heavier ash agglomerates descend through the bed.

The bed is fluidized by steam and the products of gasification. Heat requirements of the coal-steam gasification reactions cool the agglomerates to about 1600°F. Coal entrained by the downflowing agglomerates is removed by elutriation in the bottom section of the gasifier. Part of the ash agglomerates are recycled to the combustor. The remainder is rejected from the system.

Char from the top of the gasifier is removed continuously and burned with air in the combustor. The char burns at 2000°F, under conditions to form agglomerates. Hot agglomerates are recycled to the gasifier. Flue gas from the combustor is passed through waste-heat recovery, purification, and energy-recovery systems.

Raw product gas from the gasifier (1600–1800°F) passes through a heat-recovery section, scrubbing, and acid-gas removal. Purified gas is then compressed, shifted, and methanated to produce pipeline-quality gas.

Miscellaneous Processes

In addition to the projects described above that are at the pilot-plant stage, several other processes are being considered for the production of high-Btu gas from coal. A brief summary of the principal features follows.

ATGAS   Gasification takes place in a molten iron bath at 2600°F and 5 psig. Sulfur from the coal is removed in a molten slag.

HYDRANE   Gasification takes place in a stream of pure hydrogen at high pressure. The raw gas produced has a very high methane content (73%).

MOLTEN SALT   Gasification takes place in a molten sodium carbonate bath at high pressure. The salt catalyzes the reaction so that appreciable methane is formed.

GENERAL CONSIDERATIONS

The efficiency of coal conversion processes to produce high-Btu gas is in the range of 50–70%, depending on the type of process used. This range of efficiencies may be raised to 65–75% by including the heating value of by-products such as oils, tars, and sulfur.

For the production of 250 million scf/day of high-Btu gas, about 15,000–30,000 tons of total coal per day are required, depending upon the process and the rank of the coal. In addition to the coal gasified, 10–20% of the total coal used goes to supply process steam and energy for compressors, pumps, and other mechanical equipment.

The plant producing high-Btu gas from coal also produces a range of by-products. Process water streams in such a plant may contain phenols and cresols, light aromatics (for example, benzene), oils and tars, ammonia, sulfur compounds, traces of hydrogen cyanide, coal char, and ash. Gas streams may contain sulfur dioxide, hydrogen sulfide, and nitrogen oxides. Particulates arise from operations with coal, char, ash, lime, and dolomite. According to the Federal Power Commission, typical quantities of by-products within major chemical categories for 250 million cf/day of pipeline gas would be as follows:

Sulfur (primarily as H_2S)	300–450 long tons/day
Ammonia	100–150 tons/day
Hydrogen cyanide	0 to possibly 2 tons/day
Phenols	10–70 tons/day
Benzene	50–300 tons/day
Oils and tars	Traces to 400 tons/day

Plant designs will provide for recovery of the gas by-products where economical or for other environmentally satisfactory methods of disposal.

The investment cost and coal price are two major factors contributing to about 80–90% of the cost of the production of the high-Btu gas. At present, coal prices are in the range of $0.30–$1.00/million Btu. At a 60% efficiency of coal to high-Btu gas, the cost of coal could be in the range of $0.50–$1.67/million Btu of pipeline gas. Investment costs have been escalating rapidly in the last two years, making it difficult at this time to quote established figures. The estimated cost of the Lurgi WESCO project (250 million scf/day) was reported to be $447 million in the summer of 1974. In January 1975, this cost was reported to have escalated to $853 million. It is expected that the investment cost of several of the new processes now under development will be lower than that of the Lurgi process by 20–30%.

Copyright 1976. All rights reserved

CLEAN LIQUIDS AND GASEOUS FUELS FROM COAL FOR ELECTRIC POWER

✕11004

S. B. Alpert and R. M. Lundberg

Electric Power Research Institute, Palo Alto, California 94303

INTRODUCTION

The utilization of coal to produce gaseous and liquid fuels for producing power in an environmentally acceptable way is the major objective of the fossil fuel research and development program in the United States. This chapter represents a summary and appraisal of the major programs underway that will supply fuel for generating electric power.

It is certain that coal will play an increasingly important role in the future of the electricity industry. Recent estimates indicate that domestic oil production may already have peaked. The US Geological Survey forecasts indicate that perhaps one half of the oil ultimately available has already been produced. In a finite time interval of 10-20 years, US oil production can be expected to diminish, and our dependence on fossil fuels other than oil is necessary.

In the United States, generating stations use about 4 trillion cubic feet of natural gas. Many of the stations burning natural gas are concentrated in the region from Texas to the state of Washington. Before long, such installations need to switch to fuels other than natural gas. This fuel is already in short supply and is rapidly escalating in price. There is an urgent need for fuels derived from coal that are economical and that could provide energy for power to be produced in environmentally acceptable systems.

The gas industry today faces a central challenge to continue to supply customers with a premium quality fuel. Relatively high prices for synthetic natural gas, compared with those for natural gas, do not alter the requirement to provide such fuel for the pipeline. In contrast, electric utilities have other demands such as cost and environmental effects of available fuels.

Again, unlike other industries, the power industry needs a wide spectrum of fuels for base load, intermediate load, and peaking generation plants, which are integrated for the most economical power cost. The average electrical system will operate equipment at base load, which uses relatively inexpensive fuel efficiently. The requirement makes efficient base load plants high in capital investment but low in fuel cost. The lowest fuel cost in an electric system is nuclear power plants. Although specific situations can alter a generalization, the trend is toward nuclear-fueled plants for base load service.

The economical use of peaking capacity requires low-capital-cost equipment. This lower-cost equipment can utilize higher-cost fuels. The integrated solution of capital costs, fuel costs, and efficiencies for each unit results in the most economic system.

Traditionally, the electrical utilities have operated a variety of systems in which capital and fuel costs each vary by as much as 500% and efficiencies vary by a factor of three.

The electricity industry can best be characterized as an interrelated cost-effective system that uses efficient high-capacity plants to provide electricity to the public over sophisticated transmission and distribution systems. It is only in less-developed nations that electricity is a luxury and an expensive, unreliable commodity. A bench mark of an industrial state is the availability of electricity for industrial and commercial uses. One key feature of industrial growth has been the increasing size of production plants. Tabulated below is the change in scale of electricity plants in the United States.

	Average Size of New Power Stations (MW)
1953–1955	120
1956–1960	200
1961–1965	250
1966–1970	350
1971–1973	450

This change in scale summarizes a remarkable progression in plant size. It corresponds to capacity changes achieved in the petroleum and industrial chemical industries.

Today coal-fired stations produce over one half of the electricity generated and consume over 50% of the coal mined. Clean fuels from coal are needed because of environmental concerns and dwindling supplies of domestic oil and gas. High prices for oil and gas imports and uncertain sources of imports further encourage process development for producing clean gas and oil from coal. A program of development is underway to deal with this situation. The effort is funded by the federal government and the private sector, and it emphasizes processes that use coal. Coal represents 90% of the fossil fuel reserves in the United States.

COAL LIQUEFACTION

Coal has been converted to clean fuels in historic efforts. The largest plants were operated in Germany during World War II at extreme pressures of up to 600 atmospheres. In relatively small-scale plants (by US standards) the Germans converted about 25,000 tons of coal per day to fuels needed for the war effort, where cost was not a constraint. The largest system in Germany handled 600 tons/day as shown in Table 1. Because of small scale and extremely high projected costs, that technology is unacceptable today for application in the United States. Efforts

Table 1 Coal liquefaction by hydrogenation methods: historic efforts

Company	Year	Reactor Train Capacity (Tons/Day)	Pressure (psi)	Catalyst	Product	Solid Separation
IG Farben Industries[a]	1945	500–600	3500–10,000	$FeSO_4$	distillates	distillation, centrifuge
Ruhrkohle (Pott-Broche)	1945	100	1500	none	extract	batch filter
Nippon Chiso	1950	100	3500	$FeSO_4$	distillates, chemicals	centrifuge
Bureau of Mines	1953	200	10,000	$FeSO_4$, Sn, etc.	distillates	distillation, centrifuge
Union Carbide	1956	300	6000	—[b]	chemicals	distillation, centrifuge
Consolidation Coal	1969	20	500[c] 3500[d]	none[c] $CoMo$[d]	extract, distillates	hydroclones

[a] Early efforts by Standard Oil of New Jersey at Baton Rouge in cooperation with IG Farben.
[b] Not available.
[c] On extraction step.
[d] On hydrogenation, CoMo is cobalt molybdate.

are underway to better the historic methods and achieve large-scale, high-capacity plants using modern engineering at about 100 atmospheres.

Table 2 shows that the engineer has learned to construct large, efficient plants that in single-train operations process thousands of tons of raw materials per day using equipment of high reliability. In the process industries, plants have operated reliably 95% of the time. Industrial plants process many thousands of tons of raw materials to produce products needed by society.

Plants in the United States producing clean fuels for utilization by the electricity industry will need to process tens of thousands of tons of coal daily in order to have economic systems that provide fuel competitively and meet environmental standards for use in peaking and intermediate-load power stations now utilizing gas or oil. A plant producing 100,000 barrels of oil per day will need to feed about 30,000 tons of coal per day. Coal liquefaction is attractive to the electricity industry in that there is no major change in practice, presuming the petroleum industry accepts the "risk" and the opportunity of establishing a coal liquefaction industry. There is no certainty at this time that the petroleum industry will do the job. The capital cost required for a "typical" coal liquefaction plant, about $20,000 per daily barrel of capacity, must be financed. Such technology is capital-intensive, compared with oil refineries requiring about one fifth as much capital. The oil industry has simply no experience with the intensive capital required, nor is it apparent that incentives exist to justify the immense risk.

Table 3 lists the major pilot plants in the United States that produce clean liquids from coal. It is disappointing that while much has been learned about the technical nature of processes, we are just now carrying out the engineering, construction, and operation of relatively large-scale pilot plants to prove and test a first generation of coal liquefaction methods by the early 1980s.

In the development of process technology there are several guidelines that have been established, based on chemical engineering practice: (*a*) make your mistakes on a small scale and (*b*) do not continue to operate small test equipment beyond the point where commercially useful information is obtained.

In general, the pilot plant studies of coal liquefaction have been carried out in small equipment far beyond the point where important commercial data are being obtained. It is a serious mistake to continue to operate a bench-scale pilot plant to firm up mathematical models or to search for imagined or hopeful technical breakthroughs.

Scale-up of a new process is always a risky venture. However, delay in taking the technology to the larger-scale commercial test is a mistake in a technical society

Table 2 Size of US industrial plants (tons/day)

Coal-fired power stations	12,000
Oil refineries	50,000
Cement plants	8,000
Ethylene plants	2,000
Ammonia plants	2,000

Table 3 Operating pilot plants liquefying coal[a]

Process	Status	Capacity (Tons/Day)	Source of Funds
Solvent-Refined Coal	operational	6–12	Southern Company EPRI[b]
Solvent-Refined Coal	operational	50	ERDA[c]
CONSOL CSF	shut-down	26	ERDA
H-Coal (catalytic)	operational	3	ERDA, EPRI, Sun Oil, Amoco, ARCO, Ashland Oil Co.
Exxon EDS	operational	1	Exxon
Gulf CCL (catalytic)	construction	1	Gulf
Synthoil	construction	8	ERDA, Bethlehem Steel

[a] Excludes bench-scale units. Smaller-scale units at various locations are investigating alternatives in liquefaction, e.g. Lummus, Universal Oil Products, and Continental Oil Co., as well as others. In addition there are coal liquefaction programs at the University of Utah, Auburn University, University of North Dakota, and at other universities.
[b] EPRI = Electric Power Research Institute.
[c] ERDA = Energy Research and Development Administration.

that is faced with fuel shortages, high prices, and "double digit" inflation. The escalation of steel, labor, and the like can far exceed the economic benefits derived from long-duration, small-scale programs when programs needed to establish critical engineering know-how are not carried out.

Rapid expansion of capacity in commercial plants must rest on several firm bases. The objectives of a process research and development program are to provide:

1. A correlation and data summary of the commercially useful background technical data showing limits where data were obtained in small equipment.
2. A definition of operable and inoperable combinations of process variables that define safe operating areas.
3. A design handbook and an operating manual for commercial plants based on large-scale pilot plant tests. Continual checking for consistency of observations at each step is necessary.
4. A skilled team of specialists who represent an organization with detailed knowledge of the practice developed to serve as consultants for commercial plants.
5. Perhaps the most important product, an enthusiastic management with the entreprenurial insight, imagination, and resources sufficient to carry new technology to commercial reality.

Process development is not a schedule of discrete events that stretches from an inventor's concept to commercial plants employing an array of skills. Rather, it is a three-dimensional spiral of knowledge that expands with the added information from many engineering disciplines. At every step in the process of development, continual cross-checks and reconfirmation is necessary in order to achieve commercial success in the marketplace.

In some instances such as in gas phase reactions, intermediate scale operations can be skipped. In such instances the chemistry and kinetics of the reactions are relatively simple and completely understood. However, when developing a process for coal conversion, since the chemist can barely define the nature of coal and has little chemical definition to offer, the chemical engineer needs to take discrete steps at several levels of capacity before bringing the technology to the status where it is ready for commercial application. In general, factors of 10 or 20 in scale are used in each incremental increase of capacity.

In the development of a coal liquefaction technology, several aspects control the system that is developed.

1. Coal is a solid material that needs to be introduced into a pressure system as a slurry made with liquids generated in the process. These liquids dissolve the ground coal, prevent coke formation in the reactor system, and release hydrogen to effect the chemical reactions desired.
2. Hydrogen must be generated, preferably from coal or unreacted high molecular weight residues. Hydrogen in the system reacts to lower the molecular weight of the feed to a desired product and to convert sulfur and nitrogen compounds in the coal to species that can readily be removed by chemical reagents in secondary processing equipment.
3. The heat of reaction that results from hydrogen reacting with coal (which is an exothermic reaction representing a heat evolution of 100–200 Btu/scf hydrogen consumed) must be under control.
4. Ash and "unreacted" or highly condensed coal species need to be removed from the final product in equipment that ultimately can be engineered for reliable large-scale commercial practice.
5. Reactor systems must take into account the hydrodynamic, erosion, and corrosion problems associated with multiphase systems.
6. Attention must be given to materials of construction and special components that are necessary when commercial plants are constructed. Components may need to be developed in parallel with the process.
7. Particular attention must be given to heating and cooling process streams in equipment that has not been engineered for, or applied to, multiphase systems. In such heat exchange equipment, ill-defined hydrogenation reactions occur with coal or its products.

Processing schemes differ in the methods of adding hydrogen to the recycle liquids, which are rich in partially hydrogenated polycyclic aromatics. Processes using hydrogen atmosphere at elevated pressure include:

1. noncatalytic methods, which have the least flexibility to control the chemical nature of the liquids and the range of product characteristics;
2. catalytic techniques that use supported catalysts in special reactor systems; and
3. donor solvent systems that separate distillates from the extracted product and hydrogenate the distillates that can be removed by conventional vacuum fractionation.

The donor systems catalytically treat the solvent in a separate processing step to tailor the chemical species in the solvent desired for the process. Maximum flexibility is obtained in donor solvent approaches.

The noncatalytic method is being developed by the Pittsburgh and Midway Coal Mining Company in a 50-ton/day pilot plant located at Tacoma, Washington. Extensive bench-scale operations have been carried out. This work is sponsored by the federal government.

The Southern Services Company, with the support of the Electric Power Research Institute, has carried out extended solvent-refining operations on eastern bituminous coals at Wilsonville, Alabama in a 6-ton/day unit. Continuous operations lasting up to 75 days indicate that at selected operating conditions, and on selected coals, the noncatalytic solvent-refining method can produce a solid low in sulfur content and relatively ash-free. More than 90% of the coal is converted to useful products. However, at relatively low pressures, and on certain coals, the process can become inoperable because of several factors. The interrelations of the factors are presently imperfectly understood.

It is believed that in the noncatalytic solvent-refining process, the ash in the coal serves as a weak catalyst for adding hydrogen to the solvent as it passes through the reactor system. Apparently, the ash particles retained in the reactor system can grow in size by an obscure mechanism.

Gulf Oil Company, Hydrocarbon Research Inc., the US Bureau of Mines, and the Lummus Company are exploring systems that use supported catalysts. The catalyst promotes dissolution of the coal and offers flexibility in achieving a product that can range from a heavy liquid fuel to a light product analogous to a crude oil with a wide boiling range. The latter light product requires relatively more severe conditions, i.e. increased consumption of hydrogen, than in solvent refining of coal.

Hydrocarbon Research Inc. has operated a 3-ton/day pilot unit. At the conclusion of an extensive program now underway, they are planning a larger-scale pilot plant at Ashland Oil and Refining Company's refinery in Kentucky. This development is supported primarily by the Energy Research and Development Administration and a consortium of industrial organizations. The process, designated as the H-coal process, features the use of an ebullated liquid-gas-solid fluidized system that has been applied commercially for desulfurizing and hydrocracking petroleum feedstocks in commercial plants.

The donor solvent approach to liquefaction of coal is being pursued by Exxon, Continental Oil Company, and the Universal Oil Products Company. Few details have been published. However, Exxon is operating a one-ton/day pilot plant at Baytown, Texas, and has probably achieved the most progress in advancing commercial practice of the donor solvent approach.

The following are being considered as methods for removing the ash from the liquefied coal product:

1. rotating mechanical filter devices that require a coating of clay to aid filtration efficiency;
2. cyclone separators that remove an ash-rich portion from the ash-free portion of the product;

3. continuous centrifuges that use centrifugal force to effect ash separation;
4. solvent precipitation methods that depend on separation by settling;
5. distillation of the product produced; and
6. combinations of the above.

It is uncertain which of the above is best or if there is a single method that can be used by all processes over the wide range of desired product characteristics. Efforts are underway to test the different methods. The recent experience with filters at the 6-ton/day pilot plant in Wilsonville, Alabama is encouraging in that relatively high filtration rates of 100–200 lb/hr per ft^2 of filter cross-sectional area have been achieved. However, such filters require high maintenance and labor as well as an expensive filter aid to form a cake that is coated on the filter. These factors recommend continuation of programs to explore several routes for removing ash from coal-derived liquid products.

Over the long term, hydrogen that is added to coal in producing liquids for the power industry should preferably be derived from the unconverted carbonaceous residues from coal or from the gaseous hydrocarbons produced. Least desirable is the use of hydrogen from natural gas, which represents current practice in the process industries. Hydrogen consumption in processes ranges from 2% by weight of the coal in the least severe liquefaction processes to as much as about 10% where refined light synthetic liquid fuels are desired. Little work has been reported on the use of feedstocks derived from coal liquefaction processes to generate hydrogen. Such work is necessary for economic application of liquefaction technology.

Table 4 summarizes the relatively large scale plants announced. Only a few of the coal liquefaction technologies have passed the first hurdle of demonstrated sustained operations at a capacity of 1–10 tons of coal/day. Testing of first-generation process technologies is likely, according to present schedules, to be completed by the late 1970s. If prudent risks are taken, consistent with petroleum industry practice, commercial plants using these processes may be on stream in the early 1980s. A second generation of process technologies may reach commercialization by the mid- to late 1980s.

Liquefied coal from these processes can have a wide range of characteristics. The fluidity of the liquid primarily depends upon the amount of hydrogen added to coal during processing. The product can range from light distillates, equivalent to refined kerosine and diesel oils suitable for gas turbine fuel, to a solid material that

Table 4 Larger-scale plants liquefying coal in the United States

Process	Location	Capacity (Tons/Day)	Status
Solvent-Refined Coal	Sheffield, Ala.	2000	design and cost estimates
H-Coal	Catlettsburg, Ky.	600	design and cost estimate, construction in 1976
Clean Fuels-West	not determined	500–1000	evaluation underway

Table 5 Inspections on coal and solvent-refined coal products from Wilsonville, Alabama

	Feed Coal	Solid Product
Source of Coal	Kentucky	Kentucky
Inspections		
Sulfur (wt%)	3.1	0.8
Ash (wt%)	8.9	0.1
Heating value (Btu/lb)	13,060	15,700
Volatile matter (wt%)	38	71
Melting point (°F)	—	360

melts at about 400°F and is essentially free of ash and reduced in sulfur so that it is acceptable for combustion as a solid in electrical generation equipment of conventional design.

Table 5 shows the coal feed and solvent-refined coal product from the Wilsonville, Alabama plant that was operated at 120 atmospheres. The sulfur content has been reduced from 3.1% to 0.8% and the ash content lowered from 8.9% in the coal to 0.1% in the product.

In order to utilize a new fuel, large-scale tests in commercial combustion equipment used for generating power are necessary. A test program that will prove out coal-derived liquids under boilers has been initiated and will culminate with combustion tests in commercial equipment used to generate electricity. In order to commercialize completely the processes for liquefying coal, the fuel preparation technology will need to be modified and perhaps the process technology changed, should the combustion tests so dictate. Thus, in order to gain acceptance, the entire system from coal to bus-bar electricity needs to be studied. The process will not be satisfactory until the product is acceptable for commercial utilization in power plants.

COAL GASIFICATION

Historically, the gasification of coal to gaseous fuel for combustion in gas engines is a very old idea dating from the time of the creation of the electricity industry. In Europe coal was blown with air to produce a reducing gas that could be combusted. The product gases were used to fuel primitive gas engines that generated electricity. That primary technology was superseded by the success of efficient steam turbines that evolved from naval applications. Funds were available to develop steam turbines for moving ships and pumping water from coal mines. The gas engine technology in the power industry was replaced and not resurrected for some 50 years.

Over recent decades modern pulverized coal combustion evolved. There are no commercial plants owned by utilities in the United States today using gaseous fuel from coal to generate electric power. In Europe gas turbines are presently coupled with combustible gaseous fuel from coke oven operations from the steel industry to generate power. However, such plants are relatively small in size, the largest representing a capacity of some 19 MW.

The earliest large-scale application of gaseous fuel from coal used for power generation was in Germany at Lunen where in the 1920s about 20 MW of capacity was operated with use of the historic Winkler process for gasifying coal by the then-new fluid bed technology.

For power generation, relatively simple processes can be used to generate combustible gaseous fuel from coal in contrast to complex low-capacity processes needed for pipeline gas. The ideal gaseous component for power generation is carbon monoxide, whereas for pipeline gas the ideal component is methane. In the United States, a program of development is underway to provide gaseous fuel from coal by using relatively simple technologies that can process coal at a high capacity. The fuel generated can be readily cleaned of particulates and sulfur by proven methods that use simple chemistry. The fuel gas generated can be used (*a*) to retrofit plants that now use natural gas, oil, or high-sulfur coal, and (*b*) to provide fuel for a gas turbine–steam cycle system that generates power efficiently.

Retrofitting appeared attractive as an alternative to stack gas scrubbers, which are required for coal-fired plants that are not in compliance with federal or state pollution standards. However, based on the cost of available gasification methods, it is now concluded that there is no large market for using the gaseous fuel from coal to retrofit existing power stations for emission control. Stack gas cleaning of flue gas and the use of low-sulfur coal represent lower cost options than does installing a new gasifier and integrating it with an existing plant.

The commercial options for coal gasification available for the power industry are shown in Table 6. The commercially available process technology is of relatively low capacity by the standards of the electricity industry. Gasification of coal in such plants represents a loss in efficiency, in that only about 75% of the heat in the coal shows up in the product gases. In order to obtain experience in the use of integrated gasifier-turbine systems, tests with noncompetitive systems are probably desirable to obtain experience in developing advanced methods useful for power generation.

In order to develop a coal gasification system that generates power efficiently, three essential parts need to be developed:

1. reliable, efficient, high-capacity gasification process technology;
2. control systems that provide the interface with gas turbines and steam cycles; and
3. efficient high-capacity gas turbines that are reliable for service on the gaseous fuel product from coal and that operate at inlet temperatures of about 2400°F.

The loss of efficiency from generating gaseous fuel from coal represents an enormous economic penalty to an industry that uses more than 98% of the heat in fuel purchased for power generation. This loss should be reduced by new gasification methods.

The reliability of gas turbines operating on light petroleum fuels at modest temperatures is still not acceptable to the power industry. In order for gasification systems to gain acceptability, reliability of gas turbines needs to be improved, operating gas turbine temperatures need to be advanced beyond the temperature of

Table 6 Commercial options for gasifying coal

Process	Type	Size of Gasifier (in MW of Power Generated)	Operating Pressure (atm)
Lurgi	moving bed	30	25
Koppers-Totzek	entrained	50	1
Wellman, Riley Stoker	moving bed	8	1
Winkler	fluid bed	15	1

about 2000°F available in today's equipment, and a high-capacity gas turbine capable of generating over 100 MW of power needs to be developed. In addition, the entire gasification-combustion-generation system must be integrated by using controls that allow the system to be operated to follow changes in daily demand for electrical output.

When such systems are developed and installed, acceptability by the electric power industry will represent a change in practice equivalent to the introduction of nuclear generation equipment in the 1960s. Reduction in the cost of power by using advanced coal gasification systems and improvements in emissions are of similar magnitude as achieved with nuclear power plants in the 1960s.

Types of Gasifiers

Three methods are available to the chemical engineer for contacting coal with oxidants that generate gaseous fuel. These include (*a*) *fixed beds*, (*b*) *entrained beds*, and (*c*) *fluid beds*.

The preferred gasifier useful for power generation should have the following characteristics:

1. an ability to handle a large quantity of coal per gasifier (its capacity should be at least the equivalent of that required to supply gaseous fuel for a large gas turbine);
2. an ability to convert essentially all of the coal to combustible gaseous components;
3. a minimum of gas cleanup problems, thereby requiring the gaseous product to contain (*a*) no tars or phenols, (*b*) low dust loadings, and (*c*) ash in an easily disposable form;
4. an uncomplicated configuration leading to ease of operation so that cycles in electric demand can be followed;
5. an avoidance of exotic materials of construction, such as complicated refractories, so that conventional construction can be used;
6. a maximum capability to be safely operated using a reliable coal feed system; and
7. an ability to operate at moderate pressure levels consistent with compression-ratio requirements of gas turbines.

All gasifiers to be used for power systems represent an approximation to an ideal system that satisfies these requirements.

FIXED-BED PROGRAMS   In the United States and in Europe relatively large tests of fixed-bed gasification technology as power systems are underway. Table 7 summarizes the major projects and their estimated dates of completion. The two large-scale tests of the historic Lurgi process represent a first step in an evolutionary process of developing an array of efficient coal gasification technologies useful for power generation.

The Lurgi process is a dry ash coal gasification system that mechanically deals with the properties of coal. It suffers from a number of limitations. An important economic penalty is the large amount of steam needed to moderate the temperature at the bottom to about 1800°F in order to maintain the ash in a dry, nonsticking form. Tars and phenols are produced as by-products that are potential pollutants requiring equipment for cleanup. These by-products are also a severe economic penalty for the process because of the cost of the cleanup equipment.

A consortium has joined in a development program at Westfield, Scotland that is being carried out by the British Gas Corporation. This new gasification development, which was originally pioneered in England at Solihull in the 1960s, can achieve the benefits of higher capacity (4–7 times), higher thermal efficiency (10–20% increase), and fewer pollutants compared with the Lurgi process.

ENTRAINED GASIFICATION PROGRAMS   Currently available is the Koppers-Totzek gasification system. This system has been applied for generating synthesis gas (hydrogen plus carbon monoxide) from coal using molecular oxygen, at locations outside of the United States. It suffers economically from its relatively low capacity and relatively low thermal efficiency. It represents a relatively high cost option when considered for generating electric power, compared with conventional pulverized coal combustion with stack gas scrubbing of the flue gas. There are no plans known to apply the Koppers-Totzek gasification system for power generation from coal in the United States. Advanced entrained gasification technologies are being developed in several programs. Table 8 shows the projects and their estimated completion

Table 7 Major fixed-bed projects for power generation systems

Project	Source of Funds	Size (MW)	Date of Completion
Powerton (Lurgi)	EPRI and Commonwealth Edison	30	1980
Steag-Lunen (Lurgi)		150	1976
Bureau of Mines/ Tennessee Valley Authority	ERDA	1	—
British Gas Corp. (slagging gasifier)	EPRI and other organizations	120	1978[a]
General Electric	General Electric	1	1978

[a] Gasifier only, not power-generating system.

Table 8 Entrained gasifier development projects

Method	Source of Funds	Capacity (MW)	Date of Completion
Pressurized Koppers-Totzek	Shell Oil, Koppers GMBH	15	1978
Combustion Engineering	EPRI, ERDA	10/100	1978/1983
Babcock & Wilcox	EPRI	40	1980
Foster Wheeler	ERDA and utilities	40	1981/1985
Texaco	Texaco, Inc.	?	?

dates. The new advanced entrained processes promise high-capacity, cost-competitive methods relative to conventional pulverized-coal-combustion practice that could have application to electricity production in the 1980s.

FLUID-BED PROGRAMS Relative to the development of advanced moving-bed and entrained-bed coal gasification systems, the status of fluid-bed technology is less advanced. Several small pilot plant programs at about one-MW capacity are being tested by the Institute of Gas Technology, Westinghouse Electric, FMC Corporation, and Occidental Petroleum Company. It will probably be the late 1980s before such systems are developed commercially.

Advanced technologies also being explored use molten salts of alkali metals to gasify coal. Severe engineering problems need to be solved in this process. Atomics International and the M. W. Kellogg Company are pursuing a pilot plant program at a utility site.

Advanced fluidization techniques that promise high-capacity systems for gasifying coal are being studied at the City College of New York.

SUMMARY

The development of processes for coal gasification power-generating systems requires large capital expenditures. To obtain a test at an operational scale of about 100 MW, in which the developed system can be considered commercially available, usually requires a typical time schedule of 3–8 years and funding levels of 100–200 million dollars.

World events as well as technical factors will decide whether the solutions under current development will be timely. It is recognized that the successful completion of the tasks will be an achievement for the worldwide electricity community.

Both liquefaction and gasification will have an important place in the future economic production of electric power. The power industry will have a need to convert hundreds of millions of tons of coal to clean fuels.

ACKNOWLEDGMENT

The contributions of Messrs. Neville Holt, Ronald Wolk, and Bert Louks to this paper are acknowledged.

Copyright 1976. All rights reserved

THE NUCLEAR FUEL CYCLE ❖11005

E. Zebroski and M. Levenson

Electric Power Research Institute, Palo Alto, California 94303

NUCLEAR POWER CAPACITY AND GROWTH—UNITED STATES AND WORLDWIDE

As of 1976, the United States has about 55 large commercial power reactors in operation, totaling just under 40,000 MW. A total of nearly 180,000 MW is firmly committed and under construction for operation through 1985. This is the net amount after deferral or cancellation of an additional 60,000 MWe previously on order. Additional capacity will start commercial operation at a rate averaging 8000 MW/year from 1976 to 1979; additional capacity will come into operation at an average rate of 20,000 MW/year during the period 1980–1984 inclusive (1, 2). The electrical generation from this nuclear capacity will rise from a current value of about 7–8% of all thermal generation in the United States in 1975 to about 12% in 1980 and about 24% in 1985, which is the equivalent of a saving in fossil fuels of 5–6 million barrels of oil per day. For comparison, total imports in 1975 are under 4 million barrels per day.

Considering only light-water reactors (LWRs) (effectively excluding Canada, the United Kingdom, and Russia) US nuclear capacity is about 55% of world nuclear capacity, divided among twenty different countries as of 1975 (see Table 1). This

Table 1 Present and projected nuclear capacity (GWe)[a]

Region	Type	1975	1980	1985
United States	LWR	39.2	80.0[b]	178.0[b]
	others	1.2	1.2	1.5[b]
Western Europe	LWR	24.8	86.8	206.7
	others	11.5	15.8	15.8
Asia and others	LWR	7.1	27.3	63.2
	others	6.1	15.9	17.9
Total	LWR	71.1	194.1	447.9
	others	18.8	32.9	35.2
	all	89.9	227.0	483.1

[a] Source: (1, 2).
[b] Adjusted for commitment as of December 1975.

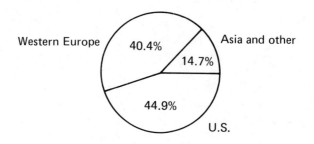

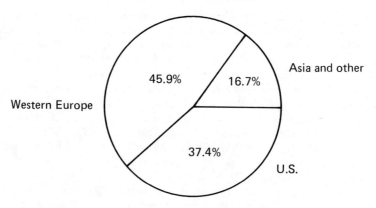

Figure 1 Present and projected worldwide nuclear capacity.

falls to about 40% of total world capacity by the start of 1985 with the commitment and construction of about 238 GWe of LWRs overseas. The overseas reactors are dominated by Japan, with 56 units totaling 49,000 MW, followed by France with 21 units and Germany with 20 units, totaling together about 37,000 MW.[1] Sweden and Spain follow with 10 and 9 reactors, respectively, totaling about 7000 MW each. A large number of countries, including Belgium, Italy, Finland, Switzerland, and Taiwan, have 6–7 units each under construction, typically totaling about 5000 MW. Iran has 4 units scheduled, and India, Korea, Mexico, and the Philippines have 2 reactors each planned or in operation.

Not counted in the totals above are reactor types other than LWRs. These predominate mainly in the countries of their origin—namely, Canada, with about 11,000 MW of heavy-water reactors operating and scheduled through 1985, and

---

[1] This text refers to firm commitments. Projections overseas are considerably higher, as shown in Figure 1 and in (2).

Britain, with about 11,000 MW of gas-cooled reactors and an initial commitment for heavy-water reactors. Some heavy-water plant capacity has been ordered by (or is operating in) India (6 units, 1300 MWe), Pakistan, Korea, and Argentina. Gas-cooled reactors are represented mainly in Britain, with small additional capacities in France and the Soviet Union, plus two prototype reactors in the United States. While there remains an active interest in advanced versions of gas-cooled reactors, including potential for high-temperature process heat in Germany, the absence of commitment activity precludes significant additions to electrical generation by gas-cooled reactors in the coming decade.

Liquid-metal-cooled breeder reactors are represented by four medium-scale prototypes totaling 1500 MW operating or near completion, none of them in the United States. In the United States a medium-sized prototype project is under way (Clinch River Breeder reactor) and, although beset with some uncertainties, mostly political, Clinch River appears likely to be completed in the coming decade. An additional medium-scale prototype breeder reactor is under construction in Germany.

Design studies for a full-scale commercial breeder reactor are under way in the United States in a joint program supported by the Electric Power Research Institute (EPRI) and the Energy Research and Development Administration (ERDA). These design efforts have the potential for maturing into one or more commercial plant commitments in the United States by about 1980. Of the large commercial plants, only the French project is reasonably likely to reach operation within the coming decade, but additional large-scale projects in Germany, Japan, Britain, and the Soviet Union are likely to follow soon thereafter. While it is technically feasible that an initial set of large-scale commercial breeder reactors could come into operation by the mid 1980s, the late 1980s appears a more likely target date in the United States, since progress is dependent on resolution of institutional, political, and regulatory obstacles (which are discussed elsewhere in connection with the LWR fuel cycle).

FUEL PERFORMANCE IN LIGHT-WATER REACTORS

Mechanical Performance

Mechanical performance of nuclear fuels in LWRs has generally been more than adequate, but with extremes that in some plants have required early displacement of certain batches of fuel, and conversely, extremes where fuel has performed flawlessly, even beyond design expectations (3, 4, 4a). The occasional instances of less-than-adequate performance in the years 1969 through 1972 have been traced to three main sources: hydriding of zirconium cladding due to residual moisture left in UO_2 pellets during fuel fabrication; cladding creepdown and cladding collapse when associated with pellet densification; and pellet-clad interaction associated with certain instances of rapid power increase of the reactor or rapid changes in local power shape. The first two causes of fuel failure have been essentially eliminated or made statistically insignificant by design and process changes evident in fuel manufactured since 1972. The third cause is now kept under control by restrictions in rate of change of total power (for pressurized water reactors, PWRs) and for rate of change of power and local power shape [primarily for boiling-water reactors (BWRs), but also for some maneuvers for PWRs]. Further design changes to more finely divided (smaller-diameter) fuel have been made for both BWRs and PWRs, which should eliminate some of the restrictions on maneuvering rates that at present cause some loss of plant productivity of both BWRs and PWRs. The performance of a particular fuel rod is influenced by (a) the interaction between the UO_2 fuel pellet column and the zirconium alloy cladding tube, and (b) the interactions of the fuel and cladding, separately, with the reactor core environment. The latter includes the effect of neutron flux, temperature, stress, and trace elements in water and fuel. The various fuel and cladding phenomena are summarized in Figures 2 and 3. Most of these phenomena have been studied in detail. Extensive research, develop-

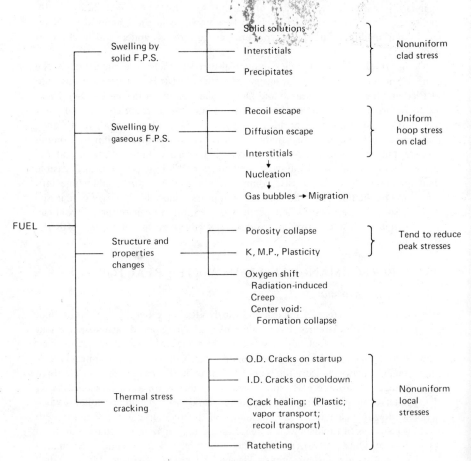

Figure 2 Phenomena occurring in fuel.

ment, and testing are under way to further optimize fuel and cladding properties with respect to irradiation performance.

Table 2 shows the main factors that have been shown to limit fuel performance—

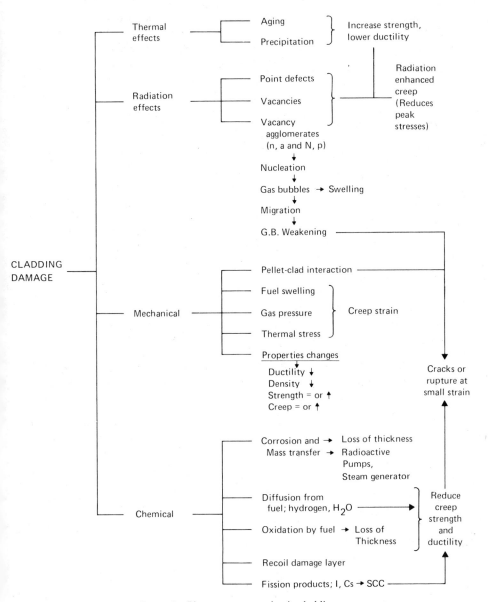

Figure 3 Phenomena occurring in cladding.

Table 2 Main factors that have limited fuel performance and remedies

Factors	PWR	BWR	Remedies
Hydriding of Zr	X	X	Elimination of moisture in fabrication
Scale deposition		X	Elimination of copper tubing from feedwater heaters
Enrichment errors	X	X	100% rod scan
Clad collapse	X		Prepressurized cladding, stable pellets
Pellet densification	X	X	Stable structure
Other manufacturing and handling defects	X	X	<5% of defects
Clad corrosion or fretting		X	Rare; control of clad quality and cleaning; spacers
Clad growth and bowing	X		Tube tolerances; axial clearances; spacer design
Channel bulging		X	Thicker wall; control of residual stress
Pellet-clad interaction on power increases	X	X	1. Slow power rise (PWR) 2. Plus local power shape control (BWR) 3. Plus fuel design changes (both)

either by causing cladding to leak, or by distortion of the cladding such that it becomes advisable to discharge the fuel prematurely because of increased possibility of leakage on further operation.

HYDRIDING AND RELATED EFFECTS By far the largest single cause of fuel failure has been so-called hydriding, which is due to the entry of excessive amounts of hydrogen into the zirconium cladding with consequent embrittlement or local blistering of the cladding (5). Zirconium alloys normally corrode slowly and uniformly in high-temperature water. A fraction—typically 10–20%—of the hydrogen produced by the corrosion of zirconium in water enters the zirconium metal and diffuses away from the surface. The hydrogen also diffuses down the temperature gradient from the hotter region of the fuel to cooler regions. Under certain conditions, the rate of hydrogen entry into the cladding locally exceeds the rate of diffusion away from the surface; then the solubility limit of hydrogen is exceeded locally and the zirconium hydride phase forms. Zirconium hydride is about 20% less dense than zirconium metal, so the precipitation of hydride is associated with the formation of local blisters and brittle regions of the cladding. On application of the stresses and strains associated with fuel operation and especially with changes in local power level, the brittle regions can crack and lead to perforation of the cladding.

The source of hydrogen on the inside of the fuel rod is generally residual moisture adsorbed in the UO_2 pellets. Traces of adsorbed hydrogen are also present for some fuel microstructures. Obviously, isolated instances of gross contamination by water or organic compounds can also be a source of hydrogen, but these occurrences are now relatively easy to guard against and apparently have been very rare relative to systematic residual traces of moisture in the UO_2 pellets.

The principal source of moisture in pellets appears to be simply absorption from the atmosphere. Pellets that have less than 7–8% porosity (greater than 93% of theoretical density) have largely only closed porosity—that is, porosity that does not communicate with the surface of the pellet. At lower densities, increasing communication with the surface occurs, and the accessible adsorption area can hold from 10 ppm to over 50 ppm of water. Moisture retention by pellets can be eliminated either by maintaining pellets in a moisture-free environment during cooldown after sintering and subsequent handling prior to sealing into the fuel rods, or by vacuum-degassing pellets at elevated temperatures either just before or after loading into cladding tubes. Routine maintenance or attainment of hydrogen and moisture content of the order of 1 ppm or less is possible by such means. An additional margin of safety against "the tail of the distribution" of retained moisture and hydrogen can be provided by the incorporation of a getter in the plenum region of the fuel rod. The getter is chosen from a class of alloys that do not form stable, coherent oxide films and that react with moisture and/or hydrogen several orders of magnitude more rapidly at operating temperatures than does the cladding itself.

A variety of mechanisms other than the simple presence of moisture have been postulated to account for or contribute to the occurrence or rate of hydriding. For example, fluoride ion, chloride, or iodine can accelerate the entry of hydrogen into zirconium, presumably by reducing the resistance of the oxide film to permeation by hydrogen. Also, concern has been expressed on the effect of scratches of the interior surface of cladding caused by loading pellets into the tubes. The condition of the surface—for example, sandblasted, electropolished, or etched by a hydrofluoric acid–nitric acid mixture—has also been postulated to influence the occurrence and rate of hydriding. None of these hypotheses has been substantiated by a majority of investigators when the levels of trace contamination and the operating conditions typically present in manufactured fuel are tested. However, since anomalous behavior of even a few thousandths of a percent of the inside surface of zirconium tubing could lead to operationally significant leakage rates, most manufacturers appear to consider it prudent both to specify and to maintain extreme vigilance against inadvertent contamination of either the pellets or the cladding. The observation of many batches of fuel that perform to nominal lifetime with cumulative rod failure rates of the order of 0.02% or less suggests that this issue is now reasonably well understood and under control, at least for fuel operated at moderate peak power levels.

PELLET DENSIFICATION AND CLADDING CREEPDOWN   During the period 1971–1973, several instances were observed in which batches of fuel developed gaps in the fuel column (6). In addition, the creepdown and collapse of the cladding under the action of external coolant pressure (2200 psi in PWRs) was observed in 7–10% of the fuel rods in certain regions of two reactors (Beznau I in Europe and Ginna in New York). Subsequently, extensive phenomenological research, as well as irradiation testing programs, found that certain microstructures of UO_2 pellets (grain-size and porosity-size distribution) were subject to further densification at relatively low temperatures in reactor operation despite the fact that the pellets had been sintered at much higher temperature (typically 1650°C) during manufacture. An intense series of investigations was conducted from 1972 to 1974 in the United States, the United Kingdom, Germany, and Canada. These investigations are now in general agreement that stable microstructures can routinely be produced at either low or high densities by specifying relatively coarse grain size, with significant amounts of large-sized porosity typically in the range of 5–20 μm, and by avoiding significant amounts of fine porosity of the order of 1 μm or less (7–9). When these conditions are satisfied, the additional densification of the fuel is typically less than 1%, either under irradiation or under extreme additional sintering out of reactor (for example, 1700°C for 24 hr). Essentially all fuel manufactured since 1973 has been stable with respect to such densification.

Even before densification and gaps in fuel columns became a significant operational issue, the creepdown of cladding (reduction in diameter under the combined influence of external pressure and radiation-induced nonthermal creep) had been widely observed in PWR fuels. Prepressurizing the cladding during manufacture to a level of several hundred pounds per square inch of helium served to reduce the collapsing stress due to external coolant pressure and reduced the rate of cladding creepdown such that nominal contact with the pellet was typically delayed to the second or third cycle of operation. The additional incentive of the operating penalties associated with pellet densification led to general adoption of prepressurization for PWR reactor fuel.

A variety of other defects have occurred in isolated instances for both PWRs and BWRs (see Table 2). With the exception of pellet-clad interaction, the basic

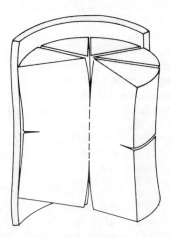

Figure 4 Schematic of pellet and clad deformations during start-up (dimensional changes exaggerated).

remedies for each of these are now well in hand and generally applied, so that normal vigilance in manufacture and quality control should reduce these sources to negligible proportions, even in the aggregate.

PELLET-CLAD INTERACTION   The most severe cladding deformation during normal operation is the localized elastic and plastic deformation of the cladding produced by UO_2-pellet interactions during power changes. The diagram in Figure 4 illustrates how the thermally induced distortion of the pellet generates local stress concentrations (and, therefore, local strains) in the cladding tube (10). A combination of such stresses sometimes augmented by chemical action by the chemical environment of the interior of the rod, namely fission product species, can result in so-called pellet-clad interaction (PCI) failures.

PCI is now controlled primarily by accepting limitations on rate of power increases, both overall and in local power shape.[2] This remedy, while workable, is expensive in terms of lost plant output from the partial capacity operation during slow increase (where otherwise the plant mechanically could be providing full power). This loss of output is estimated to range from 3% to 6% with fuels manufactured prior to 1974. For future fuel manufacture and future plants, the effect of these restrictions should be reduced—but probably not entirely eliminated—by the use of more finely divided fuel that permits a lower linear heat-generation rate (kilowatts per foot of fuel rod) while maintaining the same total power for the reactor. The effect of PCI in causing measurable strain of cladding is most pronounced at ratings above 10 kW/ft (although there is detectable elastic strain at even lower ratings). Ramps in power either overall or locally, which involve maneuvers above 10 kW/ft, must be performed slowly in order to minimize plastic strain of the cladding. Ramps above 10 kW/ft are better tolerated by the fuel during the first 20% of rod life—roughly the first half year of operation.

By proper management of fuel and power shapes, it is possible to design reactor core and fuel with initial peak ratings of 12–14 kW/ft, but which experience peak ratings of only 10–12 kW/ft after the first half-cycle, and less than 10 kW/ft after the first cycle. Such PWR and BWR fuel designs are summarized in the middle and right-hand columns of Table 3. For such fuel, the loss of capacity due to restrictions on power maneuvers should be reduced to less than 2%. For example, this is accomplished by the use of more rods of reduced diameter, per bundle of fuel, while the total number of bundles remains the same. For BWRs the modified fuel bundle consists of an 8 × 8 array of 63 fuel rods and 1 water tube, occupying the same space as previously occupied by a 7 × 7 array of 49 rods. For the PWR, a 17 × 17 array replaces former use of 14 × 14 or 15 × 15 arrays. The 9 × 9 and 16 × 16 options are also used by some vendors.

Because of the very large economic value of even a 1% gain in average capacity factor of nuclear plants (11), there remains a further incentive to reduce any loss of capacity (or premature discharge of some fuel). Accordingly, substantial research and development and analytical efforts are being directed to the fundamental

---

[2] The power increase is limited in rate only if preceded by operation at lower (local) power levels for several days.

Table 3 Comparative fuel design data—PWR/BWR options

	PWR			BWR		
Design Parameter	15 × 15	16 × 16	17 × 17	7 × 7	8 × 8	9 × 9
Fuel rod diameter (inch)	0.422	0.382	0.374	0.562	0.493	0.438
Core height (inch)	144	150	144	144	148	148
Clad wall thickness (inch)	0.024	0.025	0.023	0.032	0.034	0.030
Av. linear power (kW/ft)	7.03	5.2	5.43	7.10	6.04	4.75
Av. specific power (kW/kg UO_2)	32.7	37	32.6	18.7	21.7	21.8
Av. power density (kW/liter)	106	96	105	51	56	56
Pellet-clad gap—diametral (inch)	0.0075	0.007	0.0065	0.011	0.009	0.009

phenomena of PCI and to the possibility of basic design modifications that would essentially eliminate such phenomena completely as performance limitations on fuel at reasonable ratings. There is a further long-term incentive that the more quantitative understanding and control of interaction phenomena will eventually permit the use of higher linear heat-generation ratings. For example, increase in the average linear heat-generation rate from about 6 kW/ft to about 8 kW/ft would permit increasing reactor output from 1300 MWe to 1700 MWe without changing the size of the vessel. In addition, resulting higher specific power (kilowatts per kilogram of fuel) would have a small beneficial effect on the uranium supply required to provide inventory for a growing nuclear economy.

Effects of Fuel Defects

Fuel for LWRs typically operates for three or four annual cycles (after an initial transient in which part of the first charge of fuel resides in the core only one to two years). In the course of operation, about 3% of the total uranium atoms initially present undergoes fission with the resulting accumulation of fission products in the UO_2 ceramic matrix. Many of these fission products have short half-lives and rapidly decay to either long-life or stable isotopes. A few fission products have medium or long half-lives, and a few have physical properties such that diffusional escape from the UO_2 solid matrix is possible. These include noble gases, (isotopes of xenon and krypton, iodine, and cesium. (Detectable amounts of other elements can be found in the reactor cooling water, but these are generally a minor contribution to the total radioactivity, which is dominated first of all by activation of corrosion products of the primary system, and to a lesser extent by the volatile activities of the rare gases, iodine and cesium.)

Fuel defects typically are observed as microscopic fissures in the zirconium cladding, which eventually penetrate the cladding wall and permit the rare gas isotopes, primarily xenon and krypton, to escape to the reactor coolant. For the BWR, the xenon and krypton are removed from the circulating water by the steam separators and pass through the turbine with the steam to the condenser and to the air ejector, which maintains the condenser under partial vacuum. The gas extracted from the condenser by the air ejector is allowed to decay in a holdup line, and finally passed over a charcoal absorber bed for permanent holdup and decay of xenon and krypton.

For the PWR, xenon and krypton build up in the primary coolant during operation with defected fuel. Small amounts of these isotopes find their way to the secondary system through leaks in the steam generator (which is the heat exchanger between the primary and secondary system). For the PWR, charcoal beds also serve to trap the xenon and krypton for decay and disposal.

Early BWRs and PWRs did not trap the emitted xenon and krypton, but instead diluted them with large volumes of air and emitted them to the atmosphere in dilute form. At times, emission rates as large as $\frac{1}{2}$ C/sec were disposed of in this fashion. Because of the dilution effect and emission from a high stack, the "fencepost dose"—that is, the dose to a person or animal residing the entire year at the immediate boundary of the power plant—was typically in the range of 1–3 mrem per year, and rarely, if ever, greater than 5 mrem/year. With the advent of larger numbers of reactors, and an increased emphasis on control of emissions from both fossil and nuclear plants, all new reactors provide for adsorption of most of the rare gases.

The iodine isotopes emitted to the coolant predominantly stay in the water phase. (Small amounts do appear in the steam with the BWR and are also trapped on the charcoal bed.) The principal isotopes are iodine-131 with an 8-day half-life and iodine-133 with a 21-hr half-life. In the event of small leaks in the primary system, or when the reactor is shut down and opened up for refueling or repair, the radioiodine activities, if allowed to build up to levels greater than a few microcuries per cubic centimeter, can limit the access time for refueling or maintenance.

Because of the small amounts of radioactivity emitted by any single leaking rod, reactors typically continue to operate on a normal schedule with as many as 5–10 defective rods. With significantly larger number of defective rods—10–20 or more—the emission rate of radioactivity begins to approach either allowable license limits for coolant radioactivity, or prudent levels for allowable in-plant radiation levels. When this number of leakers occurs, it is common to employ one or more of the following remedies to limit radioactivity of the coolant:

1. reduction in power level of a local region of the core where the leakers are suspected to be, by insertion of control rods in this region;
2. reduction in total reactor power;
3. avoidance of power maneuvers (since emission rates increase after increasing or decreasing reactor power).

By use of the foregoing measures, reactors have very rarely been forced to shut down due to fuel defects, even during the early period when occasional batches of fuel showed relatively high rates of perforations. However, there has been a significant loss of output in given periods. Also, when the trend in increasing radioactivity indicated that significant reductions in output might be forced later in the cycle, it has often been advisable to reschedule an earlier refueling shutdown than would otherwise be required and to replace the normally discharged fuel—together with any additional fuel suspected of having leakers—in order to assure full-capacity operation during the following cycle.

The forced outage associated with reactor fuel or core (including control rod drives) has been historically less than 3%. However, the total effect of forced outage, added scheduled outage, and reductions in capacity have been estimated to aggregate about 10% in the 1972–1973 period, and somewhat less in 1974 (11). Of the five manufacturers currently producing LWR fuel in the United States, one has a current fuel defect rate (1974–1975) of the order of 0.02–0.03% (inferred from iodine levels in the coolant). Another, with relatively small production, has even lower fuel defect rates to date. BWR fuel, on the other hand, has shown a cumulative failure rate of 0.76% based on direct examination and somewhat lower values inferred from iodine levels (3). Considering only fuels manufactured since 1972, BWR fuel also shows a failure rate well below 0.1%. Actual failure rates may be as low as 0.01%, but this cannot be firmly established for another one or two years until the remainder of the earlier fuels has been discharged.

Behavior of Fuel in Postulated Loss-of-Coolant Accident

One of the concerns affecting both fuel design and reactor operation is the nature of fuel behavior in the event of a postulated loss-of-coolant accident (LOCA). In a LOCA it is postulated that coolant escapes from the primary system through a large rupture in the primary pressure boundary. The fission heat source is turned off by a combination of loss of moderator and insertion of control rods. However, the fuel continues to emit heat due to decay of short-lived fission products contained within the fuel, initially at about 10% of the previous full-power heat rating, then decaying with time to the -1.2 power. While the coolant is actually escaping, the fuel continues to be cooled by the flashing mixture of water and steam. Extensive analytical and test programs are under way to define this phase of "blowdown heat transfer." Part way during the blowdown, emergency core cooling systems (ECCS) are brought into play. These resupply coolant either by entry from the bottom of the reactor vessel (plenum fill) or by spray into the top of the fuel (core spray). The regulatory requirement is that a "coolable geometry" be maintained throughout this series of events, such that the possibility of an uncooled region that continues to heat up and eventually starts to melt (core meltdown) is avoided. Initially, for old reactors without modern designs of ECCSs, peak fuel temperatures of the order of 1200°C were calculated. At such temperatures, rapid and extensive oxidation of the zirconium cladding by the steam would occur, and the maintenance of a coolable geometry (with high assurance that postulated core meltdown would not occur) was in dispute.

More recent calculations of peak cladding temperatures, which take into account the heat removal that occurs during the blowdown phase of a LOCA, have been in the range of 500–1000°C, depending on the particular design, specific power of operation, and assumed effectiveness of the ECCS.

In order to assure coolable geometry, the regulatory requirement calls for no more than 17% oxidation of the cladding for the peak local power and temperature condition, including the oxidation that may occur from the internal surface of the cladding as well as the external surface. Major efforts are under way to further quantify the oxidation rates of zirconium over the range of temperature and pressure conditions involved in such calculations. Work to date tends to confirm that the Baker-Just correlation is conservative (that is, predicts higher oxidation rates than are likely to occur in reality) over most of the temperature range now of interest in plant licensing and operation.

The mechanical properties of the cladding both as a function of temperature and as a function of the degree of oxidation that has occurred for a given time-temperature history are also important in licensing calculations. Because of the nature of the safety issue involved, it is necessary to obtain not only the reasonably probable values of oxidation rates and properties but also the probability distribution of extreme values of such parameters—including the outer ranges of measurement error, inherent experimental uncertainties, and materials and environment variations. If one takes the worst case of every conceivable variation in materials and environment and uncertainty in parameters and analytical models, this results in a large degree of conservatism and leads to calculated core meltdown for cases in which the physical reality of such an event is actually extremely small.

One remedy for calculated excessive temperatures and a LOCA is to restrict the peak local power in the reactor. Significant loss of potential reactor output has occurred in the past and is likely to continue to occur, due to lack of full definition, together with estimates of statistical uncertainty, of the data and calculations on blowdown heat transfer, zirconium properties at elevated temperatures, and oxidation rates as a function of time-temperature history. All of these issues are being pursued in substantial R&D programs conducted by industry, EPRI, and the Nuclear Regulatory Commission. As the data continue to develop from these programs, added evidence indicates that early calculations indeed had large margins of safety in them; it is highly probable that actual peak temperatures in a LOCA event, when and if one ever occurs, will be substantially lower than those calculated for licensing purposes. The major remaining concerns appear likely to center on the effectiveness and statistical reliability of operation of ECCS.

NUCLEAR POWER COSTS

Economic Performance of Fuel

The incremental cost of electric power generation in nuclear plants is now (in 1974–1975) less than one half of the average incremental cost of generation using coal as fuel, and less than one fifth of the cost of generation using oil or gas (12–14). The incremental cost of nuclear generation is perhaps as low as 10–15% of the incremental cost of spot purchases of oil, and in some cases low-sulfur coal, when additions to supplies contracted in previous years must be found. At present essentially all coal available for electric power generation in the United States is consumed, and the added energy requirement is made up by oil and gas. Any displacement of fossil fuel by nuclear generation is equivalent to a reduction in the requirement for imported oil, now at $12/bbl ($13–$16/bbl delivered) but expected to rise in the next few years to $18–22/bbl. Taking the lower price, it is evident that even the relatively small fraction of electricity generation available from nuclear plants in 1974 provided a saving in the national (imported) fuel bill of the order of $1–$2 billion, and that this will rise to about $8 billion annually by 1980 and to over $20 billion annually by 1985. Taking into account probable higher oil prices and some escalation, displacement values of $40–80 billion per year are not improbable (in 1985 dollars). Only a small part of this national economic benefit will be offset by increases in the cost of uranium and separative work and increases in costs associated with reprocessing, shipping, waste disposal, and safeguards.

Table 4 shows typical ranges in the elements of cost of the nuclear fuel cycle. Since most current nuclear fuel consumption corresponds to the earlier unit costs (listed in the column showing 1965–1973 costs), fuel cycle costs ranging from well below 2 mills/kW-hr up to about 2.4 mills/kW-hr are typical for 1975 and earlier. Typical fuel cycle costs for the next few years should be in the range shown by the column headed 1975, or about 2.7 mills/kW-hr, rising to 4.5–5 mills/kW-hr by 1985. Eventually, with the reprocessing of spent fuel, a considerable credit for uranium and plutonium content may be realized. However, it is unlikely that this credit will be significantly larger than the costs of interim storage and eventual reprocessing, and in any event the present worth of these credits is small and may safely be neglected.

The long-term costs for incremental coal production are also subject to substantial

Table 4 Nuclear fuel costs 1965–1985

Factors (per unit uranium)	Range			Mills/kW-hr[a]	
	1965–1973	1974–1975	1984–1985 (estimated)	1975	1985
Ore cost ($/lb)	6–8	8–15[b]	15–25[a]	0.94	1.43
Conversion ($/lb)	1.0–1.25	1.25–1.40	1.40–1.80	0.09	0.10
Separative work ($/SWU)	28–38	38–43	50–100	0.58	1.24
Fuel fabrication ($/lb)	28–41	35–41	41–53	0.33	0.40
Shipping, reprocessing, and waste disposal ($/kg)[c]	20–40	40–80	100–150	0.15	0.28
Interest charges	12–15	12–15	15–18	0.61	1.23
Total				2.70	4.68
Coal (¢/10^6 Btu)[d]	30–45	45–150	150–220	14.10	20.70
Incremental cost advantage, nuclear over coal				11.4	16.0

[a] Using highest value in range for each time period; all figures in 1975 dollars.

[b] Spot purchases over $35/lb have been recorded, but on a very thin market.

[c] Or cost of long-term fuel storage and safeguards without reprocessing and neglecting present worth of uranium and plutonium credits.

[d] Assuming a heat rate of 9000 Btu/kW-hr for coal vs 10,000 Btu/kW-hr for nuclear. Considerably higher spot-purchase prices have been recorded for both uranium and coal, but on thin markets. Oil prices are 30–50% higher than coal.

uncertainties because of environmental, regulatory, and investment factors. The estimates in Table 4 are not meant to be predictive so much as to show the nature of the comparison over a reasonable range of assumptions. Table 4 suggests that over a fairly wide range of future possibilities in respect to both uranium and coal cycle costs, nuclear generation has a very substantial incremental cost-margin advantage over coal generation and an even larger margin over oil or gas.

Nuclear Plant Capital Costs vs Fossil Units

The discussion of fuel cycle costs indicates that for existing generating capacity there is a strong cost incentive to generate as much energy as possible with nuclear plants rather than with fossil units. In this case, the capital cost of the plant has already been expended, and the comparison is only between the incremental fueling costs of nuclear plants versus fossil plants (other operating and maintenance costs are comparable for nuclear and fossil units, and in any event amount to only about 10% of fueling costs).

On the question of commitment of new generating capacity, the relative capital costs of nuclear and fossil units must be considered, as well as effective capacity factor (which is the ratio of actual output for the year compared with the theoretical limit of output that would be obtained if the unit ran full time at full power).

The escalation in nuclear plant capital costs has been very rapid, with different elements of cost increasing between 15% and 30% per year since 1971. The overall effect is that the dollar cost of nuclear capacity coming into operation in the period 1975–1985 will average three to four times higher than the average cost of capacity installed prior to 1975. Figure 5 shows the main elements of this cost increase. While there are many hypotheses on the exact driving forces for each element of cost, the basic correlation is the cost of delay, and of slowing of the construction cycle from an average of four to five years prior to 1971 to the current range of seven to ten years. This is evidenced in the current distribution of costs, which shows that nearly half of the total costs of the plant is attributed to the cost of interest during construction and escalation. The cost of the actual equipment in the plant has fallen from 40% of the plant costs to less than 10% of plant costs. The causes of extended construction cycles are at least partly attributable to increased complexity and to indecisiveness or jurisdictional uncertainties—in the siting, licensing, and regulatory process. Other contributing factors include declining productivity of field construction labor, and the diversion of technical, engineering, and managerial talent to the ever-increasing complexity of regulatory processes.

Despite the dramatic increase in nuclear plant capital costs, the capital cost advantage of fossil generation capacity appears to be declining rather than increasing. The very low capital costs associated with gas-fired units are now unavailable due to the declining supply of natural gas and priority on its use for home heating. The lower cost of oil-fired units is unavailable in many areas where conversion to coal is being mandated by the Federal Energy Administration.

Many of the same factors that have escalated nuclear plant capital costs are now appearing also in coal plant capital costs. Extended siting and construction cycles are being experienced, as well as the slowing effect of increased complexity of environmental regulations. Under favorable circumstances, coal plant capital costs can be as much as 40% lower than nuclear plant capital costs, for example, about $450/kW for coal versus $700/kW for nuclear for plants completed about 1980. Where favorable circumstances permit realization of a capital cost advantage for coal, the maximum difference of about $250/kW is worth about 5 mills/kW-hr. This is still less than half of the 11–16 mill/kW-hr fuel cost advantage for nuclear units.

Despite the cost advantages to the consumer, a considerable number of nuclear plants have been delayed and a few have been cancelled (along with some fossil

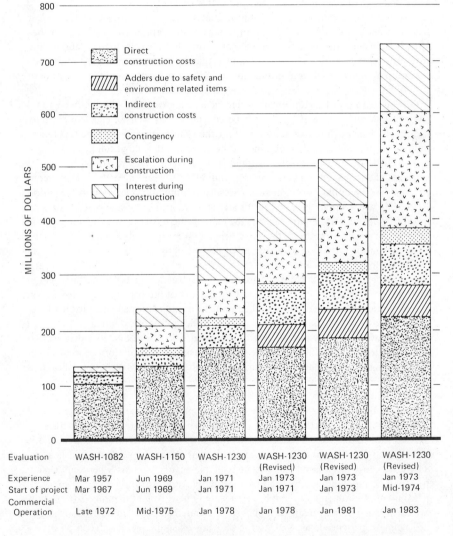

Figure 5   Elements of nuclear plant cost escalation. Source: (13); basis 100-MWe LWR.

units). Two basic reasons appear to explain such delays. First, there is only limited availability of capital to utilities. This is due in turn to declining net earnings, which are subject to extreme swings because of state and federal regulatory actions. As fossil fuel costs have escalated sharply, even small delays (or discounts) in the ability of utilities to pass on the increased cost of fuel, as a part of the cost of operation and of billing to customers, can create an almost immediate cash flow deficit. With earnings subject to such oscillations, both the stock and bond markets have become either expensive or unavailable for some utilities. The result has been to force the purchase of lowest-first-cost generating capacity even though the lifetime costs may be substantially higher. A second reason for some delays in nuclear plants is the earlier availability of fossil units due to shorter construction schedules. Coal plants can still be constructed in three to four years (although the siting time before construction is becoming longer). The quickest source of capacity is gas turbines, which can be oil- or gas-fired and can be put in place in about two years or less, but result in very high fuel costs—three to ten times higher than nuclear fuel costs.

Productivity of Nuclear and Fossil Plants

The comparison of capital costs for coal and nuclear plants is subject to the additional factor of the relative productivity of the two types as measured by the capacity factor attained. Figure 6 shows the annual and cumulative availability factor distribution for

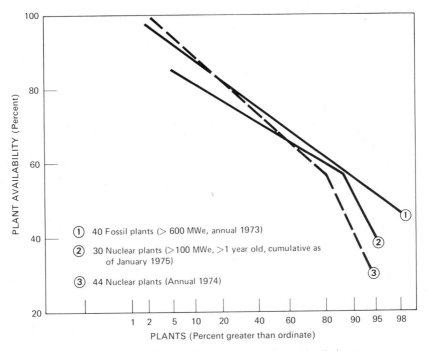

Figure 6 Availability distributions for nuclear and fossil plants.

large nuclear units and large fossil units. The average capacity factors are 58% for nuclear and 59.5% for fossil. The near coincidence of the averages and distributions for the capacity factors must be at least partly discounted since if both coal and nuclear units are available to operate on a given system, the differences in incremental fueling costs since 1972 would dictate a strong preference for operating the nuclear unit rather than the coal unit. Countering this effect, however, is the fact that many large coal units cannot be maneuvered at rates required by system load changes, so nuclear plants have sometimes been used to accommodate load swings despite the higher incremental cost of keeping the coal plant at full load.

Perhaps the most significant observation with respect to Figure 6 is that the upper halves of the distributions of fossil and nuclear plants are closely comparable on this key measure of productivity, and this suggests that a nominal 75% capacity factor used in economic evaluations in comparing plant purchases is realizable. For both types of plants the lower half of the distribution is also comparable, but shows ample opportunity for improvement.

Taking the actions that would be required to make the lower half of the distribution perform as well as the upper half for nuclear plants would be worth the equivalent of saving more than one million barrels of oil per day by 1985. Furthermore, the additional output would correspond to the productivity of over $20 billion worth of plant capital investment, which would be required to get the same output if productivity remains unchanged. Table 5 illustrates these relationships for an assumed capacity of about 200,000 MW, which are now committed or under construction for completion by 1986–1987.

While fossil and nuclear plants have roughly the same distribution of performance as measured by capacity factors to date, both distributions are subject to deterioration as well as improvement in the future. The use of lower-grade coals, or the requirement of changing from one variety of coal to another, reduces the productivity of a coal plant. The addition of increasingly complex antipollution equipment reduces the effective net generation of the plant so the overall net energy output deteriorates and in some cases becomes only about equal to nuclear units. Finally,

Table 5 Oil equivalent displaced by nuclear generation capacity of 200 GWe in 1986[a]

Measure of Displacement	Nuclear Plant Average Capacity Factor			
	0.55	0.65	0.75	0.85
Output (GWe-yr/yr)	110	130	150	170
Oil equivalent (million bbl/yr)	1480	1750	2020	2300
Oil equivalent (million bbl/day)	4.1	4.8	5.6	6.3
Oil equivalent (billion $/yr)				
Oil at $7/bbl	10.4	12.3	14.2	16.1
Oil at $11/bbl	16.3	19.3	22.3	25.2
Oil equivalent with 240 GWe (million bbl/day)[b]	4.9	5.7	6.7	7.6

[a] GWe = 10^6 kW electrical; GWe-yr = 8.76×10^9 kW-hr.
[b] Federal Energy Administration "accelerated supply" case.

Table 6 Capital requirements of coal vs a nuclear system[a]

	Nuclear	Western[b] Coal	Eastern[c] Coal
Plant	2 × 1100 MWe	3 × 700 MWe	3 × 700 MWe
Completion date	1/85	1/85	1/85
Specific investment ($/kWe)			
Power plant	950	765	805
Fuel cycle	110	180[d]	220[d]
Total	1060	945	1025

[a] Including escalation and interest during construction at 7.5%/yr and 8%/yr, respectively. Source: (13).
[b] Midwestern power plant without SO_2 cleanup.
[c] With SO_2 cleanup.
[d] Assumes existing rail system can accommodate additional traffic.

the increased complexity plus increasing reliability requirements of antipollution equipment force up the capital costs of coal plants. Table 6 shows an estimate for a coal plant with pollution control and with incremental fuel transportation and handling investment, which yields capital costs essentially equivalent to a nuclear plant (14).

CLOSING THE NUCLEAR FUEL CYCLE

Reprocessing and Recycle

Figure 7 shows the nominal nuclear fuel cycle for LWRs, starting with mining of the ore and concluding with reprocessing of the spent fuel, disposal of the radioactive wastes, and recycle of the slightly enriched uranium and plutonium. Historically, at least ten very large-scale reprocessing plants have been built and extensively operated to recover uranium and plutonium from spent fuel. Four of these plants were built in the United States for the production of plutonium for weapons purposes, and several are still in partial operation.

Plutonium production of the order of 100–200 tons has occurred in the United States, and at least a comparable amount overseas, primarily in Russia, but also in the United Kingdom, France, and China. This corresponds to the chemical reprocessing of spent fuel containing over 200,000 tons of uranium. Three smaller commercial fuel-reprocessing facilities have been built, but none are presently in operation or have a firm schedule for operation.

The reasons for nonoperation of commercial reprocessing capacity are not due to lack of established technology. The basic technology of chemical separations for weapons production has been largely published and is relatively easy to duplicate in any industrial nation. The main obstacles in the United States are the rapid change of environmental and regulatory requirements over the last five years, culminating in the GESMO (Generic Environmental Statement on Mixed Oxide), which was

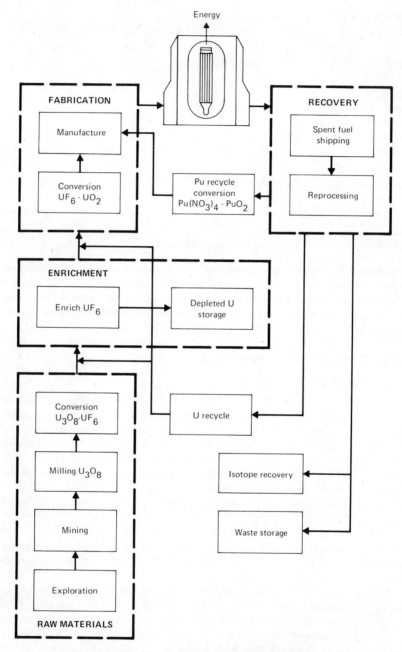

Figure 7 Nuclear fuel cycle light-water reactors.

issued in 1974 (15). At present the Nuclear Regulatory Commission is considering a delay in further resolutions of some of the questions raised in the initial draft of GESMO (16)—most notably those with respect to safeguarding fissile material such as highly enriched ^{235}U, plutonium, or ^{233}U.

In addition, there still remain substantial regulatory and political uncertainties on the shipping and handling of spent fuel and separated fissionable material. Also, the procedures, technology, and location for long-term disposal of radioactive wastes have been studied (17, 18), and the decisions for demonstration of one or more preferred approaches are beginning to be made. Assuming these issues are resolved in the next two to four years, large-scale reprocessing demonstration is likely to be under way by about 1982, and reprocessing capacity that can keep in step with the discharges of spent fuel is likely to occur by the middle to late 1980s.

Spent Fuel Storage

The near-term effect of this delay in the tail end of the fuel cycle is to raise the requirements for storage of spent fuel to accommodate a 7–10-year delay between discharge of fuel and the availability of adequate reprocessing capacity. All power reactors have a spent fuel storage pool typically capable of holding about one full core of spent fuel. For many reactors, it is possible to increase the storage pool capacity to hold two to three cores by relatively straightforward structural changes and special modifications to increase the permissible density of fuel storage. These offer the prospect of storage of three or more cores at the reactor site. Where this is possible, it will meet a 10-year storage requirement. Where this is not practical, shipping of the spent fuel to one of the existing reprocessing facilities is feasible. These facilities are expanding their spent fuel storage capacity to accommodate the anticipated storage overflow beyond that available at reactor sites.

Other options, so far not being visibly exploited, include the development of storage capacity at new sites that are potential locations of reprocessing plants in the 1980s, or adapting ERDA facilities to LWR storage needs. The large-scale reprocessing plants operated by ERDA [formerly the Atomic Energy Commission (AEC)] at Hanford, Washington, and Savannah River, Georgia, could be readily modified to provide a very large volume of secure spent fuel storage. The concern that significant numbers of reactors might have to shut down for lack of spent fuel storage capacity appears to be overstated (unless total regulatory blockage is postulated). Table 7 shows the buildup of spent fuel requiring storage (19) and some of the options for accommodating such storage. The relatively high costs shown in Table 4 for estimated costs of reprocessing (or storage plus reprocessing) are intended to include an estimate for the incremental capital costs of expanding pool storage capacity, either at reactors or at reprocessing plants, and the associated safeguards and procedural costs.

Radioactive Waste Disposal

The preferred route for long-term disposal of radioactive wastes is deep burial in a geologically stable formation, well removed from water tables. Taking naturally occurring uranium, thorium, and the radioactive daughter products as "tracers"

Table 7 Spent fuel storage capacity and requirements

Projected Fuel Storage Facilities		Projected LWR Spent Fuel Discharge	
Facility	Capacity (metric tons, uranium)	Year	Cumulative Discharge (metric tons, uranium)
On-site storage basins at reactors (unmodified)	7240[a]	1975	1659
		1976	2649
		1977	3832
		1978	5154
AGNS Reprocessing Plant, Barnwell, SC	360[b]	1979	6596
		1980	8312
		1981	10,242
GE Reprocessing Plant, Morris, Ill.	90[c]	1982	12,712
		1983	15,775
		1984	19,829
NFS Reprocessing Plant, West Valley, NY	260[d]	1985	24,633
Estimated		1990	58,685
Estimated		2000	200,913

[a] Average design accommodates about 1.33 cores.
[b] Facility complete but not licensed to store fuel.
[c] Licensed to store fuel.
[d] Licensed capability but modification required to reach full capacity.

for the likelihood of entry of heavy elements into the biosphere, several investigators have concluded that the likelihood of entry of reasonably well-buried fission product wastes into the biosphere can be made a lower order of magnitude than the interaction of natural occurring radioactivity with the biosphere (20, 21). The first meter of the earth's crust contains more uranium and radium and other daughter products than the cumulative net radioactive production of a nuclear power economy growing for a century. The disposal of the man-made radioactivity—in stable formations typically 1000–2000 m below the earth's surface—would seem to offer reasonable assurance that entry of even small amounts of this material into the biosphere is much less probable than for naturally occurring radioactivity.

One principal complication of deep disposal has been concern with the heat generation associated with the residual fission products. For the first few years of decay, these heat-generation rates are sufficient that a relatively conductive and easy-to-excavate location such as a natural salt dome has seemed to be attractive to permit conduction of the residual decay heat without requiring excessively high temperature at the source. The cooling requirement is essentially eliminated if the spent fuel is stored for a period of the order of ten years before the separated radioactive waste is encapsulated for permanent disposal.

In part because of the relatively small amount of radioactive waste produced by the civilian power economy (relative to that already produced as a by-product of weapons plutonium production), the nominal policy established by the AEC has been to use retrievable surface storage through the 1990s, presumably to be followed by permanent "deep bed" storage for ultimate disposal (22). This policy was intended to allow ample time for development, testing, and demonstration of both the technology and the regulatory requirements and procedures for both interim and ultimate disposal. However, this policy also has had the disadvantage of seeming to imply that ultimate disposal, even at the relatively small quantities required for the next two or three decades, is extremely difficult or fraught with unusual hazards for society or the environment. More recently, ERDA appears committed to pursue significant near-term demonstration projects on ultimate disposal. This involves converting the by-product radioactive waste stream to a highly insoluble mineral form (a ceramic, a glass, or concrete). (Some studies in Europe favor incorporation of some of the lower-level wastes in an asphalt or coal-like matrix for burial.) After conversion to an insoluble form, the material would be sealed into a heavy-walled cylindrical steel container, which would be sealed hermetically and placed in the geological stratum for disposal (23). Details of final placement would vary with the geological environment and the age of the fission product waste, which would determine whether or not special efforts to ensure thermal bonding to the environment were required. This general approach to disposal assumes that the overall probability of transport to the biosphere is made small compared with natural radioactivity by the product of the following factors:

1. incorporation of the fission products in insoluble glass or ceramic, which would be resistant to significant leaching over geological periods of time, even if groundwater were to obtain access to the ceramic;
2. enclosure in cannisters for convenience in handling and emplacement and additional isolation from the environment while the activity levels are high;
3. placement in regions shown by geological evidence to have been stable for tens to hundreds of millions of years, and in a location where access of groundwater is highly improbable.

A number of factors also must be considered with respect to the placement of the radioactive-waste disposal site. Foremost, of course, is the geological suitability. However, the surface operation, shipping, and handling associated with the disposal require the environment to be one that makes such operation acceptable to local citizens and require also state and regional public and political acceptance. In addition, assuming the possibility of eventual obsolescence of current conventional nuclear energy sources (for example, the postulated wide availability of fusion-based or solar-based bulk energy sources in the twenty-first century), there is the requirement that such a disposal operation be capable of being closed off in a manner that makes the probability of future interactions with the biosphere vanishingly small, even in the absence of any further custodial effort by mankind.

In addition to the geological burial, several other options have been actively explored, at least in concept. These include disposal of cannisters in sedimentary continental shelf regions where sedimentation from continental rivers would provide progressively deeper burial below the bottom surface of the ocean, and disposal of cannisters into antarctic ice sheaths, potentially reaching disposal depths greater than one mile. Other schemes—not as yet feasible, but of some conceptual interest as by-product developments of other technology—include "burning" fission products and actinides in reactors or in the excess neutrons available from hypothetically economical fusion reactors, and extraterrestrial disposal by launching payload-carrying rockets into the sun.

Consideration of multiple approaches to nuclear waste disposal, while it is likely to protract near-term resolution and adoption of a preferred set of waste management alternatives, may have the eventual benefit of assuring a high probability of developing an adequately safe and acceptable approach to nuclear waste management, and at the same time avoiding possible future delays.

Safeguards for Fissile Materials

SOURCES OF CONCERN ON SAFEGUARDS   Perhaps the knottiest public policy issue with respect to nuclear power is concerned with the safeguarding of enriched fissile material from possible diversion and use for terrorist threats. In principle, such terrorist threats could originate from any of the following:

1. capture or theft of any of the more than 80,000 nuclear weapons stored or deployed in various countries around the world;
2. capture or theft of weapons-grade materials from enrichment plants or military production plants or stockpiles;
3. theft or clandestine diversion of products from civilian reprocessing plants or fuel fabrication plants that handle either plutonium, ^{233}U, or ^{235}U of greater than 20% enrichment, or combinations of these materials;
4. capture or theft of fabricated fuel elements containing plutonium, ^{235}U, or ^{233}U enriched above 20%, which can be chemically separated from the remaining uranium (or thorium) in fuel elements;
5. capture or theft of spent fuel elements—not yet reprocessed—from which one of the foregoing materials could be separated from ordinary uranium and from fission products;
6. "pseudo theft" or alleged diversion of any of the above.

The sixth, and perhaps most probable case, is a pseudo-theft or pseudo-diversion. There is now widespread distribution of fissionable materials around the world. The allegation of a successful theft or diversion in another country (as a source of material to threaten any major city in the world) is not subject to convincing verification or disproof by the local officials and citizenry. Many bank robberies have been perpetrated by the threat of a paper bag containing a weapon-shaped object, but no weapon. Increasing preoccupation of the media and the entertainment world with terrorist activities makes such attempts almost inevitable. The presence or absence of an American reprocessing or plutonium-recycle program makes little or no difference in the probability of occurrence of such threats.

The first five options listed above are ranked in order of increasing requirements with respect to four factors involved in making a successful or credible attempt at diversion or theft. These include

1. special scientific knowledge, technological experience, and practical skills of personnel;
2. materials resources: weapons, vehicles, staging area, communications, money, special physical facilities, etc;
3. force: suitable personnel, organization, training, and discipline;
4. special inside knowledge unique to a particular location.

The most attractive target for capture or theft by a would-be terrorist is a nuclear weapon. Although this requires force and specialized inside knowledge, it obviates the need for the technical skills, resources, and physical facilities required to convert raw materials into a weapon. For all of the other options (except pseudo-theft) a much larger minimum requirement of technology, special skills, and resources is set. Willrich & Taylor (24) have postulated the relative ease with which a person or small group of persons with good basic scientific knowledge could in principle produce a crude nuclear weapon from suitable fissionable raw material. The media have dramatized the story of a graduate student who found sufficient information in the library to sketch a design for a crude nuclear weapon that was confirmed as "not being implausible" by a scientist in Sweden.

Perhaps a more skeptical view of this possibility would be by analogy to the ability of a reasonably well-informed technical person to sketch up a workable concept for a small jet-propelled airplane, or a medium-sized computer. Given access to manufactured modules for most of the critical parts, construction of such a project by a small dedicated group of artisans is conceivable. However, if the project must literally start from the raw materials in inconvenient chemical and physical form, and with very substantial hazards associated with handling and processing the materials, one obtains a rather different view of the probability of the "garage-operation weapon."

RELATED SOURCES OF CONCERN; ROLE OF TIMELY US SAFEGUARDS Perhaps the most serious obstacle to terrorist use of nuclear materials as the chosen route for performing threats or mischief is merely the basic question of the effort and resources required for various options for mischief. If greater threats or mischief are accessible by much more readily available means (which require substantially smaller assets in the form of special skills, resources, and force), one may postulate that the more readily accessible means will prove to be overwhelmingly attractive. Such options include chemical or biological agents, or non-nuclear explosives. One option for a non-nuclear explosion, which merely disperses some nuclear material, is much easier to attain than even a low-grade nuclear explosion. Actual experience with such explosions (from crashes of military aircraft carrying nuclear weapons) suggests much less-widespread consequences than those often portrayed. Dispersion of chemical or biological agents offers more readily available routes to mischief or threats. Once reasonably effective safeguards are established, further extremes of protection of domestic nuclear material appear unlikely to reduce materially the overall terrorist risks to the population, considering the other options available.

This line of reasoning requires facing the uncomfortable reality that the world is inherently a dangerous place in which to live,[3] and these dangers are not measurably reduced—and may even be increased—by restriction or continued delay of the tail end of the fuel cycle in the United States. For example, the political and technical question of safeguards may be quickly and decisively resolved. Practice and experience will be acquired in operating such a system. This can help to set a framework for world action in the next decade for safeguards procedures. For example, standards of the American Society of Mechanical Engineers (ASME), nuclear standards, and regulatory requirements and procedures used in the United States have been generally adopted (with only minor modifications) by most non-Communist countries as they begin to develop nuclear power industry and capacity. The timely establishment and practice of safeguards procedures in the United States will help to set the pattern for such systems elsewhere in the world. Even the Soviet bloc countries and third world countries have tended to adopt many US and Western European practices. For example, Soviet reactors have all been built without pressure-containment buildings capable of withstanding "design basis accidents" as used in Western practice. As the Soviet commitment to larger units and a larger number of units begins to be implemented, it appears that the Soviet Union will adopt the concept of containment as a desirable feature of population and environmental protection.

Conversely, if the policy on American practice, procedures, and regulations concerning safeguards continues to move very slowly or to be excessively perfectionist, it may take many years to resolve. In this event, the rapidly growing number of nuclear plants overseas will result in a variety of disparate practices with respect to both safeguards and waste disposal. Plutonium is already being recycled in reactors in Italy and Germany. It is hard to avoid the conclusion that exogenous systems will be less secure than those that would be established if preceded by actual large-scale practice in the United States. The direct effect of a slow or unrealistic position on safeguards (and permanent waste disposal) in the United States may well be a substantial increase in the likelihood of threatened or actual terrorist diversions.

MEASUREMENT OF RELATIVE ADEQUACY OF SAFEGUARDS SYSTEMS   Previous discussion has covered the inherent obstacles to production of a nuclear weapon even assuming access to some form of fissionable material. In addition, there are the imposed obstacles that can be provided by a combination of

1. physical security systems;
2. accountability systems;
3. detection systems for diversion or unauthorized transport;

[3] A quantitative expression of some of these risks, but not including terrorist activities, is summarized in Chapter 6 of the Reactor Safety Report (25). A seventh or "direct route" to one or more nuclear devices is available starting with uranium ore in the ground. A sustained effort over many years and resources of the order of several hundred million dollars are involved. The resources and skills required tend to define a clandestine government-level effort rather than a terrorist band. The availability of such a route elsewhere in the world is essentially unaffected by safeguards measures taken in the United States.

4. external reaction force and communications systems;
5. tagging, denaturing, or spiking of fuel material (or fabricated fuel) with sources of penetrating radiation.

It seems obvious that some balanced combination of all or most of the above will be more effective than primary reliance on any one or even two of these factors. For example, even extremes of physical security and guard forces provoke imaginative scenarios that penetrate the security and bring to mind the "Maginot Line" kind of deficiency of an excessive dependence on fixed vaults and guards. Similarly, primary reliance on any one or two of the other alternatives, including accountability systems, reaction force, denaturing, etc, suggests imaginative scenarios that thwart the defensive system. A denaturing system in which the stolen material cannot be handled without extensive facilities and elaborate purification perhaps comes closest to a single-system deterrent.

The basic conceptual difficulty in devising an adequate and publicly acceptable combination of elements to make up a safeguards system has been the lack of a plausible rationale or methodology for measuring "how safe is safe enough?" on either a relative or an absolute basis. One possible approach to this dilemma is suggested here (26). The proposed methodology consists of the following steps:

1. A given safeguard system is described in terms of its functional and operating characteristics (and in terms of probabilities that some of the more important characteristics are degraded part of the time).
2. A series of successful attempts at theft or diversion are assumed and described.
3. Each of the successful scenarios is characterized by the combination of four basic factors required, which were listed earlier: (a) skills; (b) resources; (c) force; and (d) inside knowledge. These factors constitute the required inventory of resources and skills.
4. Each assumed safeguard system is characterized by the smallest combination of resources and skills required to mount a successful diversion or theft.

It is evident that even in the absence of absolute measures of the probability of occurrence of any given diversion action, the relative measure provided by the skills and resources inventory required can be used to rank alternative designs of safeguard systems combinations.

DENATURED FUEL CYCLES A particular example of such use is to measure the relative effectiveness of different types of denaturing or spiking of fissionable material with sources of penetrating radiation as a means of reducing the likelihood of successful diversion. One has a variety of options to evaluate—for example, "natural denaturing," which results from only partial decontamination of the fuel from its associated fission products (27). One can select various levels and combinations of fission product activity associated with a given weight of fuel.

One has also the option of the addition of spiking agents that are not naturally occurring fission products. These can offer more effective half-life and radiation characteristics than those available from fission products alone. For example, fission products at a level required to produce a lethal dose in an hour or so, for example, would decay after a few hundred days so that several hours' access would be possible (albeit on a suicidal basis). The addition of long-lived radioactivity such as cobalt-60 could be used to extend the period of deterrence to five to ten years. On the other hand, relatively low levels of fission products (or spiked activities, or alpha sources of neutrons) can be used to make diverted material relatively easy to detect over considerable distances. Clandestine diversion would be easier to detect, and overt diversion or theft would be easier to track and would provide direction for reaction force. Qualitative argument on these alternatives bogs down in the absence of even relative measures of their merit (however, it is clear there would be some differences in fuel fabrication costs). The "resources and skills inventory" approach provides relative measures of the obstacles to a would-be diverter.

Some of the elements of an approach to measure effectiveness of safeguards are found in an existing study (28). This study uses an "events tree" methodology to estimate the societal costs of assumed safeguard systems. Similar events trees can be used as the outlines of postulated successful diversion scenarios, and this forms the basis for measuring the inventory of resources and skills. In principle, one could also estimate absolute probabilities of successful diversion attempts by multiplying together the judged probabilities that the resources, skills, force (and motivations), etc, required would in fact be assembled without prior detection by law enforcement agencies.

DEMONSTRATION OF SYSTEMS AND TIMING An interesting feature of several approaches to measures of safeguards is that they yield a resources and skills inventory that amounts to a fingerprint of a potential terrorist group. This opens the obvious opportunity for improved anticipation and countermeasures. It seems circumstantially probable that such studies have been conducted for much of the last 25 years in establishing the security systems for military weapons material. (Probably also for the supply chain of fissionable material associated with the large number of US Navy reactors now in operation.) The results of such studies are obviously not in the public domain since at some level of detail they could provoke attempted diversion or sabotage.

The success of such military systems over a 25-year period in safeguarding large amounts of military material is instructive. Military material is much more convenient and attractive for terrorist purposes than civilian fuel material, and it has a greater degree of geographical dispersion worldwide than fuel-reprocessing facilities are ever likely to have. This experience is encouraging to the belief that publicly and politically acceptable safeguard systems can indeed be devised and operated. It will take about 15 years before the amount of concentrated fissionable material produced and processed as a by-product of the civilian nuclear power economy approaches the amount of material already produced for military purposes. During this time, safeguards systems can be set up and demonstrated on an industrial scale with minor effects on already existing risks. Meanwhile, spent fuel can be stored without reprocessing. Even without reprocessing, the civilian power program can continue to operate and grow for at least another decade—at the expense of somewhat higher requirements for uranium mining and separative work. Fuel costs become about 10%

higher if spent fuel is merely stored (29) rather than reprocessed and recovered (see Table 4). This is still much lower in cost than fossil fuels. In 10–15 years, such spent fuel will represent a large and technically readily available energy mine. Such an energy mine is comparable in value to one of the world's major oil fields if the residual uranium and plutonium values are used for recycle fuel in LWRs. If the plutonium in this energy mine is used to fuel breeder-reactor capacity (starting in the 1990s), the energy reserve from just this small amount of plutonium is greater than all of the oil and gas so far discovered in the world. The lead time required to build practical manufacturing capacity and start the growth of capacity of appropriate reactors is 10–15 years. The cost and availability of fuel energy in the 1990s will be largely determined by policy decisions and technological options that are deployed—or not deployed—for large-scale exploitation starting in the late 1970s.

Literature Cited

1. Nuclear Assurance Corporation. 1975. *Nuclear Industry Status, Nuclear Fuel Status and Forecast, Nuclear Power Plant Status and Forecast*. Atlanta, Ga: Nuclear Assurance Corp.
2. Report on nuclear programs around the world. 1975. *Nuclear News Buyers Guide* 18(3): 39–56
3. Elkins, R. B., Williamson, H. E. June 1975. *Experience with BWR Fuel Through September 1974*. Rep. No. NEDO 20922. Nuclear Energy Div., General Electric
4. Kramer, F. W. 1975. PWR fuel performance—The Westinghouse view. In *Proc. Joint Topical Meet. Commercial Nucl. Fuel Technol. Today, April 28–30, 1975, Toronto, Canada*. Doc. No. CNS ISSN 0068-8517, 75-CNA/ANS-100. Canad. Nucl. Soc.
4a. Cherry, B. H. 1975. US operating experience. See Ref. 4
5. Proebstle, R. A., Davies, J. H., Rowland, T. C., Rutkin, D. R., Armijo, J. S. 1975. The mechanism of defection of zircaloy-clad fuel rods by internal hydriding. See Ref. 4
6. Jordan, K. R. 1975. Densification and fuel rod flattening-reliability impact. See Ref. 4
7. Robertson, J. A. L. 1975. Nuclear fuel failures, their causes and remedies. See Ref. 4
8. Chubb, W., Hott, A. C., Argall, B. M., Kilp, G. R. 1975. The influence of fuel microstructure on in-pile densification. *Nucl. Technol.* 26:486
9. Brite, D. W., Daniel, J. L., Davis, N. C., Freshley, M. D., Hart, P. E., Marshall, R. K. March 1975. *EEI/EPRI Fuel Densification Project*. Rep. No. EPRI 131, Elec. Power Res. Inst. Springfield, Va: Nat. Tech. Inf. Serv.
10. Levy, S., Wilkinson, J. P. D. 1974. A three dimensional study of nuclear fuel rod behavior during startup. *Nucl. Eng. Des.* 29: 157–66
11. Zebroski, E. L., Lapides, M. E. 1975. Evolving incentives and technical programs for attaining higher plant productivities. *Proc. Am. Power Conf., April 21–23, 1975, Chicago, Ill.*, Vol. 37
12. Budwani, R. N. 1975. Nuclear power plants: what it takes to get them built. *Power Eng.* 79(6): 38–45
13. US Atomic Energy Commission. 1974. *Power Plant Capital Costs—Current Trends and Sensitivity to Economic Parameters*. Rep. No. WASH-1345. Washington DC: US AEC
14. Davis, W. K. 1975. *Converter Reactor Alternatives*. Presented at AIF Conf. on Energy Alternatives, Feb. 19, Washinton DC
15. US Atomic Energy Commission. Aug. 1974. *Generic Environmental Statement Mixed Oxide*. Rep. No. WASH-1327 (draft). Washington DC: US AEC
16. *General Environmental Statement, Mixed Oxide: Treatment of Safeguards and Deferrals of Licensing Actions*. May 8, 1975. Federal Register Docket No. 75-12335
17. Battelle Pacific Northwest Laboratories. May 1974. *High-Level Radioactive Waste Management Alternatives*. Rep. No. BNWL-1900. Richland, Wash: Battelle Pacific Northwest Lab. 4 vols.
18. US Atomic Energy Commission. May 1974. *High Level Waste Management Alternatives*. Rep. No. WASH-1297. Washington DC: US AEC
19. Lapides, M. E. 1975. *Spent Fuel Temporary Storage*. Presented at 3rd Ann. Conf. on Energy and Environ. Assess-

ment, Univ. Calif., Berkeley, September 8-12
20. Hamstra, J. 1975. Radiotoxic hazard measured for buried solid radioactive waste. *Nucl. Saf.* 16(2):180-87
21. Cohen, B. L. July 1975. *The Hazards in Plutonium Disposal*, Univ. Pittsburgh, Pittsburgh, Pa. Submitted for publication (Summary in *Physics Today*. Jan. 1976, pp. 11-15)
22. US Atomic Energy Commission. Sept. 1974. *Environmental Statement—Management of Commercial High Level and Transuranic Contaminated Waste*. Rep. No. WASH-1539 (draft). Washington DC: US AEC
23. Battelle Pacific Northwest Laboratories. 1974. See Ref. 17, Vol. 2
24. Willrich, M., Taylor, T. B. 1974. *Nuclear Theft: Risks and Safeguards*. Cambridge, Mass: Ballinger
25. US Nuclear Regulatory Commission. 1975. *Reactor Safety Study—An Assessment of Accident Risks in U.S. Commercial Nuclear Power Plants*. Rep. No. WASH-1400 (NUREG-75/014), pp. 103-30. Washington DC: US NRC
26. Zebroski, E. L. 1975. *Overview on Plutonium Utilization and Safeguards*. Presented at 3rd Ann. Conf. on Energy and Environ. Assessment, Univ. Calif., Berkeley, Sept. 8-12
27. General Electric Company. 1975. *Deterrent Action Study—Fissile Materials: Denatured Plutonium*. Prepared for EPRI under contract RP301-1.
28. US Energy Research and Development Administration. June 1975. *Societal Risk Approach to Safeguards Design and Evaluation*. Rep. No. ERDA-7; also National Science Foundation. 1975. Unpublished work conducted by Science Applications, Inc.
29. Wolfe, B., Lambert, R. W. 195. *The Back End of the Fuel Cycle*. Presented at AIF Fuel Conf. March 20, Atlanta, Georgia

Copyright 1976. All rights reserved

SOLAR ENERGY ✖11006

Frederick H. Morse
Department of Mechanical Engineering, University of Maryland,
College Park, Maryland 20742

Melvin K. Simmons
Lawrence Berkeley Laboratory, University of California,
Berkeley, California 94720

THE USES OF SOLAR ENERGY

It is widely recognized that the inexhaustible energy of the sun is received on the earth in sufficient quantities to make major contributions to the future energy needs of the world. However it is yet uncertain and controversial whether we now have the means to economically collect and convert that solar energy into forms useful for our needs. In this article we review the approaches to solar energy conversion and assess the status and economic feasibility of the technologies involved. It is helpful in defining these various approaches to solar energy conversion to distinguish between technological and natural collection of solar energy.

Technological Collection

The approaches defined here have the common feature that the initial collection and conversion of sunlight occurs within a man-made device.

HEATING AND COOLING OF BUILDINGS The heating and cooling of buildings, and some other applications such as drying crops or heating water for industrial processes, can be achieved by collection of solar energy on "flat plate" collectors. Such collectors do not concentrate sunlight; they consist merely of a black surface directed skyward and covered by one or more transparent covers to prevent loss of heat. Such collectors easily achieve temperatures of 100°C. Heat is obtained from the collector by circulating water or air through it, and the heat is then stored for later use.

SOLAR THERMAL CONVERSION In solar thermal conversion, solar energy is collected as high-temperature heat, generally by means of mirrors or lenses that track the motion of the sun and direct a concentrated solar flux onto a receiver. Temperatures up to 500°C can be generated by this means, high enough to produce the high-pressure steam used in modern steam turbines to generate electricity.

131

PHOTOVOLTAIC CONVERSION   Photovoltaic conversion is a nonthermal process for the production of electricity directly from solar energy. Incident sunlight is used to create free charges in a semiconducting material, and the charges are then collected at the surface of the material by metallic contacts. The familiar example of this is the use of solar cell arrays to power satellites.

Natural Collection

Natural collection of solar energy occurs on the surface of the land and oceans of the earth, giving rise to wind and weather, the growth of plants, and warm surface waters in the oceans. Each of these energy forms can be further converted by man for his purposes.

WIND ENERGY   It has been estimated that the power contained in the winds over the continental United States and the arc of Aleutian islands extending from Alaska is about 10^{11} kW (1), greatly exceeding the present electricity-generating capacity of the United States of about 5×10^8 kW. The winds at many potential sites are very strong and remarkably repeatable and predictable. Large wind turbines, erected on the plains or along the continental coast, can efficiently extract the momentum in this moving air and generate electricity.

BIOCONVERSION   The energy stored in organic matter (often referred to as biomass) by the photosynthetic process can be used for the production of clean fuels. In a bioconversion process, the methods now being developed for utilization of the energy in urban waste would be applied to convert to fuel crops grown specifically for their energy content. The products of conversion processes can be either synthetic natural gas or liquid fuels, such as alcohol, that could replace gasoline. If through advanced agricultural practices, including the modification of plant genetics, energy plantations were operated to produce crops continuously through the year at a 3% conversion efficiency, then only about 2% of the US land area could provide stored solar energy equivalent to the present US electrical requirements.

OCEAN THERMAL CONVERSION   Between the Tropics of Cancer and Capricorn, where the average intensity of solar energy is at a maximum, 90% of the earth's surface is water. The incident solar energy heats this surface water to temperatures up to 85°F, while the waters at depths below 600 meters remain cold, typically 35–38°F. A heat engine can operate across this temperature difference, though only at about 2% efficiency. The amount of heat in these surface waters is so large that even at this efficiency, immense amounts of energy are available. Conversion of only 1 Btu per pound of the warm Gulf Stream would provide 700×10^{12} kW-hr annually (1).

These various approaches to solar energy conversion can be seen to span a wide range of technologies. Some are very well defined and are now entering the stage of commercial application, while others require significant research to establish their feasibility. However, they are all potential uses of solar energy, and it is the purpose of this review to introduce the reader to the many technical, economic, and other issues that surround their future development and utilization.

THE SOLAR RESOURCE

The use of solar energy adds a new consideration to the design of any system—that is, the energy source is variable. As the sun crosses the sky, its intensity at a point on the earth's surface varies due to the geometric effects associated with its position and the changing effects of attenuation and scattering by the intervening atmosphere.

The intensity of solar radiation on a surface normal to the sun's rays above the atmosphere, known as the solar constant, is 1353 W/m^2 (2). This radiation has a relatively smooth spectral distribution in the wavelength range from 0.3 to 3 microns. The solar energy incident on a surface located on the earth's surface is attenuated by water vapor and dust in the atmosphere and is also scattered, thereby forming the diffuse component of the solar energy reaching the surface. The intensity of solar energy at noon on a clear day is about 1000 W/m^2, with most of this energy in the form of direct (normal incidence) radiation. The clear-sky solar intensity (direct and diffuse) has been monitored at several locations in the United States and the data generalized for engineering design (3, 4).

While clear-day insolation values are helpful in designing a solar system, account must be taken of cloud cover. This may be accomplished by using insolation data from a previous year, or the average of several previous years, for the site being considered. The National Weather Service network has monitored the number of sunshine hours per day, total solar radiation, and, in some cases, the direct component of solar radiation at 90 sites throughout the United States (5). The basic insolation instruments are the pyranometer (for total radiation) and the pyrheliometer (for the direct component).

Ideally, the performance of thermal collection systems should be independent of spectral distribution of the insolation and should depend only on the total intensity. However, the performance of selective absorbers and cover materials is optimized only over a particular spectral region, and their degradation has a definite spectral dependence. The output of photovoltaic devices is strongly spectral-dependent, as is the natural photosynthetic process and other photochemical processes.

It is important in comparing different techniques for collecting solar energy to note that nonconcentrating collectors (such as flat-plate collectors and photovoltaic cells) can utilize diffuse as well as direct radiation. Since solar thermal systems employ a concentrating collector of some kind, only the direct component of the radiation is useful. This difference can be critical in locations where the sky is often overcast.

Solar energy arrives at the surface of the United States at an average rate of 4.76 kW-hr/m^2-day (1), so over a year a square kilometer would receive 1.7×10^9 kW-hr. In 1974, the total energy consumed by the United States for all purposes was about 2.4×10^{13} kW-hr. Accordingly, 55,000 square kilometers of land, dedicated to solar energy conversion at 10% efficiency, could meet, on the average, the entire 1974 US energy requirement.

The situation changes somewhat when one compares the energy densities of various uses to that of solar radiation. The energy arriving on the roof of a typical home over a year is several times greater than the thermal energy needs of that building. The same is true for the electrical needs of a single home. However, as the energy density of the application increases, as, for example, one moves from single family dwellings to large apartment buildings, it becomes impossible to obtain sufficient energy from the solar energy received only on the land occupied by the building. Land dedicated to the conversion of solar energy will be necessary for such high-density applications.

STATUS OF SOLAR ENERGY TECHNOLOGIES

Heating and Cooling

Of the promising applications of solar energy, heating and cooling of buildings with solar energy is generally considered to be the closest to commercialization. The history of solar energy utilization for heating and cooling of buildings in the United States began with the use of solar water heaters in Arizona, California, and Florida in the early 1900s (4). Just as with wind generators, their use declined with the availability of low-cost energy from fossil fuels. Interest in space heating with solar energy developed in the 1930s first with the utilization of large south-facing windows to admit winter sunshine and then with the application of separate collectors used in conjunction with some form of thermal energy storage. Insulated tanks of water, rock beds, and materials that undergo a change of phase with absorption of heat were all used (6).

Beginning in the 1940s, and continuing into the early 1970s, a series of experimental houses were built to test many different solar heating concepts. These include, among others, the MIT houses, Löf's houses in Colorado, the Telkes house in Massachusetts, an office in Princeton, New Jersey, the Bliss home, the Hay building, the University of Arizona solar laboratory in Arizona, the Bridgers and Paxton office building in New Mexico, the Thomason homes in Washington DC, and the University of Florida and the University of Delaware houses (4, 7). Essentially all of these buildings were financed by individuals, foundations, or universities, with the goal of demonstrating that solar energy could provide most of the heat needed by the building and that air conditioners driven by solar heat (or nocturnal radiation) could provide the cooling.

The high cost of the equipment used in these pioneering experiments prevented solar heating from moving into the marketplace. While the recent major research, development, and demonstration efforts have brought many of the technical aspects of such applications under investigation, the status of their utilization has not yet changed significantly. Solar water heaters, commercially available in Australia, Japan, India, Israel, and the USSR, are not appreciably used in the United States. Solar heating systems have been successfully demonstrated in ever growing numbers and are becoming commercially available, but for widespread use to occur, prices must be lowered further (8).

The thermal and operational performance of the major components for solar heating and cooling systems must be improved. Existing components have generally not had the adequate development and testing that would insure long-lived performance at design conditions (8). The most common collector for solar heating and cooling today is the flat-plate collector. This collector typically consists of a metal absorber plate (painted black or coated with a high-absorptivity, low-emissivity material called a selective surface) to which fluid passages are attached. One or two glass or plastic covers on top and insulation on the bottom serve to reduce thermal losses (9). Work is underway to further reduce the thermal and optical losses by means of evacuated spaces, honeycombs, special materials, concentration, and tracking, thereby making it practical to operate the collector at higher temperatures (10).

A variety of methods for solar cooling are under investigation, including absorption refrigeration, adsorption, heat pumps, Rankine/vapor compression, and nocturnal radiation. More effort is needed to identify the most promising approaches and to determine how solar cooling is to be integrated into a heating and cooling system (11).

In the solar heating and cooling systems tested to date, the storage medium for thermal energy has usually been water. Since it seems likely that water, with its numerous advantages, will continue to be widely used for thermal energy storage, better containers must be developed. The present experience with phase-change materials suggests that significant research and development is still needed before acceptable storage units using this method are available (12). Related to these major components, and in need of considerable improvements, are the heat-transfer fluids, the various heat exchangers, and the control systems.

There have been several plans proposed to promote the widespread availability and use of solar energy for meeting the thermal needs of all types of buildings throughout the United States. The present plan of the Energy Research and Development Administration (ERDA) (8), formulated in accordance with the 1974 Solar Heating and Cooling Demonstration Act, places major emphasis on a series of demonstrations. Earlier plans, such as those suggested by the NSF/NASA Solar Energy Panel (1) and the National Science Foundation, as reported in the Project Independence Report (13), also emphasized a strong demonstration effort, but placed more emphasis on the development of improved components and subsystems. The ERDA plan consists of three parallel efforts. The major effort is the demonstration of residential and commercial solar heating and cooling systems. A second effort is the development of improved components for use in the demonstration systems, and a third effort is the research and development on advanced technology for heating and cooling systems.

The relative nearness of this particular solar energy application has evoked considerable public attention, interest, and controversy. In numerous hearings, arguments have arisen concerning the roles of small business vs large industry, the importance of research vs demonstration, the role of tax incentives and subsidies, and how many demonstration projects are needed to succeed in stimulating the creation of a viable industrial and commercial capability for producing and distributing solar heating and cooling systems. The motivation is indeed great, for the use of such systems in just 1% of the buildings in the United States could save approximately 30 million barrels of oil per year (14).

Solar Thermal Conversion

The technology of solar thermal conversion dates back to an exhibition in Paris in 1878 where sunlight was focused onto a steam boiler that operated a small steam engine. A more complex system was built by Harrington in New Mexico half a century ago. He focused sunlight onto a boiler and ran a steam engine that pumped water uphill into a storage reservoir. Water drawn from the reservoir operated a turbine and generated the electricity that lighted a mine (15). Harrington's system had all the functions usually proposed for contemporary solar thermal conversion facilities: light collection and concentration, conversion to heat, storage of energy, and generation of electricity. These systems and others, built and operated during the last 100 years (16), demonstrate that an adequate technological base for solar thermal conversion has long existed.

What remains to be resolved are the answers to three fundamental questions about solar thermal conversion. Can solar thermal conversion become economically competitive with combustion of fossil fuels as a source of high-temperature heat? What are the best designs for the collection and conversion of sunlight in a solar thermal facility? What are the best uses of the high-temperature heat from solar thermal conversion? The first of these questions we address in the section on the economics of solar energy. In the present section we review the status of work on the latter two questions, regarding the technology of solar thermal conversion.

COLLECTION AND CONVERSION Two basic techniques, employed singly or in combination, are used to achieve production of high temperatures from solar energy. The temperature to which a surface is heated by a certain flux of incident solar energy is determined by the balance of incident radiation and loss by conduction, convection, and radiation. Given that measures are employed to restrict loss by conduction and convection, then achievement of high temperature at a certain solar flux results from use of a *selective surface* that absorbs visible sunlight but does not lose energy by radiation of infrared (9).

The temperature obtained from solar energy can be increased by boosting the flux of incident sunlight by use of concentrating mirrors or lenses. This method is carried to its extreme in the solar furnaces that have been built and operated to conduct materials research at temperatures as high as 4000°K (17–19). These furnaces employ a much higher concentration ratio than necessary for generation of electricity or most process heat applications, so they are not actually prototypes of solar thermal conversion facilities. However, they can be usefully employed as test beds for solar thermal components.

The techniques of selective surfaces and concentration can be employed in combination. Thus a fairly low concentration ratio, obtainable with simple optics, can be combined with a selective surface to efficiently produce temperatures high enough for electrical power generation.

We can distinguish three basic geometries for collection of sunlight for solar thermal conversion: nonconcentrating, concentrating to a line, and concentrating to a point. It is difficult to operate practical cycles for conversion to electricity at the temperatures available without any concentration (20). However, nonconcentrating collectors with selective surfaces may be useful for providing industrial process heat at moderate temperatures. With concentration to a line, concentration ratios of about 10 to 20 can be achieved. Of particular interest is the Winston collector, which achieves moderate concentration to a line without need for tracking the sun (21). Line focusing, alone or in combination with selective surfaces, provides temperatures high enough for electrical generation (22). With point focusing, concentration ratios can be as high as 1000, although lower ratios are sufficient for electrical generation (23). There is probably no need for use of selective surfaces in a point-focusing system.

Of these various configurations for collection and conversion to heat, the central tower/heliostat design (24) is now favored for the first solar thermal pilot plant (25). In this design, a field of heliostats (trackable mirrors), perhaps 15,000 individually controlled units of about 36 m^2 area each, would reflect sunlight to the top of a tower at the center of the field. Each tower would collect heat sufficient to generate about 50 MW of electricity (25). The concentration ratio in this design is quite high, and selective surfaces are not required. The high-temperature heat is collected at the top of the tower in a working fluid, perhaps water/steam or a eutectic salt (26), and piped to the ground for use in electrical generation or another high-temperature application.

APPLICATIONS   Most of the present effort in solar thermal conversion is directed toward the generation of electric power in large central plants for distribution in an electric power network. Within this application there are various options. The solar thermal plant could be run as a baseload, in which case it would require a large amount of energy storage and could provide power to the network almost constantly day and night. Or it could be run as an intermediate load plant, providing power for about 12 hours a day, with a smaller storage requirement. Finally, it could provide only peaking power and would require very little energy storage capacity. The comparative study of these possibilities by the Aerospace Corporation has identified the intermediate load plant as having the best economics (25).

There are significant opportunities for the use of solar thermal conversion in providing industrial process heat. About 18% of the fuel consumption in the United States is for generation of industrial process heat at moderate temperatures (process steam) and another 11.5% is used for high-temperature process heat (27). Solar thermal conversion facilities to provide process heat could be sited at the plant requiring the heat and could be sized to match the requirements of the plant. A demonstration of the use of low-temperature solar heat in an industrial process is now being built at the Sohio uranium mining and milling complex in Grants, New Mexico (28).

Photovoltaic Conversion

The photovoltaic effect, noted by Becquerel in 1839, is the basic process of solar cells, which were first fabricated in 1954 by workers at RCA and the Bell Laboratories. Photovoltaic conversion systems are based on the absorption of light in semiconducting materials to generate free charges that drift across a junction between two types of semiconductors and are collected at contacts applied to the exterior surfaces of the material. The theoretical limit to efficiency for conversion of solar energy to electrical energy is about 25% for a single semiconductor device operating at room temperature. While the early silicon cells achieved an efficiency of only 5%, present solar cells operate at 10–15% with efficiencies approaching 25% predicted for the near future (29). The single-crystal silicon solar cell has been, and continues to be, the mainstay of photovoltaics. Solar cells have also been developed from other materials, including most notably cadmium sulfide and gallium arsenide.

The silicon cell has been integrated into arrays and the output conditioned to meet the electrical energy demands of satellites. The 20-kW Spacelab power system is the largest of such solar cell systems built to date. Terrestrial applications have been limited to small units for remote applications such as recharging batteries on offshore drilling platforms, microwave repeater stations, and buoys and other navigational aids. Two larger experimental systems that provide on-site electric power to meet residential needs are now in operation: a combined photovoltaic and thermal system using cadmium sulfide cells (30) and a 1-kW silicon array whose electrical energy output is used to produce hydrogen by electrolysis (31). The present cost for these solar cells is now down to about $20/peak watt.

Silicon is the earth's second most abundant material, and the cost of the metallurgical-grade material from which the cells are made is only $600/ton. However, because of the way in which silicon cells are produced, they are very expensive. The basic production process used in the past incorporates about 50–60 steps with numerous heating and reheating operations and begins with the growing of a single crystal out of molten silicon by the Czochralski process. This batch process is slow and expensive and does not lend itself to mass production. A subsequent sequence of cutting and polishing steps, with considerable loss of silicon material, precedes the formation of the semiconductor junction and attachment of the electrical contacts (32).

Obviously, improvements can be, and are being, made, with process steps being consolidated or eliminated. Even without additional technical improvements, significant cost reductions can be realized by the application of mass production techniques. It has been predicted that if the present annual market of cells with a total peak capacity of 60 kW were increased to about 100 kW, the price of a silicon array would decrease to $5/peak watt (33). However, this price is still too high, and new approaches are being developed that hold the promise of further cost reductions. A goal of less than $0.50/peak watt has been set for achieving economic competitiveness of large-scale terrestrial photovoltaic systems with future fuel prices.

In order to achieve this additional cost reduction of a factor of 10 or more, it will be necessary to lower the cost of the starting material. Recent studies have shown that solar cells with good electrical characteristics can be made from a lower-grade, and accordingly less expensive, silicon. This development, coupled with a new production technique called the edge-defined, film-fed growth (EFG) process, supports an expectation that solar cells costing less than $0.50/peak watt can be produced. The basic idea of the EFG process is the placement of a die into the crucible containing the molten silicon. The liquid rises through the die and crystallizes as it is removed from the top. Ribbons of silicon several meters long and about 5 cm wide have been made with this process. Since additional silicon can be added to the melt while the process is in operation, it is a continuous process that lends itself to the production line.

Many other semiconductors besides silicon exhibit a photovoltaic effect. Of these, cadmium sulfide–copper sulfide has received the most attention. Although many of these devices have the potential for high-volume production and for costs less than that of single-crystal silicon cell arrays, they appear to require much more research and development (34). A wide variety of these devices are presently being studied.

The challenge in the development of photovoltaic technology is to produce arrays that can be used to generate electricity on an economically competitive basis. Within the next ten years the practicability of a photovoltaic system having array costs less than $0.50/peak watt should be established. The electricity cost for a residence requiring an average power of 1 kW would then be 40–50 mills/kW-hr (13). The major effort will most likely be made with the single-crystal silicon cell. A low-cost, high-volume process for producing the raw silicon starting material must be developed, along with a method for producing large silicon crystals. Finally, an automated fabrication and encapsulation method for the cell and arrays must be devised to complete the process. In order to keep the area of the arrays reasonable, the efficiency of the cell should be maintained at its present level of about 16% or higher; 20% is a reasonable goal.

A series of system experiments and demonstration projects is being planned. These systems, increasing in size from several kilowatts to several hundred kilowatts, will provide a market stimulus for the solar cell manufacturers and also will provide system performance data. An analysis is being made of the criteria to be used for selection of preferred applications within the categories of on-site generation, central-station generation, and fuel production (35). The questions of a system's scale—small to very large—and function—baseload, intermediate, or peaking power—must also be addressed.

Production of Fuels

Among technologies for large-scale production of useful fuels from solar energy, those that utilize the natural photosynthetic process in plants or algae are best established and thus potentially important for the near future. These include energy plantations on land, ponds for growth of algae or water plants, and ocean-sited kelp farms. Alternatives to the natural photosynthetic process, such as photochemical conversion or solar-thermochemical conversion, are under investigation but are still at the stage of basic and applied research. These alternative techniques are potentially of great importance in the more distant future.

ENERGY PLANTATIONS   Existing methods of land agriculture could be used now in energy plantations for the production of significant amounts of biomass to be burned directly as a low-sulfur fuel or converted to liquid or gaseous form (36). However, new crops and growth and harvesting techniques may be needed to avoid competition with food production for use of land and water. This is a severe constraint on the production of biomass for energy because of the worldwide demand for good land and water for food crops (37). Since in an energy plantation the goal is to produce the maximum amount of biomass, crops such as eucalyptus trees, rubber plants, or sunflowers may be used because of their rapid growth and high productivity as measured in energy content (38, 39). The energy efficiency of land agriculture is quite low, typically less than 0.5% of the solar energy being stored as biomass. However, the photosynthetic process is capable of higher efficiencies; up to 12% efficiency is noted in some experiments (40). The efficiency of energy plantations could be greatly improved if crops can be developed that perform photosynthesis efficiently in full sunlight and do not lose excessive amounts of energy by respiration. Recent advances in plant cell culture and genetic modification may provide the techniques needed to develop such high-energy-efficiency crops (41).

ALGAE PONDS Liquid wastes from homes, industry, and agriculture contain nutrients that can support the growth of algae in ponds. Systems have been designed that use heterotrophic bacteria to oxidize waste materials to nutrients, and algae to use the nutrients and collect solar energy to produce biomass (42). Algae cultures of nearly 10^9 liters have been established, and the potential efficiency of solar energy conversion of the cultures is better than 5%. However, algae produced in such ponds may turn out to have a higher value as a feed for animals than as an energy source.

OCEAN FARMS The substantial requirements of energy plantations for land can be avoided by establishing ocean farms of seaweed species with high growth rates, such as kelp. An investigation is now being made of the feasibility of constructing large subsurface structures in the open ocean that could support the growth of kelp. The immense areas of the ocean, now practically devoid of life, could then be used for kelp farms to produce food or energy. Initial results indicate such farms may be feasible, but the addition of fertilizers, especially nitrogen, to the ocean water will be required for good growth (43).

PHOTOCHEMICAL CONVERSION An attractive long-term alternative to the growth of biomass and subsequent conversion to gaseous or liquid fuel is the direct production of a fuel by a photochemical process. The process most widely investigated at present is photolysis, the splitting of water to produce hydrogen. The approaches to photochemical conversion can be divided into three categories by the nature of the chemical system utilized: biological, biochemical, or synthetic.

In a biological approach, whole organisms are used to conduct the splitting of water into hydrogen and oxygen; thus the term *biophotolysis* (44). However, the liberation of molecular oxygen generally inhibits the activity of hydrogenases, the biological enzymes that produce molecular hydrogen (45). Thus an important task for research in this area is to find species or mutations possessing hydrogenases that are effective in the presence of oxygen. An alternative is to produce the hydrogen and oxygen separately, as has been done with cultures of blue-green algae (46). However, in all known cases of biophotolysis the rate of hydrogen production is extremely small, and great progress will be required before practical conversion schemes can be designed.

In a biochemical approach, enzyme systems would be obtained from biological organisms and then combined in an appropriate reaction cell to perform all the steps involved in collecting energy and driving the water-splitting reactions (45). Production of hydrogen, at least at low rates for short periods of time, has been demonstrated (47, 48), but more basic research on the biochemical mechanisms of photosynthesis and inventive ideas for incorporating molecular components into systems will be required for this technique to become practical.

In a synthetic approach, a complete chemical system for photolysis would be designed and synthesized without using any components taken from plants or algae. This has a great advantage in that problems of instability of biological components can be avoided. There are some promising ideas about the form such chemical systems might take (49, 50), but this must be considered a long-term research problem. An alternative to a purely photochemical approach is a hybrid of photovoltaic conversion and electrolysis, in which light falls on a semiconductor electrode in a solution and drives a water-splitting reaction. This effect has been observed (51), but it is far from being a practical conversion device.

THERMOCHEMICAL CONVERSION   If a practical and economical technology for the collection of solar energy as high-temperature heat can be developed, then a new route for the production of fuels from solar energy is opened: thermochemical splitting of water to produce hydrogen. Processes for thermochemical production of hydrogen are being actively investigated because of their potential utilization with high-temperature nuclear reactors (52). Hundreds of possible processes that use various reactants in closed cycles have been investigated with the aid of computer programs. Many of these were reviewed at the 1974 Miami conference on the concept of hydrogen economy (53). Either point-focus solar collectors or line-focus collectors with selective surfaces could provide temperatures high enough (above 1000°K) to drive these cycles. However, these thermochemical processes are complex and might be impractical. They require at least two steps for the separate liberation of hydrogen and oxygen, and arguments based on thermodynamics suggest the need for more than two steps (54). Large amounts of reactants, high temperatures, and perhaps high pressures, and extensive mixing, reaction, and separation steps are required (55), so the design of a practical process will be a challenging task.

ELECTROLYSIS   An alternative to thermochemical production of hydrogen is to use the solar heat first to generate electricity by a solar thermal conversion scheme and then use the electricity to produce hydrogen by electrolysis of water. Electrolysis could also be used to produce hydrogen in conjunction with any solar energy technology that leads to electricity: photovoltaic, ocean thermal, or wind energy conversion. Electrolysis is a proven technology that has been used commercially in both small and large applications (53). Efficiencies are good, with only about 115 kW-hr of electricity required to produce 1000 cu ft of hydrogen, for an energy efficiency of 83% (56).

Wind Energy Conversion

Prior to and during the early 1900s, wind energy was widely used in rural areas of the United States to provide motive power for grinding crops, pumping water for irrigation, and charging electric batteries. The widespread availability of inexpensive electricity in the 1930s, brought about largely by the efforts of the Rural Electrification Administration, caused a decline in these small-scale uses of wind energy, which has not been reversed. A number of large experimental wind machines, designed for electric power production, with capacities of 100–1000 kW have been developed. The largest of these machines was the 1.25-MW Smith-Putnam wind generator, built in 1941 on Grandpa's Knob in Vermont; it proved that a practical machine could be built that would generate electricity in large quantities and feed into an electric power network. A large number of turbines for electrical power generation were also built in Europe. However, in spite of many technical improvements, these machines were unable to compete with the declining price of electricity from fossil fuels (57, 58).

The rise of fuel prices in the 1970s, coupled with three decades of technological advances, has renewed interest in the possibility that economically attractive wind energy systems can now be designed (59). For these systems to have a significant impact on the national scale will require not only improvements in the performance of major components but also development of new design concepts and applications. In addition, the wind energy resource must be assessed and techniques developed to predict the wind characteristics at potential sites for these systems.

The energy available to a wind turbine is the kinetic energy of the wind passing through the area swept by the blades of the turbine. This energy flow increases with the cube of the wind velocity, so selection of sites is of great importance. Theoretically, 59.4% of this energy would be extracted by a perfectly designed turbine (60). The efficiency actually achieved by a well-designed turbine is about 45%, and an overall system efficiency of 30–40% for generation of electricity can be obtained. Thus the blades of a turbine intended to provide 1 MW of power from a wind of 18 mph would have a diameter of about 180 ft.

At present, new concepts are being studied for improved rotors and energy storage and conversion components. The interaction of wind turbine systems with electric power networks is being investigated to assess the need for energy storage or backup generating capacity. In addition, the possibilities of on-site wind conversion for industrial or agricultural applications are being studied.

A sequence of experimental and demonstration systems are planned for construction, ranging from small units of less than 100 kW for rural applications to large multi-unit systems in the range of 10–1000 MW (61). The 100-kW experimental wind turbine generator presently under construction at the NASA Lewis Research Center and scheduled to commence tests in 1975 is the first large wind machine in this sequence (62).

Ocean Thermal Conversion

Although the concept of generation of electricity from naturally occurring differences in ocean temperatures dates from the 19th century, the only attempt to demonstrate its feasibility has been a series of experiments conducted by Claude in 1929 (63, 64). In these experiments, Claude attempted to operate a heat engine between the warm surface waters of the tropical oceans and the cold waters at depth. However, Claude used the ocean water itself as the working fluid, so the high-pressure supply to the turbine was limited to the vapor pressure of the warm surface water. Claude's experiments were of limited success for a number of reasons, including the large amounts of energy required for pumping and the problems of protecting the plant from vagaries of weather. The use of water as the working fluid cannot be ruled out on the basis of his experiments, but current proposals generally favor a secondary working fluid with a higher vapor pressure than water (65).

The concept of ocean thermal conversion has been revived and further developed by a number of advocates, including the Andersons (66) and Zener (67). A number of conceptual designs have been prepared and operating efficiencies and economics estimated (68–70). An important element of these designs is development of heat exchangers with low cost per surface area and a small temperature drop from the ocean waters to the working fluid, which may be ammonia, a fluorocarbon refrigerant, or an organic fluid such as propane. In general, these studies have reported favorably on the technical feasibility of ocean thermal conversion and have argued strongly for its economic feasibility.

Reviews of the concept of ocean thermal conversion have recently been performed by TRW Systems and Energy Group and by Lockheed (71). These reviews, intended to provide an independent assessment of the concepts being advanced by advocates of ocean thermal conversion, have generally concluded that the technical problems of the systems can be solved and that the economics of the systems are promising.

THE ECONOMICS OF SOLAR ENERGY

Although it is often said solar energy is free, this is no more true for solar energy than for any other energy resource. Oil in the ground costs us nothing: nature and time have provided it. However, to extract, transport, refine, and distribute oil to the consumer costs us in both labor and capital. The same is true of solar energy: the cost is in the labor and capital required to make it useful to our needs.

Solar energy does differ from most of our present energy sources in that it is very intensive in "first cost". Thus in comparing a solar thermal power plant with a coal-burning plant, we must decide how to compare flows of costs and benefits that have different distributions in time. This question of how to compare the present and future utility of alternatives is a subtle and difficult problem in economics. The question has ramifications at the level of the individual decision by a single firm or individual consumer (e.g. choosing between a solar and a coal-burning plant for required new generating capacity), and at the level of how a nation should influence by policy the transition from energy sources of finite extent such as petroleum to inexhaustible sources such as solar energy.

Discounted Cash Flow

First we consider how a single firm or individual will make an economic comparison of a solar energy technology with an alternative technology that requires continuing purchases of fuel in the future. For a single firm, the comparison between the present and the future is essentially determined by the interest rate the firm must pay to borrow money (the cost of capital). If the interest rate is r, to invest one dollar in capital equipment now will cost the firm $(1+r)^n$ dollars when it repays the loan n years later. Thus $(1+r)^n$ dollars n years in the future is worth one dollar now, or equivalently, one dollar n years in the future is worth only $(1+r)^{-n}$ dollars now (72). If one wishes to compare two investment alternatives that have different schedules of payment and return, then this can be done by discounting (multiplying) all future costs and income that will occur n years in the future by the factor $(1+r)^{-n}$ and summing over all future years. This is the discounted cash flow method of investment analysis. It serves to translate all future costs and payments into their present value (73).

For normal interest rates, and for the periods of service generally considered for solar energy facilities, this discount factor strongly influences economic comparisons. For example, one dollar saved in fuel 20 years from now is worth only 0.148 dollars in present value if the interest rate is 10%. Thus a company that must pay 10% interest ought not to pay more than 0.148 present dollars for a solar energy system to save one dollar of fuel 20 years from now. One sees that a high discount rate tends to discourage investment in facilities with high initial costs that have a stream of benefits extending into the future, such as solar energy systems. Thus policies that serve to provide low interest rates for solar energy facilities can be very effective in improving their economic competitiveness.

Future Energy Prices

A major source of uncertainty in comparing solar energy technologies to present technologies is the future price of fuels, including oil, natural gas, coal, and uranium. If fuel prices rise faster than the interest (discount) rate, then solar energy systems with quite high initial costs will be shown more economical in a discounted cash flow analysis. Some analyses of the economics of solar energy applications have assumed rates of price increases for fuels as high as 14%. Experience in 1974 is certainly consistent with such high rates; however, to assume that such rates of increase will continue for 20 or 30 years is unjustified. This would mean an increase in energy prices of more than an order of magnitude, so that the cost of only the energy incorporated in manufactured goods, now about 5% of total cost (74), would become comparable to their total present cost. The consequences of this on the national economy would be so severe that no reasonable deductions can be made about costs of labor or capital for solar energy devices in such a scenario.

Some resource economists would actually argue that the price of energy will decrease in the long term. This school of thought is represented by Barnett and

Morse, who see the effect of technological progress as more important in determining long-term price trends than resource depletion (75). This would mean that solar energy technologies must come down to quite low initial costs to be competitive.

Given such a wide range of opinions about future fuel prices, any economic analysis of a solar energy facility intended to provide energy for 20 or more years is very uncertain. The best we can do is use a moderate estimate of future fuel prices and always keep in mind the sensitivity of our analysis to any variation from our assumed prices. A moderate assumption about future fuel prices is perhaps represented by the Aerospace study of solar thermal conversion (73), which assumes that natural gas prices will increase at about 7% and coal prices will increase at about 3.5% between now and the year 2000. It was also assumed that future oil prices will preclude the use of oil for electrical generation.

Future Resource Use

The concept of discount analysis can also be applied to society as a whole to investigate policies that involve the use of solar energy to save finite resources for future generations. For example, we can inquire about the relative utility (benefit to society) of consuming a barrel of oil now or saving it for use at some time in the future (76). In a discount analysis approach to distributing over time the consumption of a finite amount of oil to obtain optimum utility, we would discount the value of consuming a barrel of oil n years in the future by $(1+r)^{-n}$. Thus, with a value of r about 10%, discount analysis would have us consume a barrel of oil now at a value of $10, even if it would be worth $1000 to industry 50 years from now as a chemical feedstock. This concept of discounting the future utility of consumption has been rejected by some economists, such as Ramsey, who in developing a theory of saving states "we do not discount later enjoyments in comparison with earlier ones, a practice which is ethically indefensible and arises merely from the weakness of the imagination..." (77). However, the existence of high interest rates for investment in solar energy represents, in its effect, a policy of consuming finite energy resources now in preference to conserving them for the future.

An example of a public policy that tends to favor the utilization of capital-intensive but fuel-conserving technologies, such as solar energy, is the regulation of public utilities. According to the Averch-Johnson thesis (78), the fact that the rate of return on investment in capital by a public utility is established at a fixed percentage by the regulatory commission leads to an "oversubstitution" of capital for labor or fuel. The oversubstitution is in comparison with the amount of capital that provides the best marginal return on investment with present fuel prices (78).

Heating and Cooling of Buildings

The economics of heating and cooling a building with solar energy depend strongly on the local climate and the detailed characteristics of the building loads. Because the costs of a solar system must be paid for every day, but the system returns value only when the load requirements match the presently available or stored solar energy, the economics of a system for a particular application must be estimated by an hour-by-hour analysis of the predicted loads and solar energy inputs.

Methodologies for such analyses have been developed and applied to investigate the economics of systems in various climates (79–83).

Although exact economic evaluations require the detailed analysis discussed above, we can derive a simple rule that identifies the approximate requirements that a solar energy system must meet for economic competitiveness. [This rule was suggested by M. A. Wahlig (personal communication).] Consider the type of system that employs circulating fluids in collectors and provides space heating for a building. In an average climate, about 1200 Btu fall on each square foot of collector each day. The collector is about 50% efficient, and there are about 200 days of the year when the heat can be used to meet a building load, so the amount of useful heat collected per year per square foot of collector is about 0.12 million Btu. Each year a payment must be made for the interest (about 10%) and amortization (over about 20 years). This annual payment will be about 0.12 times the initial cost of the total system. Thus, *solar energy systems for buildings provide one million Btu at a cost approximately equal to the total initial cost of the system per square foot of collector.* Of course this rule is very crude and is intended only to indicate the range of allowable costs for solar components.

The present cost of moderate-performance flat-plate collectors suitable for space heating applications is about $6/ft^2. The cost of the remainder of the system components (storage tank, plumbing, and controls) and installation may about double this cost to approximately $12/ft^2 of collector. Present (1975) prices for fossil fuels for home heating vary with region, but they are about $1.50/million Btu for natural gas and about $2.50/million Btu for fuel oil. Thus, if one considers only present prices, solar heating is not competitive with gas or oil. However, even for a solar system to be installed now, comparison should be made with the prices to be paid for oil and natural gas over the next 20 years. In addition, for solar systems to be installed in the future, mass production, improved technology, and familiarity of building contractors with solar technology should reduce the initial cost. Based on consideration of these factors, the three Phase Zero studies of solar heating and cooling estimate that solar heating systems will reach economic competitiveness with fossil fuels around 1985 (81–83).

The above estimates refer to fully active solar heating systems. In addition to these systems, there exists a wide range of passive or partially passive systems, including the use of overhangs above windows, drumwalls (84), and the Skytherm house (7). Many of these less complex systems are already economically justifiable, and while they do not provide full thermostat control of interior temperatures, they can greatly reduce requirements for fossil fuels.

Solar Thermal Conversion

In an examination of the economics of solar thermal conversion for generation of electricity, it is not sufficient to consider only the solar thermal plant itself. Instead one must consider the entire utility network that will combine solar and nonsolar energy sources. In any network, the installed capacity must exceed the expected peak loads by some margin that provides for the occasions when some of the generating plants are out of service (for either planned maintenance or breakdown).

A solar thermal plant will also suffer insolation outages, and for this reason extra backup capacity for the solar plant will be required in the network. Thus, an economic analysis of solar thermal conversion requires consideration of both the amount of fuel supplanted by the solar plant and the amount of installed capacity supplanted. Similar considerations apply to the photovoltaic and wind energy applications in electric power networks. This sort of margin analysis for solar thermal conversion has been performed by Aerospace Corporation (73).

The cost to a utility of a solar thermal plant is expected to be largely the cost of construction, and it is this parameter that has received the most attention. For valid comparison with the capital costs of fossil and nuclear plants, consistent methods of dealing with projected escalation of labor, equipment, and materials must be used (85). The cost estimates for solar thermal plants are dominated by the cost per square meter of the collectors (heliostats in a central receiver plant). Estimates of the cost of heliostats for central receiver plants range from \$3.20/ft^2 (23) to \$8.20/ft^2 (22). With an assumed cost of heliostats of \$3/ft^2, Aerospace Corporation estimates a busbar cost of electricity of 4.8¢/kW-hr in 1991 dollars (2.8¢/kW-hr in 1974 dollars). This is within range of the projected costs for electricity from fossil fuels in 1991.

Photovoltaic Conversion

For applications of photovoltaic conversion in electric power networks, the methodology of margin analysis introduced under solar thermal conversion is relevant. However, for applications involving local generation at a residence or commercial building, a different analysis is required. Much of the cost of electricity to a small consumer is the fixed cost of distribution and service rather than an incremental cost per unit of energy. Thus, the savings made possible by photovoltaic arrays that reduce but do not eliminate the need for electricity from the network are significantly limited. For the present, these detailed considerations are secondary to the crucial question of the cost per unit area of photovoltaic devices. The cost goals for economic competitiveness have been estimated to be about \$4 for central-station applications, and about \$6 for residential applications, per square meter of solar array (86).

The present costs of solar cells for terrestrial applications (used in remote sites where other energy sources are not available) is about \$2000/m^2, and the cost of the high-quality cells used in space applications is about another factor of ten higher (87). Thus, the goal in photovoltaic research and development must be to reduce costs eventually by a factor of about 500. This will require significant technological advances, in addition to the cost reductions made possible by mass production of millions of cells. Consideration of the cost of the high-purity silicon used in cells leads to the conclusion that advances in the technology for purification of silicon are also required (88).

Production of Fuels

Of the various technologies by which fuels can be produced from solar energy, only those that utilize the natural photosynthetic process in plants are well enough defined to allow economic analysis. The economics of land-based energy plantations have been examined by InterTechnology Corporation (36) and by Stanford Research Institute (38). The assumptions used in these analyses are favorable to low production costs: locale in the southwestern United States where insolation is high, adequate water supply, optimum use of fertilizers, and no significant problems of pests or disease. The conclusion reached is that biomass can be produced at a cost competitive with fossil fuels under the assumed conditions. InterTechnology Corporation arrives at an estimate of slightly more than a dollar per million Btu, and Stanford Research Institute arrives at an estimate of less than a dollar per million Btu as biomass and about $2 per million Btu as a synthetic natural gas. Such costs are quite competitive with prices of fossil fuels, but the assumptions made regarding the availability of water in the Southwest render this scenario somewhat unrealistic. The important economic competition of biomass production is with food production, and it seems that if such favorable growing conditions could be established, the highest value for the crop would be obtained if food were grown rather than energy (89).

Wind Energy

A careful economic analysis of the use of wind energy in electric power networks will require the same sort of margin analysis used in the economic analysis of solar thermal plants. Assuming that the maintenance costs of large wind turbines are found to be low, achieving economic feasibility for applications of wind turbines to electric networks will depend on a lower capital cost than has been involved in large (more than 100 kW) machines built in the past.

The large wind turbines built in the past have been primarily experimental devices that were one of a kind, so their capital costs do not provide a good guide as to the costs to be anticipated in widespread utilization. For example, the Smith-Putnam turbine built at Grandpa's Knob in the 1940s cost a total of about one million dollars over the life of the experiment for a rated capacity of 1250 kW, or about $1000/kW. At that time the capital cost of fossil fuel plants was about $125/kW, and fuel prices were low. It was estimated that production models of the Smith-Putnam turbine would have cost $191/kW, so it did not seem competitive (57, 90, 91). The 100-kW and 1-MW turbines to be built as part of the ERDA program will provide experience to allow better estimates of the construction costs possible now.

Small-scale applications of wind energy have been widely used and are economical where the costs of obtaining service for electric power networks at remote sites are prohibitive. A variety of small systems for such applications are being marketed, with prices about $5000–$8000 for a 1-kW-rated capacity (92).

Ocean Thermal Conversion

Ocean thermal power plants require no storage and provide power at all times (baseload), which tends to make their economics more favorable than those of other solar technologies. However, they must be constructed to endure the marine environment with little maintenance, so construction costs may be high. In the past, advocates for ocean thermal conversion have estimated the capital cost of large ocean thermal plants to be about $200–$300/kW, certainly competitive with fossil and nuclear energy (65). Preliminary reports of the more recent analyses by TRW and Lockheed place the cost of energy from these plants at about 2.7¢/kW-hr in 1985, higher than the previous estimates, but still potentially competitive (71).

THE ENVIRONMENTAL IMPACT OF SOLAR ENERGY

Solar energy is widely touted as a clean source of energy, and drastic comparisons are offered with fossil and nuclear energy technologies and their environmental impacts. This view is not without some foundation; indeed we agree that the environmental impacts of solar energy should be relatively benign. However, the time is long past when we ought to allow any new technology, no matter how innocuous it might appear at first, to be developed and deployed without a careful examination of the possible environmental alterations that might result. We describe here some of the possible deleterious impacts of solar energy technologies, but without any implication that these impacts are more severe than those of the technologies that solar energy might replace.

It is striking how little serious attention has been paid to these possible impacts of solar energy. One reason for this neglect might be that, until recently, few thought solar energy would ever be a serious alternative to existing energy technologies, but this reason surely no longer holds. We believe a careful examination of the possible environmental impacts of solar energy is now overdue.

Local Impacts

Solar energy facilities will have a certain impact on the local environment of their site, particularly on the heat and water balance of the region. The temperature, amount of sunlight, and humidity at the surface of the ground and in the first meter or so above the surface define a microclimate in which plants, insects, and animals live. These parameters of the microclimate are determined by a number of factors, including the albedo of the surface, the surface roughness that determines the surface wind patterns, and the degree to which the surface is shielded from radiation loss to the night sky (93, 94). Solar energy collectors for either thermal or photovoltaic conversion, which permanently cover a large fraction of the ground, will drastically alter this microclimate and thus cause major changes in the type of life that can best survive there. Large arrays of wind turbines may have similar effects if they cause significant alteration of surface wind patterns. Solar collectors for heating and cooling of buildings would have a similar effect, except that they will probably be integrated into the buildings and thus will make only a minor perturbation on the already drastic environmental alterations caused by urbanization and suburban development. Solar heating and cooling of buildings may have an important impact if its use requires widely dispersed buildings in order to avoid shading. Larger land areas would then suffer the impacts of development.

Ocean thermal conversion will have its own special impacts. The operation of the ocean thermal plant will bring up cold waters close to the surface. The most important impact of this will probably not be a change in the temperature of surface waters, although this might occur. The major impact will be the effect on the local ecosystem of the addition, by the cold waters from the depths, of high levels of nutrients not normally present at the surface. The local surface ecosystem will be drastically altered, although this impact might be considered beneficial if it leads to higher productivity of commercial fish (95).

The most drastic of the local impacts of solar energy might be those resulting from the production of biomass for fuels in energy plantations. If previously unused land is employed, then the preexisting natural ecosystem will be replaced with the forced monoculture common in agriculture. In addition, the energy plantation will be a source of some air pollutants, such as dust, pesticides, and allergenic pollens, and water pollutants, such as fertilizer runoff, pesticides, sediments, and increased salinity (96). These effects are all part of present agricultural practice, and the effect of energy plantations will be to extend them to previously unaffected regions.

Regional Impacts

With the need for regional land-use planning becoming more widely recognized, it is important that we understand how the use of solar energy will impact regional land-use problems. The most obvious of these impacts is the requirement of the solar facility itself. The amount of land required per unit of energy obtained varies widely among solar energy technologies. A 1000-MWe solar thermal plant might require as little as 14 km^2 of collector area (25), while an energy plantation that produces biomass fuel for electrical generation might require as much as 635 km^2 for the same electrical power output (38).

Water requirements and compatibility with regional supplies must also be considered. Some solar technologies, such as photovoltaic and wind energy conversion, might have no need for water and others, such as solar thermal conversion, might be able to minimize requirements by using dry cooling towers, but bioconversion for fuel production will have significant water requirements that must be included in selecting sites.

Solar energy facilities will cause demographic shifts with important regional consequences. Construction and operation of a solar facility will require a certain additional population, but potentially more important is the industry that might be attracted to a previously undeveloped region by a new and inexhaustible source of energy. With this industry would come residential areas and urbanization. Thus the existence of solar technologies would create strong pressures for development of new areas that are presently wild or used for other purposes.

National Impacts

Any energy facility causes the generation of pollutants at sites throughout the nation because of the diverse industrial activity required for the manufacture and transportation of components for the facility. In assessing the total environmental impact of solar energy, we must include in our consideration the entire economic system involved in producing the solar energy facility. Thus, in accounting for the pollution released by solar energy, we must include a certain fraction of the pollution generated by a steel plant that fabricates, among other things, supports for solar collectors; a corresponding fraction of the pollution generated by the coal-burning plant that supplies the steel plant with electricity; and so on through an infinite chain of industrial activities (97). Of course, similar considerations apply to all other energy facilities, such as nuclear plants, and with all things considered, solar energy should still be shown to be a very clean source of energy.

The material requirements of a solar energy facility, and the impact on the national economy of supplying those materials, should be included among the impacts of the facility. In some cases the requirements for certain materials that would be associated with the widespread deployment of a solar technology are large compared with the present use of that material in the national economy. For example, the amount of gallium required for a 1000-MWe photovoltaic plant using gallium arsenide solar cells would be quite large—indeed, comparable to the estimated national production of gallium from now to the year 2000 (97, 98). Even if the solar technology can afford to pay high enough prices to increase drastically the production of some needed material, we should not ignore the impact on other segments of the economy of having to match that higher price.

Among the material requirements of solar energy is one of particular importance: energy. A certain investment of energy will be required for construction of a solar energy facility, and a long period of operation may be necessary to pay back this energy debt. As an extreme example, the energy required to fabricate a 4-cm^2 silicon solar cell by the techniques used in the past has been estimated to be 2.8 kW-hr (86), so that about 40 years of operation would be required to regain the energy investment. The energy investment in a solar facility can be minimized by proper design and selection of low-energy-cost materials (99), but in an accelerated program of solar development, the net energy balance of solar facilities might be negative for a time.

Perhaps the most important among the national economic impacts of solar energy will be problems in providing the high initial investment capital required for solar facilities. Enormous amounts of capital would be required to build solar facilities rapidly enough to match exponential projections of energy demand. Even the development of practical and economically competitive solar energy technologies will not allow an unrestricted growth in energy use.

Social Impacts

Little attention has been given to the potential social impacts of solar energy utilization. Certainly part of the reason for this is that those developing the technologies have taken as their task the development of one-for-one replacements of existing energy sources. Thus, for example, in solar heating and cooling of buildings, the emphasis is on development of systems that would be installed by existing building contractors and would provide to the resident the same thermostat-controlled living environment to which he is accustomed. To the resident there would be little or no apparent evidence that the source of energy was solar, rather than gas or electricity. Thus the social impact might be negligible.

However, there are technological options, and ways of applying new technologies, that might have significant social impacts by leading to changes in life-styles. For example, the building resident might choose to install a passive solar system to temper, but not control exactly, the living environment of his residence. This is a choice of life-style that is at variance with the mainstream of US experience, but might be compatible with the economical use of solar energy.

Technologies such as solar heating and cooling of buildings, bioproduction of fuels, and harnessing of wind energy might allow individuals to establish lives completely independent of the national energy system. There are already movements in this direction, although the number of individuals involved is small. The development of viable energy options that allow independence might serve to further these social movements (92).

It is impossible now to predict what the long-term social consequences of solar energy utilization might be, but any technology that becomes as widely used as we expect solar energy to be in the long term will certainly have a major impact on our lives and our social institutions.

THE FUTURE UTILIZATION OF SOLAR ENERGY

The Near Term

The present utilization of solar energy in the United States is negligibly small. Even if the use of wood for heating is included within this category, present use provides less than half a percent of the nation's energy requirements. The factors that will determine how quickly this situation can be changed include the development of solar energy industries, the progress of the federal research and development program, and the effect of legislation intended to provide incentives for solar energy and remove institutional barriers to its use.

SOLAR ENERGY INDUSTRIES   Within the last few years interested and enterprising individuals have founded a surprisingly large number of small solar energy companies. In addition, a number of large US firms have begun development of prototype solar energy components, in anticipation of a potentially large market. This activity is greatest in the area of heating and cooling of buildings, where a large number of firms now offer components and complete systems (92, 100), and in photovoltaic conversion, where the large semiconductor companies have recently begun to explore the possibilities of developing new products (34). The Solar Energy Industries Association was formed in 1974 to aid communication between these industries and to represent them in matters relating to solar energy legislation at the federal, state, and local levels (101).

One of the first serious problems to be faced by the solar energy industries is that of maintaining their credibility and reputation. Some solar energy firms have been guilty of making unreasonable claims for the performance and economics of their components and systems for solar heating and cooling. If this practice were to spread unchecked, the credibility of all solar energy firms, and of the technology itself, would be severely damaged. The Solar Energy Industries Association has recognized this hazard and is striving to establish a code of ethical practice for its member firms. Government-established performance standards for solar components should also alleviate this problem.

SOLAR ENERGY RESEARCH AND DEVELOPMENT   Solar energy R & D programs are underway in many nations other than the United States, especially in Germany, France, Japan, USSR, and Australia. However, the US program is by far the largest, with a budget in fiscal year 1976 of about $110 million.

The ERDA solar energy R & D program is very aggressive, with a high priority on the earliest possible demonstration of feasibility of each of the approaches to solar energy conversion. We have previously referred to a number of the pilot plant and demonstration projects in this program. They include the heating and cooling of buildings demonstration program, the 10-MWe solar thermal pilot plant, large demonstration arrays of photovoltaic cells, and the 100-kW wind turbine. At the same time, longer-range research and development is underway on the technological advances that will be required for the widespread use of solar energy in the longer term (61).

SOLAR ENERGY LEGISLATION   Within the United States, a national commitment to the development and application of solar energy was expressed by the Solar Energy Research, Development, and Demonstration Act (Public Law 93-473) and the Solar Heating and Cooling Demonstration Act (Public Law 93-409). These laws are the basis of the federal research and development program and of the heating and cooling demonstration program, which seeks to aid the establishment of a solar heating and cooling industry. Further legislation to provide incentives for solar energy has been proposed. This legislation would provide low-interest federal loans for solar heating and cooling systems for residences or would provide federal insurance for loans for solar energy systems, thus lowering the interest rates that will be charged by commercial lending institutions.

Legislation has been passed in a number of states, and proposed in many others, to provide incentives for solar energy systems by not including the added value of a solar heating and cooling system in the assessed value of a home, thus lowering the property tax on solar homes.

NEAR-TERM IMPACT OF SOLAR ENERGY   The feasibility of large-scale utilization of solar energy in the near term, before 1985, is a hotly debated subject. Much of this controversy is a spillover from the debate over the hazards of nuclear energy, with critics of nuclear power often proposing an accelerated development of solar energy as an alternative to the growth of the nuclear industry. We have no doubt that in the longer term solar energy will serve to reduce significantly the need for other energy sources. However, the possibility of solar energy making a large contribution to the nation's energy requirements before 1985 is small. Reasons for this include the large investment in existing energy systems, the huge capital requirements of a massive introduction of solar energy technology, and the significant technological advances required for solar energy to become competitive economically with the presently important energy sources. These factors will probably keep the contribution of solar energy to the US energy requirements to about 1% of the energy demand in 1985.

The Far Term

Progress in dealing with the technical, economic, and environmental factors discussed in this article will, in our opinion, lead to a "coming of age" of solar energy in the years between 1985 and 2000. We expect that in the period after the year 2000 solar energy will have become one of the conventional energy sources used in many regions of the world. However, attainment of this eventual success will demand patience and a continued dedication to the advancement of solar energy technologies in the intervening years. There is some danger that unrealistic expectations of near-term widespread application of solar energy will be suddenly dashed by harsh realities and that support for the continued development of the technologies might weaken. Such an eventuality would endanger the great contributions that solar energy can make in the longer term if it is given steady and continuous support.

We have described in this chapter a wide variety of approaches to the utilization of solar energy, and limitations of space have prevented the inclusion of many other promising approaches now under investigation. Many of these approaches will probably not survive to the point of deployment and widespread utilization, but a number of them will be selected as each being best for a particular energy demand and included in future energy systems that will draw upon inexhaustible sources of energy.

Literature Cited

1. Donovan, P. et al. 1972. *An Assessment of Solar Energy as a National Energy Resource*. Rep. No. NSF/RA/N-73-001, NSF/NASA Solar Energy Panel. Springfield, Va: Nat. Tech. Inf. Serv.
2. Thekaekara, M. P. 1974. The solar constant and solar spectrum and their possible variation. In *Report and Recommendations of the Solar Energy Data Workshop*, Nov. 29–30, 1973, pp. 86–92. Rep. No. NSF/RA/N/74/062. Washington DC: Nat. Sci. Found.
3. Threlkeld, J. L. 1963. Solar irradiation of surfaces on clear days. *ASHRAE Trans.* 69:24–36
4. Yellott, J. I. 1974. Solar energy utilization for heating and cooling. In *ASHRAE Handbook and Product Directory, Applications Volume*, Chap. 59, pp. 59.1–59.20. New York: Am. Soc. Heat. Refrig. Air-Cond. Eng. Inc.
5. Jessup, E. 1974. A brief history of the solar radiation program. See Ref. 2, pp. 13–20
6. Burda, E. J. 1955. *Applied Solar Energy Research*. Menlo Park, Ca: Stanford Res. Inst.
7. Shurcliff, W. A. 1975. *Solar Heated Buildings: A Brief Survey*. 9th ed. W. A. Shurcliff, 19 Appleton St., Cambridge, Mass. 02138
8. US Energy Research and Development Administration. 1975. *A National Plan for Solar Heating and Cooling*. Rep. No. ERDA-23. Washington DC: ERDA
9. Duffie, J. A., Beckman, W. A. 1974. *Solar Energy and Thermal Processes*, p. 386. New York: Wiley-Interscience
10. Sargent, S. L., ed. May 1975. *Proc. Solar Collector Workshop*. Rep. No. NSF/RA/N/75/015. Washington DC: Nat. Sci. Found.
11. deWinter, F., ed. 1974. *Proc. Solar Cooling Workshop, Feb. 6–8, 1974, Los Angeles*. Rep. No. NSF/RA/N/74/063. Washington DC: Nat. Sci. Found.
12. National Science Foundation/US Energy Research and Development Administration. 1975. *Proc. Solar Storage Workshop, April 1975, Univ. Virginia, Charlottesville, Va*. Washington DC: ERDA
13. US Federal Energy Administration. 1975. *Project Independence: Solar*

13. *Energy.* GPO Publ. No. 4118-00012. Washington DC: GPO
14. Ray, D. L. 1973. *The Nation's Energy Future.* GPO Publ. No. 5210-00363. Washington DC: GPO
15. Daniels, F. 1964. *Direct Use of the Sun's Energy,* pp. 9–11. New Haven, Conn: Yale Univ. Press
16. Jordon, R. C. 1963. Conversion of solar to mechanical energy. In *Introduction to the Utilization of Solar Energy,* ed. A. M. Zarem, D. D. Erway, pp. 125–52. New York: McGraw-Hill
17. Trombe, F., Le Phat Vinh, A. 1973. Thousand kW solar furnace built by the National Center of Scientific Research, in Odeillo (France). *Sol. Energy* 15:57–61
18. Sakurai, T. et al. 1964. Construction of a large solar furnace. *Sol. Energy* 8:117–26
19. Davies, J. M., Cotton, E. S. 1957. Design of the Quartermaster solar furnace. *J. Sol. Energy Sci. Eng.* 1(2,3):16–22
20. Colorado State University, Solar Energy Applications Laboratory/Westinghouse Electric Corporation. 1974. *Solar Thermal Electric Power Systems.* Final Rep. under NSF/RANN grant GI-37815 to Colo. State Univ., Fort Collins, Colo., and Westinghouse Electr. Corp., Pittsburgh, Pa. Rep. No. NSF/RANN/SE/GI-37815/FR/74/3. Washington DC: Nat. Sci. Found.
21. Winston, R. 1974. Solar concentrators of a novel design. *Sol. Energy* 16:89–95
22. Powell, J. C. et al. 1974. *Dynamic Conversion of Solar Generated Heat to Electricity.* Final rep. under NASA/LEWIS contract NAS3-18014 to Honeywell Syst. Res. Center, Minneapolis, Minn. and Black & Veatch, Consult. Eng., Kansas City, Mo. Rep. No. NASA CR-134724, Vol. II. Washington DC: Nat. Sci. Found.
23. Vant-Hull, L., Easton, C. R. 1974. *Solar Thermal Power Systems Based On Optical Transmission.* Prog. rep. Jan. 1 to June 30, 1974, under NSF/RANN grant GI-39456 to Univ. Houston, Houston, Tex., and McDonnell-Douglas Astronautics West, Huntington Beach, Calif. Rep. No. NSF/RANN/SE/GI-39456/PR/74/2. Washington DC: Nat. Sci. Found.
24. Hildebrandt, A. F. et al. 1972. Large-scale concentration and conversion of solar energy. *Trans. Am. Geophys. Union* 53:684–92
25. Bos, P. B. et al. 1974. *Solar Thermal Conversion Mission Analysis: Summary.* Rep. Nov. 1974 under NSF/RANN contract NSF-C-797. Rep. No. ATR-74 (7417-16)-1. El Segundo, Calif: Energy and Resourc. Div., Aerospace Corp.
26. Skinrood, A. C. et al. 1974. *Status Report on a High Temperature Solar Energy System.* Rep. No. SAND74-8017. Livermore, Calif: Sandia Lab.
27. Ross, M. et al. 1974. *Efficient Use of Energy: A Physics Perspective,* p. 13. Rep. Summer Study on Tech. Aspects Efficient Energy Utilization. New York: Am. Phys. Soc.
28. Lawrence Livermore Laboratory, Solar Energy Group. 1975. *LLL-Sohio Solar Process Heat Project, Report No. 2, 1 May 1974.* Rep. No. UCID-16630-2. Livermore, Calif: Lawrence Livermore Lab.
29. Wolf, M. J. 1972. The fundamentals of improving silicon solar cells. In *Solar Cells: Outlook for Improved Efficiency,* p. 56. Washington DC: Nat. Acad. Sci. Also 1960. *Proc. IRE* 48:1246
30. Boer, K. W. et al. 1975. Progress report on Solar One. In *Extended Abstracts of the International Solar Energy Congress, July 1975, Los Angeles,* p. 342. Int. Solar Energy Soc., 12441 Parklawn Dr., Rockville, Md.
31. Hass, G. M., Bloom, S. 1975. Preliminary results of the operation of the MITRE photovoltaic energy system. See Ref. 30, p. 118
32. Detailed solar cell array cost analysis. See Ref. 13, p. VII-C-37
33. Jet Propulsion Laboratory. Oct. 1975. *Assessment of the Technology Required to Develop Photovoltaic Power Systems for Large-Scale National Energy Applications.* Rep. No. NSF-RA-N-74-072, JPL, Pasadena, Calif. Washington DC: Nat. Sci. Found.
34. Jet Propulsion Laboratory. Oct. 1975. *Executive Report of Workshop Conference on Photovoltaic Conversion of Solar Energy for Terrestrial Applications.* Rep. No. NSF-RA-N-74-073, JPL, Pasadena, Calif. Washington DC: Nat. Sci. Found.
35. Aerospace Corporation. April 1975. *Mission Analysis of Photovoltaic Solar Energy Systems.* Rep. No. ATR-75-(7476-01)-1,2. El Segundo, Calif: Aerospace Corp.
36. Szego, G. C., Fox, J. A., Eaton, D. R. 1973. The energy plantation. *Proc. 7th Intersoc. Energy Convers. Eng. Conf., Aug. 1972, San Diego, Calif.* Washington DC: Am. Chem. Soc.
37. Borgstrom, G. 1972. *The Hungry Planet.* New York: Macmillan
38. Alich, J. A., Inman, R. E. 1974. *Effective Utilization of Solar Energy to Produce Clean Fuel.* Final Rep. June 1974,

NSF/RANN grant GI-38723. Menlo Park, Calif: Stanford Res. Inst.
39. Calvin, M. 1974. Solar energy by photosynthesis. *Science* 184:375–81
40. Zelitch, I. 1975. Improving the efficiency of photosynthesis. *Science* 188:626–33
41. Carlson, P. S., Polacco, J. C. 1975. Plant cell cultures: genetic aspects of crop improvement. *Science* 188:622–5
42. Oswald, W. J. 1973. Productivity of algae in sewage disposal. *Sol. Energy* 15:107–17
43. North, W. J. 1975. *Evaluating Oceanic Farming of Seaweeds as Sources of Organics and Energy*. Prog. rep. June 1 to Dec. 1, 1974 under NSF/RANN grant GI-43881 to Calif. Inst. Technol., Pasadena, Calif. Rep. No. NSF/RANN/SE/GI-43881/PR/75/1. Washington DC: Nat. Sci. Found.
44. Hollaender, A., ed. 1972. *An Inquiry into Biological Energy Conversion*. Workshop held at Gatlinburg, Tenn. Oct. 12–14, 1972. Knoxville, Tenn: Univ. Tenn. Press
45. Gibbs, M., Hollaender, A., Kok, B., Krampitz, L. O., San Pietro, A., eds. 1973. *Proc. Workshop on Bio-Solar Convers., Sept. 5–6, 1973, Bethesda, Md*. Bloomington, Ind: Indiana Univ. Press
46. Benemann, J. R., Weare, N. M. 1974. Hydrogen evolution by nitrogen-fixing Anabaena cylindrica cultures. *Science* 184:174–5
47. Krampitz, L. O. 1973. *Hydrogen Production by Photosynthesis and Hydrogenase Activity*. Prog. rep. July 1, 1972 to March 1, 1973, NSF/RANN grant GI-34992X to Case Western Reserve Univ., Cleveland, Ohio. Washington DC: Nat. Sci. Found.
48. Benemann, J. R., Berenson, J. A., Kaplan, N. O., Kamen, M. D. 1973. Hydrogen evolution by a chloroplast-ferredoxin-hydrogenase system. *Proc. Nat. Acad. Sci. USA* 70:2317–20
49. Lichten, N. M. 1974. *Photochemical Conversion of Solar Energy*. Prog. rep. Jan. 1 to June 30, 1974, NSF/RANN grant 74-087 to Boston Univ., Boston, Mass. Washington DC: Nat. Sci. Found.
50. Calvin, M. 1974. Solar energy by photosynthesis. *Science* 184:375–81. Solar energy by photosynthesis: manganese complex photolysis. *Science* 185:376
51. Fujishima, A., Honda, K. 1974. Electrochemical photolysis of water at a semiconductor electrode. *Nature* 288:37–8
52. De Beni, G., Marchetti, C. 1972. *Mark-1, a Chemical Process to Decompose Water Using Nuclear Heat*. Presented at Symp. Non-Fossil Chem. Fuels, Am. Chem. Soc., Washington DC
53. Veziroglu, T. 1975. *Hydrogen Energy*. New York: Plenum. 2 vols.
54. Wentorf, R., Hanneman, R. 1974. Thermochemical hydrogen generation. *Science* 185:311–19
55. Shinnar, R. 1975. Thermochemical hydrogen production: heat requirements and costs. *Science* 188:1036–7
56. Gregory, D. P., Ng, D. Y. C., Long, G. M., 1972. In *Electrochemistry of Cleaner Environments*, ed. J. O'M. Bockris, Chap. 8, pp. 235–40. New York: Plenum
57. Putnam, P. C. 1948. *Power From the Wind*. New York: Van Nostrand
58. Savino, J. M., ed. 1973. *Proc. Wind Energy Convers. Syst. Workshop, June 11–13, 1973, Washington DC*. Rep. No. NSF/RA/W-73-006
59. See Ref. 1
60. Wilson, R. E., Lissaman, P. B. S. May 1974. *Applied Aerodynamics of Wind Power Machines*. Publ. No. PB 238-595. Springfield, Va: Nat. Tech. Inf. Serv.
61. US Energy Research and Development Administration. 1975. *The National Solar Energy Program: Program Definition*. Rep. No. ERDA-49. Washington DC: ERDA
62. Thomas, R. 1974. The Lewis Research Center wind project. In *Proc. NSF/NASA/Utility Wind Energy Conf., NASA/LRC, Dec. 17, 1974, Cleveland, Ohio*
63. Claude, G. 1930. Power from the tropical sea. *Mech. Eng.* 52:1039
64. Ley, W. 1954. *Engineer's Dreams*. New York: Viking
65. Dugger, G. L. 1975. Ocean thermal energy conversion. In *Solar Energy for Earth*, ed. H. J. Killian, G. L. Dugger, J. Grey. New York: Am. Inst. Aeronaut. Astronaut.
66. Anderson, J. H., Anderson, J. H. Jr. April 1966. Large-scale sea thermal power. *Mech. Eng.*, p. 41
67. Zener, C. 1973. Solar sea power. *Phys. Today*, Jan: 48–53, June: 11–13, Sept: 11–13
68. Lavi, A., ed. 1973. *Proc. Solar Sea Power Plant Conf. and Workshop, June 27–28, 1973, Carnegie-Mellon Univ., Pittsburgh, Pa*. NTIS Publ. No. PB-228-066. Springfield, Va: Nat. Tech. Inf. Serv.
69. Anderson, J. H. 1974. *Research Applied to Ocean Sited Power Plants*. Prog. Rep. Jan. 1 to Dec. 31, 1973, NSF/RANN grant GI-34979 to Univ. Mass., Amherst,

Mass. NTIS Publ. No. PB-223-067. Springfield, Va: Nat. Tech. Inf. Serv.
70. Zener, C. 1974. *Solar Sea Power.* Prog. Rep. Nov. 1, 1973 to Jan. 31, 1974, NSF/RANN grant GI-39114 to Carnegie-Mellon Univ., Pittsburgh, Pa. NTIS Publ. No. PB-228-068. Springfield, Va: Nat. Tech. Inf. Serv.
71. Edmundson, W. B. 1975. *Sol. Energy Dig.* 4(6): 1–2
72. Moore, B. J. 1973. *Introduction to Modern Economic Theory.* New York: Free Press
73. Bos, P. B. et al. 1974. *Solar Thermal Conversion Mission Analysis: Comparative Systems/Economics Analysis.* Vol. 4. See Ref. 25
74. Bullard, C. 1973. *Energy Conservation through Taxation.* Doc. No. 95, Center for Advanced Computation, Univ. Illinois, Urbana, Ill.
75. Barnett, H. J., Morse, C. 1963. *Scarcity and Growth: The Economics of Natural Resource Availability.* Baltimore, Md: Johns Hopkins Univ. Press
76. Koopmans, T. C. 1974. Ways of looking at future economic growth, resource and energy use. In *Proc. Conf. Energy: Demand, Conservation, and Institutional Problems, Feb. 12–14, 1974.* Cambridge, Mass: MIT Press
77. Ramsey, F. P. 1928. A mathematical theory of saving. *Econ. J.* 38: 548–50
78. Brannon, G. M. 1974. *Energy Taxes and Subsidies.* A Report to the Energy Policy Project of the Ford Foundation, pp. 111–13. Cambridge, Mass: Ballinger
79. Löf, G. O. G., Tybout, R. A. 1973. Cost of house heating with solar energy. *Sol. Energy* 14: 253–78
80. Löf, G. O. G., Tybout, R. A. 1975. The design and cost of optimized systems for residential heating and cooling by solar energy. *Sol. Energy* 16: 9–18
*81. TRW Systems Group. 1974. *Solar Heating and Cooling of Buildings, Phase 0.* Final rep. May 1974 under NSF/RANN contract NSF C-853 to TRW Systems Group, Redondo Beach, Calif. Rep. No. NSF/RA/N-74-022A. Washington DC: Nat. Sci. Found.
82. General Electric Space Division. 1974. *Solar Heating and Cooling of Buildings: Phase 0 Feasibility and Planning Study.* Final rep. May 1974 under NSF/RANN contract to General Electric Space Division, Valley Forge, Pa. Rep. No. NSF-RA-N-74-021A. Washington DC: Nat. Sci. Found.
83. Westinghouse Electric Special Systems Division. 1974. *Solar Heating and Cooling of Buildings, Phase 0.* Final rep. May 1974 under NSF/RANN contract NSF-C-854 to Westinghouse Electric Corporation, Special Energy Systems, Baltimore, Md. Rep. No. NSF-RA-N-74-023A. Washington DC: Nat. Sci. Found.
84. Balcomb, J. D. et al. 1975. *Solar Heating Handbook for Los Alamos,* pp. 66–7. Rep. No. LA-5967. Los Alamos Sci. Lab., Los Alamos, N. Mex. Also see: A breakaway house that's heated by the sun. Jan. 1974. *House and Garden*
85. Olds, F. C. Jan. 1973. Capital cost calculations for future power plants. *Power Eng.,* pp. 61–5
86. Wolf, M. 1972. Cost goals for silicon solar arrays for large scale terrestrial applications. *Proc. 9th IEEE Photovoltaic Specialists Conf., May 2–4, 1972, Silver Spring, Md,* pp. 342–50
87. Wolf, M. 1975. Photovoltaic power. In *Solar Energy for Earth.* See Ref. 65
88. Currin, C. G. et al. 1972. Feasibility of low cost silicon solar cells. *Proc. 9th IEEE Photovoltaic Specialists Conf., May 2–4, 1972, Silver Springs, Md,* pp. 363–9
89. Walsh, J. 1975. U.S. agribusiness and agricultural trends. *Science* 188: 531–4
90. Savino, J. M. 1975. Wind power. In *Solar Energy for Earth.* See Ref. 65
91. Smith, B. E. 1973. Smith-Putnam wind turbine experiment. *Proc. Workshop on Wind Energy Convers. Systems, June 11–13, 1973, Washington DC.* Sponsored by NSF/RANN and NASA/LEWIS. Rep. No. NSF/RA/W-73-006. Washington DC: Nat. Sci. Found.
92. Portola Institute. 1974. *Energy Primer: Solar, Water, Wind, and Biofuels.* Menlo Park, Calif: Portola Inst.
93. Sutton, O. G. 1953. *Micrometeorology.* New York: McGraw-Hill
94. Van Wijk, W. R. 1963. *Physics of the Plant Environment.* Amsterdam: North-Holland
95. Othmar, D. F., Roels, O. A. 1973. Power, fresh water, and food from cold, deep sea water. *Science* 180: 121–5
96. Brady, N. C. 1967. *Agriculture and the Quality of Our Environment.* Washington DC: Am. Assoc. Adv. Sci.
97. Hughes, E. E., Dickson, E. M., Schmidt, R. A. 1974. *Control of Environmental Impacts from Advanced Energy Sources.* Rep. No. EPA-600/2-74-002. Washington DC: US EPA
98. Weeks, R. A. 1973. Gallium, germanium, and indium. In *United States Mineral Resources,* ed. D. A. Brobst, W. P. Pratt, pp. 237–46. US Geol. Surv. Prof.

Pap. No. 820. Washington DC: GPO
99. Herendeen, R. A. 1973. *The Energy Cost of Goods and Services*. Rep. No. ORNL-NSF-EP-58. Oak Ridge Nat. Lab., Oak Ridge, Tenn.
100. US Energy Research and Development Administration. 1975. *Catalogue on Solar Energy Heating and Cooling Products*. Rep. No. ERDA-75. Washington DC: ERDA
101. Solar Energy Industries Association, Inc. 1975. *Solar Energy Industry Directory and Buyer's Guide*. SEIA, Inc., 1001 Connecticut Ave., NW, Washington DC

* Refs. 81–83 available from National Technical Information Service, Springfield, Va.

Copyright 1976. All rights reserved

GEOTHERMAL ENERGY ✖11007

Paul Kruger[1]

Civil Engineering Department, Stanford University, Stanford, California 94305

Geothermal energy, the natural heat of the earth, holds great promise to be one of the more abundant forms of energy. Kruger & Otte (1) note that although geothermal energy will be used primarily for generating electric power, its utilization in space or industrial heating, desalination of water, and mineral recovery may be of great value in conserving fossil and nuclear fuels. Conversion of natural dry steam to electric power has been underway in the United States since 1960, but general development of geothermal resources as a significant contribution to the nation's energy supply will require major efforts to: (*a*) establish adequate reserves of major types of geothermal resources, (*b*) develop appropriate technologies to utilize these resources in a cost-effective and environmentally acceptable manner, and (*c*) reduce the many legal and institutional constraints to development. Programs to accelerate these efforts are underway by both industry and government.

GEOTHERMAL RESOURCES

The thermal regimes of the earth are described by Lee (2); the regime of major interest for geothermal energy is the upper crust of the earth, which has a mean temperature gradient of 20–30°C/km depth and a mean emissive heat flux of about 1.5 μcal/cm^2 sec. White (3) estimates the heat stored, beyond surface temperatures, in the outer 10 km of the earth's crust to be about 3×10^{26} cal (approximately 6×10^{24} cal is beneath the United States). This resource base is equivalent to the heat content of 8×10^{14} metric tons of coal, $\sim 5 \times 10^{18}$ kW-hr, which represents about 375,000 times the forecast (4) for total electric power production in the United States for 1985. However, geothermal heat in the outer 10 km is too diffuse to be an exploitable energy resource on a worldwide basis. Resources suitable for commercial exploitation may be defined as localized geologic deposits of heat concentrated at attainable depths, in confined volumes, and at temperatures sufficient for electric or thermal energy utilization.

The major commercial geothermal resources of the world are hydrothermal convective systems that yield high-enthalpy fluids suitable for electric power

---

[1] This review was prepared during the author's leave at the National Science Foundation and the Energy Research and Development Administration, Washington, DC, during 1974–1975.

production. Elder (5) describes the physical processes occurring in a hydrothermal convective system; a model is shown in Figure 1. The characteristics of commercial resources suitable for electric power production are listed by White (6) as reservoirs with fluid temperature greater than 180°C, depth less than 3 km, volume greater than 5 km^3, and sufficient permeability to sustain productivity at an adequate flow rate. Unfortunately, resources with these characteristics have been located in only a few places in the world at present.

Major areas of geothermal energy concentrations are associated with tectonic-plate boundaries, recent volcanism and orogenesis, and relatively shallow depths to the mantle. Koenig (7) suggests the broad regions illustrated in Figure 2 as logical areas for exploration for geothermal resources. In the United States, the region encompasses 13 western states including Alaska and Hawaii.

Types of Geothermal Resource

Geothermal resources occur in many types of geologic formations and in areas with a variety of thermal, hydrologic, and chemical qualities. A convenient classification of geothermal resources is given by Hickel (8). Formations may be described as wet or dry and by temperature regime. Wet formations are generally the convective hydrothermal systems. White (6) describes variations of hydrothermal systems as vapor-dominated reservoirs, which produce dry or superheated steam, and liquid-dominated reservoirs, which produce hot water or a mixture of steam and water. The vapor-dominated hydrothermal reservoir is the one most sought after, since the steam produced with little or no water is directly usable in low-efficiency steam

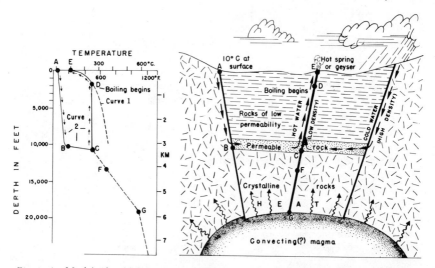

Figure 1 Model of a high-temperature hot-water geothermal system. Curve 1 is the reference curve for the boiling point of pure water. Curve 2 shows the temperature profile along a typical circulation route from recharge at point **A** to discharge at point **E**. Source: (6).

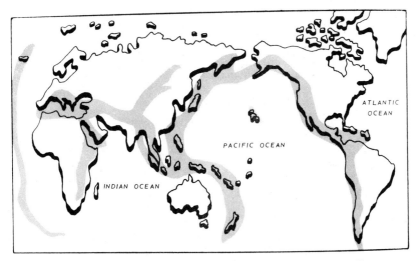

Figure 2 Regions of intense geothermal manifestations. Source: (7).

turbines. The liquid-dominated hydrothermal reservoir generally requires flashing of the hot water and separation of the steam to drive a turbine. The production of electric power from liquid-dominated systems is more difficult, since only a fraction of the hot water flashes to steam, thermal efficiencies are low, and plant operational and wastewater disposal problems are more severe. Unfortunately, liquid-dominated hydrothermal systems are believed to be many times more abundant than vapor-dominated systems (9).

Another type of wet formation not yet developed for commercial exploitation is the geopressured resource. This type of resource is described (8) as interstitial waters trapped in large, deep sedimentary basins at temperatures from 150°C to 180°C at wellhead pressures from 4000 psi to 6000 psi, and with well production rates of the order of a million gallons of fluid per day containing significant concentrations of natural gas to warrant production as a dual energy source. A description of the geopressured resources located in the Gulf Coast states is given by Jones (10), and evaluation of these resources for commercial exploitation is underway (11).

Dry formations include impermeable hot rock and magmatic deposits associated with active volcanism. Methods to extract geothermal energy from such resources with artificial circulation systems are being studied. Muffler & White (12) estimate the worldwide resources in hot dry systems to be ten times greater than the resources in hydrothermal systems.

Temperature classification is based on potential utilization. It includes high temperature ($T > 180°C$), suitable for electric power production by direct steam or flashed steam; moderate temperature ($\simeq 100 < T < 180°C$), suitable for electric power production by binary fluid conversion technology; and thermal waters ($T < 100°C$), suitable for thermal energy applications.

Exploration Methods

The objective of geothermal exploration is to locate reservoirs of sufficient thermal energy capacity to warrant economic development. Higher-quality reservoirs contain natural high-temperature geofluids and sufficient permeability to sustain adequate production rates. Lower-quality reservoirs contain lower-temperature fluids, requiring advanced conversion techniques, or have inadequate permeability or fluid flow, requiring fracture stimulation and artificial heat extraction systems.

Prospecting for underground natural resources is a difficult undertaking, but much progress has been made in locating oil, gas, and mineral deposits. Geothermal deposits are somewhat different because geothermal heat differs from material fuels in two aspects. First, it must be located as thermal energy deposits (with or without the presence of geothermal fluids). Second, as thermal energy it cannot be transported very far; it must be either used directly near the reservoir site or converted into a transportable form (primarily electricity). Therefore, the location of suitable reservoirs must be near appropriate energy markets or transmission networks.

Exploratory techniques for locating geothermal reservoirs are listed in Table 1. General descriptions of these techniques are reviewed (13) for geology and hydrology, geophysics, and geochemistry, and the philosophy of an exploration manager is given by Combs & Muffler (14).

Airborne surveys, regional reconnaissance, geologic evaluation, and geohydrologic

Table 1 Geothermal exploration methods

Exploration Surveys (Airborne):
    Aeromagnetic survey
    Thermal infrared survey
	Geochemical:
	    Chloride concentration
	    SiO_2 content

Geological:
    Tectonics and stratigraphy
    Recent faulting
    Distribution and age of volcanic rocks
    Thermal manifestations
	Na-K-Ca ratios
	    Isotopic composition of hydrogen
	        and oxygen
	Geophysical:
	    Geothermal gradient
	    Heat flow
	    Electrical conductivity
	    Seismic activity

Hydrologic:
    Surface discharge of geofluids
    Temperature of fluids
    Chemical composition of fluids
    Groundwater hydrology
    Meteorology
	Exploratory Hole Drilling:
	    Reservoir Characteristics
	        Temperature-depth profile
	        Pressure-depth profile
	        Lithology and stratigraphy
	        Permeability log
	        Porosity log
	        Fluid composition

manifestations are used to select prospective areas for further exploration. The main indicators are tectonic features, evidence of faulting, recent volcanism, and hydrothermal manifestations, such as hot springs, geysers, and fumaroles. Geochemical and geophysical methods are used to define promising areas for exploratory drilling. Geochemical evaluation prior to drilling involves analysis of surface waters and gases from hot springs and fumaroles. Significant advances have been made in chemical geothermometry with the use of silica concentrations (15), Na/K/Ca ratios (16), and isotopic analysis (17). Major geophysical techniques include heat flow and thermal gradient measurements and electrical resistivity, seismic, gravimetric, and magnetic surveys. A number of papers on specific techniques and applications are included in the proceedings of the first United Nations symposium on geothermal energy (18). More recent developments are given in the proceedings of the second United Nations symposium on the development and use of geothermal resources (19).

The final phase of geothermal exploration is the drilling of exploratory holes. At this stage, reservoir models are used to estimate the reservoir characteristics, including depth, extent, temperature, heat content, and hydrothermal properties. Drilling is generally the major part of exploration costs. The exploratory holes should be used to verify the proposed reservoir model with data from cores, downhole measurements of temperature, and, hopefully, results of production tests.

Resource Potential

Geothermal waters have been used since ancient times for bathing and local heating. The recovery of boric acid from the fumaroles in Italy became a commercial enterprise in the 1880s, and the production of electricity was demonstrated there in 1904. The first continuous source of electricity was achieved in 1913 with a 250-kW generator in Larderello, Italy. The first commercial geothermal steam turbine operated in the United States was a 12.5-MW unit installed in 1960 at The Geysers field in California. Growth at this site, the sole US geothermal power facility, has advanced in increments of 55–110 MW units. The installation of an eleventh unit in 1975 makes The Geysers field, which now has a net capacity of 502 MW, the largest geothermal electric power generating site in the world. Estimates of national geothermal power generating capacity as of 1974 are given in Table 2. Although the potential resource base for geothermal energy appears to be very large, the total world electric generating capacity is equal to that from just one modern stationary power plant.

Capacity growth at the vapor-dominated geothermal reservoir at The Geysers is expected to exceed the proven field capacity of about 750 MW (20). Plans anticipate the attainment of 900 MW at a rate of about 100 MW per year while maintaining steam production rates for existing plants.

In order to make a meaningful contribution to US energy supplies, other areas with commercial potential in the United States must be located. Over one million acres of "hot spots," areas of known geothermal resources having current potential value, were identified in 1967 by the US Department of the Interior (21) in designated federal lands in five western states. An additional 86 million acres

Table 2 Electric power generating capacity from geothermal heat, 1974

Nation	Capacity (MW)
Italy	406
United States	396
New Zealand	170
Mexico	75
Japan	40
Soviet Union	5.7
Iceland	3
Total	1096

of land in 13 states were designated as land prospectively valuable for geothermal resources. Since then several other inventories of known geothermal resource areas (KGRA) have been compiled. A general assessment of potential resources by major type is available (22).

The magnitude of US geothermal energy resources is not yet well defined, although estimates of the potential electric generating capacity are available from generalized calculations. A summary of some pessimistic and optimistic forecasts for the United States is given in Table 3, in which the capacity is also expressed in equivalent number of 1000-MWe nuclear reactors and in equivalent consumption of oil in million barrels per day (23). A national goal (24) of achieving 20,000–30,000 MW of electric and thermal power by 1985 appears to be a compromise between what is worth a national effort and what might be realistically achieved. The potential for increasing the energy supply by about 25 nuclear power plants or the conservation of one million barrels of oil per day is clearly an incentive to accelerate the development of geothermal energy in the United States.

The magnitude of hydrothermal resources required to sustain a generating capacity of 30,000 MW of electric power is estimated for liquid-dominated hydrothermal systems in Table 4. It is apparent that if hydrothermal systems are to achieve a goal of 30,000 MWe, very high priority must be given to resource exploration and assessment.

Table 3 Forecasts and national goal for development of US geothermal resources

	Pessimistic Forecasts	National Goal 1985	Optimistic Forecasts
Electric and thermal generating capacity (MW)	2,000–4,000	20,000–30,000	182,000–400,000
Equivalent in nuclear reactors (No. 1,000-MWe units)	2–4	20–30	182–400
Equivalent consumption of oil (10^6 bbl/day)	0.07–0.14	0.7–1	6–14

Table 4 Liquid-dominated resource requirements for 30,000 MWe generating capacity

Assumed parameters	
Steam flash efficiency	10%
Steam heat content	555 kcal/kg (1000 Btu/lb)
Thermal efficiency	20%
Reservoir porosity	10%
Amortization period	30 years
Mean well production	250 metric tons/hr (225 tons/hr)
Mean well spacing	0.1 km² (25 acre)
Number of wells required	9000
Fluid production rate	2.2×10^9 kg/hr (5.0×10^9 lb/hr)
Total area of resources	900 km² (2.3×10^5 acre)

RESOURCE EXTRACTION

It is apparent that the reserves of geothermal energy suitable for the production of electric power are essentially unknown. In addition to the general lack of data concerning the resources in public and private lands, the overall efficiencies with which geothermal heat can be extracted from commercial-grade reservoirs are also largely unknown. Factors involved in resource extraction include drilling technology, reservoir engineering of hydrothermal systems, development of other types of resources, and reservoir stimulation techniques.

Drilling Technology

A major factor in the economic development of geothermal fields is the techniques available for drilling and well completion in the various high-temperature geologic formations. A review of methods used for drilling wells in geothermal fields is given by Matsuo (25). Descriptions of drilling techniques at several geothermal fields are included in Armstead (13), and details of well completion practices at The Geysers field in California are described by Budd (20). Drilling is generally accomplished by the types of rotary drilling used for oil wells, with either mud or air used as the drilling fluid. However, it is expected that technical improvements will be necessary for economic drilling into deep hot-water and dry hard-rock formations. Alternate drilling methods such as turbine drilling, melting drilling, and erosion drilling are being investigated; none are commercially available yet.

Reservoir Engineering

Evaluation of reservoirs for deliverability, reserves, and longevity are required for optimum field development and production. Methods to evaluate formation characteristics include borehole measurements, production testing, and theoretical models. In some cases, economic justification for power plant construction may require production of desalinated water, commercial minerals, and process heat as by-products, or stimulation methods to increase well productivity.

The physical characteristics of the produced fluid, handling and steam-water separation facilities, turbine and generator equipment, the cooling cycle, well productivity and spacing, environmental impacts, and condensate disposal methods must be considerations in the planning and development of geothermal fields. The design of geothermal power plants varies among nations; some plants are described in the first United Nations symposium proceedings (18). A general review of electric power production from geothermal energy is given by Wood (26), and current developments are given in the second United Nations symposium proceedings (19). A detailed description of a 53-MWe power plant unit at The Geysers in California is given by Finney (27). The unit operates at an inlet rate of 1×10^6 lb/hr steam, a pressure of 114 psia, a temperature of 355°F, and an enthalpy of 1200 Btu/lb. The effluents from the plant are 800,000 lb/hr water evaporated to the atmosphere through cooling towers and 160,000 lb/hr of condensed water. However, Budd notes (20) that The Geysers cannot be considered as a typical geothermal steam field. The potential of a new field can be predicted only after the producing characteristics of the reservoir have been determined. Each geothermal field can be expected to have its own reservoir matrix and fluid properties and, thus, its own producing characteristics.

The key data required in reservoir engineering analysis involve the formation pressure, temperature, depth, thickness, permeability, porosity, thermal conductivity, fluid and rock density, viscosity, compressibility, and other such physical parameters. The estimated parameters include the magnitude of in-place geofluids, the pressure-time history of fluid withdrawal, and the fluid-heat balance in the reservoir. Several models have been proposed to describe the fluid withdrawal from a hydrothermal reservoir (19). An early model described by Whiting & Ramey (28) was applied successfully to production at Wairakei, New Zealand.

The efficiency of extracting available energy from steam and hot water hydrothermal reservoirs was compared by Ramey, Kruger, & Raghavan (29) for a reservoir with assumed 25% porosity, 1 acre foot in extent, formation temperature of 500°F, and a production history from an initial pressure of 681 psia (saturation pressure at 500°F) to an abandonment pressure of 100 psia. The data for the steam and hot water reservoirs are summarized in Table 5. It was noted that the hot water reservoir contains more energy than the steam-filled reservoir because of the greater mass of water in the former. The relative production from these reservoirs is summarized in Table 6. Although the percent of fluid mass recovery is about the same, the recovery of available energy from the hot water reservoir is about

Table 5 Relative available energy from hydrothermal reservoirs[a]

	Steam Reservoir		Hot Water Reservoir	
Component	Mass (10^6 lb)	Energy (10^6 Btu)	Mass (10^6 lb)	Energy (10^6 Btu)
Aquifer rock	5.410	233	5.410	233
Fluid	0.016	15	0.534	101
Total	5.426	248	5.944	334

[a] Source: (29).

Table 6 Relative production from hydrothermal reservoirs[a]

Parameter	Steam Reservoir	Hot Water Reservoir
Initial fluid content (lb)	16,160	533,800
Abandonment content (lb)	1,950	62,300
Production (lb) as		
steam	14,210	372,000
water	0	99,500
total	14,210	471,500
Recovery (%)		
of initial fluid mass	87.9	88.3
of total available energy	5.6	99.1

[a] Source: (29).

20 times greater than that from the steam reservoir. The recovery of energy by flashing hot pressurized water at the surface is very dependent on water temperature. The amount flashable from saturation pressure to atmospheric pressure is about 32% at a temperature of 500°F and drops to about 4% at 250°F. Table 5 indicates that most of the available heat is present in the rock. Production methods that recover the heat from the aquifer rock as well as from the geofluids would make the exploitation of hot water systems more economical. Heat recovery from the rock may be achieved by flashing the geofluid to steam within the formation or by recycling the plant condensate back through the formation.

Other Reservoir Systems

The variability of geothermal resources is perhaps the key problem in utilizing geothermal energy. Extraction and conversion technologies for dry steam reservoirs are sufficiently advanced to be commercially attractive. Hot water reservoirs, although more complex, have been successfully utilized. However, the problems associated with hot brines, geopressured basins, and hot dry rock formations are even more complex, and commercial exploitation is still further away. Since the potential of these resources may be greater than that of hydrothermal resources (12), technological advances to exploit them are clearly desirable.

Hot brine geothermal resources have been identified in the Imperial Valley in California and the Cesano field in Italy. Although temperatures in excess of 200°C make them feasible for electricity production by flashing the brines to steam, the concentration of dissolved solids, in excess of 300,000 mg/liter (about 10 times that of seawater), makes exploitation of these resources difficult. Engineering problems are especially acute because of scaling and corrosion from these hypersaline fluids. Concepts in converting the energy in hot brine geofluid include the binary and total flow systems described in the section on resource utilization.

The engineering aspects of geopressured reservoirs are not well understood, since little development has occurred to date, although many geopressured deposits have been encountered during drilling for oil and gas in the Gulf Coast area. The high pressures, often exceeding 10,000 psi, present problems in containment and energy extraction. An analysis of the potential of recovering the energy present—in the forms of thermal and kinetic energy and combustible methane—has been initiated (30). The feasibility of utilizing geopressured resources may depend on the longevity of specific resources and the presence of commercial concentrations of natural gas.

The engineering aspects of hot dry rock geothermal deposits are even less understood because the lack of heat extraction technology makes their utilization much more difficult. But, hot dry rock in the upper 10 km of the earth's crust represents an important potential resource of geothermal energy (31). The volumetric energy extractable from hot dry rock is about 1.2×10^9 kW-hr/km^3 of fractured rock. Assuming expected thermal properties and an extraction efficiency of about 10%, this represents a volumetric power extraction of 1.4 MW/km^3 for one century (23). The technical challenge is to be able to fracture such volumes of hot rock massives and achieve an artificial convective heat extraction system.

Fracture stimulation methods are useful in many types of geothermal reservoirs. In vapor-dominated systems, stimulation may restore declining pressure or connect dry holes in commercial steam fields to producing sections. In liquid-dominated systems lacking sufficient productivity for economic power generation, fracture stimulation may provide a larger wellbore diameter for increased flow rate or greater surface area for heat transfer, or it may restore porosity or permeability around wells having deposits of silica, calcite, or other precipitated minerals. In dry geothermal systems, stimulation is needed to provide large fracture volumes for heat transfer to an artificial convective extraction system.

Several fracturing methods are under study; they include hydraulic fracturing (32), thermal stressing (32), and chemical (33) and nuclear explosive (34) fracturing. Hydraulic (35) and explosive (36) fracturing methods have already been successful in stimulation of natural gas reservoirs.

Experiments to evaluate the potential for hydraulic and thermal stress fracturing for recovery of geothermal energy from hot dry rock formations are being conducted. In this method, a large-diameter vertical crack is created hydraulically at the bottom of a borehole in the geothermal formation. A second hole is drilled to intersect the upper part of the fracture, and a pump is used to initiate artificial circulation for heat extraction. It is hoped that pumping can be discontinued if a natural convective circulation is achieved. The main technical problems are the attainment of a vertical crack of about 2 km diameter with sufficient fracture area, the creation of additional fracture area by thermal stress of cold water injection, and the ability to develop a natural convective circulation without undue losses of water, especially in arid regions. Calculations indicate that under favorable conditions, the system might provide an average power of about 100 MW (thermal) for 20 years (32).

Several projects are studying the use of artificial circulation systems for heat extraction from hot rock deposits. Other projects are developing useful numerical models to determine the fluid mechanics and heat transfer properties in artificial circulation systems (23).

RESOURCE UTILIZATION

The most prevalent utilization of geothermal resources is the generation of electric power by low-pressure steam turbines. The technology for conversion by single-stage steam turbines is described in the proceedings of the several international conferences and by Finney (27) for the Geyser power station specifically. Since greater resources of non-dry steam geothermal reservoirs are expected, yielding hydrothermal fluids with temperatures less than about 200°C and with various concentrations of dissolved solids, conversion technologies other than single flash must be developed. And since many reservoirs are likely to contain large quantities of thermal water with temperatures too low for commercial production of electricity, direct utilization of the geofluids for thermal energy could result in significant savings of conventional fuels.

Electric Power Conversion Systems

In addition to improvements in the single-flash steam turbine, research is underway to develop multiple-flash steam turbines, single- and multiple-stage binary cycle systems, hybrid systems combining these two, and several total-flow concepts. A review of advanced binary cycles for generation of electric power (37) includes a case study for the hot brine resources of the Imperial Valley in California. A schematic drawing of a binary fluid power plant is illustrated in Figure 3. In this process, the geofluid flows through a heat exchanger where it heats a suitable working fluid with a lower boiling point. The superheated high-density vapor is expanded in a turbo-generator. The vapor is condensed with cooling water and recycled through the heat exchanger. The geofluid is generally reinjected into the ground.

One advantage of the binary fluid cycle is the ability to use geothermal fluids of relatively low temperature (perhaps at even less than 100°C) and geothermal waters with high concentrations of minerals and noncondensable gas. If hot, highly mineralized water can be kept under pressure to prevent flashing through the heat exchanger, it may be disposed of by reinjection, lessening considerably the environmental and operational hazards of gas releases to the atmosphere and precipitation of solids, which may plug the wells and plant equipment. The use of a pure secondary working fluid in the turbine cycle allows improved thermal and chemical efficiencies. The disadvantage of the binary cycle is primarily the added capital costs of the plant and pumping requirements (possibly in the wellbore) to keep the thermal water under pressure.

Pilot plant tests are being conducted to demonstrate the economic feasibility of the binary cycle process. A prototype vapor turbine system using isobutane as the working fluid has been described by Anderson (38). The turbine, which is a three-stage, radial flow turbine, is expected to deliver approximately 8000 kW net power at a cost of about \$350/kW.[2]

---

[2] R. Brindle, Rogers Engineers, personal communication, 1973.

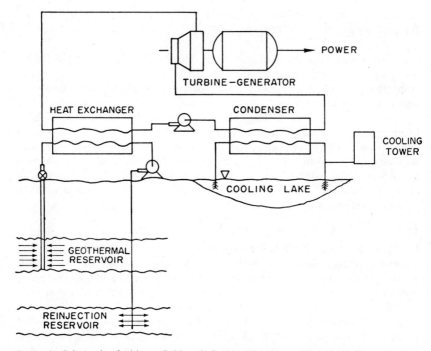

Figure 3 Schematic of a binary-fluid cycle for electric power generation from geothermal fluids.

The key components in the binary cycle are the downhole pumps and heat exchangers. Several types of downhole pumps are being developed with design concepts using: (*a*) in situ geothermal heat to operate a closed steam-generator-turbine to drive the pump, (*b*) a pump driven by a high-speed, high-temperature, high length-to-diameter electric motor, and (*c*) a hydraulically driven unit with hydraulic power from the surface. Heat exchanger concepts include fluidized sand beds to enhance the heat transfer rate and maintain a clean surface and liquid-liquid systems with direct contact of immiscible fluids, tray-tower contactors, sub-critical power cycles, or supercritical power cycles.

A binary system in the USSR for fluid energy conversion, which used an experimental 750-kWe plant to test Freon as a secondary fluid and to utilize mixed geofluids and cooling water for agricultural heating, has been described (39).

Flash and binary systems are useful in large power plants (e.g. capacity in excess of 50 MWe). They require complexes of multiple-well field development and extensive networks of fluid-gathering lines. Innovative conversion systems in which small power plants (sizes from 1 MWe to 15 MWe) are installed at individual wells are also being considered. These systems may involve a total flow concept in which both the thermal and kinetic energy of the geofluids are used for production of electricity. One of these is the impulse turbine (40), which converts the thermal energy to kinetic energy through a converging-diverging nozzle. The high-velocity output drives a hydraulic impulse turbine operated at low back pressure. An analysis was given for the utilization of the hot brine resources of the Imperial Valley, California. Research is currently being done on the design of appropriate nozzles and on the development of rugged corrosion- and erosion-resistant materials for such impact turbines.

Another concept is the helical rotary screw expander (41) based on the Lysholm gas compressor machinery developed in the 1930s. The helical screw expander can continuously expand the vapor from hot saturated liquids during pressure reduction in the machine, and, in essence, it creates an infinite series of flashing stages. A small model of a 62.5-kV prototype plant has been tested successfully with moderate salinity geofluids indicating that it can accept the total flow of untreated mineralized geofluids. Development of a prototype system of about 1250 kWe is underway.

Another concept is the bladeless turbine, which has a series of closely spaced circular disks on a shaft, housed to provide narrow channels between the disks, with nozzle introduction of geofluids. The rotary force results from viscous drag between the adjacent disks. Although the turbine seems simple and self-cleanable with hot brines, the overall efficiency is likely to be small.

Thermal Energy Utilization

Although the major emphasis on geothermal energy utilization has been for electric power production, which follows from the development of only the highest-quality steam and hot water geothermal resources, future resource development may require nonelectric, or total, utilization for economic exploitation. Thermal energy utilization appears promising for potential nonelectric applications in the United States (42). Based on 1968 data, more than 40% of the energy consumed in the United States is in forms that could be supplied by geothermal energy, namely, space heating, water heating, air conditioning, refrigeration, and process steam. An analysis of the energy cost-intensiveness of these applications in the US national economy (43) indicates that as much as 10–20% of US energy consumption could be satisfied by geothermal energy.

Non-energy interests in geothermal resources include the potential of water supply and mineral recovery. The former is the objective of a program at the East Mesa KGRA by the Bureau of Reclamation; the latter is the objective of a program of the Bureau of Mines. Both agencies are in the Department of the Interior.

It is likely that many individual geothermal reservoirs will be economically submarginal for electric power production alone. Thus, the total utilization of geothermal resources is obviously an objective for cost-effectiveness. The possibility of building a community around a geothermal resource for municipal heating, an industrial park of firms requiring process hot water and refrigeration, and a concomitant electric power station appears to be feasible in many western states. Increased production of geothermal energy can thus be achieved by stimulation of all three aspects of the technology: advanced methods of energy extraction, improved systems for energy conversion, and increased modes of resource utilization.

INSTITUTIONAL ASPECTS

Although much remains to be done to establish adequate reserves and to develop the technology to meet the goals of exploiting geothermal resources for a significant energy supply, there is confidence that the goals will be achieved (24, 44). Solutions to these problems require advances in physical research and technology. However, institutional problems also require advances. Such problems are complex; they involve public acceptance, vested interests, historical precedents, existing regulations from other resources, overlapping jurisdictions, and economic and financial factors. These problems are often more difficult to resolve than engineering problems are, and in the long run, they may be the major constraints to an accelerated and orderly development of geothermal resources. The solutions to many institutional problems may require broad public interaction, changes in regulations and legislation, and perhaps, changes in traditional investment and marketing procedures.

Economic Factors

Economic factors that affect all forms of energy supply are total capital costs per installed power unit and operational costs per unit of energy produced. For geothermal energy, both of these cost factors are very dependent on the specific characteristics of individual reservoirs and the size of the installed power plant units (13, 45, 46). Important capital costs include the investments for exploration, drilling and completion of wells, gathering lines, and waste handling systems for all utilizations. For thermal energy applications, they also include the distribution system, and for electric power production, they include the power plants and the transmission network. The production costs are influenced by the cost of capital, operations and maintenance, and plant utilization factor. In the United States, additional costs must be added for environmental pollution control.

Factual cost data for geothermal power production in the United States are available only for The Geysers field. At present, the one electric utility purchases steam from only one supplier, but it has negotiated to purchase steam for future plants from additional suppliers. In the development of future geothermal power stations, an option exists for integrated operation from exploration to power production, but the institutional problems for changing the traditional role of an electric utility purchasing "fuel" from an energy supplier may be complex. The general effect would be an increased investment cost per installed kilowatt of capacity but a decreased production cost per kilowatt-hour of energy.

Cost data from recently constructed power plant units at The Geysers are sparse, but estimates for the original plants range from about $100/kW to $150/kW (47) with an average cost of about $114/kW (48). Production costs were estimated at about 7 mill/kW-hr of which about 3.5 mill/kW-hr was the price of the purchased steam (49). These estimates included a cost of 0.5 mill/kW-hr for injection of water condensate as a disposal method. With escalation of drilling, construction, and environmental reporting costs during the past few years, these values are not useful for estimating costs of new facilities (especially for reservoirs that do not produce dry steam). Recent estimates indicate installation costs may range from $500/kW to $700/kW, and operation costs for binary conversion systems may range from 20 mill/kW-hr to 40 mill/kW-hr. Of course, such costs are hypothetical, and more precise costs will be generated as the first major reservoirs and plants are developed and operated. A great uncertainty in the total cost is the fixed exploration cost for the resource, which is independent of plant capacity, and the average drilling costs of the production, dry, and injection wells. A computer model to evaluate the relative importance of these resource and utilization costs is being developed at the Battelle Northwest Laboratories (50).

Because of large uncertainties in the technical costs of exploration, drilling, conversion efficiencies, and stimulation techniques, and because of the rapid escalation rate of these costs, it is difficult not only to estimate costs on an absolute basis but also to compare costs with those of other forms of electric power generation. Besides the costs affected by these technical factors, other factors more social in nature must be considered. Among these are public acceptance and government stimulus for accelerating the development of geothermal energy in relation to other energy sources, the interpretation of compliance with the National Environmental Protection Act of 1969 (NEPA), and the availability of investment capital for development of geothermal resources and electric and thermal power plants. These socioeconomic factors may require much public and government deliberation before general philosophies are widely accepted in practice.

Environmental Factors

Although geothermal energy is considered to be one of the least polluting of the many forms of energy available, it should be assumed that the public will insist that the environmental impact of producing geothermal energy, in all of its natural and stimulated forms, be thoroughly investigated in accordance with NEPA and any additional requirements under state and local legislation. In addition to environmental impact, it is also evident that assessment of the operational aspects of the various types of resources affecting personnel safety and plant maintenance will be required.

In the evaluation of a benefit-risk analysis, geothermal energy is expected to compare favorably with other energy resources, especially when viewed over the entire fuel cycle (51). Since geothermal energy must be utilized or converted in the vicinity of the resource, the entire "fuel cycle" from reservoir to transmission is located at one site. This cycle is different from that of material fuels whose cycle involves mining, storage, refining, transportation, reprocessing, and waste disposal, many or all of these at different locations. Furthermore, increased utilization of geothermal energy may result in a correspondingly reduced demand for material fuels in short supply such as natural gas, oil, coal, and uranium. Also, geothermal fluids may provide by-product sources of water with reduced demand for cooling water.

However, geothermal energy has its array of potentially deleterious environmental impacts. A review of the more important ones has recently been completed in a workshop sponsored by the National Science Foundation (NSF) (52) as the basis of a program to support research for baseline data and technology for monitoring potential impacts and controlling actual hazards. The major impacts include gaseous emissions, liquid waste disposal, and geophysical effects such as seismicity and subsidence. Other concerns involve thermal releases, surface water contamination, land-use planning, cooling water consumption, and visual and noise pollution.

Environmental impact may be evaluated in two ways. One way is using time duration (e.g. dividing the history of environmental impact into field development, operational, and post-operational periods). The other way is using area impact (e.g. land, water, and air). The first method recognizes that the relative importance of specific effects varies markedly during the periods. For example, noise pollution is important only during the development period; the use of silencing mufflers on operational wells eliminate this source of pollution. The release of noncondensable gases is important during the operational period of a plant, but the effects of reservoir stimulation may be important long after a geothermal field is abandoned. However, the second method is generally used to examine the overall effects of a geothermal plant on the several environmental areas.

It is anticipated that the environmental impacts of geothermal energy development will be examined for specific reservoirs by experts in many disciplines, technical and nontechnical. Many of the impacts are social in nature, and decisions regarding their acceptability will be made only after lengthy deliberations. Other impacts are technical in nature and may require reliable field information, acceptable standards, and control technologies.

The two most important physical aspects of long-term geothermal energy production are the thermal releases to the environment (their effect on cooling water and modification of local climate) and subsurface phenomena (their effect on seismicity and land subsidence). Geothermal power plants share the first aspect with fossil- and nuclear-fueled power plants, and it is expected that the environmental impact of the former can be inferred from experiences with the latter. The second aspect is more germane to geothermal energy plants. In fact, since geothermal resource areas are generally associated with regions of seismic activity, seismicity is an indicator used in prospecting for geothermal resources. On the basis of microearthquake activity measured at The Geysers geothermal area, Hamilton & Muffler (53) suggest that accurate mapping of microearthquakes can be useful in the exploration of geothermal resources. Continuous withdrawal of geofluids and reinjection of condensates may alter existing tectonic stresses and the normal patterns of earthquake activity. Although there is little evidence of land subsidence or seismic activity in existing dry steam fields, concern may exist for water-dominated fields: subsidence if there is no water reinjection or incidents of seismic activity if there is (51). Hickel (8) recommends seismic monitoring stations not only near productive geothermal areas to determine if patterns emerge related to withdrawal or injection of fluids but also near future geothermal areas prior to initial production or injection.

The major chemical releases to the atmosphere are hydrogen sulfide and boron in the noncondensable gases. Bowen (51) estimates that the release of H_2S from a 100-MW geothermal plant would be about 48.4 tons per day (with a return of

30% to the reservoir as condensate) compared with the release of 140 tons per day of SO_2 from a 1000-MW coal-fueled plant. The possibility of oxidizing H_2S to elemental sulfur with SO_2 added to the condensate is being explored as a means to reduce the emission of H_2S into the atmosphere. The colloidal sulfur would be reinjected with the waste condensate. Stoker & Kruger (54) have shown that the release of radon to the atmosphere is negligible.

Water quality is important to all forms of geothermal resources not only because of potential environmental impact but also because of the need for safe and efficient operation of the reservoir and power plant system. It is important not only for natural geothermal systems but also for systems with stimulated production. Table 7 shows a matrix of natural and introduced materials associated with the environmental and operational aspects of geothermal power plants. Contamination of surface waters will vary in extent between vapor-dominated and water-dominated fields. Condensates containing about 30% of the mass flow from the reservoir will contain higher concentrations of such contaminants as boron, ammonia, bicarbonate, sulfate, and particulate sulfur. Finney (27) notes that injection of the condensate into the ground at The Geysers plant has apparently become a successful disposal method for the condensate, since the boron and sulfate concentrations were too large for dilution disposal into surface waters. Precipitation of dissolved solids may become an important environmental and operational problem in hot water geothermal reservoirs that have high salinity. It may be a severe operational problem

Table 7 A materials matrix for natural and stimulated geothermal reservoirs and their environmental and operational impacts

	Environmental Effects	Operational Aspects
Natural reservoirs	particulate effluents: B, Hg, As, NaCl, Se, Pb, Ba	particulate effluents: SiO_2, $CaCO_3$, $CaSO_4$
	gaseous effluents: H_2S, SO_2, NH_3	reservoir precipitation: SiO_2, $CaCO_3$
	radioactive effluents: ^{226}Ra, ^{222}Rn	corrosive effluents: H_2SO_4, NaCl
Stimulated reservoirs		
Thermal fracturing	seismic activity	land subsidence
Chemical explosives	combustion products, heavy metals	ground shock
Nuclear explosives	radioactive effluents: volatile fission products soluble activation products tritiated water	radiation safety, ground shock

in geothermal waters rich in bicarbonate, which on pressure reduction may release CO_2 and result in deposition of calcium carbonate. White (6) notes the precautions required for toxic materials such as arsenic and the heavy metals. In explosion-stimulated reservoirs, additional monitoring for contamination may be required for the combustion products of chemical explosives and the radioactive products of nuclear explosives. The environmental aspects of the latter have been reviewed by Sandquist & Whan (55).

Legal Factors

An array of legal problems associated with geothermal resource development has been reviewed in another NSF workshop (56). The legal problems of geothermal resources begin with resource definition, which varies from state to state. In California, for example, geothermal resources are defined as "the natural heat of the earth, the energy—which may be extracted from naturally heated fluids—but excluding oil, hydrocarbon gas or other hydrocarbon substances." This definition leaves open the question whether geothermal resources are legally water, mineral, or gas resources, and it causes great uncertainty with respect to federal, state, and local jurisdictions. However, Hawaii considers geothermal resources to be minerals, whereas Wyoming has declared them to be water resources. As water resources, they would be subject to the very complicated set of state laws concerning water rights and regulation. As minerals, they would be subject to mining laws and such problems as ownership, depletion allowances, and write-off of intangible drilling costs. Geothermal resources have already been classified in different ways in court decisions. In one case in 1973, a US District Court in San Francisco treated The Geysers geothermal resource as "nothing more than superheated water" and therefore, not as a mineral, but in another case, the resource was determined to be a gas within the meaning of the Internal Revenue Code provisions for depletion allowance and intangible drilling costs.

Federal aspects of geothermal resources are covered under the Geothermal Steam Act of 1970, which allows for the leasing of federal lands for geothermal exploration and development. Stone (57) describes the Department of the Interior's efforts to implement the Geothermal Steam Act and divided the implementation into five general areas: 1. classification of KGRA lands for leasing purposes, 2. leasing functions, 3. supervision of operations, 4. treatment of grandfather claims, and 5. responsibility of the Department of the Interior in preparing an environmental impact statement as required under NEPA.

Ownership rights also become a serious legal problem. The federal government has given about 35×10^6 acres of land to the homesteaders, states, and railroads but has generally retained the mineral rights. However, some state grants included mineral rights, and therefore, many problems exist in the ownership aspects of federal and state lands leased for geothermal energy development. Except for resources on state or federal lands, land utilization for geothermal resources also comes under the jurisdiction of local governments.

Other institutional questions at the state level include the acreage level for commercial development, the need for long-range financial and land-use planning, and the overlapping of state regulatory agencies among themselves and with jurisdictions of local governments for permits, licenses, taxation, and especially environmental control. The latter may be affected at the federal, state, regional, county, and city government levels. For example, in some areas, authority may be divided between such agencies as a Regional Land Development Commission and a County Air Pollution Control Board.

The institutional aspects of licensing and regulating power plants are very complicated; they cover the spectrum from federal to local jurisdictions. Regulations already exist for exploration, drilling, and operation of water and mineral wells in all states. The extension to geothermal wells should be relatively simple. Yet, the need to satisfy the provisions of NEPA and any specific state environmental requirements may make geothermal resource development a slow process. For example, before the State Lands Commission in California can lease any lands under its jurisdictions, it must show at a public meeting that the lease will not have a significant detrimental environmental effect and must make an environmental impact report available to the legislature and the public. The corresponding problems of environmental impact from geothermal resources in private lands are not yet fully resolved.

Once the field is developed to the point where a utility contracts to purchase the resource and construct a power plant, other regulatory agencies such as the Federal Power Commission and corresponding state and local agencies become involved. Criteria for site selection and environmental analysis are becoming very important in power plant licensing for all types of energy resources, and their effect on geothermal energy development will probably be determined by solution of these problems on a generic basis rather than on the basis of geothermal energy alone.

Institutional problems involve many social, legal, environmental, and economic questions. The problems become more complex for land-use planning when geothermal resources span federal, state, and private lands. They involve capital investment problems for geothermal development, which may be considered to be high-risk, and involve long delays until they produce income. The problems also involve interindustrial arrangements when multiple or total utilization is needed to support economic development of electric power generation, thermal energy heating, desalination, and mineral recovery. And they involve multigovernment arrangements in the realms of regulation, licensing, and environmental control.

A NATIONAL GEOTHERMAL PROGRAM

Although significant growth of the one natural steam field in the United States has occurred since 1960, it has become apparent that a major national effort of industrial development supported by federal stimulation is needed to develop the potential of geothermal resources in its several forms as an alternative energy source. Early efforts to achieve a coordinated federal program for the support of research and development were made by an informal interagency panel for geothermal energy research. From these efforts evolved a five-year program whose objective was the rapid development of a viable geothermal industry for the utilization of geothermal resources for electric power production and other products (24). The program was evaluated under two alternate strategies. The first was "business-as-usual," which assumed continuation of current policies affecting levels of geothermal production. The second was "accelerated demand," which assumed specific changes that would result in a more rapid expansion of potential production. The major research-funding agencies that contributed to the task force program were the Atomic Energy Commission (AEC), the Department of the Interior, and the NSF, the lead federal agency. The status of the research performed with support from these agencies is described in the proceedings of a conference on research for the development of geothermal energy resources (58).

The task force estimated that under the "business-as-usual" assumptions, electric power capacity could reach 4000 MWe by 1985 and perhaps 59,000 MWe by 1990. The corresponding numbers under the "accelerated demand" assumptions were 20,000–30,000 MWe by 1985 and 100,000 MWe by 1990. Reanalysis of these values is being made under a proposed national geothermal energy research program, being directed toward (a) expanding the knowledge of recoverable resources of geothermal energy, (b) providing the necessary technological advances to improve the economics of geothermal power production, and (c) providing carefully researched policy options to assist in resolving environmental, legal, and institutional problems.

During 1974, two acts of Congress resulted in a marked change in direction for the national development of geothermal energy. The first was PL 93-410, the Geothermal Energy Research, Development, and Demonstration Act of 1974, which established a Geothermal Energy Coordination and Management Project. The Project was given responsibility for the management and coordination of a national geothermal development program, which included efforts to (a) determine and evaluate the geothermal resources of the United States, (b) support the necessary research and development for exploration, extraction, and utilization technologies, (c) provide demonstration of appropriate technologies, and (d) organize and implement the loan guaranty program authorized in Title II of the Act. The second law was PL 93-438, the Energy Reorganization Act of 1974. This Act established the Energy Research and Development Administration (ERDA) as lead federal agency with responsibility for activities related to research and development of all energy sources. The Act abolished the AEC and transferred the geothermal development function of the AEC and NSF to ERDA. On January 19, 1975, ERDA assumed responsibility for the national program of geothermal energy development. It has also assumed direction of the Geothermal Energy Coordination and Management Project, which is scheduled to submit its final report to the Congress by August 31, 1975.

In response to congressional requirements and internal needs, the Division of Geothermal Energy of ERDA prepared a comprehensive research and development plan for submission by June 30, 1975. The plan has been built upon the former plans of the Task Force for Geothermal Energy and the Geothermal Project.

The objectives being considered in the ERDA program for geothermal energy include methods to stimulate the industrial development of indigenous hydrothermal resources to provide the United States with 10,000–15,000 MW of electric and thermal power between 1985 and 1990 and to develop new and improved technologies for cost-effective and environmentally acceptable utilization of all types of geothermal resources as a long-term alternative source of energy.

The strategy of the program that might accomplish such objectives would be to accelerate industrial development of the nation's geothermal resources by (a) coordinating efforts for exploration and assessment of geothermal resources necessary to establish by 1978–1980 reserves that can support production of 20,000–30,000 MW of power, (b) demonstrating near-term and advanced technologies needed to utilize many types of geothermal resources in a cost-effective and environmentally acceptable manner, and (c) fostering rapid development of a viable geothermal industry by appropriate incentives, timely reduction of institutional impediments, and direct participation of the private sector in development and demonstration of geothermal energy technology.

Although ERDA assumes overall responsibility for management and coordination of federal geothermal activities, the scope of the federal program includes the efforts of many federal agencies. The Geothermal Steam Act of 1970 authorized the Department of the Interior to lease federal lands for geothermal resource exploration, development, and production of energy and useful by-products (such as methane, desalinated water, and valuable minerals). The leasing program is conducted by the Bureau of Land Management, which is responsible for selecting lands for lease and holding lease sales, and by the US Geological Survey, which classifies the lands by appraised value.

The US Geological Survey's geothermal research program is focused on the characterization and description of the nature and extent of the geothermal resources of the United States. The output of the program is the determination of the magnitude of the geothermal resource base on a national and regional basis. The program includes development of exploration technology, methodology for estimating energy potential of geothermal systems, environmental effects of geofluid withdrawal, and geochemical aspects of reservoir permeability.

Some of the problems that have retarded the delineation of the nation's geothermal resources through the leasing of public lands include the lack of reliable information regarding suitable resources (even on lands classified as KGRAs), insufficient requirements for early exploration and development of leased lands, and legal problems involving ownership of geothermal resources and control of their use. Under a national program, coordinated efforts by ERDA and the Department of the Interior would help to accelerate the establishment of geothermal reserves by the resource industries. Potential actions include (a) accelerated estimation of the available resources by geologic type, (b) improved technology for resource exploration and assessment and for reservoir evaluation, (c) easing of lease impediments by better methods of designating KGRAs and establishing minimum acceptable bids, (d) incentives for early development of leased lands, and (e) recommendations for legislation to resolve legal uncertainties pertaining to geothermal resources.

The second part of the strategy for the federal program would center on demonstration of near-term and advanced systems for resource utilization, development of supporting research and technology, and execution of the federal loan guaranty program. Demonstrations under the ERDA plan may be:

1. commercial-scale demonstration plants to provide the public sector with operational experience of full-scale electric power plants capable of generating energy at design production cost under pertinent environmental and institutional conditions,
2. pilot-plant facilities to prove technical feasibility, provide preliminary economic data, and provide capability for testing new and improved extraction and conversion systems for electric and thermal power production, and
3. field test facilities to improve reservoir assessment technology, evaluate reservoir characteristics and performance, test and evaluate energy extraction and conversion components and processes, evaluate material compatibility with geothermal fluids, and test environmental control technologies.

The supporting research and development program will provide for development of hardware systems, components, processes, and control techniques for installation in the demonstration facilities and field testing of reservoir evaluation technologies for the range of resource types. The supporting research and development program will also provide advanced technology for the geothermal industry, its supplier, and its support industries in order to improve productivity and utilization.

Implementation of the loan guaranty program should be coordinated with the Bureau of Land Management's geothermal leasing program and ERDA's research, development, and demonstration program. ERDA would work with companies having venture capital, reservoir developers, and leaseholders to maximize the impact of the loan program in stimulating early development of commercial electric and thermal power facilities. It is expected that the loan guaranty program will be used primarily for income-producing projects such as field development and power plant construction. Small industrial firms are likely to benefit from guaranteed loans by gaining access to necessary private capital. Regulations and procedures governing the implementation of the loan guaranty program are currently being drafted in coordination with other federal agencies such as the Small Business Administration and the Economic Development Administration. Approved regulations and operating procedures setting forth specific information required from the applicant and criteria governing the review process should be widely published as soon as possible.

Literature Cited

1. Kruger, P., Otte, C., eds. 1973. *Geothermal Energy: Resources, Production, Stimulation.* Stanford, Calif.: Stanford Univ. Press
2. Lee, W. H. K., ed. 1965. *Terrestrial Heat Flow. Am. Geophys. Union, Geophys. Monogr. Ser.* 8
3. White, D. E. 1965. *Geothermal Energy. US Geol. Surv. Circ.* No. 519
4. National Petroleum Council. 1972. *U.S. Energy Outlook: A Summary Report.* Washington DC: Nat. Petrol. Counc.
5. Elder, J. W. 1965. Physical processes in geothermal areas. See Ref. 2, Chap. 8
6. White, D. E. 1973. Characteristics of geothermal resources. See Ref. 1, Chap. 4
7. Koenig, J. B. 1973. Worldwide status of geothermal resources development. See Ref. 1, Chap. 2
8. Hickel, W. J., ed. 1972. *Geothermal Energy: A National Proposal for Geothermal Resources Research.* Univ.

Alaska, Fairbanks
9. Barnea, J. 1972. Geothermal power. *Sci. Am.* 226:70–77
10. Jones, P. H. 1970. Geothermal resources of the northern Gulf of Mexico basin. *Geothermics,* Spec. Issue No. 2, Part 1, pp. 14–26
11. Kruger, P. 1975. Geothermal resources research and technology. *Proc. Conf. Res. Dev. Geothermal Energy Resourc.* NSF Rep. No. RA-N-74-159, pp. 13–22
12. Muffler, L. J. P., White, D. E. 1972. Geothermal energy. *The Science Teacher,* Vol. 39, No. 3
13. Armstead, H. C. H., ed. 1973. *Geothermal Energy: Review of Research and Development.* Earth Sciences Series No. 12. Paris: UNESCO
14. Combs, J., Muffler, L. J. P. 1973. Exploration for geothermal resources. See Ref. 1, Chap. 5
15. Fournier, R. O., Rowe, J. J. 1966. Estimation of underground temperatures from the silica content of waters from hot springs and wet-steam wells. *Am. J. Sci.* 264:685–97
16. Fournier, R. O., Truesdell, A. H. 1973. An empirical Na-K-Ca geothermometer for natural waters. *Geochim. Cosmochim. Acta* 37:1255–75
17. Tongiorgi, E. ed. 1963. *Nuclear Geology in Geothermal Areas.* Pisa: CNR
18. UNESCO. 1970. *United Nations Symposium on the Development and Utilization of Geothermal Resources,* 2 Vols. *Geothermics,* Spec. Issue No. 2
19. *Proceedings, Second United Nations Symposium on the Development and Use of Geothermal Resources.* 1976. In press
20. Budd, C. F. Jr. 1973. Steam production at The Geysers geothermal field. See Ref. 1, Chap. 6
21. US Department of Interior. March 29, 1967. Press Release
22. White, D. E., Williams, D. L. 1975. *Assessment of Geothermal Resources of the United States. US Geol. Surv. Circ.* No. 726
23. Kruger, P. 1976. Stimulation of geothermal energy resources. *Proc. Il Fenomeno Geotermico e Sue Applicazioni.* Rome: Accademia Nazionale dei Lincei. In press (see also US Energy Research and Development Administration, Rep. No. ERDA-37)
24. Federal Energy Administration. 1974. *Project Independence Blueprint, Final Task Force Report, Geothermal Energy.* Washington DC: GPO
25. Matsuo, K. 1973. Drilling for geothermal steam and hot water. See Ref. 13
26. Wood, B. 1973. Geothermal power. See Ref. 13
27. Finney, J. P. 1973. Design and operation of The Geysers power plant. See Ref. 1, Chap. 7
28. Whiting, R. L., Ramey, H. J. 1969. Application of material and energy balances to geothermal steam production. *J. Pet. Technol.* 21:893–900
29. Ramey, H. J., Kruger, P., Raghavan, R. 1973. Explosive stimulation of hydrothermal reservoirs. See Ref. 1, Chap. 13
30. Wilson, J. S., Shepherd, B. P., Kaufman, S. 1974. *An Analysis of the Potential Use of Geothermal Energy for Power Generation along the Texas Gulf Coast.* Dow Chemical Co. Rep. to Governor's Energy Advisory Council, Texas
31. Robinson, E. R., Potter, B., McInteer, B., Rowley, J., Armstrong, D., Mills, R., Smith, M. 1971. *A Preliminary Study of the Nuclear Subterrene.* Los Alamos Sci. Lab. Rep. No. LA-4547. App. E
32. Smith, M., Potter, R., Brown, D., Aamodt, R. L. 1973. Induction and growth of fractures in hot rock. See Ref. 1, Chap. 14
33. Austin, C. F., Leonard, G. W. 1973. Chemical explosive stimulation of geothermal wells. See Ref. 1, Chap. 15
34. Burnham, J. B., Stewart, D. H. 1973. Recovery of geothermal energy from hot, dry rock with nuclear explosives. See Ref. 1, Chap. 12
35. Howard, G. C., Fast, C. R. 1970. *Hydraulic Fracturing.* Soc. Pet. Eng. of AIME, New York
36. LaRocca, S. J., McLamore, R. T., Spencer, A. M. 1974. Chemical explosive stimulation. *Pet. Eng.* 46:50–58
37. Cortez, D. H., Holt, B., Hutchinson, A. J. L. 1973. Advanced binary cycles for geothermal power generation. *Energy Sources* 1:73–94
38. Anderson, J. H. 1973. The vapor-turbine cycle for geothermal power generation. See Ref. 1, Chap. 8
39. Moskvicheva, V. N., Popov, A. E. 1970. Geothermal power plant on the Paratunka River. *Geothermics,* Spec. Issue No. 2, Part 2, pp. 1567–71
40. Austin, A. L., Higgins, G. H., Howard, J. H. 1973. *The Total Flow Concept for Recovery of Energy from Geothermal Hot Brine Deposits.* USAEC Rep. No. UCRL-51366
41. McKay, R. A., Sprankle, R. S. 1975. Helical rotary screw expander power system. *Proc. Conf. Res. Dev. Geothermal Energy Resourc.,* NSF Rep. No. RA-N-74-159
42. Stanford Research Institute. 1972. *Patterns of Energy Consumption in the US.*

Menlo Park, Calif.: Stanford Res. Inst.
43. Reistad, G. M. 1975. *Analysis of Potential Nonelectrical Applications of Geothermal Energy and Their Place in the National Economy.* USAEC Rep. No. UCRL-51747
44. Jet Propulsion Laboratory. 1975. *Status Report, Geothermal Program Definition Project.* Rep. No. 1200-205
45. Hayashida, T. 1970. Cost analysis on the geothermal power. See Ref. 18
46. Meidav, T. March 1975. Costs of geothermal steam capacity. *Oil & Gas J.*
47. Geothermal Resources Board. 1971. *The Economic Potential of Geothermal Resources in California.* Calif. Resourc. Agency
48. Kaufman, A. 1970. The economics of geothermal power in the United States. See Ref. 18
49. Stanford Research Institute. 1973. *Meeting California's Energy Requirements, 1975-2000.* Project ECC-2356
50. Bloomster, C. H. 1975. *Geocost: A Computer Program for Geothermal Cost Analysis.* Battelle PNL Rep. No. BNWL-1888. Battelle Mem. Inst., Seattle
51. Bowen, R. G. 1973. Environmental impact of geothermal development. See Ref. 1, Chap. 10
52. National Science Foundation. 1974. *Proc. Workshop Environ. Aspects Geothermal Resourc. Dev.* Rep. No. AER 75-06872
53. Hamilton, R. M., Muffler, L. J. P. 1972. Microearthquakes at The Geysers geothermal area, California. *J. Geophys. Res.* 77:2081-86
54. Stoker, A., Kruger, P. 1975. *Radon Measurements in Geothermal Systems.* Rep. No. SGP-TR-4. Stanford Univ., Stanford, Calif.
55. Sandquist, G. M., Whan, G. A. 1973. Environmental aspects of nuclear stimulation. See Ref. 1, Chap. 16
56. National Science Foundation. 1975. *Proc. Conf. Geothermal Energy and the Law.* Rep. No. NSF-RA-S-75-003
57. Stone, R. T. 1971. Implementing the Federal Geothermal Steam Act of 1970. *Proc. First Northwest Conf. Geothermal Power, Olympia, Wash.*
58. National Science Foundation. 1975. *Proc. Conf. Res. Dev. Geothermal Energy Resourc.* NSF Rep. No. RA-N-74-159

Copyright 1976. All rights reserved

OIL SHALE: The Prospects and Problems of an Emerging Energy Industry

✳11008

Stephen Rattien
Directorate for Scientific, Technological, and International Affairs,
National Science Foundation, Washington DC 20550

David Eaton
Department of Geography and Environmental Engineering, The Johns Hopkins University,
Baltimore, Maryland 21218

INTRODUCTION

This review explores technological, economic, environmental, and institutional issues relating to the development of a commercial oil shale industry. Three interrelated questions form the framework for this analysis:

1. Is oil shale a potential source of energy? Before a candidate energy source is exploited, three conditions must hold: (*a*) stored energy must exist; (*b*) a technology must exist to exploit it; and (*c*) economic incentive must exist to use the technology to produce the energy.
2. Should the United States encourage a commercial oil shale industry? Economics can be evoked to scrutinize the desirability of private and/or public investment in a commercial shale oil industry and the timing and extent of such investment. There are also noneconomic objectives that should be considered in public policy decisions concerning oil shale.
3. Assuming that a commercial shale oil industry should be encouraged, what specific actions should be taken? What are the ways in which the status quo can be altered to speed the development of an oil shale industry? Which of the many procedural, economic, environmental, and legal initiatives should be made?

THE ENERGY POTENTIAL OF OIL SHALE

The development of a commercial shale oil industry depends on the existence of sufficient stored energy, a technology to produce the fuel, the ability to overcome institutional and environmental hurdles, and economic incentives that make investment worthwhile. This section reviews some of the many studies that exist on the geology and geography of oil shale deposits, the economics of shale oil production, and the recovery technologies.

The Economic Geology and Geography of Oil Shale

Oil shale is a sedimentary rock containing hydrocarbons that can yield oil when distilled (1, 2). Three measures of the size and distribution of shale resources are of policy interest: resources in place, recoverable reserves, and commercially exploitable reserves.

1 RESOURCES IN PLACE The simplest measure of oil shale resources is how much rock contains how much oil. Such resource-in-place measurements are often widely quoted, but are not very useful for policy decisions because they do not indicate whether the resource is economically or technically recoverable. While no conclusions concerning the size of the commercially exploitable resource can be drawn from resource-in-place estimates, such estimates do permit comparisons with similar numbers developed for other fuel resources such as coal and oil. There have been at least five attempts to measure oil shale resources in place (3–7), which are shown in Table 1.

Total shale oil resources in place in the United States are estimated as high as 27 trillion bbl of oil (8). This would seem to be a large resource when compared with a similar estimate for domestic coal resources in place: 15.0 trillion bbl

Table 1 Estimates of oil shale resources in place in the Green River Formation (in billions of barrels of oil)

Shale with Yields Greater than X gal/ton	Colorado	Utah	Wyoming	Total	References
$X \geq 30$	355	50	13	418[a]	7
	311		4	315	5
				468[b]	7
$X \geq 25$	480	90	30	600	4
	607	69	60	731	3
				490	6
				1200[b]	4
$X \geq 15$	1521		260	1781	5
				860	6
$X \geq 10$	1280	320	430	2030[a]	4
				1818	7
				2468[b]	7
				4000[b]	4
$X \geq 5$				4000[a]	4
				8000[b]	4

[a] Known (only).
[b] All (including speculative).

Table 2 How technology affects oil recovery

Technology	Oil Shale Quality (gal/ton)	Recoverable Reserves (billions of barrels of oil)
Underground room-and-pillar mining plus surface retorting	35 or more	20
	30 or more	54
Open-pit mining plus surface retorting	25 or more	380
	20 or more	760
Conventional in situ retorting (after mining plus explosives)	20 or more	300
Nuclear in situ	20 or more	200

of oil equivalent.[1] Shale resources seem huge when compared with all US crude petroleum resources in place, which have been recently estimated at about 0.2 trillion bbl of oil (10).

The richer grades of shale are those containing 25, 30, or more gallons of oil per ton of shale. These high-grade shales are more limited than the total shale resource, and they occur only in a very small area. About 90% of the identified shale oil resources in the United States are located in the Green River Formation, a geological formation spanning the states of Colorado, Utah, and Wyoming. About 80% of the higher-grade shales in this formation are found in 600 square miles of the Piceance Creek Basin in Colorado. Of the two shale zones in that basin, current commercial focus is on the Mahogany Zone, a deposit of shale 30–100 feet in thickness usually lying 100–1000 feet beneath the surface (11, 12).

2 RECOVERABLE RESERVES A second measure of the oil shale resource is the quantity that could be recovered using existing technology, without considering costs. There is a wide range of opinion regarding the quantity of these recoverable reserves (3–7). The size of recoverable reserves depends strongly upon the particular recovery technique, as is shown by one such estimate in Table 2 (13). It is interesting to note that the recovery approach nearest to commercialization (room-and-pillar mining plus surface retorting) is constrained to using only the richest reserves in the Mahogany Zone.

3 COMMERCIALLY EXPLOITABLE RESERVES The most useful measure of oil shale resource is how much shale is likely to be utilized by a commercial shale oil industry, given all the constraints of geology, geography, technology, and economics. Estimates of such commercially exploitable reserves range from zero (no development) to perhaps as high as over 400 billion bbl (see Table 3). These estimates can be compared with proven petroleum reserves in the United States, including Alaska, which are estimated at only 35.3 billion bbl (14).

[1] Modified from data developed by Paul Averett, US Geological Survey, as reprinted in (9).

Table 3 Proven recoverable highest quality shale—a list of estimates (in billions of barrels of oil)

Study	Oil in Shale	Description of Resources Measured
Pessimistic	0	No commercial shale industry exists; no company irrevocably committed to entering such an industry
NPC (5)	34	Known recoverable shale, at least 30 feet thick and averaging 35 gal oil/ton
USGS (4)	80	Known recoverable resources, with 25–100 bbl oil/ton of shale
LLL (12)	270	Shale more than 25 gal/ton; more than 1000 feet thick
Interagency (3, 7)	139	Shale more than 30 gal/ton; more than 10 feet thick; known and recoverable

4 NET ENERGY AND SHALE   One conceptual limit to the use of a fuel is the possibility that it takes more energy to extract, process, and transport the fuel than is contained in the fuel itself. Although *net energy* may be defined in a number of ways, any definition that looks at energy output compared with external energy needed to operate the process will show that shale oil production is a net energy source rather than an energy sink (15).

Oil Shale Technology

The extraction of oil from shale is accomplished by retorting, for which two generic methods have been developed: (*a*) on the surface in a retort vessel and (*b*) in the earth (in situ).

Surface retorting of shale involves extraction by mining, crushing to retortable size, heating in a retort, and disposal of the waste shale. There exist many ways either to underground- or open-pit mine the shale. There are three subtly different retort designs that can be used: Paraho (Gas Combustion) (US Bureau of Mines), Union (Union Oil Company), and TOSCO (the Oil Shale Corporation) variants.

In situ retorting requires fracturing the shale in the ground to permit some fluid heat transfer agent (e.g. hot gases) to retort the shale. The released oil then flows by gravity or pressure to some collection sump from which it can be removed. There are many ways each of these tasks can be performed. The in situ mining approach, sometimes called modified in situ, seems economically most attractive at this time.

Both surface and in situ retorting will produce a heavy oil that will probably be upgraded in some way as part of the shale oil operations. Such treatment may vary in complexity from simple heat or chemical treatment to improve oil flow to complete refining into standardized liquid fuels. Ultimate products can range from a low-grade boiler fuel to gasoline.

Figure 1 is a pictorial overview of shale technology, which also describes the

OIL SHALE   187

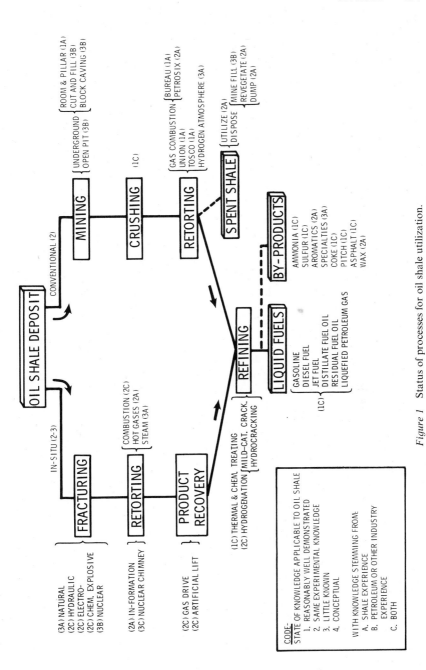

Figure 1 Status of processes for oil shale utilization.

state of and source of existing knowledge concerning each of the unit operations of shale oil operation[2] (16).

From this set of technological ideas, the first commercial shale oil operation is likely to use a surface-retorting system, composed of underground mining, shale crushing, surface retorting, and waste shale disposal. All four of the federal lease tract operations have publicly said they would use underground mining plus surface retorting.[3] Further, the Colony group would use this approach on privately held shale lands, should its project be resurrected. The research and development experience with both the surface and in situ shale recovery systems is briefly described in the following section.

1 MINING Experiments to demonstrate the feasibility of underground mining of oil shale have been made for over 30 years. The US Bureau of Mines (USBM) demonstrated mining techniques at an experimental oil shale facility from 1944 to 1956 (17). From 1955 to 1958, the Union Oil Company operated a mine to produce shale for its retort experiments. From 1964 to 1967, Mobil Oil Company operated a mine near the original USBM mine. Since 1965, the Colony Development Corporation has operated a mine near the Union mine to test mining, crushing, and retort techniques. By now there is general agreement with the feasibility of underground mining for oil shale, although there still exist some questions about the practicability of increasing the scale of operations to the expected 130,000 tons/day needed to support a retort of 100,000 bbl/day. These questions are, however, less about technical feasibility and more directed to the industrial organization aspects of managing so large a mine.

2 RETORTING There are three prime candidate retort processes that have been demonstrated at the pilot plant level: Paraho (Gas Combustion), TOSCO, and Union. All retort processes heat the shale to pyrolytic temperature (800–1000°F), which produces an oil fraction, gases containing lighter molecular weight hydrocarbons, and a solid carbonaceous residue that remains on the shale (18). The differences between the processes include the direction of solid and gas flows and the heat-transfer agent.

The Gas Combustion family of retorts was first developed by the Bureau of Mines in the 1950s (19). The retort employs a vertical, refractory-lined vessel through which crushed shale moves downward by gravity counter to the retorting gases. Recycled retort gases enter at the bottom of the retort, and air and additional gases are injected in the lower retort body to provide heat. Initial work involved pilot plants having capacities of 6, 25, and 150 tons/day (20). Cameron Engineers and Petrobras, the Brazilian governmental petroleum corporation, are operating a modified pilot plant (the Petrosix process) designed to retort 2500 tons of shale/day.

---

[2] Figure 1 is a modified version that has appeared in documents published by Lawrence Livermore Laboratory.

[3] The C-a "tract" may also utilize surface mining in addition to underground mining to extract shale.

Development Engineering, Inc. is operating another small modified retort called Paraho (20).

The Union Oil Company process employs a rock pump to move shale upward through a refractory-lined vertical retort, counter to air. Heat is supplied by the combustion of the carbonaceous residue remaining on the retorted shale. This process was tested in 1957–1958 in a pilot plant with a capacity of 1000 tons of shale/day (20). A new variant, called Union II, uses steam-gas recirculation and can produce either high- or low-Btu gas. The Union II retort has been tested in California at a small scale since 1974. Union Oil is currently designing a 1500 ton/day retort and a prototype 5000 ton/day rock pump in Colorado (20).

The Oil Shale Corporation (TOSCO) retort is a rotary kiln, using externally heated and recirculated ceramic balls and preheated shale. A pilot plant designed for about 1000 tons/day was operated by TOSCO from 1965 to 1972, as a member of the Colony Development group (20). Engineering designs have been prepared for a commercial plant to handle 61,000 tons/day (21).

There is little question that the feasibility of surface retorting of oil from shale has been demonstrated. There is uncertainty, however, regarding scale-up of retorts from prototype size and their performance reliability. A commercial retort (a facility would contain several retorts) would probably have a capacity of 10–15 thousand tons/day (21).

3 WASTE SHALE DISPOSAL Commercial oil shale operation hinges on the safe and reliable disposal of waste shale. Options for waste shale disposal have been investigated by the Colony Development Corporation since 1965 (22), and their experiences are described and analyzed in the following paragraphs.

The Colony Development Corporation has estimated that its proposed oil shale complex in Parachute Creek, Colorado, would produce about 400 million tons of waste shale over its 20-year life span (22). They considered sale of spent shale, underground disposal (backfilling), and surface disposal as options. Sale was considered infeasible because no market exists and transportation costs could be prohibitive (22). Total backfilling was rejected because the volume of the waste shale is larger than the volume of the virgin shale and thus would not fit in the mine void even with compaction (22). Further, Colony considered that backfilling would interfere with mining operations and opted for surface disposal, considering various sites. Their plan was to build a dam and diversion conduit in a gulch and to fill the gulch partially with waste shale. Waste shale would be taken from the retorts, wetted, transported to the gulch, compacted, and eventually revegetated. The dam would block the runoff from the pile and floods could be diverted via the conduits. The claimed objectives of the system would be to avoid any off-site water effluent, to dispose of the waste shale, and to permit the eventual return of the filled gulch land to "natural use."

Eleven revegetation studies have been and are being conducted by Colony or their contractors. The results from 1965 to 1973 were, according to Colony, that "... although the processed shale cannot initially support as much vegetation as native soil, it can be turned into a satisfactory growing medium with proper preparation and management" (22). The problems encountered were low fertility of waste shale due to lack of nutrients; high soluble-salt content of the processed shale; poor infiltration of waste shale piles due to "structural characteristics"; and overheating of the shale surface due to its dark coloration (22).

The authors visited the Colony test plots in 1974 and were concerned about the limited success of revegetation. Of two plots seen, one plot showed acceptable growth, but was very small and had apparently received much topsoil conditioning, fertilizer, and water. It was close to a running brook and was partially shaded by the nearby native vegetation. A second, larger plot showed little apparent growth. Perhaps its preparation was identical, but it was higher on the hill. As a result, it was without shade and not within root distance of the water table.

As part of any commercialization effort, a substantial study of the problem of waste shale disposal may have to be undertaken by interested firms, the state and federal governments, and universities to resolve these issues.

4 IN SITU TECHNOLOGY Before shale can be retorted in situ, it is necessary to prepare the rock so that heat can be efficiently applied and the oil extracted. The two methods for preparing the rock are rubblization and fracturing.

Rubblization is, as it sounds, the creation of a pile of rubble. In the in situ mining approach, a void is first created in the shale bed. Then, a column of rock over that void is blasted in such a way that a column of broken rock is created. This rubble must have sufficient space between the fragments so hot gases can pass from the top to the bottom and the oil can flow downward as well. Rubblization can be accomplished by conventional explosives following mining. The use of nuclear explosives without mining has also been proposed, showing good hypothetical economics, but suffering from many of the concerns associated with the use of nuclear devices.

In the rubblized columns, openings for air can be built into the top of the rubble mass. The top of the shale column is then ignited and supplied with compressed air to support combustion. The combustion zone heats the rock below until the kerogen liquifies, and the oil and exhaust gases flow down the column to a collection sump.

There have been two experiments conducted to verify the rubblized-column concept: one jointly by the Atomic Energy Commission and the Bureau of Mines (AEC-USBM) and the other by the Garrett Research subsidiary of Occidental Petroleum. The AEC-USBM group simulated a rubblized column by retorting shale on the surface in a specially designed retort that was used during the mid-1960s and early 1970s. Garrett has been experimenting since 1972 with actual rooms of rubblized shale 35 feet square by 70 feet high. It has plans in the near future to scale up to a rubble room 125 feet square by 310 feet high. Commercial operations would have much larger rubble columns (23).

There have been other experiments to try to verify the *fractured rock retorting* concept of shale oil production. The Bureau of Mines experimented with hydraulic fracturing and chemical explosives to provide suitable permeability for retorting (23). The shale bed was then ignited from one side, and the combustion zone passed horizontally through the shale. The Equity Oil Company and the Atlantic Richfield Oil Company used hot natural gas and steam to retort in situ during experiments between 1965 and 1971 (23). They tried to utilize existing natural fractures in the shale. From 1962 through 1972, the Shell Oil Company conducted experimental work with water injection to leach out small cavities. High-pressure steam at more than 600°F was later injected to retort the shale (23). These and other fracture/retorting in situ experiments have not led to demonstration-scale plants. In summary, the modified in situ technique has passed the proof-of-concept stage, while the non-mining fracturing/retorting and nuclear rubblization techniques have not.

The Economics of Oil Shale Technology

The existence of an oil shale resource base and a technology with which to exploit it are not sufficient conditions to insure that oil from retorted Colorado rocks will ever be sold. An additional necessary condition is the existence of an economic incentive that consists of an expectation of profitability with limited uncertainty. Uncertainties include the cost of constructing and operating an oil shale production complex, shale oil marketing factors, and the costs of controlling the substantial environmental problems associated with oil shale development. Other uncertainties relate to the price of oil and to various regulatory difficulties. If the risk-adjusted rate of return on capital invested in oil shale is equal to or better than alternative investments, there will exist an incentive to deal with the technical, environmental, and political difficulties of oil shale. The federal government is also in a position to modify marketplace decision-making through various incentives and subsidies.

Despite the recent flurry of interest, the reader should recognize that no company has yet made a commitment to build a commercial-scale shale oil plant. Colony Development, which owns private shale lands as well as a federal lease tract, recently deferred indefinitely its plans to build the first shale oil facility.

1 THE COSTS OF OIL SHALE PRODUCTION Two factors that influence the profitability of shale oil production are the capital costs and the operation and maintenance costs of the plant. There have been at least 18 separate studies of these costs of producing oil from shale.[4] Estimates have been made of most of the technologies of interest, including underground extraction plus surface retorting;

---

[4] The sources for these various economic studies were:
1. The Oil Shale Corporation, 1974, personal communication to D. Eaton;
2. (5, 6, 12, 24, 25);
3. Laurich Kennedy Associates, 1974, "A Study of Conventional Mining Methods of Fracturing Oil Shale for in-situ Retorting," unpublished report submitted to Lawrence Livermore Laboratory, University of California, Berkeley;
4. A. J. Rothman, 1974, untitled memorandum on the subject of a crash development program for oil shale, Lawrence Livermore Laboratory;
5. A series of reports by a group headed by Dr. Sidney Kattell, Office of the Bureau of Mines, Morgantown, West Virginia. Some of these reports were published; others were not. Estimates were for:

Table 4 Production cost estimates used in this study

Method of Extraction	Name of Estimate	Source of Estimates	Study Code No.
Underground extraction (U) plus surface retorting	Tosco II	The Oil Shale Corp., Braun & Co., others	1A
	NPC—audit mine	National Petroleum Council, Union Oil, others	1B
	NPC—shaft mine	National Petroleum Council, Union Oil, others	1C
	BOM—100,000 U, 30 gal/ton	Bureau of Mines, Morgantown	1D
	BOM—50,000 U, 30 gal/ton	Bureau of Mines, Morgantown	1E
	SRI—Lurgi	Stanford Research Institute	1F
	Hittman—U	Hittman Assoc., consultants	1G
Surface extraction (S) plus surface retorting	NPC—open pit	National Petroleum Council, Union Oil, others	2A
	NPC—strip	National Petroleum Council, Union Oil, others	2B
	BOM—100,000—S	Bureau of Mines, Morgantown	2C
	Hittman—S	Hittman Assoc., consultants	2D
Conventional in situ (C)	LLL—block—C	Lawrence Livermore Lab. & consultants	3A
	LLL—stoping	Lawrence Livermore Lab. & consultants	3B
	LLL—caving—C	Lawrence Livermore Lab. & consultants	3C
	BOM—15 gal/ton—C 35,000	Bureau of Mines, Morgantown	3D
	BOM—22 gal/ton—C 35,000	Bureau of Mines, Morgantown	3E
	BOM—22 gal/ton—C 50,000	Bureau of Mines, Morgantown	3F
Nuclear in situ (W)	LLL—W	Lawrence Livermore Lab.	4A

surface extraction plus surface retorting; conventional in situ; and nuclear in situ. A variety of institutions have sponsored such estimates, including governmental research laboratories, industrial groups, private consulting organizations, and a construction firm. Table 4 identifies these studies.

(a) a 50,000 bbl/day plant using 30 gal/ton of shale; underground mining plus Gas Combustion retorting assumed (February 1974),

(b) a 100,000 bbl/day plant using 30 gal/ton of shale; underground mining plus Gas Combustion retorting assumed (1974),

(c) a 35,000 bbl/day plant using 22 gal/ton of shale; in situ retorting in Wyoming used (1974),

(d) a 35,000 bbl/day plant using 15 gal/ton of shale; in situ retorting assumed (1971),

(e) a 100,000 bbl/day plant using 30 gal/ton of shale; open-pit mining plus Gas Combustion retorting assumed,

(f) a 50,000 bbl/day plant using 22 gal/ton of shale; in situ retorting assumed (1974).

Table 5 Assumptions used in estimates of oil shale profitability

Engineering
  specific processes used to extract, retort, upgrade, and transport oil from shale
  site preparation and environmental control efforts
  plant capacity, in thousands of barrels a day
Productivity
  quality of the shale, in barrels of oil per ton of shale
  plant operating reliability, in % of down time
  yield of retorting process, in % of Fischer assay
Economics
  base year dollars
  inflation factors
  interest rates
  discounting rates and periods
  expected plant lifetime
  depletion and tax (federal income, state, and local) rates
  royalties, rents, or bonus bids
  % of external financing (not generated from company)
  world oil price
Credibility
  identity and experience of persons who made estimates
  level of detail (from rough estimate to plant pre-design)
  purpose of estimate (academic exercise vs promotion, vs prelude to investment, etc)

It is difficult to compare these studies as each has a unique set of assumptions regarding engineering, productivity, and economics. Table 5 gives some of the types of assumptions normally found in these studies.

In order to compare the costs, however, certain assumptions have to be standardized, and thus the reported dollar figures in each study have been modified. The standardized assumptions are:

1. definitions of *facility* (at least extraction and upgrading),
2. two sets of base-year dollars (April 1974 and 1978),
3. plant reliability (15% down time for maintenance, repairs, etc),
4. plant lifetime (15 years expected lifetime),
5. all costs expressed in terms of cents per barrel produced.

All other assumptions in the studies were left intact. In particular, reported costs in terms of a standardized discount procedure and a discount rate were not re-evaluated. Each study utilized its own discount rates, and costs would no doubt change if these were standardized—whether to 8, 10, 12, or 16%. Despite these limitations, the resulting (pseudo)standardized estimates do represent our best estimates concerning the range of possible nominal costs of producing shale oil in facilities whose construction starts in the years 1974 and 1978. Table 6 gives the nominal costs for producing shale oil from each of the 18 studies. Nuclear in situ data are not included in Table 4, but are discussed in the next section.

Table 6 Production costs of oil shale

Process and Study Code[a]	Costs Expressed in 1974 Dollars	Costs Expressed in 1978 Dollars
Underground Mining and Surface Retorting		
Tosco—1A	$5.21	$9.57
NPC —1B	4.49	8.65
NPC —1C	4.67	8.90
BOM—1D	3.88	7.23
BOM—1E	4.25	7.71
SRI —1F	3.31	6.18
Hittman—1G	5.07	10.27
Surface Mining and Retorting		
NPC —2A	4.91	9.25
NPC —2B	4.48	8.64
BOM—2C	3.41	6.70
Hittman—2D	4.79	9.80
Conventional in situ		
LLL —3A	5.92	10.59
LLL —3B	6.74	11.99
LLL —3C	7.80	13.76
BOM—3D	8.74	15.65
BOM—3E	7.44	13.15
BOM—3F	9.55	16.73

[a] See Table 4 for descriptions of study codes.

Production cost estimates for underground extraction plus surface retorting ranged between $3.31 and $5.31/bbl in 1974 dollars. Nominal 1978 costs were calculated to fall between $6.18 and $10.27/bbl. Estimates made by government groups were lower than those made by representatives of the oil industry. The highest cost estimate (in 1974 dollars) was made by the construction consultants designing specifications for a full-scale shale facility.

Production cost estimates for surface mining plus surface retorting were between $3.41 and $4.91 in 1974 dollars and between $6.70 and $9.80 in 1978 dollars. Again, government estimates of production costs were lower than those of the National Petroleum Council, which represents the opinion of the oil industry. Estimates of costs for conventional and modified in situ systems ranged from $5.92 to $9.55 in 1974 dollars and from $10.59 to $16.73 in 1978 dollars. All six of the estimates were made by government-related groups.

Given these cost data alone, it is not possible to be definitive about the profitability of an oil shale investment. Domestic tax, shale land leasing, and royalty policies will greatly affect profitability. Of more importance, the price of world petroleum may now be at $11 a barrel, but will it remain there? Will world petroleum prices inflate with nominal dollar inflation or drop due to a weakening of OPEC? Further, the costs given in Table 6 are likely to be optimistic, reflecting a base case of a 15-year facility lifetime and an 85% plant load factor. Shale oil costs are likely to be quite sensitive to these technical assumptions. We have found that operating and maintenance costs per barrel are eight times higher with pessimistic assumptions (10-year lifetime, 50% down time) than with optimistic assumptions (30-year lifetime, 100% load factor). Overall costs per barrel were three times higher for the pessimistic case than for the optimistic case (D. Eaton, unpublished research).

2 ECONOMICS OF THE NUCLEAR IN SITU OPTION The data in the previous two sections dealt only with conventional methods of extracting oil from shale and thus have not included nuclear in situ retorting of shale. Some estimates of nuclear in situ retorting have been made by the Lawrence Livermore Laboratory. While the economics are attractive, there are a number of crucial uncertainties. For example, one base case used was a column of rubble 2000 feet high and an oil yield of 60% of Fischer assay. Forming such a large retortable column is a difficult task; achieving a yield of 60% of Fischer assay seems highly optimistic. The Bureau of Mines surface simulation of in situ retorting recovered only 60% of Fischer assay, and there was direct control of heat, air, and size of shale (6).[5]

Figure 2 shows how the production costs of nuclear in situ retorting vary as a function of the size of the rubble column and the percentage yield (of Fischer assay) of oil. These results show how very sensitive the costs of oil shale production are to modifications in technical assumptions. In 1978 dollars, the optimistic cost of $4.72/bbl is associated with technical assumptions of a 2000-foot chimney and a 60% yield. Costs rise dramatically with smaller chimneys and lower yield.

3 OIL SHALE PROFITABILITY The discussion so far has focused on costs. Profitability is of more policy interest, and results from two recent studies are reported.

The Task Force on Oil Shale for the Project Independence exercise has compiled data on the expected rates of return on oil shale investment under two alternate assumptions regarding oil prices for six different oil shale production facilities (26). The data are shown in Table 7.

Kalter and Tyner developed a simulation model for predicting profitability of shale oil facilities. They permitted investment costs, operation and maintenance costs, and the price of oil to be three stochastic variables, and performed Monte Carlo simulation for underground mining plus surface-retorting shale oil production.[6] By use of the simulation they can describe the probability of a loss from

---

[5] On the other hand, perhaps 60% is pessimistic. Surface retorts often produce yields of 90%. In a recent conference, some enthusiasts of modified in situ retorting said that 80% yields might be possible. Albert Rothman, Lawrence Livermore Laboratory, 1975, personal communication to D. Eaton.

[6] R. J. Kalter and W. E. Tyner, 1975, "An Analysis of Federal Leasing Policy for Oil Shale Lands," p. 53, report submitted to the Office of Energy R & D Policy, National Science Foundation.

the investment in shale oil facilities, with loss being defined as a rate of return below 12%. Some of their results, for a base case and with two different sets of assumptions regarding the distribution of investment and operating costs, are presented in Table 8.

The interested reader is referred to the appropriate original papers for the details of the Project Independence (14) and Kalter-Tyner calculations (see footnote 6).

4 CONCLUSIONS ABOUT COST AND PROFITABILITY   The numerical studies presented here and elsewhere seem to indicate that an incentive exists for current investment

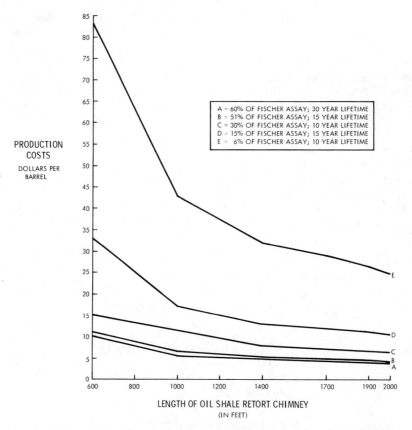

Figure 2 The sensitivity of production costs to technical assumptions for nuclear in situ oil shale. Costs are corrected for expected inflation in construction, operation, and maintenance to 1978 dollars. The technical assumptions that are modified include: 1. the depth of the in situ shale retort column, from 600 to 2000 feet; 2. the fraction of oil recovered from the shale bed, as a percentage of Fischer assay yield; and 3. the expected lifetime of the in situ operations.

Table 7 Oil shale profitability—Project Independence assessment

	Rate of Return on Investment (%)					
	Underground		Surface Mine	In situ	Modified in situ	
Shale Oil Selling Price	50,000 bbl/day	100,000 bbl/day	100,000 bbl/day 30 gal/ton	100,000 bbl/day 30 gal/ton	50,000 bbl/day 18 gal/ton	25 gal/ton
$8.35/bbl	14	16	15		11	15
$12.35/bbl	21	22	21	15	20	25

Table 8 Oil shale profitability—Kalter assessment (base case)

	Probability of a Loss (%)	
Plant Size (bbl/day)	Distribution 1	Distribution 2
50,000	15	66
100,000	4	43

in producing oil shale, particularly by the underground-mining plus surface-retorting route. However, no company has yet stepped forward to be the first such investor. Perhaps an incentive does exist, but the uncertainty of profitability is so great as to dissuade those who are risk neutral or risk averse from entering the market.

ENVIRONMENTAL AND DEVELOPMENTAL ISSUES

The development of a major shale oil industry in the Green River Formation would have many environmental effects on the lands, air, water, and natural resources of the Upper Colorado River Basin. Table 9 presents a qualitative indication of some of these impacts. The environmental issues of major concern are the effects on the quality of land, water, air resources, and regional development of shale country. In the preceding section on the energy potential of oil shale, the land impacts of shale production were discussed. Air, water, and regional development issues are considered in this part.

Air Quality

1 BACKGROUND INFORMATION A recent Supreme Court decision and subsequent US Environmental Protection Agency (EPA) regulations have structured the debate about shale development and air pollution. In affirming *Sierra Club* v.

Table 9 Major impacts of oil shale development

	Direct Effects	Indirect Effects
Physical Resources		
Land		
Surface disturbance	X	X
Erosion	X	X
Spent shale disposal	X	
Chemical wastes	X	
Trash and other solid wastes	X	X
Soil effects	X	X
Forage changes	X	X
Timber	X	X
Locatable minerals in region	X	
Other fuels in region	X	X
Aesthetic effects	X	X
Water		
Water quality	X	X
Water quantity—surface	X	X
Water quantity—ground	X	X
Air		
Plant emission	X	
Fugitive dust	X	X
Air pollutants—not plant-related	X	X
Recreation		
Wilderness		X
Fish and wildlife habitat	X	X
Fish and hunting		X
Socioeconomic Factors		
Health and Safety		
Noise	X	X
Occupational death	X	
Occupational injury	X	
Other accidents	X	X
Economic		
Jobs and income	X	
Services	X	
Taxes	X	
Housing and infrastructure	X	
Social		
Migration	X	X
Concentration	X	X
Life-style changes	X	X
Other		
Land title conflicts	X	X
Historical sites		X
Archaeological sites	X	

Table 10 Air quality—existing status and proposed standards (expressed as $\mu g/m^3$)

Pollutant	Existing Quality[a]	Permitted Incremental Air Pollution		Ambient Standards for Class III Areas	
		Class I	Class II	Federal	Colorado
Particulates					
3-month average (range)	(8.2–23.7)				
Annual average		5	10	75	45
24-hr maximum	178	10	30	150	150
Sulfur dioxide					
Annual average		2	15	80	10
24-hr maximum	136	5	100	365	55
3-hr maximum	233	25	700	1,300	
1-hr maximum	269				300
3-month average (range)	(0.2–6.0)				
Nitrogen oxides					
3-month average (range)	(6.7–7.0)				
Annual average				100	
Carbon monoxide					
8-hr maximum	14,098[b]			10,000	
1-hr maximum	14,563[b]			40,000	
Non-methane hydrocarbons					
3-hr maximum	197			160	

[a] In C-b tract (September–November 1974).
[b] Reported measurement is questionable.

Administrator of EPA,[7] the Supreme Court seemed to interpret that the 1970 Clean Air Act prohibits significant air quality degradation in nonurbanized areas. In subsequent formal rule-making procedures, the EPA put regulatory flesh upon this legal corpus by establishing three classes of air sheds (27). Class I refers to the unpolluted areas in which significant development is not to be permitted. This is enforced by permitting only very small increments in air pollutants resulting from development, as noted in Table 10. Class II is for relatively unpolluted areas in which balanced development will be permitted, allowing more substantial increments in pollutant concentrations. Class III is for primarily urban areas, where the national and state ambient standards will apply. Permitted increments in pollutant loads are not an issue here—only the gross level of ambient pollutants (28).

Table 10 summarizes the background information needed to assess the air degradation impact of a commercial shale oil industry. Existing air quality levels are noted (29) as well as the permitted incremental pollutant loads that can be

[7] *Sierra Club et al* v. *Administrator of EPA*, 344 Fed. Suppl. 253 (EDC 1973). This was the original District Court decision.

added before a state of significant degradation is reached. Finally, the concentrations in any area are limited by national and state ambient air quality standards (30). The binding environmental limit is the lower of these two ambient standards for a given pollutant.

2 THE EFFECT OF A SHALE INDUSTRY Armed with this information about the limits of air quality degradation, we can consider what may be the state of the air quality if a shale oil industry becomes a reality. There have been three significant investigations to quantify the effects on air quality resulting from an oil shale industry. Hittman Associates has developed a simplified energy-and-materials balance for several idealized shale processes. Air pollutants calculated included particulates, nitrogen oxides, sulfur oxides, hydrocarbons, carbon monoxide, and aldehydes expressed as tonnage of pollutant/10^{12} Btu of shale input. The industrial processes considered included extraction, transportation, crushing, retorting, oil refining, and pipeline use (31). The Colony Development Corporation undertook a detailed environmental assessment of its proposed 50,000 bbl/day facility at Parachute Creek, Colorado. One set of air pollution data evaluated was the maximum emission rates of sulfur oxides, nitrogen oxides, carbon monoxide, particulates, and hydrocarbons, estimated in pounds emitted from a given process per hour (32).

But perhaps the most useful study of the air impacts of shale oil development is that undertaken by Engineering-Science, Inc (33). Various data from this study are displayed in Tables 11–13. First, the expected air emissions of five critical pollutants were estimated for a representative shale-processing plant (50,000 bbl/day capacity) using any of five alternate retorting processes. The pollutants estimated (in tons per year) include particulates, sulfur dioxide, nitrogen dioxide, carbon monoxide, and hydrocarbons. The retort types used were TOSCO II, Paraho, Union Oil, Gas Combustion, and in situ. The estimated emission characteristics are shown in Table 11. Next, the emissions associated with all aspects of a shale oil complex and the related development were calculated. These numbers were

Table 11 Comparison of air pollutant emissions from oil shale processing plants[a]

	Emissions (tons/year)				
Pollutant	TOSCO II	Paraho	Union Oil	Gas Combustion	In situ
Particulates	3,245	320	425	424	1,576
Sulfur dioxide	5,835	5,380	955	1,805	8,406
Nitrogen oxides	6,412	1,655	609	609	2,256
Carbon monoxide	291	3,634	1,787	3,634	72
Hydrocarbons	1,386	2,534	4,041	2,534	971
Exhaust gas flow rate (scf/min)	265,800	375,000	378,472	376,736	1,400,000

[a] Plants of 50,000 bbl/day capacity.

scaled up to several alternate shale industry sizes. Two examples of such summary air emission statistics for a shale industry are shown in Table 12. This table shows the emissions that would result from both a 0.5 million and a 1.8 million bbl/day industry and associated secondary development by 1990.

Finally, the emissions were run through in a Gaussian-type dispersion model. Such a model can calculate an estimated air pollutant concentration giving the coordinates of imagined facilities and their emissions. Table 13 is the final output

Table 12 Total air pollution emissions in the Piceance Basin expected in 1990

	Emissions (ton/year)					
	Schedule 1 — 500,000 bbl/day in 1990			Schedule 3 — 1,800,000 bbl/day in 1990		
Pollutant	Oil Shale Industry	Secondary Development	Total	Oil Shale Industry	Secondary Development	Total
Particulates						
Point and area	16,260	1,740	18,000	59,583	2,440	62,023
Fugitive losses	2,080	—	2,080	8,228	—	8,228
Sulfur dioxide	55,071	2,304	57,375	195,283	2,338	197,621
Carbon monoxide	17,559	26,453	44,012	69,993	28,253	108,246
Nitrogen oxides	35,123	9,340	44,473	128,554	10,835	139,389
Hydrocarbons	19,185	3,753	22,938	69,315	5,435	74,750

Table 13 Summary of projected air quality in the Piceance Basin in 1990 (concentrations in mg/m^3)

	Projected Air Quality			Ambient Standards	
Pollutant	200,000 bbl/day Capacity	500,000 bbl/day Capacity	1,800,000 bbl/day Capacity	Federal	Colorado
Sulfur dioxide					
Annual average	9.9	23	82	80	10
24-hr maximum	52	104	261	365	55
3-hr maximum	240	480	1,199	1,300	—
1-hr maximum	299	597	1,494	—	300
Particulate matter					
Annual average	18	24	37	75	45
24-hr maximum	16	31	77	150	150
Nitrogen oxides					
Annual average	7	16	47	100	—
Carbon monoxide					
8-hr maximum	152	304	538	10,000	—
1-hr maximum	231	462	815	40,000	—
Non-methane hydrocarbons					
3-hr maximum	10	20	45	160	—

from such a model. Projected air quality has been calculated for three different levels of shale industry production: 0.2, 0.5, and 1.8 million bbl/day of capacity.

The important conclusion from the Engineering-Science, Inc. study is that the trade-offs between "minimize degradation" and "maximize shale oil production" perspectives are severe. The maximum permissable shale oil production that achieves even the Colorado ambient air quality standards is only 200,000 bbl/day (see Table 13). But even this level of production may be optimistic if limited degradation (Class I or Class II) is the objective.

Water Quality

1 PROCESS DISCHARGES The Congress of the United States has spoken in Public Law 92-500 regarding our national goals for clean water: (a) to eliminate discharge of pollutants to navigable waterways by 1985 (34) and (b) to limit effluents by "application of the best available technology economically achievable . . ." by July 1983 (35). The key issue is whether a shale oil industry could meet these water quality objectives. One approach is that of the Colony Development operation, which has stated: "*No Discharge Policy.* The Colony oil complex will be designed to be totally consumptive of water, and will not normally discharge any pollutant into any natural stream in the local area . . ." (36). Another perspective is that of Hittman Associates, which has done a materials-energy balance on idealized models of shale-processing complexes. It has calculated the expected pollutant emissions from the upgrading/refining end of the shale complex (37).

2 SALINITY There has been no study comparable to that of Engineering-Sciences, Inc. on the issue of effects of oil shale development on water salinity. The closest thing to a policy-oriented evaluation is the data reported by the Oil Shale Task Force for the Project Independence exercise concerning the estimated salinity increases in the Colorado River due to oil shale development. Their data, reproduced in Table 14, show the quantity of increased salinity at Imperial Dam (38). Salinity increases are traced through three different scenarios of shale development over the years 1977–1990. Although the precise significance of such an increase in salinity is unclear, there will surely be incremental economic costs associated with increased salinity that will accrue to water users in the Lower Colorado Basin. One estimate of these costs is $230,000 for each mg/liter of salinity added at Imperial Dam (38). Additional research is most desirable to estimate the severity of water quality problems of a commercial shale oil industry.

Water Supply

One additional set of issues complicating plans for shale oil development is whether sufficient water supplies exist to support the industry and its associated secondary development. The areas of uncertainty include: What constitutes a source of water supply? What is the magnitude of these supplies? How much water will a shale industry and secondary development require?

Ever since the ancient Nabateans learned to capture spring floods in the Negev desert for their use, the problem of finding water has spurred imaginative engineering

Table 14 Estimated salinity increases at Imperial Dam due to oil shale development

Level of Development Year	Schedule 1				Schedule 2				Schedule 3			
	1977	1980	1985	1990	1977	1980	1985	1990	1977	1980	1985	1990
Shale oil production (1000 bbl/day)	0	50	250	750	0	100	1,000	1,600	0	400	1,350	2,500
Water use (1000 acre feet/year)	0.9	8.7	42.9	118.0	3.3	17.4	159.0	253.0	9.6	69.6	211.0	388.0
Salt diverted at 400 mg/liter (1000 tons/year)	0.5	4.7	23.3	64.2	1.8	9.5	86.5	137.6	5.2	37.9	114.8	211.1
Increased salinity (concentrations at Imperial Dam mg/liter)												
Resulting from diversion of water	—	0.5	2.4	6.8	0.2	1.0	8.0[a]	9.3[b]	0.5	4.0	8.5[c]	10.6
Resulting from domestic return flow	—	—	0.1	0.3	—	—	0.4	0.6	—	0.2	0.5	1.0
Total	—	0.5	2.5	7.1	0.2	1.0	8.4	9.9	0.5	4.2	9.0	11.6

[a] The effects on salinity concentrations at this level were limited by the availability of noncommitted surface supplies of 133,500 acre feet/year. Approximately 69,700 tons of salt would be diverted with this water at 400 mg/liter. If supplies are obtained from some other source, such as augmentation, the salinity effects would not be the same and, therefore, were not evaluated because the source of such additional water is unknown.
[b] Footnote [a] applies, but the values are 155,400 acre feet/year and 81,100 tons of salt.
[c] Footnote [a] applies, but the values are 142,000 acre feet/year and 74,300 tons of salt.

solutions. Stream flows and springs are usually the first waters to be tapped and diverted. Wells can be dug to tap ground water and clouds seeded to enhance precipitation. Reservoirs are built to save water from season to season and from a year of abundance for a year of drought. Canals and pipelines transfer water from one river basin to another. Saline waters can be desalinated and polluted waters treated to permit reuse. To simplify this discussion, only surface flows, reservoir storage, and groundwater are considered as potential sources of water supply for a shale industry. Such a treatment should not imply either that these waters will be used to support a shale industry or that other sources will not be utilized.

1 SURFACE AND GROUNDWATER An evaluation of surface water availability to support a shale industry must include not only considerations of the stochastic nature of rainfall, but also the governing rules of water withdrawal by the western United States and Mexico from the Colorado River and its tributaries. The Colorado River Compact of 1922, the Mexican Water Treaty of 1944, and the Upper Colorado River Basin Compact of 1948 govern the interstate and international water allocations. The first two agreements require that Mexico be provided with 1.5 million acre feet and that the Lower Colorado River Basin states receive 7.5 million acre feet. The last compact specifies the percentage shares of any remaining waters among the Upper Colorado River Basin states (39).

There is no legal statement regarding the quantity of water that each state in the Upper Colorado River Basin shall receive. These states may divide the residual water after the demands of the downstream users have been satisfied. Many studies have been made to determine how much water the Upper Colorado Basin states should expect to divert. Two well-known studies are those of the Bureau of Reclamation and the Upper Colorado River Basin Commission, which estimated consumptive use of water available at 5.8 and 6.3 million acre feet, respectively (39). If the Bureau of Reclamation estimates are accepted, it is possible to compute the consumptive water available to Colorado, Wyoming, and Utah, as shown in Table 15. Table 15 also gives estimates of current (1974) water depletion, committed future depletion, and the remaining water uncommitted to use (40).

The concepts of *committed* or *uncommitted* water are used often, but they are vague terms. Remaining uncommitted water is appropriated many times over by conditional decrees, applications, and permits. However, as a practical matter, some of the committed water rights, conditional decrees, etc will never be used. Recently a task force investigating the prospects and constraints of shale oil has estimated how much surface flow is available and how much could be made available from other committed future uses to support a shale industry. These data are presented in Table 16 (40).

The total amount of water available for shale, nearly half a million acre feet per year, is not easy to interpret. One portion, 85,000 acre feet/year, may be uncommitted but under contention due to existing conditional decrees or claims. The committed water that could be diverted to shale, some 366,000 acre feet/year, includes both extant water and potential water coming from authorized, but as yet

Table 15 Surface water supply in the Upper Colorado River Basin (acre feet of water per year) (40)

State	Water Allocation	Estimated 1974 Water Depletion	Committed Future Water Depletions	Uncommitted Water for Use
Colorado	2,976,000	2,124,000	916,000	−64,000
Utah	1,322,000	825,000	381,000	+116,000
Wyoming	805,000	401,000	371,000	+33,000

Table 16 Water available for shale production (acre feet of water per year) (40)

State	Uncommitted Water	Committed Water That Could Be Diverted to Shale	Total Water Supply
Colorado	−64,000	154,000	90,000
Utah	116,000	12,000	128,000
Wyoming	33,000	200,000	223,000
Total	85,000	366,000	451,000

unbuilt, reservoirs (40). In addition to the water supplies listed in Table 15, still more water could be made available if Congress were to authorize and appropriate funds to build additional reservoir capacity. The Project Independence Task Force reported an estimate of these water supplies as 264,000 acre feet/year (40). However, these diversions might imply that certain Upper Colorado Basin states would be consuming more water than permitted in existing compact allotments.

Groundwater potential is even more ambiguous than surface water supplies. The official estimate of these supplies based on limited experimental evidence is between 2.5 and 25 million acre feet of stored groundwater (41, 42).

2 WATER DEMAND—PROCESS AND SECONDARY The uncertainty of how much water exists to support a shale complex is matched by the uncertainty regarding the demand for such water by a shale industry and its associated secondary development. There have been at least eight studies that have tried to calculate the water demands of shale industry. Some have focused only on industrial requirements; others have been concerned with shale development and associated energy and population water uses. These studies and the estimates are shown in Table 17.

The range of estimates for the total incremental demands for water for a one million bbl/day shale oil industry varies from 76,000 to 295,000 acre feet/year. The issue of where and when the demand for water will be expressed has been investigated by the Oil Shale Task Force for Project Independence (45). Table 18 gives their Schedule II estimate of the water demand resulting from oil shale development, broken down by state and by year (46).

Table 17 Estimates of water demands of shale facilities[a]

Estimate	Scope of Estimate	Estimated Requirements		Adjusted Water Demands to a 1,000,000 bbl/day Shale Oil Industry (acre feet/yr)
		Production (bbl/day)	Water Demand (acre feet/yr)	
Prien (1954)	process only	1,000,000	145,000 consumed 227,500 diverted	145,000 consumed 227,500 diverted
Cameron & Jones (1959)	process only	1,250,000	159,000 consumed 252,000 diverted	130,000 consumed 200,000 diverted
Dep. Interior (1968)	process only	1,000,000	90,800 consumed 145,000 diverted	61,800–96,000 consumed 145,000 diverted
Dep. Interior (1973)	underground mine option	50,000	consumption demand average: 8,700	
	shale complex[b]		6,060–9,600	
	associated urban[c]		740–1,000	
	surface mine option	100,000	average: 16,800	
	shale complex		12,150–18,420	
	associated urban		1,250–1,680	
	in situ	50,000	average: 4,400	
	shale complex		2,210–4,780	
	associated urban		790–920	
	technology—mixed industry	1,000,000	average: 155,000	
	shale complex		107,000–170,000	121,000–189,000 consumed
	associated urban		14,000–19,000	
Dep. Interior (1973)	shale industry & associated urban development	1,000,000	consumption demand	consumed
	lower range			76,000–82,000
	expected range			121,000–189,000
	upper range			255,000–295,000
Colony (1974)	shale complex	46,000	22 acre feet/day	175,000 consumed
Sparks (1974)	shale industry & associated urban development	1,000,000		200,000–250,000 consumed
NPC (1973)	shale complex	100,000	173,000	173,000 consumed
Interagency Task Force	shale industry & associated urban development	1,000,000		124,000–194,000 consumed

[a] Data are taken from several sources including (42–45).
[b] Shale complex includes mining and crushing, retorting, upgrading, waste shale disposal, power requirements, revegetation, and sanitary use.
[c] Associated urban uses include domestic use and domestic power.

Table 18 Water demands for oil shale development by state (acre feet/year)

State	1977	1980	1985	1990
Wyoming	0	0	8,800	13,200
Utah	0	0	34,700	52,200
Colorado	3,300	17,000	115,500	187,600
Total	3,300	17,400	159,000	253,000
Shale oil production (thousand bbl/day)	0	100	1,000	1,600

3 SHALE OIL AND WATER There is substantial uncertainty that afflicts the credibility of estimates of both water supplies and water demands associated with commercial shale oil development. The question asked should not be simply how much water, but rather what is the geographical distribution of water supplies and the timing of its availability. For example, on purely resource-availability grounds,

Table 19 Incremental population increases in shale country due to oil production

State		Year	Shale Oil Production Schedule Production (million bbl/day)	Expected Population Growth Rates (%)	
				Without Shale Oil Production	With Shale Oil Production
Colorado					
	Schedule 1	1980	50	5	9
		1985	250	5	7
		1990	750	5	7
	Schedule 2	1980	100	5	11
		1985	1,100	5	9
		1990	1,600	5	2
	Schedule 3	1980	400	5	14
		1985	1,350	5	8
		1990	2,500	5	7
Utah					
	Schedule 1	1980	50	2.5	3
		1985	250	2.5	8
		1990	750	2.5	5
	Schedule 2	1980	100	2.5	13
		1985	1,100	2.5	7
		1990	1,600	2.5	7
	Schedule 3	1980	400	2.5	16
		1985	1,350	2.5	10
		1990	2,500	2.5	4
Wyoming					
	Schedule 1	1980	50	10–15	*[a]
		1985	250	10–15	
		1990	750	10–15	
	Schedule 2	1980	100	10–15	*
		1985	1,100	10–15	
		1990	1,600	10–15	
	Schedule 3	1980	400	10–15	*
		1985	1,350	10–15	
		1990	2,500	10–15	

[a] * = insignificant differences.

the initial shale industry development is most likely in the Piceance Creek Basin in Colorado. If we look at that basin and try to imagine the existence of a million bbl/day shale oil industry, water supply becomes a problem. How can a reliable water supply of 250,000 acre feet/year for a shale industry be provided? At present, Colorado is overcommitted in its water supply by 64,000 acre feet/year. Only 78,000 acre feet/year that exist in storage could be diverted from other existing uses to support a shale industry. Another 76,000 acre feet/year could be made available if the yet-unbuilt West Divide project is constructed. Even with good intent, there appear to exist only 14,000 acre feet of assured water—not enough to supply even a single 50,000 bbl/day plant. Perhaps groundwater supplies can be tapped and perhaps they will be of a usable quality. Perhaps existing commitments to agriculture can be bought by urban and shale industry users, although the legal headaches of changing use permits and the points of water diversion could be substantial. Perhaps the Congress will permit construction of additional dams and perhaps Colorado can negotiate a modus vivendi with other Colorado River Basin states to shifting water allocations. The problem is not one of feasibility—all of these institutional and engineering alternatives are feasible. The problem is that there does exist and will exist substantial uncertainty whether sufficient water can be provided at the right time and at the right place to support a shale industry of any significant size.

Shale and Regional Development

A final objective to be considered is the attitudes and life-styles of the people in shale regions. How will they react to the changes caused by the construction and operation of a commercial shale industry?

Shale oil may directly affect six counties, including Garfield, Mesa, and Rio Blanco in Colorado, Uintah and Duchesne in Utah, and Sweetwater in Wyoming (47). The major changes in life-style that occur will result from extremely rapid population growth, which is shown for Colorado and Utah in Table 19. The significance of these rates of increase has been succinctly stated by Gilmore & Duff (47):

> From the authors' observations, a five percent annual growth rate is the maximum comfortably accommodated in semi-rural regions in the Rocky Mountains. Boom-type growth exists above a threshold in the seven percent to ten percent range. After growth rates exceed ten percent, existing institutions ... become inadequate or break down.

Some of the quality-of-life problems posed by extremely rapid urban growth are listed in Table 20.

POLICY CONSIDERATIONS AND CONCLUSIONS

A substantial resource base exists to support a shale oil industry. Several technologies have been demonstrated at a pilot plant level, although certain questions of scaling and reliability remain to be answered. An economic incentive appears to exist to encourage private investment in shale oil production. A commercial shale oil industry is clearly feasible from a resource and technology perspective.

Table 20 Quality-of-life problems associated with rapid urban growth

Housing
   The market cannot provide permanent housing in a timely fashion.
   Isolated mobile home communities offering substandard services and little community integration develop around existing towns.
   Extensive government land ownership leads to problems of monopoly or oligopoly and speculation for the limited supply of private land.
   Cost of rental housing goes up.

Employment
   Previously economically diversified communities become specialized and dependent on a single industry.
   The area labor market is unable to adequately supply skilled or trainable manpower.
   Boom conditions tend to escalate high school dropout rates.
   Employees are drawn from local crafts and professions.
   Fixed salaries or incomes do not keep pace with inflation.
   Industrial productivity and profitability suffer due to personnel problems associated with living conditions and labor undersupply.
   High employee turnover.

Services
   Public services and utilities are not provided in a timely fashion due to unwillingness of established residents to vote bonds to benefit newcomers.
   Availability and quality of services suffer, particularly health, outdoor recreation, telephone, and restaurant services.
   Health services suffer especially because of inability to provide enough facilities and inability to attract health personnel to boom areas.
   An explosion of mental health problems occurs, especially resulting from broken homes, alcoholism, etc.
   Major discontent and depression exists among wives of in-migrants.
   Lack of employment opportunities, leisure activities, and social amenities exist for women and children.

Local government
   Conventional evolutionary processes of community development and decision-making break down.
   Local government jurisdictional boundaries, tax structures, and organizations do not fit the new social and economic community patterns.
   Mechanisms for managing growth, such as zoning, may not only be inadequate, but also feared and despised by the old communities.
   Boom-inflated wages draw county and municipal employees from their jobs to construction of shale facilities.
   Tax base for property taxation and for bonding is slow to increase, while demands for facilities and services occur immediately.
   Uncertainty tied to one-industry development can limit investment.

However, substantial uncertainty exists regarding several significant factors that may have deterred investment. One set of uncertainties revolves around the impacts of a shale industry. There is some question as to the limit of development implied by the Clean Air Act of 1970 and the Amendments to the Water Pollution Control Act of 1972. There are questions about the timely availability of water for shale operations and associated secondary developments, particularly in Colorado. Problems will occur from the demographic and life-style changes wrought by a massive influx of population to the shale region.

However, the most important uncertainties are not resources, technology, environmental impacts, or regional development issues. The critical uncertainty is the economic risk. The cost of producing shale oil is increasing constantly due to high interest rates and rapid inflation in capital equipment markets, whereas the price that shale oil will fetch is as uncertain as the world price of petroleum. Massive corporate investments in shale production are likely only if shale oil could produce a return on investment substantially higher than less risky investments in alternative energy supplies.

The federal government can intervene to reduce substantially the uncertain profitability prospects of shale through leasing, fiscal, or procedural policies. If it is deemed desirable for the federal government to encourage the commercialization of shale, then there exists a wide range of policy actions, some of which are listed in Table 21. Before deciding to intervene, the government must weigh the desirability

Table 21 The range of choice in incentives to encourage oil shale commercial development

Economic incentives	Procedural incentives
Fees awarded for performance	Favorable patent policy and accelerated patent treatment
Price supports	
Concessionary loans	Simplified permit application procedures (e.g. for pipeline)
Outright grants	
Targeted financial assistance	Simplified or eliminated requirement for environmental impact statement
Technical assistance	
Special tax benefits, such as depletion, accelerated depreciation, etc	Accelerated clearance of clouded claims for shale lands and for water rights
Grants of shale lands	Supportive use of rule-making or adjudicative powers of federal agencies
Priority for leasing of shale lands	Government-advanced planning of infrastructure to support shale development
Accelerated leasing of shale lands	
Bonus-bid forgiveness	Dissemination of useful information
Sliding royalty scheme	Resolution of the nature of acceptable environmental practices, including how the Clean Air Act will be applied to pristine areas; what will be an acceptable method to dispose of waste shale; and what will be acceptable levels of saline water pollution of the Colorado River and its tributaries
Loan guarantees	
Cost-plus-profit federal contracting for shale development	
Government-funded research and development of commercially viable technologies	
Water supplies offered at low prices in sufficient quantities	
Profit levels for producers guaranteed	Resolution of the institutional delays associated with the filing of environmental impact statements and for state, federal, or other permits
Long-term contracting for shale oil by US agencies	
Charging of development costs against bonus bid	Permitting shale oil producers to get priority allocation of scarce capital equipment via classification
Accelerated forgiveness of bonus-bid debt for accelerated shale oil development	

to develop this resource in an early time frame against competing objectives, such as economic efficiency, income distribution, regional growth, and environmental quality. The issue is whether the prompt development of a new source of energy supply is worth a public subsidy and, if so, to what degree.

In summary, a shale oil industry is poised on the edge of commercialization, but no one has stepped forward to take the first plunge. The outlook for prompt and sizable production of shale oil is uncertain in the absence of federal intervention.

Literature Cited

1. US Department of Interior. 1968. *A Dictionary of Mining Mineral and Related Terms*, p. 764. Washington DC: GPO
2. Metz, W. D. 1974. Oil shale: a huge resource of low grade fuel. *Science* 184: 1271–75
3. Personal communication from W. C. Culbertson to D. Eaton. 1975. Data modified from various files including the Mineral and Water Resourc. Dep. of Colorado: W. B. Cashion's open file report; data of W. C. Culbertson; and US Geol. Surv. Prof. Pap. No. 820 (see Ref. 7)
4. Duncan, D. C., Swanson, V. E. 1965. *Organic Rich Shale of the United States and World Land Areas*, p. 9. US Geol. Surv. Circ. No. 523. 30 pp.
5. National Petroleum Council, Committee on US Energy Outlook. 1972. *U.S. Energy Outlook*. Washington DC: Nat. Petrol. Counc.
6. Lewis, A. E. 1973. *Nuclear In-Situ Recovery of Oil from Shale*. Lawrence Livermore Lab., Univ. Calif., Berkeley
7. Culbertson, W. C., Pitman, J. K. 1973. Oil shale. In *U.S. Mineral Resources*, p. 500. US Geol. Surv. Prof. Pap. No. 820
8. University of Oklahoma, Science and Public Policy Program. 1975. *Energy Alternatives: A Comparative Analysis*. Washington DC: GPO
9. National Coal Association. 1970. *Bituminous Coal Data, 1969 Edition*, pp. 97–112. Washington DC: Nat. Coal Assoc.
10. US Geological Survey, Resource Appraisal Group. 1975. *New Estimates of the Nation's Oil and Gas Reserves*. Press Release (5/7/75). Washington DC: US Geol. Surv.
11. See Ref. 8, pp. 2.4–2.7
12. Lewis, A. E. 1974. *The Outlook for Oil Shale*, pp. 4, 11. Lawrence Livermore Lab., Univ. Calif., Berkeley
13. Ibid, p. 11
14. Federal Energy Administration. 1974. *Project Independence Report*, p. 351. Washington DC: GPO
15. Penner, S. S. 1975. Report on net energy in shale oil recovery. *USCD/NSF (RANN) Workshop on In-Situ Recovery of Oil Shale*. Washington DC: Nat. Sci. Found.
16. Phillips, J. E. 1968. *Conventional and In-Situ Methods of Producing Oil from Shale*. Presented at Public Resourc. Assoc. Conf. on Orderly Dev. Oil Shale, Washington DC and Chicago, Ill.
17. East, J. H. Jr., Gardner, E. D. 1964. *Shale Mining, Rifle, Colorado, 1944–1956*. Bull. No. 611. Washington DC: Bur. Mines, US Dep. Interior
18. Sebramm, L. W. 1970. Shale oil. In *Mineral Facts and Problems, 1970 Edition*, p. 187. Bull. No. 650. Washington DC: Bur. Mines, US Dep. Interior
19. Matzick, A. R. et al. 1966. *Development of the Bureau of Mines Gas-Combustion Oil Shale Retorting Process*, p. 2. Bull. No. 635. Washington DC: Bur. Mines, US Dep. Interior
20. Federal Energy Administration. 1974. *Final Task Force Report, Potential Future Role of Oil Shale: Prospects and Constraints*, pp. 262–66. Washington DC: GPO
21. Colony Development Corp. 1974. *An Environmental Impact Analysis for a Shale Oil Complex at Parachute Creek, Colorado*, Vol. 1, pp. 115–21. Parachute Creek, Colo.: Colony Dev. Corp.
22. Ibid, pp. 140, 217, 236
23. Federal Energy Administration. See Ref. 20, pp. 279–85
24. Murray, R. G. 1972. *Liquid Recovery from Oil Shale by the Lurgi-Ruhrgas/H-Oil Process*. Menlo Park, Calif.: Stanford Res. Inst.
25. Hittman Associates, Inc. 1975. *Environmental Impacts, Efficiency, and Cost of Energy Supply and End Use*, Vol. 2. Columbia, Md: Hittman Assoc.
26. Federal Energy Administration. See Ref. 20, p. 65
27. 39 Federal Register, 42509, Dec. 5, 1974, and amendments

28. Federal Energy Administration. See Ref. 20, pp. 442–46
29. C-b Shale Oil Project. 1975. *Oil Shale Tract C-b*, pp. 56–60. Summary Rep. No. 2. Data made available by EPA
30. Federal Energy Administration. See Ref. 20
31. Hittman Associates. See Ref. 25, p. V-7
32. Colony Development Corp. See Ref. 21, pp. 283–316
33. Engineering-Science, Inc. 1974. Air quality assessment of the oil shale development program in the Piceance Creek Basin. In FEA *Final Task Force Report*. See Ref. 20, pp. 372–495
34. Federal Water Pollution Control Act Amendments of 1972. 70 Stat. 498; 84 Stat. 91; 33 USC, Sec. 1151 note. [Title I, Section 101 (a) (1)]
35. Ibid, 86 Stat. 845. [Title II, Section 301 (6) (2) (A) (c)]
36. Colony Development Corp. See Ref. 21, p. 272
37. Hittman Associates. See Ref. 25, pp. V-6–V-8
38. Federal Energy Administration. See Ref. 20, pp. 186–89
39. Water Resources Council. 1974. *Task Force Report—Water Requirements, Availabilities, Constraints and Recommended Federal Actions*, pp. 97, 98. Washington DC: Water Resourc. Counc.
40. Federal Energy Administration. See Ref. 20, pp. 163–76
41. US Department of Interior. 1973. *Final Environmental Impact Statement on the Prototype Oil Shale Leasing Program*, Vol. 1, Chap. 3, p. 46. Washington DC: GPO
42. Sparks, F. L. 1974. *Water Prospects for the Energy Oil Shale Industry*. Presented at 7th Oil Shale Symp., Colo. School of Mines, Golden, Colo.
43. US Department of Interior. See Ref. 41, Chap. 3, pp. 31–45
44. Federal Energy Administration. See Ref. 20, pp. 153–62
45. National Petroleum Council, Oil Shale Task Group. 1973. *U.S. Energy Outlook: Oil Shale Availability*, pp. 47–50. Washington DC: Nat. Petrol. Counc.
46. Federal Energy Administration. See Ref. 20, pp. 63, 158
47. Gilmore, J. S., Duff, M. 1974. Social and economic impacts of oil shale development. In FEA *Final Task Force Report*. See Ref. 20, p. 233

Copyright 1976. All rights reserved

NUCLEAR FUSION ×11009

R. F. Post
Lawrence Livermore Laboratory, University of California, Livermore,
California 94550

INTRODUCTION

Before starting this review of nuclear fusion, the reader may find it helpful to understand the context in which it was written and the viewpoint of the author in approaching the subject. The temporal context is the beginning phase of a major transitional period for the United States and other industrialized nations—a transition from a growth-oriented period, supported by inexpensive and readily available energy sources and relatively unaffected by environmental concerns, to a period in which both energy and environmental issues will assume increasing importance, and where major shifts must be made with both sources and uses of energy. The thesis to be proven herein is that nuclear fusion, a new source of energy not yet achieved, offers the best, and perhaps the only, long-term solution to the complex set of energy-related problems that man will face in the future.

This review is not addressed to the fusion specialist, but rather to those who want to understand in a more than superficial way the motivations, goals, and problems of fusion power research. It is written in a spirit of hope that a bright future is possible, and that there exists a safe and inexhaustible source of energy that can in time be tapped. Its message is at the same time sobering: the tasks ahead are exceedingly difficult, and success cannot be guaranteed on a predetermined time table. Whatever the difficulties and the uncertainties, there is no issue more important for the technical community today than the search for a source of safe and abundant energy for the future of mankind.

The monthly scientific magazine *Scientific American* has a column in which are reprinted items taken from its issues of 50 and 100 years ago. It is both amusing and informative to read these items, particularly those in which predictions were made, and then to compare with what we know today. Often today's facts do not correspond to the assertions made then. Will the same not be true 50 or 100 years hence when future readers study what we write today about the "energy crisis of the 1970s" and how we propose to solve it? Particularly, we must adopt an attitude of humility in attempting to discuss today the future role of fusion power—one of the most tantalizingly difficult and at the same time profoundly significant technical quests upon which man has yet embarked.

The first man-made release of fusion energy occurred nearly 45 years ago at the time of the discovery of fusion reactions. With the use of primitive particle accelerators, beams of deuterons (heavy hydrogen nuclei) were directed at targets also containing heavy hydrogen. Nuclear rearrangement reactions, resulting from collisions between the impinging and the target deuterons, were observed in which the total kinetic energy carried by the reaction products was much higher than the kinetic energy carried by the impinging beam particles. Exothermic nuclear combustion had been demonstrated. Fusion energy? Yes. Fusion power? No. Nor does man-made fusion power exist today, four decades later, except in uncontrolled form, in the hydrogen bomb. With the discovery of nuclear fusion reactions, the intellectual possibility of fusion power was established; the technical possibility still eludes us. Over the years, our appreciation has grown of the profound significance of achieving fusion power, a safe source of energy based on a universally available and virtually limitless fuel. Motivated by that promise, the search has now been on, in earnest, for over 20 years. Yet we must look forward to many more years of hard work before the fusion dream can come true. Why is this so?

It is easy to give a simple answer to this question; a thoughtful answer is much more difficult. The simple answer comes from recognizing the similarity between exothermic (those involving net energy release) nuclear reactions such as nuclear fusion and exothermic chemical reactions such as the chemical reaction between hydrogen and oxygen. In either case, the reaction can become energetically self-sustaining only if the circumstances in which the reaction is carried out involve a lower total energy input than that released by the reactions that subsequently occur: not all fires laid in a fireplace will light. In our example of the fusion energy released in the bombardment of a target by a beam, most of the beam particles dissipated their energy uselessly as heat and missed their nuclear targets. Only a miniscule fraction actually hit the mark and reacted. Net fusion power was not achieved, by a wide margin.

To provide a thoughtful answer to the question of why fusion power is still a hope rather than a reality, one must penetrate to the heart of the matter and assess the scientific and technological problems that have been solved and that remain to be solved before nuclear fusion can become a practical source of power. This review attempts to develop these answers and acknowledges the inevitable inexactitudes and personal biases that enter this analysis.

To relate this discussion of fusion to the research effort now being devoted to it, the following historical review may be helpful. Fusion is by now a substantial international research effort. About 1950, the first serious examinations of controlled fusion and its research problems were launched, essentially simultaneously and under the cloak of information security, in the United States, the United Kingdom, and the Soviet Union. In 1958, controlled-fusion research was declassified by international agreement. Major fusion research programs then appeared in several other nations, notably France, West Germany, Italy, and Japan. From 1962 to 1972, the US effort remained essentially static in dollars, well below its peak in 1960. Meanwhile the Soviet program grew so that by 1971 it accounted for some 38% of the world effort, at a time when the US effort had fallen to 16% of the total. Beginning in 1972, the US program began to grow again, increasing from $33 million/year to its present (1975) level of more than $100 million/year. The Japanese program, although starting from a smaller base, is now growing even more rapidly; in 1973 it represented 10% of the world effort. Since 1958, the international exchange of information in fusion research has been open and extensive, through an accelerating exchange of scientific personnel and frequent international meetings. Fusion research is indeed being taken seriously by the major industrial nations.

It is my view that the attainment of fusion power should be the keystone of the energy policy developed by the US in response to the energy problem that it faces. However, that keystone cannot be put in place if the rest of the arch falls beforehand. As we leave the age of abundant fossil energy, a transition as profound as the one we must go through must be viable at every stage. To assume that fusion power will solve our energy problems and therefore that we can pay small heed to our present predicament would probably mean both that fusion would elude us and that the United States and the other industrialized nations would soon arrive at the status of "depleted nations," i.e. at the far end of a Gaussian curve, the beginning of which is the "underdeveloped nation." Fusion power will in time provide us with all the energy we will ever need, but only if we have had the foresight and the fortitude to carry out our entire energy strategy so that fusion energy becomes the reward for a job well done. Thus, energy conservation, coal gasification, and solar heating, for example, should be as much a part of our plan for attaining a fusion-based economy as the research work in fusion itself.

However optimistic fusion scientists may be, it is clear that fusion can play no substantial quantitative role in the US energy scene for decades. Still, the question is not: Will fusion ever come about? Rather, it is: Shall we have the patience to work toward it, meanwhile adopting rational compromises, until it arrives?

THE BASICS: REWARDS AND APPROACHES

Nuclear fusion could be called nuclear combustion; that is, it is the release of energy that occurs when the nuclei of certain isotopes of the light elements collide and fuse together at ultrahigh temperatures. As we have said, these reactions were discovered in the laboratory more than 40 years ago and were soon afterwards identified as the source of energy of the sun and other stable stars, which burn ordinary hydrogen as their main fuel. Because we cannot hope to reproduce stellar fusion conditions (for which ordinary hydrogen is the fuel) on earth, we must plan to use other light isotopes as our fusion fuels. The main isotopes being considered are the heavier isotopes of hydrogen and some isotopes of helium and lithium.

The primary fuel for fusion is deuterium, sometimes called heavy hydrogen. Deuterium is a stable isotope of hydrogen that exists together with ordinary hydrogen. At the time of the creation of the solar system, deuterium was formed and was retained on the earth in the waters of the ocean. There is 1 atom of deuterium for every 6000 atoms of ordinary hydrogen in water. This energy heritage has laid in store in the oceans for countless millions of years, waiting for mankind to learn how to release it. To help to understand how enormous and virtually inexhaustible a fuel reserve deuterium represents, I would like to express the quantities involved in practical units.

From the known energy release from fusion reactions, it can be determined that the total energy content of deuterium as fusion fuel is about 100,000 kW-hr of energy per gram of deuterium. This is 4 times the energy released per gram in the fission of uranium, and is about 10 million times that released per gram in the combustion of a fossil fuel.

Today the total world energy consumption (all forms) is estimated to be about 6×10^{13} kW-hr/year. This entire amount of energy could be supplied by the fusion of only about 600 metric tons of deuterium. To obtain this amount of deuterium, a deuterium isotope separation facility would require as its feedstock the amount of water that would flow at normal water pressures through a water main only 50 cm in diameter. The amount of petroleum that would be required to satisfy the same world energy need would be 120 million barrels a day—a rate sufficient to deplete estimated present US petroleum reserves (1) in about 2 years.

We can deduce that the amount of deuterium in the world's oceans would be enough to sustain the present total world energy consumption for 100 billion years— that is, 10 times the estimated age of our universe. The development of means to generate useful energy from the fusion of deuterium will therefore permit satisfying man's energy needs, however great, for as long as civilization could be expected to exist on this planet. Furthermore, the fuel reserve on which fusion is based would be available to all nations. Fusion would indeed represent an ultimate solution to mankind's energy needs.

But what would the cost of fusion fuels be? Per unit of energy released, the cost of separating deuterium from water, using existing technology, is substantially less than 1% of the present cost of fossil fuels. In fact, the cost per unit of energy released of any of the auxiliary fuels, such as lithium, that may also be used in the first types of fusion reactors to be perfected will be comparably low. The cost of generating fusion power will therefore be essentially independent of fuel costs, instead residing virtually entirely in the capital charges and maintenance costs for the equipment. In this respect, man-made fusion power resembles natural fusion power, i.e. solar power. Unlike solar energy, however, fusion power could be employed anywhere on the globe, independently of climatological and geographic factors.

What must we do to achieve power from controlled fusion reactions? Stated concisely, we must find ways to carry out, in practical and economic ways, the following steps: (a) heat a small quantity of fusion fuel above its ignition temperature (hundreds of millions of degrees kinetic temperature); (b) confine this fuel in a heated condition long enough for the release of fusion energy to exceed the energy that was required to heat it to ignition; and (c) convert the energy thus released to useful forms, namely to electricity and process heat. All fusion power research can be understood in terms of trying to meet these three objectives. Although the problems are staggering, major progress toward their solution has been made. We base our belief that fusion will be achieved upon this progress.

The two key scientific problems to be solved to achieve fusion power, as exemplified by the first two steps above, are heating and confinement. Over the more than 20 years that fusion power research has been under way, two basically different approaches to solving these problems have evolved. Both approaches are now being vigorously pursued, with hopes of a successful scientific conclusion in a relatively few years. In the first of these, magnetic confinement, confinement is the primary problem, and the problem of heating is of secondary importance. In the second approach, pellet fusion, the critical problem is heating, and confinement is not the main issue.

The first approach, magnetic confinement, aims at solving the critical threesome of fusion power by confining a very hot, low-density, gaseous, fusion-fuel charge in specially shaped magnetic fields. As noted above, the kinetic temperature of the gaseous fuel charge would be between 100 million and 1 billion degrees;[1] the fuel particle density would be very low—only about 10^{-5} to 10^{-4} of the density of atmospheric air. In this approach, the fuel would burn rather slowly; the problem therefore is how to keep the super-hot fuel charge from being cooled prematurely by physical contact with the walls of the vacuum chamber in which it is confined. The magnetic field as employed in this approach acts as a nonmaterial furnace liner interposed between the heated fuel plasma and the chamber walls. The key scientific and technical problem then becomes how to keep the heated fuel charge from escaping too rapidly through its confining magnetic field. As I said before, in this approach heating the fuel charge to fusion temperatures is not the main problem; either various electrical means can be used or intense beams of energetic particles can be injected into the confined fuel charge gas to add kinetic energy (i.e. to heat it).

The second of the two approaches, pellet fusion, has been studied for fewer years. The basic idea is to compress and heat a tiny pellet of fusion fuel within a very short time, typically in much less than a billionth of a second (2). In this approach, the fusion energy would be released quickly, before the pellet could fly apart. The confinement principle used here, inertial confinement, would be an automatic consequence of achieving sufficiently rapid heating. Thus, heating is the critical issue for pellet fusion. Most approaches to pellet fusion are based on the use of convergent, focused light beams coming from an array of high-power lasers. In these experiments, the beams must be precisely focused; about 500 pellets of fusion fuel would be required to equal the volume of a single grain of rice.

To recapitulate, fusion research seeks to find ways to achieve the nuclear combustion of fusion fuels through heating these fuels to very high kinetic temperatures and then confining them without contact with ordinary matter. This confinement must be for a long enough time for the nuclear combustion process to release energy in excess of that required to heat the fuel to its combustion temperature. Success in fusion thus has a quantitative definition; all achievements short of this goal, however encouraging, do not of themselves constitute a definitive proof of the scientific feasibility of fusion.

[1] At such high kinetic temperatures, all atoms become totally ionized, i.e. they are broken down to electrons and positive nuclei, their elementary constituents. Such a totally ionized gas is called a plasma. The term *kinetic temperature* refers to the rapidity of motion of the particles of a gas. At a high kinetic temperature, the gas particles move about at high speeds. The heat of a gas is given by the product of its kinetic temperature and its particle density. Thus, a high-density gas at room temperature might have the same heat content as a low-density one at very high kinetic temperature.

It is very important to understand the unusual nature of the research approach through which the goal of fusion has been sought. Although the goal, the practical generation of electrical power from fusion reactions, is both applied and highly specific, the scientific phase represents an exploration in depth in a new field of physics, i.e. in the physics of high-temperature plasmas and their interaction with electromagnetic fields. In this investigation, as with any frontier field in the physical sciences, the interaction between experiment and theory is intense, indeed is crucial, to the eventual success of the endeavor. Also, the special nature of the fusion problem, the need to create extreme physical conditions (ultrahigh temperatures, super densities, etc) means that such conditions are accessible on earth only through the use of highly specialized technology. The development of this technology (high magnetic fields, intensely focused particle beams, high-power lasers, etc) thus also represents a crucial part of the pursuit of the fusion goal. Indeed, this technology has by now reached a level where it itself constitutes a considerable head start toward the next goals to be attained following the first proofs of scientific feasibility: the solution of the difficult engineering problems and the eventual demonstration of economic feasibility.

No worthwhile objective is achieved without an effort that is commensurate with its worth. Fusion power is eminently worthwhile and has been extremely difficult to achieve. But it is my firm conviction that when achieved it will offer a gift that will be deeply appreciated by future generations: an energy source that is compatible with the best stewardship of our planet Earth.

THE BASICS: FUSION REACTIONS

As I said earlier, achieving success in fusion power research, even in the scientific sense alone, is ultimately a quantitative issue. The possibility of achieving net power from fusion becomes a demonstrable reality only when a situation has been created where the recovered energy from fusion reactions exceeds the energy input, of all forms, required to initiate and sustain the fusion reaction. The implications of this requirement are severe in terms of the physical conditions they demand, that is, in terms of the combustion temperature and the combination of fuel particle density and minimum confinement time that they require. They are also obviously dependent on the particular fusion fuels that might be used.

Before quantifying these remarks, let us first consider the fusion reactions that are the main candidates for use in fusion reactors. Although in principle exothermic nuclear fusion reactions can and probably do occur in stellar interiors between many nuclear species in the lower quarter of the periodic table (for example, elements below titanium, element number 22), in practice there are fewer isotopes, those of hydrogen, helium, lithium, and possibly boron, that are candidates for fusion reactor use, even in the longest-range view. This situation comes about because, for fusion to occur, two nuclei must obviously be able to approach each other within distances of the order of nuclear dimensions. However, being positively charged, nuclei repel each other more strongly the higher their nuclear charge. Thus, fuels composed of more highly charged nuclei (those having a higher atomic number) require higher and higher kinetic temperatures to initiate fusion reactions. Beyond lithium (atomic number 3), these temperatures become so high that, considering the limited energy releases involved, it would seem that no method thus far proposed would obtain a net positive release of fusion energy.

Listed below are the four nuclear reactions, combinations of which are most often considered for use in fusion reactors. These reactions are listed in the typical balanced form used for an exothermic chemical reaction. The energy releases are given in two different units: (a) in MeV (i.e. in millions of electron volts equivalent), the kinetic energy equivalent of the energy that would be carried by a singly charged particle that had been electrically accelerated by that value of electric potential; and (b) in kilowatt-hours of energy released per gram of the fuel ions heated (where 1 kW-hr = 3.6×10^6 W·sec of energy or about 3500 Btu). The primary reactions, those thought to be the eventual primary fuel for fusion reactors, are the reactions between two colliding deuterons ($^2H_1$, here denoted by D). This reaction proceeds, with roughly equal probability, to yield either tritium ($^3H_1$, here denoted by T) or a light isotope of helium ($^3He_2$) with either a proton ($^1H_1$, here denoted by p) or a neutron (n) as the accompanying reaction product. (Both the number of nuclei and the momentum must be conserved in any nuclear reaction; this requires that there be at least two reaction products plus a balance in the total number of nuclei, before and after.) Thus, we have the reactions

$D + D \rightarrow T + p + 3.25$ MeV (22,000 kW-hr/g)   1.

or

$D + D \rightarrow {}^3He_2 + n + 4$ MeV (27,000 kW-hr/g).   2.

The reaction products T and $^3He_2$ are themselves fertile and can react with deuterons in the highly exothermic reactions

$D + T \rightarrow {}^4He_2 + n + 17.6$ MeV (94,000 kW-hr/g)   3.

or

$D + {}^3He_2 \rightarrow {}^4He_2 + p + 18.3$ MeV (98,000 kW-hr/g).   4.

It is thus possible, at such time that fusion technology permits, to conceive of a fusion fuel cycle that would use deuterium as its primary fuel, with the recovery, reinjection, and combustion of its reaction products (T and $^3He_2$). This was the cycle used for the calculation cited earlier of the total energy release from the complete combustion of deuterium, 100,000 kW-hr/g. Also, since tritium is an unstable isotope that decays with a 12-year half-life to $^3He_2$, there is the following interesting possibility: In a steady state system, the D-D-T-$^3He_2$ cycle could be preferentially weighted toward the D-$^3He_2$ branch by storing the tritium for some portion of its half-life, if this proved to be economically or environmentally advantageous. In fact, Reaction 4, the D-$^3He_2$ reaction, is itself one of the most interesting of the fusion reactions. It has both a high reaction probability and a large energy release, with all of the latter imparted to charged reaction products. This reaction thus lends itself naturally to a fuel cycle employing direct conversion of the plasma energy, a topic discussed in a later section.

While the pure D-D-T-^{3}He$_2$ cycle would in many senses represent a long-term goal for an economy based on the fusion reactor, Reaction 3, the D-T reaction, is presently being contemplated for first-generation fusion reactors. This reaction has the highest reaction probability (a cross section of about 5×10^{-24} cm^2 at 100-keV deuteron impact energy) and the second highest energy release. Thus, the D-T reaction is the easiest to initiate of all the fusion reactions. However, since tritium occurs only in trace quantities in nature, using the D-T reaction in a fusion reactor would require a source of tritium. The method most often considered would be to surround the reactor chamber with a "blanket" containing lithium or lithium-bearing alloys or compounds. This blanket would have a twofold function: (a) energy recovery, in the form of heat at high temperature, resulting from the capture of the 14.1-MeV neutrons from the D-T reaction (the neutron reaction product carries off 80% of the kinetic energy released in the D-T reaction); and (b) nuclear reactions of the neutron in natural lithium (7% ^{6}Li$_3$ and 93% ^{7}Li$_3$) leading to the generation of tritium through the reactions

$$n + {}^6Li_3 \rightarrow T + {}^4He_2 + 4.78 \text{ MeV} \qquad 5.$$

or

$$n + {}^7Li_3 \rightarrow T + {}^4He_2 + n' - 2.47 \text{ MeV}. \qquad 6.$$

Conceivably, there will in time develop the practical possibility of using fusion reactions other than those listed above. Generally, since they involve lower nuclear cross sections and higher nuclear charges, these reactions would be much more difficult to initiate. However, if they ever became usable, some would have special advantages. For example, they might offer greatly reduced neutron-induced radiation damage and lowered induced radioactivity in the reactor structure than, for instance, the D-T reaction with its 14-MeV neutron reaction product.

Two reactions deserving special mention are

$$p + {}^6Li_3 \rightarrow {}^3He_2 + {}^4He_2 + 4.0 \text{ MeV} \qquad 7.$$

and

$$p + {}^{11}B_5 \rightarrow 3({}^4He_2) + 16.0 \text{ MeV}. \qquad 8.$$

The first of these might provide a means for "breeding" ^{3}He$_2$ for reactors based on the D-^{3}He$_2$ reaction (3); the second could be of special interest from the standpoint of direct conversion and from the fact that breeding is not required.

Although some of the reactions listed above, and others that might have been listed, are at present of rather conjectural value for fusion, the totality of the possible reactions and fuel cycle combinations is quite large. This fact illustrates an important distinction between conventional (fission) nuclear power and fusion: Fission inevitably involves radioactive fission products and fissionable material with the attendant hazards. By contrast, with the progress and perfection of fusion reactor technology, fusion power should evolve toward systems in which hazards could be reduced to negligible levels and the possibility of high energy-conversion efficiencies (through direct conversion) could be realized. Although these possibilities presently represent only distant hopes, their realization cannot be ruled out a priori. They stand therefore as an additional motivation for the planned adoption of the fusion process as mankind's primary and ultimate source of energy.

QUANTITATIVE REQUIREMENTS FOR FUSION POWER

The Lawson Criterion

As stated earlier, to produce net fusion power a fusion-fuel charge must be heated to the combustion temperatures for fusion and then must be confined in this heated condition long enough for the recovered output of fusion energy to exceed, on the average, the energy inputs to the system. A succinct, if greatly oversimplified, way to state this requirement is called the Lawson criterion. This condition simply amounts to the balancing of the net fusion energy recovered during the interval of the confinement time against the kinetic energy content of the same volume of high-temperature plasma. The result is to establish a required minimum value (actually a range of values) of the product of the fuel particle density, n, and particle confinement time, τ, required to reach a break-even condition between fusion energy released and energy invested (in the form of particle kinetic energy). That the Lawson criterion can be stated in the form of such a product, $n\tau$, follows from the fact that when the fuel density is high the rate of fusion burning is correspondingly more rapid, leading to a shortened required confinement time before the break-even energy release is reached, and vice versa. Depending on various assumptions as to plasma temperature and efficiencies of handling the various energies involved, for the D-T reaction the value of the break-even $n\tau$ product ranges between about 10^{13} and 10^{15} units, with 10^{14} usually taken as a rough rule-of-thumb value. The units here are sec/cm^3, that is, with particle densities, n, measured in particles/cm^3 and time, τ, in seconds. Thus, for example, at a particle density of 3×10^{14} ions/cm^3 the break-even confinement time would be about $\frac{1}{3}$ sec. This density, which would be a typical value for a fusion reactor based on magnetic confinement, is only 10^{-5} of the particle density of air molecules at atmospheric pressure, i.e. it is practically a vacuum.

Figure 1 presents a curve of the Lawson break-even $n\tau$ value, plotted as a function of the kinetic temperature of the plasma as calculated for the D-T reaction.[2]

Note that the Lawson criterion, although approximate, is a universal one in that it applies to any fusion situation. For magnetic confinement systems, confinement times of seconds or appreciable fractions of a second are needed because of the very low fuel densities that would be employed. In pellet fusion, where particle densities are those of matter compressed to thousands of times the density of ordinary solid matter (i.e. compressed to 10^{26} particles/cm^3 or possibly higher), the Lawson break-even confinement times could be of order 10^{-11} sec. In either system, the required Lawson $n\tau$ product would be about 10^{15}.

---

[2] The curve given will differ somewhat from those found in other treatments in that only the heat content of the ions was taken into account, and the energy recovery efficiency was not included in the balancing. The curve is therefore only a rough approximation to reality.

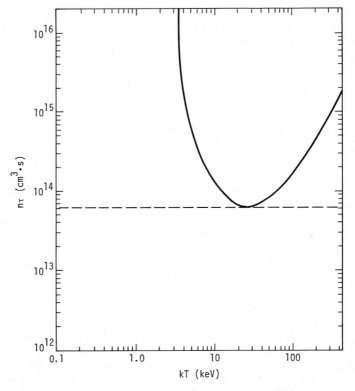

Figure 1 The Lawson criterion for the minimum product of density and burning time to produce net energy balance against plasma energy losses versus plasma temperature.

The Problems of Plasma Pressure and Fusion Power Density

What will determine the particle density at which fusion reactors must operate? Basically, it comes down to limitations arising from technological or engineering considerations. In systems based on magnetic confinement, these limitations appear from two quarters: the outward pressure exerted by the confined plasma and the fusion power released by the plasma. The outward pressure must be limited to values that can be sustained by the confining magnetic field; that is, the magnetic bottle must not burst. The pressure limit is thus ultimately set by the intensity of the confining field that can be generated. Except for highly transient cases, where inertial effects can be important, the attainable strength of the confining field will be limited by the strength of the materials from which the coil and its supporting structure are made. To provide an example of the magnitude of the pressures involved, we note that, as deduced from the laws for the pressure of a gas, a confined fusion reactor plasma at a density of 3×10^{14} particles/cm^3 (10^{-5} of atmospheric density) and at a kinetic temperature of 3×10^8 °K (10^6 times as hot as room temperature, 300°K) will exert an outward pressure of approximately $10^{-5} \times 10^6$, or 10 times normal atmospheric pressure, about 150 psi in engineering units. Depending on geometric and other factors, this may imply required magnetic field forces that are substantially larger than this value, perhaps many thousands of pounds per square inch, depending on the particular system.

To a sufficient approximation, the relationships between the magnetic field intensity and the plasma pressure being sustained by the magnetic field are given by a very simple law that resembles the law of partial pressures in ordinary gases. In pressure equilibrium, i.e. when the outward plasma pressure P_p is balanced against inward-acting magnetic forces, P_{mag}, at every point within the confinement chamber (idealized here as a long tube), the law states that

$$P_p + P_{mag} = \text{constant}.$$

Thus, where P_p is high, P_{mag} is depressed and vice versa. At the wall, P_p must become zero (no contact can be allowed between the plasma and the wall) so that the entire "pressure" is magnetic and equal to its value at the wall.

In this context, the local value of P_{mag} is numerically equal to the magnetic energy density, that is, to the amount of magnetic energy stored per unit volume. This is given by the quantity $\bar{U}_0 = B^2/8\pi$. If B is measured in gausses, U_0 is in units of ergs/cm^{-3} (i.e. 10^{-7} W·sec/cm^3). Thus our pressure balance equation becomes

$$P_p + (B^2/8\pi) = \text{constant} = (B_0^2/8\pi), \qquad 10.$$

where B_0 is the intensity of the magnetic field at the chamber wall. We now see that the effect of the confined plasma is to depress locally the value of the magnetic field below that at the wall (where $P_p = 0$). It follows that $P_{max} = B_0^2/8\pi$ is the maximum plasma pressure that could possibly be sustained by a magnetic field whose strength at the chamber wall is B_0. The plasma would then completely exclude the magnetic field, so that in this extreme case the magnetic field would act only at the boundary of the plasma. There would be no magnetic field acting on the particles inside the plasma (where $P_p = P_{max}$). We see therefore that the plasma acts as a diamagnetic material, i.e. as a material that tries to exclude a magnetic field. The physical origin of this diamagnetism is that the orbits of the charged particles in such a case produce electrical currents at the surface of the plasma that are oppositely directed to those that are flowing in the magnet coil surrounding the plasma chamber. The general situation described above is illustrated schematically in Figure 2, for the extreme case in which P_p completely excludes the external field.

In practice, it is not possible to come this close to the bursting pressure of the confining field, i.e. to where P_p actually attains the maximum value $B_0^2/8\pi$. A safety factor must be used, as in the design of any pressure vessel. In fusion research, this safety factor is denoted by the quantity β, defined as the ratio of internal plasma pressure to maximum external magnetic pressure, i.e. $\beta = (P_p)/(B_0^2/8\pi)$.

The actual value of β that may be used as an operating value in any given magnetic confinement system depends on several factors, such as the stability of the confinement. For economic reasons, high β values are generally preferred because the capital cost of a reactor may be in large part determined by the cost of the

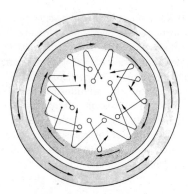

Figure 2 Plasma is contained by a magnetic field (*dotted ring*) that prevents escape of particles. Outer arrows show current in coil; inner arrows show current from deflected particles.

magnet coils, whereas the effectiveness with which the confining field is used depends on β. Since, as we shall see, the rate of fusion power generation varies as β^2, too low a value of β would imply a system with too little fusion energy release per unit volume to pay for the capital costs of the confining magnetic field. Generally speaking, β values greater than about 0.1 are believed to be required for economic reasons. For some approaches, much higher values have already been achieved in the laboratory; for them, β limitation should not be a problem. For others, experimental and theoretical results to date suggest that even a β value of 0.1 may be difficult to achieve.

In the magnetic confinement approach, two circumstances set limits on the plasma density and thus on the minimum confinement time required to satisfy the Lawson criterion: (*a*) engineering and technological limits on the strength of the confining field originating mainly from the material strength of the coil conductors and their supporting structures, and (*b*) limitations on the value of β, the magnetic utilization factor that measures the fraction of the available magnetic confining force that may be employed without failure of the confinement.

The limitations imposed by the first of these, the strength of materials, may be roughly seen from Figure 3, where the magnetic pressure ($B_0^2/8\pi$) in practical units is plotted versus the magnetic field strength in kilogausses (10 kG represents a typical value for the magnetic field intensity in commercial devices such as electric motors). Note that field values above about 250 kG imply such large mechanical stresses that steady state fields higher than this value probably cannot be achieved. The use of pulsed (inertial) techniques has allowed experimenters to reach much higher values, although usually only with accompanying destruction, or at least major deformation, of the magnet coil. Nevertheless, there are strong technical and economic reasons for attempting to reach the highest practical magnetic field values in fusion reactors based on magnetic confinement. Recent technological developments in this area, discussed in a later section, are extremely important to the future of fusion power.

What limitations are there on the plasma density in magnetic confinement systems other than those set by the magnetic field and by attainable β values? The other limitation is that of fusion power density. In a steady state (or near steady state) power-generating apparatus, there are limitations set by heat-transfer rates and other similar considerations. These limitations determine the practical maximum values of power release per unit volume of reacting fuel that can be accommodated by the reactor structure and its heat-transfer system. In fission reactors, this limit is about 100 MW/m^3 of energy-releasing volume. Power densities appreciably above this level would be very difficult to sustain in a steady state by known heat-transfer techniques. However, power densities of this level would be already reached at quite low particle densities in a fusion plasma. For example, in the D-T reaction carried out at a kinetic temperature of 3×10^8 °K, the calculated power density for a 50–50 mixture of deuterium and tritium is given by the expression

$$P_{\text{D-T}} = 5 \times 10^{-28} n^2 \text{ MW/m}^3,$$

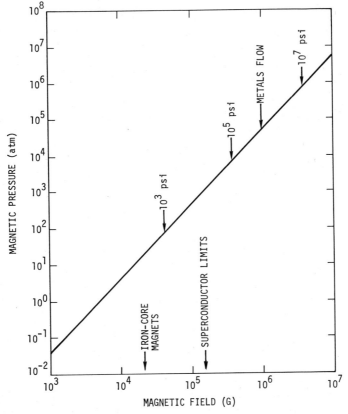

Figure 3 Limitation on strength of confining field imposed by material strength limitations.

where n is in particles/cm^3. It can be seen from this expression that a level of 100 MW/m^3 is already attained at a particle density of 4.5×10^{14} cm^{-3}. Varying as n^2, the power levels at higher densities soon reach astronomical values, far above those sustainable in steady state.

The scaling of power densities versus plasma density is shown on Figure 4. This log-log plot spans the entire range of densities contemplated for fusion reactors (either magnetic confinement or pellet fusion). Also plotted (at the side of the figure) are the minimum magnetic field values that would be required to hold the indicated densities, i.e. the values needed if $\beta = 1$.[3] Note the relatively narrow range of densities over which magnetic confinement normally would be employed, and the enormous instantaneous rates at pellet fusion densities. Between these extremes lies a "no-man's land" where it is not generally possible to contemplate the operation of fusion reactors. In this region, the power density would be either too high for steady state systems or not high enough to satisfy the strict requirements demanded by inertial confinement systems that use the pellet fusion approach. Some excursions into this forbidden region are being contemplated, although they generally involve new and difficult technological challenges such as very high, pulsed magnetic fields. However, if the technological challenges can be met, these regimes may lead to compact systems, and thus may be ones around which the most viable reactors can be designed.

From the preceding discussion it can be seen that fusion research can advance only through an intimate marriage between scientific knowledge of the high-temperature plasma state and technological ability to provide the extreme physical conditions required for the fusion process. In the next section, we review briefly the scientific problems that have had to be, and are now being, overcome en route to a viable fusion system.

THE SCIENTIFIC ISSUES FOR FUSION POWER

Magnetic Confinement

Until the scientific issues facing fusion are solved, there will remain a degree of unreality associated with speculation on the form that fusion power reactors will take. At the end of over 20 years of research toward fusion, its scientific problems are still not adequately resolved, nor will they be until at least another generation of critical experiments has been performed. I do not attempt here to describe these problems in detail, either those solved or those yet to be solved, but, instead, to try to identify them as succinctly as possible and to say what has been accomplished toward their solution and what is yet to be accomplished.

For the oldest and best understood of the approaches—magnetic confinement—the issue has been the same since the work began: What shapes, sizes, and intensity of magnetic fields are required for magnetic bottles to confine a plasma at

---

[3] In this connection, note that the relationship power $\sim n^2$ implies that the fusion power density varies as $\beta^2 B^4$, i.e. as the square of β and the fourth power of the magnetic field.

fusion reactor temperatures and densities long enough to meet the Lawson criterion? The reason for this problem is straightforward. The simplest form of magnetic confinement is confinement of a plasma column in a uniform magnetic field in a long straight tube. While this is appealing in its theoretical simplicity, it fails economically (at least by any presently known technology) simply because it is incapable of keeping the "confined" plasma from spilling out its ends. For tubes of a reasonable length (say up to several hundred meters), the confinement times would fall far short of the Lawson value. To achieve Lawson values would appear to require a tube length of order 10 km; this would lead to enormous total power outputs and correspondingly enormous capital investments. Reluctantly, we conclude that unless there is a breakthrough in the practical achievement of very high-strength magnet field coils whose structures would survive without damage for a sufficient length of time to make their capital cost acceptable, this simplest-of-all magnetic bottles cannot be used for fusion power in the form described.

Yet it is the attempt to resolve this "problem of the ends" that has resulted in virtually all of the new problems that have arisen because of the complex behavior of magnetically confined plasma. The unraveling of this behavior has been in progress for 20 years.

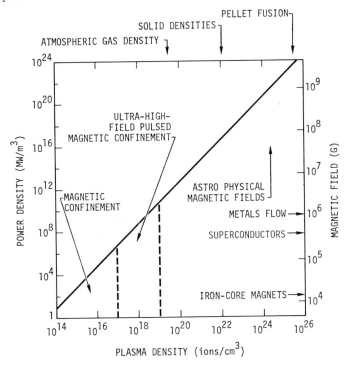

Figure 4 Limitation on plasma density in magnetic confinement systems imposed by attainable fusion power density.

In essence, there are two fundamentally different ways to attack the end problem. Both necessarily involve complicating the field shape, thereby introducing new phenomena in the behavior of the confined plasma. The two approaches go by the generic names "open" and "closed." In the open system, the magnetic field lines are allowed to leave through the ends of the confining chamber, but the plasma particles are inhibited from leaving promptly—usually by the use of magnetic mirrors (regions of intensified magnetic field at the ends of the tube). The simplest and earliest form of this so-called mirror machine is shown in Figure 5a. As shown schematically in the figure, in addition to the confinement action of the magnetic field in preventing motion across the magnetic field lines, the repelling effect of the strengthened fields can repeatedly reflect the spiraling particles as they approach the ends, thus inhibiting particle losses through the ends. However, mirror confinement is not perfect; should the particles that approach the end be moving too nearly parallel to the direction of the field lines, the mirroring action will be weakened and the particle will escape. Mirror machines therefore always have slow leaks that arise from any process—such as interparticle collisions—that can on occasion deflect a given

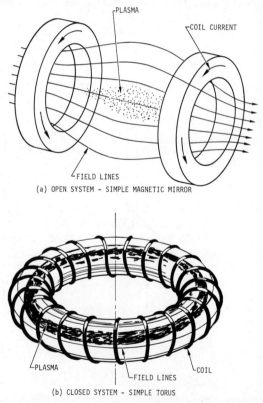

Figure 5 Open (a) and closed (b) magnetic containment configurations.

particle into an orbit that allows it to leak out the end. But theory shows that at high kinetic temperatures, deflecting collisions become less probable, whereas collisions leading to fusion reactions become generally more probable. In such regimes, mirror machines should in theory satisfy the requirements for fusion confinement and net power, provided the confined plasma does not exhibit unstable behavior.

The basic geometry of the closed magnetic confinement approach is the torus, i.e. the classic shape of the doughnut. The earliest and simplest form of such a closed system is shown schematically in Figure 5b. Here the magnetic field is produced by a magnet coil wrapped around the chamber wall as though the long tubular coil were bent into a circle, so that it closed on itself. Now, the lines of magnetic force become a family of concentric circles lying wholly inside the chamber. In principle, charged particles trapped on these lines (i.e. performing their helical orbits along them) could circle indefinitely around the chamber without touching the walls, and thus remain confined until fusion took place. According to theory, just as in the mirror machine, interparticle collisions would cause the circling particles to diffuse gradually from line to line, until in the course of time they would reach the chamber wall. Taking the simple theory at face value, however, the time for this diffusion to the walls would be more than long enough for fusion. Moreover, because the particles must diffuse across the field in order to reach the chamber walls, in principle the confinement time can always be increased simply by increasing the chamber diameter. Since elementary theory predicts confinement time increases as the square of the chamber dimensions, it would be easy to satisfy fusion requirements with reasonably sized systems.

It was upon the above considerations, seemingly solidly based on elementary but fundamental theoretical considerations that predicted success, that modern controlled-fusion research began shortly after World War II. What went wrong?

What went wrong is what goes unpredictably wrong in a mob, compared with the generally predictable behavior of individuals. In fact, the predictability of the motions of a few charged particles moving in externally imposed magnetic fields (such as those in the simple mirror machine) had been at the time well confirmed in particle accelerators and similar devices. However, to achieve significant fusion power rates, the density of confined particles, although low compared with atmospheric densities, must be very high compared with that encountered in particle accelerators. Contrary to the situation in particle accelerators, in a fusion plasma it is not possible to ignore the electric and magnetic fields that can be generated by the motions and flows of the confined particles themselves when moving in concert. Thus it is possible, indeed axiomatic, that a magnetically confined fusion plasma will at least locally influence the field that confines it. The previously cited case of a high-beta plasma that completely excluded an externally imposed magnetic field from its interior is a case in point. In this case, the influence was ostensibly benign: the effect was merely one that allowed the transfer of internal pressure from the plasma to the magnetic field, thereby allowing a state of pressure equilibrium to exist between them.

Only in the simple case just described—a straight plasma column in a uniform magnetic field—does such a simple equilibrium exist. In the two next-simplest cases described (Figure 5), no stable pressure equilibrium cases exist; the plasma finds it possible to escape confinement quickly and directly in a way resembling what happens in an aneurism in a human artery. For example, in the simple torus of Figure 5b, the magnetic field is always weakest near the outer periphery of the chamber (where the windings are spread farthest apart). If the plasma pressure is then in balance around the inner periphery, it follows that it cannot be in balance near the outer periphery. As a result the plasma will expand outward across the field, almost as rapidly as if the field were not there at all, that is, in a few millionths of a second at fusion temperatures. The plasma is said to be subject to hydromagnetically unstable motion.

Nearly the same thing can happen in the simple mirror machine (Figure 5a). Here the flow will again be sideways across the field, in the direction in which the field becomes weaker. The worst, and first-encountered, specter for all magnetic confinement systems is the hydromagnetic instability (also called magneto-hydrodynamic, or MHD, instabilities). Discovering, understanding, and finding ways to cure this kind of instability occupied fusion scientists for the first 10 to 15 years of the research period.

As of today, the hydromagnetic instability has been controlled in some approaches and ameliorated in others; however, it remains as a central problem in others, where proposed methods of controlling it have not as yet proved sufficiently effective.

The primary tool in the control of the hydromagnetic instability is the reshaping of the confining field to stop the tendency towards aneurism. The most dramatic example of the success of this approach has been demonstrated in the mirror machine.

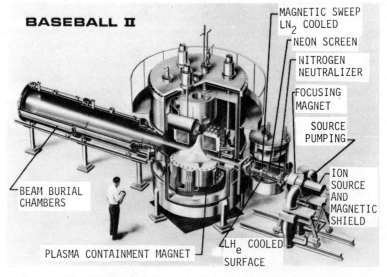

Figure 6 The LLL Baseball II superconducting, neutral injection experiment.

In a mirror machine, it is possible to reshape the confining field, while retaining the mirror effect at the ends, so that the field intensity increases in every direction away from the geometrical center of the machine. This field is called, for obvious reasons, a *magnetic well*; plasma trapped in a magnetic well cannot escape through a hydromagnetic aneurism. As a whole, it is forced to remain in the bottle, like a ball resting at the bottom of a hemispherical cup. The theory of the magnetic well was presented in 1958 (4); its efficacy was first tested in the USSR in 1961 (5); and it was soon confirmed in other experiments throughout the world. Many different forms of magnetic coils can be used to form magnetic-well mirror fields. A modern form is the *baseball coil*, shown in Figure 6, so named since the shape of the coil winding resembles the seam of a baseball.

In its pure form, the magnetic well can only be employed in open-ended systems such as the mirror machine. In a closed torus, it is not possible to reshape the field so that nowhere does it weaken outward on the periphery of the confined plasma. What can be done, however, is to arrange for this property to apply on the average as one moves around the torus. This "average magnetic well" has a stabilizing effect that is sufficient for some situations. A more powerful stabilizing effect, useful for closed systems, can in some cases be supplied by what is called magnetic shear. If the confining fields of the simple torus (Figure 5b) are supplemented by additional fields that transform the lines of force from circles to helices, and if these helices have a varying pitch angle as one moves out from the deep interior of the plasma, a basket-weave-like pattern of field lines results, as shown in Figure 7. This field-line configuration has, as would a basket made up of crossing strands, a strongly stabilizing effect on hydromagnetic instabilities. Since a hot plasma is a very good conductor of electricity, one of the simplest and most effective ways to create this field pattern in a torus is to create a strong electrical current flowing within the plasma the long way around the torus. The magnetic field produced by this flowing current, added to that produced by the external coil, produces a field with magnetic shear. This system is the one that has proved to be so effective in the closed system called a *tokamak*, pioneered in the USSR.

Unfortunately, although it greatly improves the efficiency of magnetic confinement, the control of hydromagnetic instabilities does not completely dispose of the issue of plasma instability. Although the plasma is unable to escape as a flowing whole if it cannot become hydromagnetically unstable, the particles of a magnetically

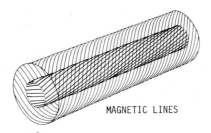

Figure 7 Basket-weave-like field-line configuration resulting from magnetic shear.

confined plasma can still find ways to escape more rapidly than would be acceptable from the standpoint of fusion requirements. This circumstance comes about because the act of magnetic confinement always carries with it the requirement for certain ordered motions or directed flow in the plasma. Examples are the induced current flowing through the plasma in the tokamak or the requirements for sufficient helicity for the orbits of the confined particles in a mirror machine. Under the right (wrong) circumstances, these flows can give rise to turbulences or oscillations within the body of the plasma that can enhance the rate of diffusion of its constituent particles out of the magnetic bottle above the rate determined by simple particle-particle collisions. Now, on a confinement time scale measured in fractions of a second, hydromagnetic instabilities can dump the plasma in a few millionths of a second. The loss rate owing to these turbulences, though typically 100–1000 times slower, could still be perhaps 10–100 times too rapid to be tolerated in a fusion reactor.

It is toward the understanding and the controlling of these residual instabilities that much of the attention in fusion research has now turned. And with this change in emphasis, the quantitative nature of the fusion quest is now most clearly seen. The control of hydromagnetic instabilities qualitatively validated the idea of magnetic confinement: it works. Now the questions are: Which system or systems can be made to work well enough quantitatively to satisfy Lawson's criterion? What shape will they take? How big will they be? What technological requirements will they pose?

Thus, although the end of the scientific quest seems to be in sight, it is still the scientific issues that are the pacing ones in the magnetic confinement approach. After a discussion of the scientific problems of pellet fusion in the next section, I describe briefly the new generation of experiments, now in operation or soon to be in operation, that may resolve these issues.

Pellet Fusion

Pellet fusion, now being most actively pursued through the route of laser irradiation of the pellet, depends for its plasma confinement entirely on inertial effects. Thus this approach would appear free from subtle instability problems that have beset magnetic confinement. Although appearing in forms different from those for magnetic confinement, plasma instability effects in pellet fusion introduce uncertainty as to its scientific feasibility. These scientific uncertainties arise from more than one quarter.

Recall that theoretical studies of pellet fusion have shown that, for it to be successful at a technically feasible level of energy release per microexplosion, the particle density at which the fusion must take place must be raised to around 10,000 times the particle densities of ordinary liquids or solids. Consider the consequences of having to satisfy Lawson's criterion while relying entirely on inertial effects. As matter is compressed spherically, its density (n in the Lawson $n\tau$ product) increases as the cube of the radial compression factor. But here confinement time (τ), measured by the time of flight of a particle out of the confinement region, clearly decreases only as the first power of the radius. Therefore, the effect is that $n\tau$ increases as the square of the radial compression factor. On the other hand, to keep the energy release per microexplosion to values that are compatible with engineering limitations, such as heat-transfer and apparatus-size limits, only a small quantity of fuel should be reacted per microexplosion. (Remember that fusion fuel releases of order 10^5 kW-hr/g consumed.) Thus, tiny pellets must be used, and this fact, taken together with the Lawson requirement, leads to the necessity for very large density-compression factors. This in turn points to the need for large compressing forces (of order 10^{12} atm, greater than the pressure at the center of the sun).

As the problem is now understood, there is only one way to achieve such enormous pressures. The surface of the fuel pellet must be irradiated at such a high power level (of order 10^{15} W/cm^2) and over such a short time that an almost instantaneous ablation (blowing away) of the outer surface of the pellet occurs before appreciable heat flow would be expected to occur into the interior. In this way, we could create a situation where the inward-acting reaction force of the outwardly ablating material creates, as in an accelerating space rocket, the reaction force that compresses the unablated material. In principle, with the use of a sufficient number of symmetrically located converging laser beams derived from ultrahigh-power lasers, the required compression forces could be generated. However, although simple in principle, the process is much more subtle than was first appreciated. For example, to compress a spherical pellet to a density 10^4 times its original density means that the elements of its volume must all be moved uniformly toward the exact geometric center of the sphere while the radius decreases by a factor 20. If this action is not accomplished with high uniformity, the compression process will go askew, leading to departure from the required compressed spherical shape. Relatively small errors in uniformity can lead to a major reduction in the achievable compression. Even starting with a near-perfect spherical ablation pattern, weak hydromagnetic instabilities during the compression process could then have a serious effect. Also, since laser irradiation of solid material at high power levels gives rise to self-generated magnetic fields, hydromagnetic as well as hydrodynamic effects may conceivably play a role in the compression process.

Another area where instability effects could act to inhibit the compression process comes from the following circumstance: To be technically feasible, the compression must proceed almost to completion without being accompanied by excessive heating of the compressing matter. If such heating occurs prematurely, the pressure (being proportional to the product of density and temperature) would rise too early in the process to permit the compression to proceed to completion. However, it is known that under certain circumstances electromagnetic waves (i.e. the light waves launched by the laser pulse) can through plasma instability effects cause the coherent acceleration of some electrons of the plasma to energies far higher than the average energies associated with the kinetic temperature of the plasma. These high-energy electrons, born early in the compression process, could then penetrate deep into the interior of the pellet, causing a preheating effect that would inhibit the compression process.

These two instability effects are but examples of why the pellet fusion approach must pass through a period of exploration of basic plasma phenomena in the presence of electromagnetic fields. This is the same fundamental issue that has had to be faced in magnetic confinement, albeit in an entirely different regime of densities, time scales, and dimensions. Out of this investigation will emerge the scientific understanding needed to specify the critical elements that will lead to success (or, conceivably, to failure) of the pellet fusion approach as it is now visualized.

FUSION EXPERIMENTS

Magnetic Confinement Experiments

Since its beginning, research in magnetic confinement fusion has been characterized by the great variety of experimental approaches to solving its problems of heating and confinement. These approaches have been essayed in hopes either that a new idea would provide an easy breakthrough to hotter, denser, or better-confined plasmas, or, as in the more systematic studies, that the experiment would provide a better understanding of the basic processes—such as plasma instabilities—that has been so desperately needed.

It would be foolish to describe here these many earlier attempts, valuable and important as many of them were. For the purposes of this review, I simply enumerate and discuss examples of the approaches that have survived the test of time, then return briefly to the as-yet-unanswerable but exceedingly important question: Will the fusion reactors of the future be the direct descendants of the present mainline approaches, or will they instead be the product of some present or future dark-horse approach, one either of a completely new genus or one representing a synthesis of features of the present approaches? In my opinion, the answer to this question is central to the future of fusion, in that it not only exposes the uncertainties inherent in our present experimental situation, but also illustrates the volatility of the fusion quest itself—in terms of what may really happen. The appearance of the pellet fusion approach is a case in point. A new technology, the laser, suddenly opened a new avenue to fusion, one that bore virtually no relationship to the previous (magnetic confinement) approach.

The three mainline experimental approaches to magnetic confinement, as exemplified in the three main components of the US program supported by the Energy Research and Development Administration (formerly the Atomic Energy Commission) are the tokamak, the theta-pinch, and the magnetic mirror. At present, the US effort on these approaches is divided in the approximate proportions 60–20–20, with the tokamak approach the major element.

The tokamak approach, pioneered in the Soviet Union, closely resembles the first-proposed form of toroidal confinement (Figure 5b), with one critical difference: In addition to the main toroidal field provided by external windings, there exists a magnetic field generated by and permeating the confined plasma itself. This field comes from a heavy current (up to 10^6 A in some experiments) that has been induced in the plasma by making the plasma the secondary winding of a transformer, the primary winding of which is excited by a rising pulse of current. In fact, this plasma current provides the basic confining force in the tokamak, through the phenomenon of the pinch effect (parallel flowing current elements tend to attract each other, i.e. they tend to pinch together).

NUCLEAR FUSION   235

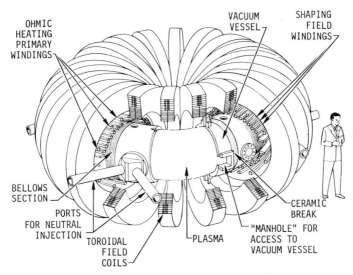

Figure 8   The Princeton large-torus PLT.

Confinement by the simple pinch effect is one of the oldest and first-essayed ideas in fusion research. But, where both the simple pinch and the simple toroid failed as a result of hydromagnetic instability, the combination of the two fields has been highly successful. Here, the role of the external toroidal field, made much stronger in magnitude than the pinch field, is to stabilize the plasma. Thus, confinement comes solely from the pinching effect; stabilization comes from the toroidal field. It follows that, as measured against the externally generated field, the tokamak is of necessity a low-beta device (where plasma pressure is small compared with external magnetic pressure). Some variations on the tokamak idea, discussed below, attempt to raise the beta limit, which in the usual tokamak is theoretically limited to a few percent.

Over the past several years, many tokamaks have been constructed in laboratories throughout the world. Their performance has on the whole certainly been better than any of the previous attempts at toroidal confinement, such as the early stellerator[4] experiments pioneered at Princeton University. Over the years, larger and more powerful tokamaks have been built, with the expectation and hope that larger size would lead to higher temperatures and longer confinement time (theoretically scaling approximately with the square of the chamber diameter). These hopes have been only partially justified. The largest of the new devices (TFR in France) does not confine plasma for times substantially in excess of the earliest successful Soviet tokamak (T-3, circa 1968). The reasons for this are only now beginning to be understood, but it is felt that their resolution must await the completion of two new and much larger tokamaks, PLT in Princeton and T-10 in Moscow. The PLT, shown as a scale model in Figure 8, is scheduled for first operation in 1975. To compare it and

[4] A stellerator is a closed toroidal device in which the shaped confining fields are provided entirely by a complex array of external windings.

T-10 with previous tokamaks, in terms of scale, Figure 9 shows the relative cross sections and major radii of several tokamaks, past, present, and future.

It is already known from tokamak experiments that the confinement of plasma in a tokamak falls short, by a substantial margin, of that theoretically predicted if the plasma is assumed quiescent. Tokamak researchers do not expect to achieve such a state. Rather, their hope is that even with instabilities present the containment margin will be a tolerable one, from the standpoint of fusion power balance, after the plasma temperature and density have been raised from their present values to those that will be required for a reactor. In maintaining this hope, they are relying on the fact that through scale-up in size and magnetic field intensity they should be able to sustain adequate confinement in the face of still further enhancement of the instability-induced losses. Such an enhancement is predicted to occur within the hotter plasmas they must attain. Only through experimentation with larger-scale apparatus, such as PLT or T-10, will it be possible to begin to resolve such questions.

Table 1 lists some of the plasma parameters achieved in present tokamaks and compares them with those thought to be needed in a tokamak reactor operated in an "ignition" mode (where plasma temperature is maintained by the fusion process).

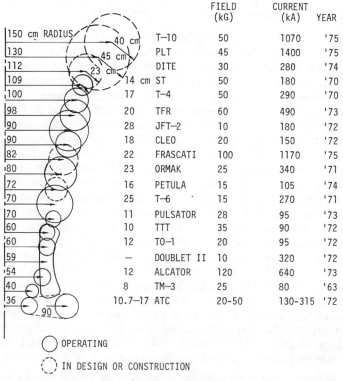

Figure 9 A comparison of tokamak parameters throughout the world.

Table 1 Plasma parameters attained by magnetic confinement experiments compared with typical values believed needed for a fusion reactor based on the same approach

Approach	Device	Experimental Status					Reactor Requirements	
		Plasma Diameter, d (cm)	Particle Density, n (cm^{-3})	Ion Temperature, T_i (keV)	Confinement Time, τ (msec)	Lawson Product, $n\tau_{\text{expt}}$ (sec/cm^3)	Ion Temperature, T_i (keV)	Lawson Product, $n\tau_{\text{reactor}}$ (sec/cm^3)
Tokamak	T-4 (1970)	30	3×10^{13}	0.4	10	3×10^{11}	15	10^{14}–10^{15}
	ST (1974)	25	4×10^{13}	0.6	10	4×10^{11}		
	TFR (1974)	40	4×10^{13}	0.8	15	6×10^{11}		
	Alcator	18	2×10^{14}	0.8	10	2×10^{12}		
Toroidal theta-pinch	Scyllac	2	2×10^{16}	0.8	0.01	2×10^{12}	10	10^{15}
Mirror machine (magnetic-well type)	PR6 (1971)	10	2×10^{12}	0.5	0.15	3×10^{8}	150	10^{13}
	2XII (1974)	20	4×10^{13}	2.0	0.5	2×10^{10}		
	Baseball II (1972)	20	2×10^{9}	2.0	1000	2×10^{9}		
Stellerator	Model C (1966)	10	10^{13}	0.06	0.2	2×10^{9}		
Mirror machine (simple mirrors)	Toy Top (1961)	5	10^{13}	3.0	0.05	5×10^{8}		

Note that some of the best results have been obtained from a small machine, Alcator, at MIT, where the high magnetic field used played an important role. It is not known, however, whether these results can be extrapolated in a practical way to reactor conditions, considering the very high magnetic fields that they seem to imply would be required.

I conclude the discussion of the tokamak by mentioning two proposed variations that, although not yet proven, give promise of increasing its probability of sucess as a reactor.

The first of these is the idea of elongating the cross section of the tokamak plasma column, giving it the general shape of a napkin ring. There are indications from theory that such a shape should permit operation at lower values of the external toroidal field (thus at higher beta values) than in the conventional tokamak with its circular plasma cross section. Several devices of this general type are in operation or under construction throughout the world. The largest such device is the Doublet III, the design and construction of which was recently initiated at General Atomic in San Diego, California. Figure 10 is a drawing of Doublet III, showing its scale and the elongated plasma column utilized.

The second variation on the tokamak is really a variation in mode of operation, one that is also being contemplated in other devices, such as mirror systems. It is called a two-component mode of operation. Recall that our description of the fusion process thus far has been entirely in terms of a thermonuclear description; that is to say, we have visualized the release of fusion energy as being the result of heating an entire fuel charge to an ignition kinetic temperature, one at which the fusion reactions occur as a result of the continued and random collisions going on within

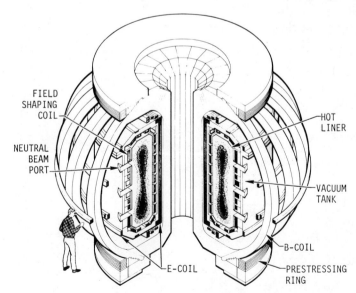

Figure 10 The General Atomic Doublet III experiment.

the heated plasma. Such a thermonuclear description is indeed appropriate for a star, for a hydrogen bomb, or for pellet fusion. But deeper reflection shows that what we really want is a net exothermic fusion energy release occurring from a system that we will then call a fusion reactor.

Recall again the circumstances, mentioned at the beginning of the article, of the discovery of fusion reactions: A beam of energetic deuterons was directed at a target containing other deuterons (or tritons, in later experiments). Fusion occurred. But the target, being ordinary matter at room temperature, rapidly damped the energy of the impinging beam particles as they penetrated the target. Only those few beam particles that could react before they were slowed down made fusion reactions. However, the situation is very different if the target is a plasma at near-thermonuclear temperatures. Now, an impinging energetic particle launched with an initial kinetic energy that is very much higher than the mean energy of the target plasma has a high probability of undergoing a fusion reaction before being eventually slowed down.

Thus, perhaps fusion research will come full circle: Fusion reactions were discovered first by beams on targets, over 40 years ago. It seems possible that this first idea, revitalized by the discovery of the means to create the right kind of target (through research in magnetic confinement) and by new beam technology, will be the best way to operate a fusion reactor. The two-component idea is being taken seriously in the United States and in many foreign programs. High-current neutral beam sources that could meet the requirements for high power are being developed at various laboratories, including the Lawrence Berkeley Laboratory (LBL) (6) and the Oak Ridge National Laboratory (7). Figure 11 is an artist's drawing of the TCT (two-

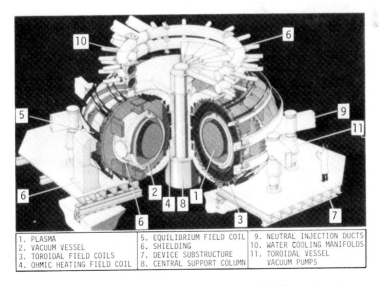

Figure 11 The Princeton two-component torus (TCT) experiment.

component torus) experiment in planning at Princeton. If successful, this experiment could begin operation about 1980 and would reach a break-even point in fusion power, certainly a milestone event for fusion research. It is no wonder that results from PLT and T-10 are being awaited with great interest, for the possibility of success with TCT will probably hinge on the results of these two experiments.

Despite hopes for the unqualified success of a front-runner, wise planning in a research field such as fusion requires that the program have breadth. The present uncertainties are too great, and the present state of understanding of plasma behavior is too limited, to guarantee the success of any one approach. Fusion is still in a research phase, however much we wish that it had by now passed to a development phase. As stated earlier, the particular examples that we discuss where breadth is being maintained in the US magnetic confinement program are the theta-pinch

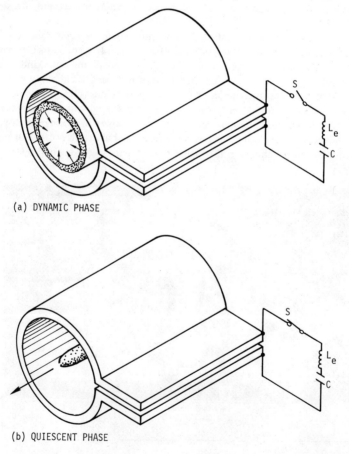

Figure 12 Basic principles of the theta-pinch magnetic confinement system.

program, centered at the Los Alamos Scientific Laboratory, and the magnetic mirror program, centered at the Lawrence Livermore Laboratory (LLL).

The theta-pinch concept, one of the earliest to be explored in magnetic confinement fusion research, is in reality a mode of plasma heating rather than a particular confinement geometry. Yet the technique has been brought to the point where, in combination with a particular geometry (for example, with a toroidal system with shaped fields, as in the stellerator) it provides a total approach in its own right. The idea of the theta-pinch process is basically simple: A plasma column is created inside a one-turn magnet coil (See Figure 12). When the magnetic field is pulsed up to high values very rapidly (in microseconds or fractions of a microsecond), shock heating and heating by magnetic compression take place (compressing any gas heats it; a familiar example is the diesel engine). The resulting hot plasma is very dense (typically 10^{16} particles/cm^3) and possesses a very high beta value (of order 0.9 in some experiments). These properties are all very desirable ingredients for an approach to fusion.

For many years, theta-pinch experiments centered on the use of straight, open-ended systems. On their microsecond time scale, hot, dense, stable, high-beta plasmas were produced. Figure 13 shows a laser interferogram "picture" of the highly compressed plasma column in such a device (Scylla IV at Los Alamos). In this experiment, the plasma beta value was about 0.9, and the plasma pressure was about 15 atm—a high value indeed. On the same microsecond time scale, the hot plasma flowed out the ends of the tube (remember "the problem of the ends"). It was readily apparent that only at tube lengths of several kilometers would it be possible to hope for net power (requiring Lawson products of order 10^{15} owing to the system involved). In recent years, therefore, most theta-pinch research has turned to the exploration of toroidal geometries, not resembling the tokamak in proportions, but being rather in the shape of large-diameter rings of small cross section (to minimize the destabilizing effects of the curvature of the tube on a high-beta plasma). The largest such device in the world is the Los Alamos Scyllac, a photograph of which is shown

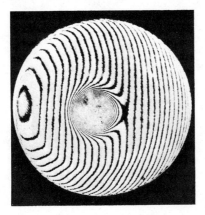

Figure 13 End-on interferogram of plasma in the Los Alamos 3-m theta-pinch experiment.

in Figure 14. The critical dimensions and other parameters for Scyllac are listed in Table 1.

It was known from the start that the central issue for devices such as Scyllac would be how to avoid hydromagnetic instabilities. It was hoped that in that particular device, proper shaping of the confining field (a mixture of a simple solenoidal field together with helical field components generated by helical grooves machined into the inner surface of the pulsed coil) would largely stabilize these instabilities, thus making them weak enough to be controlled by electronic negative-feedback currents applied to special windings. This hope did not materialize. At this point in the Scyllac experiment, the rate of growth of instabilities at full field and maximum plasma temperature was found to be too rapid to be controlled by the feedback devices presently available. As a result, operation is now being scaled back in temperature to slow down the instability growth rates to a value where the present feedback system may be able to gain control. In addition, renewed emphasis is being given to ways to enhance the shock heating process (through the achievement of faster-rising magnetic fields). If this method of heating proves to be sufficiently successful, a much smaller amount of magnetic compression would be needed to achieve the required heating, resulting in a fatter plasma column (i.e. a column with a diameter more nearly equal to that of the diameter of the chamber wall). Here,

Figure 14 The Los Alamos Scyllac experiment.

the effect of the nearby conducting wall would be to help to stabilize hydromagnetic modes so that stability might be achieved without relying on feedback methods. As with the tokamak, resolving these issues will require either a technological or a physical scale-up in the experiments.

Whereas hydromagnetic instabilities have remained on the agenda for both the tokamak and the theta-pinch approaches, for the last ten years they have not been a problem in mirror research. The introduction of the magnetic-well idea has allowed mirror researchers to avoid this particular pitfall and thus to concentrate on the central problem for mirrors—how to minimize the rate of plasma leakage through the mirrors. That this is indeed a central problem arises from the circumstance that the theoretical rate of particle leakage through the mirrors, even assuming a completely quiescent plasma, is rapid enough to make the power balance margin for mirrors relatively small. For a mirror fusion reactor to function, all operations in which energy flows are involved (i.e. plasma heating and the handling and recovery of energy from the system) must be as efficient as possible while still being consistent with economic constraints. Because the power balance margin is not large, even when these conditions are satisfied, the plasma itself must be sufficiently quiescent for its confinement time to approach the theoretical limit calculated from the effects of collisions between the particles (which set the so-called classical lower limit on the rate of mirror losses). In mirror systems, where losses occur by leakage out the ends, scaling up the size (diameter) does not improve the confinement time as it does in a closed system, where particles must diffuse across the field lines in order to be lost. In a mirror system, therefore, the route to net power lies in capitalizing on these three circumstances: (a) operation at very high plasma temperatures (thereby reducing the collisions that deflect relative to those producing fusion); (b) achieving efficient energy-handling techniques (mentioned above); and especially (c) achieving adequate plasma quiescence.

Over the several years that mirror confinement studies have been under way, the problem of achieving quiescent plasmas in mirror systems has turned out to be a very difficult one. The difficulty arises from the fact that within a dense plasma between magnetic mirrors there exists the possibility that plasma waves and oscillations will be built up by mechanisms very similar to those that cause laser action. If these oscillations are allowed to grow to large amplitudes, the plasma turbulence they produce can cause the plasma to leak through the mirrors at a much faster rate than the classical rate predicted for a quiescent plasma (that is, one in which there exists only the normal level of thermal fluctuations). To keep these waves from growing, it is necessary to prepare the plasma carefully, avoiding conditions where the particle velocities are not well randomized; and it is also necessary to damp out the unstable waves by controlling the shape and intensity of the confining magnetic field.

One of the earliest techniques for creating a high-temperature plasma in a mirror system was that of neutral beam injection, pioneered in early experiments at LLL. Many experiments have been performed in various laboratories around the world to use this technique to build a hot, mirror-confined plasma. Neutral beam injection is now recognized as one of the most promising techniques for heating plasmas to fusion temperatures (and for providing the energetic particles for two-component reactor systems). The idea involved is simple. A high-current, highly focused beam of ions is created by an ion source device similar to a particle accelerator in which a beam of ions is first extracted from a plasma produced by an electrical discharge and is then accelerated by grids resembling those used in accelerating beams of electrons. This beam of ions is then passed through a neutralizer cell, a region where a low-pressure gas is maintained. Under the proper conditions, a high percentage of these ions are converted back into atoms by the simple process of capturing an electron from the surrounding gas atoms. The energetic atoms in the beam thus formed lose none of their original energy, and they remain directed. However, the neutral atoms in the beam can freely cross through the field lines of a confining field (through an aperture in the chamber), enter the chamber, and there be broken up (reionized) by impact with any plasma that is already present. In this way, new energetic ions and their accompanying electrons can be introduced into a plasma, thereby either heating it and building up its density or maintaining the temperature and density if they are already high. Thus, neutral beam injection offers an avenue to solving the twin critical problems of heating and refueling a fusion reactor.

In early neutral beam experiments, such as the Baseball I experiment at LLL, or the Phoenix experiment in the United Kingdom, only low-current neutral beam sources were available. It was soon found that the buildup of plasma density was limited to very low values (of order 10^8 particles/cm^3). This limitation was traced to the presence of high-frequency plasma oscillations that were stimulated by the very highly peaked (very poorly randomized) nature of the plasma being produced. In later work with the Baseball II experiment at LLL (shown in Figure 6), the plasmas, albeit still at low density, were better randomized and thus more highly stable. If the density was kept low, these plasmas exhibited losses that were close to the classical value. Above a threshold density, instability effects appeared. The plasma had to be built up past the unstable density "gap" to much higher densities, where collisional and other effects had been observed in other experiments to lead to improved stability. As a consequence, Baseball II is now undergoing modifications to increase the density of its plasma. These include the use of a laser-pellet target plasma and higher-intensity neutral beam sources.

One of the most successful of the higher-density experiments has been the 2XII experiment at LLL (8). In it, highly quiescent plasmas at high densities (10^{13}–10^{14} particles/cm^3) and high temperatures (10–30×10^6 °K) have been achieved through careful control of the plasma conditions. Not only were quiescent plasmas obtained, but also high-beta values were demonstrated. In 2XII, beta values of 0.5 were demonstrated, in close agreement with theoretical limits. This same theory predicts that beta values approaching 0.9 should be obtainable in mirror reactor systems (9), a major advantage if achievable.

Despite the encouraging results, the empirically determined conditions required to maintain stability as the plasma temperature is raised are not at present well understood, so that it is not yet possible to assure that the attempt at yet higher temperatures will be successful.

In the earlier experiments, the 2X facility operated in a slow pulsed mode (a milliseconds time scale). Here, a plasma derived from "plasma guns" was injected into the evacuated plasma chamber through one of the mirrors; there it was trapped and compressed and then heated by pulsing the magnetic field (in a manner somewhat similar to the theta pinch, but on a much slower time scale). Figure 15 shows a cutaway drawing of 2X, illustrating the processes involved.

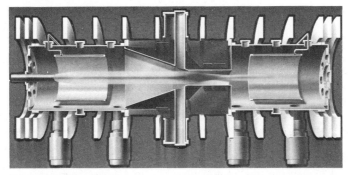

(A) PLASMA INJECTION

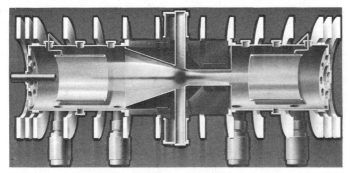

(B) TRAPPING AND COMPRESSION

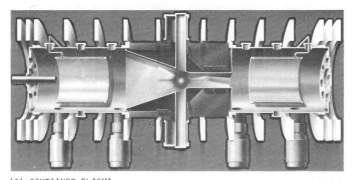

(C) CONTAINED PLASMA

Figure 15 The LLL 2XII experiment, showing the slow-pulsed mode of plasma injection, trapping, and containment.

Since the plasma-injection magnetic-compression technique is limited as to the plasma temperatures that it can achieve, and since the neutral beam technique is such a powerful method of plasma heating, 2XII has now been rebuilt (2XIIB), with the addition of a powerful array of neutral beam sources. These sources, developed at LBL as a joint program with LLL, are capable of injecting several hundred amperes equivalent of neutral atoms into the 2X plasma, which now functions as a target plasma to trap the beams. With this new facility, now being put into operation, plasma temperatures have been raised to the 10^8 °K level, at high plasma densities. The task of learning how to maintain plasma stability under these more demanding conditions is now beginning. One promising lead, already observed, is that stability is greatly enhanced in the presence of a moderate-density background plasma that permeates the mirror confinement region. This result points also to the advantages of further exploring the two-component idea for mirror system, for which the theory predicts improved stability. Again, only time will tell if these hopes can be realized.

In reviewing the progress and present status of each of the major approaches, we see that each has its own special advantages, predicted by theory and generally confirmed by experiment. Yet also each has its Achilles' heel, i.e. a source of difficulty not as yet shown experimentally to be resolvable in a practical way. For the tokamak, the primary question arises from the known presence of instability-induced losses. These losses are theoretically predicted to increase as the plasma conditions approach those required for a reactor. Will they then be such as to require a scale-up to such a huge size that the tokamak becomes impractically large? For the closed theta pinch, the achievement of high plasma temperature and high plasma beta values are both demonstrably possible. Here the question is the oldest one in fusion: How do we adequately suppress rapidly growing hydromagnetic instability modes? For a successful theta-pinch reactor, these instabilities must be suppressed over a period of time that is about 10,000 times longer than the presently observed confinement times of about 10 μsec. Can this be done? Only more experimentation can answer this question. For the mirror approach, where hydromagnetic instability has been controlled, and high-density, high-beta quiescent plasmas have been demonstrated, the key question can be succinctly stated: Will it be possible to maintain an adequate degree of quiescence as the plasma temperature is scaled up to 10–20 times the value where quiescent regimes have been demonstrated? This is a difficult requirement, one with as yet unknown scientific pitfalls.

What does it all mean? It means first that research on fusion by magnetic confinement is still very much in a scientific phase, possibly near the end of this phase, but definitely not out of it. It also means that the entire program would be well advised to keep a close watch on its dark horses, those as-yet-unproved new ideas (or variations on older ones) that give promise of bypassing the problems of the mainline approaches. The two-component idea is an example. There are many others, too numerous to mention, such as the multiple-mirror idea or the many variations on the closed systems, that could be just what is needed. I believe that it could set back the achievement of fusion by many years to close out the investigation of such options prematurely. We give ourselves credit for too much prescience when we assume that we know which particular fusion approach will turn out to be the most viable one.

Pellet Fusion Experiments

Because it is a much newer field, the experimental results in pellet fusion are much more limited. This limitation arises not only from the newness of the field but also from the fact that the key question is so much more simply posed: Will the pellet be compressed and heated to the fusion point before it disintegrates, or will it not? Because I am not an expert in the field, my critique is inadequate, but I would summarize progress in pellet fusion as follows. Extensive theoretical and computational work has shown the extreme conditions required for pellet fusion to succeed, including the need for increasing the density of order 10^4 over the normal densities of the pellets. Laser-pellet fusion experiments based on the available theory have been performed in several laboratories in the United States and abroad and have shown the presence of thermonuclear neutrons at compression ratios in the low hundreds. In this country, some pioneering experiments were done by KMS Fusion (10); these were soon followed by work at LLL and at other laboratories that confirmed and extended the KMS results. Further major progress toward laser fusion goals now awaits the construction of larger laser systems, essential for the achievement of conditions closer to fusion. No one knows whether these will be successful or, if they are successful, whether the next step beyond, requiring even greater compressions, will be successful scientifically. Members of the laser community are enthusiastic in their belief that they hold the key to fusion, but only time will tell. Also, as in the magnetic confinement approach, some dark-horse technique, for example pellet compression by focused particle beams rather than by lasers, may hold the key.

FUSION TECHNOLOGY

Although the discussion of progress toward fusion has thus far been concerned mainly with scientific issues, I have alluded to the importance of technology in fusion research. In fact, in considering the history of fusion research, it appears that even where the main issues concerned the probing of theoretical concepts and the development of understanding, the pacing element was usually the development of specialized technology. Except for early proof-of-principle experiments, or the investigation of some of the basic phenomena in plasmas, existent conventional technology has not been adequate to do the job. Thus, in parallel with the scientific quest there has been a technological one aimed at solving the very complex problems of creating and protecting the fragile entity that a fusion plasma represents. Plasma is fragile because to minimize radiation losses it must be kept free from appreciable contamination from high-atomic-number impurities (which can come from the chamber wall) and it is fragile because the quality of its confinement may depend sensitively on how it has been prepared and heated.

Together with the need for special technology for the creation of pure hot plasmas has come the need for novel noninterfering techniques for measuring the plasma properties (density, temperature, purity, level of turbulence, etc) in order to deduce what is going on. This topic is a large subject in itself. Suffice it to say that these measurement problems have been essentially overcome; satisfactory means have been found for determining all of the critical parameters.

Returning to the question of the fusion technology itself, I simply list and describe briefly the most important areas in which these developments have occurred. This cursory treatment should not be construed as indicating that fusion technology is of minor importance in fusion research and development. Quite the contrary; it is an essential part, and its importance will continue to grow as fusion power comes closer to being realized.

Magnetic Fields

Obviously a key issue in the magnetic confinement approach, the development of magnetic fields has kept pace with experimental demands. High-intensity pulsed fields can be generated up to limits imposed by materials and with pulse rise times scaled down to microseconds. Pulsed technology is essential to the theta-pinch approach. It has also been used in many other experiments, 2XII, for example, where the field coils are larger (meters in diameter) and where mechanical strength is provided by the use of advanced fiber-composite technology. The development of the new high-field superconductors,[5] proceeding at a rapid pace, now permits the generation of large volumes of high magnetic field without the expenditure of magnet power and at lower overall cost than that of ordinary conductors. This development will no doubt play a critical role in most approaches based on magnetic confinement. For an example of a fusion experiment that utilizes a superconductor coil, see Figure 6, the Baseball II experiment at LLL.

Vacuum and Wall Technology

Maintaining a high degree of vacuum and keeping contaminating atoms from the fusion plasmas have been continual necessities from the onset of fusion research in magnetic confinement. Although problems remain, for example in tokamak research, these problems are largely solved at the level needed for the scientific phase. Where they will again loom larger will be in connection with the D-T reactor, where bombardment of the chamber wall by 14-MeV neutrons will present difficult problems with materials, with regard to both contamination effects and the influence of radiation damage on the lifetime of the wall.

High-Energy Neutral Beams

The special technology concerned with high-energy neutral beams, long a part of fusion research, has been rapidly developing recently and is now viewed as one of the most promising techniques for plasma heating. It is used in the tokamak program (ATC in Princeton, ORMAK at Oak Ridge, and other experiments), and forms the backbone of the mirror program (e.g. 2XIIB and Baseball II at LLL, LITE at United Technology, and OGRA III and IV in the USSR). Pulsed neutral beam sources are

---

[5] A superconductor, generally a special alloy or an intermetallic compound, is a conductor that totally loses its electrical resistance at low-enough temperatures (that of liquid helium, for example).

now available at the megawatt power level. Twelve such sources are used in 2XIIB. Development of megawatt-level beams at higher energies (up to 200 keV), as needed for two-component tokamaks and for mirror systems, is now under way. Neutral beams will likely be one of the most important elements of both open- and closed-end reactors. Thus far, they have not been incorporated into theta-pinch systems, which use shock and compression heating methods instead.

High-Power Lasers

The laser-pellet fusion program has been able to rely on the already massive support for laser development that has been stimulated by other needs. Although lasers adequate for the next steps in the scientific program are becoming available, researchers in this field recognize that the success of laser-pellet fusion hinges on the development of yet higher-power lasers—lasers with an efficiency of conversion of electrical energy to light energy that is far higher than the fraction of 1% associated with those now in use.

Since the fusion energy release must always be debited directly for the energy required to drive the laser system, this is a critical issue, one that will no doubt require a major development effort to solve. Present lasers produce pulses of about 1000 J of energy in times less than one billionth of a second; that is, they produce a power level greater than one million MW per laser (many times the total US electrical power–generating capacity, but for a tiny time, of course). Lasers producing 100 or more times this power level are viewed by some as becoming necessary for laser fusion experiments as these experiments ascend the temperature scale to fusion reactor conditions.

Direct Electrical Conversion

I mentioned earlier the possibility that fusion reactors will someday utilize techniques for the conversion of fusion plasma energy directly into electricity without the use of a heat cycle (as in steam turbines, etc). There are two distinct areas in the operation of some types of fusion reactors where direct conversion could become important. For example, in both the theta-pinch and mirror approaches a substantial investment of electrical energy is made to bring the fuel charge to reacting temperatures. Efficient recovery of this invested energy from those charged particles that did not react would help greatly in achieving a positive power balance. Second, and of longer-range interest, is the important possibility of employing direct conversion in connection with reactions such as the D-^{3}He$_2$ reaction, in which the fusion energy is imparted entirely to charged particles. Given a good confinement scheme, so that power recirculation is minimized, such a system could, in principle, exhibit a much higher system conversion efficiency than would be possible using thermal conversion alone and possibly at a lower capital cost. In theory, direct conversion of plasma energy could be carried out at efficiencies approaching 100%; practical considerations will limit the efficiency to values that are lower but still potentially much higher than those possible in thermal plants.

Experiments carried out at LLL and checked by computer code have demonstrated direct conversion of the energy of charged particles (escaping through a mirror from a simulated confinement region) that is in excess of 85%, versus the 40–50% conversion typical of the best thermal plants today. Figure 16 shows an artist's drawing of the test experiment. A combination of magnetic fields and high-voltage electrodes was used to convert the initially randomly distributed, charged-particle energy into high-voltage direct current.

In the theta-pinch example, it is proposed to convert directly the energy of the plasma that remains confined at the end of the fusion burning cycle by allowing it to expand against the confining field. In this way, a portion of the magnet pulse energy can be recovered from the currents that are induced in the field windings. Thus far, there has been little experimental investigation of this possibility. According to theory, however, the recovery efficiency achievable by this magnetic technique would be limited to lower values than those achievable by the direct electrical technique described earlier.

Although direct conversion is not needed in solving the scientific problems that face fusion researchers today, the work being done in this area is laying the groundwork from which practical direct conversion techniques may someday evolve.

Visions of Fusion Reactors

Increasingly, as the scientific problems of fusion are being solved, attention has been turning to the appearance of fusion reactors based on the present approaches. Because the remaining scientific problems have not been solved, there is an inherent element of unreality in all such studies. Nevertheless, these studies serve at least one useful purpose: They help prevent compounding the element of scientific uncertainty with an additional unreality that arises from not appreciating the major engineering problems, problems associated with bringing fusion out of the laboratory and into the electrical utility industry.

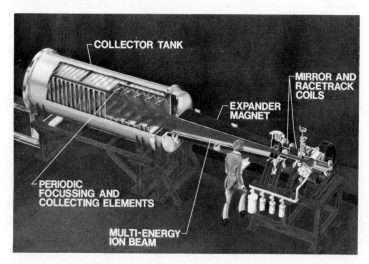

Figure 16 A direct conversion test facility.

One result seems clear from these studies. With the possible exception of pellet fusion systems, the present approaches will lead to large central-station units; there will probably be no viable miniature fusion reactors (11). For the D-T cycle, the required thickness of the neutron-absorbing blanket (of order 1 m) will already impose a lower size limit.

Although mirror systems might be reasonably small, toroidal systems such as the tokamak would be quite large because a large size would probably be needed to achieve adequate confinement (theoretically, confinement time varies as the square of the diameter of the plasma column). Because of reduced confinement time allowed, the two-component concept may help reduce the size. Mirror systems appear to offer the possibility of smaller units, especially if the two-component idea can be implemented practically. Although in the theta-pinch the chamber diameter is small, the overall size and power output of both theta-pinch and tokamak reactors may be quite large, with power outputs of thousands of megawatts.

An example of a fusion reactor is shown in Figure 17, which depicts a 1000-MWe mirror reactor with a direct converter attached.

One feature common to all the reactor studies carried out so far is somewhat disturbing. Each system involves complex and novel technology, often substantially more complex than that encountered in fission reactors. Considering the technical difficulties that have beset the fission program, it is worrisome that even more complex problems may occur for fusion, at least for those approaches studied so far.

This situation puts additional importance on the search for smaller, better, and especially simpler approaches to fusion—the arena of the dark horses. In my opinion, the day when fusion power will first be put to practical use may possibly depend more on finding a "new invention" (or a new twist on an old invention) than on the patient pursuit of present approaches, important though that pursuit is to maintain the present momentum. One possible approach to fusion power has not even been mentioned thus far in this review. Paper studies have been made of the feasibility of generating fusion power by what are essentially the same techniques used to tap geothermal energy. A huge cavity would be excavated deep in solid bedrock, and with small hydrogen explosives periodically detonated within this cavity, high-temperature steam could be generated and piped to the surface to produce power (12). Will this be the fusion reactor system of the future? It could be.

I conclude this brief discussion of speculations on fusion reactors with a suggestion, one that is being studied seriously, for introducing fusion technology into the economy at an earlier date than seems likely through the pure fusion route. The idea would be to use a driven D-T fusion plasma as a source of neutrons for breeding and/or energy multiplication in a blanket containing a subcritical lattice of fertile fission material (depleted uranium, for example) (13). As an alternative to the breeder reactor, such a system might offer economic and other advantages. Certainly the demands for fusion power gain and reliability that would have to be satisfied would be far less than those required for a pure fusion reactor. Fusion-fission might therefore be the door through which fusion would first enter the power scene. The pros and cons of such a possible course are just now being debated vigorously in the fusion community.

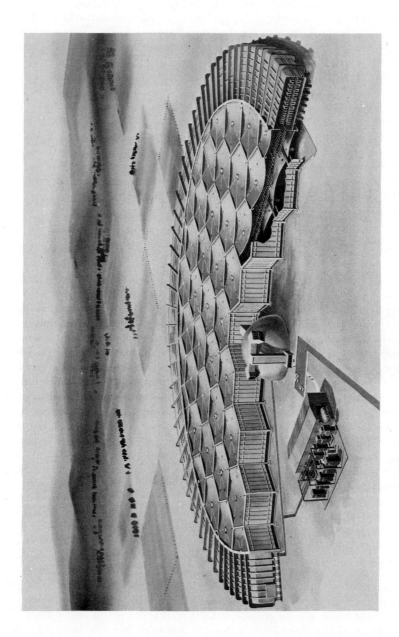

Figure 17 A fusion reactor with an attached direct conversion facility.

Fusion Power and the Environment

Fusion, being a nuclear process, involves hazardous radiation, i.e. neutrons, gamma rays, and radioactive materials. Fusion reactors must therefore operate in shielded enclosures, as is true for any device involving radiation. Studies have been made of the potential external hazards of fusion reactor systems, as well as of other aspects of their environmental impact. The consensus is that vis-à-vis conventional nuclear sources of power, i.e. the fission reactor, fusion reactors could be designed to have a much lower hazard potential. But these advantages are not automatic; they will require both care and ingenuity on the part of the reactor designer to be sure that they are achieved.

In brief, the points made in one recent study (14) are the following:

1. The total inventory of volatile radioactive isotopes (tritium) is of such a magnitude that even its instantaneous release would not represent a catastrophic event in terms of prompt fatalities. Dilution and dissipation would be expected to be relatively rapid, and the critical areas involved (possible prompt fatalities) would be of the order of the site size. Routine releases can apparently be kept to levels now required of light-water reactors (5 mrem/hr at the plant boundary), with tritium control techniques now under development.

2. The inventory of nonvolatile radioactive isotopes resulting from neutron-induced activity in the inner structure of a pure-fusion reactor is subject to control by the designer. With clever choice and arrangement of materials, both the total inventory and the problem of radioactive afterheat can be minimized.

3. Except for fusion-fission hybrid systems, the problems of diversion of nuclear materials for the clandestine production of nuclear weapons is obviously greatly reduced. However, even in a pure-fusion reactor, the presence of neutrons would mean that security precautions would be necessary to prevent the introduction of uranium in the blanket for the purpose of breeding plutonium for use in weapons. The fusion-fission hybrid would of course possess the same kinds of problems relating to plutonium diversion that are of concern in fission reactors.

4. The environmental impact of obtaining fusion fuels (deuterium and lithium) should be essentially negligible. That of obtaining the structural materials for the reactor itself would appear to be entirely acceptable. In the long term, obtaining some of these materials could tax known reserves. Again, design ingenuity may circumvent reliance on particularly scarce materials.

5. Fusion systems offer improved thermal efficiencies (through higher-temperature operation and, possibly, direct conversion). If this proves possible, the local problems of heat removal could be alleviated.

In summary, although one cannot be definite in discussing a nonexistent power-generation system, it nevertheless appears likely that fusion reactors, particularly as they evolve, will present a smaller hazard to man and to the environment than any other major power source with the exception of solar power.

FUSION: THE UNCERTAIN CERTAINTY

For the title of the closing section I have borrowed the title of a talk given several years ago by the late Theos Thompson, then a Commissioner of the Atomic Energy Commission. I believe that his title reflects very well the present nature of the fusion quest. At this point, most scientists and engineers who have seriously considered the fusion issue would agree that fusion power is in fact achievable. I believe all will agree to its desirability as an ultimate energy source. The difficulty comes from trying to predict when fusion power can begin to make a real difference in the world energy picture.

We have the realities of the time and effort required to transfer an as-yet-untested scientific and technological development into hardware that must be reliable and that must compete economically with alternatives. Traditionally this is a long process. Nor can it be expedited and accelerated in the same way as the first Apollo moon landing. There, the basic scientific knowledge was entirely in hand, cost was not an issue, and all effort was concentrated on achieving one product—one manned lunar module to the moon and back. Putting together a complete new energy system is quite another matter. This is not to deny that an intensified effort on fusion research will yield a shortened timetable to fusion power—it certainly will. What it does say to me is that, given that intensified effort, there are two areas that will yield the highest probability of early success: (a) consolidating the scientific basis for fusion, including pushing the special technology needed; and (b) cultivating the development of new insights and new inventions that can result in simpler and smaller-scale fusion reactor systems. These systems could be brought on-line with less complexity on a lower buy-in price than any of these conceptual designs proposed so far. I believe that the challenge can be met, and I believe that future generations will thank this generation for rising to meet it, thereby achieving safe and abundant energy for the future of mankind.

ACKNOWLEDGMENT

This work was performed under the auspices of the United States Energy Research and Development Administration, under contract no. W-7405-Eng-48.

Literature Cited

1. Hubbert, M. K. 1975. Survey of world energy resources (1973). In *Perspectives on Energy*, ed. L. C. Ruedisili, M. W. Firebaugh. London: Oxford Univ. Press. 527 pp.
2. Emmett, J. L., Nuckolls, J. H., Wood, L. L. 1974. Fusion power by laser implosion. *Sci. Am.* 230:24–37
3. Post, R. F. 1962. Critical conditions for self-sustaining reactions in the mirror machine. *Nucl. Fusion*, Suppl. Pt. 1:99–123
4. Berkowitz, J. et al. 1958. Cusped geometries. *Proc. 2nd UN Int. Conf. Peaceful Uses At. Energy* 31:171–76
5. Ioffe, M. S., Yushmanov, E. E. 1962. Eksperimental'now issledovanie neustoychivosti plazmy v lovushke s magnitnymi pro"kami. *Nucl. Fusion*, Suppl. Pt. 1:177–82
6. Berkner, K. H. et al. 1974. Performance of LBL 20-kV, 10-A and 50-A neutral beam injectors. *Proc. 2nd Symp. Ion Sources Formation Ion Beams.* Pap. VI-9

7. Jernigan, T. C. et al. 1974. Plasma properties and performance of the 10-cm DuoPIGatron ion source. *Proc. 2nd Symp. Ion Sources Formation Ion Beams.* Pap. VI-9
8. Coensgen, F. C. 1974. Plasma containment in 2XII. *Proc. 5th Conf. Plasma Contr. Nucl. Fusion* 1:323–28
9. Rensink, M. E. 1974. Theoretical studies of plasma confinement in magnetic mirrors. See Ref. 8, 1:311–19
10. Charatis, G. et al. 1974. Experimental study of laser-driven compression of spherical glass shells. See Ref. 8, 2: 317–33
11. Post, R. F., Ribe, F. L. 1974. Fusion reactors as future energy sources. *Science* 186:397–407
12. R&D Associates. 1974. *Fusion Power in Ten Years,* Rep. No. RDA-JTR-4100-002; *Project PACER Final Rep.,* No. RDA-TR-4100-003
13. Moir, R. W. et al. 1975. *Progress on the Conceptual Design of a Mirror Hybrid Fusion-Fission Reactor.* Rep. No. UCRL-51797. Livermore, Calif.: Lawrence Livermore Lab.
14. Holdren, J. P., Fowler, T. K., Post, R. F. 1976. Fusion power and the environment. *Proc. Short Course: Energy and the Environment—Cost-Benefit Analysis.* Georgia Inst. Technol., Atlanta. In press

Copyright 1976. All rights reserved

WASTE MATERIALS ✵11010

Clarence G. Golueke and P. H. McGauhey
Sanitary Engineering Research Laboratory, University of California,
Richmond, California 94804

INTRODUCTION

Status of the Literature

Concern with solid waste per se is not new to the chronicler of public affairs. A Roman signpost unearthed within this century and dating back more than 1500 years warns the citizens, "Take your refuse farther or you will be fined!" This ancient warning underscores the fact that of the three basic problems of refuse management—collection, transportation, and disposal—disposal has been the most imperfectly solved for a very long time and has, therefore, commanded most attention. For that reason, perhaps, the literature dealing with waste materials in the context of energy recovery dates back less than one decade—to the rise of conservation with a special concern for resource recovery, recycling, and similar reuse concepts. Energy appeared conceptually as a logical and useful spinoff of any waste-recycling program but without the simple and obvious merits of a direct recycling of aluminum cans and waste paper.

Published writings identifying waste materials as a possible source of energy to meet a long-predicted but little-heeded national deficiency of energy date back only about two years—to the Middle East oil embargo. Consequently, the literature that critically evaluates waste materials as a source of energy, by analyzing such factors as amount, physical and economic recoverability, and long-term role in the energy picture, is mostly in manuscript (unpublished) form. Published literature on the subject in early 1975 is still confined largely to gross general comparisons between the calorific value of major classes of solid waste and the estimated energy shortage the nation may anticipate, with little regard for the feasibility of recovery. Thus, it tends to fall into the same category as those writings which proclaim the vastness of our wealth of energy in oil shale and coal, at the same time that equally facile pens warn darkly of rationing and other drastic actions needed to scale national energy consumption forever to current petroleum supplies.

A considerable amount of the literature on energy recovery from waste materials is fragmentary or nonspecific—that is, a statement here and a comment there, often by knowledgeable individuals, state the rationale for investigative work and indicate where and how to begin, but offer too little in terms of data or evaluation of fact to justify citation in any review of literature.

As the reviewer moves further back into the literature on burning of refuse, energy loss appears as a secondary concept, as one more bit of evidence in support of some writer's particular viewpoint. Basically the interest is in reduction of the overall cost of disposal and decrease in air pollution potential, dust and paper blown about by wind in the vicinity of the incinerators, visible smoke, concentration of collection vehicles at the incinerator site, and zoning problems that unhinge any rational relationship between incinerator sites and refuse collection routes.

The principal source of literature on the utilization of waste materials is the Environmental Protection Agency (EPA). Most of the research and development, and the process demonstration involving energy from waste materials, is proceeding under grants from the EPA. These result in reports to the Agency, the availability of which is announced monthly in an EPA publication (1). In addition, technical papers based on current and completed studies are published by the investigators themselves in technical journals of such societies as the American Society of Civil Engineers, American Society of Mechanical Engineers, American Institute of Chemical Engineers, American Institute of Agricultural Engineers, etc, and in magazines such as *Compost Science, Resource Recovery, Waste Age, Soil Science,* and others concerned directly with the phenomena of incineration or bioconversion of wastes containing available energy. States with highly organized waste-control programs such as those of Connecticut (2), Pennsylvania (3), Wisconsin (4), California (5), and New Jersey (6) periodically publish accounts of their activities, as do research units of several universities, for example University of California (7–10), Purdue University (11), and the University of Illinois (12).

From the foregoing and other significant literature are drawn references appropriate to each specific subject discussed in the sections that follow. However, the reader will find that a considerable fraction of the material presented in this review is necessarily drawn from personal evaluation of the subject of energy from waste materials by the authors on the basis of their own experience. Consequently, the results may reflect more of the personal philosophy and biases of the authors than they would wish were it their task to analyze a field in which a rich literature abounds.

Finally, in citing literature to inform the reader of the status of any approach to energy recovery from waste materials, we can give no assurance that the author cited relative to that literature did not draw his inspiration from the work of some other source to which applause more rightfully belongs. It is hoped, however, that the reader may acquire an accurate evaluation of the current status and prospects for energy production from waste materials by each of a variety of processes, and that the subject may be revealed in an accurate perspective.

Transience of Definition

If one studies the general literature on solid wastes in any detail, he cannot fail to be impressed by the inordinate amount of human energy that has gone into attempts to define solid wastes precisely. It is as though a resolution of the vexing problems of collection and disposal of the ever-changing mixture comprising man's refuse depended upon his ability to analyze it with a precision analogous to the neutralization reaction in a pure acid/base solution. Obviously in a world of changing technology, rooted both in scientific discovery and in social custom and national and world economy, wastes can be fixed neither in amount nor in composition. Nor can competition between energy and higher resource values of the fiber content of solid waste be made a constant. This does not mean, however, that the absurd will not be tried and reported in the literature. If the past is any indicator, one school of thought will hold that a fixed fraction of primary power should come from waste materials, with petroleum stored underground serving to take up the variable differences between firm and total power. In contrast, the realist will contend that wastes should be put to their highest use. Recycle paper back to paper if the result is economically preferable to converting it to energy; the important thing is that nothing be wasted. Thus, conceptually, power production becomes the variable that makes for good conservation, rather than a procedure that renders America independent of the Middle East. Whether or not the whole idea of energy from waste materials is a mere "tempest in a teapot" depends upon the volume of economically reclaimable energy in comparison to energy needs.

In summarizing the matter of definition, it seems useless for us to go back to the era before World War II. In about 1950, society began an especial effort to regain control of the problem of wastes, which had grown immeasurably by the exigencies of war and the impaction of cities formerly separated by open space in which refuse dumps and hog farms were tolerated, if not particularly applauded. In those days municipal refuse was essentially the only category of solid wastes receiving public attention; hence waste materials were usually defined as components of domestic (household) wastes, plus other debris collected in urban areas.

Typically, municipal refuse was defined (13, 14) as pounds per capita per day, or percentage of daily total of:

garbage (food residues)	commercial swill	plastics
paper	miscellaneous rubbish	street sweepings
textiles	dirt	dead animals
glass	metal (ferrous)	ashes
wood	metal (nonferrous)	broken bricks and concrete
yard trimmings	rubber and leather	

With some variations, the foregoing is defined in the literature as *municipal refuse* or, minus certain inorganic fractions, as *domestic refuse*. Popularly and categorically the whole assemblage was called "garbage" by the public and by the media. The reporter for the local newspaper has been especially prone to class all municipal refuse as garbage. Sometimes this was a matter of ignorance shared with the public. Sometimes it reflected resentment of having to take a city hall assignment loathsome to his fellow journalists. But many times he misused the term garbage because its repulsiveness to the public might be effective in shocking citizens into action to improve collection service or into investing money in abating the nuisance of a burning, odorous, vermin-infested city dump. One unfortunate effect of the public's disorientation to words such as garbage has been to place the refuse handler at the bottom of the social scale (or caste); to make the refuse-handling division of the city's public works department a "Siberia" for its least talented or least popular management personnel; and so to complicate further the problem of refuse management for which no one wants to invest money.

After World War II, and especially within the past decade, the term garbage or less objectionable words no longer describe society's waste material. Municipal refuse continued at most to include residues from the household, commercial establishments, street maintenance, and light industry, but industrial wastes became a whole spectrum of residues not manageable by municipal routines. Housing expansion, urban renewal, and numerous construction programs added another class of wastes best described as *demolition debris*. Agricultural wastes from food processing and manures from penned-up animals in dairies, feed lots, and egg and poultry production became a special class of considerable volume. Water pollution laws forced changes in methods of sewage sludge disposal. Air pollution control agencies became concerned over the organic matter left in fields and burned. Conservationists, environmentalists, foresters, and wood-using industries became concerned over the percentage of any harvested tree that rotted in the forest or caught fire and burned. Aesthetic considerations, heightened by at least two major disasters from mine tailings, excited those who saw the results of large-scale strip mining or long-term operation of deep shafts. They too identified another type of wastes.

The main literature of solid waste management abounds with instances such as those described above. However, when disposal other than dumping was employed, an additional set of limited definitions has been employed. For example:

| combustible | compostable | digestible |
| noncombustible | noncompostable | nondigestible |

It is in such literature as is cited in (15, 16) that the first notes on energy transfer appear. In addition to the numerous sources cited in (1), the literature of the American Society of Mechanical Engineers and countless other sources deal with incineration. Although the objective of such literature is to define the economies, the hardware, and the air pollution/fly ash problems of refuse incineration, the reader interested in energy production from wastes in the context of national energy needs will find reading worthwhile—if only to bring a degree of sobriety to his own thinking about the "irrationality" of wasting heat from refuse incinerators. References of more direct pertinence to the chapter's subject are introduced in a later section.

The principles of bio-incineration to reclaim energy as fertilizer and soil conditioners are fully detailed in such literature as (17, 18). Energy recovery by gas production of municipal refuse is related in early news items and reports (8, 19–21). As is the case with incineration, literature directly concerned with alleviating a national energy shortage is cited in an appropriate subsequent section.

With the foregoing expansion of concern for solid waste now classified by type of origin, the old labels no longer applied. Not even the greenest reporter or the most excitable environmentalist could gain credibility by applying the term garbage to the total wastes that could be identified in such a broad list as:

municipal wastes
industrial wastes
demolition debris
animal manures

agricultural wastes
mining wastes
junk and old automobiles
litter

A search for a generic term by regulatory agencies and engineers in recent years turned up several possible new definitions for solid wastes:

1. The entire mass of residues, unwanted and obsolete materials or products, and refuse which for disposal purposes or other wastes-management purposes must be handled in the solid state (regardless of whether or not disposal is currently practical).
2. Residues of resource uses.
3. Resources we have not yet learned to use.

The last two, both of which are self-explanatory and true enough, had a certain amount of appeal but were too esoteric for general use. However, as concepts they did help to further the concept of concentration of wastes for management purposes. Especially did they help draw attention to the fact that many materials existing as scrap, empty metal cans, old tires, agricultural machinery, old automobiles, were so widely scattered under such conditions and circumstances that they were irretrievably lost to man, and, moreover, that the total so scattered is significant.

The term *solid wastes* is both generic and descriptive and has come into common use.

ENERGY IN WASTE MATERIALS

Amount of Wastes

Various estimates of the amount of solid wastes have been published. For many years the common estimates of the domestic waste component of solid wastes was 2.5 lb/capita-day, including 0.5 lb garbage. The advent of the garbage grinder, the introduction of synthetic fabrics, and the packaging of frozen foods were among the developments that disturbed these historic values. Increasing environmental standards likewise had an effect, and for some time the annual estimate of municipal refuse varied from 3 to 6 lb/capita-day. At present the total is estimated by EPA (22) to average 3.32 lb/capita-day. From this same source estimates of wastes for 1971 are the following:

Municipal waste . . . . 250×10^6 tons/day
(Domestic fraction of municipal wastes . . . . 175×10^6 tons/day)
Industrial wastes . . . . 110×10^6 tons/day
Agricultural wastes . . . . 2280×10^6 tons/day
Mining . . . . 1700 tons/day

Availability of Energy

The writer concerned with estimating the amount of energy available in various amounts of waste, as tabulated in the preceding section, is tempted to estimate the organic content of each type of waste and to multiply each by some appropriate Btu content/unit for the type of waste concerned. Such an approach is misleading for a number of reasons, including the low-density scatter of agricultural crop wastes and the low assay of burnable material in some mining and demolition wastes.

The fraction of municipal wastes that is burnable fiber (paper, plastics, rubber, etc) has been estimated at 65% (20, 23–26). The National Center for Resource Recovery (NCRR) reports (24) that a plant processing 750 tons of municipal waste per day consumes 52,000 kW-hr and is capable of recovering 250,000 kW-hr energy, 24 tons of paper, plus variable amounts of glass and metals. Moreover, it is possible to produce 127×10^6 Btu/hr in addition to the electricity.

Data on industrial wastes that can be related to the tonnage of wastes shown in the preceding section are not available. Presumably the organic industrial wastes from the food-processing industry are included in the agricultural waste category, as are agricultural wastes presently left in the field.

Various experiments have been run on the conversion of municipal refuse and animal wastes to methane by anaerobic digestion. Klein (27) found municipal refuse readily degradable by anaerobic digestion, producing 6–9 cu ft gas/lb of volatile solids introduced after a period of acclimation of the organisms.

Pennsylvania State University has estimated (26) that if all of the burnable solid waste in the United States were incinerated with 40% heat recovery, it would generate enough Btu's to heat 2.4 million homes a year in a typical midwest climate. Similarly, the EPA (22) has estimated that if wastes from all Standard Metropolitan Statistical Areas in the United States (70% of population) were collected, about one quadrillion Btu's could be generated. This is estimated as 1.5% of the nation's 1970 consumption and is equivalent to the energy in 400,000–500,000 barrels of oil per day. Moreover, it is equal to the total national needs for commercial and residential lighting.

Obviously these figures do not all reflect the energy input necessary to provide and operate the hardware necessary to produce the energy. Nor do they reflect the economics and energy ramifications involved in dedicating all organic fiber to energy production. For example, suppose that paper recycling were practiced to a much greater degree, releasing the present tree farms used to maintain tree growth for energy. If this were done by the federal government operating as effectively as private industry, the effect of the trade-off of paper for standing trees might be zero or only slightly positive. On the other hand, if the trees were left unharvested and eventually decomposed, a very great loss in usable energy would result. Moreover, the public expense of maintaining forest land for no other purpose than recreation would result in a very significant negative value in terms of energy production.

According to estimates reported by Anderson (28), approximately 880 million tons (dry weight) of organic waste were generated in the United States in 1971. A much lower total, namely 571 million tons, is reported by the International Research and Technology Corporation (29). The true number probably lies somewhere between the two extremes. At 16 million Btu/ton dry matter, the energy content of the total organic wastes would amount to about 9.1×10^{15} to 1.4×10^{16} Btu. Of this total, the larger amounts are in the form of animal manure (200 million tons) and crop residues (160–390 million tons). Estimates of the total energy content of urban wastes range from 8.3×10^{14} Btu (30) to 1.08×10^{15} Btu (31). (The difference between estimates is due to differing estimates of the total waste produced.) The significance of waste on energy sources to meet the national total energy needs is shown by comparing the estimates listed with the estimate of the total energy consumption, which in 1973 amounted to 75×10^{15} Btu (32). The lower energy estimate for urban refuse is nearly one third of the energy expected from the Alaskan pipeline.

While the amount of energy contained in the organic waste stream seems, and is, impressive, only a fraction of it is available for conversion to practical uses, as we stated earlier. For example, the largest fraction of the total potential energy is in the form of crop residues. A considerable amount of optimism is needed to convince one that it would be economically and energetically feasible to convert even a small fraction of crop residues to usable energy. Yet, projections for such conversion have been seriously made (23, 31). Yeck (31) states that 23 million tons/yr of crop residues are already being collected and that if this amount were added to that of collected animal wastes and logging and wood-manufacturing residues, the total amount would be 54 million tons/yr. At 5000 Btu/lb, the energy content of the presently collected agricultural wastes would add up to 5.4×10^{14} Btu/yr.

Proponents for the conversion of these scattered residues to energy base their hopes for the fruition of their dreams on the establishment of small-scale energy conversion facilities located near the residues. The output from these facilities could be either used locally or fed into a gridwork to be channeled to large-scale consumers of energy (23).

The feasibility of using manures from feedlots and other confined animal operations as a source of energy rests on the fact that the wastes are concentrated at one location and are readily collected. Moreover, the operators of such facilities are confronted with the need to dispose of the wastes in a manner that is at once aesthetically acceptable and safe in terms of public health and yet economically feasible. Methods exist for meeting both requirements. The energy yield can be used in powering not only the disposal operation but also the entire animal rearing facility, thus reducing the dependence on external sources. The imposition of increasingly stricter standards by public regulatory agencies plus the rising costs of energy combine to place these methods within the bounds of economic feasibility.

The feasibility of utilizing municipal wastes as an energy source is more likely and substantial progress has been made in the attainment of that goal in recent years.

TECHNOLOGY

Types of Energy Conversion Systems

The technology available for converting wastes into energy can be subdivided into a number of categories based on the principles involved in the design and application of each. A general classification could be: (*a*) thermal processes; (*b*) biological or fermentation processes; and (*c*) solar energy processes. Thermal processes can be subdivided into those which involve either combustion or pyrolysis, or a combination of the two, or wet oxidation. Biological or fermentation processes can be subdivided into those designed to produce a combustible gas (e.g. CH_4) by anaerobic digestion, or a combustible liquid (e.g. ethanol) by a series of aerobic and anaerobic fermentations. Systems involving solar energy do so indirectly in that wastes are used as a nutrient source for green plants, which in turn are digested to produce methane. On the basis of the type of plant cultivated, solar energy systems can be divided into algal production systems and higher plant systems. The overall efficiencies of energy recapture by the systems are listed in Table 1, and representative capital and annual costs in Table 2. An excellent survey of thermal energy recovery systems can be found in the feasibility study report "Energy Recovery Systems" (33).

Thermal Systems

COMBUSTION The likely candidates for thermal conversion into energy by way of combustion are those which exist in a relatively dry state. This normally would rule out most cannery wastes and manures. On the other hand, it makes municipal wastes and certain industrial wastes suitable substrates. This fact has been an important factor in the impressive amount of work done with the utilization of municipal wastes as an energy source.

The heat value per dry ton of municipal refuse delivered to the disposal site is on the order of 10 million Btu's. As a source of heat energy, the ton of refuse can be compared to 70 gal of fuel oil or 800 lb of coal. Combusting a ton of refuse can produce enough heat to generate 6500 lb of steam—compared to 7000 lb by burning the 70 gal of fuel oil. It has been estimated that the incineration of all of the municipal wastes in the United States could produce enough energy to satisfy 2.5% of the nation's prime energy requirements, or 10% of the total energy needed to heat or cool its buildings (34).

The approaches in the development of combustion systems for municipal wastes have followed three general lines: (*a*) conventional incineration; (*b*) utilization of

Table 1 Process efficiencies (33)

Process System	Overall Recapture Efficiency (%) of Primary Product		
	Electricity	Steam	Gas
Power boiler	27	77	
Incinerator		67	
Monsanto Landgard		54	
Pyrotek pyrolysis			39
Union Carbide Purox			64
Bioconversion (anaerobic digestion)			33
+incineration of organic residues (20)		63	

Table 2 Costs of systems (33)

Process System	Capital ($1000)[a]	Annual ($1000)[b]
Pyrolysis	8,500	1,692
Steam generation with processed refuse	7,400	1,388
Steam generation with unprocessed refuse	6,900	1,283
Electrical generation with processed refuse	10,000	1,776
Fuel preparation	3,000	925

[a] Based on 275 tons unsorted refuse per day.
[b] 1978 constant dollars.

suitably processed wastes as a fuel in a conventional public utility boiler; and (c) direct utilization of hot combustion gases to drive a gas turbine to generate electricity. As of late 1974, nine refuse-to-energy facilities have been in operation in the United States and Canada, four were under construction, and twenty others were in stages of planning ranging from "under study" to the awarding of contracts (35).

Incineration Incineration has always been regarded as one of the two main options in the disposal of municipal wastes. (Landfill is the other option.) Moreover, the idea of using the heat generated in the incineration process to generate steam and eventually electricity ranks historically with the concept of incineration itself. For example, in a text published in 1901 (36), reference is made to incineration facilities in Great Britain that included the installation of a boiler for the production of steam. A 15-ton/hr incinerator equipped with a boiler to generate steam for heating purposes was operated in Palo Alto, California from 1910 to 1915 (37). The rationale for the conversion of heat energy to steam and then to electricity prior to the time when energy conservation became significant in the public mind was that since the temperature of the stack gases had to be reduced from as much as 1800°F to 500°F to permit air pollution control devices to function, part of the cost of so doing could be recouped by converting the heat energy otherwise wasted to a usable form. This energy could be recovered by inserting a boiler or its equivalent between the combustion chamber and the stack. The boiler serves as the heat-exchange device. Overall efficiency of energy recovery in terms of steam produced is about 67% (33).

Generation of power with the waste heat in an incinerator is by no means without its disadvantages. A major problem is that of corrosion, which may be one or all of three types: (*a*) corrosion set off by a reducing environment; (*b*) halogen corrosion; and (*c*) low-temperature corrosion. Tube fouling is another difficulty for which no easy remedy is available (38). It results from the deposition of slag and ash on heat-transfer surfaces. It can be minimized by suitable spacing of the heat-transfer surfaces and correct use of boiler cleaning equipment.

The practice of generation of steam and electricity as part of an incineration operation has been in vogue in Europe, in the larger cities especially, since the early

1960s (39, 40). Few installations have been built in the western hemisphere, although within the last few years interest has begun to develop in the United States. A major deterrent to its widespread usage, especially on the West Coast, springs more from the incineration aspect rather than the power-generation phase. The deterrent is the fear that impossible-to-meet air quality standards might force the closure of the incinerator or require endless expensive process additions. Probably the largest installation in the United States is in Chicago, Ill. (41). Among the other large plants in operation are those in Montreal and Quebec in Canada (42); Harrisburg, Pa.; Braintree, Maine; US Navy Public Works Center in Norfolk, Va.; the thermal transfer plant in Nashville, Tenn. (35); and East Hamilton, Ontario (43).

Refuse as a supplementary fuel One of the more promising approaches to recovering energy from solid wastes, particularly urban refuse and a variety of industrial wastes, is through the use of the fiber in the waste stream either as a supplementary or as the sole fuel in a conventional utility to generate power or steam. A very excellent series of papers on this subject has been published in the magazine *Power* in the form of a special report consisting of four parts (35). Another excellent source of information is the EPA publication *Energy from Waste* (25). Recourse to these reports is recommended for one especially interested in this phase of the subject, inasmuch as it can only be touched upon in this chapter.

The use of refuse as a fuel is an avenue for cooperation between the municipality and private enterprise because it provides a means of disposing of a portion of the wastes from the former. Until the specter of a fuel shortage began to face private enterprise, it was not inclined to become involved in the garbage-disposal business. However, conditions have arisen which have changed the outlook of both the public and the private sectors. In many cases it has become impossibly difficult for municipalities to operate incinerators because of increasingly rigid environmentally oriented regulations and the unavailability of needed highly skilled personnel at the low pay scales traditionally characteristic of municipalities. An incentive to cooperation in the form of the municipality's providing fuel to a utility is the possibility of cheaper financing through the city and the possibility of attracting federal assistance to the enterprise.

The first large-scale experiments on the use of refuse as a supplementary fuel took place in St. Louis, Missouri at the Union Electric Company's Merrimac plant (44, 45). The method of utilization involves burning milled refuse along with powdered coal in a suspension-fired boiler. The Merrimac boiler is tangentially fitted and one port is installed in each corner of the boiler, between the two middle coal burners. Combustion controls automatically regulate the rate of coal firing to maintain heat output when refuse is being fed into the unit. Experience gained at the St. Louis plant indicates that refuse firing rates equal to about 20% of the total heat input appear to be suitable for coal-fired units, and perhaps about 10% for oil-burning plants. The reason for the lower fraction with oil burning is that not enough ash is produced to absorb corrosive elements in the flue gas. A slight increase in boiler emissions may result, but no final conclusions have been reached as yet. Small amounts of glass and grit not removed in processing the refuse result in considerable abrasion at bends in the pneumatic conveyer systems. Not unexpectedly, the volume of bottom ash is increased considerably. The amount of bottom ash could be decreased by milling the refuse to a smaller particle size or by combining suspension firing with a bottom grate. The latter approach has been followed at the Hamilton, Ontario facility (16).

Before refuse can be burned in suspension, it must be suitably processed. Processing consists of milling the refuse and then air-classifying the milled material to remove noncombustibles (heavy fraction) from the combustibles (light fraction). A variety of size-reduction equipment is available for milling refuse, but all types are subject to extensive wear and tear because of the abrasive nature of refuse, and all require a considerable amount of energy to operate (10). Ideally, the processing operation should be made a part of an overall resource-recovery activity. This is being done at Ames, Iowa where a $5.5 million facility is being constructed and will convert the combustible fraction of about 1000 tons refuse/week into a supplementary fuel for three existing boilers in the city's municipal power plant (35).

The onset of the energy shortage as manifested by higher fuel costs has given an impetus to the practice of recovering energy through the combustion of process wastes in industry. The impetus is amplified by: (a) the difficulty of finding land for waste disposal and the increasing severity of regulations on such disposal; (b) the need to provide air pollution control devices on existing incinerators; and (c) the low sulfur content of process wastes. The need to provide air pollution control devices necessitates, as stated earlier, a lowering of stack gas temperatures to enable the control equipment to function. The heat lost in lowering the temperature can be utilized to generate steam. The low sulfur content of most process wastes makes it possible to burn them along with more readily available high-sulfur fossil fuels and thus bring the overall sulfur concentration of the stack emissions within permissible limits. A wide variety of by-product fuels can be burned with existing boilers and ancillary equipment. Of especial usefulness are the methods used in the petroleum and paper and sugar making industries to burn their wastes. Examples of the energy value per pound of certain industrial solid wastes are bagasse, 3600–6500 Btu; bark, 4500–5200 Btu; coffee grounds, 4900–6500 Btu; rice hulls, 5200–6500 Btu; and corn cobs, 8000–8300 Btu (35).

Direct utilization of hot gases Theoretically, high energy-recovery efficiencies in terms of electric power production can be attained if the boiler is bypassed and, instead, the hot gases are used directly to drive a gas turbine to produce electricity. The concept has found its manifestation in the form of the CPU-400 system developed at Menlo Park, California by the Combustion Power Corporation (46). In the system, refuse is milled and its organic content removed by air classification. The milled organic matter is burned in a fluidized bed reactor. The hot gases from the combustion chamber are cleaned and then used to drive a gas turbine.

Considerable difficulty is being encountered in bringing the system up to the degree of reliability required for municipal operation. The first problem is the loss of medium from the fluidized bed reactor. The latest problem, perhaps an insuperable one, is plating of the turbine blades with submicron particles, principally aluminum, in the combustion gases. This plating can result in failure of the turbine after only a few hours of operation. Unfortunately, present technology can offer no remedy for the problem.

PYROLYSIS AND PYROLYSIS-COMBUSTION SYSTEMS Among the several publications on pyrolysis that could be recommended as a source of information on the subject are "Pyrolysis of Municipal Solid Waste" by S. J. Levy (47) and "Pyrolysis/ State of the Art" by N. J. Weinstein and Charanjit Rai (48). Much of the information given in this section comes from the two papers.

In the strict sense of the term, pyrolysis is the physical and chemical decomposition of organic matter accomplished by the application of heat in the absence of oxygen. Many systems have been developed in recent years that have been labeled "pyrolysis," but are in reality hybrid systems because they involve the introduction of oxygen (combustion) in one or another phase of the operation. True pyrolysis differs further from incineration or combustion by being endothermic, i.e. requiring the application of heat. In the hybrid processes the heat is supplied by the combustion of part of the combustible pyrolysis products. Another difference between pyrolysis and combustion is in their respective products. True pyrolysis results in the production of a complex mixture of combustible gases, liquids, and solid residues. On the other hand, combustion produces primarily CO_2 and water. The hybrid processes produce products of a nature that ranges somewhere between the extremes of those from true pyrolysis and those from combustion.

About 10–12 systems involving pyrolysis in one form or another were under development at the time of this writing. The five systems most advanced in terms of development at this time are: 1. an oil pyrolysis process promoted by the Garrett Research and Development Corporation (Flash Pyrolysis system); 2. a process involving gas production developed by the Union Carbide Corporation (Purox system); 3. a gas pyrolysis system developed by Monsanto Enviro-Chem Systems (Landgard system); 4. a process involving gas pyrolysis developed by the Carborundum Company (Torrax system); and 5. the Pyroteck process developed by Pyrotek Corporation. Of these, only one, the Landgard system, is near commercial operation (35, 47).

In the Garrett process, the organic fraction of refuse is finely milled and then subjected to a temperature of 1000°F in the complete absence of O_2. The two principal problems with the process are the required fineness to which the organic wastes must be ground and the high viscosity and corrosiveness of the oils. On the average the oils must be heated to 160°F in order to flow. They also have a lower Btu value than that of petroleum oils—on the order of 10,500 Btu/lb, compared to 18,200 Btu/lb of typical No. 6 fuel oil. Incidentally, the problem of a product having a high viscosity besets other systems in which organic wastes are converted to oil (49). Annual operating and maintenance costs including amortization are estimated to be on the order of $17/ton refuse (1974 dollars), and net costs, $11/ton (47).

As with many processes, the Monsanto gas pyrolysis process was originally designed mainly as a waste-treatment process and, in this case, as an alternative to incineration. Its application to energy recovery comes from the fact that it can be modified to produce steam or low-Btu gas. The reaction temperature is approximately 1400°F, and gas and small quantities of a mixture of char and inorganic solids are produced. The oxygen used in the partial oxidation phase comes from air. An important feature is that although shredding is required, pre-sortment of inorganics from organics is not. The overall efficiency of the process is 54% (33). About 5000 lb of steam (6 million Btu) are produced per ton of organic refuse charged into the system. The total cost per ton of throughput is $13.15 (1974 dollars), and the net operating cost, $0.02 (30).

Neither front-end separation nor shredding is required in the Union Carbide system. The process involves the application of temperatures as high as 2600–3000°F. These high temperatures are attained by using pure oxygen in the partial oxidation phase. The primary product of the system is gas (47). About 22 cu ft of 300-Btu gas are produced per ton of waste processed. The overall efficiency is 64%. The total operating costs, inclusive of amortization and based on 1974 dollars, would be on the order of $14/ton refuse processed, and the net cost, $5.87/ton (47).

The Torrax system developed by the Carborundum Company resembles the Union Carbide Purox system, except that air preheated to 2000°F is used instead of pure oxygen. The preheated air is provided by burning natural gas in a ceramic heat exchanger. A 75-ton/day pilot plant is being evaluated in Erie County, New York. Gas produced in the process has a Btu of only 170/scf. The overall efficiency is about 65%. Capital cost/ton of capacity is about $11,000 (1974 dollars).

The Pyrotek system involves the use of an indirectly fired reactor. Refuse is continuously fed through the reactor on a moving grate. Heat from a separate combustion chamber adjacent to the reactor drives the process. A low-Btu gas (375 Btu/scf) is produced, as are char, oils, and tars. The overall efficiency is on the order of 39%. Gas produced per ton amounts to about 10,670 cu ft. Approximately 0.004 ton of oils and 0.3 ton of solids are produced per ton of refuse charged into the unit (33).

In evaluating the pyrolysis or pyrolysis-related systems described in the preceding paragraphs, one should remember that a primary reason for selecting pyrolysis is to produce a storable, transportable fuel that can be utilized as a substitute for fossil fuels without the need for special handling facilities. The Union Carbide and the Garrett systems come closest to attaining this objective at present. Gas produced in the Union Carbide system can be used in conventional furnaces with only minor modifications. Short-term storage and limited transportation are feasible. The Garrett system's liquid fuel can be stored for only about two weeks; primarily because of its corrosive properties, special facilities are needed to use it. The difficulty in handling the Garrett system's product plus the energy and dollar costs to mill the wastes to the required degree of fineness combine to detract materially from the attractiveness of the process.

The Monsanto system has the advantage of being the furthest developed of the pyrolysis-related systems because of the full-size 91,000-ton/day plant soon to be in operation near Baltimore. While it does not meet the objective of producing a storable transportable fuel, it does have a significant energy recovery potential if it can be located near a customer for the steam produced by it.

Biological Systems

METHANE-PRODUCING SYSTEMS Methane is produced through the anaerobic digestion of organic wastes. Anaerobic digestion of wastes is a biological process whereby facultative and obligate anaerobes decompose organic wastes in the absence of molecular oxygen. Inasmuch as the methane-forming bacteria are obligate anaerobes, the fermentation must be carried on in the complete absence of atmospheric oxygen. The process proceeds in two stages—the acid production and the methane production stages—which are usually in dynamic equilibrium and proceed simultaneously in a continuous culture. The acids formed in the first stage serve as the substrate for the methane formers. Gases formed and their percentages are as follows: methane (CH_4), 54–70%; carbon dioxide (CO_2), 27–45%; nitrogen (N_2), 0.6–3%; hydrogen (H_2), $\simeq 1\%$; carbon monoxide (CO), $\simeq 1\%$; and hydrogen sulfide (H_2S), a trace. Almost all of the fuel value of the gas comes from its methane content. Pure methane has a heat value of 994.7 Btu/scf.

The anaerobic digestion of wastes is regarded today as one means of offsetting the increasing shortage of natural gas. According to McCarty (50) digestion of sludge resulting from the treatment of wastewaters produced by the population of the United States could provide some 70 billion cu ft of methane/yr. If municipal refuse were added to the wastes, another 2000 billion cu ft of gas would be theoretically possible. If, in addition, all animal wastes and crop residues were fermented, the resulting methane would double and add up to about 4000 billion cu ft/yr or about 20% of current natural gas consumption (50). Although potential is great, its realization is a long way off, and before it is attained many obstacles must be overcome. As will be seen later, its greatest utility probably rests in the treatment of concentrated animal wastes. However, anaerobic digestion of municipal refuse has its staunch proponents (12).

Anaerobic digestion has been utilized in the treatment of human wastes for well over 100 years, and the CH_4 produced by the digestion of the wastes has been put to useful purposes for almost an equal length of time. In the literature available to the authors of this review, one of the earliest reports on the utilization of gas from the digestion of garbage is that of an operation in progress in 1941 in Goshen, Indiana (27). According to the report, all of the village's garbage was ground and digested along with sewage sludge. The gas produced was used to supply the electrical needs of the community with the exception of street lights. A somewhat larger operation was reported in 1953 for Richmond, Indiana, where the plant had been in operation since 1951 (20). Again, garbage (207 tons/month) and sewage sludge were digested together. The savings in energy cost for the town were substantial.

The utilization of digestion to treat animal manures has a history almost as long as that for treating human wastes. Since the early 1900s, it has received especial attention in India (51) because it is a means of obtaining a useful fuel (CH_4) and yet conserves the plant nutrients in the manure.

At present, no consensus exists with respect to optimal digester design, or even optimal digester operation. Inasmuch as the digester structure is the most costly aspect of digestion, the motivation to simplify digester design is strong. Successful developmental work on increasing the permissible loading rate to digesters has resulted in lower volume requirements and thereby a reduction in capital costs. An important factor that widens the scope of application of digestion is that it has been reported to be rather insensitive to scale (52). The economics of most of the other energy processes respond very sharply to the scale factor, with the cost curve flattening only at productions well above 500 tons/day.

Digestion of organic refuse One of the first research institutions to investigate the feasibility of digestion of the organic fraction of municipal refuse was the Sanitary Engineering Research Laboratory (SERL) of the University of California, Berkeley (8, 14, 21). At the time of the study, the present concern about an energy shortage had not as yet arisen, and consequently, the principal interest was in assessing the feasibility of digestion as a treatment process. The study demonstrated that the organic fraction could indeed be digested to produce CH_4 at a rate and amount comparable to that from digestion of sewage sludge. Gas production per pound of dry solids ranged from 3 to 5 cu ft, of which 55–65% was CH_4. The study also brought to light an important operational problem, namely the excessive formation of a scum layer and the consequent need for mechanical mixing. (Another drawback in digestion of refuse is the need to slurry the material to 5–10% solids in water. This implies water usage and pollution.)

At a later date, when the awareness of the national energy shortage had become acute, interest in the digestion of refuse as an energy recovery method became pronounced. Perhaps the leading proponent of the approach is J. T. Pfeffer of the University of Illinois (12, 53). In a recent paper he proposes that, in addition to utilizing the CH_4 as an energy source, the organic residue (sludge) be incinerated to produce steam. By so doing, he takes cognizance of the substantial amount of energy that would otherwise remain unreclaimed in the residue. [Volatile (i.e. combustible) solids reduction ranges to a maximum of 62%.] In a mathematical simulation of such a combined system (gas production plus incineration of residue) involving a plant processing 908 metric tons/day, he concluded that the capital costs would be $14,289,000 (1973 dollars). It would produce 3905 m^3 methane/hr. With a selling price of CH_4 at $3.53/100 m^3, the system would operate at zero profit if the dump fee for refuse disposal were $5.34/metric ton. Sale of the steam from incineration of the sludge would reduce the dump fee by about $1.00. As shown in Table 1, if only methane were recovered, the efficiency of energy recovery would be about 33%. It would be increased to 63.4% if the steam from incinerating the sludge could be sold.

Gas from landfills While on the subject of producing CH_4 by way of anaerobic digestion of organic refuse, we should mention an ancillary development, namely tapping landfills for the CH_4 formed in them. Organic matter deposited in a landfill decomposes largely by way of anaerobic digestion with a resultant formation and accumulation of CH_4. In the past, the CH_4 accumulation was regarded solely as a hazard to be encountered in the use of the completed fill. Within the past year or two, the idea of tapping potential stores of CH_4 in the fills has materialized in the form of exploratory investigations. One such investigation is being conducted in Los Angeles County, California (54). As with many predictions by enthusiastic proponents of favorite energy schemes, one should be very conservative in accepting the numbers extrapolated for landfill gas stores. In this case, the investigators base their enthusiasm on the huge amounts of wastes put in landfills in Los Angeles County—about 5 million tons solid wastes/yr. They hope to draw gas from the landfills in quantities that would represent an energy recovery equal to 10–15% of that available from burning wastes. The gas is collected by sinking extraction wells into the fill and connecting the wells to a suction blower. A fill presently being investigated contains about 6 million tons of material and is expected to yield 600 million scf of 500–525 Btu gas/yr over a 10-yr period.

Digestion of animal manures Anaerobic digestion, as stated earlier, holds greatest promise in the recovery of energy from animal manures because it seems also to be the most economical and yet environmentally satisfactory method of treating such wastes, especially those from confined rearing operations. Manures are readily decomposable and in a state readily amenable to being slurried.

Amounts of average gas production to be expected from the digestion of manures from representative livestock are listed in Table 3 [data from Taiganides et al (55)]. The amount of gas produced per unit amount of manure is significantly influenced by the food intake of the animal. Thus gas production per pound of manure from animals raised in the United States is greater than that from animals raised in India.

At present, the number of anaerobic digestion installations is increasing on farms and feedlots in the United States. Gas produced is used to supply at least part of the energy requirements of the farms and feedlots (56). The largest agricultural installation reported thus far is one being constructed by America's largest beef cattle feed operator, Monfort of Colorado, Inc. The plant will cost $4 million (57).

ETHANOL-PRODUCING SYSTEMS Systems in which the end product is ethanol may also be classified as cellulose hydrolysis systems because their usual substrate is cellulose (e.g. paper) and the cellulose is hydrolyzed to glucose either enzymatically (i.e. biologically) or chemically. (Lignin may be included and perhaps wood sugars.) The glucose, in turn, serves as a substrate for yeast, which ferment it to produce ethanol as one of the end products (58). The ethanol is used as a fuel. Before the

Table 3 Gas produced from manure of humans and typical livestock (55)

Animal	Gas Production (cu ft/day)	CH_4 (%)	Heat Value (Btu/day)	Digester Size (cu ft/animal)
Human	1.3	66–75	900	1.00–4.00
Swine	6.3	55–75	3,600	1.00
Dairy cow	42.0	60–80	25,000	37.00
Poultry	0.4	60–80	250	0.37

advent of the great interest in energy, the glucose was made to serve as a substrate for the yeast *Torula,* which was then utilized as a livestock feedstuff (59–61). Reference to such feedstuff production systems can serve as a useful source of information with respect to energy production because steps prior to the actual utilization of the glucose are common to both types of systems.

Thus far, research on energy production by way of cellulose hydrolysis is almost entirely on a laboratory research scale. Two of the research centers currently engaged in such research are the US Army Natick Laboratories (58) and the Lawrence Berkeley Laboratory at the University of California, Berkeley (62).

In the Natick Laboratories, sorted cellulose (paper) is milled very finely. The milled wastes are then mixed with an enzyme (cellulase) culture fluid in a reactor. The enzyme hydrolyzes the cellulose to glucose and a crude glucose syrup results. The enzymes are produced by a strain of fungi selected on the basis of its ability to synthesize cellulase. The glucose is fermented to ethanol or used as a substrate for single-cell yeast (i.e. protein) production. According to Natick estimates, hydrolyzing one ton of waste paper will produce a half-ton of glucose, and fermenting the glucose will produce 68 gallons of ethanol.

Wilke & Mitra (62) estimate a 34% conversion efficiency (based on heat of combustion of feed) in the waste-to-sugar step, and a 27% efficiency in the overall waste-to-ethanol process. According to their cost analysis (1973 dollars), producing the glucose would come to about 1 ¢/lb and a total of 31.96 ¢ to produce a gallon of ethanol (83% yield from glucose). A criticism of this cost analysis is that it is unrealistically low with respect to preparation and handling of materials.

Solar Energy Waste Conversion Systems

The role of wastes in solar energy conversion systems is to serve as a nutrient source in the culture of photosynthetic organisms. Through photosynthesis the photosynthetic organisms convert visible light energy into plant cellular material. The energy fixed as plant cellular material is converted to the chemical energy of CH_4 by way of anaerobic digestion. The type of photosynthetic organisms employed in the process can range from single-cell algae to higher plants. Inasmuch as the subject of fixation of solar energy is covered in another chapter, reference at this point is limited to an algal conversion system developed by Oswald and Golueke in the 1950s (63, 64). In their system, single-celled algae are grown on wastewaters.

Table 4 Summary of system sizes and cost evaluations[a]

Capacity (MW)	1	10	100	1,000	10,000
Population served (at 30 kW-hr/capita)	800	8,000	80,000	800.000	8,000,000
Pond area					
(acres)	250	2,400	23,500	225,000	2,200,000
(sq miles)	0.390	4	36.7	351	3,437
Cost of power (mills/kW-hr)	21.48	18.88	16.29	14.81	13.51

[a] No cerdit taken for waste disposal.

The algae are harvested and then introduced into an anaerobic digester in which they are transformed into gas and an organic residue (sludge). The gas is from 55 to 70% methane.

While the two researchers have conducted no large-scale studies on the anaerobic digestion of algae, they have done so on the production of algae on sewage. The largest scale thus far involved a one million-liter pond (2/3 acre) capable of treating the sewage from 1000 people. Thus their predictions in terms of required pond area and algae production costs would most likely be confirmed in a full-scale operation. A summary of the projected economics of the process is given in Table 4 (65).

CONCLUSION

The problem of disposing of residues and worn-out goods is as old as the cultures of man that involve his living in a fixed site. Disposal of some fractions of waste into fire, with or without benefit from the heat generated, is certainly one of his oldest customs. In modern times, disposal of the refuse and unwanted goods of urban man—municipal refuse—has been a vexing and unsatisfactorily resolved problem, sometimes involving incinerators that have recovered little and destroyed much, including some aspects of the surrounding environment. Municipal refuse, often carelessly labeled "garbage" by the public servant, the citizen, and the press, is by weight about 65% combustible and has a heat value of some 6000 Btu per pound (wet weight).

Within the past decade the magnitude of resources wasted by disposal practices applied to demolition debris, animal manures, agricultural residues, and industrial residues, as well as to municipal wastes, has come to appall both the conservationist and the representative of the public charged with solving waste problems. During that time the generic term *solid waste* was adopted to describe all wastes handled in the solid state and to identify institutions for dealing with wastes by reclamation, recycling, reuse, and disposal. From the obvious need for resource conservation, the US Environmental Protection Agency (EPA) emerged with a program of support for research and development, manpower training, and construction of demonstration projects. Thus man was enabled for the first time to look to the breaking of an ancient routine that involved only collection, haul, and deposit of wastes.

The Arab oil embargo produced a flurry of publicity in which there was no limit to estimates of the energy presumed to be resident in nearly all types of solid waste, with the possible exception of metal mining. These estimates were projected against the estimated national energy shortage, usually with the overoptimism that makes good copy. However, the numbers presented by the most sober of analysts were impressive, although they did not generally reflect the energy input necessary to provide and operate the hardware for extracting energy from waste materials. Nor did they reflect the economics and the energy trade-offs involved in alternative uses for the materials, e.g. the utilization of forests saved by paper salvage for recreation versus energy production. Finally, they included only the roughest of estimates of the concentration of animal manures and agricultural wastes in practical amounts and the increased demand for petroleum-based fertilizers if field residues could be collected for energy conversion. Nevertheless, such items as the EPA estimate that the refuse from 70% of our population contained energy equal to that in half a million barrels of oil per day served to underscore the importance of investigating carefully the technical and economic feasibility of its recovery.

Most literature on energy generation from waste materials appears in the context of resource conservation or resource recovery in which energy is but one of the items recovered. Thermal processes (including incineration and pyrolysis), biological fermentation, and solar energy processes are all in the prototype or advanced pilot-plant stages of construction. Among the best known is the burning of fiber milled from municipal refuse at St. Louis as 10%, or more, of the fuel in a pulverized-coal-burning installation of a public utility. Fermentation of animal manure to produce CH_4 is an already established system, the most striking example being in Colorado. Solar energy conversion systems utilizing the nutrients in wastes to culture photosynthetic plants and so produce cellular plant material have been developed through the pilot stage.

For details on processes, costs, and prospects for successful application, the reader is referred to the body of this chapter. It now seems safe to conclude that energy from waste materials is due to develop, as it began, as one aspect of a resource recovery system. As such, it will contribute to the nation's energy needs in varying amounts as the volume and the constituents of solid waste change with time and as materials seek their highest economic use in the marketplace. As an answer to the threat of oil boycotts, however, it is the authors' judgment that the national pile of solid waste would be ignored in favor of our coal and oil shale deposits were it not for the unpleasant fact that the waste pile is renewed each day and that one of the ways eternally to prepare for tomorrow is to convert a part of the waste pile into energy.

Literature Cited

1. US Environmental Protection Agency. 1975. *Solid Waste Management Available Literature.* Rep. No. SW-58.23. Washington DC: US EPA
2. Hopper, R. E. 1975. *A Nationwide Survey of Resource Recovery Activities.* Rep. No. SW-142. Washington DC: US EPA
3. Penn. Dept. Health, Bureau Housing, and Environmental Control. 1970. *A Plan for Solid Waste Management in Pennsylvania.* Solid Waste Sect. Publ. No. 3
4. McGauhey, P. H., Koerper, E. C., Wisely, F. E. 1973. *Wisconsin Solid Waste Recycling Predesign Report.* Prepared for State of Wisconsin. 176 pp.
5. US Environmental Protection Agency. 1971. *California Solid Waste Management Study (1968) and Plan (1970).* Rep. No. SW-2 tsg. Washington DC: US EPA
6. New Jersey Environmental Protection Agency. 1970. *New Jersey State Solid Waste Management Plan.* Prepared by Planners Associates, Inc. 47 pp.
7. *News Quarterly.* Published quarterly by Sanit. Eng. Res. Lab., Univ. Calif., Berkeley
8. McFarland, J. M. et al 1972. *Comprehensive Studies of Solid Wastes Management, Final Report.* SERL Rep. No. 72-3. Sanit. Eng. Res. Lab., Univ. Calif., Berkeley
9. Golueke, C. G. 1972. *Abstracts, Excerpts, and Reviews of the Solid Waste Literature,* Vol. V. Sanit. Eng. Res. Lab., Univ. Calif., Berkeley
10. Trezek, G. J., Savage, G. 1974. *Size Reduction in Solid Waste Processing.*

Progr. Rep. 1973–74. Dep. Mech. Eng. Lab., Univ. Calif., Berkeley
11. *Environmental Engineering News*. Published monthly by Sch. Civil Eng., Purdue Univ., Lafayette, Ind.
12. Pfeffer, J. T., Liebman, J. C. 1975. *Energy from Refuse by Bioconversion-Fermentation and Residue Disposal Processes,* 27–28. Presented at Energy Recovery from Solid Waste Symposium, Univ. Maryland, College Park, Md.
13. University of California. 1952. *An Analysis of Refuse Collection and Sanitary Landfill.* Tech. Bull. No. 8. Sanit. Eng. Res. Lab., Univ. Calif., Berkeley
14. McGauhey, P. H., Golueke, C. G. 1969. *Comprehensive Studies of Solid Wastes Management, 2nd Annual Report.* SERL Rep. No. 69-1. Sanit. Eng. Res. Lab., Univ. Calif., Berkeley
15. University of California. 1951. *Municipal Incineration.* Tech. Bull. No. 5. Sanit. Eng. Res. Lab., Univ. Calif., Berkeley
16. Fryling, G. R., ed. 1966. *Combustion Engineering.* New York: Combustion Engineering, Inc. 809 pp. Rev. ed.
17. McGauhey, P. H., Golueke, C. G. 1973. *Reclamation of Municipal Refuse by Composting.* Tech. Bull. No. 9. Sanit. Eng. Res. Lab., Univ. Calif., Berkeley
18. Golueke, C. G. 1972. *Composting: A Study of the Process and Its Principles.* Emmaus, Pa: Rodale. 110 pp.
19. Taylor, H. 1941. Garbage grinding at Goshen. *Eng. News Rec.* 127:441
20. Ross, W. E., Tolman, S. F. 1953. Garbage grinding pays its way. *Public Works* 84:70
21. Chan, Deh Bin, Pearson, E. A. 1970. *Comprehensive Studies of Solid Wastes Management: Hydrolysis Rate of Cellulose in Anaerobic Fermentation.* SERL Rep. No. 70-3. Sanit. Eng. Res. Lab., Univ. Calif., Berkeley
22. US Environmental Protection Agency. 1974. *Resource Recovery and Source Reduction. Second Report to Congress.* Rep. No. SW-122. Washington DC: US EPA
23. National Center for Resource Recovery. 1975. Refuse derived fuel. *NCRR Bull.* 5(15):2–10
24. National Center for Resource Recovery. 1974. Resource recovery... questions and answers. *NCRR Solid Waste Manage. Briefs*
25. Lowe, R. A. 1973. *Energy Recovery from Waste.* 2nd Interim Rep. No. SW-36 I. ii. Work performed under Grant No. S-802255 to City of St. Louis. Washington DC: US EPA
26. Institute for Research on Land and Water Resources. 1975. Report issued on solid waste recycling. *Newsletter* 5(15):2. University Park, Pa: Penn. State Univ.
27. Klein, S. A. 1970. Anaerobic digestion. In *Comprehensive Studies Solid Wastes Management, 3rd Annual Report.* SERL Rep. No. 70-2. Sanit. Eng. Res. Lab., Univ. Calif., Berkeley
28. Anderson, L. L. 1972. *Energy Potential from Organic Wastes: A Review of the Quantities and Sources.* Inform. Circ. No. 8549. Washington DC: Bur. Mines, US Dep. Interior
29. International Research and Technology Corp. 1972. *Problems and Opportunities in Management of Combustible Wastes.* Prepared for US EPA, Contract No. 68-03-0060. Washington DC: US EPA
30. Sussman, D. B. 1975. *Baltimore Demonstrates Gas Pyrolysis.* Interim Rep. No. SW-75 d. i. Washington DC: US EPA
31. Yeck, R. G. 1973. Agricultural biomass byproducts and their effect on the environment. In *Proc. Int. Biomass Energy Conf.,* Winnipeg, Manitoba, Canada, pp. XI-1 to XI-8
32. Ford Foundation Energy Policy Project. 1974. *A Time to Choose: America's Energy Future.* Cambridge, Mass: Ballinger. 511 pp.
33. Jones, H. E., Dehn, W. T. 1975. *Energy Recovery Systems.* Feasibility Study Rep. Prepared by CH$_2$M Hill Co. for Gray Harbor County, Washington
34. Kearley, J. 1973. Resource and energy conservation. In *Proc. from a Two-Day Program on Central Heating, Cooling and Electric Generating Plants.* Inst. Public Serv., Univ. Tenn., Nashville
35. Schwieger, R. G. 1975. Steam generation from refuse, process and manufacturing wastes. *Power* 119(2):5.2–5.24
36. Goodrich, W. F. 1901. *The Economic Disposal of Towns' Refuse.* London: King. New York: Wiley. 340 pp.
37. Gray, H. F. 1932. Symposium on garbage and refuse disposal. *West. Constr. and Highway Builder,* 479–82
38. Defeche, J. 1969. Corrosions in refuse incineration. In *Waste Disposal. Proc. 4th Int. Congr. Int. Res. Group Refuse Disposal.* In *Schweiz. Z. Hydrol.* 3(2):498–508
39. Jensen, M. E. 1969. *Observations of Continental European Solid Waste Management Practices.* Publ. Health Serv. Publ. No. 1880. Washington DC: Bur. Solid Wastes Manage., US Dep. HEW
40. Stephenson, J. W. 1970. Incineration

today and tomorrow. *Waste Age* 1(2): 2–5, 14, 15, 18–22
41. Plants burn garbage, produce steam. 1971. *Environ. Sci. Technol.* 5(3): 207–9
42. Northwest incinerator plant is largest complex of kind in western hemisphere. 1971. *Solid Waste Manage. Refuse Removal J.* 14(5): 74, 150–52
43. The East Hamilton solid waste reduction unit—SWARV. 1975. *Waste Age* 6(3): 22–27
44. Sutterfield, G. W. 1974. *Refuse as a Supplementary Fuel for Power Plants.* Interim Progr. Rep. No. SW-36d. iii. Washington DC: US EPA
45. Shannon, T. J., Schrag, M. P., Honea, F. I., Bendersky, D. 1974. *St. Louis/Union Electric Refuse Firing Demonstration Air Pollution Test Report.* Rep. on Contract No. 68-02-1324. Washington DC: Office Res. Develop., US EPA
46. Combustion Power Corp. 1975. *Combustion Power Unit—400 System.* Interim Rep. on Contract No. 68-03-0143, US EPA
47. Levy, S. J. 1974. Pyrolysis of municipal solid waste. *Waste Age* 5(7): 14–18
48. Weinstein, N. J., Rai, C. 1975. Pyrolysis/state of the art. *Public Works* 106(1): 83–86
49. Appell, H. R., Fu, Y. C., Illig, E. G., Steffgen, F. W. 1975. *Conversion of Cellulosic Wastes to Oil.* Rep. Investigation No. 801.3. Washington DC: Bur. Mines, US Dep. Interior
50. McCarty, P. L. 1973. Methane fermentation—future promise or relic of the past? In *Proc. Bioconversion Energy Res. Conf. Univ. Mass., Amherst, Mass,* 1–5
51. Singh, Ram Bux. 1971. *Bio-Gas Plant.* Gobar Res. Sta., Ajitmal, Etawah, India
52. Ghosh, S., Klass, D. L. 1974. *Conversion of Urban Refuse to Substitute Natural Gas by the Biogas® Process.* Presented at 4th Miner. Waste Util. Symp., Chicago, Ill.
53. Pfeffer, J. T. 1973. Anaerobic processing of organic refuse. In *Proc. Bioconversion Energy Res. Conf., Univ. Mass., Amherst, Mass.,* 31–37
54. Dair, F. R., Schwegler, R. E. 1974. Energy recovery from landfills. *Waste Age* 5(2): 6–10
55. Taiganides, E. P., Bauman, E. R., Hazen, T. E. 1963. Sludge digestion of farm animal wastes. *Compost Sci.* 4(2): 25–28
56. Stoner, C., ed. 1974. *Producing Your Own Power.* Emmaus, Pa: Rodale. 320 pp.
57. *Turning Waste Into Wealth.* Feb. 25, 1975. *Oakland Tribune,* California
58. Wilson, D. G. June 1974. Alcohol from cellulose—an energy breakthrough. *Environ. Action Bull.,* 3–6
59. Miller, F. B. 1969. *Conversion of Organic Solid Wastes into Yeasts.* Publ. Health Serv. Publ. No. 1909. Washington DC: Bur. Solid Wastes Manage., US Dep. HEW
60. Rosenbluth, R. F., Wilke, C. R. 1970. *Comprehensive Studies Solid Wastes Management: Enzymatic Hydrolysis of Cellulose.* SERL Rep. No. 71-1. Sanit. Eng. Res. Lab., Univ. Calif., Berkeley
61. Callihan, C. B., Dunlap, C. E. 1971. *Construction of a Chemical-Microbial Pilot Plant for Production of Single-Cell Protein from Cellulosic Wastes.* Rep. No. SW-24c. Washington DC: US EPA
62. Wilke, C. R., Mitra, G. 1974. *Process Development Studies on the Enzymatic Hydrolysis of Cellulose.* Presented at NSF Special Seminar *Cellulose as a Chemical and Energy Resource,* Univ. Calif., Berkeley
63. Golueke, C. G., Oswald, W. J., Gotaas, H. B. 1957. Anaerobic digestion of algae. *Appl. Microbiol.* 5: 47–55
64. Golueke, C. G., Oswald, W. J. 1959. Biological conversion of light energy to the chemical energy of methane. *Appl. Microbiol.* 7(4): 219–27
65. Uziel, M., Oswald, W. J., Golueke, C. G. 1975. *Solar Energy Fixation and Conversion with Algal-Bacterial Systems.* Final Rep. for NSF Grant No. GI-39216. Sanit. Eng. Res. Lab., Univ. Calif., Berkeley

Copyright 1976. All rights reserved

HYDROGEN ENERGY ✖11011

D. P. Gregory and J. B. Pangborn
Institute of Gas Technology, 3424 S. State Street, Chicago, Illinois 60616

INTRODUCTION

As this country's energy supply moves toward a system based on nonfossil sources, we must consider how these new energy sources can best be delivered to the customer. Nuclear technology is currently used to provide electric power, and most research and development applied to the bulk conversion of energy from solar, geothermal, wind, tide, and biological sources is also associated with electric power.

In the future, nonfossil energy sources must be used in vastly increased amounts, for this is the only way to provide the United States with an adequate energy supply not dependent upon imports that will be politically and economically detrimental during the twenty-first century. It is reasonable to assume that (*a*) for economic reasons these energy sources must be harnessed or converted to usable forms in large central conversion stations located remotely from the cities or load centers, and (*b*) for technical reasons, these sources will produce useful energy either at a constant rate or at a regularly cycling rate, neither of which matches the fluctuating demands for energy, either daily, weekly, or annually. We conclude that a high-capacity storage and transmission system will be required to link the consumer with the generating station.

A further consideration is the desirability of making as few basic changes as possible in the way energy is actually used by the customer, so that his manufacturing techniques do not have to be redeveloped and his energy-consuming appliances or equipment do not have to be completely replaced.

Although electric energy can be produced from a wide variety of energy sources and is extremely clean in end use, it is not ideally suited as a universal energy form. When compared with the mixed system that we have today—a mixture of chemical fuels and electric power—an all-electric economy has the following disadvantages:

1. Unless transmission technology for electric power is vastly improved, bulk transmission of power will require considerable land area, will be excessively expensive, and will present serious instability problems as the system grows.
2. Complete electrification of residential and industrial energy needs would require enormous capital expenditures to replace current oil-, gas-, and coal-fired equipment.

3. Unless unprecedented breakthroughs occur, electric power cannot be used as a fuel for aircraft or for conventional road vehicles.
4. Electrification of many manufacturing processes would require completely new process technology to be developed.
5. Unless an unprecedented breakthrough in electricity storage occurs, a huge power-generating capacity will have to be installed, nearly equal to peak demand, and this capacity will be only partially used most of the year.

For these reasons it is wise to consider, in addition to electricity, an alternative synthetic fuel that could be used in much the same way as our present fossil fuels, but could be made from the new energy sources and an unlimited supply of materials. Chemical fuels made from substances found in air and water are the only ones that can be considered from the viewpoint of pollution, availability of raw materials, and balance of materials.

Water can be split into hydrogen and oxygen by an input of energy. The hydrogen can be used as a fuel directly, or it might be used as a raw material to produce methanol, ammonia, or hydrocarbons by using either carbon dioxide or nitrogen from the atmosphere. If hydrogen itself can be used as a fuel, it is the simplest to make, requiring less wasted energy than the others, and the cleanest to use. The concept of the hydrogen energy system has been reviewed by Gregory (1), De Beni & Marchetti (2), and Gregory et al (3).

Hydrogen does not occur naturally in the earth's crust in the uncombined state, and thus is not a potential energy source that could be used to relieve the energy crisis or to achieve US energy independence. Hydrogen is merely a link between the new energy sources and the multiple users of energy, and it can thus ease and hasten the introduction of the new energy sources. Hydrogen energy is really an additional option to electric energy, filling the same function as an energy carrier but having advantages in storability, portability, and perhaps efficiency.

HYDROGEN PRODUCTION

Commercial production of hydrogen today is carried out by the steam reformation or partial oxidation of hydrocarbons—natural gas, naphthas, or crude oil—depending on supplies and economics. Clearly, with gas and oil in diminishing supply, these sources for hydrogen will become less attractive as a basis for an energy delivery system.

Hydrogen made from coal is of questionable importance as a fuel gas. On one hand, it is relatively inexpensive (compared with the sources we discuss next), but on the other hand, it would not be as conveniently transported or utilized in transmission and combustion systems as would a synthetic conventional fuel such as methane or liquid hydrocarbons (which are probably as easy to make from coal as is pure hydrogen). Hydrogen made from coal can probably only be justified as a fuel for special applications where the unique characteristics of hydrogen can be put to advantage, such as its weight or its nonpolluting characteristics.

Small quantities of hydrogen are commercially produced today by the electrolysis of water—usually in situations where electric power is cheap, or where reliable, unattended operation is required. Electrolysis is the only presently available technology by which hydrogen can be made from nonfossil energy (nuclear, solar, geothermal, or wind), and it is this aspect that makes electrolysis so important in the context of a hydrogen energy system.

Electrolysis suffers from a disadvantage as an energy conversion system because its raw energy requirement must be supplied as a high-quality energy form: electricity. Considerable research work is under way to develop a thermal cycle that would utilize heat to achieve the chemical splitting of water to its elements without the need for intermediate electricity generation, and without the need to use the extremely high temperature of 2500°C or more (which would be required to dissociate water directly) (4). This so-called thermochemical hydrogen production concept is an important building block of a hydrogen energy system because it offers the potential for nonfossil hydrogen production at a higher efficiency and lower cost than electricity generation plus electrolysis. At present, research is at the stage where these potentials have not been realized in practice, but progress is sufficiently encouraging to continue.

Electrolysis

The technology of hydrogen production by electrolysis has been reviewed recently by Stuart (5) and Konopka & Gregory (6). An electrolytic cell operates with essentially no moving parts, can be designed to produce no by-products, and offers the physical separation of hydrogen and oxygen as well as the initial decomposition of water. A perfectly efficient cell would require 94 kW-hr of electrical energy for each 1000 scf[1] hydrogen produced. Of these 94 kW-hr, only 79 kW-hr need to be supplied as electrical energy; the remainder is needed as heat.

Several large electrolytic hydrogen plants, consuming over 100 MW, have been operated successfully, while thousands of smaller units are in use for special applications. The electrolytic process is normally operated at 60–75% efficiency and at capital costs only a fraction (about one third) of the cost of the power station needed to drive it. As power costs increase, more research is needed to develop the high-efficiency potential of electrolysis.

Three major factors determine the usefulness of an electrochemical cell for hydrogen production. One is the energy efficiency, related to the cell's operating voltage; another is the capital cost of the plant, related to the rate of hydrogen production from a cell of a given size. These two factors are closely interrelated. The third factor is the lifetime of the cell and its maintenance requirements, which involve the materials used in its construction and the operating conditions selected.

A number of advantages can be gained from operating an electrolyzer at higher pressures, including (a) reduction in specific power consumption; (b) delivery of gas at pressure, thus reducing or eliminating the cost of gas compressors; and

---

[1] All cubic feet measurements given in this chapter are at standard conditions, 68°F and 14.7 psi.

(c) reduction in the size of electrolysis cells. It can be shown theoretically that the reversible cell voltage increases with pressure. However, as decreased gas volume and higher operating pressures result in a reduced overpotential, there is usually a small overall reduction in the cell voltage. This real gain in efficiency is offset by increases in the costs of pressure vessels or stronger components.

Operating voltages can be lowered for a given current by using electrodes that carry precious-metal catalysts or incorporate sophisticated metallurgical structures, both of which are expensive but increase efficiency. Catalysts speed up the electrode reaction at the surface. The structures increase the actual physical surface area of the electrode without increasing overall cell size; a roughened surface or porous electrode with a high internal surface will also achieve this objective. Some electrodes incorporate both approaches, applying a catalyst to a highly developed surface. When expensive electrodes are used, the cell must operate at higher current densities, so that the capital cost per unit of hydrogen production does not rise beyond the economic limits.

Diaphragms prevent electronic contact between adjacent electrodes and passage of dissolved gas or gas bubbles from one electrode compartment to another (leading to a decrease in current efficiency and possibly to explosions), without themselves offering an appreciable resistance to the passage of current within the electrolyte. Dissolved-gas crossover is serious only in pressure operations; to prevent the passage of gas bubbles, the diaphragm must consist of small pores whose capillary pressure is greater than the maximum differential pressure applied across the cell.

Asbestos is the most common material for cell diaphragms. At atmospheric pressure, woven asbestos cloth is used, sometimes with fine nickel wire to support the structure. Pressure electrolyzers usually have a mat made of woven or felted asbestos fibers that produces a fine pore structure, giving a higher resistance to the penetration of gases. This mat is sometimes supported by the electrodes.

The first applications of water electrolysis in industry used the tank electrolyzer, in which a series of electrodes—alternating anodes and cathodes—are suspended

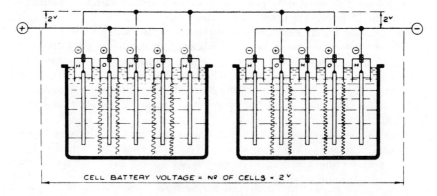

Figure 1   Schematic diagram of unipolar (tank-type) electrolyzer.

vertically and parallel to one another in a tank filled (most commonly) with a 20–30% solution of potassium hydroxide in demineralized water. Alternate electrodes, usually the cathodes, are surrounded by diaphragms, impermeable to gas but permeable to the cell's electrolyte, that prevent the passage of gas from one electrode compartment to another. The whole assembly is hung from a series of gas collectors. A single tank-type cell usually contains a number of electrodes, and all similar electrodes of the same polarity are connected in parallel electrically, as pictured in Figure 1. This arrangement allows an individual tank to operate across a 1.9–2.5 V dc supply. In general, the costs of electrical conductors rise as the current load rises, but the cost of ac/dc rectification equipment per unit of output falls as the output voltage rises.

The major advantages of tank-type electrolyzers are twofold: Relatively few parts are required and those needed are relatively inexpensive; and individual cells may be isolated for repair or replacement simply by short-circuiting the two adjacent cells with a temporary busbar connection. Some disadvantages of the tank electrolyzers are (a) inability to handle high current densities because of cheaper component parts and (b) inability to operate at high temperatures because of heat losses from the large surface areas of connected cells.

An alternative to the tank electrolyzer, in which a single electrode is either an anode or a cathode, is the bipolar electrolyzer, in which one side of each electrode is used as an anode in one cell, and the other side, as the cathode of the next cell. Figure 2 indicates the difference in the layout of electrodes in bipolar cell construction. The bipolar arrangement is also known as the filter-press electrolyzer because of its superficial resemblance to a filter press. The cells are connected in series, and individual cell voltages are additive within a battery; because the cells can be made relatively thin, a large gas output is achieved from a relatively small volume.

It is usually desirable to circulate electrolyte through the cells, thereby separating the gas and the electrolyte, and in many designs this is accomplished in a separating drum mounted on top of the electrolyzer. The electrolyte, free of gas, is recirculated through the cells, and the circulation is maintained by gas lift of the generated oxygen and hydrogen.

Although filter-press electrolyzers can operate at higher current densities and

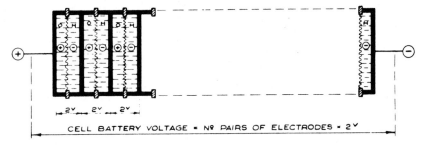

Figure 2 Filter-press (bipolar) cell construction.

appear to occupy relatively less space than tank types, they require a much closer tolerance in construction and are more difficult to maintain. Breakdowns in filter-press electrolyzers are rare, but when they occur, rejuvenation is difficult and repair may take considerable time. If an individual asbestos diaphragm is damaged, the entire battery must be dismantled. The greater capital costs of bipolar electrolyzers are offset because these electrolyzers can operate at higher current densities (producing more hydrogen per area of electrode) with virtually the same operating voltages as the tank-type unit.

Most commercially available electrolyzers use the filter-press design, although one manufacturer offers the tank-type cell (5). They all use potassium hydroxide solution as electrolytes and high-surface-area metals as electrodes. Electrodes made of perforated steel, steel wire gauze, and highly developed corrugated surfaces all find application (7, 8). All electrodes are heavily plated with nickel. In one more advanced type of cell (9), sintered porous nickel electrodes are used.

General Electric Company of Lynn, Massachusetts, has been developing a water-electrolysis system based on solid polymer electrolyte (SPE) fuel-cell technology (10). SPE fuel cells were first used in space exploration during the Gemini program, where they provided primary on-board power for seven of the spacecraft flights.

The SPE is a thin, solid, plastic sheet of perfluorinated sulfonic acid polymer having many of the physical characteristics of Teflon. Unlike Teflon, however, when a thin sheet of this material is saturated with water, it is an excellent ionic conductor, providing low electrical resistance. Used in an electrolysis cell, it is the

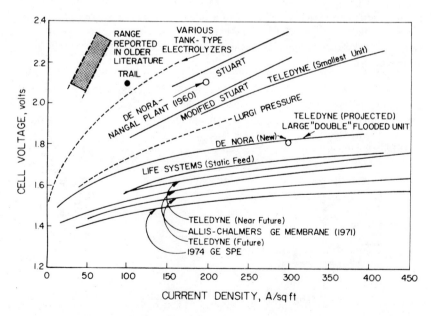

Figure 3 Comparative performances of electrolyzer systems.

only electrolyte required; there are no liquid acids or alkalies in the system. Hydrated hydrogen ions ($H^+ \cdot xH_2O$) move through the sheet of electrolyte.

Because the electrolyte is a solid, the catalytic electrodes are not required either to retain or to support the electrolyte, and they can therefore be optimized for catalytic activity at minimum cost. Currently, a thin layer of high-catalytic-activity platinum black is attached to the SPE surface to form the hydrogen electrode. A similar layer of proprietary precious-metal alloy catalyst forms the oxygen electrode. Additional metal current collectors are pressed against the catalytic layers. To date, the system has incorporated the use of niobium or titanium as the current collector and separator sheet materials.

Figure 3 (and Table 1) compares the cell operating performance of various advanced electrolyzers. These data are meant to give only a technological comparison of cell types and not a comparison of economics. A cell comparison based on voltage-current relationships is meaningless, however, unless cell cost is included.

Thermochemical Hydrogen Production

In terms of presently available technology, nuclear energy can be converted to hydrogen only by conventional electricity generation and the subsequent use of the electricity for the electrolysis of water. Thermal decomposition of water is an alternative concept that merits technological development. Because of the temperature limitations of nuclear reactors and conventional process equipment, direct single-step water decomposition cannot be achieved. However, a sequential chemical reaction series can be devised in which hydrogen and oxygen are produced, water is consumed, and all other chemical products are recycled. This multistep thermochemical method offers the potential for processes that could use high-temperature nuclear heat and be contained in conventional or developable chemical-process equipment.

At present, no commercial process for the thermal conversion of water to hydrogen and oxygen is in operation. The direct thermal splitting of water in one step requires a temperature in excess of 2500°C to obtain reasonable dissociation yields. This temperature is much higher than that which can be expected from a nuclear reactor. Several workers have proposed many multistep reaction sequences that thermally decompose water at lower overall temperatures. An example of such a chemical reaction sequence is as follows:

$$
\begin{array}{ll}
2CrCl_2 + 2HCl \rightarrow 2CrCl_3 + H_2 & 50°C \\
2CrCl_3 \rightarrow 2CrCl_2 + Cl_2 & 900°C \\
H_2O + Cl_2 \rightarrow 2HCl + \tfrac{1}{2}O_2 & 750°C \\
\hline
H_2O \rightarrow H_2 + \tfrac{1}{2}O_2 &
\end{array}
$$

In this reaction sequence, only water is split; all other materials are completely recycled.

The maximum temperatures available from newly designed high-temperature gas-cooled reactors (HTGRs) are about 800–900°C. Somewhat higher temperatures may be available in the future from the conceptual, ultrahigh-temperature, gas-cooled nuclear reactors. In the United States, an experimental model has operated for

Table 1 Present and future comparison of overall efficiencies of various electrolyzer cells

Manufacturer	Presently Available Potential			2–5 Year Projection of Potential			Future (Ultimate) Potential		
	(kW-hr/lb)	(V)	(%)[a]	(kW-hr/lb)	(V)	(%)	(kW-hr/lb)	(V)	(%)
Teledyne	32	2.1	75	22–19	1.8–1.6	82–92	15	1.24	118
General Electric	22	2.0	74	18–22	1.5–1.8	98–82	15	1.24	118
Stuart	24.5	2.04	72	22.8	1.9	77	—	—	—
Life Systems	20.5	1.7	87	18.1	1.5	98	14.96	1.24	119.8
De Nora	—	1.85	80	—	1.8	82	—	—	—

[a] Percentage of efficiency defined by $\dfrac{\text{higher heating value of hydrogen produced}}{\text{electrical energy consumed}} \times 100$.

30 days at 1100°C (2012°F). In Japan and Germany, test loops for HTGR coolants are now in operation at 950–1000°C (1740–1830°F).

In research programs now in progress, chemical reaction sequences for splitting water are derived by combining thermodynamics and chemistry to determine

Table 2 Thermochemical cycles

System	Reactions	Temperatures (°C)
DeBeni	$C + H_2O \rightarrow CO + H_2$	725
	$CO + 2Fe_3O_4 \rightarrow C + 3Fe_2O_3$	225
	$3Fe_2O_3 \rightarrow 2Fe_3O_4 + \frac{1}{2}O_2$	1425
IGT L-1	$Cd + H_2O \rightarrow CdO + H_2$	125
	$CdO \rightarrow Cd + \frac{1}{2}O_2$	1225
IGT C-5	$Fe_3O_4 + 2H_2O + 3SO_2 \rightarrow 3FeSO_4 + 2H_2$	125
	$3FeSO_4 \rightarrow \frac{3}{2}Fe_2O_3 + \frac{3}{2}SO_2 + \frac{3}{2}SO_3$	725
	$\frac{3}{2}Fe_2O_3 + \frac{1}{2}SO_2 \rightarrow Fe_3O_4 + \frac{1}{2}SO_3$	925
	$2SO_3 \rightarrow 2SO_2 + O_2$	925
IGT J-1	$2CrCl_2 + 2HCl \rightarrow 2CrCl_3 + H_2$	325
	$2CrCl_3 \rightarrow 2CrCl_2 + Cl_2$	875
	$H_2O + Cl_2 \rightarrow 2HCl + \frac{1}{2}O_2$	850
EURATOM Mark-7	$3FeCl_2 + 4H_2O \rightarrow Fe_3O_4 + 6HCl + H_2$	600
	$Fe_3O_4 + \frac{1}{4}O_2 \rightarrow \frac{3}{2}Fe_2O_3$	400
	$\frac{3}{2}Fe_2O_3 + 9HCl \rightarrow 3FeCl_3 + \frac{9}{2}H_2O$	100
	$3FeCl_3 \rightarrow 3FeCl_2 + \frac{3}{2}Cl_2$	400
	$\frac{3}{2}H_2O + \frac{3}{2}Cl_2 \rightarrow 3HCl + \frac{3}{4}O_2$	800
EURATOM Mark-1	$Hg + 2HBr \rightarrow HgBr_2 + H_2$	250
	$HgBr_2 + Ca(OH)_2 \rightarrow CaBr_2 + HgO + H_2O$	200
	$CaBr_2 + 2H_2O \rightarrow Ca(OH)_2 + 2HBr$	725
	$HgO \rightarrow Hg + \frac{1}{2}O_2$	600
EURATOM Mark-9	$3FeCl_2 + 4H_2O \rightarrow Fe_3O_4 + 6HCl + H_2$	650
	$Fe_3O_4 + \frac{3}{2}Cl_2 + 6HCl \rightarrow 3FeCl_3 + 3H_2O + \frac{1}{2}O_2$	120
	$3FeCl_3 \rightarrow 3FeCl_2 + \frac{3}{2}Cl_2$	420
G E "Beulah"	$2Cu + 2HCl \rightarrow 2CuCl + H_2$	100
	$4CuCl \rightarrow 2CuCl_2 + 2Cu$	100
	$2CuCl_2 \rightarrow 2CuCl + Cl_2$	600
	$Cl_2 + Mg(OH) \rightarrow MgCl_2 + H_2O + \frac{1}{2}O_2$	80
	$MgCl_2 + 2H_2O \rightarrow Mg(OH)_2 + 2HCl$	350
G E "Agnes"	$3FeCl_2 + 4H_2O \rightarrow Fe_3O_4 + 6HCl + H_2$	550
	$Fe_3O_4 + 8HCl \rightarrow FeCl_2 + 2FeCl_3 + 4H_2O$	110
	$2FeCl_3 \rightarrow 2FeCl_2 + Cl_2$	300
	$Cl_2 + Mg(OH)_2 \rightarrow MgCl_2 + \frac{1}{2}O_2 + H_2O$	80
	$MgCl_2 + 2H_2O \rightarrow Mg(OH)_2 + 2HCl$	350

probable reaction steps using desirable materials. Laboratory work is required to prove the workability of and the operating conditions for the reaction steps. Eventually, recycled materials must be used as reactants in the laboratory, and reaction product separations must be performed to "demonstrate" a cycle. If a cycle is operable in the laboratory and if calculations with laboratory data show good energy efficiency in the cycle, experimental work must be continued to determine reaction rates and heat-transfer requirements. Then process flow sheets can be generated and estimates of the production cost of hydrogen by a thermochemical process can be made.

The principles of thermochemical water splitting have been reviewed by Funk (11), Wentorf & Hanneman (12), Pangborn & Sharer (13), and Marchetti (14). Pangborn has considered the energy requirements for thermochemical water splitting in comparison with the energy requirements needed for electrolysis. Since the technology of thermochemical water splitting is too young to allow realistic estimates of plant costs and thus the economics of these processes, the comparison of the overall production efficiencies of electrolysis with thermochemical processes is the next best thing that can be done to review the two processes in comparison. Pangborn & Gregory (15) discussed nine published thermochemical cycles shown in Table 2. Water-splitting efficiencies were calculated by incorporating a selection process based on thermodynamics for reaction temperatures, material flow patterns, an enthalpy balance with heat exchange between material streams in the cycle, an estimate of work requirements within the cycle, and hence a calculation of an attainable energy efficiency. The calculations were based on perfect heat transfer and reaction equilibriums, and on the assumption that work requirements for materials transport within the cycle are negligible. The efficiency of the thermo-

Table 3 Comparison of efficiency of thermochemical cycles

Thermochemical Cycle[a]	Highest Heat Input Temp. (°C)	Thermochemical Efficiency (%)[b]	Electricity Generating Efficiency (%)	Thermoelectrochemical (Electrolysis) Efficiency (%)[b]	Water-Split Ratio[c]
1	1425	69	39	37	1.86
2	1225	66	37	35	1.88
3	925	57	34	32	1.78
4	875	46	33	31	1.48
5	800	36	32	30	1.20
6	725	39	31	29	1.34
7	650	41	28	27	1.52
8	600	27	27	26	1.04
9	550	29	26	25	1.16

[a] See Table 2.
[b] Efficiency is based on the high heating value of hydrogen.
[c] Ratio of moles of water split, thermochemistry to electrolysis, by the same energy input.

chemical cycle is defined by dividing the high heating value of the hydrogen produced by the sum of all the heat inputs to the cycle. The efficiencies of the corresponding electrolysis processes were calculated by assuming an electrolyzer operating at 95% efficiency, which derives its electricity from a heat engine operating at 50% (or higher depending on technology level) of the theoretical Carnot efficiency using a heat input at an equal temperature to that assumed in the thermochemical process.

Table 3 shows a comparison of the efficiency of thermochemical cycles with the corresponding electrochemical processes for the nine cycles, which operate at temperatures from 1425°C down to 550°C. Pangborn & Gregory (15) show that unless temperatures in excess of about 600°C are achieved, the efficiency of the thermochemical process is only barely greater than that of the corresponding electrolysis process. For future technology, electricity generation followed by electrolysis will become potentially more efficient, and about 700°C is the temperature of equivalent efficiency.[2] From this the authors conclude that there is a tremendous incentive for the development of thermochemical water-splitting processes to produce hydrogen. However, this incentive is based on the potential for superior efficiency of conversion of nuclear energy, and it is only valid with the arrival of high-temperature nuclear reactors capable of delivering heat at temperatures above 700°C and preferably in the 900–1000°C range.

Hydrogen from Coal

Although the production of hydrogen from coal is probably of less importance in the ultimate long range than its production from nonfossil sources, this review would not be complete without a brief discussion of the technology that is already available, and that which is under development, for hydrogen-from-coal processes. There are many plants operating that produce from coal a gas mixture containing hydrogen, and there is no reason why conventional technology could not be used to produce pure hydrogen if required.

A coal gasification plant consists essentially of two basic components. The first is the gasifier in which suitably prepared coal is reacted with steam, oxygen, or air, or a combination of all three, to produce the raw gas. Gasification is an energy-consuming operation, and the gasifier obtains its energy from the combustion of some of the coal. The second operation is a gas purification train, which takes the raw gas and converts it to a product of the desired purity. If pure hydrogen is required, a purification train can be designed and constructed to do this, by relying on fairly conventional and available technology. The design of the purification train depends markedly upon the composition of the raw gas and the pressure at which hydrogen must be delivered. The raw gas composition in turn is determined by the design, operating conditions, and coal input of the gasifier itself.

Gasifiers form into three general classes: The entrained-bed gasifier operates by the injection of coal dust or powder into an air or oxygen stream in which

---

[2] This comparison and temperature of efficiency equivalency does not necessarily hold for hybrid thermochemical cycles—those involving an electrochemical step or those requiring significant electrical inputs for cycle operation.

some combustion and some gasification are carried out in a flame-type environment. The fixed-bed gasifier operates by passage of a current of oxygen, air, or steam through a bed of coal either in a batchwise process or in a system in which the bed is moving from one end of the reactor to the other in a plug-flow mode. The fluidized-bed gasifier operates by fluidizing fine-particle coal with a gas stream so that the gasification reaction occurs fairly uniformly throughout the bed.

Some of the coal gasification processes currently under development operate by direct hydrogenation of coal to produce a high-Btu gas (900 Btu/scf or higher). In these processes, hydrogen and coal are reacted to produce a methane-rich gas stream. The hydrogen for this reaction is in turn obtained from a second gasifier, which operates on the char and coke residue left over from the primary gasification process. These hydrogen production units, which are under development in conjunction with substitute natural gas (SNG) plants, are of potential importance as pure hydrogen producers.

Some processes are being developed for the gasification of coal to produce a low-Btu gas usable as a boiler fuel. These low-Btu gasifiers are normally operated under conditions that promote the formation of hydrogen rather than methane; thus the addition of an appropriate purification train to yield pure hydrogen is an attractive option.

Table 4 shows some characteristics of several gasifiers that are either commercially available or under development for various purposes. This table is not meant to be complete, but is intended to illustrate some of the principal characteristics of the major types of gasifier.

Table 5 indicates the efficiency of the production of pure hydrogen, delivered at 1000 psi, from coal by three different gasification processes together with the appropriate purification trains. The overall process efficiencies are in the region of

Table 4 Coal gasifiers

Type	Process	Pressure (psia)	Status	Use
Suspension	Koppers-Totzek	0	commercial, 16 plants operating	ammonia
	Texaco	700	commercial (oil), 63 plants operating; pilot (coal)	ammonia
Moving-bed	Lurgi	435	commercial, 58 plants built	fuel gas
	Wellman-Galusha	0	commercial, 2 units operated	synthesis gas
Fluidized-bed	Steam-Iron(IGT)	1000	pilot under construction	H_2 feed for SNG synthesis
	U-Gas(IGT)	335	pilot	fuel gas
	Winkler	0	commercial, 16 plants built	synthesis gas

Table 5 Coal-to-hydrogen processes

Process	Coal Feed Limitations	Efficiency (%)	Remarks
Koppers-Totzek	none	56.8	commercially available
U-Gas	caking coal needs pretreatment	66.2	not intended as a hydrogen producer
Steam-Iron	caking coal needs pretreatment	44.6 58.7[a]	no oxygen plant

[a] With credit for electric generation.

45–66%. It should be noted that typical efficiencies for the production of a methane-rich, high-Btu gas from coal are in the range of 60–67%.

HYDROGEN TRANSMISSION

At the present time, the long-distance pipelining of hydrogen is an operation that is carried out by only a few specialized companies in different parts of the world, but it is a concept that is receiving much attention as the world moves into an age in which the transportation and use of unconventional fuels are likely to become of major importance. Transmission possibilities for hydrogen have been reviewed by Gregory (16) and Konopka & Wurm (17).

Hydrogen Pipeline Design

It is of interest to compare the design requirements of a pipeline for hydrogen with those of a pipeline for natural gas (for which much experience and data are available). One might expect the capacity of a given pipeline to be far lower for hydrogen than it is for natural gas, since the heating value of hydrogen is only 325 Btu/cu ft compared with about 1000 Btu/cu ft for natural gas. This implies that to deliver the same quantity of energy, three times the volume of hydrogen must be transmitted. On closer inspection, however, one finds that the capacity of a given pipe depends upon the square root of the density of the gas (17), and because the density of hydrogen is about one ninth of that of natural gas, there is a compensating factor of one third that results in the given pipe having essentially the same energy-carrying capacity for natural gas as for hydrogen. In other words, a given length of pipe of a given diameter operating in turbulent flow with the same pressure drop across its length will carry the same number of Btu's per hour when working with hydrogen as it will with natural gas.

This rather fortuitous compensation applies only at atmospheric pressure. As the pressure increases to typical pipeline operating pressures of 750 psi or so, the compressibility factor for hydrogen is somewhat different than that for natural gas, and this results in a slightly unfavorable carrying capacity for hydrogen. At 750 psi the ratio of heating values for a given compressed volume of hydrogen and natural gas has changed from 3:1 to 3.83:1.

Long-distance gas transmission lines of lengths greater than about 60 miles must be supplied with pipeline compressors at fairly regular intervals. When the compressor requirements for hydrogen are examined, we find significant differences from those for natural gas. First, hydrogen compressors must handle a considerably greater volume of gas—somewhere between three and four times the number of cubic feet for the same energy capacity. Second, the horsepower required to drive a hydrogen compressor is considerably greater than that needed to drive a natural gas compressor for the same gas energy throughput. Third, the design of rotary compressors commonly used for natural gas lines appears to be inadequate for hydrogen operation. Table 6 indicates the requirements for the conversion of an existing natural gas line to operate on hydrogen.

It is possible to estimate the cost of transmitting hydrogen by pipeline from a knowledge of the required pipeline diameter, compressor capacity and horsepower, and energy throughput required. For natural gas lines these cost data optimize at operating pressure levels of about 750 psia, with pressure ratios across the compressors of about 1.3:1. However, if we begin a complete redesign of a pipeline specifically for hydrogen, we must take into account not only the different physical properties of hydrogen, but also the cost of fuel in powering the engines used to drive the compressors. Several published studies (18, 19) on pipeline transmission costs for hydrogen differ in their conclusions quite markedly because they have assumed rather different fuel costs for their compressors. Because the compression energy is so much higher than that for natural gas, the costs for hydrogen transmission are more sensitive to this factor.

Konopka & Wurm (17) carried out a detailed hydrogen-pipeline optimization calculation in which hydrogen fuel costs are varied from $1/million Btu to $4/million Btu. They found that for a cost of $3/million Btu and a compressor-station spacing of 100 miles, optimum pipeline costs are obtained at operating pressures in the region of 1500–1800 psi, with pressure ratios across the compressors of only 1.1:1 or less. Figure 4 illustrates some optimized energy-transmission costs

Table 6 Relative capacity of transmission line on hydrogen and natural gas at 750 psia[a]

Gas	Compressor Capacity	Compressor Horsepower	Energy Delivery Rate
Natural gas	1.0	1.0	1.0
Hydrogen	1.0	0.1	0.26
	2.1	1.0	0.56
	3.8	5.5	1.0

[a] Note that these figures relate to the assumption that compressor efficiency is unchanged, an assumption valid only for reciprocating compressors—not for radial turbocompressors. The table also indicates that modifications in both the compressor and the engine must be made in order to recover the original capacity of the line. In this case, turbocompressors would be replaced with custom-designed equipment intended for hydrogen service. Considerable benefit would result from operating the line at a higher pressure, but this is considered impracticable for existing pipelines.

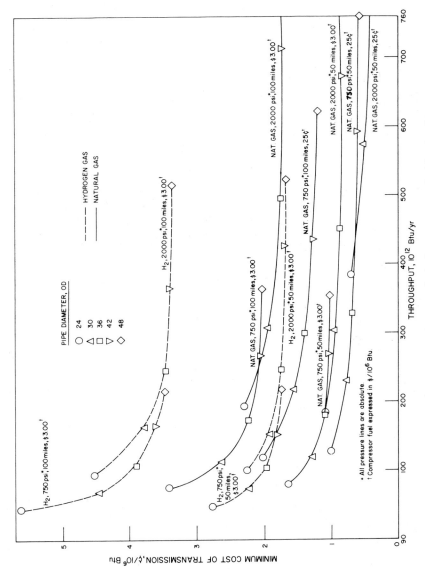

Figure 4 Optimized transmission costs for hydrogen compared with natural gas.

as a function of throughput for hydrogen at two different pressures, and these are compared with typical costs for natural gas transmission that were derived by the same methodology as that used for hydrogen. It may be seen that hydrogen transmission at typical pipeline pressures of 750 psi will cost about 250% more than natural gas transmission, but operating pipelines at 1800 psi will cost about 50% more than natural gas.

The cost of moving energy by hydrogen pipeline is significantly less than that of moving the same amount of energy by overhead electric transmission line. Electric transmission costs vary widely with voltage load factor and terrain, but a 230-kV overhead line operates at a cost of about $1/million Btu per 100 miles, between 18 and 30 times more than hydrogen transmission.

New Requirements for Hydrogen Transmission

One of the principal concerns about hydrogen transmission is the fear of hydrogen embrittlement of the pipeline materials. Operating experience with common pipeline steels at pressures up to about 500 psia has shown no problems of consequence (20), although there is some evidence from laboratory studies that a certain amount of embrittlement might occur. However, when pressures up to 1800 psia are considered, there is far less certainty about the behavior of materials. Very little conrolled testing of normal pipeline steels has been carried out in hydrogen at such high pressures. High-strength materials certainly do exhibit severe embrittlement behavior at considerably higher pressures—in the region of 10,000 psi—but there has been no extensive testing at the 2000-psi level. One mode of failure, caused by the so-called hydrogen-environment embrittlement, appears to occur only when steels undergo yielding in the presence of high-pressure hydrogen, but there is a marked discrepancy between reported behavior under laboratory conditions and actual field experience, where failures have been less frequent than would have been predicted from the laboratory data. Clearly, more experimental work needs to be carried out, but at the present time there seems to be ample evidence for confidence that normal pipeline steels are satisfactory for pressures up to typical operating values of 750 psia.

Experience with Hydrogen Pipelines

There are many miles of hydrogen "pipeline" of moderate diameters operating in every major oil refinery, but these are not in the category of long-distance transmission. Some of the industrial gas manufacturers operate small-diameter hydrogen delivery pipes, several miles in length, between their production plants and their customers. One such line in the United States is 8 inches in diameter and 12 miles long, and operates at 200 psi. Its operators say that hydrogen is handled in the same manner as oxygen, nitrogen, and carbon monoxide, and no special precautions, unique to hydrogen, are taken.

The most comprehensive hydrogen pipeline system operates in the Ruhr area of West Germany (20). It is owned and operated by Chemische Werke Huls AG (CWH) and has a total length of about 130 miles. The original system was constructed in 1938 and was expanded after 1954. The purpose of the pipeline network is to distribute by-product hydrogen. This industrial-grade hydrogen is transmitted at a nominal pressure of 15 atm (225 psia), the operating pressure at which hydrogen is produced. Electrolytic hydrogen is compressed to this level in reciprocating units. There are four separate injection points and nine separate users. A map of the pipeline system is shown in Figure 5.

Isting (20) summarizes his company's experience with this network as follows:

> It may be said that the long years of operation of an integrated hydrogen network and the fact that any detonations, explosions, hazardous situations, or major upsets have not occurred during that period provide an answer to the much discussed question of how safe the construction and operation of hydrogen pipelines is. On the other hand, hydrogen pipelines do involve problems. Safe operation calls for careful design and construction, varied experience being an essential basis for engineering hydrogen pipelines.

HYDROGEN STORAGE

The Need for Storage

In an energy system there is a need to be able to store energy somewhere between the production point and the utilization point. This need for storage is due to the almost inevitable mismatch between the optimum production rate of energy and the fluctuations in demand for energy by the users.

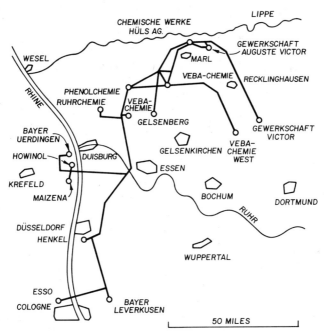

Figure 5 Hydrogen pipeline system in West Germany.

In the electric energy system, storage presents considerable difficulty because electricity itself is not readily storable. For this reason the fluctuations in demand are usually met by varying the generating rate. This is carried out by the installation of peaking generators, which tend to have high operating costs, if coupled with low capital costs, such as gas turbines or diesel generators. The operation of peaking generators requires the use of soon-to-be-scarce fossil fuels and lowers the overall efficiency of the electrical system.

In contrast to the electric energy system, both the gas energy and the oil energy systems are endowed with the capability to store very large quantities of energy. In the oil industry, liquid fuels are stored in above- and belowground tanks at the refinery, at the distribution centers, and even on the consumer's premises. In the gas industry, storage is provided to a small extent in the pipelines themselves, and to a larger extent in underground systems, which can be either depleted gas or oil fields or so-called aquifer systems: rock formations resembling oil fields but containing water. This very large storage capacity allows the gas industry to operate its production wells and its main transmission lines at a constant output and still meet the very large fluctuations in demand that exist between summer and winter.

The location of energy storage systems is very important. If a large storage system can be installed very close to the customer, the load factor on the transmission system is automatically raised, and therefore the transmission cost becomes lower. On the other hand, if storage can only be provided close to the generating station, then the transmission and distribution systems must bear the problems of fluctuating demand.

One of the advantages often claimed for a hydrogen energy system is that hydrogen is storable. Although this is true, it must be realized that the storage of hydrogen is not an easy problem compared with the storage of liquid fuels such as gasoline or oil. It is only when it is compared with electricity that storage of energy as hydrogen seems relatively easy. Indeed, some studies have been carried out in which the storage of electric energy as hydrogen has been investigated as a possible means of easing the load factor problems in the electricity system. In these studies, the use of off-peak electricity to produce hydrogen by electrolysis has been considered. The hydrogen would be stored by one of several methods to be described shortly, and then during periods of peak electricity demand, the hydrogen would be used as a fuel for either a fuel cell or some form of engine generator. The technology and economics of electric energy storage systems using hydrogen as a storage medium have been reviewed by Burger et al (21), Fernandez (22), and Gregory (23). All three reviews indicate that the round-trip efficiency of energy storage that starts with electricity and finishes with electricity is very low if present-day technology is used. However, future projections for the efficiency of both electrolyzers and fuel cells, and more efficient storage systems such as metal hydrides, could yield round-trip efficiencies above 50%, which could compare reasonably well with the only known way of bulk storage of electric power at the present time: pumped hydroelectric systems. Fernandez (22) shows that the economics of a hydrogen energy system could in fact be more attractive than competitive storage systems.

It is when hydrogen is considered as a replacement fuel in applications currently met by natural gas and oil that bulk energy storage becomes very important. Hydrogen does have the potential of being stored in large quantity and, therefore, of being able to take over some of the seasonal applications of oil and gas by using either solar or nuclear energy as an energy source.

There are five principal methods that have been considered for hydrogen storage. These are 1. line pack storage (allowing the pressure in the transmission of distribution systems to vary); 2. compressed gas storage (in aboveground or belowground pressure tanks); 3. underground storage (in depleted oil and gas fields or in aquifer systems); 4. cryogenic storage (liquid storage in vacuum-insulated or -superinsulated storage tanks); and 5. storage as metal hydrides (such as those of magnesium or iron-titanium alloys). Other chemical forms of hydrogen storage have also been considered in the past, and these include ammonia and methanol.

Line Packing

The use of line pack storage in the natural gas industry provides a relatively small-capacity storage system, but one with a very fast response time that can take care of minute-by-minute or hour-by-hour variations in demand. A hydrogen transmission and distribution system running on hydrogen would have a similar capability, although the capacity would be reduced by a factor of about 3 because of the reduced heating value of hydrogen, compared with natural gas.

Compressed Gas

Hydrogen is conventionally stored for many applications in high-pressure cylinders. This method of storage is rather expensive and very bulky because very large quantities of steel are needed to contain quite small amounts of hydrogen. In the conventional industrial hydrogen system, compressed gas is used to supply relatively small amounts of hydrogen, but when hydrogen is considered as a fuel, it is soon realized that tank storage of hydrogen is not really a practical proposition.

Underground Storage

Underground storage represents an extremely low-cost method of storing energy, and there seems to be no reason why this technique should not be applied to hydrogen. When a natural gas field or an oil field has been largely depleted of its contents, there remains a sealed geologic formation, usually partly filled with water. Its gas tightness is well guaranteed since the original fossil fuel, oil or gas, has been retained in the structure for many millions of years. In normal storage operations, natural gas is injected under pressure down the wells previously used for withdrawal, and this serves to force back the gas and water interface into the porous rock and therefore creates a new reservoir of gas. As long as the original discovery pressure of the gas is not exceeded, there seems to be no reason to expect leakage to occur because the tendency to leak is limited by the capillarity of water trapped in the pores of very fine rock. The pressure at which this water is expelled is dependent only upon the surface tension of the water and the capillary size of the porous rock and is not a function of the type of gas trapped.

It is unlikely that small underground storage systems will be found or that they will be worthwhile developing. Underground gas storage represents a very inexpensive method of storing extremely large quantities of hydrogen. The method is restricted to geographical locations where the correct geological conditions occur, and these are generally in areas known to have oil- or gas-bearing formations. It is noteworthy that the gas industry uses virtually no underground storage in the eastern part of the United States, but most of the large storage fields are in the Midwest and in the South and Southwest.

Cryogenic Storage

By analogy with the gas industry, the storage of hydrogen as a cryogenic liquid appears to be an attractive possibility. The seasonal storage of liquefied natural gas has achieved very considerable use in those areas of the country that do not have the correct geology for underground gas storage.

The aerospace industry has developed competence in the production, storage, and use of liquid hydrogen, which was selected as a rocket fuel for the Apollo space program. Very large facilities for hydrogen liquefaction have been designed and built, and large storage tanks have also been constructed. One major difference exists between handling liquid natural gas and liquid hydrogen—the storage temperature. Liquid hydrogen boils at $-423°F$ and therefore must be maintained at or below this temperature in storage unless pressure buildup can be tolerated. It is commonly regarded as necessary to use vacuum-insulated storage vessels, whereas liquid natural gas can be maintained in the liquid state at a considerably higher temperature by using superinsulation, but without the need for vacuum jackets. The principal need for vacuum-jacketed containers is that the liquid hydrogen is below the temperature at which air condenses on the surface, and thus any air in contact with the cold walls of the hydrogen container will condense and in doing so will allow more rapid heat transfer into the container. There is also a flammability danger from the fact that liquefied atmospheric gases (rich in oxygen) would concentrate in the vicinity of the hydrogen tank. Another problem concerning storage of liquid hydrogen is the considerable amount of energy required to convert hydrogen gas into the liquid phase. Not only must the latent heat of the phase change be removed from the system, but some additional precooling is required to bring the gaseous hydrogen down below the Joule-Thompson temperature, above which liquid hydrogen heats up on expansion. Thus, a liquid hydrogen plant normally requires some kind of primary refrigeration, such as a liquid nitrogen plant, to precool hydrogen. The net result is that about 25–30% of the heating value of hydrogen is required to liquefy hydrogen.

Metal Hydrides

Considerable interest has been shown recently in the possibility of storage of hydrogen in the form of a metal hydride (24). In principle, many metals form hydrides by direct reaction with hydrogen gas, which can then be decomposed by heating to high temperatures, yielding pure hydrogen and the metal, which upon cooling is in a form ready for hydriding with a fresh charge of hydrogen.

A unique property of hydride storage is that the equilibrium pressure of a hydride remains constant over a wide range of composition. This means that as a hydride tank is emptied, its pressure remains constant during the withdrawal of most of its capacity. The equilibrium pressure of the metal hydride at the given temperature can be altered very significantly with the addition of a second or third metal so as to form an alloy system.

An alloy system that has received a great deal of attention recently is iron-titanium. This system is of interest because the heat of formation of the hydride is relatively small, only about 10% of the fuel value of the hydrogen stored. Another attribute is that the equilibrium pressure of an iron-titanium hydride system at room temperature is about 3 atm and rises to about 10 atm when heated to only about 70°C. Thus the hydride system can easily be charged and discharged without changing its temperature significantly from ambient. The main disadvantage of the iron-titanium system is that the iron-titanium alloy weighs more than most other metal hydride systems, and more than systems that would contain alternative (synthetic) liquid fuels. Thus the iron-titanium system cannot reasonably be selected for portable-fuel applications such as for use in vehicles.

HYDROGEN UTILIZATION

Combustion Characteristics of Hydrogen

The combustion of hydrogen is probably the best-known and best-researched topic in the field of fundamental combustion research. However, there are very few pieces of commercially available equipment capable of burning hydrogen because its burning characteristics are somewhat different from those of more conventional fuels for residential and industrial applications.

When compared with natural gas and hydrocarbons, there are five very significant differences in the combustion properties of hydrogen. One difference is that the energy required to initiate combustion in air is about 10 times lower for hydrogen (0.02 mJ) than for natural gas and other hydrocarbons (0.3 mJ). Another difference is that the flame speed of hydrogen (9.4 ft/sec) is about 10 times greater than the flame speed of natural gas (1.0 ft/sec). These combustion properties are extremely important in the design of ignition and combustion equipment, and in particular, the high flame speed means that burner ports on conventional appliances will be far too big for hydrogen and the flame can propagate back through the port and into the body of the burner. There is no reason why a burner cannot be designed specially to operate on hydrogen; sufficient information is known about the combustion characteristics to do this.

A third major difference in the atmospheric combustion of hydrogen is that the primary combustion product is water. No hydrocarbons, carbon oxides, sulfur compounds, or other pollutants can be produced. This means that hydrogen is potentially an extremely nonpolluting fuel and might be considered for combustion in unventilated or unflued appliances. However, as with all combustion processes, which occur in air, it is possible to produce nitrogen oxides when the nitrogen and the oxygen in the air are heated in a flame to high temperatures. Flame temperatures for hydrogen combustion are expected to be somewhat higher than flame temperatures for hydrocarbon combustion, so that the production of nitrogen oxides is expected to be greater. On the positive side, however, since there are no products of combustion or pollutants other than water and nitrogen oxides, the suppression or removal of nitrogen oxides should be easier in a hydrogen combustion system than in a hydrocarbon system.

Another difference in the combustion of hydrogen compared with conventional fuels, which results from its low ignition energy, is the fact that the initiation and maintenance of catalytic combustion at relatively low temperatures on a suitable catalytic surface is far easier with hydrogen than it is with hydrocarbon fuels. It is possible to sustain the catalytic oxidation of hydrogen in air at temperatures little higher than ambient room conditions, so that flameless combustion of hydrogen to produce low-grade heat is a distinct possibility. Thus, catalytic combustion of hydrogen is an important area and is discussed in more detail later.

Finally, the fifth difference between the combustion characteristics of hydrogen and conventional hydrocarbons is to be found in the range over which mixtures of hydrogen and air are flammable. Hydrogen-air mixtures are flammable in the range of 4–75% hydrogen, compared to natural gas, for example, which is flammable in the range of 5–15%. While the lower limit is of concern as a safety feature, and these are very similar for hydrogen and natural gas, it is the extent of the flammable air-fuel composition that is of concern in engine applications. If hydrogen, rather than hydrocarbon, is used in an internal combustion engine, it is almost impossible to provide the engine with an overrich mixture that would result in an inability to run or to start. A hydrogen-fueled engine is far more tolerant to maladjustments of fuel mixtures than any other type of engine, and the broad range of flammability allows "quality-governed" (unthrottled) operation.

Industrial Feedstocks

At present hydrogen finds major application as an industrial feedstock or chemical intermediate, mainly in the production of ammonia and methanol. It is used for hydrotreating operations in refineries in order to upgrade heavy fuels into lighter fractions, for hydrogenations of many organic chemicals, and to a minor extent for the production of soaps, fats, special metallurgical atmospheres, and for other purposes in minor amounts. Normally, hydrogen used in the larger applications is produced from either natural gas or oil on site by the user himself, and is normally referred to as "captive hydrogen." This hydrogen is not normally sold or traded between companies but is produced on demand within a chemical plant. It is possible to consider the substitution of these internal sources of captive hydrogen by pipelined hydrogen from external nonfossil sources. The production of ammonia would be a case in point. Although much of an existing ammonia plant serves to produce the hydrogen-nitrogen synthesis gas from natural gas and air, this could be replaced ultimately with a pipelined supply of hydrogen and an air separation plant to provide the nitrogen. The provision of an external pipelined supply of hydrogen to oil refining operations is somewhat of academic interest, since it would appear that hydrogen made from oil on a refinery site would always be more convenient and cheaper than hydrogen from an exotic source. However, in the long term, hydrogen produced from nonfossil energy could be used to upgrade both oil and coal into lighter, cleaner fuels such as gasoline and SNG, resulting in a lower consumption of oil and coal.

A major potential growth area for chemical feedstock hydrogen is in the production of iron and steel. Currently, iron ore is reduced to iron by coal or carbon monoxide derived from coal. Processes exist, however, in which the direct reduction of iron ore by hydrogen is used to produce metallic "sponge iron," which in turn is delivered to a steelmaking process. In the long run, the use of hydrogen for direct reduction of iron ore could be the route by which the iron and steel industry makes the transition from the use of fossil fuel energy to the use of nuclear and solar energy. At the present time, the economics of such processes are unattractive.

Industrial Fuel

Vast quantities of gaseous fuels, in particular natural gas, are used in industry today as sources of process heat and process steam, and for inert or reducing atmospheres. There are huge numbers of industrial burners installed in long-life equipment that will apparently outlive the supply of the fuel on which they are now run. There are many processes such as the calcination of limestone, firing of pottery, heat-treating of metals, and formation of glass, which are far more readily carried out by gas-flame-heated systems than by electrically heated systems. Hence, there is considerable incentive for the development of a gaseous fuel made from nuclear energy to supplement and replace natural gas (and coal-based SNG), rather than the development of electrical systems that would require the complete replacement of the consumers' equipment at the time of transition from natural gas to electricity.

It is interesting to note how much energy is consumed in the United States as an industrial fuel. In particular, the huge amount of energy used to raise steam in industry is equivalent, on a Btu basis, to the energy consumed by all the automobiles in the country (each sector accounts for about 17% of the US energy demand). In the advent of a serious shortage of fossil fuels and a transition to a nuclear or solar energy economy, this very large industrial market for energy must be provided with an adequate energy supply. Hydrogen appears to be the most appropriate way to do this.

Residential Appliances

Most residential appliances use what are known as atmospheric burners. An atmospheric burner is one in which all air for combustion is taken from the surrounding atmosphere at atmospheric pressure and gas is supplied at a comparatively low pressure, normally 3–15 inches of water column. Usually the combustion air is applied to the flame in two stages. Gas is passed through an orifice, which serves to meter flow and accelerate the velocity of the gas. When this gas enters the mixing tube, it entrains air drawn in through a series of orifices near the base of the burner. In the mixing tube, gas and air are mixed together and then passed through small holes called burner ports. When the gas mixture issues from the port, it is ignited and a flame appears. Secondary air, supplied from the atmosphere, diffuses into the flame and serves to complete the combustion. The flame does not propagate back through the burner port because the speed of the issuing gas mixture is greater than the flame speed of the natural gas/air mixture.

If hydrogen were substituted for natural gas in a conventional burner, the following differences would be apparent. First, the orifice that serves to meter the flow of gas would be slightly undersized for hydrogen. The rate of flow of hydrogen through an orifice is about 10% less on a Btu basis than that of natural gas. While this would not be disastrous, it would derate the heating capacity of the burner below the stated and guaranteed performance. Second, when the flame issuing from the burner port is ignited, it would immediately propagate back through the port, because of the high flame speed of hydrogen. The flame would propagate all the way back to the mixing tube and would stabilize at the orifice. The resulting high temperature inside the mixing tube could lead to melting of the burner itself. This problem can be remedied in one of two ways: One is to make the burner ports very small, so that the gas velocity is above the flame speed of hydrogen. This could lead to problems in blockage at the burner ports in the event of food spills or other contamination. Also, when the burner is extinguished, flame propagation backwards through the port would still occur, leading to a noisy extinction within the mixing chamber. Another approach is to eliminate the primary air feed. In this case a stable flame would still be obtained at the burner port because hydrogen does not appear to require the provision of primary air. However, a conventional burner has a significant volume of gas enclosed within the mixing chamber, and upon extinction of the burner, back diffusion of air occurs into the mixing chamber before some of the flames at the burner ports have completely gone out. When sufficient air has diffused into the mixing chamber to reach the upper flammable limit, rapid flame propagation occurs with very loud report. Although this is not dangerous, the noise would presumably be unacceptable in domestic use. Thus, it seems necessary to consider the replacement of conventional gas burners with modified burners that use no primary air; have properly sized ports; use a metering orifice of slightly larger size; and have a volume between the metering orifice and the burner ports that is reduced to an absolute minimum.

Most residential gas appliances are equipped with ignition devices. In general, pilots are used, but in some cases electric spark or glow coil ignition is used. Similar ignition devices would be more than adequate for the ignition of hydrogen. Because of its extremely low ignition energy, only a weak spark is needed, or the pilot could be replaced by a suitable catalytic surface, which would suffice to initiate the flame.

Catalytic Combustion

The potential of catalytic combustion with hydrogen is very great. The primary advantage of catalytic combustion is that combustion, and the resulting release of useful heat, can be caried out without a flame at lower temperatures than in a conventional flame burner. At these lower temperatures, the production of nitrogen oxides from the air is less likely to occur and can be negligible at temperatures below 1000°F. This phenomenon has been demonstrated quantitatively by Sharer & Pangborn (25), who operated catalytic combustors that achieved greater than 99% complete combustion of hydrogen, producing nitrogen oxide levels below 0.1 ppm. For indoor ventless operation, this would result in ambient nitrogen oxide levels well below those deemed to be hazardous. Sharer and Pangborn state:

> One of the major problems encountered in the catalytic combustion of hydrogen is the maintenance of or stability of combustion on a catalytic surface. Because of the physical properties of hydrogen, flame propagation rather than catalytic combustion takes place in atmospheric burners unless one of three criteria is met:
> 1) If the laminar velocity of the combustion mixture of hydrogen and oxygen or hydrogen and air is above the flame velocity of hydrogen, flame propagation cannot exist. This principle is used in conventional flame-type burners when operated with natural gas. Laminar gas velocity through the burner ports is faster than the flame velocity of natural gas (1.0 ft/sec). The flame velocity of hydrogen is 9.4 ft/sec, or almost a factor of 10 greater. It is certainly not practical (and we did not find it possible) to operate a catalytic, atmospheric burner with this high flow rate over the entire catalyzed surface and achieve complete combustion.
> 2) If a fuel-air mixture is outside the flammability limits, flame propagation cannot occur. The lower (lean) flammability limit of hydrogen in air is 4%, and the upper (rich) limit is about 75%. To maintain a mixture that is above the rich limit, complete combustion will never occur. Therefore, rich mixtures are undesirable. However, a system that stays below the lower limit will not support flame combustion. In properly designed systems, catalytic combustion can proceed below this lower limit so that essentially 100% of the fuel is consumed. Greater surface areas or catalyst loadings per unit of fuel consumed are required for systems that operate this lean, and such systems are less desirable. A lean mixture is difficult to obtain in atmospheric burners, and we found no way to obtain this condition in a practical burner.
> 3) If all points on the catalyst surface or in the system that have contact with the combustion mixture are kept below the autoignition temperature of the fuel gas, flame combustion cannot be initiated. The autoignition temperature of hydrogen in air was determined experimentally to be 1085 ± 5°F.

The catalytic burner surface must be designed so that the rate of heat generation is equal to the rate at which heat is removed when surface temperatures are below 1085°F. To use catalytic burners in appliances, a heat sink (load) must be applied to remove heat. For example, water is the load in a water heater, and air is the load in a space heater.

High-activity catalyst surfaces have been developed that permit hydrogen-air mixtures to self-ignite at room temperatures. This will allow catalytic appliances to operate without standing pilots or electric ignition systems. Determination of the lifetime of high-activity catalysts is a parameter that is being studied at IGT.

Attainable Efficiencies

Two factors cause inefficiencies in appliances. One is unburned fuel, which is intolerable when hydrogen is considered for an unvented household appliance. The other is the heat losses that occur as unused heat is lost to the room and as waste heat is vented to the environment through chimneys.

Conventional appliances that incorporate flame combustion of natural gas require venting and usually lose about 25% of the input energy as waste heat out the vents. However, catalyst surfaces that directly heat fluids (air, water, etc., as described above) give high efficiencies automatically. Appliances can be controlled to maintain combustion-surface temperatures below 200°F. These need not be vented because the hazard of hot flue gas does not exist. . . .

Another approach to catalytic combustion of hydrogen has been suggested by Baker (26). He found that a conventional gas burner, when surrounded by stainless steel wool, could be ignited in the conventional way with flame combustion. As the steel wool became heated, however, combustion rapidly became predominantly catalytic on the heated steel surface and open flame combustion decreased to an insignificant level. This type of catalyst, a relatively high-surface stainless steel wool, is potentially cheaper than the platinum catalyst employed by Sharer et al (25) and demonstrates a significant reduction in nitrogen oxide production (2–5 ppm) when compared with the conventional flame-type burner (200–300 ppm). However, the work of Sharer et al indicates that the amount of platinum used in their catalytic combustion equipment was very small and that it would not be a limiting or significant factor in the total cost of a catalytic appliance. Baker's approach would be more amenable to the conventional design of domestic or industrial appliance, while the Sharer and Pangborn approach would require complete redesign and replacement of conventional equipment with catalytic combustion appliances.

Hydrogen as a Fuel for Land Vehicles

Considerable interest has been shown in the potential use of hydrogen as a fuel for automobiles, trucks, and buses. The main incentive for the use of hydrogen is its inherently clean combustion and thus elimination of many of the pollutant problems associated with vehicle engines. Several research teams, mainly associated with universities rather than industry, have investigated the conversion of conventional automobile engines to run on hydrogen. Contemporary projects in the United States on hydrogen-fueled internal combustion engines have recently been reviewed in detail by Escher (27). He discusses 15 organizations that have converted engines and vehicles to operate on hydrogen. In general the conversion of an automobile piston engine to run on hydrogen is relatively simple, requiring a major modification in the way in which fuel is admitted to the engine, but few other major changes. There are three ways in which fuel may be introduced into the engine:

1. The conventional gasoline carburetor can be exchanged for one designed to run on a gaseous fuel. A simple conversion of this type results in satisfactory low-speed operation but can produce severe backfires and erratic performance at high throttle openings. Some researchers have found that alteration of the ignition timing has assisted engines operating in this mode, but more stable operation can be obtained by the injection of massive quantities of water into the fuel-air mixture. This can easily be accomplished by arranging a considerable amount of exhaust recirculation, since the exhaust gases contain large amounts of water vapor.
2. Hydrogen can be injected directly into the fuel cylinder by a third valve; this somewhat resembles a fuel injection engine or a diesel engine configuration.

This technique is considered by its proponents to be much safer than the mixing of hydrogen-air mixtures in a carburetor and passing them through a hot inlet manifold. However, this requires very considerable modifications to the engine itself.
3. The carburetor and inlet manifold can be removed completely and hydrogen injected directly into the inlet ports. Control of the engine is carried out by varying the amount of hydrogen supplied, instead of throttling the air, which is the more normal method of controlling the speed of gasoline engines. This method works with hydrogen because the flammability limits of the hydrogen-air mixture are much wider than those of gasoline.

Escher reviews performance and emission aspects of the different hydrogen engines tested by the various research teams. He reports that engines can operate in an ultralean regime when running on hydrogen because hydrogen engines are able to operate considerably below an equivalence ratio of about 0.8, where gasoline engines reach their lean misfire limits. Such lean burning operation accomplishes two major gains: increased thermal efficiencies and very low levels of oxides of nitrogen emissions.

Escher points out that the increased thermal efficiency is obtained during lean-out of the engine mixture only at the expense of power levels. Thus the rated horsepower of the engine is significantly reduced by running it in this lean condition.

In several studies of emissions from hydrogen-operated automobile engines, wide variations in results of nitrogen oxide emissions have been reported. In general, Escher concludes that a hydrogen engine can produce very high levels of oxides of nitrogen, considerably higher than gasoline in certain instances. For hydrogen engines the peak generation of nitrogen oxide occurs in the vicinity of 0.8 equivalence ratio. However, at equivalence ratios of 0.4 and lower, the level of this emission product falls to an acceptable level in accordance with the projected federal standards of 0.4 g/mile. In addition, charge diluent techniques such as extensive exhaust-gas recirculation and water injection can also maintain acceptable levels of nitrogen oxide emissions.

Thus it appears that the conversion of existing gasoline engines to operate on hydrogen is relatively straightforward and can give rise to improved operation in terms of efficiency and level of emissions.

The most significant problem in using hydrogen as an automotive fuel is that of storage of hydrogen on board the vehicle. The storage of hydrogen as a compressed gas appears impractical, since the weight of the cylinders is almost 100 times greater than the weight of fuel. For example, the standard-sized car with a 20-gal gasoline tank stores about 116 lb of gasoline. An amount of energy equivalent to this would weigh only 40 lb in the case of hydrogen, but might require several thousand pounds of steel tank to contain it. The alternatives of metal hydrides and cryogenic liquid hydrogen have been considered in some detail (28). A metal hydride system that yields reasonable efficiency, the iron-titanium system, is also extremely heavy and a corresponding weight of hydride in a 20-gal gasoline tank would be about 1500 lb. The liquid hydrogen tank is considered the lightest of all, but the cost of a liquid hydrogen tank is thought to be much higher than what the average automobile owner would wish to afford, together with the need for complex handling and servicing equipment at the filling station.

The dispensing of hydrogen to a hydrogen-fueled automobile also involves severe problems. The provision of either high-pressure hydrogen or liquid hydrogen to the refueling station site in an urban location cannot be considered simple or safe, although the technology for it certainly exists. The possible choices for alternative automobile fuel in the future have been examined by Gillis et al (29). They conclude that hydrogen is an almost ideal fuel for vehicles but that it can only be considered in the relatively long-range future, i.e. after the year 2000, because it will take this long for hydrogen distribution and vehicle storage systems to be developed in a satisfactory way and for economic factors to make it a competitive fuel.

Hydrogen as a Fuel for Aircraft

The case for hydrogen-fueled transport aircraft has been discussed by Brewer (30). He maintains that hydrogen is a candidate alternative fuel that offers significant potential advantages (in performance, pollution, noise, and cost) when used in aircraft. Accordingly, it is recommended for initial and early use in aircraft with its use in other energy sectors growing as its availability increases.

The use of hydrogen as an aircraft fuel has been considered with enthusiasm because of the extremely light weight of hydrogen on an energy-per-pound basis. Hydrogen has an energy content of 52,000 Btu/lb, which is about two and a half times that of (conventional) liquid hydrocarbon fuel. The conversion of aircraft gas turbine engines to operate on hydrogen appears to be a relatively simple problem and was demonstrated during the 1950s (31).

Table 7 Comparison of conventional jet-fueled vs liquid-hydrogen-fueled subsonic transport aircraft (current technology)

	Subsonic Transport		Change (%)
	Jet-Fueled	Hydrogen-Fueled	
Payload[a] (lb)	56,000	56,000	
Range (n mi)	3,400	3,400	
Cruise speed (mach)	0.820	0.820	
Gross weight (lb)	430,000	318,000	−26
Operating empty weight (lb)	239,200	215,350	−10
Fuel weight capacity (lb)	137,000	46,650	−65.9
Fuel volume (cu ft)	2,920	11,050	278
Wing area (sq ft)	3,460	2,830	−18.2
Wing span (ft)	155	140	
Wing loading (lb/sq ft)	124	113	
Specific fuel consumption [(lb/hr)/lb]	0.677	0.216	

[a] 272 passengers plus 1600 lb cargo.

Brewer reports a preliminary conceptual study of the subsonic hydrogen-fueled commercial transport aircraft following the general design of the Lockheed L-1011 Tristar. A direct comparison was drawn between two aircraft based strictly on differences imposed by the fuel.

In one hydrogen version of the aircraft, the fuel is carried in wing-tipped tanks, compared with the conventional jet-fueled version where fuel is carried within the wings. This difference is required because although hydrogen is very lightweight, it is also very bulky, requiring about three times the volume for the same amount of energy. Table 7 shows the comparison of the principal design features of the two versions of the aircraft. It is interesting that not only does the gross takeoff weight decrease by a factor of 26%, mainly due to the reduction of fuel load, but the operating empty weight is also 10% lower in the hydrogen version. Another study, carried out by Kirkham & Driver (32), indicates data for a large cargo transport aircraft. The hydrogen version was shown to reduce fuel consumption by more than one third in the energy required per ton-mile when compared with the conventional jet fuel version.

Hydrogen has long been considered as the only fuel possible for supersonic and hypersonic transports because at very high speeds fuel consumption by an aircraft is very high, and thus the saving in weight by using a lightweight fuel becomes of major importance. In addition, a supersonic aircraft requires cooling of the aerodynamic surfaces, which would otherwise become overheated by friction of the air. Liquid hydrogen can be used to cool these surfaces while it is being vaporized before being fed to the engine.

Hydrogen for Electricity Generation

Hydrogen can be used as a fuel for electricity generation in several different ways. However, the high cost of hydrogen is likely to make its conventional use in power stations unattractive. The development of the fuel cell is receiving very considerable attention at the present time in the context of the decentralized power generator. The prospects for fuel cells in dispersed power generation have been reviewed by Lueckel et al (33). They show that by locating generating units within a distribution network, pollution sources would be dispersed, transmission line requirements lessened, existing substation and generating sites utilized, area protection provided, and reserve requirements reduced. Considerable effort has gone into the development of fuel-cell-dispersed generators of approximately 20 MW in unit size. Fuels such as natural gas and low-sulfur oil have been considered, and both of these fuels require reforming or preliminary reaction to produce hydrogen-rich gas before being acceptable to the fuel cell itself. If a pure hydrogen supply was available by pipeline to these dispersed power generators, considerable improvements could be made in both the efficiency and the capital cost of the units. It is estimated that hydrocarbon-air fuel-cell systems incorporating some form of fuel reformer will have overall efficiencies of about 40%, while the equivalent hydrogen-air unit would have an efficiency of 50% or greater, because of the elimination of the fuel reformer. Similarly, elimination of this unit would reduce the capital cost of the fuel-cell system, perhaps by 20%. If, in addition, an oxygen supply is also made available to the electric generator, the efficiencies of fuel-cell units can be raised probably to 55% or 60%.

Another potential for generating electricity using hydrogen as a fuel opens up if we consider the possibility of a supply of oxygen along with the hydrogen. Escher (34) has proposed use of a hydrogen-oxygen closed-cycle steam turbine system in which hydrogen and oxygen are burned together in a burner that is quenched with water to produce steam as its only product. The temperature of the steam is determined by the amount of water used in the quenching circuit. This high-temperature, high-pressure steam is then expanded through a conventional steam turbine and condensed, and most of the water is returned to the quenching system. Such a generating unit would be completely nonpolluting and could have efficiencies as high as 50% or greater. The possibility of hydrogen cooling in the steam turbine blades, analogous to the air-cooling techniques used in gas turbine blades, could result in the use of even higher steam turbine temperatures than are presently attainable in conventional steam systems, since the hydrogen used in cooling could be recovered and recycled through the system without considerable penalty.

CONCLUSION

Hydrogen offers much promise for the future for energy delivery in a mixed electricity-chemical fuel economy based on nonfossil sources. There seem to be no major roadblocks to its introduction other than sheer economics. At the present time, it is cheaper to use more conventional fuels derived from fossil sources and to use electricity.

There are gaps in technology and areas for technological improvements that constitute topics for hydrogen energy research. Principal among these are nonfossil-based methods for hydrogen production (other than electrolysis). Interest in hydrogen energy concepts has been mounting rapidly in the last few years and several major symposia (35–40) on the subject have been held. An important conclusion from the work surveyed here is that hydrogen energy has as great a potential for application as does electrical energy, with which it should usually be compared as an alternative. It appears likely that a mixed hydrogen-electric energy system will develop sometime after the year 2000.

ACKNOWLEDGMENTS

The authors wish to acknowledge the cooperation of many of their colleagues at the Institute of Gas Technology in preparing the papers and reports from which most of this material was drawn. In particular, the contributions of the following staff are noted with thanks: T. D. Donakowski, K. G. Darrow, Jr., W. J. D. Escher, J. C. Gillis, D. Johnson, A. J. Konopka, J. C. Sharer, and J. Wurm.

Literature Cited

1. Gregory, D. P. 1973. The hydrogen economy. *Sci. Am.* 228:13–21
2. DeBeni, G., Marchetti, C. 1970. Hydrogen, key to the energy market. *Euro Spectra* 9:46–50
3. Gregory, D. P. et al. 1972. *A Hydrogen*

Energy System. Am. Gas Assoc. Cat. No. L21173, Arlington, Va.
4. Bilgen, E. Feb. 1975. *On the Feasibility of Direct Dissociation of Water Using Solar Energy.* Publ. No. EP75-R-10, Ecole Polytech. Montreal
5. Stuart, A. K. 1973. *Modern Electrolyser Technology.* Presented at Am. Chem. Soc. Symp. Non-Fossil Fuels, April 1972, Boston
6. Konopka, A. J., Gregory, D. P. 1975. Hydrogen production by electrolysis: present and future. *Proc. 10th Intersoc. Energy Convers. Eng. Conf., Aug. 1975, Newark, Del.*
7. *De Nora Water Electrolyser.* 1973. Oronzio de Nora Implanti Elettrochimici SpA, Milano, Italy
8. *Hydrogen From Water.* 1973. Lurgi Express Information T1084/6.73, Lurgi Apparate-Technik GmbH., Frankfurt
9. Kincaide, W. C., Williams, C. F. 1973. *Storage of Electrical Energy Through Electrolysis.* Timonium, Md: Teledyne Isotopes Co.
10. Titterington, W. A. 1974. *Status of GE Company SPE Water Electrolysis for Hydrogen/Oxygen Production.* Presented at World Energy Syst. Conf., June 1974, Hurst, Texas
11. Funk, J. E. 1972. *Thermodynamics of Multi-Step Water Decomposition Processes.* Presented at Am. Chem. Soc. Symp. Non-Fossil Fuels, April 1972, Boston
12. Wentorf, R. H., Hanneman, R. E. 1974. Thermochemical hydrogen generation. *Science* 185:311–19
13. Pangborn, J. B., Sharer, J. C. 1974. Analysis of thermochemical water-splitting cycles. See Ref. 38, pp. 499–516
14. Marchetti, C. 1973. Hydrogen and energy. *Chem. Econ. Eng. Rev.* 5 (1): 7–25
15. Pangborn, J. B., Gregory, D. P. 1974. Nuclear energy requirements for hydrogen production from water. *Proc. 9th Intersoc. Energy Convers. Eng. Conf., Sept. 1974, San Francisco*
16. Gregory, D. P. *Hydrogen Pipelines.* 1975. Presented at 3rd Int. Pipeline Technol. Conv., Jan. 1975, Houston, Texas
17. Konopka, A., Wurm, J. 1974. Transmission of gaseous hydrogen. *Proc. 9th IECEC.* See Ref. 15
18. Johnson, J. E. 1973. *The Economics of Liquid Hydrogen Supply for Air Transportation.* Presented at Cryog. Eng. Conf., Aug. 1973, Atlanta, Ga
19. Wurm, J., Pasteris, R. F. 1973. *The Transmission of Gaseous Hydrogen.* SPE Pap. No. 4526. Presented at 48th Ann. Fall Meet. Soc. Pet. Eng. AIME, Las Vegas
20. Isting, C. 1974. *Experience with a Hydrogen Pipeline Network.* Presented at EUROCON '74, April 1974, Amsterdam
21. Burger, J. M. et al. 1974. Energy storage for utilities via hydrogen systems. *Proc. 9th IECEC.* See Ref. 15, pp. 428–34
22. Fernandes, R. A. 1974. Hydrogen cycle peak-shaving for electric utilities. *Proc. 9th IECEC.* See Ref. 15
23. Gregory, D. P. 1974. *The Use of Hydrogen for Energy Storage.* Presented at 168th Natl. Meet. Am. Chem. Soc., Sept. 1974, Atlantic City, NJ
24. Reilly, J. J., Hoffman, K. C., Strickland, G., Wiswall, R. H. 1975. Iron titanium hydride as a source of hydrogen for stationary and automotive applications. *Proc. 20th Power Sources Symp., May 1974, Atlantic City, NJ,* pp. 11–17
25. Sharer, J. C., Pangborn, J. B. 1974. Utilization of hydrogen as an appliance fuel. See Ref. 38, pp. 875–88
26. Baker, N. R. 1974. Oxides of nitrogen control techniques for appliance conversion to hydrogen fuel. *Proc. 9th IECEC.* See Ref. 15
27. Escher, W. J. D. 1975. Survey and assessment of contemporary U.S. hydrogen-fueled internal combustion engine projects. *Proc. 10th IECEC.* See Ref. 6
28. Billings, R. E. 1974. *Hydrogen's Potential as an Automobile Fuel.* Publ. No. 74004. Billings Energy Res. Inc., Provo, Utah
29. Gillis, J. C., Pangborn, J. B., Vyas, K. C. 1975. The technical and economic feasibility of some alternative fuels for automotive transportation. *Proc. 10th IECEC.* See Ref. 6
30. Brewer, G. D. 1973. *The Case for Hydrogen-Fueled Aircraft.* Presented at 9th Propulsion Conf., AIAA, Nov. 1973, Las Vegas
31. Witcofski, R. 1972. Potentials and problems of hydrogen-fueled supersonic and hypersonic aircraft. *Proc. 7th Intersoc. Energy Convers. Eng. Conf., Sept. 1972, San Diego,* pp. 1349–54
32. Kirkham, F. S., Driver, C. 1973. *Liquid Hydrogen-Fueled Aircraft Prospects and Design Issues.* Presented at 5th Aircraft Design Flight Test Oper. Meet., AIAA, Aug. 1973, St. Louis, Mo. Pap. No. 73-809
33. Lueckel, W. J., Eklund, L. G., Law, S. H. 1972. *Fuel Cells for Dispersed Power Generation.* Presented at IEEE Meet., Feb. 192, New York. Pap. No. T72-235-5

34. Escher, W. J. D. 1975. Hydrogen-oxygen utilization devices. *Proc. Semin. Key Technologies for The Hydrogen Energy System, Tokyo.* Washington DC: Nat. Sci. Found.
35. American Chemical Society. 1972, 1973. Symp. Nonfossil Fuels, April 1972, Boston; Symp. Hydrogen Fuel, Aug. 1973, Chicago
36. Intersociety Energy Conversion Engineering Conference. 1972–1975. Symposia on hydrogen, at the 7th–10th meetings. Am. Chem. Soc., Washington DC
37. Cornell University. Aug. 1973. Int. Symp. Workshop on Hydrogen Energy, Ithaca, New York
38. Veziroglu, T. N., ed. 1975. *Hydrogen Energy: Proc. The Hydrogen Econ. Miami Conf. (THEME), March 1974, Miami Beach.* New York: Plenum. 2 vols.
39. University of Miami. March 1975. Symposium–Course on Hydrogen Energy. Clean Energy Res. Inst., Univ. Miami
40. University of Miami. March 1976. 1st World Hydrogen Energy Conf., Miami Beach. Clean Energy Res. Inst., Univ. Miami

Copyright 1976. All rights reserved

ENERGY STORAGE ×11012

Fritz R. Kalhammer
Electric Power Research Institute, Palo Alto, California 94303

Thomas R. Schneider
Public Service Electric and Gas Company, Newark, New Jersey 07101

INTRODUCTION

In modern societies, energy is extracted, supplied, and "consumed" (converted into heat and useful work) at different times, in different locations, and in different ways. To reconcile these differences, energy must be stored in large amounts. Despite today's complexity of energy supply and use, energy storage remains largely limited to the familiar forms of fossil fuel storage: in the oil tanks, gas reservoirs, and coal piles of utilities; in the fuel oil tanks of industrial, commercial, and residential consumers; and in the gasoline tanks of transportation modes.

Only a few new forms of energy storage have been introduced during the past 50 years. Among these, electrically heated water tanks and hydroelectric pumped storage have resulted in significant economic and operational benefits to consumers and utilities alike. At a time when society has become acutely aware of the problems and constraints surrounding the key issue of adequate energy supply, the national expectation is that energy storage will eventually provide major conservation, environmental, and economic benefits.

Most importantly, large-scale conservation of irreplaceable petroleum and natural gas resources would become possible if advanced energy storage devices and systems were to power electric vehicles, replace (in conjunction with additional coal or nuclear baseload plants) the combustion turbines used by electric utilities, allow space heating and cooling to be done with off-peak electric power, and permit a more effective utilization of solar energy.

Major environmental benefits can be expected from future, large-scale use of energy storage: electric vehicles would result in greatly reduced urban air and noise pollution and traffic congestion, and dispersed energy storage in utility systems would defer or eliminate the need for unsightly electric transmission lines. Finally, significantly lower costs of energy—for transportation, electric power, and heating/cooling—should be achievable if cost-competitive energy storage methods could be introduced into the transportation, power generation, and residential/commercial sectors. How, and to what extent, energy storage is likely to be used will depend critically on the success of current efforts to develop advanced energy storage devices and systems.

This article begins with a review of those energy storage applications that have potential for major societal benefits. Subsequently, status and prospects of the major energy storage technologies are examined, with a view toward identifying the more promising approaches. This information is followed by a summary and some comments on the outlook for energy storage.

APPLICATIONS OF ENERGY STORAGE

Opportunities for significant new applications of advanced energy storage methods in modern industrial societies are readily identified by considering the flow of energy, from its extraction to its ultimate use. Figure 1 shows this flow, including the locations and forms of energy storage as currently used. As noted before, existing energy storage is limited almost entirely to chemical energy (fossil fuels). Other noteworthy points are that (*a*) with the exception of hydroelectric pumped storage, the lack of storage methods for electric energy reduces the opportunities to use this versatile energy form; (*b*) with the exception of hot water and brick storage, only fuel energy is presently stored sufficiently close to the major end uses to be fully effective in reconciling supply and demand; and (*c*) the biomass (vegetation) is the only—and rather inefficient—medium for conversion and storage of solar energy.

Together with the potential benefits previously presented, the foregoing points indicate that the major opportunities for new applications of energy storage are in electric utility systems, transportation, the residential/commercial sector, and the utilization of solar energy.

Electric Utility Systems

Much of the current interest in energy storage—and many of the ongoing efforts to develop advanced storage technologies—originated from the realization that energy storage can improve the operation and economics of electric power systems.

The role of utility energy storage becomes apparent by considering the daily and weekly variations in electricity demand, as expressed by the typical weekly load curve shown in Figure 2 (*top*). The steady demand base throughout the day is met by baseload plants, the units with the highest efficiency and lowest operating costs— typically, modern fossil and nuclear steam-generating units. The broad daily peak is served by intermediate generating equipment, which comprises a system's less modern fossil steam units and, where necessary, combustion turbines. Peak demands are satisfied by older fossil steam units, combustion turbines, and diesels. Although this "generation mix" approach has worked quite well in the past, it is becoming increasingly costly and restrictive. Sharply increased fuel prices create a heavy cost penalty for older, less efficient equipment, and the limited supplies and high costs of natural gas and distillate fuels make the use of low-capital-cost combustion turbines less attractive than in the past. Thus there is a clear conservation and economic incentive to use baseload plants as a source of the power now generated by peaking and intermediate equipment. In a simplified way[1] this use is shown in Figure 2

---

[1] In the illustrative example, a combined daily and weekly cycle of the energy storage system is shown. In actual applications, specific cycles and discharge scheduling will be chosen on the basis of economic optimization studies.

ENERGY STORAGE   313

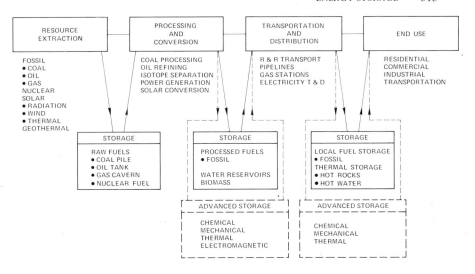

Figure 1 Energy flow and storage.

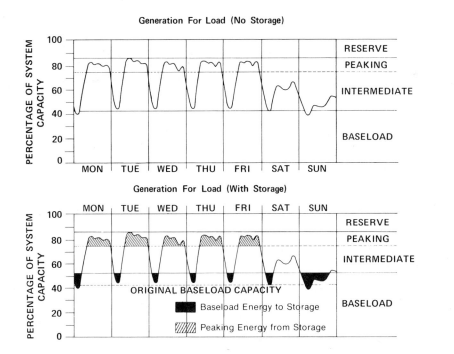

Figure 2 Typical weekly load curve of an electric utility.

(*bottom*): the more efficient and economical baseload capacity of a system is increased, and the excess capacity during night and weekend periods is used to charge the energy storage system. Discharge of that system provides for the daily peak demands, thus replacing oil- and gas-burning generating capacity. From published projections (1) it is readily estimated[2] that replacement of peaking combustion turbines would result in the conservation of nearly 100 million barrels of petroleum per year by 1985.

The favorable experience with hydroelectric pumped storage—which in the United States goes back to a 1929 installation at the Connecticut Light Company—has already shown that energy storage on electric power systems can yield substantial operating and economic benefits. Significant expansion of pumped hydro storage from 8100 MW in May 1974 to about 40,000 MW in 1993 can be expected (2). However, this capacity represents less than 4% of the generating capacity projected for that year (3). Geographic, geologic, and environmental constraints are likely to limit pumped hydro storage to a steadily decreasing percentage role thereafter. This potential must be compared with recent estimates [see, for example, (4)], which indicate that between 10% and 20% of an average utility system's generating capacity could be installed in the form of energy storage. This represents a very large ultimate market for competitive new energy storage methods.

For the near term, energy storage can effectively increase the utilization of existing baseload equipment. In the longer term, energy storage will permit an increase in the percentage of baseload capacity. In both instances, availability of low-cost energy for charging the storage device and acceptably low capital costs of storage systems are necessary to make energy storage a truly attractive option for utilities.

Besides the availability and cost of charging energy and of energy storage devices, several other factors will determine to what extent and how electric utilities will use energy storage. Several preliminary analyses of these questions have been carried out (4–7). These studies indicate that 10–12 MW-hr of storage capacity should be available for each megawatt of power-generating capacity from storage. Operationally attractive energy storage systems would be expected to provide power for 2000–2500 hours per year, depending somewhat upon the generation mix, actual incremental operating costs of generation, and the power system's available storage capacity (8–10). Simple sensitivity analyses (7, 9, 11) can be used to approximate the influence of key parameters on the economic competitiveness of energy storage methods for utilities. For example, as shown in Figure 3, the economic break-even cost of energy storage systems may be expressed as a function of peaking equipment cost, with parametric dependence on the costs of combustion turbine fuel and of charging energy for the storage device.

The detailed determination of maximum energy storage capacity and storage system power capacity requires study of utility system load data. Such studies are done routinely by utilities considering specific installations of pumped hydro storage. Methodology for describing characteristics of energy storage systems and representative electric utility load data is currently being developed by the Public

---

[2] The estimate is based on assuming the following combustion turbine usage by 1985: (*a*) total capacity, 66,000 MW; (*b*) average load factor, 0.075; and (*c*) average heat rate, 12,000 Btu/kW-hr.

Service Electric and Gas Co. in a study for the Electric Power Research Institute and the US Energy Research and Development Administration. A useful approach to the economic description of energy storage systems is to employ a cost function that separates the capital cost elements into power-related and storage capacity–related capital costs (12). This approach tends to differentiate between the different storage methods in terms of their suitability for shorter or longer periods of energy storage.

Certain advanced energy storage devices, such as batteries and flywheels, appear capable of being sited closer to the load than conventional generating equipment for peaking and intermediate cycling power. Dispersed siting of energy storage could result in significant savings or deferments of capital costs for new transmission and distribution facilities; dispersed storage could also increase system reliability.

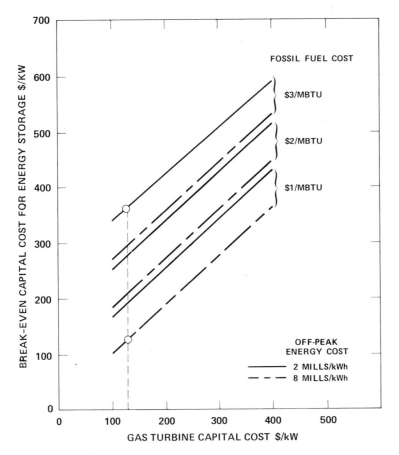

Figure 3 Break-even capital costs for energy storage vs gas turbine capital costs for intermediate load operation. Source: (7).

Preliminary analyses of these benefits have been made (8, 9, 13). Inasmuch as capital cost credits could materially affect the economic competitiveness of batteries and other energy storage devices (14), detailed analyses of specific applications need to be carried out in the coming years. Realistic analyses must include comparisons with generation equipment such as fuel cells (possibly also combustion turbines), which also could be sited close to load centers. Another factor to be considered is that installation costs per unit of storage system capacity tend to be higher for smaller installations, even for basically modular devices such as batteries.

Energy storage is almost invariably associated with inefficiencies that increase storage system operating costs. Devices with low efficiencies will be economical only for large ratios of a utility system's peak power costs to the off-peak power costs. In practice, efficiency is not considered a problem area because most of the proposed advanced energy storage methods appear capable of efficiencies above 70% (see Table 2, in Summary, near the end of this chapter) and the economic benefit of efficiencies above 80–85% is quite small.

Lifetime of a storage system is a more critical factor. The annual carrying charges can become economically prohibitive for devices with a short life, for example less than 10–15 years. To achieve utility-type service life, some energy storage systems may require upgrading or replacement of life-limiting components or subsystems as part of their scheduled maintenance.

Overall, the anticipated conservation and operational benefits constitute a large incentive toward development and introduction of new methods for utility energy storage. Current uncertainties regarding applicable technical and economic feasibility criteria—and the prospects for advanced technology to meet these—will receive

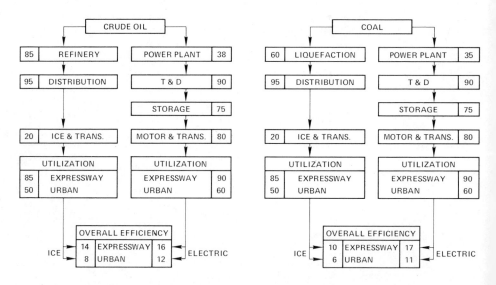

Figure 4 Energy conversion efficiencies in transportation. Source: (12).

clarification in the study by the Public Service Electric and Gas Co. and through the refined analyses expected over the coming few years.

Transportation

The propulsion of electric vehicles could become one of the most important applications for advanced energy storage devices. The transportation sector currently accounts for more than half of the US petroleum consumption, or approximately 25% of total energy use; highway vehicles consume 75% of that fraction (15). This huge amount of energy is used quite inefficiently: the average US mileage of 13.5 miles/gallon (16) corresponds to an overall efficiency of only 8%, based on a road energy requirement of 0.16 kW-hr per ton-mile of a typical urban driving cycle (17). A major shift of transportation requirements to electric vehicles could lead to a substantially more efficient transportation system and result in large-scale conservation of irreplaceable petroleum resources.

The expected, higher efficiency of electric vehicles for two different driving modes is compared with the internal combustion engine (ICE) in Figure 4. Similar data have been derived by other authors (18, 19). Because the efficiency of batteries (and other advanced power sources such as flywheels and fuel cells) increases with decreasing load, the efficiency advantage of electric vehicles is greatest in urban driving. For a coal-based fuel energy system—the inevitable trend for the future—this advantage could reach 50%. Accordingly, very significant petroleum savings can be predicted even for limited populations of electric vehicles (12, 19).

Significant operating-cost savings of battery-powered vehicles are established for industrial trucks (20), anticipated for US postal delivery vehicles (21), and projected for commercial vans in Europe (22). These savings are expected to increase and apply to broader classes of electric vehicles in future years if and when fuel costs increase and low cost off-peak electric energy is available for battery charging. Also, a significant population of advanced electric vehicles could improve load factor and power generation economics of electric utility systems (23). Yet another benefit from widespread use of electric vehicles would be a major reduction of air and noise pollution, especially in urban areas (24). In the United States, the economic benefits from reduced air pollution could reach several hundred million dollars per year (25).

Clearly, major incentives exist for the introduction of practical—that is, technically and economically feasible—electric vehicles. It is generally agreed that the development of the vehicle itself and the required power conversion subsystems are primarily design and production engineering tasks [see, for example, (26)]. The status of the relatively more advanced European efforts in this area has recently been summarized (27). Also, provision of the needed electric energy supply is likely to be well within the capability of future electric utilities (12, 19, 20). Thus, the pacing step in the realization of broadly useful electric vehicles will be the development of improved energy storage devices. In Table 1, the situation is put in perspective by comparing storage device criteria for several classes of electric vehicles with the capability of the lead-acid battery. While this comparison illustrates the basic limitations of current battery technology, it also indicates that a twofold or threefold increase of energy density to perhaps 30–35 W-hr/lb would be sufficient to make

Table 1 Minimum energy and power densities for electric vehicle power sources[a]

	Family Car	Commuter Car	Utility Car	Delivery Van	City Taxi	City Bus	Lead-Acid Battery
Conventional vehicle construction							
Energy density (W-hr/lb)	135	41	26	50	96	81	12–15
Power density (W/lb)	94	46	40	55	45	36	20–30
Lightweight vehicle construction							
Energy density (W-hr/lb)	87	28	18	33	64	55	
Power density (W/lb)	60	31	28	36	30	25	

[a] Adapted from (28).

batteries and other energy storage devices far more competitive power sources for vehicle propulsion.

This conclusion did not attract enough attention in the past when the focus was almost exclusively on matching the performance and range of the standard US car with electrically propelled vehicles (29). The events during and following the oil crisis have increased public interest in electric vehicles of more modest performance. This interest is reflected in the recent introduction in the US Congress of a bill (H.R. 8800, September 8, 1975) that would authorize a $160 million, four-year federal program for research, development, and demonstration of commercially feasible electric vehicles. The text of that bill includes a particularly well stated and complete summary of the incentives for electric vehicle development.

A vigorous research and development program to develop advanced batteries and other energy storage devices suitable for electric vehicle propulsion should result in significantly improved vehicle capabilities within 10 years, with good prospects for further improvements if technically and economically feasible batteries of high energy density (≥ 100 W-hr/lb) can be developed. The status and prospects of the more promising technologies for vehicular energy storage are reviewed further below.

Residential, Commercial, and Industrial Sectors

At present, the only residential and commercial applications of energy storage are based on thermal storage in hot water or ceramic brick storage heaters. The significance of these applications for the energy system as a whole lies in a partial shift of the electric energy demand to off-peak hours when it can be met with baseload generation operating on coal or nuclear energy. For the electric utilities, this shift tends to result in improved load factors for generation, transmission, and distribution equipment, and hence, improved economics of operation.

Hot water heaters controlled by timing devices have been used in parts of the United States and in Europe for decades, generally with positive but limited impacts on utility load curves. A considerably larger impact can be expected from the use of electric storage heaters because the amounts of energy involved are several times larger. In the United States, there has been little incentive for customers to install storage heaters because, historically, most utilities have not offered reduced off-peak rates (30). This may change if current experiments (31) with storage heaters prove technically and economically successful. In Europe, storage heaters are already in widespread use. The incentive for customers to make investments of about $100–200/kW is provided by off-peak rates that are only a fraction (for example, 25–30%) of the daytime rates. For RWE, Germany's largest utility, installations of storage heaters have reached an aggregate demand of nearly 25% of the system load, with a significant improvement of the system load factor in the peak (winter) season (32). The RWE system is essentially saturated with storage heaters since further increases would result in undesirable effects, including new demand peaks during originally off-peak hours, and the need to reinforce the distribution network. An important aspect of extensive use of storage heaters is the high level of control needed by utilities if a positive impact on their load curves is to be assured. Information obtained from RWE indicates that the associated development efforts and capital investments were substantial.

For many US utilities, electric air conditioners represent a large part of the daily demand during the summer peak periods. Shifting this demand to off-peak hours through use of "coolness" storage would be very desirable, but only a few experimental installations have been described so far (33). A limitation is that coolness storage systems will always be more bulky and costly than storage heaters since heaters have much higher specific energy storage capacities.

Thermal or other energy storage is of less significance in the industrial sector. Many modern industrial processes are continuous and operate at high load (plant) factors, with little opportunity for energy storage. Some processes, such as production of steel, glass, and aluminum, have inherent thermal storage in the molten materials. This type of storage could become more important if socioeconomic forces were to shift working hours increasingly toward the daytime period. An interesting application of energy storage is for certain industries to shift part of their energy-intensive chemical processing to daily, weekly, and seasonal periods of low demand, with the processed chemical intermediate serving as a form of chemical energy storage. Because industrial consumers of electricity are usually subject to a demand charge that is determined by the maximum power used, energy storage to shave sharp demand peaks might have some merit. Case-by-case analyses of energy storage economics and the comparison with alternate techniques of demand management (such as manual or computer-assisted equipment scheduling) are needed to establish opportunities for industrial applications of energy storage.

Utilization of New Energy Sources

UTILIZATION OF SOLAR ENERGY Solar energy has potential for becoming a major new source of energy. An NSF/NASA Panel Report (34) projected that by the year 2000, 10% of the heating and cooling demand and 10% of the demand for electric energy in the United States could be met by utilizing solar energy. Although appreciably lower utilization percentages have been extimated more recently (35), there remains a substantial incentive to develop the solar energy resource.

The main problems associated with efficient and cost-effective utilization of solar energy derive from the diffuse and intermittent nature of the resource. Use of energy storage in solar energy systems can solve or mitigate these problems. Most importantly, energy storage bridges the gap between solar energy input during sunny days and energy demands at night or during cloudy daytime periods. The general approach in designing solar energy utilization systems is therefore to provide for substantial storage capacity (36–43).

For solar heating and cooling of buildings, only thermal energy storage is appropriate (38–40). The larger the installation for solar energy collection and storage, the smaller the dependence of buildings on supplementary electric or thermal energy. If little or no storage is provided, solar energy utilization systems can only "displace energy," that is, replace some of the energy supplied by conventional sources. On the other hand, if the system includes sufficient storage, it can "displace capacity," that is, reduce the demand for generating capacity placed on the conventional sources. The important consequences of this factor for the cost-effectiveness of solar energy utilization have been emphasized recently (35). Detailed analyses of overall energy systems will be required to estabilsh the most economical storage capacities of systems for residential and commercial utilization of solar energy.

Energy storage is also a practical necessity for solar-thermal generation of electric power. Depending on the type of solar-thermal system and the location of storage within the system, various kinds of thermal storage will be more appropriate than batteries, which tend to result in higher costs of electricity (35). On the other hand, batteries are the logical complement of photovoltaic solar conversion systems (36) unless these systems, which are currently prohibitively expensive, are aimed almost exclusively at energy displacement. Some credit for displacement of capacity may accrue to solar-assisted cooling systems because of a significant coincidence of high solar energy input levels with maximum requirements for air conditioning.

Conversion of wind energy is an indirect way of utilizing solar energy. Wind energy conversion systems feeding power directly into the electric network have been proposed. Again, such systems would primarily displace energy, although some generating capacity would be displaceable in geographic areas where a significant coincidence exists between wind levels and heating requirements (44). However, wind conversion systems would be more broadly useful if they were coupled to energy storage, either mechanically to hydro pumped storage, flywheels, and compressed air installations (37), or electrically to electrolytic hydrogen production and storage systems (45).

ADVANCED UTILIZATION CONCEPTS   Energy storage could become an important factor for the optimal utilization not only of solar energy but also of other energy resources. For example, closed-loop chemical energy conversion and storage systems (see section on chemical energy storage) could be coupled thermally to nuclear fission reactors. Nuclear heat would thus become economically transportable over significant distances (46), and the efficiency of energy end use as heat could be higher,[3] compared with systems using electric energy as the intermediate energy carrier. High-temperature nuclear reactors might become efficient producers of hydrogen if practical, closed cycles for thermochemical splitting of water can be developed. This hydrogen could be used flexibly, as part of systems for utility energy storage and peak power generation (9), as a key chemical, or as an important fuel.

Because of the dispersed and frequently rather remote geographic location of geothermal energy resources, it has been proposed (47) to couple this resource to the US energy system via hydrogen production and storage subsystems.

In general, energy storage appears to offer considerable potential for more flexible and efficient utilization, reduced environmental impacts, and improved economics of future energy resources.

---

[3] Detailed comparisons must take into account the transportation losses in electric and chemical energy transport, the impact of heat pumps on overall thermal efficiency, and so forth.

METHODS AND TECHNOLOGIES OF ENERGY STORAGE

In the following sections, we review energy storage technologies with potential for application to nationally significant uses. On the basis of the stored energy form, these technologies fall under four broad groups: mechanical, thermal, chemical, and electromagnetic energy storage. Each storage method has specific technical and cost characteristics. These are discussed briefly; a comparison table is included with the Summary section.

Mechanical Energy Storage

HYDROELECTRIC PUMPED STORAGE ("PUMPED HYDRO") This method is based on pumping water from a lower to a higher level with expenditure of pumping energy. Much of this energy is recovered by allowing the water to return to the lower level through a turbine driving an electric generator. The water may be pumped between two dedicated reservoirs, or the lower (upper) reservoir may be a naturally occurring body of water such as a lake, river, or ocean.

Hydro pumped storage is a technology already in a mature state of development, and an extensive body of knowledge exists (48, 49). Continuing improvements in performance are being made, but the basic energy conversion devices are well developed and available from major equipment supplies. Many plants have been built and more are planned, and new plant costs can be estimated with considerable reliability. Properly designed plants operate efficiently and with low maintenance costs (50). Research and development problems relate mainly to the proposed underground siting of lower reservoirs (51, 52) and to overcoming environmental objections to siting (53).

In underground pumped storage the lower reservoir and power plant are located in deep underground caverns and the upper reservoir is at the surface. By being free of surface topographical restrictions, the siting of these underground plants should be considerably easier than the siting of conventional pumped storage facilities. The underground reservoir and power plant could use naturally occurring caverns, abandoned mines, or a mined-out cavern consisting of a tunnel labyrinth excavated specifically for the pumped storage reservoir. In existing mine sites, firsthand knowledge of the subsurface rock formations is available, and existing shafts can be used.

With the elevation difference (head) between the upper and lower reservoirs now a variable parameter, the major design restrictions are equipment capability and rock conditions. If very high heads are used, there may be cost penalties associated with very deep mining; nevertheless, mining costs per unit of energy storage capacity should decrease with depth because of proportional reductions in the volume of the reservoir.

Depending upon the head selected, there are several choices of equipment and powerhouse arrangements and also the option of using an intermediate reservoir and powerhouse. The latter would permit the use of conventional equipment despite heads with a very high total pressure.

Several engineering cost estimates have been made and at least one utility has filed for a license from the Federal Power Commission for a potential underground hydro pumped stage installation (GPU Service Corporation). Without taking credit for mined rock, the existing cost estimates are in the neighborhood of $200/kW for a sufficiently large head. The installation of at least one underground pumped hydro plant should occur by 1992 (53).

Equipment needed for underground pumped hydro is commercially available and in widespread use for conventional hydroelectric and pumped storage facilities. Several manufacturers are developing equipment capable of handling very high heads in a single lift. This equipment should improve the economics of underground pumped hydro.

Underground construction and mining technologies are available and can be adapted for this system. The largest uncertainty is the underground reservoir: its cost, durability with pressure cycling, and the rate of water leakage into the lower reservoir. Costs are heavily dependent upon suitability of the site and local labor conditions. Given a specific site and knowledge of the geological locations, good accuracy in the cost estimates is expected. Since the economics of scale in pumped hydro dictates sizes in the range of 200–2000 MW, these units require substantial transmission facilities unless sites can be developed within or near large urban load centers.

COMPRESSED AIR–GAS TURBINE SYSTEMS Compressed air storage, as currently conceived, uses a modified combustion turbine (split Brayton cycle), uncoupling the compressor and turbine so that they can operate at different times and incorporating the intermediate storage of compressed air. During off-peak load periods, the turbine is disengaged and the compressor is driven by the generator, which is now used as a motor and takes its power from other generating units through the system's interconnections. The stored compressed air is subsequently used during peak load periods when it is mixed with fuel in the combustion chamber,[4] burned, and expanded through the turbine. During that period, the compressor clutch is disengaged and the entire output of the turbine is used to drive the generator, which feeds power to the electrical system.

Since in normal operation the compressor consumes about two thirds of the power output of the turbine, the rating of the gas turbine operating from the stored compressed air is increased roughly by a factor of 3. This permits redesign of the compressors, the combustion process, and the combustion turbine, free from the allodynamic and thermodynamic restrictions inherent in designs of conventional combustion turbines. Current estimates for the heat rate of the combustion turbine operating from stored compressed air are in the range of 4000 Btu/kW-hr to 5400 Btu/kW-hr. A compression/generation energy ratio in the range of 0.65–0.75

[4] In an advanced concept proposed by I. Glendenning of the Central Electricity Generating Board Engineering Laboratories, Marchwood, United Kingdom, the heat liberated during compression would be stored and used to reheat the air during expansion (discharge). No fuel would be required in this mode of compressed energy storage.

should be readily available. The variable maintenance cost should not be any greater than that for a conventional combustion turbine.

The compressed air may be stored in naturally occurring reservoirs (caverns, porous ground reservoirs, and depleted gas or oil fields) or man-made caverns (dissolved-out salt caverns, abandoned mines, or mined hard-rock caverns). Air storage may be accomplished either at variable pressure or, through the use of a hydrostatic leg, at constant pressure. Each approach has its advantages and all are applicable to different underground reservoirs.

Compressed air storage is an old concept. According to a personal communication from R. A. Huse of Public Service Gas and Electric Co., conceptual design work was underway in the late 1930s and the earliest public work relating to this concept is Gay's patent (54). However, a more extensive literature on the subject was published only in the past five years (55–64). Considerable interest in this concept has been expressed in Sweden, Finland, Denmark, Yugoslavia, France, and the United States, and one commercial unit is under construction in Huntorf, West Germany (65).

The required development work relates mainly to the physical-mechanical characteristics of the storage reservoirs and the adaption and modification of existing equipment for use with suitable couplings and possibly unique valving. Some research is required to investigate: (a) geological conditions for underground storage, (b) new approaches to underground cavern construction, (c) energy losses in storing and moving air, (d) alternative concepts of air storage, and (e) corrosion effects on turbines from air contamination. The major uncertainties include the cost of the air storage facilities, the performance and durability of the storage facility with pressure and thermal cycling, and leakage from the storage reservoir. Also, additional geological survey work to identify the availability and number of possible sites is necessary before the future role of compressed air storage may be assessed.

FLYWHEELS   Energy storage in form of the kinetic energy of a rotating mass has been used almost since the beginning of the industrial age. The technological advances in rotating machinery and high-strength, lightweight materials achieved since then—especially in the past few decades—hold out promise for longer periods and greater specific capacity of energy storage, which raises the possibility of new applications.

Most of the proposed, advanced applications of flywheel systems have been directed at either vehicular propulsion or electric utility energy storage. For vehicles, several modern applications are reported (66–69), while utility system applications have been restricted to special-purpose uses for smoothing pulsed power (70, 71). The proposed superflywheel designs deal primarily with the wheel itself without treating the full energy storage system in sufficient detail (72–74). While extensive design experience exists for spacecraft momentum wheels (75) and for small vehicular flywheels, the large wheels proposed for utility applications appear to be outside the size of current, state-of-the-art, cost-effective designs.

Previous applications have used steel wheels and direct mechanical drives. Advanced wheel designs have proposed the optimal use of essentially anisotropic materials such as glass-fiber reinforced plastic. A well-designed flywheel made from advanced composite materials should be capable of storing in excess of 30 W-hr/lb.

Basic approaches to wheel design that have been discussed in the literature include (a) a brushlike design that uses thin rods or filaments of high-strength materials (74), (b) multi-rim constant-thickness wheels (72), and (c) constant-stress wheels (75). Innovative designs will be necessary to achieve high working stress levels in advanced composites and must take into account realistic stress levels in fabricated wheels, practical methods and acceptable costs of fabrication and quality assurance, and the fracture tolerance of the wheel material. Wheel configurations of high energy density and potentially low cost will not by themselves insure feasibility of compact, economical flywheel systems inasmuch as auxiliary and safety subsystems will contribute materially to the size and cost of a flywheel system.

Thermal Energy Storage

Thermal energy storage may be defined as storage of energy in the form of (a) sensible heat and (b) the latent heat associated with phase changes. The major technical parameters for thermal energy storage include the storage medium, the operating temperature range, and the mode of heat exchange between the storage subsystem and the heat source/sink. Any practical system must include not only the thermal energy storage and transfer subsystems but also provisions for control and insulation.

Thermal energy storage can be useful in a wide spectrum of applications, including (a) hot water heating, (b) heating and air conditioning of buildings with off-peak (or solar) energy, (c) low-temperature process steam storage, (d) central-station thermal storage (for conventional or solar-thermal power plants), and (e) industrial process heat storage. Depending largely on the temperature of the storage medium, these uses may be grouped into applications of low-grade (relatively low-temperature) and high-grade (high-temperature) heat.

STORAGE OF LOW-GRADE HEAT   Storage of sensible heat in hot water reservoirs is established commercial practice and is not reviewed here. Heat storage in the ceramic bricks of storage heaters[5] has gained commercial acceptance in Europe. Among the key tasks in making storage heaters practical were the design and refinement of control methods (31, 76). Development continues on improved ceramic materials of high specific heat capacity and good thermal cycling ability, new approaches to high-quality thermal insulation, and extension of the concept from individual room units to central systems.

The operation of air conditioning systems with off-peak power requires coolness storage. Two prototypes of coolness storage systems were constructed and tested at the University of Pennsylvania (77, 78). Although coolness storage has not yet found significant applications, it would be a useful option in the service areas of summer-peaking electric utilities, and commercial applications appear to be economically attractive.

Storage of relatively low-temperature heat will be a key requirement for the residential and commercial utilization of solar energy, and appropriate approaches

---

[5] Although storage heaters, strictly speaking, are not low-temperature devices, they are discussed in this category because the stored energy is ultimately used as low-temperature heat.

have been explored experimentally (79). Thermal energy storage systems based on storage of latent (phase-change) heat are attractive in principle because of their high specific storage capacity. However, despite their greater bulk, systems using sensible-heat storage in liquids are almost certainly more practical and economical because they avoid the problems and costs associated with heat exchangers.

Storage of waste heat from power plants or industrial processes for later use is another possible application of low-grade heat storage. For example, municipal heating systems can be integrated with power plants via hot water storage and transport (80). This energy system approach could improve the economics of generating equipment such as fuel cells that permit ready recovery of their waste heat. A rather more speculative possibility is the long-term, seasonal storage of waste heat in thermal wells (81). Detailed analyses from the standpoint of a total energy system wound be necessary to establish the merits and economic trade-offs of waste-heat storage and utilization. The comparatively high cost of transporting low-quality thermal energy is one of the major factors to be considered in such analyses.

STORAGE OF HIGH-GRADE HEAT The advantage of storing high-temperature heat is twofold: (*a*) the specific storage capacity of a high-grade heat storage system tends to be high (comparable for example, with chemical energy storage; see Summary, Table 2) and (*b*) high-temperature heat can be converted with good efficiency into other forms of energy, especially work. Storage of high-temperature, high-pressure steam-water mixtures is the prime example in this category. The basic technology of steam storage is well understood, and thermal storage has been in service since 1929 in the power plant of Berlin-Charlottenburg, which has a power rating of 50 MW and a storage capacity of 67 MW-hr (82). Steam can be stored in steel tanks (83, 84), which is expensive, or in underground reservoirs (85). To be economical, innovative engineering designs must be applied toward reduction of the costs of pressure vessels.

While the basic technologies associated with thermal storage via hot water and steam are straightforward, the large size and high costs of the storage vessels suggest consideration of latent-heat thermal storage systems because of their higher specific storage capacity. The problems with this approach include corrosion by high-temperature working fluids, such as fused salts, and the complex heat transfer to and from the two-phase system. Novel approaches might prove attractive, such as a direct-contact heat exchanger (86) or the modification of steam-water storage in Ruth accumulators by addition of suitably shaped phase-change material. Whether latent-heat storage will find future applications depends primarily upon the development of novel heat exchanger designs and the identification of phase-change materials that combine high thermal conductivity and stability under cycling with a small change of density upon melting and freezing.

In power plant applications of high-grade heat storage, steam storage would probably be used together with a separate peaking turbine. In another approach, hot feedwater storage would be integrated into the design of the station; this would require a rather sophisticated main turbine. Although a thorough analysis of probable capital costs has not been made, preliminary information suggests that such systems could be economically attractive (84) if costs of storage tanks can be reduced.

Similar storage schemes can be developed around working fluids other than water; sodium would be particularly appropriate in conjunction with sodium-cooled nuclear reactors. However, safety and regulatory considerations could impede the acceptance of central, thermal energy storage because of the required interfaces with the working fluids of nuclear power plants.

As discussed previously, practical solar-thermal power-generating systems will require thermal energy storage, and the entire range of possibilities under consideration for central power plant thermal storage can be considered for this application of the future.

Chemical Energy Storage

In the broadest sense, chemical energy storage is the storage of energy as the chemical potential of metastable reactants that can be made to react with a net release of energy. Storage of energy in chemical form has two inherent advantages. The high energy density of a chemical system results in compact, generally low-cost storage and ready transportability of energy, and chemical energy is readily converted into other useful energy forms by a variety of methods and devices. These advantages are responsible for the almost exclusive use of conventional chemical fuels as today's energy storage media.

The chemical energy storage methods and systems reviewed in this article meet a second criterion that excludes conventional fuel storage: the reactant systems containing the stored energy must be re-formed readily from their reacted (the discharged) state upon addition of energy in a suitable form. This more restrictive definition focuses on advanced methods that seek to retain the basic advantages of chemical energy storage while adding new characteristics that increase its usefulness. All chemical energy storage systems may be described functionally by the general scheme shown in Figure 5, but widely differing subsystem configurations and conversion techniques are used in the specific storage devices and systems reviewed in the following.

BATTERIES In secondary or "storage" batteries, the conversion from electrical to chemical energy (charging) and the reverse process (discharging) is performed by way of electrochemical reactions. The electric form of input and output energy, compactness, and the modular characteristics common to electrochemical devices make batteries potentially the most useful among advanced energy storage methods. A number of different electrochemical systems have been developed, or offer prospects for development, into practical storage batteries. We review the more promising of these briefly. Key characteristics of advanced battery technologies are discussed more thoroughly in several recent reviews (87–92).

The lead-acid battery is the only chemical energy storage system with near-term prospects for application in vehicular propulsion and possibly also for utility energy storage. For application to vehicles, reductions in weight and cost of the battery (per unit of stored energy) are highly desirable. The key to lower specific weight and cost is a better utilization of lead; this is being achieved in advanced designs using thin grids (93). The use of experimental lightweight lead-acid batteries has already

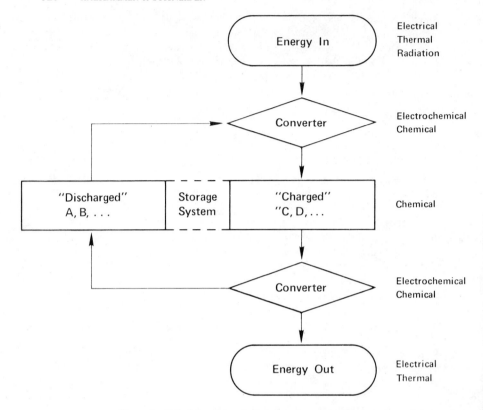

Figure 5 Principle of chemical energy storage systems.

resulted in a significantly improved driving range of prototype vehicles (94). It is not yet clear whether the combination of low cost and long cycle life needed for utility energy storage can be met with state-of-the-art lead-acid batteries. Several US battery manufacturers[6] are currently examining the feasibility of developing suitable designs.

If the technical goals for mobile and stationary applications can be met, inexpensive mass-production techniques remain to be developed before lead-acid batteries can capture extensive new markets over the next five or ten years.

Iron/nickel oxide ("nickel/iron") and zinc/nickel oxide ("nickel/zinc") batteries are under development for vehicular applications (95, 96) where they might introduce the first significant improvement in energy density over the lead-acid battery, possibly by 1980. The major technical shortcomings—gassing and a significant self-discharge rate for the nickel/iron battery, limited cycle life of the zinc electrode

---

[6] ESB, Inc., Gould Inc., and C&D Batteries Division of Eltra Corp., under contracts respectively with the Electric Power Research Institute, the US Energy Research and Development Administration, and the International Lead-Zinc Research Organization.

in the nickel/zinc battery—are likely to preclude their use for utility energy storage.

Metal-gas batteries are functional hybrids between conventional batteries and fuel cells. Because of their potential for good energy density and reasonably low cost, the zinc-air, iron-air, and zinc-chlorine systems are the most interesting representatives of this group. The zinc-air battery has been under development for many years, but problems with zinc electrode rechargeability have forced developers into increasingly complex engineering approaches to cell design and operation; to date, none of these has been commercialized. The key problems of the iron-air battery are its gassing and self-discharge tendency. Zinc-air and iron-air batteries are handicapped with respect to efficiency by the poor reversibility (high voltage losses) of the air electrode and with respect to energy density and cost by the need to remove carbon dioxide from the ambient air used for discharge. For these reasons alone, application of metal-air batteries to utility energy storage does not appear attractive. The outlook for vehicular applications is still uncertain.

The zinc-chlorine battery is attracting considerable interest, primarily because an apparently practical solution to the problem of chlorine storage has been developed (97) at EDA (Energy Development Associates, Madison Heights, Michigan). Potential for high efficiency, good cycle life, and low cost are claimed for this battery, and a vehicle test (using a mechanically charged battery) was successful (98). A substantial program to develop a 20-kW-hr vehicle battery is in progress at EDA, and the merits of the zinc-chlorine battery for utility energy storage are currently being explored. Areas of uncertainty include the cycle life of zinc and chlorine electrodes and the ultimately achievable efficiency and cost of completely engineered batteries. If these questions can be resolved positively within the next two or three years, a commercial zinc-chlorine battery could become available within another five years.

Redox batteries—in which the positive and/or negative active materials are dissolved in the electrolyte—have been proposed (99, 100) for large-scale energy storage. The potential advantage of this approach (compared with more conventional battery designs) is that external reactant storage in tanks tends to result in relatively low capital costs for the storage-related part of capital costs. This characteristic might qualify redox batteries for accumulating and storing energy over longer periods— for example, weekends—than can be handled economically by conventional batteries. Because the development of redox batteries is at an early stage, research and development over three to five years are likely to be required before the true potential of this new battery type for large-scale energy storage can be assessed.

Sodium/sulfur batteries were first described (101) by workers at the Ford Scientific Research Laboratories who introduced the battery's key component, the solid β-alumina electrolyte. At the battery operating temperature of about 250–330°C, the sodium and sulfur as well as most of the discharge products are liquids. Liquids have no structural memory; hence, there is no possibility for the development of the cumulative, deleterious changes of the electrodes that tend to limit the life of more conventional batteries.

Work on sodium/sulfur batteries is currently underway in a number of laboratories around the world [for example, see (102)]. Recent results at Ford and at the

Laboratoires de Marcoussis of CGE in France indicate that there is probably no inherent limit to the cycle life of the β-alumina electrolyte (103). This potential for practical cycle life needs to be verified for batteries constructed with commercially feasible materials and fabrication techniques. Available cost estimates range from \$15/kW-hr to \$30/kW-hr, but they are rather speculative at the present state of development. Assuming continued technical advances, a commercial sodium/sulfur battery might be established by approximately 1985.

A battery using a sodium negative, a β-alumina electrolyte, and antimony chloride positive is being developed at ESB, Inc. (104). Although similar in concept to the sodium/sulfur battery, the ESB battery operates at somewhat lower temperature, with less demanding requirements for seals and containers.

Lithium/iron sulfide batteries are being developed at Atomics International (AI) (105) and the Argonne National Laboratory (ANL) (106). Both programs emphasize utility energy storage, but design of a vehicle battery is also addressed at ANL. Considerable progress has been made through development of solid iron sulfide positive electrodes and solid or liquid lithium-alloy negative electrodes. For either approach, the capability for long cycle life and potential for low cost are yet to be established. A recent engineering cost analysis (107) indicates a probable manufacturing cost of between \$20 and \$25 per kilowatt-hour of storage capacity. Extensive development and engineering efforts will be required over the coming three to five years to verify the battery's technical and cost potential. Even if current approaches prove to be basically successful, the establishment of a commercial technology is likely to require at least another four to six years.

Conventional batteries represent a special case of chemical energy storage in which all functions of the energy storage system (see Figure 5) are combined into a single device. In redox batteries, this close coupling of energy conversion and storage is relaxed to some extent because the energy-carrying reactants are stored outside the energy converter, although close by because the same converter is used in the reconversion step.

In other chemical energy storage systems, such a close coupling between the major functions does not usually exist. This results in several potential advantages: (*a*) possibility of using a variety of initial conversion steps to couple energy storage with various primary energy sources, (*b*) low cost of bulk chemical energy storage, (*c*) transportability of the stored energy, and (*d*) added flexibility through use of different modes for reconversion of the stored energy. The obvious disadvantage (compared, for example, with batteries) associated with the physical separation of functions is the need for separate devices and generally higher capital costs for the system and lower efficiencies. A case-by-case analysis of chemical energy storage systems is required to establish whether they offer net advantages over other forms of energy storage.

HYDROGEN ENERGY STORAGE SYSTEMS Hydrogen energy storage represents the best-known example of advanced chemical energy storage. Several approaches have been proposed and explored for each of the required subsystems—hydrogen generation, storage, and reconversion—which can be combined in various ways into overall energy conversion and storage systems. More detailed surveys are given in available bibliographies, literature surveys, and conference proceedings (108–110).

For hydrogen generation from water and energy,[7] electrolysis is the only established industrial process. Current electrolysis technology [for a complete review, see (111)] is handicapped by modest efficiency and high capital costs, but considerable potential appears to exist (112) for development of more efficient, lower-cost electrolyzers. Realistic targets for advanced technology might include efficiencies up to 100% and electrolysis equipment costs between $100/kW and $150/kW. At present there is little commercial incentive to develop such technology, but with a significant research and development commitment, these targets might be achieved within four to six years.

Closed-cycle thermochemical processes are being proposed for hydrogen production via water splitting, but current work is still in the conceptual and early laboratory stages [see (113–116) for several recent accounts]. The incentive to develop such processes derives from the potential for efficiencies and economics that might be superior to those offered by electrolysis, particularly if sources of fairly high-temperature heat—such as high-temperature, gas-cooled reactors (117, 118) or perhaps focused solar heat (119)—become available. The theoretical and laboratory studies currently underway in several organizations should result in a more realistic assessment of practically achievable efficiencies and capital costs for thermal water splitting within the next few years. The development of practical processes for thermal splitting of water will undoubtedly require extensive laboratory and engineering efforts for at least ten years. Integration of these processes with the nuclear heat source and commercialization of the entire hydrogen production system are likely to require another ten years and large capital investments.

Hydrogen storage, the second major subsystem of hydrogen energy storage systems, can take several different forms. Storage of compressed hydrogen is technically feasible now, but the economics are not fully established because of uncertainties regarding suitable containment methods and materials. Storing hydrogen in more concentrated forms—as a cryogenic liquid (120) or chemically bound in metal hydrides (121, 122)—is technically feasible and logistically attractive. However, cryogenic storage of hydrogen carries a significant efficiency penalty that is unacceptable for large-scale energy storage on utility systems, and capital cost, logistics, and safety are likely to present severe problems for mobile applications. The outlook is better for metal hydride storage, but development efforts over a three- to five-year period are still required to establish the probable technical and economic characteristics of this method for hydrogen storage.

Reconversion of hydrogen to electric energy can be done in fuel cells or in combustion-based devices (gas-fired boilers or gas turbines). The fuel cell approach offers potential for high efficiency, with 60% as a realistic target for pure hydrogen fuel. Although a commercial technology is not now available, much of the fuel cell technology currently being developed at Pratt & Whitney for electric power generation (123) will be applicable to fuel cell systems operating on pure hydrogen.

---

[7] The scope of this review excludes hydrogen production from fossil fuels and water.

Probable technical and economic characteristics, and the first generation of a commercial fuel cell technology, are expected to be established within the next three to four years.

Advanced combustion technology such as "hotshot" hydrogen-oxygen gas turbines also appear to offer potential for high energy conversion efficiencies—on the order of 50% and higher for a combined combustion turbine-steam cycle (124).

Complete hydrogen energy storage systems consist of various combinations of hydrogen generation, storage, and reconversion. An experimental electrolyzer/ hydride storage/fuel cell system of 12.5-kW electric power output has recently been described (125). A much larger prototype test facility is being designed at Brookhaven National Laboratory, with completion planned for 1977. At a projected cost of \$300/kW (126), future electrolyzer/hydride storage/fuel cell systems might be economically justifiable for some electric utility systems. Other combinations of hydrogen energy storage subsystems have been examined, with generally similar conclusions (9). However, the uncertainties of available cost projections must be recognized. Current, simplified analyses of hydrogen storage systems have not considered credits for special characteristics, such as benefits from dispersed energy storage (7, 9) or the high operational flexibility of hydrogen systems (127). On the other hand, the overall efficiencies of hydrogen energy storage systems are likely to be lower and the capital costs higher, compared with several other energy storage methods. Accordingly, it is imperative that the unique advantages of hydrogen energy systems be adequately identified and, if possible, quantified to guide development and applications of hydrogen energy storage technology.

CLOSED-LOOP CHEMICAL SYSTEMS Other recently proposed concepts for chemical conversion and storage of energy are based on closed-loop chemical reaction systems. Such systems would be thermally coupled to nuclear (or solar) heat sources to achieve an energy-absorbing chemical change. The absorbed energy, now in chemical form, would be storable and transportable, possibly over significant distances. At the point of consumption, the reaction would be allowed to proceed in the reverse direction, with evolution of heat at a somewhat lower temperature. To be suitable, the forward and reverse reactions must be readily reversible and must occur at useful temperatures. One of the most promising concepts appears to be the ADAM-EVA system ($CH_4 + H_2O \rightleftarrows CO + 3H_2$) that is currently under development in Germany (128) for efficient and economic transport of nuclear heat to industrial and residential users. The very similar HYCO concept ($CH_4 + CO_2 \rightleftarrows 2CO + 2H_2$) of General Electric appears to have promise for off-peak energy storage and for transport of heat (129). Several other reversible chemical reactions can be considered for these applications (129).

The function of the chemical reaction system in the applications just described is to convert energy into a stable, readily handled form—a function that is as much stabilization as it is storage per se. Going one step further, one may include under chemical energy storage those reactions (or reaction systems) that can be used to directly convert solar or nuclear radiation into the chemical energy of metastable reaction products. The general problems with such photolytic and radiolytic reactions appear to be lack of selectivity and the difficulty of stabilizing the energetic primary products [for a review, see (130)]. If these problems can be solved, chemical energy conversion and storage might eventually offer more efficient and less costly routes for the utilization of solar and fusion energy—the energy sources of the future.

Superconducting Magnetic Energy Storage (SMES)

The application of superconductivity to power systems is a technology in a very early stage of development (131, 132). The proposed use of a superconducting inductor for energy storage takes advantage of the principle that energy can be stored in an inductor of zero resistance for, theoretically, an infinite amount of time. An inductor made from a superconducting material and stabilized with a normal conductor is housed in a large Dewar. The superconducting magnet is charged using off-peak energy, and during peak periods, energy is fed back to the system through an inverter, which is used as a rectifier during charging. Work on superconducting systems for energy storage is being carried out at the Los Alamos Scientific Laboratory (133, 134) and the University of Wisconsin (135, 136). Much of the present effort is concentrated on small-scale experimental work and estimation of the economics of large SMES systems using current technology.

Current cost estimates—based on upward scaling of known technology by several orders of magnitude in size and storage capacity—must be examined carefully (134, 135) before conclusions can be reached about eventual economics. So far, all data indicate that this will be a rather expensive method for energy storage. Further development of superconducting materials with high critical currents and high fields at reasonable cryogenic temperatures could reduce projected system costs. Since total stored energy increases with the volume, and the quantity of superconductor and support structure increases with the surface area, the relative costs of storage should decrease as the magnet size increases. Small (100 MW-hr to 1000 MW-hr) magnets would be too costly; only when sizes exceed 10,000 MW-hr of storage capacity do the economics begin to appear attractive. This minimum economical size requires a very large magnet that must be buried in bedrock to reduce the costs of support structures.

Many detailed technical problems remain to be solved before SMES can be considered feasible, and substantial reductions in costs will be necessary before it can be applied to bulk energy storage in utility systems. Other applications do exist: SMES makes a very attractive pulsed power supply device (137), and its use in conjunction with the proposed magnet systems for controlled thermonuclear fusion is of considerable interest. The feasibility of SMES for utility energy storage might eventually benefit from development of superconducting technology for these applications.

SUMMARY

Storage of energy is basic to the functioning of all societies, even simple ones. Modern industrial societies depend vitally on storage of very large amounts of energy, primarily in the form of fossil fuels. The multiplicity and complexity of energy-dependent functions and services in a modern society open up broad opportunities and potential benefits from storage of energy in new forms.

In electric utility systems, pumped hydroelectric storage has already demonstrated significant benefits. The full potential of utility energy storage can be achieved only through development of advanced, more broadly applicable storage methods and systems of competitive technical and economic characteristics. Development of such systems could result in reduced costs and higher reliability of electric service. The conservation goals suggested by President Ford for the electric utilities could eventually be met by a shift—with the aid of energy storage—from petroleum to a mix of coal and nuclear fuel for generation of peaking and a part of the intermediate cycling power.

Energy storage (in lead-acid batteries) is currently used only on a small scale for propulsion of electric vehicles. Major societal benefits could result from the development of advanced energy storage devices meeting the power source criteria for larger populations of practical electric vehicles. The shift from a petroleum fuel base for urban vehicles to electric power generated centrally from coal and/or nuclear fuel, and the higher efficiency of electric vehicles in urban driving, offer potential for large conservation benefits and a drastic decrease of urban air pollution.

Thermal energy storage in the residential sector is a commercial reality in Europe today. Although climatic and economic conditions are substantially different in the United States, it appears likely that residential and commercial thermal storage installations, including storage of coolness, could generate economic and petroleum conservation benefits for the US utilities and the public. Industrial opportunities for energy storage are less well defined and are likely to be limited to specific industries that depend on batch-type operations. Social forces could lead to shift of industrial operations more toward daytime hours. In conjunction with increased electrification, this could result in an increased need for industrial or, alternatively, utility energy storage.

Use of thermal energy storage will be a key to the more efficient and economical utilization of solar energy for residential and commercial heating and cooling and for electric power generation. Energy storage and transport via advanced chemical reaction systems appear to have potential for efficient and economical utilization of nuclear or solar high-temperature heat.

The broad range of possible uses of advanced energy storage methods gives rise to an equally broad range of desirable storage device and system characteristics. In response to the opportunities for large-scale applications, a variety of energy storage technologies are currently under study and development. These efforts are funded by industry, the electric utilities (primarily through the Electric Power Research Institute), and the federal government.

Information on characteristics and status of candidate energy storage methods is summarized in Table 2. Several changes made in adapting this table from (12) reflect the better understanding (mainly of cost factors) developed over the past year. Additional changes of technical and cost parameters must be expected, especially for those technologies that are in an early stage of development.

Pumped hydroelectric storage is an established, operationally and economically viable technology. Because underground pumped hydro will have similar techno-economic characteristics but fewer siting constraints, a number of electric utilities are likely to turn to this storage method. The major barriers to widespread use are geologic factors, environmental and space constraints because of the large size of commercially feasible installations, and long construction times.

Compressed air–gas turbine energy storage systems are close to commercial feasibility, with the first demonstration by a German electric utility expected in 1977. If existing uncertainties regarding geology and costs of caverns and future availability of combustion turbine fuel can be resolved, compressed air storage could become attractive to US utilities.

Kinetic energy storage in flywheels is attractive in terms of storage efficiency and prospective low costs of power handling capacity. However, even the advanced flywheel designs currently under study and exploratory development are unlikely to meet the storage capital cost criteria that apply to utility energy storage and electric vehicles. This high cost is likely to restrict the application of flywheels to uses where high power capability over short discharge periods is of prime importance.

Storage of thermal energy in storage heater systems is technically rather simple and has demonstrated operational and economical benefits in European residential and commercial applications. Similarly, technical and economic feasibility may be anticipated for the thermal storage systems required for residential/commercial utilization of solar energy. Higher-grade heat can be stored as steam and pressurized hot water. This type of thermal energy storage appears to have good technical potential for integration with the steam supply systems of electric utility plants. The major uncertainties are economic (especially regarding the feasibility of developing storage vessels of acceptably low cost) and regulatory (primarily regarding the required integration of the energy storage system with the nuclear steam supply).

Batteries qualify for the broadest range of energy storage applications on the basis of prospective performance and operational characteristics. Their large-scale use will be contingent on successful development of advanced batteries with longer life, lower costs, and (especially for electric vehicles) higher energy density than the lead-acid battery.

Hydrogen and other chemical energy storage systems have unique potential for utilization of primary energy sources and flexible use of the stored energy. However, unless major advances result from current research and development on hydrogen production, storage, and conversion technologies, relatively low efficiency and high capital costs will be major barriers to the introduction of hydrogen energy storage systems. The status of other chemical energy conversion and storage systems is not sufficiently advanced to permit projections of their ultimate applicability and cost; however, their basic potential appears to justify increased conceptual and experimental exploration.

OUTLOOK

It is reasonable to expect introduction of the first compressed air and underground pumped hydro storage installations into service in US utilities by 1990. During the

Table 2 Projected characteristics and status of some energy storage systems[a]

Type	Round Trip Efficiency (%)	Capital Costs[b] C_p($/kW)	Capital Costs[b] C_s($/kW-hr)	Energy Density (kW-hr/ft³)	Development Stage	Potential Application
Mechanical						
Pumped hydro	67–75	100–140	2[c]–15	0.04[d]	Existing application; engineering studies for underground	Central energy storage for peak shaving and load leveling
Compressed air–gas turbine system	65–75[e]	120–150	3–10	0.1–0.5	First commercial demonstration 1977	Central energy storage for peak shaving and load leveling; reserve generating capacity
Flywheels	70–85	80–120	50–100[f]	0.5–2	Initial development	Distributed energy storage; power factor correction; emergency generating capacity
Thermal						
Steam (pressure vessel)	70–80	150–250	15–25	up to 1	Historical installations; engineering studies of modern systems	Central energy storage, integrated with baseload steam generation
Hot oil	65–80	150–250	10–50			
Batteries						
Lead-acid	60–75	60–100	25–50[g]	1–2	State-of-the-art	Distributed energy storage for daily peak shaving; stand-by and emergency generating capacity; vehicle propulsion; energy storage in solar energy systems
Advanced aqueous	60–75	60–100	15–50[g]	1–3	Small prototypes	
High-temperature	70–80	60–100	15–35[g]	2–5	Laboratory cells	
Redox	60–70	100–200	5–15[g]	0.5–2	Conceptual and laboratory studies	

System					Status	Application
Chemical						
Hydrogen (electrolysis plus fuel cell)	35–55	300–400	5–30	N.A.[h]	Advanced development of subsystems	Central energy storage with distributed generation; combined gas/electric energy systems
Reaction systems (closed loop) $CH_4 + H_2O \rightleftharpoons CO + 3H_2$	?	?	?	N.A.	Conceptual studies and initial development	Conversion, storage, and transport of nuclear and solar energy
Electromagnetic						
Superconducting magnets	80–90	40–50	35–200	0.5–1	Concept; key components under development	Central energy storage and system stabilization (large-scale only)

[a] Adapted from (12).
[b] Total storage system capital cost is given by $C_t = C_p + t_{max} \times C_s$, where t_{max} is the maximum period for which the storage system can be discharged at its rated power.
[c] Assuming one existing reservoir (lake).
[d] Assuming 3000-ft head.
[e] Efficiency with respect to recovery of stored energy.
[f] Not including subsurface vault (estimated at $20–50/kW-hr).
[g] Not including installation (estimated at $3–10/kW-hr).
[h] Not applicable.

same period, advanced lead-acid batteries should find increasing use in electric vehicles, with some possibility that their usefulness can be demonstrated also for utility energy storage. Storage of low-temperature heat will also increase as residential and commercial utilization of off-peak power and solar energy increase. All of these uses require only evolutionary changes and adaptation of existing technology.

The introduction of more advanced energy storage technologies and systems will depend greatly on the success of ongoing research, development, and engineering efforts. The next two to five years should resolve basic questions regarding the technoeconomic potential of advanced batteries, hydrogen and other chemical energy storage systems, flywheels, and high-temperature thermal storage. Some of these technologies might begin to find commercial applications in the mid- to late 1980s.

The realization of large-scale applications of new energy storage methods will depend not only on the success of research and engineering development but also on a number of institutional factors. These include the large costs of commercializing new technologies, future costs and availability of energy sources and conversion equipment, and regulatory strategies affecting competitive situations in the energy sector.

Some of the more important strategy options are in (a) pricing and tax policies with respect to oil and natural gas—the major fuels used for transportation, home heating, and electric peak power generation; (b) restrictions and priorities in fuel allocations; (c) electric power pricing policies to achieve better electric load management; (d) tax and other financial incentives to promote use of energy storage devices; and (e) the extent of national commitments to the nuclear option and the utilization of solar energy.

Despite the considerable uncertainty, the potential benefits and technological possibilities of energy storage appear sufficiently large to insure for it an important role in the energy systems of the future.

Literature Cited

1. Federal Power Commission. Aug. 1973. *Report of the Task Force: Utility Fuels Requirements (1974–2000)*. Washington DC: FPC
2. Federal Power Commission. May 1974. *Staff Report on the Role of Hydroelectric Developments in the Nation's Power Supply*. Washington DC: FPC
3. Federal Power Commission. Sept. 1974. *Staff Report on Electric Utility Expansion Plans for 1984–93*. Washington DC: FPC
4. Rosengarten, W. E., Kelleher, A. J., Gildersleeve, O. D. Oct. 1972. *Wanted: Load-Leveling Storage Batteries*. Presented at Fall Meet. Electrochem. Soc., Miami Beach, Fla.
5. Ku, W. S., Sulzberger, C. L. July 1960. *Determination of Pumped Storage Requirements and Limitations on a Long Range System Basis*. Presented at IEEE Summer Power Meet., New Orleans, La. Pap. No. 31PP66-496
6. Ley, R. D., Loane, E. S. July 1962. General planning of pumped storage. *J. Power Div. Proc. ASCE*, p. 21
7. El-Badry, Y. Z., Zemkoski, J. Aug. 1974. The potential for rechargeable storage batteries in electric power systems. *Proc. 9th Intersoc. Energy Convers. Eng. Conf., San Francisco*, p. 896
8. Lewis, P. A., Zemkoski, J. March 1973. *Prospects for Applying Electrochemical Energy Storage in Future Electric Power Systems*. Presented at IEEE Int. Conv., New York

9. Fernandes, R. A. Aug. 1974. Hydrogen cycle peak-shaving for electric utilities. See Ref. 7, p. 413
10. Fernandes, R. A., Gildersleeve, O. D., Schneider, T. R. Sept. 1974. Assessment of advanced concepts in energy storage and their application on electric utility systems. *Trans. 9th World Energy Conf., Detroit, Mich.*
11. Casazza, J. A., Huse, R. A., Sulzberger, V., Salzano, F. J. Aug. 1974. *Possibilities for the Integration of Electric, Gas and Hydrogen Energy Systems. CIGRE 1974 Session, Paris.* Pap. No. 31-07. Available from Public Service Electric & Gas Co., Newark, NJ
12. Kalhammer, F. R. Sept. 1974. *Energy Storage: Incentives and Prospects for Its Development.* Presented at Symp. on Energy Storage, Div. Fuel Chem., Am. Chem. Soc. Meet., Atlantic City, NJ
13. Berkowitz, D. G., Brown, J. T. March 1974. *Batteries for Peaking Generation.* Westinghouse 1974 Electr. Util. Eng. Conf., East Pittsburgh, Pa.
14. Kalhammer, F. R., Zygielbaum, P. Nov. 1974. *Potential for Large-Scale Energy Storage in Electric Utility Systems.* Presented at Am. Soc. Mech. Eng. Winter Meet., New York. ASME Reprint No. 74-WA/Ener-9
15. US Department of Transportation, Transportation Energy R & D Goals Panel. 1972. *Research and Development Opportunities for Improved Transportation Energy Usage.* Summary Tech. Rep. NTIS Publ. No. PB 22 612. Springfield, Va: Nat. Tech. Inf. Serv.
16. Strombotne, R. L., Yang, B. Feb. 1974. *Evolution and Effects of Energy Conservation in the Transportation Sector.* Presented at Meet. Am. Assoc. Adv. Sci., San Francisco
17. Salihi, J. T. 1973. Energy requirements for electric cars and their impact on electric power generation and distribution systems. *IEEE Trans. Ind. Appl.* 1A-9(5):516
18. Salihi, J. T. March 1975. Kilowatthours vs. liters. *IEEE Spectrum,* p. 62
19. Nelson, P. A., Chilenskas, A. A., Steunenberg, R. K. Nov. 1974. *The Need for Development of High-Energy Batteries for Electric Automobiles.* Rep. No. ANL-8075. Argonne Nat. Lab., Argonne, Ill.
20. Ferrell, D. T., Jr., Salkind, A. J. 1967. Battery-powered electric vehicles. In *Power Systems for Electric Vehicles,* p. 61. US Public Health Serv. Publ. No. 999-AP-37
21. The U.S. postal service. 1974. *Adv. Battery Technol.* 10:2

22. Plust, H. G. Dec. 1974. *Elektrostrassenfahrzeuge—Technischer Stand und Entwicklungstendenzen.* Presented at Kraftfahrtechnisches Seminar, Institut für Kraftfahrzeugwesen, Technische Hochschule, Aachen, Germany
23. Oehms, K.-J., Busch, H. Feb. 1974. The supply of propulsion energy to electrically powered vehicles and its incorporation in the load curves of public utilities. *Proc. 3rd Int. Electr. Vehicle Symp., Washington DC.* Pap. No. 7411
24. Mencher, S. K., Ellis, N. M. Nov. 1971. Environmental impact of air pollution. *Proc. 2nd Int. Electr. Vehicle Symp., Atlantic City, NJ*
25. Ahern, W. R. 1973. Measuring the value of emissions reduction. In *Clearing the Air: Federal Policy on Automotive Emissions Control,* ed. H. D. Jacoby, J. D. Steinbrenner, and others. Cambridge, Mass: Ballinger
26. Vasant, C. A. 1967. The mechanical design of electric automobiles. In *Power Systems for Electric Vehicles,* p. 143. US Public Health Serv. Publ. No. 999-AP-37
27. Busi, J. D., Turner, L. R. June 1975. *Electric Vehicle Research, Development and Technology—Foreign.* Unclassified Rep. No. DST-1850S-403-75. Defense Intelligence Agency, Washington DC
28. George, J. H. B., Stratton, L. J., Acton, R. G. 1968. *Prospects for Electric Vehicles: A Study of Low-Pollution—Potential Vehicles—Electric.* Summary Rep. to US Dep. Health, Education, Welfare. HEW Rep. No. C-69260. Washington DC: US Dep. HEW
29. Starkman, E. S. 1968. *Prospects of Electrical Power for Vehicles.* Presented at Mid-Year Meet. Soc. Automot. Eng., Detroit, Mich. SAE Pap. No. 680541
30. Gordian Associates. July 1974. *Minutes and Commentary of the Utility Load Management Conference, Paris,* p. 45. New York: Gordian Assoc.
31. Van Cleave, D. Nov. 1974. Storage heaters and ripple control: a way to shape demand to fit capacity. *Electr. Light Power* 52:27
32. Biasin, K., Lange-Hüsken, M. 1974. Aufladeregelung und -steuerung elektrischer Speicherheizungen—Entwicklung und Tendenzen. *Elektrowarme Int.* 32(A6):A273
33. Dudley, J. C., Freedman, S. I. 1973. Power reduction in air conditioning by means of off-peak operation and coolness storage. *Trans. ASME, J. Eng. Power,* Pap. No. 73-WA/Ener-1
34. National Science Foundation/National

Aeronautics and Space Administration, Solar Energy Panel. Dec. 1972. *An Assessment of Solar Energy as a National Energy Resource.* Univ. Maryland, College Park, Md.
35. Spencer, D. F. April 1975. *Solar Energy: A View from an Electric Utility Standpoint.* Presented at Am. Power Conf., Chicago, Ill.
36. California Institute of Technology, Jet Propulsion Laboratories. Oct. 1973. *Workshop Proceedings: Photovoltaic Conversion of Solar Energy for Terrestrial Applications, Volumes I & II.* Rep. No. NSF-RA-N-74-013. Washington DC: Nat. Sci. Found.
37. Savino, J. M., ed. Dec. 1973. *Workshop Proceedings: Wind Energy Conversion Systems.* Rep. No. NSF/RA/W-73-006. NASA Lewis Research Center, Cleveland, Ohio
38. General Electric Co. May 1974. *Solar Heating and Cooling of Buildings: Phase 0. Feasibility and Planning Study Final Report.* Rep. No. NSF-RA-N-74-021 (A-E). Washington DC: Nat. Sci. Found.
39. TRW Systems Group. May 1974. *Solar Heating and Cooling of Buildings: Phase 0.* Rep. No. NSF-RA-N-74-022A and TRW Rep. No. 25168.002. Washington DC: Nat. Sci. Found.
40. Westinghouse Electric Corp. May 1974. *Solar Heating and Cooling of Buildings: Phase 0.* Rep. No. NSF-74-CS84-1-(1-3). Washington DC: Nat. Sci. Found.
41. Aerospace Corp. Jan. 1974. *Solar Thermal Conversion Mission Analysis.* Rep. No. ATR-74 (7417-05)-1. Vols. 1–4
42. Aerospace Corp. Jan. 1975. *Solar Thermal Conversion Mission Analysis: Southwestern United States.* Rep. No. ATR-74 (7417-16)-2. Vols. 1–4
43. Aerospace Corp. Jan. 1975. *Mission Analysis of Photovoltaic Solar Energy Systems.* Rep. No. ATR-75 (7476-01)-1
44. Galanis, N., DeLisle, A. Aug. 1973. Performance of wind driven heating systems. *Proc. 8th Intersoc. Energy Convers. Eng. Conf., Univ. Penn., Philadelphia, Pa,* p. 376
45. Dambolena, I. G., Kaminsky, F. C., Rikkers, R. F. Aug. 1974. A planning methodology for the analysis and design of wind-power systems. See Ref. 7, p. 281
46. *Nukleare Fernenergie: Feasibility-Studie.* Dec. 1974. Prepared by Kernforschungsanlage Jülich, Gmbh, and Rheinisches Braunkohlenwerk A. G., Köln, Germany
47. Maslan, F., Gordon, T. J. March 1974. Geothermal energy as a primary source in the hydrogen economy. *Proc. The Hydrogen Economy Miami Energy (THEME) Conf., Univ. Miami, Miami Beach, Fla.,* p. S1-25
48. Garvey, T., Karadi, G. 1971. Pumped storage bibliography. In *Pumped Storage Development and Its Environmental Effects,* ed. R. Krieck, G. Karadi, S. Csallany. Am. Water Resourc. Assoc.
49. Schlimmelbusch, J. S. June 1971. *Pumped Storage, A Bibliography (1961–1970).* Bonneville Power Administration Library, Portland, Oreg.
50. Ferreira, A. 1973. Initial operating and testing experience: Northfield Mountain pumped-storage project. *Proc. Am. Power Conf.* 35:955
51. McCreath, D. R., Willett, D. C. 1973. Underground reservoirs for pumped storage. *Bull. Assoc. Eng. Geol.* Vol. 10, No. 1
52. Warnoch, J. G., Willett, D. C. March 1973. Underground reservoirs for high headpumped storage stations. *Water Power,* p. 81
53. O'Brien, T. May 4, 1975. JCP & L asks clearance for plant under lake. *The Star Ledger,* Newark, NJ, p. 6
54. Gay, F. W. Jan. 1948. Means for storing fluid for power generation. *US Patent* No. 2,433,296
55. Whitehouse, G. D., Council, M. E., Martinez, J. D. 1968. Peaking power with air. *Power Eng.* 72 (1):50
56. Olsson, E. K. A. May 1970. 220-MW air storage power plant. *ASME Pap.* No. 70-GT-34
57. Korsmeyer, R. B. Feb. 1972. *Underground Air Storage and Electrical Energy Production.* Rep. No. ORNL-NSF-EP-11. Oak Ridge Nat. Lab., Oak Ridge, Tenn.
58. Harboe, H. 1971. Economical aspects of air-storage power. *Proc. Am. Power Conf.* 33:503
59. Ayers, D. L., Hoover, D. Q. 1974. Gas turbine systems using underground compressed air storage. *Proc. Am. Power Conf.* 36:379
60. Rogers, F. C., Larson, W. E. 1974. Underground energy storage. *Proc. Am. Power Conf.* 36:369
61. Giramonti, A. J., Lessard, R. D. 1974. Exploratory evaluation of compressed air storage peak-power systems. *Energy Sources* 1 (3):283
62. Rudisel, D. A. June 1974. *Energy Storage for Electric Peaking Power.* Northern States Power Co. Res. Dep., Minneapolis, Minn.
63. Day, W. H., Alff, R. K., Jarvis, P. M. July 1974. *Pumped Air Storage for Electric Power Generations.* Presented

at IEEE 1974 Summer Meet. and Energy Resour. Conf.
64. Glendenning, I. May 1975. Compressed air storage. *Cent. Electr. Generat. Bd. Symp. Long-Term Studies, Oct. 1974. CEGB Research,* No. 2. p. 21. Leatherhead, UK: Central Electricity Generating Board
65. Herbst, H-C. Oct. 1974. Air storage-gas turbine, a new possibility of peak current production. *Proc. Tech. Conf. Storage Syst. for Secondary Energy, Stuttgart, FRG.* Transl. from German by D. K. Dreyer
66. Dann, R. T. May 17, 1973. The revolution in flywheels. *Mach. Des.,* pp. 130–35
67. Biggs, F. Nov. 1974. *Flywheel Energy Systems.* Rep. No. SAND 74-0113. Sandia Lab., Livermore, Ca.
68. The Oerlikon Electrogyro. Dec. 1955. *Automob. Eng.* (London), p. 559
69. Kalra, P. May 1975. Dynamic braking. *IEEE Spectrum,* p. 63
70. Ringland, W. C., Little, C. W. Feb. 1958. *Specifications for CFSF and DF Power Supply for the C-Stellorator Facility at Project Matterhorn.* Rep. No. CTR-2. Princeton Univ. Plasma Lab., Princeton, NJ
71. Kilgore, L. A., Washburn, D. C., Jr. Nov. 1970. Energy storage at site permits use of large excavators on small power systems. *Westinghouse Engineer,* p. 162
72. Post, R. F., Post, S. F. 1973. Flywheels. *Sci. Am.* 229 (6): 17
73. Rabenhorst, D. W. Aug. 1974. *The Multirim Superflywheel.* Tech. Memo. TG 1240. Appl. Phys. Lab., Johns Hopkins Univ., Baltimore, Md.
74. Rabenhorst, D. W., Taylor, R. J. Dec. 1973. *Design Considerations for a 100-Megajoule/500-Megawatt Superflywheel.* Tech. Memo. TG 1229. Appl. Phys. Lab., Johns Hopkins Univ., Baltimore, Md.
75. Notti, J. E., Cormack, A., III., Schmill, W. C. April 1974. *Integrated Power/ Attitude Control System (IPACS) Study.* NASA Rep. No. CR-2383. Vols. 1 and 2. Washington DC: NASA
76. *Elektrowärme Int.* Nov. 1974. Vol. 32 (Edition A, No. 6), pp. A273, 276, 279, 284, 288, 292, 296
77. Dudley, J. D., Freedman, S. I. Nov. 1973. *Power Reduction in Air Conditioning by Means of Off-Peak Operation and Coolness Storage.* Presented at Am. Soc. Mech. Eng. Winter Ann. Meet. ASME Reprint No. 73-WA/Ener-1
78. Lorsch, H. G. 1972. *Solar Heating Systems Analyses.* Rep. No. NSF/RANN/SE/ GI27970/TR72/19. Nat. Center for Energy Manage. Power, Univ. Penn., Philadelphia, Pa.
79. Lorsch, H. G. Aug. 1974. Thermal energy storage devices suitable for solar heating. See Ref. 7, p. 572
80. Haal, S., et al. 1971. *Various Types of Heat Storage for Use in Conjunction with Electric Heating of Buildings Having Existing Water Radiator System.* Presented at 8th World Energy Conf., Bucharest, Rumania. Pap. No. 2.1-58
81. Meyer, C. F., Todd, D. K. Aug. 1973. Heat storage wells for conserving energy and reducing thermal pllution. See Ref. 44, p. 000
82. Beckman, G., Fritz, K., Gilli, P. V. Oct. 1974. *Thermische Energiespeicherung zur Wirtschaftlichen Spitzenstromerzegung aus Kernkraftwerken.* VOR Conf., Stuttgart, Germany
83. Gilli, P. V., Beckman, G. Aug. 1973. Steam storage adds peaking capacity to nuclear plants. *Energy Int.,* p. 16
84. Gilli, P. V., Beckman, G. Sept. 1974. *The Nuclear Steam Storage Plant: An Economic Method of Peak Power Generation. Trans.* 9th World Energy Conf., Detroit, Mich. 5:162
85. Margen, P. H. Oct. 1971. Thermal energy storage in caves. *Energy Int.,* p. 22
86. Bundy, F. P. Nov. 1974. Power density generating plant with nuclear reactor/ heat storage system combination. *US Patent* No. 3,848,416
87. Schwartz, H. J. Oct. 1973. *Electric Vehicle Battery Research and Development.* Presented at Fall Meet. Electrochem. Soc., Boston, Mass. Tech. Memo. NASA TMX-71471. Washington DC: NASA
88. George, J. H. B. Feb. 1974. *The Development of Electrochemical Power Sources for Electric Vehicles.* Presented at 3rd Int. Electr. Vehicle Symp., Washington DC. Pap. No. 7428
89. Cook, A. R. 1974. A review of battery research and development. *Proc. Am. Power Conf.* 36:1046
90. Douglas, D. L. Sept. 1974. *Batteries for Energy Storage.* Presented at Symp. Energy Storage, Div. Fuel Chem., Am. Chem. Soc. Meet., Atlantic City, NJ
91. Binder, H., Sandstede, G. 1975. Forschungsrichtungen bei Batterien und Brennstoffzellen im Hinblick auf die zukünftige Energieversorgung. *Chem.-Ing.-Tech.* 47:51
92. Yao, N. P., Birk, J. R. Aug 1975. Battery storage for utility load leveling and electric vehicles: a review of advanced batteries. *Proc. 10th Intersoc. Energy Convers. Eng. Conf., Univ. Delaware,*

Newark, Del., p. 1107
93. Takagaki, T. Feb. 1974. *On Advances of On-Road Electric Vehicles and Those Batteries in Japan*. Inf. released by Japan Storage Battery Co., Kyoto, Japan, at 3rd Int. Electr. Vehicle Symp., Washington DC
94. Joseph Lucas (Industries) Ltd., Birmingham, UK. Feb. 28. 1975. News release
95. Feduska, W., et al. Feb. 1975. *Performance Characteristics of a New Iron-Nickel Cell and Battery for Electric Vehicles*. Presented at SAE Automot. Eng. Congr., Detroit, Mich.
96. Kucera, G., Plust, H. G., Schneider, C. Feb. 1975. *Nickel-Zinc Storage Batteries as Energy Sources for Electric Vehicles*. Presented at SAE Automot. Eng. Congr., Detroit, Mich.
97. Symons, P. C. Jan. 1975. Process for Electrical Energy Using Solid Halogen Hydrates. *US Patent No. 3,713,888*
98. Symons, P. C. 1973. *Batteries for Practical Electrical Cars*. SAE Pap. No. 730248. New York: Soc. Automot. Eng.
99. Beccu, K. D. 1974. Zur Deckung des Spitzenbedarfs elektrischer Energie mittels elektrochemischer Energiespeicher. *Chem.-Ing-Tech.* 46:95
100. Warshay, M., Wrights, L. O. Feb. 1975. *Cost and Size Estimates for an Electrochemical Bulk Storage Concept*. Tech. Memo. NASA TMX-3192. Washington DC: NASA
101. Kummer, J. T., Weber, N. May 1967. A sodium-sulfur secondary battery. *Proc. 21st Ann. Power Sources Conf., Atlantic City, NJ*
102. Kyle, M. L., Cairns, E. J., Webster, D. S. March 1973. *Lithium-Sulfur Batteries for Off-Peak Energy Storage: A Preliminary Comparison of Energy Storage and Peak Power Generation Systems*. Rep. No. ANL-7958. Argonne Nat. Lab., Argonne, Ill.
103. Lacennec, Y., et al. 1975. Factors influencing the lifetime of pure beta-alumina electrolyte. *J. Electrochem. Soc.* 122:734
104. Werth, J. J. April 1975. Alkali-Metal Chloride Battery. *US Patent No. 3,877,984*
105. Heredy, L. A., Parkins, W. E. 1972. Lithium-Sulfur Battery Plant for Power Peaking. Presented at IEEE Winter Power Meet., New York, NY. Pap. No. C 72-234-8
106. Nelson, P. A., et al. March 1974. *High-Performance Batteries for Off-Peak Energy Storage*. Rep. No. ANL-8038. Argonne Nat. Lab., Argonne, Ill.
107. Ivins, R. O., Chilenskas, A. A., Kolba, V. M., Towle, W. L., Nelson, P. A. April 1975. Design of a lithium/sulfur battery for load leveling on utility networks. *Proc. 1975 IEEE Southeastcon, Charlotte, NC*
108. Cox, K. E., ed. Jan. 1974. *Hydrogen Energy: A Bibliography with Abstracts*. Energy Inf. Center, Univ. N. Mex., Albuquerque, N. Mex.
109. Olien, N. A., Schiffmacher, S. A. Feb. 1974. *Hydrogen—Future Fuel*. NBS Tech. Note 664. Boulder, Colo: US Nat. Bur. Stand.
110. *The Hydrogen Economy Miami Energy (THEME) Conference Proceedings, Miami Beach, Fla.* March 1974. Univ. Miami, Coral Gables, Fla.
111. Smith, D. H. 1971. Industrial water electrolysis. In *Industrial Electrochemical Processes*, ed. A. T. Kuhn, p. 127. New York: Elsevier
112. Gregory, D. P. Aug. 1972. *A Hydrogen Energy Distribution System*. Inst. Gas Technol. Final Rep. IU-4-6 to Am. Gas Assoc.
113. DeBeni, G. March 1974. Considerations on iron-chlorine-oxygen reactions in relation to thermochemical water splitting. See Ref. 47, p. S11-13
114. Russell, J. L., Jr., Porter, J. T. March 1974. A search for thermochemical water splitting cycles. See Ref. 47, p. S11-49
115. Funk, J. E. Aug. 1974. The generation of hydrogen by the thermal decomposition of water. See Ref. 7, p. 394
116. Knoche, K. F., et al. Nov. 1974. Water splitting processes of the iron-chlorine family. *Proc. III/19, BNES Conf. on High Temp. and Process Appl., London*
117. Quade, R. N., McMain, A. T. March 1974. Hydrogen production with a high-temperature gas-cooled reactor (HTGR). See Ref. 47, p. S3-21
118. Barnert, H., Schulten, R. March 1974. Nuclear water splitting and high temperature reactors. See Ref. 47, p. S3-1
119. Swet, C. J. March 1974. Thermochemical water cracking using solar heat. See Ref. 47, p. S7-21
120. Newton, C. L. 1967. Hydrogen production, liquefaction and use. *Cryog. Eng. News* 9: No. 8, p. 50 (Pt. 1); No. 9, p. 24 (Pt. 2)
121. Wiswall, R. H. Reilly, J. J. Sept. 1972. Metal hydrides for energy storage. *Proc. 7th Intersoc. Energy Convers. Conf., San Diego, Calif.*, p. 1342
122. Vanvucht, J. H. 1970. Reversible room temperature absorption of large quantities of hydrogen by intermetallic

compounds. *Philips Res. Rep.* 25:133–40
123. Lueckel, W. J., Farris, P. J. July 1974. *The FCG-1 Fuel Cell Power Plant for Electric Utility Use.* Presented at Summer Meet., IEEE Power Eng. Soc., Anaheim, Calif.
124. Hausz, W., et al. 1972. *Hydrogen Systems for Electric Energy.* Rep. No. 72TMP-15. Gen. Electr. Co. Center for Adv. Studies (TEMPO)
125. Burger, J. M., et al. Aug. 1974. Energy storage for utilities via hydrogen systems. See Ref. 7, p. 428
126. Salzano, F. J., Cherniavsky, E. A., Isler, R. J., Hoffman, K. C. March 1974. On the role of hydrogen in electric energy storage. See Ref. 47, p. S14-19
127. Casazza, J. A., Huse, R. A., Sulzberger, V. T. Aug. 1974. *Possibilities for Integration of Electric, Gas, and Hydrogen Energy Systems.* Presented at Int. Conf. on Large High Tension Syst., (CIGRE), Paris. Available from Public Service Electric & Gas Co., Newark, NJ
128. Häfele, W. 1974. Energy choices that Europe faces: a European view of energy. *Science* 184:360
129. Hanneman, R. E., Vakil, H., Wentorf, R. H., Jr. Aug. 1974. Closed-loop chemical systems for energy transmission, conversion and storage. See Ref. 7, p. 435
130. Paleocrassas, S. N. March 1974. Photolysis of water as a solar energy conversion process: an assessment. See Ref. 47, p. S7-33
131. Hein, R. A. 1974. Superconductivity large-scale applications. *Science* 185:211
132. Hadlow, M. E. G., Baylis, J. A., Lindley, B. C. Aug. 1972. Superconductivity and its applications to power engineering. *Proc. IEE, IEE Rev.* 119 (8R):1003
133. Hassenzahl, W. V., et al. March 1973. *Magnetic Energy Storage and Its Application in Electric Power Systems.* Presented at 1973 IEEE Int. Conv. New York. See also Rep. No. LA-5258-PR, Los Alamos Sci. Lab., Los Alamos, N. Mex.
134. Hassenzahl, W. V., et al. Aug. 1973. *The Economics of Superconducting Magnetic Energy Storage Systems for Load Leveling.* Rep. No. LA-5377-MS. Los Alamos Sci. Lab., Los Alamos, N. Mex.
135. Peterson, H. A., et al. April 1975. *Wisconsin Superconductive Energy Storage Project.* Presented at Am. Power Conf., Chicago, Ill.
136. Boom, R. W., et al. July 1974. *Wisconsin Superconductive Energy Storage Project.* Feasibility Study Rep., Vol. 1. Coll. Eng., Univ. Wis., Madison, Wis.
137. Gilbert, J. S., Kern, E. A. Feb. 1975. *Analysis of Homopolar Generators and Superconductive Inductive Energy Storage Systems as Power Supplies for High Energy, Space-Based Lasers.* Rep. No. LA-5837-MS. Los Alamos Sci. Lab., Los Alamos, N. Mex.

Copyright 1976. All rights reserved

ADVANCED ENERGY CONVERSION

×11013

Edward V. Somers and Daniel Berg
Westinghouse Electric Corporation, Pittsburgh, Pennsylvania 15235

Arnold P. Fickett
Electric Power Research Institute, Palo Alto, California 94303

INTRODUCTION

Recent events have greatly expanded the importance of and the need for developments in energy conversion and conservation. Advanced energy conversion methods for the generation of electricity are being adapted to the near-term and long-term fuel crisis with emphasis being placed on the aspects of the fuel/power plant interconnection. First of these aspects is the adaptability of any energy conversion process to its use with coal and/or its use with nuclear fuel in thermal and breeder reactors. One views coal usage to include energy conversion processes coupled with both gaseous and liquid coal-derived fuels. Within the United States coal and nuclear fuel are the abundant available fuels of the next 30 years, and energy conversion patterns should shift from oil and natural gas to coal and nuclear fuel. Second is the aspect of greatly improved thermal efficiency. With the departure of the era of low-cost fossil fuels, new energy conversion processes will be designed to conserve fuel usage and to use the thermal values from waste streams. And third is the aspect of better technological energy conversion design leading to improved resource utilization and to improved environmental impact. Materials of construction are important, and the power plant design of the future must increasingly consider the drain on these resources. The use of air, water, and land resources also must be factored into the design of all new energy conversion processes. Renewable fuel resources from the sun and from the heat of the earth as distinct from the coal and nuclear resources will be available for man's use indefinitely. New energy conversion developments are being adapted to solar and geothermal sources, and this emphasizes those energy conversion developments applicable to the low-temperature end of coal- and nuclear-fired power plants.

The conversion of fuel to electrical and to mechanical energy consumes about one half of the 80 quadrillion (10^{15}) Btu of fuel energy used today (1). That the conversion to electrical energy will shift to coal or nuclear fuel is predictable. Most electrical energy is generated in large central-station-type electric power plants, and most mechanical energy is generated in transportation engines—the diesel engine, the aircraft gas turbine, and the automobile engine. Of the approximate 20 quadrillion Btu of fuel energy used to produce electricity, about 45% is derived from coal, and the balance somewhat equally from nuclear, oil, natural gas, and hydro. Predictions for future additions of electric power plants indicate that future fueling for most of these plants will be either coal or nuclear (2). Advanced energy conversion concepts for large power-plant generation of electricity are in accord with this fueling prediction. A large part of oil-derived fuel energy is used for transportation engines. Of this 20 quadrillion Btu used for transportation, nearly 60% is consumed by automobiles, about 20% by trucks, about 10% by airplanes, and about 4% by railroads (3). Although the quantitative picture is unclear, large changes will occur in the amount and use pattern of transportation fuel because of a restricted utilization of oil in the future. Advanced energy conversion concepts for transportation include developments for fuel-fired automobile engine cycles, with improved efficiency and greatly reduced pollution emissions, and developments for the electrification of mass transit and urban vehicles. The former includes many variations in the Otto cycle and in the Rankine cycle. The latter includes direct electrically driven trolleys and trains and energy storage in automobiles and buses containing batteries or small flywheels. Although automotive transport will adapt to coal-derived oil products, electrical transportation will shift to coal- or nuclear-generated electricity with a more efficient use of the fuel energy. We do not discuss the automotive engine developments other than as they intertwine with the advanced energy conversion concepts for large power-plant generation of electricity, nor do we discuss electrified transportation.

There are three main paths for obtaining high thermal efficiency in the advanced energy conversion developments. 1. The top cycle temperatures are being increased to make available more of the thermal energy for conversion to mechanical and/or electrical energy. 2. Cycle fluids and their accompanying energy conversion processes are being adapted to their best range of operating temperatures so that combined cycles of all types are being developed using two or three different cycles thermally in series. 3. The fuel-cell development is aimed at highly efficient devices. An example of the first and second paths is that of the combined-cycle power plants using gas-turbine and steam-turbine power plants. An efficient magnetohydrodynamic (MHD) power plant also will be a combined-cycle plant in that it will include a steam bottoming plant to use the heat rejected from the MHD generator. An example of the third path is the fuel-cell development for megawatt-size efficient power plants. The fuel cell conforms to energy conversion rules unlike those used in thermal cycle design; it is a power device that by selection of the electrolyte type can operate efficiently at temperatures near room temperature or at temperatures up to 1800°F.

An important facet in the construction of future advanced energy conversion systems is the availability of materials of construction and their environmental impact. Both the abundance of the material (4) and the price of the material as affected by international political and economic control of the resource are important (5). Abundance of materials is usually quantified in conceptual design work of a cycle application. Since a power plant affects the air, water, and land resources surrounding it, important design constraints are introduced to meet existing and future governmental regulations and guidelines on plant emissions and their environmental impact. For example, control of SO_2 emissions to meet various governmental regulations affects power plant design and operation.

The advanced energy conversion techniques with potential for coal and nuclear-fuel usage, for implementing high thermal efficiency, and for resource/environmental conformance are those using open- and closed-cycle gas turbines, MHD, potassium topping cycles, low-temperature closed cycles, and fuel cells. These selections are similar to those of NASA for its ECAS program (Energy Conversion Alternative Studies) being worked on by the General Electric Company and Westinghouse Electric Corporation (5a). The low-temperature cycle work encompasses power plants adapted to solar and geothermal resources.

Gas turbines and combined cycles are being developed for open-cycle firing using coal as a fuel. Control of sulfur emissions from coal-burning plants is generally required to meet stack-emission regulations. The coal also can be burned in a desulfurizing, pressurized, fluidized bed or can be converted to a desulfurized gas or liquid burnable in the gas-turbine combustor. Fluid-bed combustion limits the top firing temperature of the power plant. It does have design options for in-bed boiler tubes to generate steam, and these produce another version of an efficient combined cycle. Coal-derived fuels, such as low-Btu gas, can be used in combined-cycle gas turbines that are being developed to reach inlet temperatures up to 3000°F with efficiencies in excess of 50%. Closed-cycle gas turbines with moderate inlet temperatures up to 1600°F are being developed for use in high-temperature nuclear reactors or in coal-fired combustion units. Closed-cycle developments in the lower temperature range include the use of waste heat from the gas-turbine cycle for industrial process and district heating steam.

Magnetohydrodynamics in two of its developments, open-cycle coal-fired plasma and closed-cycle nuclear or coal-fired plasma, aims to produce large amounts of power at ultrahigh temperatures from 3000°F to 4500°F. Coupled with a bottoming cycle, MHD shows calculated thermal efficiencies in excess of 50%. Liquid-metal MHD (LMMHD) operates in a closed cycle at top temperatures below 1800°F, and with a bottoming plant will attain potential efficiencies in excess of 40%. The aspect of good plant reliability at these lowest of MHD-operating temperatures is the goal.

The potassium-steam binary cycle best fits the coal/nuclear fuel requirement at top temperatures of 1600°F. This cycle has potential efficiency in excess of 50%. The low top pressure of about 2 atmospheres in the potassium cycle lessens material demands on the coal-fired potassium boiler.

Low-temperature closed cycles are being developed for waste heat recovery systems, for automotive and mobile power applications, and for use with low-temperature heat sources. The latter usage includes the renewable fuel resources of the solar power plants, thermocline power plants, and geothermal power plants.

The fuel cell converts chemical energy directly to electrical energy, avoiding the

Carnot cycle limitation of thermal machines. This results in the potential of an overall efficiency above 50%. In addition, the fuel cell is quiet, nonpolluting, and compatible with a wide variety of fuels, including clean coal fuels as well as petroleum, natural gas, and hydrogen. The fuel-cell types under active development are the low-temperature acid and alkaline fuel cells and the intermediate-temperature molten carbonate fuel cell.

OPEN COMBINED-CYCLE GAS-TURBINE POWER PLANTS

The coal-fired combined cycle using a gas turbine topping a steam turbine is probably the most promising power plant for short-term and long-term development payoff. Extensive gas turbine development during the past decade has brought more efficient simple- and combined-cycle units using petroleum fuel into the marketplace (6). State-of-the-art combined-cycle efficiencies are in the vicinity of 38–42% of higher heating value, equal to or better than efficiencies of the best steam plants. These high efficiencies are due to the well-developed cooling design of the gas turbines, which permits them to be operated at inlet temperatures up to 2100°F, and to the use of well-designed steam bottoming plants.

The inlet temperatures are limited by acceptable blade stresses and by the effects of erosion, corrosion, and deposition arising from contaminants in the combustion gas. Much past work has gone into specifying fuel contaminants to reduce corrosion of and deposition on the gas turbine blades to acceptable levels.

With the new need to fire gas turbines with coal, development work on coal-derived clean fuels and an acceptable coupling of coal-burning units to gas tur-

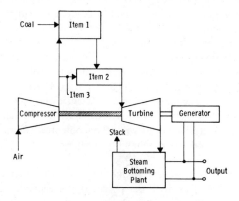

Plant Type	Gasifier C. C.	Fluidized Bed C. C.
Item 1	Gasifier	Fluidized Bed
Item 2	Combustor	Particulate Cleanup
Item 3	Air Duct	No Duct

Figure 1 Coal-fired open combined-cycle (CC) plants.

bines is being accented. Coal-derived clean fuels may include low- and medium-Btu gas, solvent-refined and distilled coal products, and methanol. The processes under development to produce these fuels (7, 8) are discussed elsewhere in this volume. Objectives in the clean-fuel programs for gas turbines must include plant design for control of fuel contaminants at low cost. Wherever these fuels are used as direct replacements for petroleum fuel, the gas turbine specifications on fuel contaminants (9) will apply. Coupling of coal-burning equipment directly to the gas turbines will alter these specifications somewhat. Either of two routes for direct coupling will be used: a combined-cycle plant will be integrated with a low-Btu gasifier; or a fluidized-bed will replace the usual oil- or gas-fired combustor in the gas turbine (10, 11). These are shown in Figure 1.

Development of the gas turbine using coal-derived clean fuels takes the path toward higher temperature (and higher pressure ratio), more efficient machines with structural cooling sufficient to meet mechanical stress, and controlled gas-side corrosion and oxidation. Turbine inlet temperatures are expected to reach 2500°F and to proceed above 3000°F. The power plant efficiency with coal-derived liquids would be greater than 50%. Efficiency of a low-Btu gasifier/power plant would be near 45%. Turbine cooling with compressor air, steam, and liquid water are methods for reaching successively higher temperatures in the turbine development path (12, 13). Ceramic blades to be used uncooled are under development (14).

Low-Btu gasification with the coal gasifier integrated into the power plant is one path for direct coal-fired gas-turbine development. A commercial pressurized air-blown fixed-bed gasifier with gas cleanup by wet scrubbing has been integrated into a pressurized boiler/gas turbine plant operated by Steag (15) in Germany. Development of the fixed-bed gasifier with a combined cycle continues (16). The fluidized-bed gasifier with in-bed desulfurization and with hot gas cleanup of particulates is under investigation under sponsorship of the US Energy Research and Development Administration (ERDA) (17). One objective of this investigation is improvement in efficiency of a coupled gasifier/combined cycle by cleaning the hot gas without cooling. The combined-cycle power plant development used with either the cold or hot gasifier product would produce turbine inlet temperatures as high as 3000°F and efficiencies up to the 45% level.

Another development path for direct coal-fired gas turbines replaces the standard combustor with a coal-burning desulfurizing fluidized bed. This is part of the work on a fluidized-bed boiler being pursued throughout the world (18, 19). Two options are available with the pressurized fluidized bed used either with or without in-bed boiler tubes. The option with in-bed boiler tubes is a counterpart of the pressurized boiler/gas turbine plant of Steag (15); the option without in-bed tubes is being pursued by the Combustion Power Company, Inc. (20). Such a combustor coupled into a combined-cycle power plant would produce efficiencies of about 40%, since the fluidized-bed temperatures are limited to below 1800°F. The lower turbine inlet temperatures permit reasonable gas turbine sizes of 25–75 MW with accompanying combined-cycle power plant sizes of 70–300 MW. The plant appears to have near-term commercial potential with capital costs relatively close to those of the smaller-sized modular combined-cycle power plants. As such, it may be more than competitive in electric energy cost with other coal-burning power plants. As cited earlier, the pressurized fluidized-bed boiler option is aimed at greatly reduced boiler cost. Development problems exist in the area of turbine erosion, corrosion, and deposition. This is reflected in coupling the turbine design limitations on fluidized-bed gas contamination with the associated hot gas cleanup. Environmental effects from plant emissions and waste streams, and their control to meet legal regulations, are also under investigation.

Past experience on burning of pulverized coal in gas turbines showed that ash carry-over from the combustor into the turbine severely eroded the blades. Removal of particulates in the hot gas to low concentrations (less than 10 grains/10^3 standard cubic foot of gas) was required to reduce the erosion to tolerable amounts (21). Erosion and deposition effects were investigated on a small scale, and environmental effects, e.g. SO_2, were not considered. Recent work on the burning of coal in a desulfurizing fluidized bed has evaluated the particulate, SO_2, and NO_x emissions and the fouling of turbine blades by erosion, corrosion, and deposition. Experimental facilities at the British Coal Utilization Research Association Ltd. were used under a contract from the Office of Coal Research to investigate these effects (22). Data from the work indicate relatively low values of SO_2 and NO_x compared to coal-fired steam plant emissions. With gas temperatures of 1600–1650°F, erosion, corrosion, and deposition effects on gas-turbine bladings appear to be similar to those experienced with firing with GT2-fuel, a distillate. This improvement in contamination effects from combustion gas apparently arises from the large difference in combustion temperatures between the standard combustion process and the fluidized-bed combustion process.

The Combustion Power Company, Inc. is evaluating fluidized-bed combustion of coal without in-bed boiler tubes in its 1-MW Process Development Unit containing a Ruston gas turbine (20). The test interval is set for 1000 hours. This project, also under ERDA sponsorship, will derive design information on combustion, emissions, and blade fouling for scale-up to the larger units.

CLOSED-CYCLE GAS-TURBINE POWER PLANTS

High-power closed-cycle gas turbine systems have been under development since 1939. These systems work with a highly pressurized gaseous working fluid that can include noble gases, CO_2, N_2, and other gases (23) and that leads to compaction of both the turbine and heat exchange apparatus (24). Partial load power can be obtained by pressure control in the system, and this permits operation of the unit at relatively constant temperatures and with relatively constant efficiency (24). Heat rejection to ambient occurs over a range of relatively high temperatures that permits the use of either compact dry cooling towers (25) or a subposed low-temperature power cycle (26). Use of dry cooling towers permits environmental protection of water courses and conservation of water in water-deficient areas (27). Subposed power cycles will increase the thermal efficiency of the power plant but with increased plant investment in the low-temperature power plant.

A schematic of a closed-cycle gas-turbine power plant is shown in Figure 2. The heater can be either nuclear or coal fired, and the precooler and the dry cooling tower can be replaced with a vapor-cycle bottoming plant. The heater is necessarily either a high-temperature gas-cooled reactor (HTGR) or a heat exchanger fitted to a coal-fired combustor. The HTGR will heat the working fluid directly from its nuclear fuel elements (25). Reactor outlet temperatures of existing HTGR plants are near 1500°F, and this extends the closed-cycle gas-turbine temperature to 1500°F. The coal-fired combustor provides heat to the combustion gas that is heat exchanged to the turbine working fluid. Working-fluid outlet temperatures of coal-fired plants range from 1300°F to 1382°F and are limited to this range by fireside corrosion of existing heat exchangers (28).

The application of the closed-cycle gas turbine to the HTGR is proceeding in the United States (25) and in Europe (29). Small high-power gas turbines are to be placed within the nuclear reactor cooling loop, thus eliminating the heat exchanger, so that the reactor coolant and turbine working fluid will be one and the same—in this case, helium. The turbomachinery and recuperators are to be located in the prestressed concrete reactor vessel. The attainable gas-turbine inlet temperature corresponds to that defined by the maximum reactor outlet temperature of 1500°F with present core and fuel element designs. With nuclear reactor thermal levels centering at about 3000 MW, the US and European designs are aiming at 1000–1200 MW electrical.

The US design for a HTGR gas turbine, by Gulf General Atomic (GGA), calls for use of four 260-MW electric gas turbines. The European designs of Hoch-Temperatur-Reaktorbau GmbH (HRB) will use one 1000-MW electric gas turbine. The GGA gas-turbine inlet temperature is 1500°F, and the HRB is 1560°F. The selection of power unit level in the US design is cited to rest on multiple units for emergency core cooling and on better scale-up of the development units from 30 to 50 MW to commercial size. The largest operating closed-cycle unit is a 30-MW electrical unit recently built in Vienna, and this has a turbine inlet temperature of about 1380°F.

Aimed toward scaling-up the gas turbine design to needed HTGR power levels is

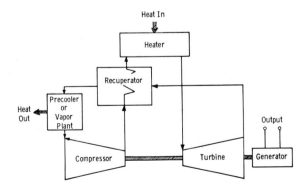

Figure 2 Coal-fired closed-cycle plant.

the Oberhausen gas-turbine power plant, being built by Energieversorgung Oberhausen AG (EVO) and scheduled to produce power at a 50-MW electrical level in late 1974 or early 1975 (30). This plant is a first in two design aspects: it is the largest closed-cycle power plant to be built, and it is the first application using helium as a working fluid in a commercial plant. The unit, as with the previous European units, couples electric power with district heating. The Oberhausen unit will deliver a maximum of 53.5 MW thermal to the heat distribution network. The unit is to be fired by coke-oven gas, a clean fuel. It faces a 750°C (1382°F) limit on turbine inlet temperature imposed by fossil fuel fireside corrosion.

The US development on the 275-MW electrical level is proceeding (31, 32). This incorporates fundamental investigations on the effects of helium impurities on materials and on compressor and turbine aerodynamics. Microamounts of hydrogen and carbon monoxide are cited to produce catastrophic effects by decarburizing of gas turbine blade materials at turbine operating temperatures, and the mechanism of attack and its rate and methods of control require investigation (33). Highly efficient compressors and turbines require good aerodynamic design to produce an efficient power cycle because helium machinery requires small blade weights and large hub-to-tip ratios. The Oberhausen gas turbine is part of the European helium development.

Several small closed-cycle electric power plants that are also used for district heating have been built in Europe and successfully operated using bituminous coal, oil, and gas. The closed-cycle working fluid is air. The design and operation of four such units built by Gutehoffnungshütte Sterkrade AG (GHH) was recently surveyed (28). These units were located in Germany at Oberhausen, Haus Aden, Coburg, and Gelsenkirchen. The limiting feature on their turbine inlet temperature was that of fireside corrosion of the air heater tubes. The four units were all operated at temperatures below 710°C (1310°F). The necessary use of much austenitic alloy in the air heaters poses an economic impediment to the closed cycle. Similar design and operation was noted for the 10-MW electrical unit built by Escher Wyss and installed in the USSR using Moscow district coal (34). This unit operated at turbine inlet temperatures of 680°C (1256°F) and experienced fireside corrosion that was eventually corrected. Bammert and Deuster cite a turbine inlet design temperature of 750°C (1382°F) for the 50-MW electrical Oberhausen helium unit (30).

The major limitation placed on coal-fired closed-cycle systems is that of permissible maximum air-heater temperature. With the technical and cost limitations on temperature arising from metallic air-heater structures, raising these temperatures may follow the successful development of ceramic air heaters being pursued within the United States.

Commercial realization of a coal-fired closed-cycle power plant will require a combination of increased plant efficiency and reduced costs of the compressor, turbine, recuperator, and precooler to compensate for the cost of the coal-fired heater. The immediate efficiency goal of such a unit would be near 40% with a turbine inlet temperature of 1500°F.

MAGNETOHYDRODYNAMIC POWER PLANTS

The three MHD systems currently under development are the coal-fired open cycle, the closed cycle (nuclear or coal-fired), and the liquid-metal cycle (nuclear or coal-fired). All three systems coupled to a bottoming steam plant have potential thermal efficiencies in excess of 40%, and the two plasma types, in excess of 50%. Because of this, both fuel consumption and thermal pollution would be reduced for a given power output. Potentially, large savings in electric energy costs are projected for central-station MHD power plants.

Open-cycle MHD combines the electric generator with the gas expander. It eliminates the need for high-temperature moving parts. The combustion unit is internal to the cycle, and there is no need for a heat exchanger. High cycle temperatures leading to high thermal efficiencies are required. Large US programs are aimed at demonstration units with supporting materials and cycle design work. These will burn clean coal-derived gas or liquids (35), or will burn coal directly with accompanying ash and sulfur carryover into the generator (36), or will employ in-plant combustion deashing and gasification (37). A version of the latter type of unit is schematically shown in Figure 3.

Two MHD workshops were held in 1974 to establish an integrated MHD program for the United States (38, 39). The EPRI/OCR (Electric Power Research Institute/Office of Coal Research) Workshop defined a program for staged implementation of MHD that embodies systems analysis, subsystems development, construction of a 50–100-thermal MW Engineering Test Facility, and eventual construction of a 250–500-thermal MW Engineering Demonstration Plant. The NSF (National Science Foundation)/OCR Workshop effected interchange of information on critical materials developments among the various MHD investigators, particularly those working on the NSF and OCR programs.

All high-temperature components of an open-cycle MHD generator—the combustor, the generator duct, and the air preheater—require designs of proven performance and reliable operating life. Fabricating techniques for large superconducting magnets must be developed. Environmental problems arising from high-temperature formation of NO_x and from SO_2 formation from the coal must be adequately solved. Satisfactory seed recovery must be attained.

High-temperature combustors are being developed with different controls on slag and sulfur carry-over into the generator. Single-stage combustors experiencing temperatures to 2700°K have been built with water-cooled steel walls that operate with a slag coating. Multistage combustors have been proposed and are being developed to greatly reduce slag carry-over into the generator. These contain stages using preliminary pyrolysis or gasification and combustion followed by the final stage of high-temperature combustion fired with the in-process derived, relatively ash-free clean gas (37, 40–42). A single-stage combustor adapted to high slag carry-over is exemplified by the University of Tennessee's design concept for a MHD generator (36, 43). The multistage combustors with low slag carry-over are repre-

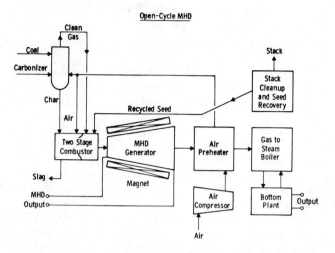
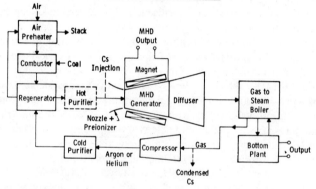
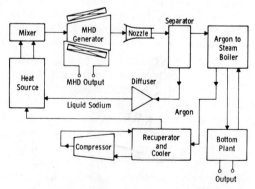

Figure 3 Magnetohydrodynamic power plants.

sented by design concepts from AVCO Everett (37), US Bureau of Mines (41), and Westinghouse (42).

Generator-duct design involves the duct material and the electrical characteristics of the flow/electrode/insulator interactions for sustained electrical performance of the generator. Much development effort is aimed at long-life tests of duct structures employing different geometrical electrode/insulator arrangements for the Hall and Faraday voltage takeoffs. The direct combustion of coal under development at the University of Tennessee involves high slag loadings within the generator duct, and slag layers deposited on the metallic electrode surfaces protect them thermally while permitting satisfactory electrical current transmission (37). Multistage combustion with low slag carry-over is simulated in testing with relatively clean fuel-fired generators by AVCO (44) with water-cooled duct walls and by Westinghouse (45) with uncooled walls. A US/USSR cooperative program dealing with MHD materials has been established by the Fossil Fuel Division of ERDA, and US electrode-insulator test models will be tested in the USSR U-02 MHD unit (38). Evaluations in real and simulated MHD environments with different amounts of coal slag present are to be conducted to define electrode and insulator configurations displaying satisfactory life, current uniformity, thermal and mechanical stability, and low interelectrode current leakage. MHD duct performance is being evaluated with combined analysis and model generator experiments to establish flow and electrical design correlations scalable to the larger generators needed for prototype and commercial operation (36, 43, 46).

MHD air preheaters have formidable design problems arising from the operating environment of high-temperature and highly reactive seed material. Preheat temperatures of 2000°F and 3000°F are best obtained with regenerative (cyclic) units. Refractory matrices are being tested for thermal shock resistance, high mechanical loadings, and corrosion resistance to slag/seed liquids impacting on the heat exchange surfaces (47–49). Work is centering on compounds of alumina, magnesia, and titania. Previous analysis of the MHD cycle has found regenerative air preheaters to be acceptable thermodynamically.

Superconducting magnets are needed for best operation of the MHD generator. Magnet systems for a 250-MW thermal test facility have been designed but not yet constructed (50). Work on seed recovery and control of NO_x emissions continues.

The current status of open-cycle MHD is advanced experimental. The physical basis of the concept has been amply demonstrated. However, there is considerable imprecision in the design correlations for the electrical parameters of the gas and electrodes. The materials needed for the components in the various concept designs are not defined. The infusion of large-scale funding will accelerate the progress of the past two years, and the construction of large test facilities for longtime operation is expected to provide answers to the commercial viability of the various MHD concepts.

The closed-cycle MHD system operates at temperatures of 2500–3500°F, considerably below those of the open-cycle MHD. Conceptually, the cycle is adaptable to either very high temperature gas-cooled nuclear reactors or to coal firing with a heat exchanger coupling the combustion stream with a closed-cycle MHD loop.

Although the lower peak temperatures mitigate the material and component construction problem for a coal-fired system, they are considerably above expected operating temperatures for commercial gas-cooled reactors. Additionally, the coal-fired/closed MHD loop heat exchanger would be a nonmetallic type, and present concepts center on ceramic regenerative types (51). These pose difficulty in maintaining the purity of the working fluid in the MHD loop.

Considerable progress has been claimed recently in closed-cycle MHD, and nonequilibrium gas conductivities appear to be sufficiently high to permit good power extraction with high thermal efficiency from reasonable design concepts of power plants (52, 53). Recently proposed conceptual power plants are coal-fired, since nuclear-fired plants await the far-future developments of 2500–3500°F reactors. The recent EPRI/OCR Workshop confined its program and recommendations to open-cycle MHD but did cite intended similar treatment of closed-cycle MHD in the near future (38).

A schematic of a coal-fired closed-cycle MHD power plant is shown in Figure 3. This plant concept heats the MHD working fluid with a regenerator and rejects heat from the MHD exhaust to a steam bottoming plant. Compared to the open-cycle concept, three new and essential pieces of equipment are shown here: the regenerator and two purifiers.

Development work on closed-cycle MHD during the past two years has been centered on supersonic MHD duct performance in short-time tests and on extension of these results to power plant concepts and their associated plant performance (53). Alkali-seeded helium or argon working at high flow velocities through a strong magnetic field of 4–6 tesla magnetically induce a high nonequilibrium gas conductivity. This gives rise to large power extraction per unit volume. This method of providing nonequilibrium ionization is superior to the older method of using external electron guns or irradiation devices to produce the ionization. MHD duct experiments are being conducted on blowdown apparatus operating in power-producing modes for several seconds. Many key experimental parameters under investigation include working fluid type, pressure and temperature levels, impurity content of the working fluid, electrode configurations, high working fluid expansion rates, and magnetic field levels up to 4 tesla (54). Thermodynamic analyses of nuclear-heated and coal-heated power plant concepts have been conducted in parallel, and these show potential thermal efficiencies in excess of 50%.

The coal-fired power plant requires the use of regenerative ceramic heat exchangers heated with slightly oxygen-deficient combustion to acceptable MHD inlet temperatures of 3000°F or higher. The temperatures are in excess of those reached by the air preheaters of the open-cycle MHD. These regenerators are operated on a rotating cyclic schedule, somewhat like that used in blast furnace operation, with heating of the regenerator by combustion gas followed by heating of the argon gas by the regenerator. Since the argon working fluid must be pure, purge systems must be employed to maintain the purity of the gas by removing combustion gas and recovering noble gas from the regenerators when the switch is made from heating to cooling and vice versa. Two types of purge systems can be employed for hot purification: a hot low-pressure purge and a vacuum-pumping purge. The cold purifier removes particulates, CO, CO_2, N_2, O_2, and H_2O, all of which are considered impurities in a nonequilibrium plasma MHD generator. These purge systems coupled with the cyclic operation of the regenerator require successful operation under difficult corrosion/temperature conditions.

The current status of closed-cycle MHD lags behind that of open-cycle MHD. Because of the purity requirements placed on the working fluid, experimental models are more difficult to implement. With continued accumulation of successful experimental data and establishment of economic feasibility of conceptual power plants, closed-cycle MHD will next move into longer testing of small power plants.

Liquid-metal MHD has been under investigation for many years. Two different cycles are being investigated. The NASA Technology Utilization Office and the Jet Propulsion Laboratory evaluated LMMHD cycles, selected a potassium working fluid and a cesium-lithium working fluid for experimental and analytical work, and fitted multistaged LMMHD separator cycles to a conceptual bottoming steam plant to obtain an estimated operational efficiency of 40% and 45% (55). These LMMHD separator cycles first mechanically separate the vapor and liquid and then generate MHD power from the kinetic energy of the separated liquid. The Argonne National Laboratory, funded primarily by the Office of Naval Research (56, 57), has investigated two component flows in LMMHD generators. These generators use a mixture of either argon/sodium or helium/sodium in the MHD duct, and separation of the liquid metal and gas follows the MHD generator, as shown in Figure 3. Here, the inert gas provides the thermodynamic working fluid and the liquid metal provides the electrodynamic conversion fluid in the MHD duct.

Problems appear in three areas: in efficiently coupling the LMMHD generators to bottoming steam plants with both nuclear and coal heat sources; in establishing MHD duct performance; and in designing efficient vapor/liquid and gas/liquid separators. Plant studies coupling LMMHD generators to steam bottoming plants are being continued to establish the potential of these plants. To date, LMMHD inlet temperatures of 1600°F and higher appear necessary for plant thermal efficiencies above 40%. Experimental LMMHD generator-duct studies are needed to establish design data on duct efficiency and duct operation for either the mixed or the liquid-metal flow types. The mixed-flow type is more complex, since the two-phase flow is coupled to the electrodynamic performance. Scaling information for larger ducts is needed. Chemical interaction of the liquid metal and the materials of construction continues to be evaluated to established long-time materials design. Efficient separator designs must be established, especially for the separator cycles. In the mixed-flow cycle, metal carry-over into the low-temperature (compressor) part of the cycle equipment must be virtually eliminated.

LMMHD holds promise of lower plant thermal efficiency than do the other two MHD types. Experimental work continues on empirical evaluation of the MHD duct flow and electrical parameters for the mixed-flow generator. The goal of a nuclear-reactor-heated power plant awaits the development of a reactor with outlet temperatures of 1600–2000°F. The goal of coal firing for similar temperatures would require fluidized-bed combustors with in-bed heating tubes. This may be difficult to accomplish because of fireside corrosion control required with coal firing (28).

POTASSIUM-STEAM BINARY-CYCLE POWER PLANTS

The potassium Rankine cycle was developed in the 1960s as part of the auxiliary power supply program for space vehicles. A potassium power plant with an output of about 200 kW, with the potassium boiler fired with natural gas, was operated for nearly 5000 hr at a turbine inlet temperature of 1500°F (58). The use of the potassium cycle to top a steam cycle forms the conceptual basis of a large power plant, which was originally proposed for high-temperature nuclear reactor plants (59). Recently, the concept was directed toward fossil fuel firing of the potassium boiler (60). Development of the fossil fueled power plant concept has since followed two parallel paths: 1. apparatus component testing of a module of the potassium boiler-furnace combustor and a module of a potassium condenser-steam boiler, and 2. evaluation of conceptual power plants.

The first power plant concept, proposed with a pressurized potassium boiler (60), is illustrated in Figure 4. The pressurized boiler operates isothermally about 200F° above the boiling temperature of the potassium with continuous combustion throughout. The gas turbine recovers the energy from the hot pressurized gas discharging from the boiler, and the gas turbine exhaust heats the feedwater of the steam cycle. This concept puts about 85% of the heat from the fuel through the potassium boiler and about 15% through the gas turbine. With a potassium turbine inlet temperature of 1540°F, the boiler pressure of the potassium is 2 atmospheres.

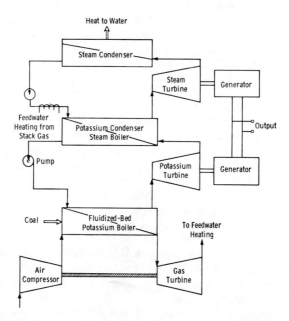

Figure 4 Potassium-steam binary-cycle power plant with fluidized-bed combustor.

Since the boiler is pressurized, the boiler tubes would be designed for small negative-pressure difference across their relatively thin walls while still maintaining the outside wall temperature near 1600°F (61). This requires smaller metal allowance for fireside corrosion. This can be compared to the European coal-fired experience of boiler tubes designed for positive-pressure differences of 40 bars and outside wall temperatures of about 1500°F (28). Leakage of combustion gas, air, or water into the potassium loop (the boiler, turbine, and condenser) must be reduced to tractable levels by careful design and operation of the unit. Control of combustion hot spots is conceptually afforded by replacing the pressurized furnace boiler with a fluidized bed. This may permit direct coal firing of the boiler. The Oak Ridge National Laboratory (ORNL) continued the design study through 1974, and based on their modular design of a 600-MWe unit, they have built a boiler module to full scale (62). Their burner concepts include a variety of burners designed to insure against flame impingement on the tube surfaces, and these include several types of gas and oil burners and a coal-fired fluidized-bed combustion chamber, all producing continuous combustion throughout the tube bundle. The low potassium pressure in the power loop requires turbines capable of handling large volumetric flows. Preliminary design of these machines is indicated in a NASA/OCR document (63). The ORNL and NASA efforts are aimed at providing a sound design basis for the potassium-steam power plant.

Several conceptual power plants have been evaluated technically and economically. These include evaluation of four different potassium boiler-furnace-combustor concepts, two atmospheric furnaces, and two pressurized furnaces under a NASA contract in 1973 (63). Of the four concepts, the pressurized boiler-furnace-combustor and a pressurized fluidized-bed boiler-combustor were examined in detail for technical design and cost. The plant with the pressurized boiler was conceptually adapted to clean fuel firing, while that with the pressurized, fluidized-bed was conceptually adapted to coal. Both plants were estimated to produce electricity at the same cost. Costing of these potassium-steam cycles continues in studies under OCR sponsorship (64). Expected follow-up contracts for the NASA and ORNL efforts include further preliminary engineering design and continued costing studies.

LOW-TEMPERATURE CLOSED-CYCLE POWER PLANTS

There has been considerable development of low-temperature closed-cycle power plants. This work has involved both organic and inorganic working fluids with much development on organic working fluids being confined to small reciprocating and turbine engines for transportation applications (65, 66), and with work on the inorganic working fluids confined to larger-power turbine machinery, e.g. ammonia bottoming of large steam plants (67) and ammonia power plants for thermocline power plants (68). Working fluids such as SO_2 and CO_2 are also of interest. The work on very low temperature power plants includes that of thermocline power plants using propane (69, 70) and several hydrocarbon and fluorocarbon working fluids (71). Rankine cycles with either subcritical or supercritical pressure are also useful for extracting waste heat from the exhausts of gas turbine, diesel, and gas engines (72, 73); geothermal power plants extracting heat from geothermal hot water (74); a low-temperature solar power plant (75); and a suposed power plant replacing the precooler of an HTGR gas-turbine power plant.

Because of different properties of the fluids and of mixtures (65), selection of the fluid best fitting an application involves a detailed evaluation of the conceptual design to fit a specific application. This includes the type of heat source, e.g. waste heat stream, a geothermal stream, a solar-heated stream, or even the lower-temperature combustion streams, particularly with respect to maximum extraction of available energy from it. Much future attention will be given to developments applying the low-temperature closed cycle to power plants with large power ratings deriving their energy from nuclear or fossil sources and to the renewable sources such as geothermal energy.

FUEL-CELL POWER PLANTS

Conceived in the 1800s, fuel cells became practical devices in the 1960s when they powered the Gemini and Apollo space vehicles. Since that time, efforts have focused on developing low-cost, efficient fuel-cell power plants for terrestrial applications (Figure 5). As an electrochemical device, the fuel cell avoids the Carnot cycle efficiency limitation of thermal machines, as well as the NO_x and CO pollution associated with combustion. Thus, the fuel cell offers the potential of high efficiency over a wide range of loads, negligible pollution, and water conservation. In addition fuel cells are modular devices and can be fabricated and installed in response to demand, thereby minimizing escalation and interest costs during construction. For these reasons fuel cells offer substantial benefits to gas or electric utilities if used as dispersed or on-site generators within the utility network (76–78). Potential benefits include deferred transmission costs, reduced transmission losses, efficient load-following capability, cost-effective capacity expansion, and reduced construction costs.

The major commercial fuel-cell developments are the following:

1. TARGET (79) was initiated in 1967 to develop a 40–200-kW natural gas fuel cell for use at shopping centers, apartment buildings, or other on-site applications. Over $50 million has been expended to date by United Aircraft and 36 gas/electric utilities (80).
2. FCG-1 was initiated in 1972 by United Aircraft and nine electric utilities to

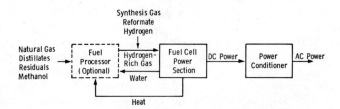

Figure 5 Fuel-cell power plant for utility application.

develop a 26-MW power plant for commercial service by the end of the decade (81). Approximately $50 million has been committed to date for research and development efforts (80).
3. Alsthom-Exxon initiated a joint $10–20-million program in 1970 aimed at vehicular applications (82, 83).

In addition to the primary electric/gas utility applications, the fuel cell must also be considered for commercial and industrial applications and for transportation. The waste heat of the fuel cell could be available as steam for heating/cooling (heat pumps) in condominiums, shopping centers, or industrial parks. This "total energy" concept employing both electric and thermal output is an ideal application for the fuel cell. There are industrial processes that require dc power and produce hydrogen as a by-product that is presently wasted. Recovery of that hydrogen and conversion to dc power via a fuel cell would increase the overall process efficiency, e.g. the production of chlorine by electrolysis of brine. A fuel cell should be considered as the power generator for transportation vehicles, if either methyl fuel or hydrogen become candidates for transportation fuels. The former fuel appears to be a more viable candidate than the latter. Thus, methanol fuel cells should be seriously considered for intermediate-term development.

Efforts toward fuel-cell development can be considered in three time frames.

1. Near-term efforts are principally aimed at fuel-cell power plants that could be in commercial service by 1980. These power plants would use natural gas or petroleum fuels and a phosphoric acid fuel-cell power section. The primary objective of these programs is improved efficiency with existing fuels. The power plant goal is 37–40% efficiency (hydrocarbon, HHV, to ac power).

2. Intermediate-term or second-generation fuel-cell programs would expand the application of fuel cells by broadening fuel capabilities to include coal-derived fuels and increasing the overall efficiency to 45%. Commercial introduction would be in 1985. Advanced technologies such as the molten carbonate fuel cell are being considered in addition to the aqueous electrolyte technologies.

3. Long-term programs are generally considered in terms of high-temperature fuel-cell systems such as the solid oxide. However, highly efficient lower-temperature concepts compatible with a wide range of fuels should be considered. The long-term goal would be the efficient direct use of a hydrocarbon fuel or, preferably, coal. Efficiencies as high as 60% have been cited for fuel-cell concepts operating directly on coal (84). Commercialization of such systems is aimed at the 1990s.

The three fuel-cell types under development include the low-temperature aqueous fuel cell, the molten electrolyte fuel cell, and the solid-oxide electrolyte fuel cell.

Low-Temperature Aqueous Systems

Low-temperature aqueous systems operate below 200°C and include phosphoric, sulfuric, and sulfonic acids; ion-exchange membranes; and alkaline electrolytes.

The major constraints in the development of these technologies are the need for precious-metal catalysts in acid and intermediate pH fuel-cell systems and the reaction of alkaline electrolytes with CO_2 from the fuel or air to form carbonates.

The phosphoric acid technology has been the focus of the major efforts toward terrestrial fuel-cell development over the past seven years. In addition to the TARGET and FCG-1 efforts, significant phosphoric acid programs have been supported by the US Army MERDC (85–87) and Engelhard Industries (88, 89). Phosphoric acid is favored by its ability to operate at relatively high temperatures (up to 200°C) without decomposition and with low vapor pressure of water. This allows the electrocatalysts to operate at a high rate and, in addition, simplifies the integration with a fuel-processing subsystem, i.e. steam reformer.

The major thrust in phosphoric acid research is toward the development of carbon electrode structures with minimal platinum catalyst content and long-term stability ($\geq 40{,}000$ hr life). The most serious limitation is the growth of high-surface-area electrocatalyst particles with time (90). This results in a loss of performance over the life of the fuel cell.

The phosphoric acid technology is well developed. Under the TARGET program, over 60 experimental 12.5-kW phosphoric acid prototype units were fabricated and installed in some 35 sites. Under the FCG-1 program, a megawatt-size demonstrator is scheduled to be tested in 1977.

The phosphoric acid technology will likely meet the near-term (1976) fuel-cell technical goals of the TARGET and FCG-1 programs (Table 1). However, major improvements will be required to meet the intermediate-term (1980) technical goals of 45% efficiency on a wide range of fuels.

Sulfuric acid systems have historically suffered from electrolyte instability at the desired range of 150–200°C. High-water-partial pressure and the tendency for formation of sulfur and hydrogen sulfide have been the major limitations.

Recent results (91) have suggested that electrocatalysis in a sulfonic acid medium may be superior to that in a phosphoric acid medium. However, the trifluoromethane sulfonic acid monohydrate used in that study was limited to $\leq 150°C$. These results indicate the need for further research toward developing a higher temperature sulfonic acid capability.

The ion exchange membrane is essentially a sulfonic acid wherein the acid is attached to a polymeric Teflon-like backbone. As with the other sulfonic acids, the ion exchange membrane or solid polymer electrolyte (SPE) is limited by the vapor pressure of the water. This limitation is exaggerated as the SPE provides negligible suppression of the vapor pressure. The SPE fuel cell does offer potential long-life capability (92). However, extension to operation at 150–200°C is problematic.

Alkaline fuel cells offer high performance. They are well developed and do not require precious-metal catalysts. However, they are limited by the need either to remove CO_2 from the fuel and air or to process the electrolyte to regenerate hydroxide from carbonate formed during operation. Both approaches are being considered and offer promise.

Molten Electrolyte Fuel-Cell Concepts

The two molten electrolyte fuel-cell concepts are the alkali metal hydroxide (400–500°C) concept developed by Prototech (93) and the molten carbonate (600–700°C)

Table 1 Past and projected fuel-cell progress (based on engineering subsystems)

	Past Progress		Goals	
	1968	1972	1976[a]	1980[b]
Precious-metal content (g/kW)	200–800	100	10	0
Fuel/oxidant	C_3H_8/air	NG/air	naphtha/air	*[c]/air
Life on air (hr)	1,000	1,000	40,000	40,000
Efficiency on air/hydrocarbon (%)	17	31	37	45
Capital cost (1972 $/kW)	100,000	1,000	200	150

[a] For commercial service by 1980.
[b] For commercial service by 1985.
[c] * = wide range of available utility fuels including coal and coal-derived fuels.

fuel-cell technology researched by Texas Instruments, General Electric, Institute of Gas Technology, and others.

The molten hydroxide avoids the problem of carbonate formation by use of a palladium diffuser to separate the hydrogen from the reformed fuel and prescrubbing the air. The technology development program has been modest to date and limited by air-inlet plugging and material- and component-life problems such as diffuser failures.

The molten carbonate technology has also been limited by component-life capability. However, recent progress has been somewhat encouraging (94, 95). Small cells have operated in excess of 7000 hr. Efficiency projections of 45% (hydrocarbon, HHV, to ac power) at capital costs of $210/kW can be made, based on present state-of-the-art technology. A series of six cell stacks are currently undergoing testing.

The molten carbonate technology is compatible with CO_2, operating at a temperature where reaction rates are high and electrocatalysts are not poisoned by species such as CO. Considerable efforts will be required to extend the life capability to the desired 40,000 hr. Material-life limitations can be expected to be the major concern.

Solid-Oxide Fuel-Cell Technology

Perhaps the ultimate in fuel-cell technology is an electrochemical device that can operate directly on coal. The solid-oxide fuel cell operating at 1000°C offers this possibility (96). However, a major long-term research and development commitment of $5–10 million per year for 15 years is likely to be required if this technology is to be commercialized in the 1990s. As a result, there is presently no industrial support of this technology. Possibly if the federal government becomes involved in a National Fuel Cell Program, a significant solid-oxide effort would be initiated.

Recent investigations into lower-temperature (600–800°C) solid-oxide systems (94, 97) may avoid some of the material problems present at 1000°C. However, these efforts are too preliminary to be significant at this time.

The past and projected fuel-cell progress is shown in Table 1.

The phosphoric acid technology is the only near-term approach that could be commercial by 1980. In the intermediate term, the molten carbonate fuel cell appears promising, although the requirement of $\geq 40,000$ hr of life will remain a challenge. In addition, advanced concepts for phosphoric acid electrolytes and new concepts for alkaline electrolytes must be considered as candidates for commercialization by 1985. In the longer term, direct-fueled concepts such as solid oxide should be developed to take full advantage of the ultimate potential of this versatile power generator.

CONCLUSION

The potential for coal and nuclear fueling of power plants using advanced energy conversion techniques with high-temperature efficiency and/or lower energy conversion cost appears to be significant with combined-cycle conversion using high-temperature energy conversion with a gas turbine, magnetohydrodynamic generator, or potassium turbine coupled to a lower-temperature energy conversion using a Rankine working fluid such as water, the halocarbons and hydrocarbons, and the inorganic working fluids such as ammonia, SO_2, and CO_2. Much work is being done on fuel-cell power plants that would use either coal directly as in the solid electrolyte cell or coal-derived fuels such as distillates, methane, methanol, or hydrogen.

Much of the work in the areas discussed lies in establishing viable technology. The need for viable technology that also is economically competitive in the "then existing" costs of energy conversion is apparent. Cost studies of conceptual systems using advanced energy conversion techniques are constantly being revised to include the latest technical developments. To this end, NASA is funding two studies—one with the Westinghouse Electric Corporation and the other with the General Electric Company on the technology and costs of advanced energy conversion systems under its ECAS (Energy Conversion Alternative Studies) program (5a). The results of these in-progress studies should be available in late 1975.

Literature Cited

1. National Academy of Engineering, Task Force on Energy. 1974. *U.S. Energy Prospects, An Engineering Viewpoint.* Washington DC: Nat. Acad. Sci.
2. Olmsted, L. Sept. 15, 1974. 25th annual electrical industry forecast. *Electr. World*, pp. 43–48
3. Hirst, E. March 1972. *Energy Consumption for Transportation in the US.* Rep. No. ORNL-NSF-EP15. Oak Ridge Nat. Lab., Oak Ridge, Tenn.
4. Klaff, J. L. June 1973. *Final Report on the National Commission on Materials Policy.* Library Congress Card No. 73-600202
5. Tilton, J. E. March 1975. Cartels in metal industries. *Earth and Mineral Sciences*, pp. 6, 41–44. Penn. State Univ., University Park, Pa.
5a. National Aeronautics and Space Administration. March 1974. Energy-related research and development. *Committee on Aeronautical and Space Sciences, US Senate, 93rd Congress, 2nd Session,* p. 28; July 1975. *Gas Turbine World*, p. 33
6. Wood, B. April 1972. Combined cycles: a general review of achievements. *Combustion*, pp. 12–22
7. Bodle, W. W., Vyas, K. C. August 1974. Clean fuels from coal. *Oil Gas J.*, pp. 73–88
8. Vyas, K. C., Bodle, W. W. March 1975. Coal and oil-shale conversion looks better. *Oil Gas J.*, pp. 45–54
9. Young, W. E., Lee, S. Y. 1975. Gas turbine hot stage parts in aggressive atmospheres. *Joint ASME/ASTM Gas Turbine Panel, Int. Gas Turbine Conf., Houston, Texas*
10. Archer, D. H., Somers, E. V. 1974. Fluidized bed gasification and combustion for power generation. *9th World Energy Conf., Detroit.* Pap. 4–17
11. Archer, D. H. Sept. 1973. In *Clean Fuels from Coal Symp., Inst. Gas Technol., Chicago*
12. Dundas, R. E. 1972. Design of the gas turbine engine. *Sawyer's Gas Turbine Eng. Handb.* 2:79–145
13. Kydd, P. H., Day, W. H. 1975. An ultra high temperature turbine for maximum performance and fuels flexibility. *Am. Soc. Mech. Eng. (ASME) Pap.* No. 75-GT-81
14. Bratton, R. A., Holden, A. N., Mumford, S. E. 1974. Testing ceramic stator vanes for industrial gas turbines. *Soc. Automot. Eng. Tech. Pap.* No. 740236.
15. Rudolph, P. F. H. May 1970. New fossil-fueled power plant process based on Lurgi pressure gasification of coal. *Am. Chem. Soc., Div. Fuel Chem., Prepr.* 14
16. Ahner, D. J., May, T., Sheldon, R. C. Sept. 1974. Low B.T.U. gasification combined cycle power generation. *ASME/IEEE Joint Power Conf.*
17. US Department of Interior, Office of Coal Research. 1973. *Clean Energy from Coal Technology*
18. Ehrlich, S., McCurdy, W. A. 1974.

Developing a fluidized-bed boiler. *Proc. 9th Annual IECEC, San Francisco.* Pap. 749133
19. Keairns, D. L., et al. Dec. 1973. *Evaluation of Fluidized Bed Combustion Process.* Rep. No. EPA-650/2-73-048 a, b, c, and d. Washington DC: US EPA
20. US Department of Interior, Office of Coal Research. 1975. *Energy Conversion from Coal Utilizing CPU-400 Technology, TR-74-105.* R&D Rep. No. 94
21. Smith, J. 1967. US Bur. Mines Rep. No. R16920.
22. US Department of Interior, Office of Coal Research. 1974. *Pressurized Fluidized-Bed Combustion.* R&D Rep. No. 85, Interim Rep. No. 1
23. Bammert, K., Klein, R. 1975. The influence of He-Ne, He-N_2, and He-CO_2 gas mixtures on closed-cycle gas turbines. *ASME Pap.* No. 75-GT-124
24. Pietsch, A. 1974. Potential applications for closed gas turbines. *Proc. 9th Annual IECEC.* San Francisco. Pap. No. 749087
25. Adams, R. G., Bell, F. R., McDonald, C. F., Morse, D. C. 1973. HTGR gas turbine power plant configuration studies. *ASME Pap.* No. 73-WA/Pwr-7
26. Barber, R. E. Nov. 1974. Rankine cycle systems for waste heat recovery. *Chem. Eng.*, pp. 101–106
27. Manning, G. B., Stone, L. K. 1974. The closed cycle gas turbine and the 1985 US water pollution abatement goals. *ASME Pap.* No. 74-GT-123
28. Bammert, K., Rurik, J., Griepentrop, H. 1974. Highlights and future development of closed-cycle gas turbines. *ASME Pap.* No. 74-GT-7
29. Haselbacher, H., Eiermann, A. 1974. Development of helium gas turbine systems in the nuclear field. *ASME Pap.* No. 74-GT-123
30. Bammert, K., Deuster, G. 1974. Layout and present status of the closed-cycle helium turbine plant Oberhausen. *ASME Pap.* No. 74-GT-132
31. Gulf-GA Project Staff. 1973. *HTGR Gas Turbine Power Plant Control and Safety Studies.* Rep. No. GA-A 12865
32. Jordan, D. J., Mayer, P. M. 1973. *HTGR-Gas Turbine Power Plant Study Phase II Final Report.* Rep. No. PWATM-4759
33. English, R. E., et al. 1973. *Conversion Techniques R&D Program FY 75–79.* Rep. No. WASH 1281-4
34. Kaganovich, S. A., et al. 1972. Features of operation of the air heater in a closed-cycle gas turbine plant. *Teploenergetika,* pp. 19, 41–44

35. Sonju, O. K., Teno, J. April 1974. Experimental and analytical research on a two-megawatt high performance MHD generator. *14th Symp. Eng. Aspects Magnetohydrodynamics, University of Tennessee.* Pap. No. I.3
36. Wu, Y. C. L., et al. 1974. On direct coal fired MHD generator. See Ref. 35, Pap. 1.2
37. Gannon, R. E., Stickler, D. B., Kobayashi, H. 1974. Coal processing employing rapid devolatilization reactions in an MHD power cycle. See Ref. 35, Paper II.2
38. Electric Power Research Institute/Office of Coal Research. May 1974. *EPRI/OCR Workshop on the Development of Open-Cycle MHD, Palo Alto, Calif.*
39. National Science Foundation/Office of Coal Research. November 1974. *NSF/OCR Engineering Workshop on MHD Materials, MIT Energy Laboratory, Cambridge, Mass.*
40. Henry, J. M., Kurtzrock, R. C., Bienstock, D. Nov. 1974. MHD combustors: design and materials of construction. See Ref. 39, Session VA Pap.
41. Lacey, J. J., Demeter, J. J., Bienstock, D. March 1972. Production of a clean working fluid for coal-burning open-cycle MHD power generation. *12th Symp. Eng. Aspects Magnetohydrodynamics, Argonne, Ill.* Paper VI.2
42. Way, S. 1967. US Patent No. 3,358,624
43. Dicks, J. B., et al. 1974. The direct-coal-fired MHD generator system. See Ref. 35, Pap. II.1
44. Rosa, R. J., Petty, S. W., Enos, G. R. 1974. Long duration testing in the Mark VI facility. See Ref. 35, Pap. I.5
45. Young, W. E., et al. 1974–1975. *Magnetohydrodynamic Investigation for Coal-Fired Open-Cycle Systems.* Monthly Prog. Rep. 1–20. OCR Contract No. 14-32-0001-1540. Office Coal Res., Washington, DC
46. Zankl, G., et al. 1974. Results of the IPP test generator and the design of a 10 MW/el short-time combustion MHD generator. See Ref. 35, Pap. I.1
47. Hals, F. A., Gannon, R. E., Kivel, B. 1974. Auxiliary component development work at AVCO Everett Research Laboratory, Inc., See Ref. 35, Pap. II.3
48. Bates, J. L. 1974. Material development goals for open-cycle MHD air preheaters. See Ref. 39, Session IV.A Pap.
49. Uhlmann, J. R. 1974. Thermal shock behavior of preheater materials. See Ref. 39, Session IV.A Pap.
50. Hatch, A. M., et al. 1974. A super-

51. US Department of Interior, Office of Coal Research. 1974. *Evaluation of a Fossil Fired Ceramic Regenerative Heat Exchanger.* R&D Rep. No. 92, Interim Rep. No. 1
52. Williams, J. R., et al. 1972. Exploratory investigation of an electric power plant utilizing a gaseous core reactor with MHD conversion. *Proc. 7th Annual IECEC, San Diego,* pp. 1305–11
53. Zauderer, B. March 1974. Closed cycle MHD potential impressive. *Electr. World,* pp. 94–95
54. Zauderer, B. 1972–1974. *Investigation of a Non-Equilibrium MHD Generator.* Office Naval Res. Contracts with General Electric, Contract No. N00014-73-C-0039, Project Code 9800, and associated Annual Rep.; and p. 1039 of Ref. 57
55. Phen, R. L. 1974. *Liquid Metal Magnetohydrodynamics (LMMHD) Technology Transfer Feasibility Study,* Vols. I, II. JPL Rep. N74-13466 and 13467
56. Petrick, M. 1973–1974. Office Naval Res. Contracts with ANL
57. Subcommittee on Energy, Committee on Science and Astronautics, US House of Representatives, 93rd Congress. 1974. *Inventory of Current Energy Research and Development,* Vol. 1, p. 1039. (Committee print)
58. Moor, B. L., Schnetzer, E. 1971. *Three Stage Potassium Vapor Turbine Test.* NASA Contract. Rep. 1815. Nat. Aeronaut. Space Admin.
59. Fraas, A. P. 1966. A potassium-steam binary vapor cycle for a molten salt reactor power unit. *J. Eng. Power, ASME Trans.,* pp. 88, 355–66
60. Fraas, A. P. January 1973. A potassium-steam binary vapor cycle for better fuel economy and reduced thermal pollution. *J. Eng. Power, ASME Trans.,* pp. 53–63
61. Fraas, A. P. 1974. *Operational, Maintenance, and Environmental Problems Associated with a Fossil Fuel-Fired Potassium Steam Binary Vapor Cycle.* Rep. No. ORNL-NSF-EP-30. Oak Ridge Nat. Lab., Oak Ridge, Tenn.
62. Fraas, A. P., MacPherson, R. E. 1974. *Survey of Gas and Oil Burners for Use with USF/RANN-ORNL Potassium Boiler.* Rep. No. ORNL-NSF-EP-45. Oak Ridge, Nat. Lab., Oak Ridge, Tenn.
63. Rossbach, R. J. 1973. *Final Report of Joint NASA/OCR Study of Potassium Topping Cycles for Stationary Power.* Rep. No. GESP 741, under NAS3-17354
64. Ralph M. Parsons Co. 1974. Assessment of potassium-steam binary cycle (Objective 9.2). In *Clean Liquid and/or Solid Fuels from Coal.* R&D Rep. No. 82, Interim Rep. No. 4, Tech. Eval. Serv. US Dep. Interior, Office Coal Res.
65. Miller, D. R., Mull, H. R., Thompson, Q. E. 1973. *Optimum Working Fluids for Automotive Rankine Engines,* Vols. II, III. Rep. No. USEPA-APTD1564 & 1565
66. Sindermann, F. 1973. *Optimum Working Fluids for Automotive Rankine Engines,* Vol. IV. Rep. No. USEPA-APTD 1566
67. Slusarek, Z. M. 1973. *The Economic Feasibility of the Steam-Ammonia Power Cycle.* Rep. No. PB-184331
68. Connell, J. W., Gowan, J. G. 1973. *Condenser Requirements for an Ocean Thermal Gradient Power Plant.* Rep. No. NSF/RANN/SE/GI-34979/TR/73/3
69. Lessard, R. D. 1973. *Technical and Economic Evaluation of Ocean Thermal Difference Power Plant Turbomachinery.* Rep. No. NSF/RANN/SE/GI-34979/TR/73/18
70. Marshall, J., Ambs, L. 1974. *Evaluation of the Major and Support Fluid System Necessary for the Operation of a Rankine Cycle Ocean Thermal Difference Machine.* Rep. No. NSF/RANN/SE/GI-38979/TR/74/5
71. Ambs, L., Marshall, J. 1973. *Ocean Thermal Difference Power Plant Turbine Design.* Rep. No. NSF/RANN/SE/GI-38979/TR/73/17
72. Angelino, G., Moroni, V. April 1973. Perspectives for waste heat recovery by means of organic fluid cycles. *J. Eng. Power, ASME Trans.,* pp. 75–83
73. Sternlicht, B. June 1975. The equipment side of low level heat recovery. *Power,* pp. 71–77
74. Brewer, W. A., et al. September 1972. Discovery and development of geothermal resources. *Joint Power Conf. IEEE/ASME/ASCE*
75. Prigmore, D. R., Barber, R. E. 1974. A prototype solar powered Rankine cycle system providing residential air conditioning and electricity. *Proc. 9th Annual IECEC, San Francisco.* Pap. No. 749026
76. Lueckel, W., Eklund, L., Law, S. 1973. Fuel cells for dispersed power generation. *IEEE Trans. Power Appar. Syst.,* PAS-92: 230–36
77. Fernandes, R. A. 1974. Hydrogen cycle peak-shaving for electric utilities. *Proc. 9th Annual IECEC, San Francisco.* Pap. No. 749032

78. Fickett, A. P. 1975. An electric utility fuel cell: dream or reality. *Proc. 37th Am. Power Conf., Chicago*
79. Podolny, W. H. 1973. *Progress in the TARGET Fuel Cell Power Plant Development Program.* Presented at 12th World Gas Conf., Nice, France
80. Podolny, W. H. 1974. *Statement to the Subcommittee on Energy Research, Development and Demonstration, Committee on Science and Technology, US House of Representatives, 93rd Congress*
81. Lueckel, W. J., Farris, P. J. 1974. *The FCG-1 Fuel Cell Power Plant for Electric Utility Use.* Presented at IEEE Power Eng. Soc., Anaheim, Calif.
82. Wall Street Journal. December 1970. *Fuel Cell Development Set by French Firm, Jersey Standard Unit*
83. Heath, C. E. 1973. Fuel cells. *Science* 180:542
84. Sverdrup, E. F., Warde, C. J., Glasser, A. D. 1972. A fuel cell power system for control power station generation using gas. In *Electrocatalysis to Fuel Cells*, 255–77. Seattle: Battelle Mem. Inst. and Univ. Wash. Press
85. Huff, J. R. 1972. *A Status Report on Fuel Cells.* Rep. 2039. USA-MERDC, Fort Belvoir, Va
86. Kunz, H. R., Gruver, G. A. 1974. *The Catalytic Activity of Platinum Supported on Carbon for Electrochemical Oxygen Reduction in Phosphoric Acid.* Presented at Fall Meeting, Electrochem. Soc., New York
87. Camp, R. N., Baker, B. A. 1972. Low cost air cathodes. *Proc. 7th Annual IECEC, San Diego, Calif.* Pap. No. 729002
88. Adlhart, O. J. 1972. The phosphoric acid fuel cell, a long life power source for the low to medium wattage range. *Proc. 7th Annual IECEC, San Diego, Calif.* Pap. No. 729163
89. Adlhart, O. J. 1972. The air-cooled matrix-type phosphoric acid cell. See Ref. 84, pp. 181–88
90. Stonehart, P. 1974. Fuel cells. *Energy Workshop on Electrochemistry,* Rep. UNC-100. Chapel Hill: Univ. N. Carolina Press
91. Adams, A. A., Barger, H. J. 1974. A new electrolyte for hydrocarbon air fuel cells. *J. Electrochem. Soc.* 121:987–90
92. Nuttall, L. J. 1975. Solid polymer electrolyte fuel cell states. *Proc. 10th Annual IECEC, Newark, Del.*
93. *Fuel Cell Process Using Peroxide and Superoxide and Apparatus.* October 1969. *US Patent No. 3,471,334*
94. Electric Power Research Institute. 1975. *Advanced Technology Fuel Cell Program.* Interim Rep. EPRI 114
95. King, J. 1975. Advanced fuel cell technology for utility applications. *Proc. 10th Annual IECEC, Newark, Del.*
96. Westinghouse Electric Corporation. 1973. *Development of a High Temperature Solid-Electrolyte Fuel Cell Power System.* Proposal to Office Coal Res.
97. Tuller, H. L., Norwick, A. S. 1975. Doped ceria as a solid oxide electrolyte. *J. Electrochem. Soc.* 122:255–59

Copyright 1976. All rights reserved

THE ENERGY INDUSTRY AND THE CAPITAL MARKET

×11014

William E. Pelley, Richard W. Constable, and Herbert W. Krupp
Bankers Trust Company, PO Box 318, New York, NY 10015

STATEMENT OF PURPOSE

As the United States attempts to restructure its energy supplies, there has developed an associated concern over the ability of the capital markets to supply the financing that will be required. This study, which is based on a more detailed report[1] by Bankers Trust Company, develops the possible funding needs of the domestic energy industries through 1990 and projects a possible parallel development of the US capital markets. The relationship of the supply of funds to the demand is then investigated.

In presenting this study, it is not our intent to make forecasts of either the necessary route our energy policies should take or the actual factors that will influence the structure of our capital markets. Rather than trying to predict the future, this analysis attempts to establish reasonable assumptions of both economic and energy developments. If there is adequate capital indicated under this set of assumptions, such adequacy should also exist for other reasonable combinations of energy development.

METHODOLOGY

The approach taken in this analysis was to accept an overall national requirement of British thermal units (Btu's) on a yearly basis and break down this demand into its various components. The supply from the various energy sources was then compiled and related to the demand sectors to assure that there were no incompatible assumptions of utilization, such as having to run the transportation sector on nuclear power. The analysis of these demand/supply factors is given in detail in the following sections of this chapter.

The data used in studying the supply components are either taken directly from various expert sources or modified to conform to the authors' interpretations. Frequently it has been necessary to choose one case out of several offered, such as in the National Petroleum Council studies or the Federal Energy Administration

---

[1] A copy of the more detailed report is available from Public Relations, Bankers Trust Company, PO Box 318, New York, NY 10015.

reviews. There are no major areas of energy supply where we have presumed to perform original analysis. Where modifications have been necessary, they conform to a conservative attitude toward the development of domestic energy supplies.

Of at least equal significance, and involving original effort by Dr. Edward Campbell and some of his associates in the Economics Department of Bankers Trust, is the projection of the possible capital market available through 1990. The work relates to a flow-of-funds analysis and is discussed later in this article.

ASSUMPTIONS

The underlying premises of this study are based on the belief that the fundamental institutions of the United States will not change radically in the next 16 years. They are summarized as the following.

(a) An average 3.6% annual growth rate of the Gross National Product (GNP) and a 3.1% rate for domestic energy consumption will occur.
(b) No major wars or Mideast conflicts will affect world energy supplies on a sustained basis.
(c) Recovery from the present business recession will continue on a gradual basis through the rest of this decade.
(d) Government deficits will continue into the 1980s.
(e) Environmental constraints will not impose new and unforeseen capital-intensive requirements on the energy industries or increase existing time delays.
(f) Imports of crude oil will be available and tolerated.
(g) Foreign sources of capital will at least equal the demands of foreign energy development.
(h) Either free market conditions will tend to prevail or governmental regulations and policies will stimulate energy development as effectively and efficiently as would free market conditions.

ENERGY DEMAND BY CONSUMING SECTOR

During the 15-year period prior to the Arab oil embargo of October 1973, the United States demand for energy advanced at an average rate of slightly over 4% per annum. Then in 1974 this growth was altered by the impact of the embargo and the recession, and demand actually fell by 2%.

Energy consumption will unquestionably begin to grow again once the US economy resumes its upward path. But because of higher energy prices and greater governmental intervention, the rate of growth is expected to fall below the historical rate of 4%. The Bankers Trust Company study analyzes the individual energy-consuming sectors and assigns probable growth rates to each (see Table 1 for details). In the aggregate this approach indicates a future growth rate of roughly 3%. In this study the consumption of electricity has not been included in the household and commercial, industrial, and transportation sectors, for the fuel sources required to generate electricity for these sectors have already been accounted for in the electrical generation market.

Table 1 United States energy supply and demand (quadrillion of Btu)

	1974	1977	1980	1985	1990
DEMAND					
Domestic					
Household and commercial	13.8	14.8	15.9	17.3	18.9
Industrial	21.1	21.7	22.1	22.7	23.6
Transportation	18.3	19.8	21.2	24.0	26.8
Electrical generation	19.7	23.2	27.6	36.9	49.3
Total	72.9	79.5	86.8	100.9	118.6
Exports	2.3	2.5	2.7	3.1	3.6
Synthetic conversion loss	—	0.1	0.2	0.9	3.5
Total Demand	75.2	82.1	89.7	104.9	125.7
SUPPLY					
Domestic					
Coal	14.9	17.2	20.1	24.7	32.1
Petroleum	20.6	19.0	24.2	26.0	27.4
Gas	21.1	19.6	18.5	17.5	16.5
Nuclear power	1.2	2.8	5.0	12.2	23.8
Hydropower	3.0	3.2	3.3	3.8	4.3
Shale	—	—	0.1	1.4	4.2
Other	—	—	0.1	0.3	0.7
Total	60.8	61.8	71.3	85.9	109.0
Imports					
Petroleum	13.4	19.0	16.3	16.7	13.9
Gas	1.0	1.3	2.1	2.3	2.8
Total	14.4	20.3	18.4	19.0	16.7
Total Supply	75.2	82.1	89.7	104.9	125.7

Household and Commercial

Most fossil fuel energy in the household and commercial sector is used for heating, cooking, and hot water purposes. Since 1960 consumption of fossil fuel energy in this sector has advanced by 3% per annum. The study projects such consumption to grow at only half that rate through 1990.

The reduction reflects in part conservation practices by consumers to reduce their markedly higher energy costs, and in part the increased growth of electric power for home heating. Conservation practices will include new building standards designed to increase insulation and reduce heat loss as well as the more immediate action of setting the thermostats lower in winter and higher in summer. The use of electricity in place of the direct use of fossil fuel will increase because of its convenience and its developing price competitiveness. In addition, the supply of electricity may be more secure than that of other fuels.

Although shortages of natural gas may occasionally develop in some regional markets and although the total US supply of gas will likely decline within the next decade and a half, the availability of supply for the residential market in most regions will continue to expand. Gas is ideally suited for the high-priced residential market, and the Federal Power Commission (FPC) has accordingly given this market the highest priority rating. This means that supplies to industrial and utility customers will be curtailed to provide additional gas for a growing number of residential customers.

Thus, despite an expected decline in total gas supply through 1990, residential and commercial customers will increase their consumption of gas. The study has assigned a 2.5% per annum growth rate for gas to this sector, far less than the 6% rate experienced in the 1960s, but still one percentage point greater than the rate for petroleum. It is felt that consumption of petroleum will be restricted by the uncertainty of supply and the recent rise in price relative to natural gas and electricity.

Industrial

Industrial demand for fossil fuels rose by 3% per annum during the 1960s, but then declined to under 2% during the early 1970s. In 1974 consumption sagged by 1%, mainly in response to higher energy prices and a depressed economy.

In the future the industrial growth rate of fossil fuel consumption will likely fall well below the post-1960 rate of 2% or 3%. The study projects a 0.6–0.7% rate through 1990. Higher prices and slower economic growth are partially responsible for the projected slowdown, although another important factor will be the growth of electricity at the expense of fossil fuel consumption. Manufacturing firms within the industrial market cannot afford the risk of energy supply interruptions. Therefore, an increased number of industrial users will switch to electricity or coal where possible to avoid interrupted gas or embargoed oil supplies.

Despite the high price and vulnerability of petroleum, its demand is projected to grow at the robust rate of 5% per annum through 1980. This growth in the next few years is expected because the growth rate for natural gas is negative and its shortage will force many industrial customers to switch, at least temporarily, to residual fuels until longer-range plans to switch to electricity are completed. The actual growth in petroleum consumption in the next few years will depend on the magnitude of the gas curtailments, which in turn depends principally on the severity of the upcoming winters, the growth of the economy, and the decline in natural gas production.

Transportation

Virtually all energy consumed within the transportation sector is petroleum-based. During the 1960s petroleum demand in this sector rose at an average annual rate of 3.5%. In the 1970s the rate jumped to over 4%. The four main factors accounting for the increase were (*a*) larger cars and government-required safety features raising the weight of vehicles; (b) the Environmental Protection Agency (EPA) requirement of emission control devices in new cars; (c) increasing popularity of air-conditioned automobiles; and (d) more rapid growth of the vehicle population as consumers enjoyed increasing affluence. The first three factors reduced mileage per gallon, while the last increased mileage driven.

All of these factors will be influenced by higher fuel costs and governmental requirements for efficiencies and conservation. Cars will become smaller and lighter and engine efficiencies will be improved, all of which will give more mileage per gallon. Not only will governmental standards move in this direction, but the auto industry will support these changes for competitive reasons as their material and labor costs increase.

Bearing these factors in mind, the study projected an average annual growth rate of 2.5% through 1990, somewhat higher in the early years and somewhat lower in the later years.

Electrical Generation

During the 1960s and through 1973, the demand for electricity rose at an average rate of just under 7%, nearly double the rate of the other energy-consuming sectors. This study projects electric energy demand through 1990 to increase 6% per annum, just one percentage point below the recent historical growth rate and still more than double the rate of the other sectors.

Despite the recent upsurge in electricity prices and the strong likelihood of further increases, the projection of a 6% growth rate seems reasonable. As noted earlier, the reliability and availability of electricity will make it even more of a premium energy source than before. In the household and commercial market, electric heating will become increasingly acceptable. In the industrial market, natural gas shortages and potential petroleum embargoes will force shifts to electricity consumption. In the transportation market, the role of electricity will increase as public transportation systems are developed in and between our urban areas in the 1980s.

NUCLEAR POWER   Within the electrical generation sector, nuclear fuel consumption should exhibit the greatest surge, rising from 1.2 quadrillion Btu's in 1974 to 23.8 quadrillion in 1990. Whereas in 1974 nuclear power provided just 6% of the utility industry's requirements and less than 2% of the nation's total energy needs, by 1990 it will provide utilities with almost 50% of their requirements and 20% of the nation's needs.

HYDROPOWER   Hydropower will show little growth during the forecast period, mainly because the best hydro sites have already been developed. In 1974 hydropower accounted for 15% of all electricity generated; by 1990 it will account for under 10%.

NATURAL GAS   During the late 1960s orders for gas-fired utility plants slumped to near zero. As a result the growth rate for consumption of natural gas by utilities fell from 6% per annum to just 1% by the early 1970s. In 1974 it actually declined by 10%. In the future, utility gas consumption will continue to decline as older plants are retired or forced to convert to coal by order of the federal government.

The study projects gas consumption for electricity to fall to 1.8 quadrillion Btu's in 1990 from 3.3 quadrillion in 1974. In the process the gas share of the electricity market will slump to 4% from its current level of over 16%.

PETROLEUM Consumption of fuel oil by utilities rose 12% per year during the 1960s and accelerated to 25% during the early 1970s. The rapid growth in the last few years has been largely caused by some utilities shifting out of coal and into low-sulfur oil in order to meet environmental standards for air quality. This demand by utilities for residual fuel oil will continue to grow at a rate of 2.5% per year through 1980 despite the four- to fivefold increase in price and the vulnerability of the foreign supply. Additional consumption will result from the start-up of new oil-fired plants ordered before the 1973 embargo. After 1980 virtually no new oil-fired plants will be put on line. Demand will decline as older oil plants are retired and replaced by coal-burning or nuclear-powered plants. By 1990 the market share of oil for utilities is projected to fall to less than 4% from its current level of 17%.

COAL As the remaining fuel source for utilities, coal accounts for about 45% of the electrical generation market. In the 1960s coal consumption within this sector grew about 6% per annum. By the 1970s the rate had declined to 4.5%, mainly because governmental air quality regulations discouraged the construction of coal plants in certain regions. For the next five years or so, the coal consumption by utilities is projected to resume growing at 6% per annum. This accelerated rate reflects the industry's response to the delay or curtailment of nuclear plant construction, and to curtailment of natural gas supplies, and in some cases as an effort to move out of higher-priced oil. Sometime after 1980 this rate of growth will ease as nuclear power becomes more important, but nevertheless, coal demand will rise by 3–3.5% a year during the 1980s. By 1990 it will still be supplying almost one third of the energy needs of the utility industry.

Other Areas of Demand

EXPORTS This study assumes that the only significant export of an energy material will be metallurgical coal. Although this is not used domestically, it must be mined as part of the fuel supply and treated as such in the coal supply analysis. It may seem illogical to assume fuel exports at a time when fuel imports are undesirable, but the rationale used in accepting it is that this coal is in reasonable supply and is a strong contributor to our balance of payments. For these reasons coal exports are expected to increase during the 1975–1990 period.

ADJUSTMENT FOR SYNTHETIC FUELS The synthetic fuel adjustment is a minor factor in the total study. It represents a fuel loss in converting a primary fuel to a more desirable form, and it is included as a necessary item to balance out the demand/supply relationship shown in Table 1.

Summary of Demand

With each sector of consumption analyzed, the resulting demands by sectors for particular time intervals are shown in the top part of Table 1. Although the cumulative totals are not given here, the total Btu demand of all sectors by years through 1990 does conform to the initial assumption that the growth of domestic demand will average some 3% a year over this time span.

THE SUPPLY OF ENERGY AND ITS CAPITAL COSTS

The US energy supply must be divided into two sources: domestic production and imports.

Imports

Imports are virtually restricted to petroleum and natural gas, with the latter fuel coming in by pipeline from Canada or by liquefied natural gas (LNG) tanker from overseas, mainly from Algeria and Indonesia. Petroleum imports will be the balancing fuel in the sense that where domestic production of any energy source falls short of requirements, oil imports will have to increase. For example, if nuclear plant construction continues to be delayed, then the utility demand for oil will be higher than presently assumed. Since the US capability for crude oil production is projected at maximum capacity, petroleum imports would have to increase to meet this higher demand for residual fuel. Similarly a gas shortage would increase the demand for refined products, which could only be supplied by more oil imports.

Since this study assumes that all readily available sources of domestic supply will be utilized to their maximum and new sources of conventional and synthetic fuels will be developed for commercial application as rapidly as possible, there will be no domestic capacity in reserve to call on if domestic demand for energy exceeds domestic supply—which it presently does and will continue to do. In order to meet this estimated excess demand, the United States must import the necessary extra supplies. Although it is not included in this report, a review by Bankers Trust Company indicates that foreign sources are capable of supplying such crude oil and gas to the United States through 1990 even in the face of growing demands in the rest of the world. Therefore, the addition of imports to balance out domestic requirements is necessary and can be achieved.

The amount of gas to be imported is largely restricted by the economics of liquefaction, transportation, storage, and regasification. At present these economic factors are joined by a concern by some critics over the safety of moving and storing the fuel. Both of these forces will discourage the future growth of this fuel source. This study gives credit, therefore, only to those projects underway that can be expected to be finalized.

The figures for petroleum imports are purely derived as the necessary addition to balance the supply with demand. In this study the demand for imports continues to rise to over 9 million barrels per day (Mb/d) through 1977; then it decreases to a more stable level of 6.5 Mb/d. In the latter part of the 1980s it again begins to decrease as nuclear power and synthetic fuels become more significant.

Coal

SUPPLY Coal is projected to provide this country by 1990 with more energy than any other domestic fuel source, including nuclear power. By then its share of the energy market will have climbed to 25%, up from 20% in 1974, when both natural gas and petroleum provided more Btu's than did coal. In this study about one sixth of all coal in 1990 is assumed to be used as a base in the manufacture of synthetic gas and oil.

To meet the huge requirements projected in the study, the supply of coal will have to increase by almost 5% per annum through 1990. Looked at another way, production will have to more than double in just 16 years. The realization of this goal will be difficult. Many environmental problems are presently unresolved. A strong government position will be required to achieve the necessary development.

Under normal conditions it takes two to four years to develop a strip mine and three to five years for an underground one. But the sudden surge in demand for coal mining equipment and materials has created shortages that may considerably extend the lead times. In addition, 60,000 to 80,000 additional miners, possibly more, will have to be hired and trained within this 16-year period. Many thousands of hopper cars and engines will have to be manufactured and many thousands of miles of railroad tracks will need to be repaired and upgraded.

It is clear that unless the government acts soon to stimulate and encourage this development, the goal will not be realized. In that case more oil for industry and the utilities will be used and less coal-derived synthetic fuel will be produced. This would mean that petroleum imports will have to be in excess of those projected in this study, but the additional requirements could then possibly exceed available foreign supply.

In order to provide the supply of coal represented by the Btu's in the projection, two principal areas of capital cost must be satisfied: mining of the coal itself and transportation to the consumer. It was anticipated that the trend toward a greater proportion of western coal in total production would continue and that this coal would be principally surface mined. The overall ratio of US surface-mined coal to deep-mined coal is projected to change from approximately 1:1 at present to 2:1 by 1990 (1), while concurrently the average Btu content of surface-mined coal will decrease as western coal becomes predominant in the mix. This factor will result in more tons being mined than would otherwise be required to provide the same Btu supply, although at a lower capital cost per ton. The projection is for total annual coal tonnage to increase from the current level of about 600 million tons per year to approximately 1.1 billion tons in 1985 and 1.5 billion tons by 1990. The rate of increase in the later years is augmented somewhat to reflect the needs of an expanding synthetic natural gas (SNG) industry.

CAPITAL COSTS The capital costs for opening new coal production, as well as replacing depleted mines, were estimated to be $25 per ton of new capacity for deep mines (2) and $12 per ton for surface mines (3). These figures are considered sufficiently high to include the capital portion of surface mine reclamation. The resulting total capital requirement for coal production over the 16-year period is $19.6 billion.

TRANSPORTATION Investment in coal transportation required to handle the projected tonnage was found to be virtually as great as the cost of opening new mine production. The total through 1990 was projected to be $19.4 billion. Because of the increasing average distances from mines to consumers, as well as increasing coal production, the equivalent of nearly 3000 additional unit trains of four locomotives and 100 hopper cars each will be required by 1990 to handle this increase. To an important extent, barge transportation may replace some of the projected unit trains, but still the overall transportation cost would not be greatly changed. The additional cost of replacing worn trackage was not included, but replacements of hopper cars were, averaging over 4% per year due to increased wear inherent in unit train operations.

Petroleum and Natural Gas

OIL SUPPLY Spurred on by a tripling of new domestic crude oil prices, drilling rates within the United States have soared above their depressed levels of the early 1970s. Nevertheless, because of a decreasing discovery rate and the long lead time required between discovery and production, output of petroleum liquids in the lower 48 states will rise only slightly through 1980. Including Alaska, however, US production of petroleum liquids is expected to climb by almost 2 Mb/d to over 12 Mb/d in 1980. Thereafter it will climb a modest 2% per annum with most new output coming from Alaskan and offshore wells. By 1990, production is projected to climb to 14 Mb/d, up by 35% from 1974 (4).

Despite the 35% rise in output by 1990, a gap of 9 Mb/d between domestic demand and supply is projected. Of this gap, 6.5 Mb/d will be filled by imports, and the rest from syncrude. Thus, dependence on foreign sources of petroleum will still be high in 1990, with the United States importing 30% of its petroleum requirements. Nevertheless, such dependence would be less than during the peak year of 1977, when the United States is projected to import 9 Mb/d, or 50% of its petroleum supply. Dependence will drop sharply in 1978 (possibly even by late 1977) when Alaskan crude oil begins to flow into the lower 48 states.

GAS SUPPLY Natural gas production will have an unavoidable decline during the next decade. A recent FPC report (5) found that the elimination of price controls, while desirable, will only retard this decline, not prevent it. Recent production figures reflect these slumping rates. Whereas gas production advanced by 7% per annum in the 1960s, it showed no growth during the early 1970s, and actually declined by 4% in 1974. Bearing in mind these statistics, the Bankers Trust Company study projects a 22% decline in natural gas production through 1990. In that year natural gas will supply just one sixth of US energy requirements, down from one third today.

Supplemental gas, both imports and syngas, will play a much more important role in the future than it does today. Imports via pipelines from Canada will grow only slightly and not until the late 1980s, when Canada's northern fields are finally developed. However, imports of LNG are projected to rise from the near-zero level in 1974 to 1 quadrillion Btu's by 1980. Thereafter, LNG imports are projected to remain level.

OIL AND GAS CAPITAL COSTS Of the numerous studies done to date on estimating the capital costs involved in domestic oil and gas exploration and development, Bankers Trust Company, in its review, has preferred the work done by the National Petroleum Council (NPC) in its 1972 report (6). That study covers six different cases involving various combinations of rates of drilling and discovery. Bankers Trust Company found information conforming to its assumption of a moderate drilling effort with a moderate rate of discovery for oil and a low rate for gas. The capital costs calculated for this situation were adjusted from the NPC 1970 dollar base to a 1974 base and were extrapolated from 1985 to 1990 based on the projected market for petroleum products. With this procedure, it was estimated that capital expenditures for exploration and development would increase from under $10 billion per year during the mid-1970s to over $16 billion per year by the late 1980s. Over the projected 16-year period, these capital costs in 1974 dollars are estimated to total $220 billion, which is 70% of all expenditures in the oil and gas category.

REFINERY CAPACITY The determination of refinery capital costs was based on the projected availability of all additional new crude supplies, whether domestic, foreign, or synthetic. From these crude supplies, required annual additions to refining capacity in barrels per day were calculated, after allowance for retirements and necessary reserve capacity. Total capital expenditures were then derived by multiplying the figures for barrel-per-day additions by $2500, which is the assumed capital cost per barrel per day. Implicit in the figures for capital costs are the assumptions that refiners will enjoy all the cost advantages of scale and that they will tend to emphasize, somewhat more than in the past, the simpler and less expensive processing of residual fuels rather than the more complex refining processes directed toward increasing recovery of the lighter petroleum fractions.

PIPELINES AND LIQUEFIED NATURAL GAS FACILITIES Construction of oil and gas pipelines during the projected period is related to the expected discovery rates for oil and gas and includes the necessary transmission and distribution investments for the gas industry. The total capital costs of all these facilities are estimated to be $22.3 billion. In addition, LNG terminal facilities, exclusive of the associated gas pipelines and LNG tankers, were projected to add $2.8 billion. The total, therefore, for the oil and gas transportation sector, not including tankers, is $25.1 billion. To keep the industry analysis within a domestic frame of reference, the projection assumes that prospective Alaskan gas is delivered through a trans-Alaska line with associated LNG terminal facilities.

TANKERS For a projection of tanker requirements for that portion of the US petroleum supply still to be imported during the period 1975–1990, it was assumed that financing of these vessels, when required, generally would be handled overseas. On the other hand, LNG ships for the African and Alaskan routes were expected to be financed and built domestically.

The current oversupply of world crude tankers precludes the necessity of much additional construction of these vessels for several years, while at the same time tankage for expected Alaskan production is already available or under construction. Overall requirements for financing tankers were projected to total only $6.4 billion during the study period. About one quarter of these funds represents petroleum-carrying capacity, while the remainder is forecast for LNG vessels in the late 1970s and early 1980s.

MARKETING   At $600 million per annum, marketing capital costs are projected to fall below expenditures of recent years, mainly because the growth in petroleum demand will be lower. Moreover, with domestic hydrocarbons in short supply, the petroleum companies will emphasize exploration and development rather than marketing.

DEEPWATER PORTS   On the basis of discussions within the industry it was estimated that three deepwater ports will be built between 1976 and 1980 at a total cost of $1.5 billion.

Synthetic Fuels

DEVELOPMENT   By 1990 the production of synthetic petroleum products is estimated to reach 2–2.5 Mb/d. This will be some 10% of all US petroleum requirements. The projection assumes implicitly that a favorable economic, environmental, and political climate will prevail during most of the forecast period. That is, the study assumes that the environmental problems of water scarcity, waste disposal, and strip mining will be solved within the next few years. It further assumes that the price of petroleum will remain high, or will even rise somewhat, and that the government will provide some kind of financial guarantee to spur development of synthetic fuels.

If some or all of these assumptions prove to be wrong, then synthetic production of crude in 1990 could conceivably approach zero. If so, imports would have to be at least 2 Mb/d greater than projected in the study. Dependence on foreign crude would be 45% of all requirements, not 30% as shown in the study. Since such a high level of dependence will probably be unacceptable to the US government for both military and economic reasons, it appears likely that the political and environmental problems will have to be resolved for policy reasons.

As with syncrude, the development of syngas depends heavily on a favorable environmental, political, and economic climate. If the favorable climate fails to materialize, the projection of 1.8 quadrillion Btu's in the study by 1990 will greatly overstate actual production. If underlying conditions are favorable, syngas production would provide the United States with 10% of its gas requirements.

CAPITAL COSTS   With the exception of syngas from petroleum, capital expenditures for synthetic fuel facilities were determined by calculating the total possible capacity for each fuel. Total capital requirements were calculated by multiplying annual additions by capital costs per unit. By this method, total capital expenditures were calculated at $35 billion through 1990. Two thirds of these expenditures are expected to occur during the last five-year period, during which average annual expenditures are expected to rise to $4.5 billion.

SYNGAS FROM PETROLEUM   Because of the Federal Energy Administration (FEA) order prohibiting petroleum feedstock allocations for facilities started after May 1, 1974, no new proposals for syngas from petroleum are likely during the forecast period. Thus, capital expenditures will be made only for those two plants currently under construction. These expenditures will amount to $85 million in 1975 and 1976 (7).

SYNGAS FROM COAL   Information supplied by competent sources indicates that capital costs of syngas plants are in the neighborhood of $2.5 million for each million cubic feet per day (Mcf/d) of capacity, excluding the capital costs associated with the mining of coal (8). This means that a typical 250 Mcf/d plant will cost over $600 million. Sixteen such plants are proposed by 1990. Thus, capital expenditures on syngas plants will approximate $10 billion during the forecast period.

SYNCRUDE FROM COAL   Although good estimates on capital costs for syncrude plants are unavailable, the study assumed that the per-barrel capital cost for these plants, which includes the mining of the coal, will be equal to that of shale plants, or $10,000/bbl, an assumption that seems reasonable. Since the coal mining capital cost of $2500/bbl per day is assigned to the coal sector in this study, it is subtracted from the $10,000 to obtain the coal syncrude plant cost of $7500/bbl per day. Multiplying this figure by the capital additions results in projected capital requirements of $8 billion, with most expenditures occurring after 1985.

SYNCRUDE FROM SHALE   Capital cost estimates for a 100,000 bbl/day plant vary widely, ranging from the FEA's low estimate of $520 million (9) to Union Oil Company's high estimate of $1.5 billion (10), or $5200–$15,000/bbl per day. Since most estimates seem to center around $10,000, Bankers Trust Company used this figure as a per-barrel capital cost, which indicates that over $15 billion will have to be invested through 1990 to achieve the projected output levels.

Electric Utilities

FOSSIL FUEL   Fossil fuel facilities generated 1455 billion kW-hr in 1974. This production is expected to increase to over 2000 billion kW-hr in 1990, an increase of over 40% in 16 years. However, fossil fuel facilities will drop from 79% of total deliveries in 1974 to 44% in 1990 as nuclear power becomes the dominant source of electricity.

The mix among the various fossil fuels used for electric generation in 1974 was coal 59%, oil 20%, and gas 21%. In 1990 these percentages are expected to change to 84%, 8%, and 8%, respectively. These changes have been referred to in part in the discussions of fuel supplies.

NUCLEAR   The supply of nuclear power is assumed to be constrained by regulatory and engineering factors, rather than by a shortage of basic nuclear fuel. For the next ten years the projection includes only 187 new facilities that are under construction or have been announced. In the late 1980s approximately 140 additional facilities of 1000 MW each are presumed to be constructed. A five- to nine-year lead time is expected in all new construction for conventional nuclear facilities, and no attempt has been made to include breeder plants.

HYDROPOWER   The generation of power by this low-cost fuel supply is limited to only a few significant locations. This study accepts the projection by Olmsted (11) of the future expansion of this source.

GENERATING CAPITAL COSTS   In projecting the number of kilowatt-hours generated by each energy source, the Bank relied heavily on the forecast made by Olmsted in *Electrical World* (11) for nuclear, hydro, and fossil fuel additions to generating capacity. Furthermore, the Bank's projection of kilowatt-hour demand through 1990 closely approximates Olmsted's. Consequently, it seems reasonable for the Bank to use his forecast of capital expenditures.

Total capital spending including the nuclear fuel cycle and stack gas desulfurization between 1975 and 1990 will exceed $417 billion. This accounts for more than half of all capital expenditures by the energy industry and exceeds expenditures of the oil and gas industry by over 40%.

NUCLEAR FUEL CYCLE   In addition to construction of nuclear generating plants and transmission and distribution facilities for the electric power produced, provision must also be made for the capital costs of the nuclear fuel cycle. This includes the total supply spectrum from uranium mining and milling, conversion, enrichment, fabrication, and transportation to reprocessing and waste storage. The capital requirements used in this study—$13.3 billion for the fuel cycle as a whole—were obtained by relating the projected new generating capacity required by this study to the total capital costs determined in a similar earlier study by the NPC (12).

FLUE GAS DESULFURIZATION   Present indications are that the technology for removal of sulfur oxides from emissions of coal-fired power plants is becoming more reliable and more widely accepted. One of the principal objections remaining is the considerable capital cost of installing desulfurization equipment in new or existing plants. For the purpose of the study it was assumed that 60,000 MW of present capacity, identified by the EPA (13), would be retrofitted. The capital cost of this undertaking, including provision for the handling of solid waste, was found to be $5.2 billion. In addition, one third of new coal-fired plants constructed during the 16-year projection period were estimated to require desulfurization equipment at a cost of $3.4 billion. Therefore, the total capital requirement for this purpose totals $8.6 billion.

BALANCING FUEL SUPPLY AND DEMAND

Table 1 gives a summary of the balance between the fuel demand sectors and their supplies. Although this table presents only figures for selected years through 1990, it is based on our projection by years for each factor.

It should be noted that any increase in the demand sectors or any decrease in the domestic fuel supply will require an increase in imports. To restrict imports would mean a significant change in our total energy consumption and therefore in our style of living. Such a change lies outside of the basic assumptions used in this study.

EXTERNAL FINANCING REQUIREMENTS

Although much of this study is devoted to determining the total capital expenditures of the energy industry, the main purpose is to analyze the ability of the capital market to meet the external portion of these expenditures. As a basis for projecting the ratio of internally generated funds to those generated externally, Bankers Trust Company reviewed the available historical data for the fossil fuel and electric utility industries and the factors that influenced them. From these, the Bank estimated the future percentage requirements of both industries for external financing.

The oil and gas component of the fossil fuel industry accounts for about 90% of that industry's capital expenditures. It is a high-risk, capital-intensive business. It is composed of some of the largest corporations in the United States along with thousands of smaller producers, refiners, and marketers. In the early 1960s oil and gas firms on the average raised about 25% of their capital requirements externally with the smaller firms financing a much larger share than the majors. By the late 1960s and early 1970s the average rate had climbed to approximately 40% (14). Since the financial requirements of the rest of the industry, principally the coal sector, are relatively insignificant, it can be assumed that the external share for the entire fossil fuel industry has approximated 40% in recent years.

In looking forward, we believe that the industry's external share will not differ significantly from recent years. Admittedly the penalizing effect of losing federal tax incentives will have a tendency to increase the industry's reliance on external markets. Moreover, massive new capital spending programs as projected in this study would similarly tend to force heavier dependence on the capital markets. However, these factors will be more or less offset by the increased cash flow resulting from the ultimate decontrol of all crude and new gas prices, or at least from the establishment of higher and more realistic price ceilings. Bearing in mind these offsetting trends, we feel that the industry will continue to call on the capital markets for about 40% of its annual requirements through 1990.

The electric utility industry holds a different position in the capital market. The publicly owned utilities that account for 20% of all capital expenditures in the utility industry rely almost entirely on government appropriations or credit for their external requirements. Consequently they do not call on the private business capital market as defined. The investor-owned companies make up most of the rest of the industry. Because they have been regarded as being in a low-risk business, at least until recently, the highly regulated investor-owned utilities have externally financed a much larger proportion of their capital requirements than the oil and gas firms. For example during the late 1960s the utility industry financed in this manner about 50% of its capital outlays, up from about 35% in the early 1960s. In the last two years this percentage has climbed to approximately 70%.

During 1973 and 1974 the industry was probably close to the limit of its capability to raise funds. Investors were dissatisfied with the safety of and return on their investments. Consumers were resisting higher prices, while environmental pressures were increasing costs, and the inflation-prone economy was forcing the payment of extremely high interest rates to the banks and in the money markets. Probably no industry has come closer to disaster before being saved by the concern of regulators, lower interest rates, and a declining rate of demand, which permitted a curtailment of the construction of new facilities.

Having found the limit of its market at 70% of requirements, the investor-owned industry is assumed in this study to finance externally at the 65% level. At such a high level, the industry will still be very sensitive to interest rates, regulatory lags and delays, and increasing costs of construction. However, it is unlikely that managements will be able to retreat to lower external financing as we move ahead into a more electrical age.

CAPITAL MARKET ANALYSIS

Method

The development of a projection of the capital market from 1975 through 1990 involves considerable subjective analysis. As a starting point, however, the relationship of flow of funds to the GNP was analyzed on an historic basis. This approach established ratios between the funds raised in various sectors of the economy and the GNP. These ratios were modified to adjust historic averages to the immediate business cycle and its development through the remainder of this decade. Finally, the GNP was forecast on an average annual growth rate of 3.6%. From this base the total capital market available to private industry was calculated as the total funds supplied from the corporate debt, equity, and bank loan sectors. These projections are presented in Table 2. All amounts are expressed as 1974 constant dollars.

The most significant assumption in the projection is probably the acceptance of a 3.6% overall rate of growth in the GNP. This is lower than the 4.5% experienced in recent years, and it reflects the opinion of Bankers Trust Company that several constraints have developed recently in our economic climate:

1. The cost of goods with relatively inelastic demand has risen sharply. This is particularly true of food and energy, and there is little prospect that either will experience significant future decline in price.
2. Higher service content of total US output means higher cost and therefore less disposable income for durable goods. This conforms to the assumption in the analysis of energy requirements that the rate of material output will slacken.
3. The rate of growth of the labor force will decline from 1.7% per year during the 1970s to 1% in the 1980s.

Offsetting these factors are possible increases in productivity and more efficient use of resources. These positive forces are unlikely to overcome the effects of the constraints so that the 4.5% growth rate of the past will give way to the 3.6% projected rate.

Table 2 Projection of gross national product and related capital markets (billions of 1974 dollars)

	Actual	Estimated		Projected				
	1974	1975	1976	1977	1978	1979	1980	1981	1982
Gross National Product	1397	1337	1396	1466	1532	1599	1671	1745	1820
Funds Raised For:									
US government	13.0	70.9	65.6	41.0	23.0	11.2	8.4	7.0	5.5
State and local governments	16.3	16.0	16.8	16.1	16.9	16.0	16.7	17.4	18.2
Consumer credit	9.6	4.0	14.0	16.1	18.4	20.8	21.7	22.7	23.7
Home mortgages	34.0	28.1	34.9	35.2	36.8	36.8	36.8	36.6	36.4
Foreign bonds	2.0	2.7	1.4	1.5	1.5	1.6	1.7	1.7	1.8
Business Sector									
Corporate equity	3.5	5.3	7.0	8.8	10.7	11.2	13.4	14.0	14.6
Corporate bonds	19.7	24.1	15.4	17.6	19.9	22.4	25.1	26.2	27.3
Mortgages	22.5	21.4	20.9	20.5	21.4	22.4	23.4	24.4	25.5
Business loans	56.9	6.7	20.9	29.3	32.2	33.6	35.1	36.6	38.2
Total Business	102.6	57.5	64.2	76.2	84.2	89.6	97.0	101.2	105.6
Total Funds Raised	177.5	179.2	196.9	186.1	180.8	176.0	182.3	186.6	191.2

Table 2 (*Continued*)

	1983	1984	1985	1986	1987	1988	1989	1990
				Projected				
Gross National Product	1896	1974	2053	2133	2214	2296	2379	2462
Funds Raised For:								
US government	3.8	2.0	20.5	21.3	22.1	23.0	23.8	24.6
State and local governments	19.0	19.7	26.7	27.7	28.8	29.8	30.9	32.0
Consumer credit	24.6	25.7	34.9	34.1	33.2	34.4	35.7	36.9
Home mortgages	36.0	35.5	2.1	2.1	2.2	2.3	2.4	2.5
Foreign bonds	1.9	2.0						
Business Sector								
Corporate equity	15.2	15.8	16.4	17.1	17.7	18.4	19.0	19.7
Corporate bonds	28.4	29.6	30.8	32.0	33.2	34.4	35.7	36.9
Mortgages	26.5	27.6	28.7	29.9	31.0	32.1	33.3	34.5
Business loans	39.8	41.5	43.1	44.8	46.5	48.2	50.0	51.7
Total Business	109.9	114.5	119.0	123.8	128.4	133.1	138.0	142.8
Total Funds Raised	195.2	199.4	203.2	209.0	214.7	222.6	230.8	238.8

Economic Sector Analysis

To assist in understanding the modifications of the historic capital flows, a few comments are helpful on each economic sector involved.

FEDERAL GOVERNMENT   Government policy must cope with an unemployment rate that will not decline to 5% until the mid-1980s, although a steady reduction in unemployment is expected up to that time. As a consequence the government will remain a net borrower of funds through 1984. Since it is impossible to predict the time of another business cycle after that, the future is averaged out to a noncyclical projection.

STATE AND LOCAL GOVERNMENTS   There will probably be a shift in capital spending away from highways and schools and toward projects for improving the environment such as sewage treatment and the rebuilding of central cities. The expected increase in services such as welfare and health should be financed from tax receipts rather than long-term borrowing. The present fiscal problems facing many communities will abate as local governments adjust their budgets to decreased revenues or market resistance to their debt.

CONSUMERS   The credit requirements of this sector are closely related to spending on large-item durables such as automobiles and housing. Such demands will be smaller in the future because less personal income can be devoted to them. Consequently, there will be a trend to smaller, less costly homes and automobiles, and the expansion rate of consumer credit will be lower than that in the early 1970s.

BUSINESS   The breakdown of this sector in Table 2 is largely self-explanatory. The immediate call on the equity, bond, and bank markets is low as business works its way out of the present recession. By the early 1980s this sector is expected to be back to historic levels.

Observations on the Capital Market Projection

Table 2 shows that the capital market is expected to decline from $102.6 billion in 1974 to $57.5 billion in 1975, a 44% slump. Thereafter through 1980, during the post-recession surge in the economy, the market is expected to grow at an average annual rate of over 10%. Despite the high growth rate, the capital market projected for 1980 is still $5.6 billion smaller than the inflation-expanded market of 1974.

In Table 3 the capital requirements of the energy industry are related to the total capital market. As can be seen, the industry's share remains rather stable throughout the period with the exception of 1975, when it is much larger. This phenomenon occurs, however, not because energy requirements are so much greater in 1975 than in other years, but because the demand for external funds by other economically depressed industries is estimated to be much smaller in 1975 than in other years. In other words, the high share of energy in 1975 primarily reflects the shrinking of the capital market and not the rising requirements of the energy industry. With the normalization of the capital market after 1975, the energy share will decline by

Table 3 Energy industry share of capital market (billions of dollars)

	Actual Dollars			Projected 1974 Dollars							
	1961–1965	1966–1970	1971	1975	1976	1977	1978	1979	1980	1981	
Fossil Fuel Industry											
Capital outlays	37.9[a]	53.0[a]	9.9[a]	16.5	15.8	17.1	19.2	20.6	21.4	22.0	
Financed externally (40%)	9.2[a]	20.9[a]	5.2[a]	6.6	6.3	6.8	7.7	8.2	8.6	8.8	
Electric Utilities											
Capital outlays (100%)	23.4[a]	46.2[a]	14.6[a]	18.7	17.3	18.1	18.7	19.4	21.0	23.3	
Investor-owned share (80%)	17.3[b]	36.6[b]	11.9[b]	14.9	13.9	14.5	15.0	15.5	16.8	18.6	
Financed externally (65%)	8.5[a]	23.7[a]	9.8[a]	9.7	9.0	9.4	9.7	10.1	10.9	12.1	
Total Financed Externally	17.7[a]	44.6[a]	15.0[a]	16.3	15.3	16.2	17.4	18.3	19.5	20.9	
Total Capital Market	102.4[c]	211.8[c]	63.0[c]	57.5	64.2	76.2	84.2	89.6	97.0	101.2	
Energy Share of Market (%)	17.3	21.1	23.8	28.3	23.8	21.2	20.7	20.4	20.1	20.6	

	1982	1983	1984	Projected 1974 Dollars							
				1985	1986	1987	1988	1989	1990	1975–1990[d]	
Fossil Fuel Industry											
Capital outlays	23.2	23.5	24.7	26.1	26.7	27.7	28.6	29.4	30.0	372.6	
Financed externally (40%)	9.3	9.4	9.9	10.4	10.7	11.1	11.4	11.8	12.0	149.0	
Electric Utilities											
Capital outlays (100%)	25.5	26.9	27.9	28.0	29.6	31.9	34.3	37.0	39.9	417.4	
Investor-owned share (80%)	20.4	21.5	22.3	22.4	23.7	25.5	27.4	29.6	31.9	333.9	
Financed externally (65%)	13.2	14.0	14.5	14.6	15.4	16.6	17.8	19.2	20.7	217.0	
Total Financed Externally	22.5	23.4	24.4	25.0	26.1	27.7	29.2	31.0	32.7	366.0	
Total Capital Market	105.6	109.9	114.5	119.0	123.8	128.4	133.1	138.0	142.8	1685.0	
Energy Share of Market (%)	21.3	21.3	21.3	21.0	21.1	21.6	21.9	22.5	22.9	21.7	

[a] See Ref. 14, pp. 104, 107.
[b] Edison Electric Institute. *Statistical Year Book.* Various issues.
[c] *Federal Reserve Bulletin.* Various issues.
[d] Totals may not add because of rounding off of numbers.

1980 to an historic 20%. Thereafter it gradually rises to almost 23% in 1990. Clearly no capital crunch develops based on the assumptions used in this study.

Naturally, in a long-range projection of this nature some of the estimates used will have a wide range of possible values. Yet even if some of the principal underlying variables are modified significantly, the overall conclusion still holds true that no radical rise in the energy industry's share of the capital market is expected.

To test this premise, the 6.25% annual growth rate of the capital market could be reduced 25% to 4.6%. This leads to a capital market projection of only $114 billion by 1990. In this smaller capital market the energy industry would account for a 28% or 29% share. Although it is presumptuous to predetermine the level of dedication at which financing becomes impossible, it does seem reasonable to conclude that a percentage approaching 30% would not be unthinkable in a free market provided it offered enough return to the investor.

To test a different variable, assume the energy growth rate exceeds 3% a year. The increased demand on the limited supply will bid up the price of energy. The higher prices would discourage energy consumption and thereby reduce capital requirements for expansion. In addition, profits should rise. This would provide the industry with more internally generated capital and would limit its demands on the external capital markets. Finally, the higher profits would provide the industry with the necessary base to attract both equity and debt capital in the amounts necessary to meet the nation's energy requirements as set by the free interaction of supply and demand.

CONCLUSION

The total capital needs of the energy industry through 1990 are estimated to be about $790 billion. Of this amount some $423 billion are likely to be generated internally, while $367 billion will be raised externally. Impressive as these numbers are, the total availability in the capital markets from 1975 through 1990 will probably reach $1685 billion. On this basis the energy industry will require only 21.7% of that supply. This dedication falls within its historic share.

It should be noted that from the early 1960s on there is a gradual increase in yearly dedications of capital to energy. Although it is small and is distorted by the current recession, it reflects the underlying movement toward a more electric-oriented economy that is fundamentally less efficient in its use of primary fuels. It is also indicative of our moving away from cheaper sources of energy as they are depleted and into other long-range supplies that are even more capital intensive but that will carry us into the next century.

Of major importance to these conclusions, however, is the problem of whether the energy industry can command its historic share of capital. This study only indicates that the market will be capable of supplying the required funding. The ability of industry to obtain its share of the market is dependent on industry's attractiveness to the investing groups that compose that market. Capital will only be available to the extent that industry can offer it a satisfactory rate of return in the competitive marketplace. At the present time the federal government, and many state and local governments, are promoting policies, laws, and regulations that impede the ability of the energy industry to generate enough profits to be attractive in the capital markets. If this punitive attitude is maintained, the present stress in our energy structure will turn into an overwhelming crisis as industry strangles under the resulting curtailment of its supply of capital.

Literature Cited

1. National Academy of Engineering, Task Force on Energy. 1974. *U.S. Energy Prospects: An Engineering Viewpoint*, Chap. 4. Washington DC: Nat. Acad. Eng.
2. Norris, T. G. May 1975. *Coal Outlook for Conoco—Industry Perspective*. Presented to New York Society of Security Analysts
3. Katell, S., Hemingway, E. L. 1974. *Basic Estimated Capital Investment and Operating Costs for Coal Strip Mines*. US Bur. Mines Inf. Circ. 8661
4. National Petroleum Council. Sept. 1974. *Emergency Preparedness for Interruption of Petroleum Imports*, p. 59. Washington DC: NPC
5. Bureau of Natural Gas, Federal Power Commission. Dec. 1974. *A Realistic View of U.S. Natural Gas Supply*
6. National Petroleum Council. 1972. *U.S. Energy Outlook*, Chap. 4. Washington DC: NPC
7. Brumm, J. Apr. 23, 1975. Transco feels SNG from naphtha is answer. *Oil Daily*, pp. 1, 3
8. Summary of coal conversion plants. March 1975. *Coal Age*, pp. 94–96
9. Federal Energy Administration. Nov. 1974. *Project Independence Report*, p. 133. Washington DC: GPO
10. Shale roadblocks draw fire from Union. Nov. 1974. *Oil & Gas*, pp. 42–43
11. Olmsted, L. M. Sept. 1974. 25th annual electrical industry forecast. *Electr. World*, pp. 43–58
12. See Ref. 4, Chap. 6
13. US Environmental Protection Agency. 1974. *National Strategy for Control of Sulfur Oxides from Electric Power Plants*
14. Hass, J. E., Mitchell, E. J., Stone, B. K. 1974. *Financing the Energy Industry*. Cambridge, Mass: Ballinger. 138 pp.

Copyright 1976. All rights reserved

THE PRICE OF ENERGY ✗11015

Diana E. Sander[1]
Public Service Commission, State of New York, Albany, NY 12208

INTRODUCTION

Price is essentially a device to allocate supply. Its rationality in eliciting and distributing efficiently the supply of energy or any other group of commodities, services, or resources hinges directly on the extent to which price reflects marginal or incremental costs—the cost of producing another increment of supply or the savings permitted by producing one less unit of output—in a market where production is free to respond to consumer demand without artificial restraints or subsidies. It is worthwhile to restate this concept of efficient economic price before proceeding to examine summarily the nature of current prices in key sectors of the energy market as these prices reflect historic government policy adopted for a variety of goals, as well as the influence of industry structure and regulatory practice on price determination in different parts of the market. Such a review may help to flush out altered price relationships and desirable policy redirections. Proposals to rely more heavily on market forces in confronting current energy problems have received enormous focus in recent months vis-à-vis administrative actions such as rationing and price controls or other regulatory actions that need substantial surveillance. This renewed interest in classical price theory is becoming apparent not only for petroleum and gas production and transmission, for example, but also in the distributive industries traditionally regulated as natural monopolies, i.e. the electric power and gas utility industries.

This article does not have as its chief objective a detailed statistical analysis of trends in production, price, reserves, or future levels of demands among the components of the energy market. Others have covered these subjects in great detail in other chapters of this book and in a great mass of literature (1–5). Rather, the intent is to present an overview of how price has evolved in these past years as an effective or ineffective tool to signal the consumer and help him choose those volumes and energy sources that will yield the greatest possible satisfaction from

---

[1] The author is a principal economist with the New York State Public Service Commission. (Publication does not necessarily imply acceptance by the Commission of the views expressed.) She wishes to acknowledge with appreciation the encouragement and constructive comments received from Chairman Alfred E. Kahn, whose permission to draw freely from his writings has expedited the completion of this chapter.

society's limited productive resources as well as to permit society's environmental and other goals to be met.

In approaching this discussion of energy price there is a need to disentangle a number of elements that impinge on what the price of each of the major fuels is and should be—for each fuel and in relation to each other. Among these elements are the physical scarcity of energy resources nationally and/or internationally; the prognosis of the duration of OPEC supply restrictions; the extent of distortions in the level of demand for combined energy sources because of historically low price accomplished via subsidies to the industry on one hand and/or to the consumer as opposed to the industry on the other; the reversal of historic attitude toward external costs as these relate to a clean environment and the attendant dislocations; and last, but no less urgent, the consistent thread of governmental policy, which has adopted domestic energy price as a means of financing a measure of national security in energy and stimulating a dependable independent supply. None of these is discussed in depth, but all are referred to, since their influence is greater or less in the price of specific fuels or in the objectives of government in influencing those prices over time.

While deviations from optimal pricing and resource allocation make it impossible to conclude as a general proposition that application in any single sector of the normative rules is desirable, we assume that an attempt at a conscious policy is nevertheless better than none or one that can result in inconsistent, contradictory action. The theory of second-best emphasizes the need for considering the implications of any particular policy in the presence of suboptimal conditions elsewhere in the economy (and more particularly within the sector in which we are operating). As a practical matter, we can maintain that it is possible to make informed judgments about the ways in which directly relevant imperfections might suggest modifications of the rules (6-8). How to modify these rules to take account of deviations from first-best conditions within and outside the energy sector will depend on whether these must be taken as given or whether these are in fact subject to change. This latter opportunity to change many of the distortions that have marked segments of the energy market has been accompanied by a rising mood to do just that in the period since the dramatic OPEC initiatives in 1973.

In the energy sector, we have some prices that are markedly above and others that are below their respective marginal costs. The burden of taxes and the effects of regulation differ sharply from one energy form to another, and major externalities associated with competing energy forms are dissimilar. The significance of these differences has escalated rapidly. What prior to 1973 appeared to be long-run problems of energy supply and demand were made short-run and critical by the end of that year. These problems revolved around the prospects that the United States could be cut off from Arab oil sources and made susceptible to higher prices and political coercion through our growing energy demand and lagging domestic supply.

Energy problems have been especially difficult to adjust to because of their rapidly changing character. The growth in nuclear generation of electricity has been slower than expected (see Table 1). The effectiveness of environmental pressures has, on the other hand, sharply increased. This has had many consequences—among them a sudden, enforced shift of large numbers of electric generating stations from

Table 1 Gross consumption of energy by major source, 1965–1974[a]

Year	Total Energy Consumption		Percentage of Total Energy Consumption[b]					
	Amount (quadrillion Btu's)	Change (%)[c]	Total	Petroleum	Natural Gas	Coal	Hydro-power	Nuclear
1965	53.3	4.1	100.0	43.6	30.2	22.3	3.9	0.1
1966	56.4	5.8	100.0	43.2	30.9	22.1	3.7	0.1
1967	58.3	3.3	100.0	43.5	31.3	21.0	4.0	0.1
1968	61.8	6.0	100.0	43.8	31.7	20.5	3.8	0.2
1969	65.0	5.2	100.0	43.8	32.3	19.6	4.1	0.2
1970	67.1	3.3	100.0	44.0	32.8	18.9	4.0	0.3
1971	68.7	2.3	100.0	44.5	33.2	17.6	4.1	0.6
1972	71.9	4.7	100.0	45.8	32.0	17.3	4.1	0.8
1973	74.7	3.9	100.0	46.4	30.6	17.9	3.9	1.2
1974[d]	72.7	−2.7	100.0	46.0	30.1	18.3	4.2	1.4

[a] Source: (9).
[b] Detail may not add to totals because of rounding.
[c] Based on unrounded data.
[d] Preliminary estimate by Council of Economic Advisers.

coal to oil, a sharp increase in demand for low-sulfur fuels, a loss of energy efficiency in internal combustion engines, and a slowing of electric utility expansions and oil refinery construction, through delays in site approvals and postponement of the delivery of Alaskan petroleum resources (10, p. 6). Total energy demand was meanwhile rising sharply[2] at least up to 1973, reflecting among other things preferences for big cars, increased market saturation of air conditioning, and the enforcement of environmental controls on motors.

HISTORIC DEPARTURES FROM EFFICIENT PRICES— A TRADITION

Some degree of general price control for the energy industries had been in effect from 1971 to 1973 as part of the price control program for the general economy.[3] Aside from this, however, regulation of the price of natural gas in interstate markets went

[2] While total energy use per capita only doubled over the past century, electric energy use and the use of aviation fuels rose far more rapidly. Total (rather than per capita) electric power usage doubled every 10 years; natural gas consumption doubled in 14 years (1958–1972), while petroleum consumption doubled in 18 years (1954–1972). Meantime, coal consumption dropped sharply over the post–World War II period, particularly in transportation and as a fuel for the generation of electricity. In contrast, petroleum use in electricity generation grew very fast. [See testimony of Dr. James R. Nelson in (11).]

[3] For a chronology of price controls affecting the energy sector see (12).

back two decades. In the states, regulation of electricity and gas distribution had a much longer history.

Governmental policy has been contradictory with regard to the various fuels, and the nature and extent of governmental interjection has ranged over the decades from promotion and protection of domestic energy production on the one hand to greater emphasis on conservation on the other. The most prominent purpose of our various public policies as they bear upon energy has been to promote its use, via tax subsidy and other devices, in order to make energy cheaper and more abundant than it would be in a laissez-faire market (10, p. 11). The assumption has been that cheap and abundant energy is necessary for economic growth and that some measure of economic growth is required to attain social and economic goals without sharp dislocations.

Promotion

It was the fear of shortage in the early 1920s that spurred the development of oil tax policy in subsequent years. The two major preferences that the oil industry has enjoyed under the federal income tax laws—"percentage depletion" and the privilege of expensing intangibles—have had the purpose and effect of encouraging an enormous flow of capital into the oil industry, thereby producing a lower price and a greater abundance of domestic oil and gas than would otherwise have been available, at least for several decades.[4] Percentage depletion, which was essentially revoked by 1975 tax revisions,[5] allowed natural resource producers the privilege of treating a part of their income as tax-exempt even if it far exceeded the costs of the assets being depleted. The government has long felt that tax incentives for the production of minerals were necessary to ensure adequate exploration and production of exhaustible resources. The concept of percentage depletion was the reward to the producer as he extracted the resources from the ground.

The statutory percentage depletion that was allowed varied from 22% of gross income for oil and gas (reduced from $27\frac{1}{2}$% in 1969[6]), to 15% for oil shale, and to only 10% for coal. The structural differences among the fuel industries that permitted them to take advantage of this provision translated to net benefits of

---

[4] These tax arrangements, along with other government policies, encouraged a more rapid consumption of domestic oil and gas resources than would otherwise have taken place. At some point, which we may or may not have reached, this tendency would be expected to make oil and gas less abundant and, as we move along the diminishing returns function, higher in cost than they would have been but for those policies of the preceding decades.

[5] Percentage depletion for domestic and foreign production of major companies was repealed on March 26, 1975, effective as of January 1, 1975. Independents, however, may still take a revised depletion allowance, which will be at the rate of 22% in 1975–1980 but will decrease (except on secondary and tertiary recovery) to 15% by 1984. Gas under fixed-price contracts made prior to February 1, 1975, and regulated gas under FPC control produced and sold before July 1, 1976, will also be eligible for the allowance (13).

[6] Public Law 91-172 amending Section 613 of the Internal Revenues Code of 1954 (14).

about 15% of gross income for oil and gas, 10% for uranium, and about 5% for coal (15, p. 26).

The tax provision that allows current deductions for intangible drilling and development costs for oil and gas, and certain exploration and development costs for other minerals, benefits the entrepreneur by permitting him to take his deductions all at once instead of depreciating the capitalized expense over time, therefore letting him derive the time value of money. (A rule of thumb equates the current deduction as twice as valuable as the usual depreciation procedure.) The special tax benefits of percentage depletion combined with the deduction of intangibles and the normal investment credit for tangible drilling costs are estimated to have been equivalent to a relatively uniform investment credit of about 50% of producer's net income in the oil, gas, and coal industries. It was the combination of the two benefits—depletion allowance and deduction for intangibles—that made them so valuable. The privilege of expensing the major portion of an oil producer's investment costs would have been (partially) offset by the correspondingly smaller depreciation charges against taxable income in future years, since the expenses that he would have capitalized and been permitted to depreciate for tax purposes would have been correspondingly smaller. But this offset was destroyed by percentage depletion since it could be taken as a percentage of sales, no matter how small the remaining capitalized costs subject to depreciation or depletion.

It has been calculated that the differential effect of these provisions on delivered price—if these were in fact all passed on to a utility generating electricity—would vary substantially, amounting to 12–13% for use of oil and gas, but only about 3% for coal and uranium.[7] The differential spread in the relative final impact of these tax provisions on the prices of different fuels is tied to the capital-labor rates in these industries. As energy incentives, these tax provisions have therefore worked unevenly among the various energy resources.[8]

Cartelization

The successful, if uneconomic, flow of excess capital funds into oil exploration and production as a result of tax incentives caused overcapacity in the domestic oil industry for years. Domestic excess capacity was paralleled by excess capacity abroad over much of the post–World War II period (16–18). In the name of conservation and national defense, the excess oil capacity generated by special tax measures was made tolerable to the industry within the United States by a policy

---

[7] If the petroleum industry does pass on to the electric utility the full amount of its increase in tax caused by the loss of percentage depletion, many states, particularly along the Atlantic seaboard, whose utilities are more heavily dependent on oil than other fuels, will bear a much larger than average increase in fuel expense.

[8] Estimates of the tax revenue loss to the US Treasury from oil depletion in 1974 were in the vicinity of $2.4 billion for domestic oil (priced at an average of $7.00/bbl), $350 million for gas, $130 million for coal, and $10 million for uranium. [See (15, pp. 32–33).] The forecasted revenue loss to the industry in 1975 from revision of domestic and foreign percentage depletion is estimated at $1.7 billion in 1975, and it is expected to rise to $1.9 billion by 1984 (13).

of state production controls—prorationing—and eventually by instituting import controls in 1959, which in effect organized the industry into a cartel.[9] (Not all states adopted prorationing, although the major producing states accounting for most oil production followed such programs.)

Prorationing to market demand was instituted in the 1930s in an attempt to put some limitations on the "rule of capture," a system of property rights over the ownership and exploitation of subsurface minerals that allowed anybody who owned surface land to have title to whatever he could capture from below it. This led to unrestricted drilling of wells by competitors of a single oil reservoir, instead of a unified development approach that would preserve the high underground pressure and hence permit more efficient development of oil and gas flows. Prorationing policy, which started out with conservation goals, also served, however, to permit sharp production cutbacks that insulated the market price from the increasing surplus capacity—almost completely until 1958, and with a high degree of success in the 1960s.

In a free market the price would have begun to fall to stop excessive inflow of capital. Instead, price was actually increased several times in the 1950s, despite growing excess capacity. The industry justified these price actions with evidence of rising costs of finding, developing, and producing oil. These were, however, the inevitable consequence of the tax incentives and production control systems under which the industry operated. The greater the tax preference and the higher the price of oil, the further the search could be carried to marginal, high-cost reservoirs and the larger the bonuses and royalties that explorers could offer leaseholders for the privilege of looking for oil on their land. Holding the price far above the cost of efficient producers and raising after-tax returns on investment above those in other industries will attract a rapid inflow of capital if entry is free. Production cutbacks that maintain price and protect profits also keep capital coming in, until the cost burden of excess capacity is just sufficient to eliminate the artificial stimulus to investment that was created by supernormal profits in the first place. The higher the price, the more excess capacity the producers will tolerate before a new equilibrium is produced (22).

The burden of excess capacity fell most heavily on the efficient producers in the domestic industry as the domestic cartel tried to keep everyone in business. The numerous small, comparatively inefficient firms were given quotas larger than could be justified if the intent was to produce the maximum output allowed by each state at minimum cost. (Hundreds of thousands of low-output oil wells were completely free of control so long as their output remained below the state's maximum production cutoff point.) The cartel also protected the volume of profits to the higher-cost firms by varying production allowables with the depth of the producing reservoir

---

[9] While classic private monopoly was infeasible in domestic oil, given the relatively low concentration of production and the rule of capture, government cartelization lent force to the possibility of private price fixing, given the important role of large vertically integrated oil companies, which controlled pipeline access and the posting of prices in the field and had a strong incentive to take profits at the crude oil production level. [See (19–21).]

because the costs of production are much greater at greater depths. "It was thus reasonable to people brought up in the medieval tradition of just price to give the high-cost producer a larger share of the production quota. It was another way in which costs were raised to predetermined price, rather than price reduced to marginal cost of the most efficient firms able to supply the quantities demanded" (22). The soaring figures for well completions in the first half of the post–World War II period were a sign not of economic health but of inefficiency of the industry's governing institutions, which encouraged overinvestment and hence higher costs of production.

To the extent, then, that tax subsidies were reflected at all in retail prices, the high price of domestic oil was subsidized, but the American consumers were paying on net prices for domestic oil that were much higher than the price at which international oil was available. Furthermore, high costs provided a direct incentive to US companies themselves to look abroad both for supplies for their refineries and for production opportunities.

Restriction of Foreign Oil Imports

It was inevitable that imports would have to be brought under control too, first by voluntary controls (1954) and then by mandated quotas (1959), if oil prices were to be maintained and the domestic oil industry fully protected. These quotas brought foreign production into the "domestic" cartel as far as the American market was concerned. The protection of the domestic industry from foreign imports denied to the American consumer the fruits of the subsidy that our tax law extended also to foreign exploration and production. The cost/benefit of tolerating higher domestic oil prices in relation to long-run national defense requirements has not been fully explored in relation to alternative, and perhaps less costly, defense programs that could have been devised in the energy area, but the OPEC embargo in 1973 gave a remarkable push to the value of the programs in effect. There is little doubt that the tax program initially spurred drilling of wells.[10]

Natural Gas

Gas production benefitted from the same promotional federal tax subsidies as oil. It was, in fact, difficult to sort out the impact of these provisions on gas versus oil since gas was often found along with oil or instead of the oil sought, probably as often as the result of a direct search for dry gas. State prorationing policies for gas, in contrast to those for oil, had little effect on market supplies and price. Emphasis was on the aspect of prorationing concerned with production to ensure maximum efficient rate of recovery from a reservoir and on equities among individual owners in the management of a reservoir. [Oil prorationing restrictions necessarily limited

---

[10] The depletion allowance stimulated the drilling of production wells, however, rather than exploratory wells for establishing reserves, thus using up (except for the intervention of prorationing) natural resources. Such alternative policies as developing new energy technologies or building stockpiles may, in the long run, have been more efficient in terms of societal cost.

the supply of casinghead gas found in association with oil, versus gas found by itself (nonassociated gas).][11]

In addition to promoting supply via tax preferences, the governmental regulatory policy of setting maximum prices for interstate wellhead gas served to stimulate demand. Promotion of demand outweighed the promotional subsidies for production, and the gap between the supply and demand at the regulated price increased steadily.

The spread of interstate pipelines after World War II was the most significant spur to natural gas production and nationwide markets. The purchase from the War Assets Administration of the Big Inch and Little Inch pipelines (originally built during World War II for the transmission of oil and products) by Texas Eastern Transmission Corporation and the conversion of these lines to gas transmission was a big stimulus. (The long-distance movement of natural gas began with the completion of the Tennessee Gas Transmission Line in 1944, which transported natural gas 1265 miles from Texas to West Virginia.) Shortly thereafter, Transcontinental Gas Pipeline Corporation obtained authority to construct a natural gas pipeline from Texas to the New York City area; meanwhile the East Coast utilities began to convert the installations of their customers from manufactured gas to natural gas.

The economic justification for the adoption of a gas regulatory policy was based on the assumption that gas supply from existing wells would not be responsive to high prices and that rising gas prices would merely provide economic rents or windfalls to producers (24). This assumption was also premised on the belief that capital irretrievably sunk in an enterprise has a lower opportunity cost than incremental capital costs and should be rewarded accordingly. The latter was an underlying consideration in the 1965 decision by the Federal Power Commission (FPC) to introduce a two-price system for natural gas, with a lower price for old gas supplies already committed for sale under long-run contracts and a higher price for new additional supplies in new contracts. Maximum gas prices were viewed in some sense as a competitive dampening influence on the price of oil, already well beyond its efficient price.

Presumably, the price of new gas should have been adjusted upward fast enough to track costs of additional supplies, or inevitably the exploration and production of new gas supplies was likely to be dampened. Regulation failed to exercise adequate surveillance procedures for assuring an adequate long-term supply of natural gas while permitting the extension of pipelines to new markets. Thus demand burgeoned, unconstrained by price or findings requirements. Instead, reserve and deliverability requirements in relation to sales were repeatedly reduced. Unanticipated were the surges in demand for natural gas over the post–World War II period and perhaps a greater responsiveness of supply to regulatory price policy than anticipated. How much higher gas prices would have needed to rise before investment in gas production in the United States competed effectively with high-

---

[11] At the beginning of the 1970s, four fifths of total gas production came from nonassociated gas wells compared with 72% in 1960 and 66% in 1950. The decline in oil well drilling necessarily impacted these proportions (23).

return foreign investment opportunities in oil and gas has not been explored in detail. Also uncertain is the net benefit or detriment that might have flowed from a policy of permitting gas prices to rise inefficiently in a cartel setting for domestic oil over the last ten years compared with having opened the US market to cheap petroleum products from abroad. This is a policy dilemma that the country faces even more intensely in 1975. Since the price of oil was artificially held above competitive levels in the 1950s and 1960s, the advocates of gas price regulation could argue that it was undesirable in effect to let the price of gas rise to that cartel level. On the other hand, regulation undoubtedly encouraged the growth of demand, while limiting the incentive of producers to find and supply gas (25). A similar situation exists today, with the international oil price being the one sustained by the cartel, and continued regulation of the domestic oil price on the one hand advocated to deny domestic producers the benefit of that monopolistic price and on the other hand criticized as encouraging demand and discouraging the replacement of imports with domestic production.

Be that as it may, the promotion of gas demand at prices below incremental cost led to a deepening shortage of natural gas during the 1970s. Historically below prices of competing fuels, low gas prices caused widespread conversions from coal and oil. It also encouraged wasteful consumption of gas. Meantime, according to experts (26), a gradual exhaustion of lower-cost oil and gas resources occurred in the lower 48 states. Prices, which were held well below a level that would clear the market in the 1970s, generated a huge excess demand, all channeled overseas.

It may be worth digressing to review briefly the strong governmental influences in the separate segments of the natural gas industry. At the production stage, the FPC regulates the field portion of natural gas production consigned to interstate pipelines. In contrast, sales to users within the producing states are unregulated by the FPC (although subject to regulation by state or municipal commissions). Interstate sales by pipelines to local gas distributors ("city gate sales") are also regulated by the FPC. With the growing gas shortage, prices unregulated by the FPC have soared to three and four times the interstate level and the flow of new onshore gas to the interstate pipelines has slowed to a trickle.

Although major petroleum firms are also large producers of natural gas, most of them do not own large pipeline transmission companies, and most are not directly engaged in significant retail sales of natural gas.[12] (Because FPC regulates integrated companies more stringently than independents, most large petroleum companies have preferred to remain independent producers of natural gas and avoid cost-of-service regulation, preferring instead maximum price regulation.)

Promotion of natural gas demand was made more effective by the adoption by the FPC of the Atlantic Seaboard formula which, by allotting 50% of capacity costs[13]

[12] The degree of vertical integration has fallen sharply over the last 30 years—from about 40% in 1947 to under 10% in more recent years (27).

[13] In order to make gas more competitive with coal in selected markets, some modification in the application of the Atlantic Seaboard formula was made so that 60% of costs were to be recovered through the demand charges and 40% through the commodity component of gas rates.

to the commodity (or energy) charge, was in effect subsidizing the seasonal peak users of the interstate pipeline system—primarily residential customers of the northern regions of the country—by charging them less of the capital costs of the pipeline, which peak responsibility pricing would require. Thus the market for residential heating held by oil or coal was penetrated more competitively. Gas heating was responsible for using about 25% of natural gas production in the early 1970s. The largest share of natural gas production (over two fifths) was going to the industrial sector.

At least part of the successful expansion of the industrial market for gas resulted from the interruptible concept of pricing. The unused delivery capability in transmission and distribution systems during summer months provided incentive to find off-peak markets. The relatively low-cost gas became especially attractive to industrial customers. In addition to price, demand was spurred by new industrial applications of natural gas, by improvements in gas-activated equipment, and by increased use of gas as a petrochemical feedstock (e.g. fertilizers, ammonia) and in prime motors for compressors and pumps, particularly in the gas transmission and oil pipeline industries.

The off-peak sale of gas to industrial customers was economically efficient in terms of capital investment of pipelines and wells. Indeed, to the extent that the pipelines, under the Atlantic Seaboard formula, failed to follow peak responsibility principles, these off-peak sales were overpriced and thus inefficiently discouraged. Nevertheless, these sales of gas, particularly as a boiler fuel, became more and more questionable as the price of the gas itself came to be held farther and farther below the market clearing level. In addition to the off-peak gas used in manufacture, electricity generation also provided a large market, consuming about one fifth of natural gas production in the early 1970s. The adequacy of the overall level of gas prices became more serious as the expansion of natural gas production began to slow and turn down.[14] Because natural gas is traditionally sold under long-term contracts, its price responded slowly to increased demand.

Coal

In contrast to petroleum products, the pricing of coal in the United States was much less subject to governmental regulation since World War II. By the same token, the industry benefitted much more modestly from the tax provisions discussed earlier. Also, the domestic market for coal was not concerned with imports and did not need protection from foreign supplies. The United States is a net exporter of coal.[15] Of all the fossil fuels, coal was, however, impacted earlier by social concerns and external costs—first via mine-safety regulations, which tended to increase coal costs and price, and also by the emphasis on coal early in the environmental awakening to air pollution. These legislative restrictions caused a sharp turnaround in the demand

---

[14] Total natural gas production peaked out in 1972–1973 and then began declining (28).

[15] The United States exports about one tenth of its total production, three quarters of which is coal of coking quality (metallurgical coal) used in steel production. There are a significant number of captive coal mines owned by US steel producers.

for coal on the North Atlantic seaboard, especially from the utility sector of the coal market.[16]

While only coal among the fossil fuels showed a substantial price rise in the 1968–1972 period, the increased revenues per ton apparently did not serve as an effective incentive to expand production. In part, some of these additional revenues were absorbed by increased labor costs related to changes in health and safety standards under the 1969 Coal Mine Health and Safety Act. While some mines undoubtedly expanded production, others that could not profitably comply with new standards were abandoned. Although the stagnant market for coal was over by the mid-1960s, as both domestic and foreign demand rose, predictions that nuclear reactors would displace coal in electric power generation and uncertainty about the stringency of rules on sulfur emissions in metropolitan areas, which might make coal noncompetitive in those utility markets, were other factors that held back investment in new mines. More recently, concerns over legislation related to strip-mining reclamation standards as well as to modification of industrial pollution standards that may prove temporary have been additional negative factors in soliciting an increase in coal supply.

Considerable interest in Congress has also been directed to the question of whether horizontal acquisitions of coal reserve and production facilities by major oil companies were a contributory factor in coal price rises before 1972 and more recently (29, p. 7). (Similar fears are held about the market control of uranium and the eventual formation of a cartel because of the high concentration of control over production and reserves in the United States, again involving oil companies.) Economic factors other than conspiracy to reduce production and raise prices are said by some experts to explain the price rises at the turn of the decade.

Over the last 20 years a number of large petroleum companies have begun production of coal and uranium and many of the 20 largest petroleum companies appear to have acquired substantial reserves (27, 30). In addition to growing horizontal concentration, a narrowing of control within the coal industry per se has resulted from the advantages of using unit trains introduced in the early 1960s. Use of unit trains in transporting about half of coal shipped by rail, terms of contracts specifying rapid loading and unloading, and minimum weight restrictions designed to minimize operating costs all gave the large producer a significant competitive advantage over producers who could not meet the volume levels required by unit trains. The trend toward mine-mouth electric generating plants (with ultrahigh voltage transmission of power to market) will also improve the position of large suppliers compared with smaller ones. Large-scale economies in strip mining may be another accelerating force to concentration of the coal industry that will affect the market price of coal in the next few years. The pattern of utility purchases of coal from limited numbers of suppliers under long-term contracts may also serve to increase concentration of production. It is estimated that nationwide, electric power generation accounts for 70% of the coal market. This figure may

---

[16] Coal usage along the North Atlantic seaboard dropped 73% in the 1968–1973 period, from 21 million tons in 1968 to 5.8 million tons in 1973 (29, p. 7).

well expand as the limited push for coal conversion of electric utility plants by the Federal Energy Administration continues and/or is supplanted by congressional action requiring conversions on a massive scale by 1985 (31, May 26, p. 1).

Distribution of Electricity and Gas

Turning next to the electricity and gas distribution industry, we find a history of governmental regulation going back to the beginning of this century. The basic right to regulate in the public interest rests with the states under the concept of police power. (The federal government regulates under the constitutional power over interstate commerce.)[17] The purpose of state regulation historically has been to limit the earnings of utilities to a fair return on the fair value of the property under the Fourteenth Amendment. In general, regulatory commissions showed little interest in returns from individual rates. The rate structures that evolved embodied various internal subsidizations of use and classes of customer. Regulatory agencies normally took note of rates for small users, particularly the residential group, assuming the large user could protect himself. Company rate structures in the main were designed to be promotional. For follow-on rates to be low enough to be competitive with other fuels for large appliances, it was necessary that the rates be free of the greater portion of customer and demand charges. The initial blocks of consumption carried these charges. This practice also had the effect of stabilizing company revenues in the event of economic recessions.

Rates were designed to subsidize the customer with elastic demand while maintaining the overall stability of the revenue required. The opportunities for taking advantage of technological advances and economies of scale spurred the individual utilities to stimulate growth.[18]

When legislatures or the federal government felt that regulation was insufficiently effective in forcing prices down, they proceeded with other policies—low-cost interest loans to rural electrification cooperatives; exemption of public-owned generation and distribution companies from taxes, enabling them to raise capital more cheaply; etc. The federal government embarked on many river-basin projects, and major public power projects connected with these served as models of achievement in lowering electricity prices. The Tennessee Valley Authority was an outstanding example of large systems benefitting from economies of scale, low rates, and rapid growth (34, 35).

The final major illustration of subsidy of energy use is the general failure until recently (perhaps with the exception of coal noted earlier) to impose on fuel and power production the costs of environmental pollution and/or congestion, treating clean air and water and space as completely free goods. Since the production of energy and fuels and the consumption of energy (e.g. in transportation) have in general been major polluters of air, water, and space, this market failure has meant

---

[17] The Public Utility Act of 1935 subjected electric utilities to selected financial provisions of the Act. The second part of the Act called the Federal Power Act provides for regulation by FPC of interstate business of the utility companies (32).

[18] See, for example, a description of the objective rate plan introduced in the 1930s by private utilities as promotional tariff (33).

that price has not reflected all the costs imposed on society by the consumption of energy and that these costs have in fact been shifted to society as a whole.[19]

PRICE ISSUES IN THE BALANCE
Deregulation
Although general economic controls were essentially phased out by April 30, 1974, the Petroleum Allocation Act, passed after the OPEC nations imposed their oil embargo on the United States, authorized price controls on petroleum products until August 31, 1975. In anticipation of that date, and in the face of unrelenting pressure from OPEC and inadequate domestic production levels to permit major cutbacks in foreign imports, there has been intense administrative and legislative effort to formulate a more rational, coordinated US energy policy. This policy would, in the short term, synthesize the place of the fossil fuels, nuclear power, solar power, synthetic fuels, and imports as sources of energy supply and would, in the long term, encourage conservation through prices and taxes.

The legislative proposals have wrestled with the competing solutions of deregulation (phased, selective, total), import fees versus import quotas, special use taxes on gasoline, taxes on fuel used in industrial production, taxes on windfall profits, government stockpiling powers, etc. The implications for at least some of the alternative actions under current legislative consideration have been implicitly addressed in the historical discussion in the first section.

For petroleum, the issues revolve around the desirability of bridging the price gap (see Table 2) between regulated "old" oil on the one hand and domestic oil from new fields and from imports on the other, to allow higher price and market forces to operate in the expansion of domestic supply (from new and old fields via more costly technology). Advocates of quotas believe market operations to be ineffective, however, because of what they conceive as structural defects in the energy industry, which make the free market incapable of effecting national goals. Advocates of quotas versus import fees feel, in addition, that the effectiveness of price in curtailing demand may fall short of expectations in curtailing imports to a desirable proportion of domestic consumption. They prefer, therefore, the administrative control route (with or without some continued measure of price control). Each view holds paramount a different aspect of energy policy: the stimulus to production, the distributional equities, or the national defense concerns.

While the issue of deregulation at this writing revolves more tightly around oil policy, the issue of deregulating natural gas is probably not far behind. (Deregulation could take the form of deregulation of new gas and continuing antitrust surveillance for anticompetitive behavior, or complete deregulation of new and old gas fields.) The question with respect to natural gas is apparently impacted more by the presumed lack of reserves than is the case for oil. Thus the question is asked: How much more will the consumer have to pay for a current gas supply in order

---

[19] No attempt is made here to evaluate the cost/benefit of each departure on net to society except to note that price has not functioned or been permitted to function as an efficient allocator of energy. For a discussion of issues and alternatives, see (36).

Table 2 Crude petroleum[a]

Month	Wellhead Cost ($/bbl)		Refiner Acquisition Cost ($/bbl)		
	Old	New	Domestic	Import	Composite
1974					
Jan.	5.25	9.82	6.72	9.59	7.46
Feb.	5.25	9.87	7.08	12.45	8.57
Mar.	5.25	9.88	7.05	12.73	8.68
Apr.	5.25	9.88	7.21	12.72	9.13
May	5.25	9.88	7.26	13.02	9.44
June	5.25	9.95	7.20	13.06	9.45
July	5.25	9.95	7.19	12.75	9.30
Aug.	5.25	9.98	7.20	12.68	9.17
Sept.	5.25	10.10	7.18	12.53	9.13
Oct.	5.25	10.74	7.26	12.44	9.22
Nov.	5.25	10.90	7.46	12.53	9.41
Dec.	5.25	11.08	7.39	12.82	9.28
1975					
Jan.	5.25	11.28	7.78	12.77	9.58
Feb.	5.25	11.39	8.29	13.05	10.16
Mar.	5.25	11.46p[b]	8.29p	13.17p	n.a.[c]

[a] Source: (37).
[b] p: preliminary.
[c] n.a.: not available.

to solicit presumably only a small incremental supply of gas from limited reserves? (See Table 3.) If indeed the potential gas discoveries are very small, the sharp dislocations in existing patterns of gas supply and price and the attendant equity dislocations may seem, at least to some, not worth the cost of the extra stimulus to supply. If continued regulation of natural gas is presumed necessary, the issue of what form it should take is addressed below. Deregulation would eliminate the dichotomy of price and priority of use between intra- and interstate markets, which is currently skewing market supply and demand.

According to prevailing opinion of production and reserve levels of natural gas in the United States, higher gas prices are inevitable, either through higher unit delivery costs resulting from underutilization of pipeline facilities, or as a result of higher cost increments of gas supply. The deepening gas shortage in recent years has impelled the FPC to experiment with innovative changes in the interstate pipeline rate structure (38) in an attempt to direct the available gas supplies away from "low-priority" uses.[20]

Although the FPC has rejected a full-scale application of marginal cost pricing in the interstate gas pipeline industry, it has proposed a variation that gropes for

[20] "Low-priority" use administratively determined may or may not coincide with priorities arrived at in a free market.

Table 3 Wholesale prices of natural gas (¢/10^3cf)[a]

Year	Field Price[b]	Border Price[c]	Resale Price[d]
1960	15.53	22.90	n.a.[e]
1961	16.30	25.00	n.a.
1962	16.54	24.98	n.a.
1963	16.63	25.06	n.a.
1964	16.55	24.43	n.a.
1965	16.66	24.21	n.a.
1966	16.70	23.96	n.a.
1967	16.96	24.12	35.28
1968	17.20	23.87	35.17
1969	17.48	24.57	36.09
1970	17.93	25.79	38.07
1971	19.13	28.00	42.23
1972	20.54	30.61	45.87
Jan. 1973	21.37	31.43	46.40
Nov. 1973	23.85	35.17	51.82
Mar. 1974	25.66	43.12	56.90
Mar. 1975	31.57	102.52	77.83

[a] Source: (12) and selected press releases from the Federal Power Commission.
[b] Average field price of natural gas sold to interstate pipelines.
[c] Average Canadian border price of natural gas sold to major pipelines.
[d] Average price of natural gas sold by interstate pipelines for resale.
[e] n.a.: not available.

some of the same objectives—namely, signaling the consumer with the current cost information of new supplies. Recent incremental pricing proposed by the FPC would be limited to new high-priced gas supplies derived from imports of liquefied natural gas, for example. The cost of these high-priced supplements would not be rolled in and averaged with the cost of wellhead gas but sold under a separate schedule to test the market.[21] (In the event of pipeline curtailments, however, beneficiaries of supplemental supplies from nonconventional sources would become the lowest priority customers.) The large price spread between price of "old" and "new" wellhead gas[22] from traditional sources has been cited as a major reason for reluc-

---

[21] The FPC imposed an incremental pricing requirement on the importers of the first large baseload supply of liquefied natural gas (LNG) from Algeria in June 28, 1972. To assure that incremental LNG prices would be passed through to ultimate customers, the FPC prohibited the importing pipelines from selling to distributors that did not have separate rate schedules for reselling the LNG at incremental rates. (Op. No. 622) (39). Overwhelming opposition led to elimination of the requirement that distributors sell to retail customers on an incremental basis (Op. No. 622-A) (39).

[22] "New" wellhead gas is defined as gas produced from wells commenced on or after January 1, 1973, and new dedications of natural gas to interstate commerce on or after January 1, 1973 (Op. 699 and 699-H) (39). In these opinions, the FPC abandoned area rate making for a uniform national rate that would be reviewed biannually and rolled in for all new gas as defined above and for expiring contracts.

tance to go all the way with marginal cost pricing of interstate gas. However, the FPC wants to have the Natural Gas Act amended to permit it to go beyond "current costs" in cost-of-service procedures to determine prices so that it can take account of the price of substitutes and alternative supplies or second-best considerations.

Conservation has become a much more important factor in recent FPC rate reform. In interstate markets for natural gas, the principal discrepancy of current economic significance between incremental cost and average revenue requirements revolves around the cost of the commodity rather than capacity.[23] The greatest threat of economic waste is therefore not excessive construction of new transmission capacity but wasteful use of the gas itself. The latter has caused a turnaround in current attitudes of economists and regulators alike toward the Atlantic Seaboard formula's failure to follow peak-load principles in charging for pipeline capacity and hence discouraging economically efficient construction of storage capacity.[24] Recognizing that peak responsibility related to pipeline capacity deserves less attention and conservation of gas supplies more attention, the FPC issued a rate decision in 1973 in which 75% of fixed costs, instead of 50%, was assigned to the commodity charge.[25] This allows more of the revenue requirement (based on rolled-in historic costs) to be derived from that charge.

By making gas more expensive, the new "rate-tilting" of regulated gas should encourage customers with dual-fuel equipment to burn other fuels and might set a floor for the rates charged by gas distributors to their interruptible customers within their respective states. It would also eliminate the promotion of gas demand for uses considered "nonpriority" under the Commission's curtailment end-use policy.[26] The FPC has avoided a complete shift to volumetric rates from two-part rates because of its concern over revenue stability for the pipeline companies, as gas supply falls below capacity.[27] (Volumetric rates would also eliminate the rate

---

[23] Underutilization of many transmission facilities does not eliminate completely new pipeline construction since "gathering" facilities, as well as offshore pipelines and increased storage, is still an ongoing requirement of pipeline operation.

[24] Underground storage capacity for natural gas is currently estimated at 6 trillion cubic feet according to the American Gas Association, which represents about $2 billion in investment in 367 underground reservoirs in 26 states (29, p. 9).

[25] See FPC Op. No. 600-A, No. 671, and, more recently, Docket No. RM 75-19 (39). The latter is exploring abandonment of two-part rate schedules and adoption of volumetric rates to be designed by end use, which would increase the price relatively more for industrial than for other categories of customers, with the goal of discouraging industrial use of natural gas.

[26] See FPC Docket Nos. RP 74-42 et al, announced proceedings scheduled for the period of July 15, 1975 to August 15, 1975 to review priority of service categories for 14 pipelines that will be curtailed during winter 1975–1976 (39).

[27] It has been observed, however, that the shift of demand charges to the commodity portion of the structure is also a means of preserving the full return to stockholders of an investment only partially delivering a service. Innovative incentives would be preserved if differential returns on investment were allowed when excess capacity is substantial vs inconsequential. For example, the return on excess sunk capital costs might be reduced for

incentive for distribution utilities to store gas for peak use in winter or to install peak-shaving equipment for the peak-day requirements during the winter heating season.)

If legislative preference precludes deregulation and market allocation of supplies, the alternative is administrative allocations. In fact, the "incremental proposal" put forth by the FPC continues to rely on direct allocations and priorities set by administrative decisions. The incremental cost price would be charged only to those users whose demand is deemed to be of lower priority or more price-elastic than that of others. Under such circumstances the function of price is considerably undermined and may contribute negligibly to proper rationing of total supply. Nevertheless, to abandon completely the function of price in this area ignores the issue of considering second-best solutions not only within the energy sector of the economy but in all sectors simultaneously. The sharper the deviations from efficient economic price for any single energy commodity, the greater the impact of the residual or deficient purchasing power for nonenergy products, as well as for other types of energy.

There is little doubt that interstate competition among industrial firms in the various regions of the country will sharpen according to whether companies gain access to unregulated, intrastate natural gas. [The issue is under consideration; see (39).] (The characteristic pattern of natural gas use in manufacturing establishments for the major industrial states in 1971 is shown in Table 4.) In addition, the inability of the FPC to require that incremental prices paid by distributors for selected incremental supplies be passed along in tariffs to final customers undermines the net potential of embarking on such a pricing policy at all.

The state regulator may currently see his choice (as presented by the incremental policy of the FPC) as either rolling in the price of new and old gas to the final customer, thus providing a declining subsidy to existing customers as the availability of old gas diminishes, or carrying through the differentiated rates for incremental gas to only selected groups of customers, thereby giving unwarranted lower prices to existing customers who have captured the right to old gas. His choice then is not in terms of economic efficiency but on the political-social grounds of avoiding large short-run dislocations. The state regulator is also faced with another problem in rolling in new and old supplies and letting price rise as it may—that of protecting existing gas customers from overzealous attempts by gas distribution utilities to expand markets by means of synthetic gas supplies that may or may not be economic. Rolled in with field gas, the new exotic supplies may open the opportunity for the promotion of gas more effectively in new markets and yield more favorable returns to stockholders, perhaps at the expense of existing customers with sunk costs in gas equipment. Of course, if alternative forms of energy are barred by the current state of technology, the zero incremental cost of transmission and distribution may take on new interest. (What the most efficient developmental choice should be

---

equity toward the level of prevailing interest rates for bonded indebtedness, in order to prod management of pipeline companies to secure new supplies of field gas via exploration. To shift to the consumer the full risk of loss of supply is an inefficient way to deal with the issue of returns on sunk costs vs new investment.

Table 4 Reliance on electricity and purchased fuels by type, manufacturing establishments in the United States, and leading industrial states in 1971 (percentage of total kilowatt-hour equivalents)[a]

State	Total	Purchased Electricity[b]	All Other Purchased Fuel[c]	Type of Fuel			
				Fuel Oil	Distillate	Residual	Natural Gas	Coal
United States	100.0	15.3	86.0	11.3	4.6	6.7	50.5	14.9
Leading industrial states	100.0	13.3	86.6	10.8	4.7	6.1	49.7	15.6
California	100.0	19.1	81.1	3.6	2.5	1.1	66.5	0.7
Connecticut	100.0	15.6	85.0	54.2	14.1	40.1	13.5	<0.1
Illinois	100.0	13.2	87.2	7.9	4.2	3.7	46.9	21.7
Indiana	100.0	13.3	89.5	12.0	4.9	7.2	51.5	18.3
Massachusetts	100.0	15.6	85.2	42.9	15.8	27.1	15.2	0.6
Michigan	100.0	15.1	85.8	4.6	1.9	2.7	40.3	27.5
New Jersey	100.0	11.6	76.9	38.0	19.1	18.8	18.9	3.6
New York	100.0	18.3	82.2	24.4	10.8	13.6	22.0	16.3
No. Carolina	100.0	19.6	80.6	24.1	6.9	17.2	23.8	16.1
Ohio	100.0	15.5	84.8	3.5	2.5	1.0	39.3	35.4
Pennsylvania	100.0	13.1	87.2	16.0	6.4	9.6	39.1	25.5
Texas	100.0	7.4	92.8	7.8	0.2	0.6	82.2	2.0
Wisconsin	100.0	12.7	87.6	5.7	3.1	2.6	42.1	29.2

[a] Estimated in the New York State Public Service Commission based on data published in (40, 41).
[b] Data are based on sales to large light and power customers by the total electric utility industry (investor-owned and public) and are therefore only rough approximations of industrial sales.
[c] Includes fossil fuels unspecified by type.

must receive consideration independently from the sunk costs of old forms.) The break-even point between adjusting regulated price upward to reflect new high-price sources of gas and allowing prices to rise to reflect higher per-unit costs of excess pipeline capacity will become an ever more critical piece of policy information.

Incremental Pricing of Electricity

We turn now to electricity, where deregulation is not an issue, and set forth a position for more rational pricing within the regulatory context. Electricity prices began to attract substantial attention in the early 1970s. A long era of steadily declining prices for electricity had just about come to an end and rates were either stabilizing or turning upward. But it was not costs of electric power inputs per se that set off the interest in prices, although inflation had reared its ugly head; rather it was new environmental issues linked to air and water pollution and the intense division of opinion on whether zero economic growth (42, 43) or continued economic growth should be the preferred social goal. It was environmental intervention in the up-to-then rather mundane rate case that helped to sharpen the focus on electricity price as a means of achieving new objectives, quite unrelated to the traditional function of raising adequate utility revenues. Thus, prices of primary fossil fuels were under careful scrutiny during the 1960s (and in fact during previous decades) as old concerns with monopolistic control of industry supply and price (44) were intermittently activated, as extension of federal control to interstate gas was initiated, and as national defense and protection of irreplaceable resources continued to receive attention. At the same time, however, the electricity and gas distribution companies, which had been granted monopoly status and were under state regulation, received much less public interest. (The issue of public power and federal development of river basins had had their day in an earlier era.)

Utility rate structures had worked well enough for the electricity and gas distribution industry and the consumer in the 1950s and 1960s to raise adequate revenue and to track declining average costs reflecting technological advances and dramatic economies of scale. Theoretical concern with market efficiency and the desirability of optimizing societal welfare via marginal cost pricing, while debated seriously here and abroad, was not implemented in the US energy sector.[28]

The relatively cheap price of fuels that prevailed in the 1960s tended to discourage a keener interest in fuel efficiency of power generation if it meant higher capital costs because interest rates were rising. A great expectation that nuclear-fueled plants would be built rapidly to supplement fossil fuels for electricity generation was another factor. Fuels came to the fore as an issue for electric utilities with the passage of environmental legislation that spurred a switch to low-sulfur oil from coal and residual oil, particularly on the East Coast.[29]

---

[28] In contrast, practical applications of marginal cost pricing to electric tariffs were first adopted in France (1957) in the so-called Green Tariff and in the United Kingdom (1962). [For descriptions of European applications see (8, 45–49).]

[29] The sharply rising cost of low-sulfur fuels in recent years reflected shortages of refinery capacity to produce low-sulfur fuels. These were caused by past government policies in the

The initial intervention by the environmental groups in rate cases was intended to raise rates by recognizing externalities associated with new generation of electricity and to slow demand and economic expansion in the light of new concerns about economic growth and the exhaustion of natural resources. The issues that they addressed shifted, however, as utilities scrambled to keep up with rising costs. Rising cost factors facing the industry accelerated the pace of rate increases in a crisis-like succession of rate cases all over the country. The switch from creeping inflation to more rapid price increases especially affected the capital-intensive expansion programs of the utilities. Delays in siting new plants caused monetary costs per kilowatt of new construction to escalate sharply because of rapidly rising interest rates as well as increasing construction costs. Some utilities began to face blackouts, while for others operating costs rose to parallel the inefficiencies that beset their systems.

The push for reform of the tariff structure in the early 1970s was directed toward demand management via inverted rates[30] and higher rate levels, particularly for large industrial customers, rather than toward cost-based rates. Subsequent escalation of utility costs caused intervenor policies to switch first to flat rates (51) and finally to marginal cost pricing, reflecting peak and off-peak differences, implemented through time-of-day metering. The presumption that marginal costs were above average costs at least for generation of electricity meant that the goal of the consumer to keep rates down could be assisted by a restricted building program for electricity generation—also the objective of the environmentalist. Efficient economic prices differentiating peak and off-peak usage would eliminate a subsidy to peak users by off-peak users that was stimulating an inefficient expansion of consumer demand and hence generation of electricity that was damaging to the environment. Slowing system growth by pricing above historic costs was proposed to permit stabilization or even a decline of the rates to small users.

Not all rate reformers are equally sanguine or vigorous in asserting their belief in the ability to achieve simultaneously many different energy goals in the long

---

petroleum market for crude vs refined products as well as uncertainty of future government policy on environmental issues, as these were likely to affect the future profitability of refinery expansion or conversion to low-sulfur products. These factors served to exacerbate the financial plight of utilities, as did the vigor and success of environmental groups in delaying new refinery expansion and the building of the Alaskan pipeline. The growing shortage of natural gas and the FPC policy toward the use of gas under utility boilers also posed new fuel problems for the electric utilities. The impact of these several policies differed significantly among the regions of the United States.

[30] Simply stated, the proposal was to invert the declining block structure, in wide use, thus charging more per kilowatt-hour for larger users than for smaller ones. The assumption was that charging a higher, rather than a lower, unit price as consumption grew would inhibit demand; there was also an implicit assumption that the change would serve an egalitarian purpose, by eliminating and reversing any existing discrimination against smaller consumers, who might also be poorer (50).

run merely by reforming the cost calculation of electricity tariffs.[31] There is, however, widespread agreement at least that moving tariff design closer to efficient pricing can have substantial unrealized benefits in the allocation of societal resources. The elimination of subsidies to users of electricity, which are counterproductive to achieving national goals of conservation, national security, and general health and welfare, would be one such obvious benefit.

Thus one of the key issues in the formulation of electricity price before the state regulatory agencies in 1975 is the consideration of a shift from existing costing methodologies, particularly from fully allocated cost, to incremental cost pricing. The state of New York is currently engaged in a generic hearing on rate structure whose purpose is to explore the entire problem of determining a suitable design for electricity rates along marginal cost-pricing principles outside a specific rate proceeding (53).[32] Beginning with the proposition that the principal purpose of rate making is economic efficiency and proper economic signals to buyers, the hearing is designed to explore the enormously difficult but soluble problems of definition, measurement, and administration. The first qualifying consideration is the cost of administering such a pricing system, since economic efficiency requires that the additional cost of developing and administering a closer approximation to marginal cost pricing be exceeded by the incremental benefits. The second is the principle of second-best. In setting rates it will be essential to take into account the extent and direction in which prices in other markets may diverge from that standard. In addition, a conscientious effort to estimate the impact of rate structure alterations on the total revenue flow to companies will require moving prudently but expeditiously. Through opportunities for users to avoid higher costs that are imposed on the system by peak consumption, many may escape the burden of increasing costs (54).

The Wisconsin Commission carried out a comprehensive review of rate principles in the Madison Gas and Electric rate case in 1974 (55, 56) and accepted long-run incremental cost, a practical variant of marginal cost, as the appropriate measurement of cost and basis of pricing. It also ordered the company to implement full peak-load pricing for large customers and to undertake experimental work on the feasibility of various forms of time-of-day metering for small residential and commercial customers.

The Federal Energy Administration recently announced it will spend $1.2 million to finance six time-of-day pricing studies. These were awarded to cooperating agencies in Arizona, California, New Jersey, Ohio, Connecticut, and Arkansas, and will extend over several years.[33]

National Economic Research Associates has developed a model for time-of-day

[31] Supply problems must be addressed also through stimuli to greater exploration, research, and development of innovative technology even in the face of reductions in demand achieved by conservation practices (52).

[32] Several other states—including Florida, California, Massachusetts, and Arizona—have been reviewing their rate policies outside of a specific rate case.

[33] See (31, May 5, p. 2) for greater detail.

marginal costing and has been testing it with data from Wisconsin Electric Power. Pacific Gas and Electric, in conjunction with the California Public Utilities Commission, is undertaking two studies, one for residential customers and another for large industrial customers.[34]

The Edison Electric Institute (EEI) and the Electric Power Research Institute (EPRI) have formulated a joint plan of study to determine on a national scale the potential for peak load and time-of-day pricing. The study is to be completed in a year, once approved by all of its joint sponsors: National Association of Regulatory Utility Commissions, EEI, and EPRI.[35]

The Vermont Public Service Board ordered the introduction of optional time-of-day rates for residential consumers in 1974 to stimulate the installation of heat-storage equipment.

Marginal cost pricing is the controlling economic principle for the efficient allocation of resources. While its implementation by regulation requires judgment and difficult determinations of how to measure and apply it efficiently, it is to be hoped these difficulties can be overcome. Current deliberations over the application of the theory also include consideration of whether short-run or long-run costs should be used as the basis of tariff structure, in the light of goals of stability of price, optimum efficiency, and administrative difficulties. The greatest economic efficiency theoretically flows from using short-run incremental cost as a basis of pricing from a given system of plant and equipment, rather than from the cost consequences of a continuing incremental flow of business after plant and equipment have been adapted to that change in volume. Restated, long-run incremental costs measure how costs will change after a span of time sufficiently long for the system planners to adapt the supplying system to the change in demand (all measured in current[36] costs). It was once desirable to try to reduce rates closer to marginal costs in markets responsive to price in order to minimize waste arising from not producing more (even though the revenue requirement made it impossible to charge everyone these low rates without incurring a deficit). Likewise, it becomes desirable at present to raise rates closer to marginal costs for more elastic demands to accomplish the reverse objective—provided that marginal costs now do exceed average revenue requirements in general because of inflation.

The most important effect of pursuing incremental costing, which recognizes peak responsibility differentiation, will be on the measurement of capital or capacity costs (54). Note that the issue in pricing electricity is therefore not exclusively, or even

---

[34] See (31, Mar. 17, pp. 5, 8).

[35] See (31, June 2, p. 3) for details of topics to be studied.

[36] Some clarification is required in discussing incremental vs average costs, average historical costs, and economies of scale. The presence or absence of economies of scale can be measured only in the same dollars, in the same time period, and with a given technology. It is therefore possible to have marginal costs below average costs even after a long period of inflation. Inflation can move whole cost functions upward, without necessarily diminishing their slope. Once, however, marginal costs exceed historic average costs (the revenue requirement), the traditional justification for price discrimination to encourage demand disappears (57).

primarily, conservation of fossil fuels, but the efficient allocation of capital. On the other hand, peak–off-peak pricing would presumably produce energy savings as well, by tending to shift consumption from times when energy can be supplied only by bringing into service high-marginal-cost generating facilities to times when it can be supplied wholly with much more energy-efficient, baseload generating plants.

As long as declining block-rate structures stated in terms of energy consumption ignore the question of whether or not demands coincide with a system peak, they will inevitably produce economically inefficient results.

In the electricity sector, as in any other, marginal cost-pricing methodology, under whatever name, will have the task of making a proper determination of pricing periods, peak versus off-peak, and measuring their respective incremental costs because different kinds of capacity are constructed to serve different kinds of load. Short-run compared with long-run incremental costing would have the advantage of requiring much simpler data inputs for computation (primarily fuel costs).[37]

Pricing based on short-run incremental costs (SRIC) would automatically embody peak responsibility principles since short-run marginal costs vary directly with the level of system demand. Systematic discrimination among the customer classes required to adjust revenues to budgetary constraints would be less disparate because adjustment would presumably be made in inverse proportion to short-run rather than long-run price elasticities of the classes. Prevailing views based on available studies indicate that price elasticities for the short run tend to be smaller than those for the long run. Because of the state of the arts in this area of research, dependence on short-run elasticities might therefore be a far more comfortable choice for the administrator. (There is no presumption that the differential between the revenue requirement and SRIC in dollars or percentages would be consistently smaller or larger than if LRIC were the basis of pricing.) SRIC might also avoid the dilemma of how to meet undesirable dislocation of current market equilibria if future costs (adjusted for inflationary price increases) are conceivably higher or lower than current costs because of nonoptimalities in respective electric power systems. (Such dislocations could produce either shortages or excess capacity during selected seasons or times of day, creating the additional issue of how to allocate the cost between consumers and equity.)

Long-run incremental costing and pricing on the other hand, by using a time period long enough to permit readjusting the current imperfect power system to an optimum system for future demand, may be better able to cope with existing inefficiencies, although it would depend more heavily on forecasting skills.

While marginal cost-pricing studies and separate price elasticity studies for peak and off-peak demands should eventually provide information that would permit demand management and price setting beyond the economically efficient level in order to achieve some other goal, a theoretical marginal cost-pricing approach

---

[37] The smooth upward-sloping short-run marginal cost curve is presumed to permit the recovery of capital costs by charging the fuel costs of the least efficient plant; the revenue differential between the least efficient over the fuel costs of more efficient plants should be equal to the fixed costs of those plants in an optimum system.

would not deny to the consumer the free choice of buying or not buying. He would merely be charged the economic price. This does imply, however, that production capacity would not be restricted, but the supply would be allowed to respond to effective demand at economic price according to the classical concept of that theory. An attempt to restrict supply by exploiting demand price information and charging prices above marginal costs would be a distortion of the economic pricing system.

The economist is not in a position, however, to rule out for the community the privilege to deviate from charging marginal cost prices—whether it is for social goals[38] linked to redistribution of incomes, environmental goals, or national defense. What his methodology permits in case of such deviation is an estimate of the true costs of the undertaking. It would, of course, also be important to measure the benefits derived rather than to view such costs in isolation.

The enormous capital requirement projected for the electric utility industry, even after revision in the last year of projected building requirements in the face of a slowing of the demand growth rate,[39] has led to various suggestions for government insurance and/or subsidy in order to assure that the industry obtains adequate funding at moderate cost. In the last analysis this might be yet another example of departure from efficient pricing for a social goal that needs very careful scrutiny, unless of course the lower interest charges were a direct reflection of lower risks associated with a change in institutional arrangements.

Adjustment Clauses

Outside the basic costing methodology employed in rate cases, there have been several important procedures for expediting the recovery of costs by utilities, which have made electricity prices the focus of public outcry in the last 18 months. The most important has been the automatic adjustment clause, the most common of which is the fuel adjustment clause, used widely among the states. The ability of the utilities to pass along their escalating fuel costs during 1973–1974 permitted them a prompt recovery of one of the largest components of their sharply increasing costs at a time when general inflation, on the one hand, and regulatory lag in translating these costs into higher rates, on the other, subjected them to severe financial stringency.

While the fuel adjustment clauses have raised questions about the utility companies' diligence in fuel contract negotiation because higher costs are passed along completely (though typically with a lag), there is little doubt that, in contrast to delays of general rate cases, an automatic fuel adjustment tends to promote economic

---

[38] "Lifeline rates" is a social rate-making proposal that has had considerable publicity in recent months. It would provide some minimum amount of electricity (and/or gas) at a low cost differentiated from other consumption. For a critique of the issue see (58).

[39] The consumption of electric power historically has risen at a rate of about 7% a year. No increase in demand for electric energy at all was recorded in the United States in 1974 over 1973. A moderate growth rate of 5–6% is being forecast or at least judged desirable for the rest of this century by some studies. See summary of Edison Electric Institute study, *Economic Growth in the Future,* in (31, June 2, p. 7). See also (59, 60).

efficiency by translating variable costs into price, promptly affecting consumer demand for power. The fuel clause acts as a crude surrogate for short-run marginal costs insofar as these vary from month to month. The automatic fuel adjustment clause has, in a period of sharply rising fuel costs (see Table 5), also produced a substantial flattening of the rate designs in effect in the electric power industry, particularly among large industrial and commercial customers. The problem with these clauses, however, is that it is average fuel costs that form the basis of the adjustment rather than differentiated costs based on time-of-day and varying fuel efficiencies of the generators used, or, often, variation in the Btu content of fuel purchases.[40] In addition, there is some lag (30–90 days) in registering on the customer's bill the cost of fuel incurred by the utility.

There has been a substantial push in the utility industry for broader automatic adjustment clauses that would cover other major cost items such as taxes, interest, wages, and conservation. In isolated instances such modified clauses have been adopted. These have less merit than the fuel adjustment clause. Fuel cost is a major part of short-run marginal cost and so it is especially efficient to have such charges reflected quickly in price. In contrast, adjustment clauses designed to recover charges related to historic sunk costs do not have similar justification. If regulation were to use its rate-setting authority effectively for both the consumer and the utility by monitoring efficient management, as well as recognizing legitimate costs, a plethora of automatic adjustments could be avoided.

The increasing use by regulatory agencies of future test years in calculating revenue requirements and hence price levels is another step to efficient pricing since tariffs should reflect the costs of the period in which they are to be in effect rather than historic costs and distorted signals for choice. Such a procedure also reduces the utility company's need for a rate case determination almost on the heels of the previous decision. By preserving their discretion over the use of regulatory lag, the regulatory agencies are currently grappling with the need to meet the unusual inflationary pressures on the utilities, while maintaining utility incentive for improved productivity. Price structure that is economically efficient is more likely to discourage any tendency of the utility to push for growth beyond its optimum level[41] although it would not necessarily ensure against a nonoptimal system marked by an inefficiently high capital/labor ratio.

[40] In 1974, four fifths of the states had such clauses in effect. Among the larger utilities, 65% have fuel adjustment clauses in their residential schedules, 77% had such clauses in their commercial schedules, and 83% in their industrial schedules according to an FPC survey. (Comparable figures for 1970 were 35%, 58%, and 72%, respectively.) See (63) for a critique of these clauses.

[41] The Averch-Johnson effect—i.e. the postulated tendency for a firm to add to its rate base as long as the cost of capital is below the rate of return allowed, defraying its loss on peak service meantime by overcharging off-peak customers—is less likely to occur, however, when a firm is faced with such severe inflationary spiral that new capitalization will eventually lead to a rate increase rather than a rate decrease. In such a case, the incentive to overinvestment is likely to disappear and theoretically the utility should try to avoid capital investment. Overcapitalization is more likely to occur when a firm is able to enjoy technological advances of such magnitude or such large economies of scale in the face of growth of demand that its unit costs fall in spite of inflationary forces. See (64).

Table 5 Utility fossil fuel prices, 1974–1975 (cents per million Btu)[a]

Region	All Fossil Fuels			Natural Gas			Residual Fuel Oil			Coal		
	February 1975p[b]	June 1974	February 1974	February 1975p	June 1974	February 1974	February 1975p	June 1974	February 1974	February 1975p	June 1974	February 1974
National average	106.4	87.7	81.6	65.2	47.9	39.8	202.7	194.9	185.9	81.7	69.5	56.9
New England	198.8	184.7	175.7	113.2	124.7	73.3	204.1	201.1	190.5	134.8	95.9	114.2
Middle Atlantic	147.1	137.6	129.0	84.5	77.3	72.7	204.9	207.7	208.1	104.7	88.6	69.5
East north central	85.6	76.9	57.0	92.7	76.1	62.4	180.3	198.2	127.2	78.4	71.7	52.4
West north central	69.1	47.2	40.5	43.8	41.7	38.0	184.7	179.3	154.8	57.9	42.0	36.3
South Atlantic	120.2	119.0	100.6	68.5	59.8	57.3	184.6	181.5	167.3	97.0	90.2	76.7
East south central	83.1	62.5	52.4	79.8	52.8	48.1	184.3	171.5	132.2	79.3	57.9	49.8
West south central	67.3	50.0	46.2	63.0	43.6	35.2	183.1	161.1	126.8	21.0	17.7	13.6
Mountain	62.9	40.3	48.1	66.7	49.2	54.5	197.4	199.2	174.9	30.8	25.7	26.8
Pacific	194.7	117.9	160.3	83.6	50.7	47.6	234.8	202.5	191.2	57.7	35.5	n.a.[c]

[a] Source: (61, 62).
[b] p: preliminary.
[c] n.a.: not available.

SUMMARY

Under the pressure of rising energy prices, governmental policy has been groping for greater rationality in both the legislative and the regulatory arenas in place of taking piecemeal actions.[42] While supply has the lion's share of attention, demand management and rate structure have attracted renewed interest. Policy reevaluations have led to an attempt to formulate clear goals, reconcile them where they conflict, and design means of reaching them. The issue of national security and its relation to domestic versus international energy supplies has been a major focus that shapes the short-run as well as long-run considerations.

The role of efficient price is probably receiving more serious consideration vis-à-vis direct administrative controls than it has in some time. Deregulation to varying degrees and over wide time frames[43] is gaining new advocates in the oil and gas sectors of the energy market. With regard to state regulatory functions, discussion emphasizes a review of the cost basis of the old rate structure and a move closer to incremental costing and theoretical efficient price via peak pricing and time-of-day metering.

The structure of the oil, gas, and coal industries continues to be viewed with a jaundiced eye. Expressions of willingness to deregulate are couched in terms of continuing antitrust surveillance and excess profits taxes in order to avoid windfall profits accompanied with little compensation in the way of a supply response.

The backdrop of soaring prices of nonregulated oil via the influence of the OPEC cartel[44] on the domestic market encouraged Congress at long last to eliminate the percentage depletion allowance, one of the tax subsidies with a long history of stimulating inefficiency in the petroleum industry. The question remains whether the elimination of this almost $2-billion tax subsidy, if accompanied by decontrolled oil prices, will accomplish the goal of increased oil exploration and production, now that the goal is to cut back overdependence on foreign oils dramatically. Will high prices via decontrol and the elimination of percentage depletion work to permit the flow of capital for drilling only exploratory wells rather than production wells because inflationary expectations provide an incentive to keep

---

[42] Congress had not yet passed its major legislation as of the beginning of June. Therefore, President Ford followed his earlier announced intention and added a second dollar to the import fee on foreign oil. (The first dollar of import duty went into effect on February 1, 1975.)

[43] Deregulation of old oil prices in 39 months was suggested by the Administration (in contrast to as much as four and five years, suggested by various congressional factions). The Emergency Petroleum Allocation Act of 1973 is scheduled to expire August 31, 1975. A simple extension is viewed unlikely as the struggle continues to find some plan for phasing out controls on oil.

[44] Although OPEC promised in January 1975 to freeze oil prices until September 1975, a move to sever the link with the dollar and start quoting prices in special drawing rights, an artificial money made up of a combination of 16 currencies including the dollar (currently valued at $1.40) may raise the cost of a barrel of oil somewhat before then.

oil in the ground rather than flowing?[45] Moderate revisions made in computation of the overseas tax credit for foreign production were logical, but it remains to be seen whether these went far enough in view of the stated intent of Project Independence.

If current ceiling prices on petroleum do not restrict price below marginal production costs, they are not in effect causing domestic shortages by making capital difficult to obtain.[46] And since the price of newly produced oil (in contrast to gas) is unregulated, the only way in which regulation could be discouraging production is by defining as old oil and subjecting to price control those incremental supplies that might be forthcoming at higher prices from fields producing prior to 1973. Rationing, instead of higher decontrolled prices, would be justified if higher price could not be counted on to induce an expanded supply either because of industry structure or because short-run inabilities to start new production would only provide economic rents to current producers in the interim period. To the extent that producers are sufficiently free of the law of capture to do so, they would presumably hold old oil in the ground rather than develop or produce it, so long as there was reason to expect the price, now regulated, to rise to the world level, no matter what the relation of its regulated price to marginal production costs. The question to which we do not have the answer is to what extent regulation confined in principle to old oil is discouraging production. The distributional consequences of alternative policies are hard to predict.

Industry spokesmen contend that regulation applied only to old oil or gas, even if correctly applied, will still retard the efficient adjustment of supply because it reduces the flow of internally generated funds available for developing new supplies of fuel. As a matter of economic principle, however, it would suffice to provide the correct incentive in the form of an unregulated price for new supplies. Efficient prices should not be padded further in order for the industry to generate more of its own capital. The avoidance of capital markets as the efficient allocator of scarce capital by government subsidy or by permitting suppliers to retain (inframarginal) economic rents represents in the latter case a deviation from efficient capital funds allocation, and in both cases an undue burden on the consumer. A key market factor in oil and gas supply and resultant equilibrium in price has been uncertainty about government policy. This is likely to gain some clarification in 1975.

The implementation of efficient electricity price as an allocator of fuel, capital, and environmental resources in the next decade is of enormous significance, particularly since electricity competes with other forms of energy over large areas of use and utilities exercise such a major influence in the fuels markets. Their influence stems from the proportion of their consumption to total energy consumption, their use of individual fossil fuels, and their unique ability to use coal and nuclear fuels. For these reasons, rate reform of electric power may create one of the

---

[45] Recent statistics for the first quarter of 1975 show the drilling rate at an all-time high (data from the Federal Energy Administration).

[46] Nordhaus claims that prices prevailing even before the OPEC actions were too high in relation to production costs (65).

few opportunities for regulation to influence energy market prices positively and to allow electricity price to give some competition to cartel-influenced petroleum prices so that the international cartel does not succeed in setting a floor for domestic energy prices.

The issues of energy price in the next few years will be heavily influenced by international political policy, but the extent to which the resultant economic conditions, whatever they are, will be influenced by the priority of economic efficiency versus other goals will be a crucial determinant of the general welfare. The choices will revolve around the competing forces of deregulation and administrative controls, or antitrust surveillance; regulated markets utilizing or superseding market forces; short- or long-run priorities; the responsiveness of energy consumption to conservation; and efforts to subsidize price. In any case, the era of cheap energy at least for the next ten years seems to have come to an end.

Literature Cited

1. Ford Foundation, Energy Policy Project. 1974. *Energy Consumption in Manufacturing.* A report by the Conference Board. Cambridge, Mass: Ballinger
2. Federal Energy Administration. Nov. 1974. *Project Independence, A Summary.* Washington DC: GPO
3. Schurr, S. H., ed. 1972. *Energy, Economic Growth, and the Environment.* See also Appendix, Energy consumption: trends and patterns, by J. Darmstadter, pp. 220ff. Baltimore: Johns Hopkins Univ. Press
4. National Petroleum Council, Industry Advisory Council to the US Dept. of the Interior. 1974. *Energy Conservation in the U.S., Short-Term Potential 1974–1978.* Washington DC: Nat. Pet. Counc.
5. National Petroleum Council, Industry Advisory Council to the US Dept. of the Interior. 1972. *U.S. Energy Outlook, A Summary Report of the National Petroleum Council.* Washington DC: Nat. Pet. Counc.
6. Baumol, W. J. 1967. Reasonable rules for rate regulation: Plausible policies for an imperfect world. In *Prices: Issues in Theory, Practice, and Public Policy,* ed. A. Phillips, O. E. Williamson. Philadelphia: Univ. Penn. Press
7. Kahn, A. E. 1970, 1971. *The Economics of Regulation: Principles and Institutions,* Vols. 1, 2. New York: Wiley
8. Turvey, R. 1967. Practical problems of marginal cost pricing in public enterprise: England. See Ref. 6
9. Council of Economic Advisors. 1975. *Economic Report of the President,* p. 80. Transmitted to the Congress Feb. 1975. Washington DC: GPO
10. Kahn, A. E. 1974. *Our National Energy Policy.* Lecture at Univ. Delaware, March 4, 1974. Unpublished
11. US Congress, Senate, Committee on Interior and Insular Affairs. Jan. 1974. *Market Performance and Competition in the Petroleum Industry,* Pursuant to S. Res. 45, A national fuels and energy policy study, 93rd Congress, Serial No. 93-24, Pt. V, pp. 92–95
12. Foster Associates, Inc. 1974. *Energy Prices, 1960–1973,* A report to the Energy Policy Project of the Ford Foundation. Cambridge, Mass: Ballinger
13. Petroleum Publishing Company. 1975. *Oil Gas J.,* Vol. 73, various issues
14. Commerce Clearing House. 1975. *Standard Federal Tax Reporter,* 755CCH, Para. 3554, pp. 42098ff
15. Brannon, G. M. 1974. *Energy Taxes and Subsidies,* A report to the Energy Policy Project of the Ford Foundation. Cambridge, Mass: Ballinger
16. Adelman, M. A. 1972. *The World Petroleum Industry.* Baltimore: Johns Hopkins Univ. Press
17. Organization for Economic Cooperation and Development. 1969. *Impact of Natural Gas on the Consumption of Energy in the OECD European Member Countries.* Paris: OEDC
18. Organization for European Economic Cooperation. Jan. 1961. Trends in economic sectors. In *Oil: Recent Developments in the OEEC Area, 1960.* Paris: OEEC
19. DeChazeau, M. G., Kahn, A. E. 1959. *Integration and Competition in the Petroleum Industry.* New Haven, Conn: Yale Univ. Press

20. Jacoby, N. H. 1974. *Multinational Oil: A Study in Industrial Dynamics*, Studies of the Modern Corporation, Graduate School of Business, Columbia Univ. New York: Macmillan
21. Kahn, A. E. 1964. The depletion allowance in the context of cartelization. *Am. Econ. Rev.* 54: 286–314
22. Kahn, A. E. 1970. The combined effects of prorationing, the depletion allowance, and import quotas on the cost of producing crude oil in the U.S. *Nat. Resour. J.* 10(1): 53–61
23. American Gas Association. 1974. *1973 Gas Facts*. Arlington, Va: Am. Gas. Assoc.
24. Kahn, A. E. 1960. Economic issues in regulating the field price of natural gas. *Am. Econ. Rev.* 50(2): 506–17
25. MacAvoy, P. W. 1972. The regulation induced shortage of natural gas. In *Regulation of the Natural Gas Producing Industry*, ed. K. C. Brown, pp. 169–91. Baltimore: Johns Hopkins Univ. Press
26. Adelman, M. A. Winter 1972–1973. Is the shortage real? Oil companies as OPEC tax collectors. *Foreign Affairs* 51(2): 69–107
27. Mulholland, J. P., Webbink, D. W. 1974. *Concentration Levels and Trends in the Energy Sector of the U.S. Economy*, Federal Trade Commission, US Econ. Rep., p. 57. Washington DC: GPO
28. Federal Power Commission. 1975. A time for decision and action. *Natural Gas Survey*, Vol. I, Chap. 1. Preliminary draft
29. *Weekly Energy Report*, Vol. 3, No. 22. 1975. New York: Llewellyn King
30. Duchesneau, T. D. 1975. *Competition in the U.S. Energy Industry*, A report to the Energy Policy Project of the Ford Foundation. Cambridge, Mass: Ballinger
31. *Electr. Week*. 1975. Various issues. New York: McGraw-Hill
32. Caywood, R. E. 1972. *Electric Utility Rate Economics*, sixth printing. Originally published 1956. New York: McGraw-Hill
33. Kennedy, W. F. 1937. *The Objective Rate Plan for Reducing the Price of Residential Electricity*. New York: Columbia Univ. Press
34. Hellman, R. 1972. *Government Competition in the Electric Utility Industry: A Theoretical and Empirical Study*. New York: Praeger
35. Christian, V. L. Jr., Vaughan, C. M. 1971. Some aspects of cost and demand in the pricing of electric power. *Land Econ.* 47: 281–88
36. Kneese, A. V., Schultze, C. L. 1975. *Pollution, Prices, and Public Policy*, A study sponsored jointly by Resources for the Future and Brookings Institute. Washington DC: Brookings Inst.
37. Federal Energy Administration. 1975. *Petroleum Situation Reports*. Washington DC: Nat. Energy Inf. Center. Various issues
38. Wald, H. 1975. Rate reform for electric utilities and gas pipeline companies. *Transp. J.* 14(3): 30–41
39. Federal Power Commission. June 11, 1975. *Order Reopening Proceedings, Providing for Hearings and Establishing Procedures*, Docket Nos. RP 74-42 et al; April 4, 1975. *Policy with Respect to Certification of Pipeline Transportation Agreements*, Docket No. RM 75-25; Feb. 20, 1975. *End Use Rate Schedules*, Docket No. RM 75-19; Dec. 4, 1974. *Opinion and Order Modifying Opinion No. 699*, Opinion No. 699-H, Docket No. R-389-B; June 21, 1974. *Opinion and Order Prescribing Uniform National Rate for Sales of Natural Gas Produced from Wells Commenced on or after Jan. 1, 1973, and New Dedications of Natural Gas to Interstate Commerce on or After Jan. 1, 1973*, Opinion No. 699, Docket No. R-389-B; Oct. 31, 1973. *United Gas Pipeline Co.*, Opinion No. 671, Docket No. RP 72-75; Oct. 5, 1972. *Columbia LNG Corp. et al*, Opinion No. 622-A, Docket No. CP 71-68 et al; June 28, 1972. *Columbia LNG Corp. et al*, Docket No. CP 71-68 et al; May 8, 1972. *El Paso Natural Gas Co.*, Opinion No. 600-A, Docket No. RP 69-6 (Phase II)
40. Edison Electric Institute. 1972. *Statistical Year Book*. New York: Edison Electr. Inst.
41. US Bureau of the Census. 1972. *Fuels and Electric Energy Consumed*, 1972 Census of Manufacturers, Spec. Rep. Ser. MC 72 (SR 7). Washington DC: GPO
42. Gordon, S. 1973. Natural resources as a constraint on economic growth: today's apocalypses and yesterday's. *Am. Econ. Rev.* 63(2): 106–10
43. Rosenberg, N. 1973. Innovative responses to materials shortages. *Am. Econ. Rev.* 63(2): 111–18
44. Federal Trade Commission Staff Report. Aug. 22, 1952. *The International Petroleum Cartel*. Submitted to the Senate Subcommittee on Monopoly of the Select Committee on Small Business, 82nd Congress, 2nd sess. Washington DC: GPO
45. Boiteaux, M. 1964. Marginal cost pricing; Peak load pricing; The "tarif

vert" of Electricité de France; Boiteaux, M., Stasi, P. The determination of costs of expansion of an interconnected system of production and distribution of electricity. In *Marginal Cost Pricing in Practice,* ed. J. R. Nelson. Englewood Cliffs, NJ: Prentice-Hall
46. Caillé, P. 1971. Marginal cost pricing in a random future as applied to the tariff for electrical energy by Electricité de France. In *Essays on Public Utility Pricing and Regulation,* ed. H. M. Trebing, MSU Public Utilities Studies, Inst. Public Utilities, Mich. State Univ.
47. Balasko, Y. 1975. *Optimal Forms of Electricity Tariffs.* Electricité de France, Conf. on Electricity Tariffs, Madrid, April 1975. Paris: Int. Union of Producers and Distributors of Electric Energy
48. Turvey, R. 1968. *Optimal Pricing and Investment in Electricity Supply, An Essay in Applied Welfare Economics.* London: Allen & Urwin
49. Anderson, D., Turvey, R. Dec. 1975. *Electricity Economics: Theory and Practice.* Baltimore, Md: Johns Hopkins Univ. Press. In press
50. Sander, D. Feb. 1972. *The Inverted Rate Structure—An Appraisal, Part 1, Residential Usage.* OER IX, New York State Dept. Public Service
51. New York State Public Service Commission. April 28, 1975. *Consolidated Edison Co. of New York—Electric Rates.* Opinion No. 75-9, Case No. 26538
52. Ford Foundation, Energy Policy Project. 1974. Advisory board comments. In *A Time to Choose: America's Energy Future,* Final Rep., pp. 349ff. Cambridge, Mass: Ballinger
53. New York State Public Service Commission. April 30, 1975. *Proceedings on Motion of Commission as to Rate Design for Electric Corporations,* Case No. 26806. Minutes of prehearing conference
54. Kahn, A. E. 1975. *Efficient Rate Design: The Transition from Theory to Practice.* Presented at the Symposium on Rate Design Problems of Regulated Industries, Kansas City, Mo., Feb. 24, 1975
55. Public Service Commission of Wisconsin. Aug. 8, 1974. *Application of Madison Gas and Electric Company, Findings of Fact and Order,* Docket No. 2-U-7423
56. Berlin, E., Cicchetti, C., Gillen, W. J. 1974. *Perspective on Power, A Study of the Regulation and Pricing of Electric Power,* A report to the Energy Policy Project of the Ford Foundation. Cambridge, Mass: Ballinger
57. Sander, D. 1974. Rate structure concepts—A comment. *Papers and Presentations, Regulatory Information Systems Conf., St. Louis, Mo., Oct. 1973*
58. Pace, J. D. 1975. The poor, the elderly and the rising cost of energy. *Public Util. Fortnightly* 95:26–30
59. Stelzer, I. 1975. Electric price for the next decade. *Conf. Board Rec.* 12:53–56
60. Hass, J. E., Mitchell, E. J., Stone, B. K. 1974. *Financing the Energy Industry,* A report to the Energy Policy Project of the Ford Foundation. Cambridge, Mass: Ballinger
61. Federal Energy Administration. Feb. 1975. *Monthly Energy Review,* pp. 44–46. Washington DC: Nat. Energy Inf. Center
62. Federal Power Commission. June 17, 1975. *Monthly Fuel Cost and Quality Information,* News release, No. 21479, pp. 11, 17, 21
63. National Association of Regulatory Utility Commissions. July 8, 1974. *Automatic Adjustment Clauses Revisited,* Econ. Pap. No. 1R. Washington DC: NARUC
64. Johnson, L. 1974. The Averch-Johnson hypothesis after ten years. In *Regulation in Further Perspective: The Little Engine That Might,* ed. W. G. Shepherd, T. G. Gies. Cambridge, Mass: Ballinger
65. Nordhaus, W. D. 1973. The allocation of energy resources. *Brookings Pap. Econ. Activity* 3:529–70

Copyright 1976. All rights reserved

ENERGY SYSTEM MODELING AND FORECASTING

✵11016

Kenneth C. Hoffman
Brookhaven National Laboratory, Upton, New York 11973

David O. Wood
Energy Laboratory, Massachusetts Institute of Technology,
Cambridge, Massachusetts 02139

INTRODUCTION

The energy system consists of an integrated set of technical and economic activities operating within a complex societal framework. Energy is a vital component in the economic and social well-being of a nation and must be considered explicitly in the formulation of regional, national, and international policy. As the importance of energy in policymaking has become apparent, research and analysis in the field of energy system modeling and forecasting have grown rapidly. The field has evolved from one almost exclusively the domain of government regulatory agencies and planning groups in the major sectors of the energy industry into one in which many federal and state agencies are active in the development and application of energy models and forecasts. Energy system models are now used extensively for regional, national, and international forecasting and for policy formulation and analysis.

Energy system models are formulated using theoretical and analytical methods from several disciplines including engineering, economics, operations research, and management science. Techniques of applied mathematics and statistics used to implement these models include mathematical programming (especially linear programming), econometrics and related methods of statistical analysis, and network analysis.

The purpose of this review is to provide an introduction to the scope, applications, methodology, and content of energy system models, particularly those developed and used in the United States. In the following sections, we discuss the purpose, scope, and applications of energy system models. The important methodologies used to implement these models are surveyed, a classification of models is provided, and representative models are reviewed. The review is not intended to be exhaustive or to provide a comparative evaluation of models designed for similar purposes. Rather, the models are reviewed to illustrate the structure of recent and current efforts by energy system modelers to provide useful and constructive analytical tools for understanding and solving energy planning and policy problems.

SCOPE OF ENERGY SYSTEM MODELS

The concept of a model usually evokes an image of a complex, computerized system of mathematical equations providing detailed information concerning the operation of the process being modeled. In fact, models may be simple or complex depending upon the purposes for which the model is intended. Simple judgmental models may be most appropriate when monitoring the overall performance of a process. When more detailed information about the process is required and/or when the model is used for planning of complex decision steps, such as the choice of an optimal generation mix for an electric utility, then more complicated models employing theoretical specifications from relevant disciplines and techniques of applied mathematics are appropriate. The choice of theoretical structure and implementation methods and the level of detail required to satisfy the purpose for which the model is being designed represent the art of modeling as distinct from the science of modeling. The first order of business in evaluating any model is to determine the appropriateness of the detail, theory, and implementation methods vis à vis the purposes for which the model was designed.

In addition to the theoretical structure and the implementation methods, energy system models may be characterized by the level of detail and number of processes and activities modeled, whether the model is intended primarily for predictive or normative purposes, the appropriate geographical detail, and the treatment of uncertainty.

The scope of energy system modeling ranges from engineering models of energy conversion processes (e.g. nuclear reactors) or components of such processes to comprehensive system models of the nation's economy in which the energy system is identified as a sector. Although engineering or physical models of conversion processes, ecosystems, etc are energy-related, they are not included in this review. The models or forecasts that are included are best characterized by the coverage of various fuel supplies and demand for them and by the methodology employed. Thus, the scope of the models reviewed includes addressing the supply and/or demand for specific energy forms such as natural gas and electricity, analysis of interfuel substitution and competition in a more complete energy system framework, and analysis of the interrelationships between energy, the economy, and the environment.

Energy system models are employed for both normative or descriptive analysis and predictive purposes. In normative analysis, the primary objective is to measure the impact on the system of changing some element or process that is an exogenous, or independent, event in the model. Predictive models are used to forecast energy supply and/or demand and attendant effects over a particular time horizon. Most models have both normative and predictive capability, and a partition of models into these classes can be misleading. Whenever such a classification is used here, it is intended only to identify the primary objective of the model.

Geographical detail appropriate for a given model again depends upon the purposes for which the model is designed. A model of energy flow in a particular production process is specifically related to the plants in which that process operates. Such a model has no geographical dimension. However, a model of utility electricity distribution has a very explicit regional dimension, which is defined by the market area of the utilities being modeled.

Treatment of uncertainty in a model is an important distinguishing characteristic and is closely related to the implementation methods chosen. Uncertainty may arise because certain elements of the process to be modeled are characterized by randomness, because the process is measured with uncertainty, or because certain variables used as inputs to the model may, themselves, be forecasted with uncertainty. The methods for dealing with these problems are important in evaluating the predictive capability and validating the model. The validation of normative models is quite different from that of predictive models. Since normative models deal with how the energy system should develop, given an objective, the issues of validation deal more with the representation of the structure of the energy system and the accuracy of input parameters. For predictive models, validation includes evaluation of both the model's logical structure and its predictive power. Three levels of predictive capability may be identified. First, there is the ability to predict the direction of a response to some perturbing factor (e.g. a decrease in GNP due to a fuel supply curtailment). A second level of capability involves the ability to predict the relative magnitude of a response to alternative policy action or perturbing factors, and the third level involves the prediction of the absolute magnitude of the response to a perturbing factor. Validation against the requirement of the first two levels is a minimum requirement, and a model may be quite useful even if it cannot be validated at the third level. At both the second and third level, validation of a conditional form is usual, and restrictions on the perturbing factors and their range of availability must be specified. Perturbing events outside the scope of the model, such as acts of God, must, of course, be taken into consideration in evaluating predictive capability.

APPLICATION OF ENERGY SYSTEM MODELS

Energy system models are developed and applied in a wide variety of energy planning and policymaking activities. Before we review specific models, it is useful to classify the types of planning functions and the requirements that must be imposed upon models if they are to be useful in supporting these planning activities.

Ayres (1) provides a useful classification of modes and levels of planning. He defines three levels of planning: policy planning, strategic planning, and tactical or operational planning. Policy planning involves the formulation of goals or objectives and may be accomplished with little regard to technology as long as technical factors do not constrain the selection among alternative goals. Strategic planning concentrates on the development of a set of alternative paths to the desired goals and generally includes the establishment of criteria by which alternative strategies may be evaluated and ranked. Lastly, tactical planning deals with the determination of the steps necessary to implement a particular strategy.

Energy system models provide support at all three planning levels, for regulatory agencies; for industrial planning; for planning, management, and evaluation of R & D programs; and for national energy policy and strategy planning. The objectives of these planning activities and the requirements imposed upon the models are discussed below.

Regulatory Planning

State and federal regulatory agencies are engaged in both operational and strategic planning involving issues related to the regulation of natural monopolies (e.g. electric utilities and gas pipelines, the siting of energy facilities, and public safety). Typically, forecasts of variables such as demand for the regulated product are required for a 5–20-year time horizon corresponding to the life of the facility as a basis of justification of the need for expansion of capacity and new facilities. With the current interest in modification of rate structure as a potential means of controlling energy demand, the role of forecasting models that include price effects will become increasingly important. This feature is also important in the prediction of the response on the supply side to higher prices. For example, the regulation of natural gas prices is now an important energy strategy issue and has stimulated considerable activity on price-dependent supply and demand models for this resource (2).

State regulatory agencies have primary responsibility in the siting of new energy facilities. Although forecasting or predictive models are employed to analyze the need for capacity expansion, the siting problem also requires analysis of a normative or descriptive nature. Analytical models have been developed to evaluate the effect of a given plant on the air and water quality and ecology at a proposed site and have been applied to the problem of optimal plant location, although location is frequently constrained by political and other factors. The latter question involves both physical models of the ecosystem and energy/economic models that permit quantification of the trade-offs among the many attributes of a particular site.

Industrial Planning

The primary planning activity in the energy industries is at the tactical level and involves the scheduling of production levels and the routing of energy products. Petroleum companies, in particular, are faced with a complex transportation problem involving many supply and demand centers with associated storage facilities. Most companies use large optimization models to assist in solving the scheduling and routing problem and to optimize production scheduling.

The energy industry is also active in developing models for forecasting future demand for products and in planning and siting new facilities. In these areas, their energy modeling activity is similar to that outlined for the regulatory agencies. These forecasting and analysis activities are of great importance to the industries in view of the crucial role that they play in the future development of individual companies.

Research and Development Planning

Industry and the federal government are deeply involved in planning, managing, and evaluating energy R & D. Industry-sponsored R & D is generally directed toward near-term applications in response to corporate goals and objectives. The energy R & D sponsored by the federal government, however, is much longer range and involves quite advanced and innovative technologies such as fusion, solar energy, and breeder reactors. The formulation of energy R & D policy at the federal level requires a broad assessment of the technical, economic, and environmental characteristics of new technologies and their potential role in the energy system. Forecasts of the demand for and the supply of energy from established technologies are required with longer time horizons than are necessary for siting and regulatory policy. In addition to long-range forecasting models, descriptive and normative models are required to estimate implementation rates and the competitiveness of new technologies with existing ones. Sophisticated management methods are employed to manage and evaluate the technical programs and involve network planning and scheduling tools.

Strategic Planning and Policy Analysis

The formulation of an integrated national energy policy is currently in progress and is supported by comprehensive modeling and forecasting activity. The nation's energy policy is closely interrelated with economic and social policy and with international developments. Questions of economic growth, balance of trade, and protection of the environment must be considered in a balanced way, and complex trade-offs must be made among these and other national objectives. Once formulated and adopted, a national energy policy must be adaptive and must evolve as conditions and objectives change.

The Project Independence study performed by the Federal Energy Administration (FEA) involved the development and application of a large-scale forecasting energy model (3). Similar policy-oriented studies and associated modeling research efforts are in progress at many research centers. These efforts at modeling and the associated development of data involve the integration of energy system models with macroeconomic and environmental models in order to measure and evaluate the important interactions between economic growth, societal goals as reflected in environmental policies and national goals of energy independence, and future energy market conditions.

METHODOLOGIES

Energy system models are formulated and implemented by the theoretical and analytical methods of several disciplines including engineering, economics, operations research, and management science. Models based primarily on economic theory tend to emphasize behavioral characteristics of decisions to produce and/or utilize energy, whereas models derived from engineering concepts tend to emphasize the technical aspects of these processes. Behavioral models are usually oriented toward forecasting uses, whereas process models tend to be normative. Recent modeling efforts such as the FEA Project Independence model evidence a trend toward combining the behavioral and process approaches to energy modeling in order to provide a more comprehensive framework in which to forecast the condition of future energy markets under alternative assumptions concerning emergence of new production, conversion, and utilization technologies. In part, this trend is the result of recognizing that formulating and evaluating alternative national eneigy policies and strategies require an explicit recognition of technical constraints.

Methods for implementing energy models include mathematical programming (especially linear programming), activity analysis, econometrics, and related methods of statistical analysis. Process models are usually implemented using the programming techniques and/or methods of network and activity analysis, whereas the behavioral models use statistical methods. The remainder of this section provides an introduction to these techniques and is intended to provide background information for the review of specific models presented in the next section.

Mathematical programming has been used in energy system modeling to capture the technical or engineering details of specific energy supply and utilization processes in a framework that is rich in economic interpretation. In mathematical programming, series of activity variables are defined representing the levels of activity in specific processes. These are arranged in a series of simultaneous equations representing, for example, demand requirements, supply constraints, and any other special relationships that must be defined to typify technical reality or other physical constraints that must be satisfied. An objective function to be minimized or maximized must be specified (usually cost, revenue, or profit); there are many algorithms available to solve very large problems (up to approximately 10,000 variables). The methodology of linear and nonlinear programming and numerous practical applications are described by Dantzig (4) and Wagner (5).

The linear programming (LP) technique has been used far more than other mathematical programming methods because of its efficiency in solving large problems. Nonlinear relationships may be captured in such models by using piecewise linear or step function approximations. Nonlinear and dynamic programming techniques are also used for special purposes.

The mathematical programming methodology has especially interesting and useful economic interpretations. A dual problem formulated in terms of prices is associated with any LP problem formulated in quantities. The solution to the quantity optimization problem yields both the optimal activity levels in physical terms and the prices that reflect the proper valuation of physical inputs to the real process represented by the model. Important information concerning the economic interpretation of the solution is provided. Thus the LP technique provides a natural link between process and economic analysis.

Mathematical programming models and related optimization techniques such as calculus of variations and LaGrange multipliers are generally classified as normative techniques since they presume the existence of an overall objective such as cost minimization or profit maximization. It is possible to reflect multi-objective criteria as some weighted combination of objectives, and, indeed, some objectives such as environmental control can be expressed through special constraint equations in the model. Nevertheless, the validity of this technique as a predictive tool depends on the ability to capture and represent the objectives of the players in various sectors of the energy system and in those sectors of the economy and society that affect the energy sectors. The technique is normative in that it determines optimal strategies to achieve a specific objective with a given set of constraints.

Interindustry techniques are frequently employed in energy modeling, primarily for descriptive purposes.[1] Primary data for constructing interindustry sale and purchase accounts are collected by the government on the transactions between various sectors of the economy (agriculture, ferrous metals, electricity, oil, retail trade, etc). These data are expressed in terms of a common unit, the dollar, and are available for all census years for the period 1947–1967 (6).

The input-output approach has been adapted to energy studies by converting the inputs from the energy sector to other industry sectors from dollar flows into energy units such as the British thermal unit (7, 8). In this format, the direct energy inputs from the oil sector to the agriculture sector, for example, are specified in the input-output matrix. The interindustry flow table may be converted into a coefficient table measuring the quantity of input required from one sector per unit of output for another sector. The coefficient matrix represents a model of the production process.

Important assumptions of the model include fixed technology and zero price elasticity since input proportions are assumed independent of relative prices. With these assumptions, the model may be used to estimate the total direct and indirect energy requirements necessary to produce a given level of final demand.

The input-output approach is limited by the difficulties and time delays in assembling the interindustry flow data and by the apparent restrictiveness of the key assumptions. As noted, the table currently available for the United States is based upon 1967 data. Recently, the Department of Commerce, the agency responsible for assembling and publishing interindustry accounts, has initiated a program to develop and publish annual updates to the input-output table in order to provide more recent information. Regarding restrictiveness of assumptions, Hudson & Jorgenson (9) have used econometric techniques to implement an interindustry energy model for which the input-output coefficients are explicitly a function of the relative prices of all inputs. These important developments will significantly increase the utility of input-output analysis for energy analysis.

Econometrics is concerned with the empirical representation and validation of economic theories and laws.[2] Econometric methods involve the application of statistical techniques to estimate the structural parameters of one or more equations derived from economic theory and to test hypotheses concerning these parameters. The appropriate method for a particular estimation problem will depend on the assumptions concerning the statistical properties of the process generating the observed data.

The principal method of econometrics is regression analysis. The regression model combines the economic model derived from theory with a statistical model of the process from which the observed data are assumed to be generated. Statistical methods may be used to test hypotheses concerning the assumptions of the statis-

---

[1] Input-output analysis originated with Leontief, for which he was recently awarded the Nobel prize. See (5a) for the original development and (5b) for a recent compendium of research in input-output analysis.

[2] There are many excellent econometrics textbooks including (10, 11). An advanced treatment is given in (12).

tical model as well as hypotheses concerning the economic model. Examples include testing the hypotheses that a particular parameter is not significantly different from zero, that parameters in different equations of the model are not significantly different, or that combinations of parameters are equal to some specific value.

Econometric methods are used in modeling two types of energy processes: behavioral and technical processes. Behavioral processes are characterized by a decision-making agent hypothesized to adjust behavior in response to changes in variables outside his direct control. For example, one could hypothesize that a household would distribute its expenditures between energy and other types of goods and services on the basis of its income and wealth and the relative prices of energy and the other products and that the distribution would be consistent with some household objective function.

Technical processes are characterized by purely technical relations. An example would be the production function of a firm in which maximum potential output is a function of the quantities of inputs available such as capital, labor, energy, and other material inputs. Given a suitable functional form for this relationship and observations on capacity output and associated inputs, econometric methods could be used to estimate the parameters of the function. Alternatively, the technical relation might be used to derive behavioral relations concerning the firm's demand for input factors at a given output level.

Econometric and engineering/process methods are sometimes alternative approaches to modeling technical processes. An example of the two approaches to modeling the supply of electricity in the United States is provided by the work of Griffin (13) and Baughman & Joskow (14); Griffin used an econometric approach and Baughman & Joskow employed an engineering/process approach. These two models are reviewed in the next section and are illustrative of the contrasting characteristics of each approach.

The system dynamics approach (15) evolved from studies of specific industrial operations to global applications. It involves simultaneous linear and nonlinear equations used to represent functional relationships between parameters of interest in a problem. Both flow and stock variables may be represented, and feedback relationships may be taken into account. The technique was developed in the engineering field and is quite powerful. A major difficulty in large-scale models has been the development and verification of the functional relationships used and the interrelationships and feedback mechanisms represented by the model structure.

The techniques of game theory (16) are receiving increased attention in energy modeling. This is particularly true when decisions to be modeled are made on a basis other than optimization or market equilibrium. This methodology has promise for modeling some aspects of international trade of energy resources, fuels, and energy-intensive products. Game theory deals with the quantification of the outcomes of interactions between two or more players with each player having options that can affect the outcome. With a given uncertainty of the strategy to be used by another, the selection of a best strategy includes consideration of the relative payoffs and risk aversion by the players.

REVIEW OF ENERGY MODELS

This section includes a representative sample of models that have been developed and applied to analysis of the energy system and to the development of forecasts for planning purposes. The objective is to emphasize the assumptions and methodologies of selected models rather than to provide an exhaustive review of all models. The selection of models is somewhat arbitrary and does not imply any superior capabilities in comparison with other models of the same generic class that are not discussed.[3]

The energy models are discussed in several groups according to their scope, and they range from supply-oriented models of a single fuel to models encompassing the overall energy system coupled to the economy. The four major groups of models and forecasts reviewed here are

1. *sectoral models*, covering the supply or demand for specific fuels or energy forms,
2. *industry market models*, which include both supply and demand relationships for individual or related fuels,
3. *energy system models*, which encompass supply and demand relationships for all energy sources,
4. *energy/economic models*, which model the relationships between the energy system and the overall economy.

No single categorization of energy models can represent all of the important characteristics; this classification is intended to highlight the scope of particular models. In each of the above groups, both activity- or engineering/process-oriented models and econometric models are used extensively. These approaches are frequently combined to capture the strengths of the process technique in representing technical detail and the ability of econometric methods to represent aspects of behavior.

Sectoral Models

Sectoral models are defined as relating to some specific energy process or activity forming a part of a specific energy industry market. Typically, models in this category focus on either the supply or the demand side of the market. Process models are used most often for characterizing energy supply and capacity expansion, whereas econometric models are used to characterize demand.

---

[3] Several general reviews of energy models have been completed and others are being compiled. Anderson (17) surveyed mathematical programming approaches to analysis of the electric sector, and Decision Sciences Inc. (18) conducted a survey of energy models under a contract with the Council on Environmental Policy. Broad surveys of energy system models are now underway at the International Institute of Applied Systems Analysis in Laxenburg, Austria (19). Several major conferences have been held on energy modeling, and the proceedings (20-22) of these conferences include detailed descriptions of a variety of energy models. Taylor (23) has completed a review and evaluation of demand models for electricity.

Process-oriented supply models have been developed and applied most extensively to the analysis of oil refining and transportation operations. The refining of crude oil involves a series of unit operations including simple distillation, cracking of heavy molecules into lighter fractions by a number of techniques, hydrogenation, and desulfurization. The yield of lighter fractions such as naphtha and gasoline can vary from 30% to over 80%, depending on the process employed and the characteristics of the crude. Oil refineries are generally designed to handle a specific type of crude oil; however, the demand for major oil products such as naphtha for petrochemicals, gasoline motor fuel, light distillate fuel for turbines and heating, and residual oil for power generation varies on a seasonal and shorter-term basis.

The international scope of the transportation and allocation of crude oil and refined products provides an important application for energy modeling. The characteristics of crude oils vary by gravity (density) and by the level of sulfur and other contaminants they contain. The allocation of this crude to the appropriate refineries and of the refined products to storage and demand centers is optimized by many oil companies using the LP technique. Although many of these are proprietary, the Energy Research Unit at Queen Mary College under the direction of R. I. Deam has published its model (5c). This model is global in scope and includes 25 discrete geographical areas. Fifty-two types of crude oil and 22 refining centers are represented along with 6 types of tankers that may be selected for transport.

The model includes the following refinery processes and products:

Processes
1. Crude distillation unit
2. Vacuum distillation unit
3. Alkylation
4. Catalytic reforming
5. Desulfurization
6. Hydrocracking
7. Catalytic cracking
8. Desulfurization
9. Coking
10. LNG regasification
11. SNG production

Products
1. Liquid petroleum gas
2. Motor spirits (gasoline)
3. Petrochemical feedstocks (naphtha)
4. Kerosene
5. Gas oil
6. Residual fuel oil
7. Bitumen
8. Coke

The LP matrix for this model is quite large (about 3500 rows and 13,500 columns). The exogenous inputs to the model include future demand for products by region, refinery technology, costs of product refining, and transport of specific crudes and products. The model is solved to determine the optimal allocation and routing of crude oil and products between sources, refineries, and demand centers at some future target date. The requirements for new refineries, tankers, and production facilities to satisfy the projected level and distribution of demands are also determined. Because the model includes the transport and refining costs for crude from specific sources, it provides a basis for analyzing the relative price of these crudes in a competitive market or in a controlled market where relative prices are set to reflect the differences in transportation and refining costs among the many sources.

Most sectoral econometric modeling efforts in the energy area have focused upon the demand for a single energy input in one particular use. Such models are used principally to provide an analysis of the determinants of demand and to forecast demand with given estimates of the variables that are exogenous to the model, including price and other variables measuring the market size for the energy inputs (population, GNP, income, etc). These models have been designed to focus on specific policy issues such as gasoline tax policy. Since they are limited in scope, they generally do not have broad policy applicability.

Taylor (23) recently surveyed and evaluated econometric models of the short- and long-term demand for electricity in the residential and commercial sectors. The models surveyed are classified according to regional detail and the measure of electricity price used, and short- and long-term prices and income elasticities are summarized. Taylor reviewed the special problems associated with modeling the demand for electricity, including the fact that such demands are derived demands depending on the stock and utilization rates of equipment, the characteristic of fluctuating utilization rates for the equipment (peak demands), and the effects of the regulatory process on the pricing schedules. Taylor concluded that, to varying degrees, modeling efforts have not yet dealt adequately with these problems (especially, the problem of incorporating the appropriate price schedule). Table 1 summarizes classification information and elasticities for the models surveyed by Taylor and for other electricity demand models published since his review was completed.

Sweeney (35) has developed a model of the demand for gasoline in order to support analysis of conservation policies affecting automobiles. Vehicular gasoline consumption for any time period is a derived demand that depends on the total number of miles driven and the average number of miles per gallon (mpg) for the fleet in operation during the period. The demand for vehicle miles is estimated by a function of real disposable income per capita, the unemployment rate, and the cost per mile of automobile travel, including the cost of gasoline and time (permitting introduction of speed limits). The average mpg for the fleet is first estimated by prediction of new car purchases per capita as a function of lagged automobile purchases per capita, total vehicle miles per capita, real disposable income per capita, and the unemployment rate. A sales-weighted average mpg of new cars is estimated by a function of automobile efficiency and the price of gasoline. The mpg for the fleet is then estimated by formation of a weighted harmonic mean of the mpg estimates for new cars and each vintage of old cars where the weights are the shares of each vintage in the total vehicle miles demanded.

Other models of the demand for gasoline have been developed by Lady (36), Verleger (37), McGillivray (38), Windhorn et al (39), and Adams, Graham & Griffin (40). The long- and short-term price and income elasticities for each of these models, the measure of gasoline prices used, and the type of data are summarized in Table 2.

Table 1 Summary of price and income elasticities for models of electricity demand[a]

Type of Demand	Type of Price	Price Elasticity Short-Run	Price Elasticity Long-Run	Income Elasticity Short-Run	Income Elasticity Long-Run	Type of Data
Residential						
Houthakker (24)	M	−0.089	NE	1.16	NE	CS: cities (United Kingdom)
Fisher & Kaysen (25)	A	≈ −0.15	≈ 0	≈ 0.10	small	CS, TS: states
Houthakker & Taylor (26)	A	−0.13	−1.89	0.13	1.94	TS: aggregate United States
Wilson (27)	A*	NE	−2.00	NE	≈ 0	CS: SMSAs[b]
Mount, Chapman & Tyrrell (28)	A	−0.14	−1.20	0.02	0.20	CS, TS: states
Anderson (29)	A*	NE	−1.12	NE	0.80	CS: states
Lyman (30)	A	(≈ −0.90)		(≈ −0.20)		CS, TS: areas served by utilities
Houthakker, Verleger & Sheehan (31)	M	−0.09	−1.02	0.14	1.64	CS, TS: states
Griffin (13)	A	−0.06	−0.52	0.06	0.88	TS: aggregate United States
Commercial						
Mount, Chapman & Tyrrell (28)	A	−0.17	−1.36	0.11	0.86	CS, TS: states
Lyman (30)	A	(≈ −2.10)				CS, TS: areas served by utilities
Industrial						
Fisher & Kaysen (25)	A	NE	−1.25			CS: states
Baxter & Rees (32)	A	NE	−1.50			TS: industries (United Kingdom)
Anderson (33)	A	NE	−1.94			CS: states
Mount, Chapman & Tyrrell (28)	A	−0.22	−1.82			CS, TS: states
Lyman (30)	A	(−1.40)				CS, TS: areas served by utilities
Baughman & Zerhoot (43a)	A	−0.11	−1.28			CS, TS: states
FEA (3)	A	NE	−1.33			TS: aggregate United States

Table 1 continued

Residential and Commercial					
Baughman & Joskow (34)	A	−0.13	−1.31	0.08	0.52	CS, TS: states
FEA (3)		NE	−0.44		1.90	TS: aggregate United States
Industrial and Commercial					
Griffin (13)	A	−0.04	−0.51			TS: aggregate United States

[a] Based on Taylor (23, Table 4), with the additional entries for Griffin (13), Baughman & Joskow (34, p. 24), and FEA (3). Abbreviations used are NE, not estimated; TS, time-series; A, ex post average price; CS, cross-section; M, marginal price; A*, average price for a fixed amount of electricity consumed per month.
[b] Standard Metropolitan Statistical Areas.

Table 2 Summary of price and income elasticities for selected models of gasoline demand[a]

Model	Price Elasticity		Income Elasticity		Measure of Gasoline Price	Data
	Short	Long	Short	Long		
Sweeney (35)	−0.12	−0.72	0.85	0.78	retail excise taxes	annual TS: United States
Lady (36)	−0.16	—	0.48	—	retail excise taxes	monthly TS: United States
Verleger (37)	−0.16	−0.54	0.32	1.06	retail excise taxes	quarterly TS, CS: states
McGillivray (38)	−0.22	−0.69	—	—	retail including taxes	annual TS: United States
Windhorn et al (39)	−0.14	−0.93	—	—	CPI, gas and oil	annual TS: United States
Adams et al (40)	−0.90	−1.5	0.5	1.0	retail including taxes[b]	

[a] See Table 1, footnote a, for abbreviations. CPI, consumer price index.
[b] Cross section for 20 OECD countries for the year 1969.

Industry Market Models

Models for energy industry markets include process and econometric models as well as integrated process/econometric models, which characterize both the supply and the demand for a specific or related set of energy products. Such models are very useful and are applicable to all energy-use categories, although the greatest utility is in providing a consistent framework for planning industrial expansion and studying the effects of regulatory policy on the industry. Much of the modeling work in this area involves the coupling of process and econometric techniques to exploit their strength in representing supply and demand relationships, respectively.

Adams & Griffin (41) combined an LP model of the US refining industry with econometric equations determining endogenously the prices, quantities demand, and inventory adjustments for the major petroleum products. Exogenous inputs to the econometric/LP model are the refining process configurations, product quality specifications, factor input prices (crude oil, etc) economic activity, and the stocks of petroleum-consuming equipment. In the first step, the requirements for the various petroleum products are determined in the demand equations. Using these requirements as output constraints, the solution to the LP model indicates the volume of crude oil required, process capacity utilization, operating costs, and outputs of by-products such as residual oil. In turn, capacity utilization, inventory levels, and crude oil prices determine the product prices. The structure of the model indicating the relationship of the macroeconomic model with the linear program is shown in Figure 1. The LP model has 227 equations and 334 variables. The combined econometric/LP model was applied to a sample period, 1955–1968, and it traced the development of the industry over that period with good accuracy.

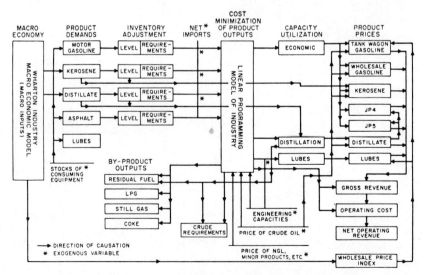

Figure 1 Structure of Adams-Griffin petroleum industry model (41).

No provision was made to reflect technological change in the LP production functions, but statistically estimated adjustments were made to the crude oil inputs to account for the implementation of more advanced refineries.

Mathematical programming has been used extensively in the analysis of electric utility operations and expansion plans. The review by Anderson (17) includes a description of over 50 models used in that industry. Methodologies discussed in the Anderson review cover marginal analysis and simulation and global models using dynamic programming, linear programming, and nonlinear programming. In the electric utility optimization models, the demands for electricity and prices of fuels and facilities are usually exogenous inputs, and the models are used to select the least-cost investments to satisfy increased demands. The output generally includes specification of the type of plants to be built (nuclear, coal, oil, hydro, gas turbine, etc), location of plants for models that include regional definition, replacement, and the scheduling of plants on a weekly and/or seasonal basis.

A systems dynamics model of the coal industry is being developed by Naill, Miller & Meadows (42) to study the role of coal in the transition of the US energy system from oil and gas to renewable resources up to the year 2100. The interrelationships in the coal production sector between demand, investment, labor, and production are modeled along with the oil and gas sector and the electric sector. Time delays associated with R & D and plant construction are included in the synthetic fuels sector where liquid and gaseous fuels are produced from coal. The demand for energy and the market shares of various fuels are determined endogenously as a function of price, GNP, and population. These variables are exogenous to this model, although in more comprehensive systems dynamics models they are also determined endogenously.

MacAvoy & Pindyck (2) developed an econometric policy simulation model of the natural gas industry. The model has been used extensively to analyze the effect on the industry of current and proposed federal regulation of the wellhead price of gas and permissible rates of return for pipeline companies purchasing and selling natural gas in interstate markets (2, 43). The model focuses on the supply of reserve additions and the demand for gas by pipeline companies for sale in wholesale markets. The supply of additions to gas reserves in any period is the sum of new reserves discovered and extensions and additions to reserves. New reserves discovered in a producing region are the product of wells drilled, the proportion of successful wells, and the average size of find. New discoveries of both oil and gas are estimated since, in exploration and development activity, oil and gas are joint products.

An important feature of the MacAvoy-Pindyck model is that the drilling projects initiated depend on driller choices between the intensive and extensive margin. The extensive margin refers to projects in new fields with lower probabilities of success but larger expected size of find if the project is successful. The intensive margin refers to projects in known fields with higher probabilities of success and correspondingly smaller expected size of find (since, presumably, the better projects will have been drilled first). This choice is modeled as a function of economic costs and as a measure of the risk averseness of the producer. The average success ratio for projects initiated is a function of this drilling choice. The size of discovery incorporates the effects of geological depletion by depending negatively on the number of previous successful wells in the region, since better prospects are likely to be drilled first, and positively on higher gas prices, since this shifts the producer's drilling portfolio toward the extensive margin. The model also estimates changes in reserves due to extensions and revisions, thereby providing a complete reserve-accounting framework.

Production from reserves depends on the reserve base and the field price. The marginal cost of production depends on the reserve level relative to the production level. Lower reserve-to-production ratios imply higher marginal costs, with the regulated price setting an upper bound for marginal cost and, therefore, possible production levels.

Demand for natural gas by industrial, residential, and commercial customers depends on the wholesale price of gas, the prices of alternative fuels, and market-size variables such as population, income, and investment levels. The wholesale price of gas is a function of the wellhead price and a pipeline markup that depends on operating and capital costs and the regulated profits of the pipeline companies. The wholesale markets are also defined on a regional basis. The flows of natural gas between producing and consuming regions are estimated by using a network model characterized by an input-output table of flow coefficients between each of the producing and consuming regions. The difference between the production flows and demand levels in the consuming regions is a measure of the excess demand for natural gas in each region.

Griffin (13) has developed an econometric model of the supply and demand for electricity. The model is estimated by using national time-series data. Major variables determined by the model include the demand for electricity in the residential, industrial, and commercial sectors; nuclear capacity expansion; distribution of generation requirements between nuclear, oil, gas, and coal; and the price of electricity.[4] Important exogenous variables include various measures of market size such as population, real disposable income, GNP, the price of oil, gas, and coal, the GNP deflator, total generating capacity, construction costs, and other operation costs. The model is simultaneous because the average price of electricity, a determinant of demand, depends on the generating mix. The model has been used to conduct simulation studies of the impact on demand and the generating mix of alternative projections of relative fuel prices.

Baughman & Joskow (14) have developed an engineering/econometric model of electricity supply and demand. The model combines an engineering supply model with an econometric demand model and links the two with an explicit model of the regulatory process by which the price of electricity is determined. The supply model for electricity is regional, encompassing the nine census regions. Each region is assumed to have eight potential plant types available, and a ninth type, hydroelectric, is treated as exogenous. The plant types are

1. Gas turbines and internal combustion
2. Coal-fired thermal
3. Gas-fired thermal
4. Oil-fired thermal

[4] The income and price elasticities for the Griffin model are presented in Table 1.

5. Light-water uranium reactors 7. Plutonium recycle reactors
6. High-temperature gas reactors 8. Liquid-metal fast breeder reactors

The model characterizes the decision process by which operation and expansion of the electricity supply system takes place based on cost minimization techniques employed by the industry. The econometric demand model is based on a state classification of data. Demands for electricity, natural gas, coal, and oil are estimated for the residential, commercial, and industrial sectors by functions of fuel prices and various market-size variables.[5] The price of electricity is controlled by state regulatory agencies. Transmission and distribution requirements and costs are estimated by using an econometric approach. The procedure is first to estimate requirements for five types of transmission and distribution equipment and then to estimate the maintenance and operating costs as a function of the installed capacity for these five types of equipment. The Baughman-Joskow model simulates the process by which electricity prices are determined based on calculations of the rate base derived from inputs from their supply model and assumptions about the rate of return permitted by the regulatory agency, the rate of depreciation, and the effective tax rate.

The model takes as exogenous fuel and other operating costs, as well as construction costs and plant-operating characteristics. The electric power supply industry is assumed to expand to meet expected demand based on an exponentially weighted moving average with time adjustment of recent actual demands. This expected demand projection will, of course, differ from the actual consumption in any given period. Adjustments in operating capacity due to differences between projected and actual demand are assumed to take place in future optimizing decisions. Generation requirements by plant type are calculated by using an estimate of the load duration curve (percentage of time that load equals or exceeds a given output level) and a merit-order ranking of plant types by operating and fuel costs. Since the load duration curve and merit-order ranking are independent of the projected demand, the fact that projected and actual demand may differ will not affect the order in which capacity is utilized to meet actual demands.

An international, multi-commodity model of the world oil market has been developed by Kennedy (44). Activities in the model include production of crude oil, transportation, refining, and consumption of refinery products. Refinery products include gasoline, kerosene, distillate fuel, and residual fuel. A seven-region classification of crude oil production and product consumption is used, including the United States, Canada, Latin America, Europe, the Middle East, Africa, and Asia. The exogenous variables in the model are the regional crude oil supply equations, the product demand equations, refining technology, and governmental policy. Potential governmental policies that might be reflected in the model include tariffs, excise taxes, environmental restrictions, refinery subsidies, and policies intended to shift the supply curves or demand curves through tax measures or direct regulation. For example, the OPEC tax on production can be analyzed within the

---

[5] The income and price elasticities for the Baughman & Joskow model are presented in Table 1. The model for the residential and commercial sector is described in Baughman & Joskow (34), while the industrial sector model is discussed in Baughman & Zerhoot (43a).

framework of this model. The model determines consumption of products, production of crude oil, equilibrium prices, refinery capital structure, and refinery output by region, as well as world trade flows of crude oil. The demand and supply functions are static functions. The author suggests that the model is most useful in analyzing equilibrium conditions in a 5- to 15-year time period. The world oil market has been used to simulate the effects of alternative Persian Gulf royalty fees for the year 1980. The model provides a convenient framework for simulating other governmental policy actions, both within the Persian Gulf and by other regions reflected in the model.

Energy System Models

Analysis and modeling of the overall energy system, including supply and demand sectors as well as all fuels and energy forms, were stimulated largely by the need to develop forecasts of total energy demand. Much of the initial work in this area involved the development of overall energy balances for the United States in which forecasts for individual fuels were assembled. These forecasts highlighted many problems involving such factors as resource definition and interfuel substitution, which must be handled in a consistent manner for all fuel types and sectors and which led to increased modeling of the entire energy system.

One of the first systematic attempts to account for all energy flows in a consistent manner was that of Barnett (45). Barnett's approach involves obtaining a national energy balance of energy supplies and demands by type. The emphasis was on quantity flows expressed in physical units and a common unit, the Btu. This approach has been extended and refined by Morrison & Readling (46) and by DuPree & West (47). As an accounting approach, the energy balance system focuses attention on a complete accounting of energy flows from original supply sources through conversion processes to end-use demands. The approach accounts for intermediate consumption of energy during conversion processes as well as efficiencies at various points in the energy supply system.

The energy balance methodology has been employed in forecasting studies in the following way. Independent estimates of demand by each of the major end-use sectors for each of the detailed energy types are developed by relating demand to aggregate economic activity and trends in energy consumption. Independent estimates of supply of major energy types are developed and compared with the demand estimates. Differences are resolved, usually in a judgmental way, by assuming that one energy type is available to fill any gap that may exist between supply and demand. This energy type is normally assumed to be imported petroleum, including crude oil and refined petroleum products. The DuPree-West (47) study provides an excellent example of the execution of a forecast employing this methodology.

The National Petroleum Council (NPC) also employs the energy balance approach in developing forecasts of expected energy consumption. The NPC models (48, 49) use econometric techniques to forecast energy demands and engineering and judgment models to forecast supplies. However, the forecasts from the models are substantially modified by judgmental information provided from the various working groups of the NPC. The energy balance framework is used to ensure the consistency of the various component forecasts. An important feature of the NPC approach is that it permits incorporation of subjective, specialized industrial information into an energy balance framework and thus provides an important source of industrial expectations about future energy markets.

The process type of energy system model encompasses all alternative fuels and energy sources and frequently employs network analysis in order to represent technical detail and to capture the interfuel substitution possibilities. The network is used to represent the spatial or interregional flows of energy as well as the alternative processes and fuels that may be used in specific demand sectors. This representation of the energy system may be augmented with optimization or simulation techniques or used simply as a framework to exhibit information and options.

The model developed by Baughman (50) to study interfuel competition uses systems dynamics to simulate the flow of resources (coal, oil, natural gas, nuclear fuel) to the various demand sectors (residential, commercial, industrial, transportation, and electricity). The model has been applied at the national level but might also be formulated at the regional level. The model includes representation of the economic cost structure of the energy system along with investment decisions and physical constraints on the supply of coal, oil, natural gas, and nuclear fuels. Demands are developed in two components, a base demand that is not sensitive to price, and a market-sensitive demand that includes incremental and replacement demands. The model is used to simulate interfuel competition and to develop the quantities and prices of fuels and energy sources that are used over time as demands for, and the availability and cost of, resources change. The model has been used to develop projections of oil and gas use as relative prices change; it is being extended to address regional analyses.

The *reference energy system* approach was developed by Hoffman (51) and was applied to the assessment of new energy technologies and policies. This approach gives a network description of the energy system in which the technical, economic, and environmental characteristics of all processes involved in the supply and utilization of resource and fuels are identified. All steps in the supply chain (the extraction, refining, conversion, storage, transmission, and distribution activities) are included along with the utilizing device (combustor, air conditioner, internal combustion engine, etc). A reference energy system representing a detailed projection of historical growth to the year 1985 is shown in Figure 2. Each link in the network corresponds to a physical process and is characterized by a conversion efficiency, capital and operating cost, and emissions of air and water pollutants per unit of energy input. The system is used to evaluate the role of new technologies and the possibilities of interfuel substitution. Substitution is heavily dependent on the characteristics of utilizing devices, and these devices are represented in the network for all functional end uses such as space heating, air conditioning, and automotive transport. The resource, economic, and environmental impacts of new energy technologies are determined by inserting them into the reference system at appropriate levels and efficiency and recalculating the energy flows, cost, and emissions.

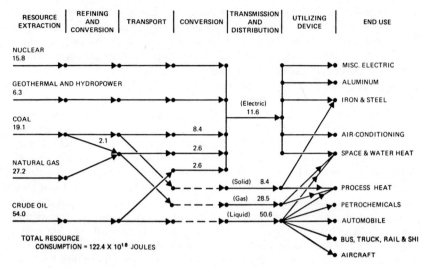

Figure 2 Reference energy system—historical growth projection to 1985 (52). Flows in 10^{18} joules. Solid line indicates real process.

The reference energy system is supported by an energy model data base (53), which includes data elements for approximately 600 supply processes and 200 functional end uses. Both simulation and optimization models may be used to develop and quantify the reference energy system.

Cazalet (54) of Stanford Research Institute developed a network model of the US energy system, which is solved by a successive approximation algorithm using decomposition techniques. The model does not use an explicit optimization approach but looks for market equilibrium between energy supplies and demands using the marginal cost of resources and cost of supply and conversion technologies as the basis for energy prices. The network includes 30 supply regions and 8 demand regions and covers 17 time periods of varying length with a time horizon to the year 2025. The demand sectors are defined on a functional basis as in the reference energy system. The input parameters include supply and demand curves for all regions where the quantities of fuels that would be forthcoming are specified as a function of price. The characteristics of conversion and delivery technologies are represented in the model, and interregional transportation costs are identified. The important output information provided by the model includes regional prices for fuel by region and time period, resource production levels, interregional flows, and demands for fuels. The model has been applied by Gulf Oil Company to the analysis of synthetic fuel strategies, specifically, an analysis of the coal gasification option in the Powder River Basin of Montana and Wyoming. The model is also being used by the Council on Environmental Quality for analysis of western energy resource economics.

Debanne (55) has developed a network model of a North American energy supply and distribution system. It accounts for physical interregional flows where the nodes in the network may represent oil, gas, coal, hydro, and nuclear conversion centers. The arcs represent pipelines and other appropriate transport facilities. The model determines optimal locations and expansion of capacity to satisfy increased regional demands. The model takes interfuel substitution into account and includes the interactions of price with supply and demand.

A number of LP models that are similar to those employed for optimization of the generating mix in the electric sector have been developed for the analysis of the complete energy system including both the electric and the nonelectric sectors. The Brookhaven Energy System Optimization Model (BESOM) (52, 56), developed by Hoffman & Cherniavsky, was designed to determine the optimal allocation of resources and conversion technologies to end uses in the format of the reference energy system. This model focuses on the technical structure of the energy system including the conversion efficiencies and environmental effects of supply and utilizing technologies. It is currently applied at the national level. The model may also be formulated for regional or interregional analysis. A wide range of interfuel substitutability is incorporated in the model, and the load-duration structure of electrical demands may be expressed. The model is quantified for a future point in time. The energy sources compete in the optimization process to serve specific functional demands such as space heat, petrochemicals, and automotive transport. The energy demands to be satisfied and the constraints on specific energy sources and environmental effects are specified exogenously. These may be inputted as either fixed or price-sensitive constraints. The energy sources provided in the model include a number of alternative central-station electric systems, general-purpose fuels delivered directly to the consumer, and special systems such as solar energy and decentralized electric generators. The optimization may be performed with respect to dollar cost, social cost, environmental effects, resource consumption, or some combination of these factors. The model has been applied to study the optimal implementation mode for new energy technologies, break-even costs for new technologies, and strategies of interfuel substitution to conserve scarce resources.

A time-phased LP version of the BESOM model has been assembled by Marcuse & Bodin (57), and it incorporates the same technical detail and constraints but treats plant expansion and capital requirements in an explicit manner. The inputs necessary to drive the model include energy demand requirements in each future time period, initial plant capacity (existing capital stock) and maximum permissible growth rate for each conversion process, maximum permissible growth for the extraction and supply of each energy resource, and a discount rate. Solution of the model determines resource usage, activity level of each conversion process, and new capital facility requirements in each time period. It has been applied to cost-benefit analysis of new energy technologies and to the determination of the optimal use of scarce resources over time.

Time-phased LP models have also been developed by Nordhaus (58) and Manne (59). The Nordhaus model covers five regions of the world and nine time periods and includes all major competing resources. The backstop technology that is introduced provides a long-term substitute of possibly higher cost but almost infinite availability, which can be used to replace scarce or finite resources when they run out. This backstop technology has been taken to be the nuclear breeder reactor producing electricity and hydrogen for electric and nonelectric demands, respectively. The cost and efficiencies of all resources and technologies are reflected in the model along with demand and resource constraints. This model has been used to study the optimal allocation of scarce resources over time and, specifically, to evaluate current fuel production costs and the scarcity cost premium associated with the requirement that a more costly form of energy must be substituted at some future time for any scarce resources that are used at an earlier date.

The Manne model (59) is formulated as a single region model. For the problem of resource exhaustion, there are only two energy demand sectors, electric and nonelectric. Coal is viewed as a source of both electric energy and synthetic fuels. Several options for nuclear reactors are represented including the light-water reactor, the high-temperature gas-cooled reactor (HTR), and the fast breeder reactor. The fuel cycles for these systems are coupled, and the high-temperature reactor is viewed as a source of process heat (e.g. for thermochemical hydrogen). This model is also time-phased covering the period 1970–2030 in three-year time steps. It has been applied to determine the benefits of the fast breeder reactor as a source of electricity and of bred nuclear fuel for the high-temperature gas-cooled reactor, which in turn is used to

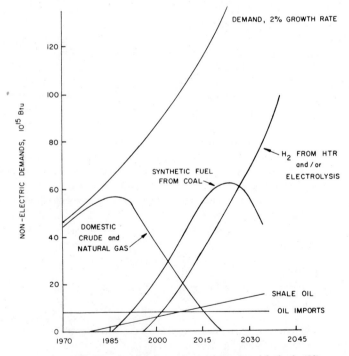

Figure 3 Nonelectric energy demand trend and fuel mix (59).

produce hydrogen as a substitute for scarce oil and gas resources. The benefits of this technology were evaluated under various assumptions regarding the availability of petroleum imports and domestic sources. An example of the output from the model is shown in Figure 3, and it illustrates the time-phasing of various fuels used to satisfy the demands for nonelectric end uses. These end uses are specified as either exogenous or price-responsive. Similar curves are generated for the fuel mix in the electric sector.

Integrated Energy/Economic Models

There is increasing research activity in the coupling of energy system models with models of the overall economy such as macroeconomic and input-output models. Many of the sectoral and energy system models discussed previously require that the energy demands be specified exogenously as input parameters. Of course, such demands must be related to trends in society (households, transportation pattern, etc) and the economy (population, GNP, industrial production, and so on). This requirement has led to the extensive use of regression analysis and other macroeconomic modeling techniques in order to generate demand levels and other inputs to the process-oriented models. Such coupling is relatively straightforward, however, and the major research activity in combined models is more fundamental in nature. The coupled energy/economic models reviewed here involve those that are used for analysis of the role of energy as a driving force and constraint on economic development. They involve a more integral relationship between the energy sector and the economy. In particular, the recognition that the cost and availability of natural resources has significant near- and long-term implications for the economy has stimulated this modeling activity.

The process-oriented energy/economic modeling has emphasized the use of input-output techniques. Herendeen (7) developed energy coefficients in physical units for coal, crude oil and gas extraction, refined oil, electricity, and gas sales in the 367-sector input-output matrix of the Bureau of Economic Analysis. The direct energy coefficients represent the Btu inputs per dollar of total output of each sector of the economy. This model is operational at the University of Illinois and has been used to analyze the energy inputs, both direct and indirect, to different products and activities.

Just (60) has developed a two-period, dynamic, 104-sector, input-output model. In this model, technological coefficients were developed for new energy technologies to analyze the expansion required in specific industries to support the implementation and expansion of new technologies such as coal gasification plants and gas turbine cycles.

Input-output analyses of the type developed for energy studies provide the basis for energy accounting. Many calculations have been performed of the energy inputs to capital projects, including the construction of nuclear power plants, shale oil facilities, and solar systems to determine how long the facilities must operate to return the energy invested in their construction. There are many approaches to such calculations, and the basic problem involves the definition of the boundaries on the energy inputs to be considered. It is clearly valid to account for the energy required to produce the physical inputs such as steel and concrete as well as the energy required to fabricate the materials into components and install them. It is less clear that the accounting should include the energy used in everyday activities at home by those who work on such projects, although some analyses include this energy as an input.

The fixed nature of the technological coefficients in the input-output matrix raises some problems in using this methodology for future-oriented studies. The fuel requirements corresponding to a projected GNP do not necessarily correspond to the quantities that may be available at that future time. Some interfuel substitution will take place in response to such limitations on specific fuels, and provision must be made to revise the technological coefficients accordingly. A combined energy system/input-output model (61) was developed by Brookhaven National Laboratory and the University of Illinois' Center for Advanced Computation to resolve this problem. This model combines the University of Illinois input-output model (62) with BESOM (52, 56). Constraints may be placed on the availability of fuels and resources in BESOM, and the required fuel substitutions are determined. Coefficients in the input-output model are revised to reflect the new fuel mix, and the input-output model is again solved with the revised mix. Several iterations are required between the two models in order to obtain a solution in which the energy demands and fuel mix are consistent in the two models.

Hudson & Jorgenson (9) have developed a macroeconomic energy model providing a new and innovative integration of traditional techniques of econometrics and input-output analysis. The model consists of a macroeconometric growth model of the US economy integrated with an interindustry energy model. The growth model consists of submodels of the household and producing sectors with the government and foreign sectors taken to be exogenous, and it determines the levels and distribution of output valued in constant and current dollars. The model determines the demand for consumption and investment goods, the supplies of capital and labor necessary to produce this level of output, and the equilibrium relative prices of goods and factors. The model is dynamic and has links between investment and changes in capital stock and between capital service prices and changes in prices of investment goods.

The macroeconometric growth model is linked to an interindustry energy model by estimates of demand for consumption and investment goods and the relative prices of capital and labor. The Hudson & Jorgenson interindustry model is based on a nine-sector classification of US industrial activity. The sectors are

Energy
1. Coal mining
2. Crude petroleum and natural gas
3. Petroleum refining
4. Electric utilities
5. Gas utilities

Non-Energy
6. Agriculture
7. Manufacturing (excluding petroleum refining)
8. Transportation
9. Communications, trade, and services

Production submodels are developed for each sector. These submodels treat as exogenous the prices of capital and labor services determined in the growth model and the prices of competitive imports and, for each sector, determine simultaneously the sector output prices and the input-output coefficients. Making the input-output coefficients endogenous is unique to this model and represents an important advance in input-output analysis.

The sector output prices and the demand for consumption goods from the growth model are used as inputs to a model of consumer behavior that determines the distribution of total consumer demand to the nine producing sectors. The distribution of private investment, government and foreign, is determined exogenously, and it completes the final demand portion of the model. Given final demands, the input-output coefficients may be used to determine the industry production levels required to support a given level and distribution of real demand.

The Hudson & Jorgenson model has been used to forecast long-term developments in energy markets within the framework of a consistent forecast of macroeconomic and interindustry activity. The model has also been used to analyze the impact on energy demands of alternative tax policies, including a uniform Btu tax, a uniform energy sales tax, and a sales tax on petroleum products (63).

The FEA has developed an integrated econometric/process model to assist in analyzing alternative strategies for achieving energy independence (3). The FEA model, the Project Independence Evaluation System (PIES), is summarized in Figure 4.[6]

There are four basic input submodels to PIES including a macroeconomic model, an industrial production model, an annual demand model, and a supply model for oil and gas production. Associated input data include estimates of coal production at alternative prices and a major data base of resource input requirements per unit of activity output. The macroeconomic and industrial production models generate estimates of the level and distribution of real output in the economy as inputs to an econometric energy demand model. The demand model is a dynamic econometric model that forecasts demands for 47 primary and derived energy products that are conditional upon assumed energy prices, industrial activity levels, the level and distribution of real output, and certain data for energy consumption technology. The model distinguishes the demand for fossil fuels between uses of fuel and power in each of three major consuming sectors (residential, industrial, and transportation) and industrial uses of raw materials. The model disaggregates the national forecasts to the census-region level of detail by means of regional energy prices and various measures of regional market size such as population, GNP, and industrial output levels.[7]

---

[6] This description of PIES draws heavily on the review of the Project Independence Report prepared by the MIT Energy Policy Study Group for the National Science Foundation (64).

[7] See Appendix II of the Project Independence Report (3) for a detailed discussion of the FEA demand simulation model.

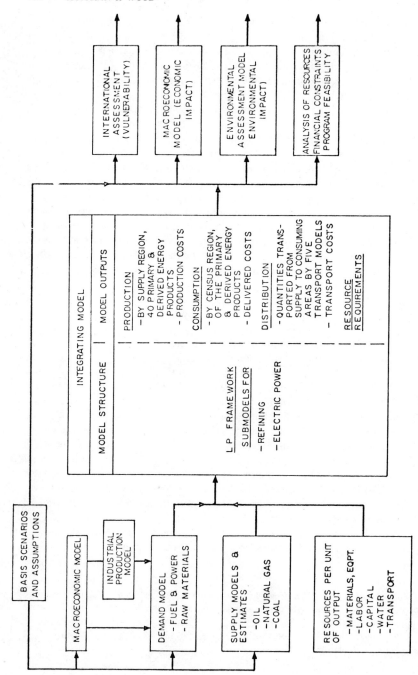

Figure 4 Blueprint of the Project Independence evaluation system (PIES).

The oil and gas supply model is an adaptation of an engineering/process model developed by the NPC. The model estimates additions to reserves and production levels for 12 supply regions, given assumptions about crude oil prices, regional drilling programs, required rate of return on investment, the expected success ratio per foot drilled, and the projected reserve/production rate.

The heart of PIES is the integrating model, an LP model that uses given estimates of regional demands, prices and elasticities, regional supply schedules, and resource input requirements to calculate an energy market equilibrium. The relation between the demand model and the LP submodel, which incorporates the supply schedules and conversion processes, may be summarized as follows: The demand model is used to calculate a price/quantity coordinate on the demand curve for each of the primary and derived energy products in the system. Associated with each of these coordinates are measures of the sensitivity of the quantities demanded to small changes in each of the prices in the demand model (own- and cross-price elasticities). In the first iteration of the integrating model, an LP problem is solved in which the minimum cost schedule of production, distribution, and transportation necessary to satisfy the given demand levels is calculated. Associated with the calculated supply quantities are implicit prices. If these supply prices differ from the original demand prices, then the solution is unstable and a new problem must be structured and solved. The procedure is to calculate new demand prices that equal one half the difference between the last iteration's supply and demand price, to use the own- and cross-price elasticities to calculate the new demand quantities, and, finally, to solve a new LP problem for the new production, distribution, and transportation schedules and the supply prices. This process is continued until the demand and supply prices are equal, at which point the energy market is assumed to be in equilibrium.

The outputs of the integrating model are then used as inputs to certain interpretive models including a macroeconomic model, an environmental assessment model, and an international assessment model. In addition, the integrating model outputs are analyzed to determine if potential limitations exist on the availability of the necessary resource inputs.

Another type of energy/economic, or energy/societal, modeling work involves global modeling of the type described in *Limits to Growth* (65). The energy sector is not described in sufficient detail in that model to warrant discussion in this review. Efforts are being made to develop more detailed models of the energy system that may be embedded in global models. The most significant example of this is the global model of Mesarovich & Pestel (66). This model encompasses energy, resources, economics, the environment, and population. The energy submodels consist of an energy resource model, a demand model, and an energy supply model. The resource model includes statistical information on energy resources allowing for the uncertainty of the resource and the feasibility of recovery. It also incorporates a simulation of the production of the resources. The demand model describes the demand for energy as a function of GNP, and the supply model links these demands to resources. The supply model covers 13 primary and 7 secondary forms of energy along with 27 conversion processes in a simulation framework.

SUMMARY

This review provides an introduction to the scope, application, methodology, and content of energy models used for forecasting future energy market conditions and in the formation and analysis of energy policy. The review of models is not exhaustive, but rather is intended to illustrate the structure of recent and current efforts by energy system modelers to provide constructive tools for energy system forecasting, planning, and policy analysis.

Energy models may be simple or complex, depending upon the purpose for which they were designed. Models range from detailed engineering/process models of particular energy conversion processes to econometric models of energy markets to total energy/economic system models, which include energy production, conversion, and utilization as one component of aggregate economic activity. The engineering/process models tend to be descriptive in character, whereas the econometric efforts emphasize the behavioral aspects of the agents making production, conversion, and utilization decisions. The more comprehensive efforts include both types of models as components, characterizing the behavior of the decision-making agents subject to the technical constraints imposed by the existing and emerging technologies.

Energy models may be classified as sector models for the supply or demand of a specific energy type; industry market models for the supply and demand of a specific form; energy system models, which include supply and demand for all energy types; and energy/economic models, which encompass the interrelations between the energy sector and the overall economy. Whatever the model type, energy models are used in forecasting and in planning and policy analysis. Energy planning activities include policy planning, or the formation of policy goals that are unconstrained by technology; strategic planning, which involves finding the most suitable path to obtaining a particular policy objective; and tactical or operational planning, which involves the determination of the appropriate steps necessary to implement a particular strategy. All three types of planning and policy analysis are utilized by the various groups employing energy system models, including regulatory agencies, at both the federal and state government levels; industry product planning and R & D planning groups; and federal agencies concerned with formulating and implementing national energy policy and planning, managing, and evaluating national programs of energy R & D.

The variety of models reviewed in this chapter suggests that a broad-ranging capability exists for supporting energy forecasting, planning, and analysis studies. The direction of current research is toward developing models that integrate engineering/process models with more behavioral models to form energy/economic systems that treat the demand for and supply of energy types simultaneously with those for other factors of production. The recent formation of the Energy Research and Development Administration, which is responsible for formulating, implementing, and managing a national program of energy R & D, will almost surely accelerate the development of these integrated models. This movement toward integration of the engineering/process approach and the behavioral approach, for modeling the energy system, seems extremely enlightened and should result in a substantial improvement in both the forecasting and the descriptive power of the resulting models.

Literature Cited

1. Ayres, R. U. 1969. *Technological Forecasting and Long-Range Planning.* New York: McGraw-Hill
2. MacAvoy, P. W., Pindyck, R. S. 1975. *The Economics of the Natural Gas Shortage 1960–1980.* Amsterdam: North-Holland
3. Federal Energy Administration. 1974. *Project Independence Report.* Washington DC: GPO
4. Dantzig, G. B. 1963. *Linear Programming and Extensions.* Princeton, NJ: Princeton Univ. Press
5. Wagner, H. M. 1969. *Principles of Operations Research.* Englewood Cliffs, NJ: Prentice-Hall
5a. Leontief, W. W. 1951. *The Structure of the American Economy 1919–1939.* New York: Oxford Univ. Press
5b. Carter, A. P., Brody, A. 1970. *Contributions to Input-Output Analysis.* Amsterdam: North-Holland
5c. Deam, R. J. et al. 1974. World energy modelling. In *Energy Modelling* (special issue of *Energy Policy*). Guildford, Surrey, UK: IPC Sci. Technol. Press
6. US Department of Commerce. 1974. *Input-Output Structure of the U.S. Economy, 1967, Volume 3: Total Requirements for Detailed Industries.* Washington DC: GPO
7. Herendeen, R. A. 1973. *The Energy Cost of Goods and Services.* Rep. No. ORNL-NSF-EP-58. Oak Ridge Nat. Lab., Oak Ridge, Tenn.
8. Reardon, W. A. June 1972. *An Input/Output Analysis of Energy Use Changes from 1947 to 1958 and 1958 to 1963.* Rep. submitted to Office of Science and Technology, Executive Office of the President, by Battelle Mem. Inst.
9. Hudson, E. A., Jorgenson, D. W. 1974. U.S. energy policy and economic growth, 1975–2000. *Bell J. Econ. Manage. Sci.* 5(2): 461–514
10. Johnston, J. 1972. *Econometric Methods.* New York: McGraw-Hill. 2nd ed.
11. Theil, H. 1971. *Principles of Econometrics.* New York: Wiley
12. Malinvaud, E. 1966. *Statistical Methods of Econometrics.* Chicago: Rand McNally
13. Griffin, J. M. 1974. The effects of higher prices on electricity consumption. *Bell J. Econ. Manage. Sci.* 5(2): 515–639
14. Baughman, M. L., Joskow, P. L. Dec. 1974. *A Regionalized Electricity Model.* Rep. No. MIT-EL 75-005. MIT Energy Lab., Cambridge, Mass.
15. Forrester, J. W. 1961. *Industrial Dynamics.* Cambridge, Mass: MIT Press
16. von Neumann, J., Morgenstern, O. 1944. *Theory of Games and Economic Behavior.* Princeton, NJ: Princeton Univ. Press
17. Anderson, D. 1972. Models for determining least-cost investments in electricity supply. *Bell J. Econ. Manage. Sci.* 3(1): 267–99
18. Limaye, D. R., Ciliano, R., Sharko, J. R. 1972. *Quantitative Energy Studies and Models: A State of the Art Review.* Rep. submitted to Office of Science and Technology by Decision Sciences Corp.
19. Charpentier, J. P. July 1974. *A Review of Energy Models.* Pap. No. RR-74-10. Laxenburg, Austria: Inst. Appl. Syst. Anal.
20. Searl, M. F., ed. 1973. *Energy Modeling.* Washington DC: Resources for the Future Inc.
21. National Science Foundation and Economic Research Unit, Queen Mary College. 1974. *Energy Modelling* (special issue of *Energy Policy*). Guildford, Surrey, UK: IPC Sci. Technol. Press
22. Benenson, P., ed. 1974. *Proc. Conf. Energy Modeling Forecasting, June 28–29, 1974.* Lawrence Berkeley Lab., Berkeley, Calif.
23. Taylor, L. D. 1975. The demand for electricity: a survey. *Bell J. Econ. Manage. Sci.* 6(1): 74–110
24. Houthakker, H. S. 1951. Some calculations of electricity consumption in Great Britain. *J. R. Stat. Soc. Ser. A,* Part 3, 114: 351–71
25. Fisher, F. M., Kaysen, C. 1962. *A Study in Econometrics: The Demand For Electricity in the United States.* Amsterdam: North-Holland
26. Houthakker, H. S., Taylor, L. D. 1970. *Consumer Demand in the United States.* Cambridge, Mass: Harvard Univ. Press. 2nd ed.
27. Wilson, J. W. 1971. Residential demand for electricity. *Q. Rev. Econ. Bus.* 11(1): 7–22

28. Mount, T. D., Chapman, L. D., Tyrrell, T. J. 1973. *Electricity Demand in the United States: An Econometric Analysis.* Rep. No. ORNL-NSF-9. Oak Ridge Nat. Lab., Oak Ridge, Tenn.
29. Anderson, K. P. 1973. *Residential Energy Use: An Econometric Analysis.* Rep. No. R-1297-NSF. Santa Monica, Calif: Rand Corp.
30. Lyman, R. A. Oct. 1973. *Price Elasticities in the Electric Power Industry.* Dep. Econ., Univ. Arizona, Tucson, Ariz.
31. Houthakker, H. S., Verleger, P. K., Sheehan, D. P. 1973. *Dynamic Demand Analysis for Gasoline and Residential Electricity.* Lexington, Mass: Data Resources Inc.
32. Baxter, R. E., Rees, R. 1968. Analysis of the industrial demand for electricity. *Econ. J.* 78:277-98
33. Anderson, K. P. 1971. *Toward Econometric Estimation of Industrial Energy Demand: An Experimental Application to the Primary Metals Industry.* Rep. No. R-719-NSF. Santa Monica, Calif: Rand Corp.
34. Baughman, M. L., Joskow, P. L. July 1974. *Energy Consumption and Fuel Choice by Residential and Commercial Consumers in the United States.* MIT Energy Lab., Cambridge, Mass.
35. Sweeney, J. 1975. *Passenger Car Use of Gasoline: An Analysis of Policy Options.* Washington DC: FEA
36. Lady, G. 1974. *National Petroleum Product Supply and Demand.* Tech. Rep. No. 74-5, FEA. Washington DC: GPO
37. Verleger, P. Dec. 1973. *A Study of the Quarterly Demand for Gasoline and Impacts of Alternative Gasoline Taxes.* Rep. submitted to Counc. Environ. Qual. by Data Resources Inc., Lexington, Mass.
38. McGillivray, R. G. 1974. *Gasoline Use by Automobile.* Working Pap. No. 1216-2. The Urban Institute, Washington DC
39. Windhorn, S., Burright, B., Enns, J., Kirkwood, T. F. 1974. *How to Save Gasoline: Public Policy Alternatives for the Automobile.* Rep. No. R-1560-NSF. Santa Monica, Calif: Rand Corp.
40. Adams, G. R., Graham, H., Griffin, J. M. June 1974. *Demand Elasticities for Gasoline: Another View.* Disc. Pap. No. 279. Dep. Econ., Univ. Penn., Philadelphia, Pa.
41. Adams, F. G., Griffin, J. M. 1972. An economic-linear programming model of the U.S. petroleum refining industry. *J. Am. Stat. Assoc.* 67(339): 542-51
42. Naill, R. F., Miller, J. S., Meadows, D. L. Nov. 1974. *The Transition to Coal.* Rep. No. DSD-18. Systems Dynamics Group, Thayer School of Engineering, Dartmouth College, Hanover, NH
43. MacAvoy, P. W., Pindyck, R. S. 1975. *Price Controls and the Natural Gas Shortage.* Washington DC: Am. Enterpr. Inst. Public Policy Res.
43a. Baughman, M. L., Zerhoot, F. S. 1975. *Interfuel Substitution in the Consumption of Energy in the United States, Part II: Industrial Sector.* Rep. No. MIT-EL 75-007. MIT Energy Lab., Cambridge, Mass.
44. Kennedy, M. 1974. An economic model of the world oil market. *Bell J. Econ. Manage. Sci.* 5(2): 540-77
45. Barnett, H. J. 1950. *Energy Uses and Supplies, 1939, 1947, 1965.* Inf. Circ. No. 7582. Washington DC: Bur. Mines, US Dep. Interior
46. Morrison, W. E., Readling, C. L. 1968. *An Energy Model for the United States, Featuring Energy Balances for the Years 1947 to 1965 and Projections and Forecasts to the Years 1980 and 2000.* Inf. Circ. No. 8384. Washington DC: Bur. Mines, US Dep. Interior
47. DuPree, W. G. Jr., West, J. A. Dec. 1972. *United States Energy Through the Year 2000.* Washington DC: Bur. Mines, US Dep. Interior
48. National Petroleum Council. 1972. *Guide to National Petroleum Council Report on U.S. Energy Outlook.* Washington DC: Nat. Pet. Counc.
49. National Petroleum Council. 1974. *Emergency Preparedness for Interruption of Petroleum Imports into the United States.* Washington DC: Nat. Pet. Counc.
50. Baughman, M. L. 1972. *Dynamic Energy System Modeling—Interfuel Competition.* Rep. No. 72-1. MIT Energy Analysis and Planning Group, Cambridge, Mass.
51. Hoffman, K. C., Palmedo, P. F. 1972. *Reference Energy Systems and Resources Data for Use in the Assessment of Energy Technologies.* Rep. No. AET-8. Associated Universities Inc., Upton, NY
52. Hoffman, K. C. 1974. A unified framework for energy system planning. In *Energy Modelling.* See Ref. 21
53. Brookhaven National Laboratory. 1974. *Energy Model Data Base, User Manual.* Rep. No. BNL 19200. Associated Universities Inc., Upton, NY
54. Cazalet, E. J. 1975. *SRI-Gulf Energy Model: Overview of Methodology.* Menlo Park, Calif: Stanford Res. Inst.
55. Debanne, J. G. 1975. *A Techno-Economic*

and *Environmental Energy Model for North America*. Presented at Summer Computer Simulation Conf., San Francisco, Calif., July 21–23, 1975
56. Cherniavsky, E. A. 1974. *Brookhaven Energy System Optimization Model*. Brookhaven Nat. Lab. Topical Rep. No. BNL 19569. Associated Universities Inc., Upton, NY
57. Marcuse, W., Bodin, L. 1975. *A Dynamic Time Dependent Model for the Analysis of Alternative Energy Policies*. See Ref. 55
58. Nordhaus, W. 1973. The allocation of energy resources. *Brookings Papers on Economic Activity* (3). Washington DC: Brookings Inst.
59. Manne, A. S. July-Aug. 1975. What happens when our oil and gas run out. *Harvard Bus. Rev.* 53(4):123–37
60. Just, J. E. 1973. *Impacts of New Energy Technology Using Generalized Input-Output Analysis*. Rep. No. 73–1. MIT Energy Analysis and Planning Group, Cambridge, Mass.
61. Behling, D. J., Marcuse, W., Swift, M., Tessmer, R. Jr. 1975. *A Two-Level Iterative Model for Estimating Inter-Fuel Substitution Effects*. See Ref. 55
62. Bullard, C. W., Sebald, A. V. 1975. *A Model for Analyzing Energy Impact of Technological Change*. See Ref. 55
63. Hudson, E. A., Jorgenson, D. W. Sept. 1974. *U.S. Energy Policy and Economic Growth, 1975–2000*. Rep. to FEA by Data Resources Inc., Lexington, Mass.
64. MIT Energy Policy Study Group. June 1975. *The FEA Project Independence Report: An Analytical Review and Evaluation*. Rep. No. MIT-EL 75-017. Submitted to Office of Energy Research and Development Policy, NSF, Washington DC
65. Meadows, D. L. et al. 1972. *The Limits to Growth: A Report to the Club of Rome's Project on the Predicament of Mankind*. New York: Universe
66. Mesarovich, M., Pestel, E. 1974. Multi-level computer model of world development system. *Proc. Symp. Int. Inst. Appl. Syst. Anal., April 29–May 3, 1974, Laxenburg, Austria*

Copyright 1976. All rights reserved

RAISING THE PRODUCTIVITY OF ENERGY UTILIZATION

✷11017

Lee Schipper
Energy and Environment Division, Lawrence Berkeley Laboratory,
University of California, Berkeley, California 94720

INTRODUCTION

Energy planners in the late 1960s, projecting growth in energy use on the basis of past experience, compared the expected future demand for energy with possible supplies and generally concluded that a gap (Figure 1) would appear between demand and supply from domestic energy sources. It was suggested that the gap be filled by a variety of solutions: expanded oil imports; accelerated use of nuclear power or .

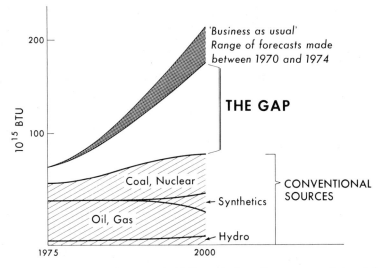

Figure 1 The energy gap. In conventional forecasts domestic energy supply lags behind demand that grows at the historical rate of about 4–5% per year. Compare with forecasts in (1–3).

coal; enhanced harvesting of domestic oil and gas; or development of synthetic fuels, solar energy, geothermal energy, or some other new form of fuel, heat, or work.[1] Today all of these supply options are still potentially important, but they are far more expensive than the oil, gas, and electricity that were available in the late 1960s. This suggests that a more productive use of energy can more economically supply much of the well-being that otherwise would be made available by using more energy. This higher efficiency, or higher productivity, of energy use is called *energy conservation*.

Until recently conservation was virtually ignored or dismissed in work dealing with energy (2, 3). A typical view was expressed in the prognosis of the Chase Manhattan Bank (5, p. 52), which asserted that

> analysis of the uses of energy reveals little scope for major reduction without harm to the nation's economy and its standard of living. The great bulk of the energy is utilized for essential purposes—as much as two thirds is for business related reasons. And most of the remaining third serves essential private needs. Conceivably, the use of energy for such recreational purposes as vacation travel and the viewing of television might be reduced—but not without widespread economic and political repercussions. There are some minor uses of energy that could be regarded as strictly non-essential—but their elimination would not permit any significant savings.

More informed studies of energy use contradict this analysis. Especially misleading is the subjective phrase "essential purposes," which obscures the whole question of efficiency. Careful analysis of energy use has revealed an enormous potential for energy conservation (6–15). The most recent forecasts from the Energy Research and Development Administration (ERDA) (16) suggest that US energy needs in the 1990s could be 20–40% below what was previously expected, as higher energy prices and new end-use technologies help Americans squeeze more economic and personal well-being from every Btu. This review deals with some of the implications of more productive energy utilization.

THE NATURE OF ENERGY CONSERVATION

Insights from Physical Sciences and Economics

All energy systems and energy use must obey the laws of physics, in particular the laws of thermodynamics (17–19)[2] (see Table 1). "Using" or "consuming" energy really means converting high-quality energy, stored in fuels or as falling water, into heat and work and ultimately into low-quality heat near the temperature of the environment and no longer available to do work. The second law of thermodynamics assures that fuel, the ability to do work, or the quality of energy, is indeed consumed when energy is "used" by society. Ideally economics guides energy consumers (and producers of goods and services) in the choices of how and how well to convert

---

[1] For reviews see other articles in this volume or the April 19, 1974 issue of *Science*. For "alternative" source see (4).

[2] An excellent introduction to the physics of energy can be found in (19).

Table 1 The laws of thermodynamics[a]

First Law of Thermodynamics
Energy can be neither created nor destroyed, but it can change form. ("You can't get something for nothing, you can only break even.")

Second Law of Thermodynamics
It is impossible to convert a given quantity of heat completely into work. In any macroscopic process involving energy conversion, some energy is always degraded in quality, so that ability to do work is lessened. ("You can't break even, you can only lose.")

Quality of energy describes the degree to which energy can be converted into work, which is the application of force through a distance. The quality of thermal energy increases with the temperature *difference* between the body of heat and the background environment. Work, electricity, and gravitational energy are of the highest quality; chemical energy stored in fuels is also of high quality, although not the highest. The first law merely states that the total quantity of energy in a closed system is conserved. The second law, however, asserts that the quality of energy can be consumed in physical processes.

[a] See (17) or chapter by C. A. Berg in this volume.

energy use into goods and services or other forms of utility that increase human welfare. It is widely recognized, however, that energy use is a complicated social phenomenon, the full understanding of which demands interdisciplinary analysis far beyond ordinary economics or physical science (20, 21).[3]

Energy is used in the economy, along with other resources, to produce goods, services, transportation, environmental conditions (heating, cooling), or other life-support systems, and conveniences. Economic resources (factors of production) include capital (with design and know-how), labor, land, energy, and the environment, which absorbs pollution. These factors are compared and evaluated for economic decisions by attaching prices, or dollar values, to them as well as to the output produced by their use. Because energy is but one input to processes, minimizing energy use alone does not always equate with minimizing total costs. What has aroused the scientific community, however, is the fact that careful energy conservation does reduce total costs. This suggests the following definition of energy conservation:

> the strategy of adjusting and optimizing energy-using systems and procedures so as to reduce energy requirements per unit of output (or "well-being") while holding constant or reducing total costs or providing the output from these systems.

Conservation techniques will (*a*) improve the delivery of energy and reduce the specific energy requirements of processes or systems, or (*b*) modify the tasks or goals of energy use. Improving the efficiency of air conditioners reduces the energy required to pump each unit of heat out of a room, while redesigning the house can reduce the amount of heat that needs to be pumped out in the first place.

[3] An excellent introduction to economics of resources and pollution can be found in (22).

From an economic point of view, conservation strategies substitute other economic resources for energy. The most important of these is capital. Conservation can be viewed as an investment, with a certain rate of return.

Conserving has always been identified by economists as the practice of saving something that might be more valuable in the future (22, 23). Furthermore, conservation as increased efficiency allows energy to be saved at no sacrifice in the goals of energy use. Since most consumers have a definite time preference toward the present, few will automatically save for the future without rewards. The rewards for conserving energy are the economic savings that conservation strategies yield.

There are other rewards for conserving energy not measured by direct monetary savings. Some benefits are difficult to perceive by the saver as, for example, lower future prices for fuels in the United States and abroad, fuels saved for use by future generations, lessened dependence on foreign sources of fuel, lessened environmental burdens, and other "hidden benefits" (14). I believe, however, that the largest stimulus to more efficient energy utilization will occur in reponse to direct economic incentives and governmental policies designed to aid those incentives.

Some discussions of conservation implicitly or explicitly equate conservation with sacrifice of desired life-styles, acceptance of lower standards of living, or denial of

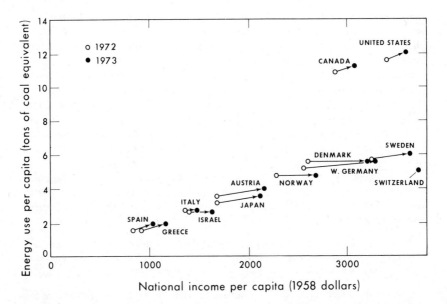

Figure 2 Per capita energy use and national income of some important industrial and emerging nations, 1972 and 1973. Note the wide variation in energy use among the nations with highest per capita income, measured in 1958 US dollars at current exchange rates. 1972 and 1973 data were compiled by A. Rosenfeld, Lawrence Berkeley Laboratory, from US and UN statistical abstracts. Data for Switzerland are based on author's estimate. National income is closely related to GNP.

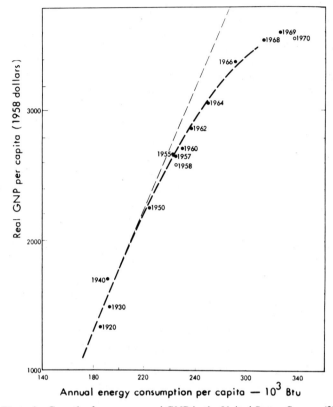

Figure 3 Growth of energy use and GNP in the United States. Source: (27).

economic opportunity to low-income groups (24, 25).[4] It will be argued here repeatedly that the strategies that bring about the largest savings in energy use need have few or none of the above effects. The difference between shortages, either short-term or long-term, and true reduction of specific energy needs brought about by more effective utilization must be borne in mind when energy policies are discussed.

Energy Use and Standard of Living

For many years it has been common practice to investigate the relation between energy use and gross national product (GNP) (26), often referred to as affluence or standard of living (Figure 2). Such relationships help to distinguish between wealthy and underdeveloped countries. Historical data also show correlations between the rise in GNP and the rise in energy use in a single country (27, 28) (Figure 3).

[4] See the general discussion of future demand in (24). Cook & Vassell (25) largely ignore conservation techniques.

These analyses, however, ignore many important factors that influence actual energy use and the well-being derived therefrom, such as

1. geographic, demographic, and meteorological differences among countries or regions;
2. cultural differences among peoples: advertising, personal habits, and values;
3. differences in economic conditions, including energy prices; the breakdown of inputs and outputs in the GNP; the pace of economic growth; and pollution;
4. physical differences in the structure of the GNP, as well as in the technical efficiency of energy use.

The present scatter of countries with high GNP and varied rates of energy use (Figure 2) suggests that the energy-GNP relationship is much more flexible than previously thought,[5] the level of GNP no longer dictating energy requirements of the economy. Additionally it has been suggested that the rate of growth in the GNP can be uncoupled from the rate of growth in energy use, especially as energy costs rise (29).

Careful comparisons among countries (30–34) often reveal energy conservation strategies and allow the side effects of these strategies to be examined at the same time. In Sweden, for example, a significant fraction of all energy consumed in thermal power plants is utilized as low-temperature process or space heating, as the details of Figure 4 illustrate. An evaluation taking into account the slightly smaller size of Swedish dwellings (compared with the United States), the higher percentage of apartments, the higher efficiencies of the well-maintained apartment heating systems in Sweden, and the Swedish climate shows that space heating requirements in Sweden are 30–40% lower per square meter of space in homes (and commercial buildings) than in the United States. Since indoor temperatures in Sweden are in the range of 23–24°C (73–75°F), the differences must be ascribed to generally more energy-efficient structures in Sweden (33, 35).

In the transportation sector Swedish automobiles are considerably lighter than those in the United States, averaging around 1100 kg. A larger fraction of short intra-city trips in Sweden are made via mass transit. More passenger trips in Sweden are made by train, with higher load factors, than in the United States, or by chartered jet, again with higher load factors than commercial aviation in the United States. Swedes travel nearly as much as Americans but with far less energy (33, 35).

In the industrial sector Sweden also uses less energy for each ton of steel, paper, cement, and most other industrial products, including allowance for the difference in heat rates. Sweden's import/export statistics indicate that 9–10% of all energy consumed in Sweden is embodied in exported products, while imports of refined petroleum products embody about 5% of Sweden's total energy consumption, this energy being consumed by refineries outside Sweden (33, 35, 36).

On balance Sweden requires about half as much energy per dollar of GNP as the United States. Some of this is certainly due to the large share of hydropower in

---

[5] See review by Craig, Darmstadter & Rattien in this volume.

the Swedish energy economy and the larger fraction of Sweden's GNP that goes to services, particularly social welfare. But most of the difference arises out of the strikingly more efficient ways in which Swedes convert fuel into comfort, transportation, and industrial output. Similar conclusions were reached in a study of West Germany (32).

Energy use per dollar of GNP has changed with time in the United States, reflecting changes in the goals of energy use as well as in efficiencies. And there are variations from place to place within the United States today in the amount of insulation in buildings (6, 9, 13), the use of public transportation (37–39), weight and other factors in automobiles, and the prices of fuels and electricity (40, 41). The differences in energy consumed for heating can be striking. Table 2 gives natural

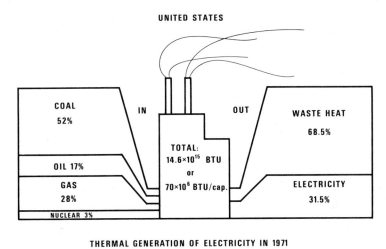

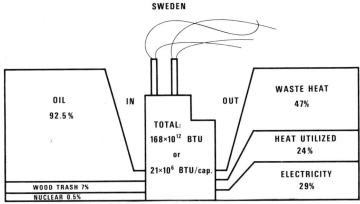

Figure 4 Comparison of utilization of output from thermal electric power plants in Sweden and the United States in 1971. Source: (14, 33).

Table 2 Residential energy consumption of the largest metropolitan areas in the United States[a]

Rank	Metropolitan Area	Thousands of Btu per Degree-Day	Rank	Metropolitan Area	Thousands of Btu per Degree-Day
1.	San Diego	72	21.	Detroit	33
2.	Los Angeles	63	22.	Dallas-Ft. Worth	31
3.	New Orleans	60	23.	Providence	31
4.	Phoenix	55	24.	Oklahoma City	31
5.	Houston	55	25.	Dayton, Ohio	30
6.	New York City	55	26.	Buffalo	29
7.	Chicago	44	27.	Columbus, Ohio	27
8.	Newark	39	28.	Baltimore	26
9.	Washington DC	37	29.	Denver	26
10.	San Antonio	37	30.	Rochester	25
11.	Louisville	36	31.	Boston	25
12.	San Francisco	36	32.	Philadelphia	25
13.	Indianapolis	36	33.	Portland, Oregon	24
14.	Pittsburgh	36	34.	Kansas City	24
15.	Cleveland	36	35.	Minneapolis	24
16.	Memphis	35	36.	Seattle	23
17.	Atlanta	34	37.	Milwaukee	20
18.	Cincinnati	34	38.	Birmingham	20
19.	St. Louis	33	39.	Hartford	17
20.	Norfolk	33[b]			

[a] Source: (42).
[b] Median consumption.

gas consumption per degree-day for the largest US metropolitan areas. The variation is enormous; in the warm cities, or where natural gas is relatively inexpensive, builders and homeowners understandably ignore energy in designing and living in homes, since total heating costs are low.

It is clear from the foregoing discussion that energy use alone is an insufficient measure of how well people live. Variations in energy use and efficiency among different countries, different regions of the same country, or during different time periods in the same country or region indicate that the energy requirements of tasks vary considerably. High rates of energy use per dollar of GNP may indicate inefficient use of energy as well as high standard of living. Understanding the efficiency of energy use will allow us to see how much more well-being can be generated from each unit of energy used.

Efficiency of Energy Utilization

PHYSICAL EFFICIENCY Physical measures of efficiency are important indicators of the potential for conservation. Traditionally the first law of thermodynamics (see

Table 1) has been used to express efficiency as

$$\frac{\text{(energy provided or transferred in desired form and place)}}{\text{(energy input to system)}}.$$

First-law efficiencies follow the flow of work and heat in systems, accounting for all uses and losses at the boundaries in and out of the system. Figure 5 shows first-law efficiencies for air conditioners (43). Motors (44) and other devices (45) have been carefully studied. First-law efficiencies are numerically less than one, except in the case of refrigerators, air conditioners, or heat pumps, where the amount of heat moved is usually greater than the amount of work (usually electricity) consumed. The "spaghetti bowl" chart, drawn first by Cook (Figure 6), shows an approximate first-law accounting of US energy use in 1971 [see also (45) and for other "bowls" from a variety of years, see (46)].

It is important to note, however, that first-law efficiencies can be misleading. As suggested by the study done by the American Physical Society (45), the second law of thermodynamics gives a more relevant measure of physical efficiency in cases

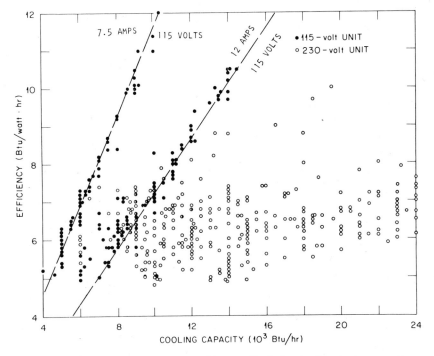

Figure 5 First-law efficiency of air conditioners. Vertical axis gives heat removed per unit of energy consumed; horizontal axis, the size of the unit. Note the wide variations in efficiency. The models that lie near the two straight lines are those constrained to operate on either 7.5-amp or 12-amp circuits. Source: (43).

where temperature changes are involved, by comparing the energy (as work or heat) theoretically required to perform a task with that actually used. Note that 60% of all fuel is consumed in order to change the temperature of environments or substances and that 85% of electricity and nearly all of transportation is provided by heat engines. Second-law efficiency is measured by

$$\frac{\text{(theoretical minimum energy required by second law)}}{\text{(energy actually consumed)}}$$

Theoretical energy requirements of tasks are based on properties of materials, environments, and heat engines that convert heat to work and vice versa. The amount of work needed to pump heat across the small temperature differences common to most household or building situations and industrial tasks is small compared with the amount of work that could be extracted from combustion of the fuels that are used in these applications—hence the low efficiencies in Table 3. By contrast, modern power plants or industries using high-temperature heat utilize a larger fraction of the ability to do work that is stored in fuel. An ever greater efficiency can be obtained by modification of industrial procedures to allow use of the exhaust heat from these high-temperature processes in other applications, including generation of electricity (47). Locating power plants near communities permits utilization of power plant waste heat (via cooling water) for district heating, as is done in Sweden (33). Putting small power plants into factories allows economic cogeneration of both heat and electricity in a mix optimized for the temperature and horsepower requirements of the factory (47).

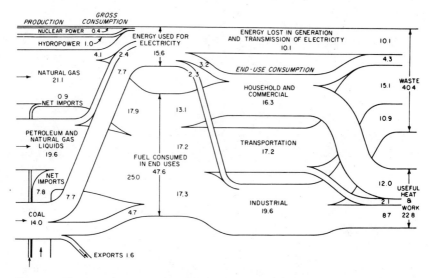

Figure 6 First-law efficiencies of energy use in the United States in 1971. Estimates by E. Cook (21). In Cook's earlier "spaghetti bowls," drawn for a variety of scenarios in (45), overall first-law efficiency was assumed to be higher, around 50%.

Table 3 American Physical Society estimates of efficiency of energy use[a]

Use	Relative Thermodynamic Quality	Percent of US Fuel Consumption (1968)	Estimated Overall Second-Law Efficiency
Space heating	lowest	18	0.06
Water heating	low	4	0.03
Cooking	low	1.3	—
Air conditioning	lowest	2.5	0.05
Refrigeration	lowest	2	0.04
Industrial uses			
Process steam	low	17	0.25
Direct heat	high	11	0.3
Electric drive[b]	(work) high	8	0.3
Electrolytic processes	high	1.2	—
Transportation	(work) highest		
Automobile		13	0.1
Truck		5	0.1
Bus		0.2	—
Train		1	—
Airplane		2	—
Military and other		4	—
Feedstock		5	—
Other		5	—
Total		100	

[a] Source: (45).
[b] *Work* is defined as *infinite-temperature* energy by the APS study.

Some processes are best understood by examining both first- and second-law efficiency. Improving heat-transfer properties from flame to boiler in a power plant, which is a first-law procedure, also increases temperature differences that the plant utilizes and therefore improves second-law efficiency (45). Often, too, redefinition of the tasks involved in energy use allows a higher physical efficiency. Use of recycled scrap lowers process heat and electricity requirements for most metals. Heat pumps might utilize groundwater, the temperature of which varies little during the year, instead of outdoor air, as a heat source for space heating. The temperature differences between the source and the indoor space is then reduced, so less work is required (45). While physical analysis of efficiency gives an important measure of possibilities for energy conservation, the tasks in question should also be evaluated.

DESIGN AND MAINTENANCE EFFICIENCY Sometimes it is useful to measure efficiency by relating the energy requirements of a task to the physical or economic output of that task. Design intensity is expressed this way, employing units such as Btu/passenger mile (for passenger transportation, see Figure 7), Btu/degree-day (for

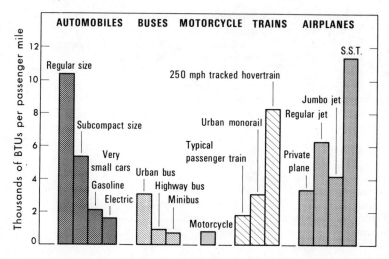

Figure 7 Design intensities of some modes of transportation, in Btu per passenger mile, using common load factors. Data from R. Rice, presented in (7). Note the wide variation in design intensities, which are far lower than possible in most cases because of low load factors.

space heating), or Btu/ton (for industrial output). Design efficiencies are the inverses of these intensities. But whether a task is carried out at its rated design efficiency usually depends on how well the system or process used is maintained. Maintenance efficiency compares rated design efficiency with operating design efficiency. The design efficiency may also depend on other factors; the seasonal dependence of the design efficiency of space heating systems (48) is a good example.

Some systems, such as mass transit and automobiles, are designed to provide far more output (passenger miles) per unit of energy consumed than is usually used (Figure 7). Uninsulated, poorly designed structures require more energy than insulated, carefully built ones (a design efficiency), and poorly maintained heating systems deliver less comfort, per unit of energy used, than well-maintained systems (a maintenance efficiency). By modifying systems, energy conservation procedures improve design efficiency, even before physical efficiencies are changed.

ECONOMIC EFFICIENCY   Economic efficiency is measured by comparing the total cost of using energy in various systems at various levels of physical efficiency. "Cost" is usually, but not always, measured as the total direct cost of providing useful output or services from energy, measured over the lifetime of the system. Cost includes purchase price, interest or opportunity cost on the investment involved, taxes, and maintenance costs (49).[6] Thus economic efficiency considers both the cost of the energy and the cost of the energy system. Ideally, although rarely in practice, economic efficiency includes environmental or other external costs associated with

---

[6] See also the Appendix in (43) for sample calculations.

energy systems (22, 23), as well as the cost of the risk of the fuel or system being unavailable in the future and the prospects for changes in fuel costs. That system which provides the desired output for the lowest cost is the most efficient. Unless otherwise indicated, this review uses the economic definition of efficiency in terms of life-cycle direct costs, although compelling reasons have been suggested (22, 23) for including environmental costs and other non-market costs or exigencies such as the reduction of oil imports or threats to national security.

Economic efficiency can also be identified with total resource productivity. For many years economists usually concentrated their studies on labor productivity, since resource costs were stable. Now, however, energy costs have risen faster than labor costs, and in some cases energy, rather than labor unions, has gone on strike, as in the case of the 1973–1974 oil embargo. If energy prices continue to lead all other factors in inflation, including wages, then it can be expected that energy users will accelerate efforts to increase the economic and physical efficiency of energy use. Part of the reason for the relatively high efficiency of energy use in Sweden can probably be ascribed to the high cost of energy there. It has been noted, however, that over the past 25 years the physical and economic efficiency of energy use in US industry has increased in spite of falling real energy prices (29). In this sense energy conservation leads to a more productive use of resources.

Physical or monetary values are not always sufficient for analyzing energy use because people often make decisions on the basis of variables such as exertion, luxury, convenience, time, risk, pleasure, or nuisance. Urban design, employment, and life-style are other social aspects of energy use that influence energy choices. For example, I use taxis when I am in Washington DC (but rarely elsewhere) because the value of time there is usually worth the expense of taxis and the loss of exercise. Important values like these must enter into energy planning and considerations of conservation strategies. The object of conservation is not to deny ourselves conveniences, preferences, necessities, or other aspects of life-styles, but instead to make these activities more economically efficient. The next sections discuss how this might come about through energy conservation.

ENERGY USE AND CONSERVATION

Kinds of Energy Use

One way to display kinds of energy use is the traditional breakdown of energy use by economic sector and task shown in Figure 8, which can be compared to Figure 6, in which the flow of fuels from source to economic sector was illustrated. Another possible description of energy use is by task and quality, as the American Physical Society study (45) suggested (Table 3).

Often it is desirable to learn the energy requirements of individual economic or physical activities. Energy analysis allows these requirements to be evaluated through accounts of purchases of energy by firms, by using input-output techniques (50–52), or measurement of fuel consumption in production processes (53–55) (Table 4). Energy-intensive activities are those for which the specific energy requirements per unit of physical or economic output are significantly higher than for the

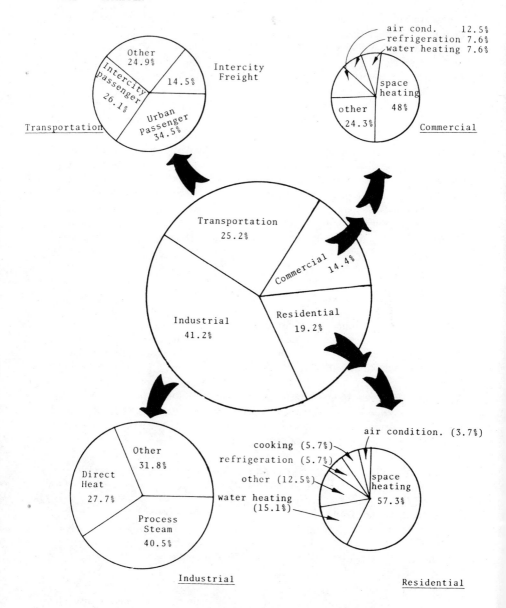

Figure 8 How does the United States use its energy? Schematic pie representation of energy uses in the United States. From Oak Ridge Associated Universities, 1974, *Citizens Energy Workshop Handbook,* Oak Ridge, Tenn. Compare with Table 3 and Figure 6.

Table 4 The energy cost of some common goods and services[a] and the energy costs associated with the production and recycling of metals[b]

1967 Example Energy Intensities[c]		Summary of the Energy Requirements for the Production and Recycle of Metals[c]		
Product	(10³ Btu/$)	Metal	Main Source	Equivalent Coal Energy (kW-hr/ton)[d]
Primary aluminum	388	Magnesium	sea water	90,821[103,739]
Fertilizers	174		mg scrap recycle	1,395[1,875]
Airlines	192			
Glass	103	Aluminum	bauxite 50% alumina	51,379[63,892]
Motor vehicles	67		aluminum scrap recycle	1,300–2,000
Cheese	73			
Apparel	50	Iron	high-grade hematite	4,270[4,289]
Hospitals	51		iron laterites	6,268[6,327]
Computing machinery	36		iron and steel scrap recycle	1,240[1,666]
Banking	18			
US average		Copper	1% sulfide ore	13,532[15,193]
Including energy	80		0.3% sulfide ore	24,759[29,766]
Non-energy goods	52		98% scrap recycle	635[853]

[a] Source: (52). These are direct plus indirect energy requirements.
[b] Source: (87). These are direct requirements only.
[c] Electricity has been counted at 40% conversion from fuel to electricity.
[d] Figures in brackets count electricity at 29% conversion efficiency.

Table 5 Energy cost of energy: efficiency of the US economy in delivering energy, 1967[a]

Sector	Efficiency (%)
Coal	99.3
Refined petroleum	82.8
Electricity	26.3
Natural gas	90.9

[a] Source: (52). Percentage of total Btu's harvested by the energy industries that are delivered to the US economy.

economy as a whole. Energy analysis also shows how much energy must be consumed to build and run energy production technologies (Table 5). Such analysis is particularly important for nuclear power programs or advanced fossil fuel conversion schemes (56–60b).[7] Energy analysis provides a reminder that investing dollars in almost everything requires investing energy.

Energy analysis can also be used to predict the energy cost of implementing conservation strategies, but consideration of the economics involved suggests that the periods for energy payoff will be short (61). Let us examine why this so. Expenditures for energy require between 2×10^5 and 10×10^5 Btu per dollar. The same dollars spent on conservation investments or other non-energy purchases require only about $6-8 \times 10^4$ Btu per dollar (50–52). Even if the present value of the energy saved from a strategy is somewhat less than the cost of the strategy, energy will be saved. However, in the case of energy technologies that are materials-intensive, such as solar photovoltaic and central-station solar power plants, large energy investments may be required, and these should be carefully evaluated.

These objective descriptions about energy use and cost avoid important judgements about the tasks involved. Must cars weigh 6000 lbs? Must a refrigerator be so poorly insulated as to "require" an electric resistance heater in its skin to drive off condensation? Questions like these suggest that while physical science and economics answer many of the technical questions about efficiencies of energy use alternatives, political and sociological analysis is required to explain why energy use patterns have developed (20, 21). Understanding the quantitative aspects of energy use is, however, essential in order to be able to evaluate energy conservation strategies. These strategies are discussed next.

Conservation Strategies and Their Applications

In Table 6 a classification of conservation strategies is suggested. The most important strategies for existing systems are leak plugging, management, and mode mixing, and for future systems, input juggling, thrifty technology, and output

---

[7] The matter of net energy during periods of growth in the nuclear program has not been settled.

Table 6 Kinds of energy conservation[a]

IN EXISTING SYSTEMS

Leak Plugging

Reducing heat and cooling losses in life-support systems, adjusting energy systems that are not running at design efficiency, and eliminating unutilized or underutilized energy by retrofit in all energy systems. Examples include insulation in buildings, heat recovery in industry. Leak plugging techniques are generally implemented once, at little initial cost, and then remain passively effective.

Mode Mixing

Changing the mix of transportation to utilize modes requiring less energy per passenger- or ton-mile.

Energy Management

Turning off lights, heat, or cooling; changing thermostat settings; improving maintenance; driving more slowly; car-pooling; increasing load factors in public transportation. Involves small but important changes in energy use. May cause minor changes in life-style and habits. Energy management, unlike leak plugging, must be actively pursued by individuals or firms. Capital costs are usually small. Also called belt tightening.

IN NEW SYSTEM

Thrifty Technology

Introduction of innovative technology not in common usage in any energy system to increase the useful output of the system per unit of energy consumed. Examples include gas heat pumps for space and industrial heat, electric ignition of gas water heaters, or new propulsion systems in transportation.

Input Juggling

Change in the mix of existing economic or physical inputs to a given kind of output. Substitutions can be among energy forms or among economic variables such as labor, capital, design (a form of capital), and machines. Solar energy substitutes capital and labor for heating; returnable bottles substitute labor for the extra energy and materials requirements of throwaways. Some leak plugging also substitutes investment capital, design, and, indirectly, labor for energy expenditures.

Output Juggling

Changes in life-style, consumer preferences, investment practices, or shifts from manufacturing to services in the economy, which lead directly and indirectly to lower (or higher) energy requirements. Shifting to throwaway containers raises energy requirements per unit of beverages. Gardening at home instead of taking a Sunday drive lowers energy use. Smaller cars, changing urban housing patterns, increased vacationing closer to home are all examples of output juggling.

[a] Modified from (14).

juggling. If strategies like these are applied to important end uses of energy, savings of 10–50% of specific energy requirements result (Table 7). Applying all of these savings to the 1975 US energy economy would have the effect of reducing total

Table 7 Applying some energy conservation strategies: possible savings[a]

Area of Strategy	Potential Savings[b]	Notes	References
Homes, Buildings			
Space heating	5–8%	Insulation; heat pumps cut electric heating needs by 50%; gas heat pump is a possibility	51, 62–67
Air conditioners	>1%	Save peak power, fewer brownouts; insulation, design, window improvements reduce heat load	43, 62, 65, 66, 68, 79
Home appliances	2%	Fluorescent lights, better motors, and insulation in refrigerators; insulation in water heaters, electric igniters replace pilot lights	44, 69–73
Design of buildings	5%	Includes redefining lighting levels and tasks; total energy systems, conserving window systems; orientations that reduce energy needs	67, 73–80
Solar heating/cooling	10%	If 40% of today's heating and cooling were solar; economics depends on cost of glass, storage, and alternative fuels	62, 75, 81, 82

Total 1973 US demand: 30×10^{15} Btu. Hypothetical demand with maximum savings: 18×10^{15} Btu (62).

Area of Strategy	Potential Savings[b]	Notes	References
Industry			
Process heat	5–12%	Once through the facility is sufficient, with insulation, leak plugging; more sophisticated treatment requires redesign, pipes, cascading high-temperature processes with lower temperature demands	68, 83–86
Total energy: cogeneration of electricity and heat at factory	3–5%	Energy independence to factories; siting communities near industry is a possibility	45, 47, 62
Returnable bottles, use of recycled materials	1–3%	Many institutional problems as "no deposit, no return" becomes ingrained	54, 87–90

Total 1973 US demand: 26×10^{15} Btu. Hypothetical demand with maximum savings: 14.5×10^{15} Btu (62).

Area of Strategy	Potential Savings[b]	Notes	References
Transportation			
100% shift to 40% lighter cars	5%	Appreciable savings in energy cost of building car, refining oil; less congestion, less wear and tear, less pollution, with less traffic	37–39, 45, 91, 92
More careful driving cycle	1%		
Improved technical efficiency of autos	5%		
Switch one half of urban passenger miles to bus	2%		
Improved load factors in rail, bus, plane, mass transit	2%	Savings in freight and other passenger modes come mainly from fuller utilization of existing routes and higher load factors	6, 37–39, 91–94
Freight mode mix improved technical efficiency	2%		

Total US 1973 demand: 19×10^{15} Btu. Hypothetical demand with maximum savings: 10×10^{15} Btu. Note that transportation is nearly 100% dependent on liquid fuels.

Table 7 continued

Area of Strategy	Potential Savings[b]	Notes	References
Other			
More durable, repairable, and recyclable goods	?	Substitutes quality work for endless throwing away	
Urban design	?	Live near work, district heating, etc	13, 33, 50, 95, 96
Changes in consumer preferences	?	"Output Juggling"—vacation near home, ride a bike, work in the garden	

[a] Source: (14, 15, 62) and references cited.
[b] Given in percentage savings of total use (early 1970s). Savings figured as optimum achievable at 1980 energy prices. Individual savings do not add. See also (6–16, 62).

energy consumption by approximately 33%, as is illustrated in Figure 9 (6–13, 45, 62). Since these savings require months to decades to be achieved, they appear as slower growth in energy use. This scenario is often referred to as the "technical fix" (see 13). Listed in order of the amount of energy saved, the three most important energy conservation applications are (a) better heating and cooling of buildings, (b) use of the second law as a guide in industrial heat treatment, and (c) reduction of the weight of automobiles. The order of implementation time is probably the reverse.

Most of the potential for energy saving in homes and buildings comes from leak plugging and input juggling, as owners or builders invest in technology and design options (listed in Table 7) that lower the requirements for energy. Energy savings

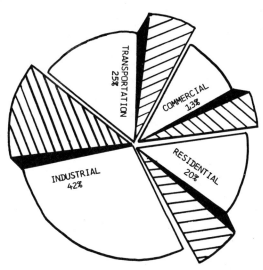

Figure 9 The energy pie of Figure 8, with savings summarized in the four sectors. Percentages given are today's breakdown of energy consumption. Shaded areas give savings. Source: (14) and Table 7.

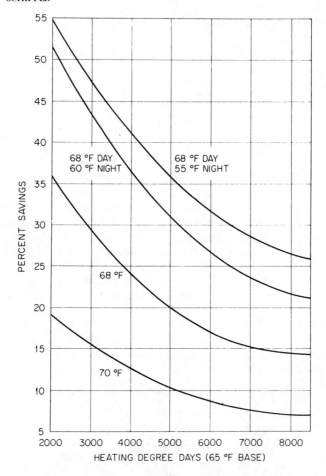

Figure 10 Management: predicted energy savings for several thermostat settings. (72°F is the reference setting, and night setback is from 10 PM to 6 AM.) Source: (97).

from management of systems also can be considerable, as Figure 10 demonstrates for thermostats [(97); for results of leak plugging and management, see (98)]. With new homes, input juggling can reduce energy consumption further with little or no change in the home environment (74, 75, 99), (Figure 11). The savings become larger as higher fuel prices justify increased use of insulation and other techniques that reduce heat loss. This illustrates well the degree to which input juggling might take place in construction of new systems.

Experiments with thrifty technology indicate, however, that even greater residential energy savings can be effected when the total home environment is considered. The Pennsylvania Power and Light Company's experimental low-energy

RAISING THE PRODUCTIVITY OF ENERGY USE   475

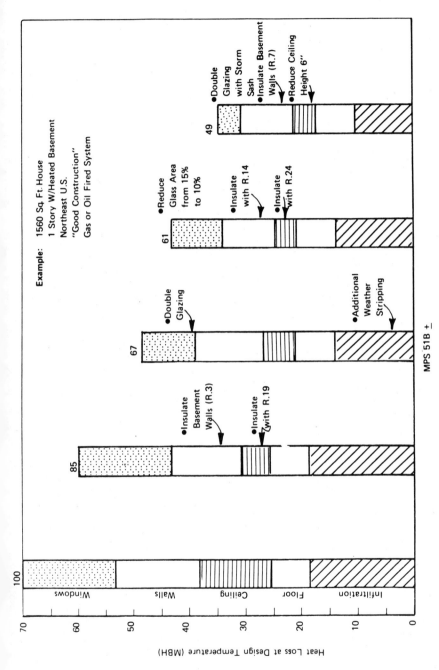

Figure 11 Leak plugging and input juggling: effect of more elaborate modifications of design and construction of a house on energy use for heating. Note the progressive savings. Source: (67).

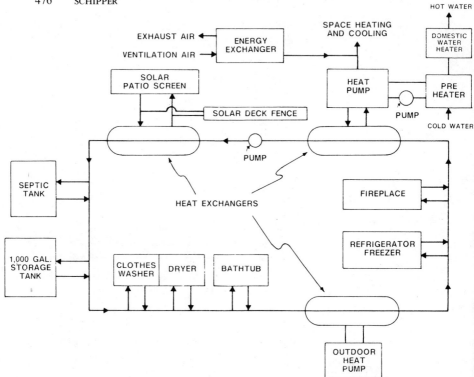

Figure 12 Input juggling and thrifty technology: schematic of the house of the future. This system is installed in the experimental low-energy house built by Pennsylvania Power and Light Co. in Allentown, Pa. Heat exchangers recapture as much heat as possible.

house, shown schematically in Figure 12, would reduce energy use by as much as two thirds compared with ordinary homes, by using solar collectors, heat pumps, heat recovery schemes, and conveniences such as automatic devices that close curtains at sunset to minimize heat losses through windows at night.[8] Other studies of actual and proposed buildings indicate that energy requirements in existing structures can be reduced by 20% in the short run and more over longer periods (76–78, 98), and consumption in new buildings can be reduced by nearly 50%, or more if solar technology is used[9] (78, 81, 82). Figure 13 illustrates the savings achieved in a new office building in Manchester, New Hampshire, by combining most of the conservation strategies discussed here.

In transportation, improvements in the physical efficiency of engines, such as greater use of diesel motors or development of advanced propulsion systems in cars (thrifty technology), would save a good deal of energy,[10] but reducing the

---

[8] Information released by Pennsylvania Power and Light Company, Allentown, Pa.
[9] See footnote 8.
[10] See discussion and references in (45). Also see (93).

Table 8 Identified savings potential of eight industrial plants[a]

Plant Type	Total Annual Energy Bill ($)[b]	Identified Savings ($)[b]	Percentage
Basic chemicals	5.5	2.39	43.4
Textiles	0.9	0.29	32.0
Agricultural chemicals	1.7	0.28	16.7
Oil refinery	10.3	1.12	10.8
Chemical intermediates	13.2	1.87	14.2
Food processing	1.1	0.33	30.1
Pulp and paper	5.3	1.70	31.5
Rubber and tires	2.9	0.47	16.4
Average	5.1	1.05	20.6

[a] Source: E. I. du Pont 1973 Energy Management Client List (83).
[b] In millions of dollars.

weight of cars (output juggling) would save even more (45, 94) (Figure 14). However, the degree to which people might reduce travel (output juggling) is difficult to estimate. Mode mixing depends on the availability and cost of alternative forms of passenger and freight transport, although increased load factors in public transit, including air transport, would save energy (38, 100). Improved freight handling procedures (management), including permission for interstate trucks to be fully loaded on return trips, would greatly increase energy efficiency in the movement of goods (6, 91).

Many of the savings in transportation are not dependent on technological or economic breakthroughs but instead await changes in socioeconomic patterns, habits, and laws. Trends can work in the wrong direction also; both a continuation of population dispersal into suburbia and exurbia[11] (101) and the development of complicated or extremely fast vehicles, such as SSTs, high-speed rail vehicles, or short-takeoff and short-landing aircraft, would increase the energy requirements per passenger-mile and the per capita miles of travel. These trends are an example of how output juggling increases energy use.

In energy-intensive industries, engineers and computers have been employed to effect immediate savings of 10–20% through leak plugging (6, 12, 68, 83–86). Table 8 shows some of the results of the energy conservation programs of E. I. du Pont Company (83). The costs of these programs and the actual equipment adjustments required were far less than the savings in fuel expenditures. Similar programs have been developed by I.B.M., Honeywell Corporation, and Johnson Controls.

In the longer run, greater energy savings in industry are realizable as input juggling and thrifty technology help reduce process energy requirements of materials

[11] See (101). The BART electric train system in the San Francisco Bay Area underscores one dilemma in the interaction of urban planning, geography, and energy use: if a rapid rail system stimulates commuters to live farther from work than they otherwise would, some or all of the energy conservation benefits of this system would be lost.

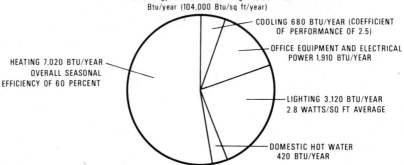
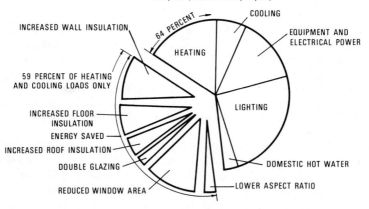
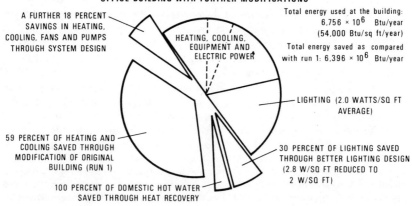

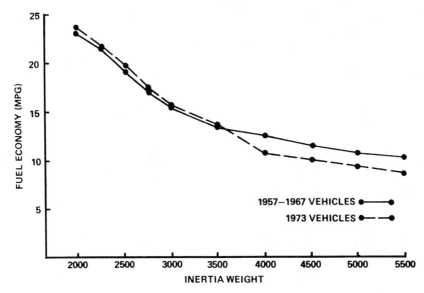

Figure 14 Design efficiencies: fuel economy vs inertial weight. The design efficiency of lighter cars is obvious. Source: (94).

closer to thermodynamic minima (45, 86; see also article by C. A. Berg in this volume) or as tasks are modified, especially to allow utilization of waste heat (45, 62). The economically optimum level of energy use will be sensitive to the price of fuel (86), but most process energy requirements declined during the past 25 years even as fuel prices fell (29).

Figure 15 shows this decline in energy consumed for an important energy-intensive product, cement (102). Input juggling has allowed each of the two processes to become more efficient, and at the same time the manufacturers have shifted to the more efficient dry process for future production. Similar short- and long-term trends have been observed in nearly every energy-intensive industry in the United States (103) and also in Sweden, where process energy requirements are generally lower (33).

← *Figure 13* All strategies: effect of progressive design modifications of an office building. Actual predictions of building being built for Government Services Administration, in Manchester, NH. Source: (77). Figures show equivalent energy units of 10^6 Btu/year.
  (a) In New England; 126,000 sq. ft.; design based on "typical" New England design criteria; weather data from Manchester, NH; wall U value = 0.3 Btu/°F-hr-ft; floor U value = 0.25; roof U value = 0.2; single glazing, 50% window/wall area ratio; shading coefficient = 0.5 (year round); 6 stories tall; 2:1 aspect ratio (length:width); long axis, north-south.
  (b) Wall, floor, roof U values = 0.06; double glazing, 10% window/wall area ratio; shading coefficient = 0.5 (year round).

The discussion of energy conservation strategies and their applications illustrates the different kinds of approaches to conservation that energy users can employ. Certainly research and development of techniques that promise even greater energy savings will play an important role in increasing the overall impact and pace of conservation. Understanding the barriers to more efficient energy use (see below) will also aid society and individuals in implementation of conservation options. But the economics of conservation will doubtless determine how fast new technologies will be adopted and how well society will work to circumvent difficulties in implementing them. Some aspects of economics are discussed in the next section.

Economics of Conservation

Economic efficiency is, in theory, the basis for decisions made by energy consumers. Economics determines how many extra dollars can be spent on improving the heat-transfer properties of an air conditioner, how much extra investment might be justified in choosing a diesel engine over a conventional one in an auto, or how much insulation should be added to a structure. All of these decisions affect physical efficiencies. In theory an informed energy consumer will invest in higher physical efficiency as long as the next increment of investment is less than the present value of extra energy saved [see especially (49)].

Most of the conservation options cited in Table 7 have been analyzed with respect to first cost, interest and taxes, and so forth. Fortunately the strategies that raise the physical efficiencies of energy systems raise the economic efficiencies as well. Some energy conservation strategies, such as insulation in refrigerators and

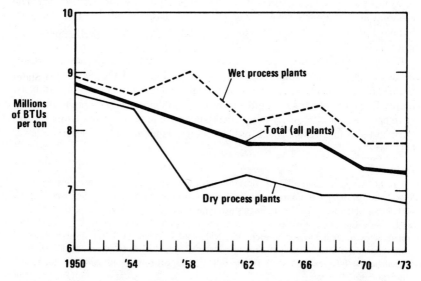

Figure 15 Historic trend in the cement industry's improvement in design efficiency. Source: (102).

water heaters, pay back the initial incremental investments in months (69–71), while others, such as industrial techniques, building insulation, and more efficient heating/cooling systems, require longer periods. Some options save money from the beginning, such as smaller cars and innovative designs of structures, while others, such as mass transit systems, are difficult to evaluate because much of the payoff accrues to society in general.

A few examples are worth noting here. Moyers (43) showed how much money a consumer could economically invest in a more efficient room air conditioner, based on the first-law efficiencies of models shown in Figure 5. His results (Figure 16) depend on number of hours of use per year, price of electricity, cost of money, and inside temperature desired. The calculations of Berg (68) (Table 9) compare yearly operating costs of different air conditioners and show clearly the savings achieved by higher efficiency. As Ross & Williams noted (62), it is usually less expensive for the user to invest in higher efficiency than for the local electricity utility to invest in extra peak capacity.

Other calculations show that proper insulation in buildings saves 20–50% of total energy and 10–30% of the expense required for heating or cooling (63) and that heat pumps save at least 30% of total energy and 20–30% of the total costs of using electric resistance heat (45, 62, 64, 104, 105). The alternative of providing extra energy by building more oil refineries, pipelines, and power plants would be far more expensive both in initial investment and in life-cycle costs (62).

Table 9 Efficiency and the cost of air conditioning[a]

Rated Cooling Capacity (Btu/hr)	Rated Current Demand (A)	Retail Price ($)	First-Law Efficiency (Btu/W-hr)[b]	Ten-Year Total (dollars/1000 Btu)
4000	8.8	100	3.96	84
	7.5	110	4.65	77.70
	7.5	125	4.65	81.45
	5.0	135	6.96	67.25
5000	9.5	120	4.58	74.90
	7.5	150	5.80	68.20
	7.5	150	5.80	70.20
	5.0	165	8.70	59.80
6000	9.1	160	5.34	67.30
	9.1	170	5.24	68.90
	7.5	170	6.96	61.80
	7.5	180	6.96	63.50
8000	12	200	5.80	67.30
	12	220	5.80	67.80

[a] Cost of air conditioning is inversely proportional to first-law efficiency. In each class of air conditioner, the model with the highest first-law efficiency has the lowest yearly cost. Examples worked out by Berg (68).

[b] Also called energy efficiency ratio (EER).

It is also important to consider the economic and energy impact of input juggling applied to all energy-using systems in a single structure. For example, Dubin et al evaluated a large number of energy conservation options that apply to electric homes in Florida (74). These are listed in Table 10, along with expected savings from each technology in kilowatt-hours per year consumed in the home (74). The incremental capital costs of each option can be compared with the present value of the electricity saved over the lifetime of the option. For a 30-year payoff time on these houses (15-year payoff time on appliances) the homeowner could invest at least 75¢ (38¢ on appliances) on conservation measures per kilowatt-hour saved yearly, assuming a present cost of 2.5¢/kW-hr and inflation in electricity cost at or greater than the general rate of interest. In these houses the cost of adding an overhang is not justified by energy savings alone. Solar heating is also difficult to justify on economic grounds because heating requirements in Florida are low, but a combined water heating/space conditioning system might be economically feasible.

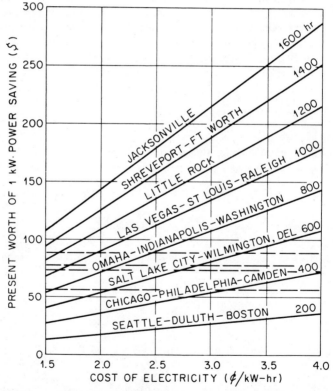

Figure 16 Present worth of 1-kW power savings as a function of annual hours and energy cost, for selected locations, 18% interest, 10-year amortization. If environmental costs were included, or if electricity was more costly during peak usage periods, consumers would be justified in paying much higher prices for more efficient air conditioners.

Table 10 Cost and energy savings of key design elements in all-electric homes[a]

Item	Extra Cost ($)	Annual Savings (kW-hr)	Investment Saved per kW-hr/yr
Styrofoam roof insulation	930		
5-ft overhang	1026		
Styrofoam wall insulation	375	2000	(65¢, insulation only)
Casement windows and french doors	800		
Door closers	80		
Single 3-ton water-to-air heat pump (design B)	1000	5420	20¢
Hot-water heater	300	3140	10¢
Refrigerator	100	650	15¢
Range	400	300	Self-cleaning, no charge
Dishwasher			
Air heater	25	110	12¢
Short wash		210	
Clothes washer	70	340	20¢
Dryer and clothes line	50	300	16¢
Freezer	50	300	16¢
Trash bins	(Save 130)	50	—
Fluorescent lighting	(Save 430)	840	—
Total	4646	13,700	

[a] Source: (74). Figures based on an all-electric home in Florida.

The kinds of analyses reviewed here for buildings have been performed on many other aspects of energy use (6, 68, 86). Such evaluations of the economic and energy savings potentials of conservation strategies allow planners to project future energy needs based on the economically optimal needs for energy. This is explored in the next section.

Conservation and the Energy Future

It has been emphasized in this review that the tasks for which energy is consumed can be carried out in different ways that require widely varying amounts of energy. Figure 17 illustrates a general relationship between economic efficiency (cost) and design efficiency. Some combination of energy with other inputs will provide the most economic use of all resources. This is shown symbolically by the optimal zero point at the minimum of the cost-energy curve in Figure 17. Environmental costs tend to push the optimum toward physically more efficient energy use, a fact rarely considered seriously in evaluations of the economics of solar energy or other supply or conservation options (22, 23, 106, 107; chapter by Budnitz and Holdren in this volume; also section on conservation and pollution, below).

That the energy use in most systems in 1970 or 1975 was economically inefficient is indicated by the position in Figure 18 of *actual* for those years. Here *projected* gives an estimate of the energy use that would occur if energy consumption grew at historical growth rates in spite of price increases. If energy costs continue to rise, however, then the economically optimal amount of specific energy consumption for tasks will fall, as is indicated by O_{90} in Figure 18.

The discussion of "barriers" (below) presents reasons why energy use today is not optimal and why certain governmental actions may be necessary if energy use is to approach the optimal point in the future. Energy conservation policy can be considered as that group of laws, standards, incentives, taxes, or other governmental, institutional, or private actions that aid this approach to the optimal energy utilization.

A curve like those in Figure 18 can be drawn for most energy-using activities or systems, as is done for space comfort in buildings in Figure 19. The data presented in Tables 8 or 9 would fit on a curve of this shape. The costs of providing comfort fall with conservation, first through retrofitting of insulation, then through more sophisticated designs and increased efficiency of new structures, as long as marginal benefits exceed costs. Some options, like solar heating and cooling, may not be the most economic in every case, as is symbolized by the upturn of the cost curve as energy use falls further. As fuel costs rise, however, solar heating and

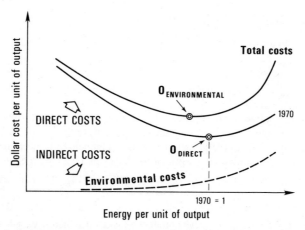

Figure 17 Optimizing energy use. There are many combinations of energy use and other economic inputs that provide a given output. This figure shows the total cost of that output for different amounts of energy use. As energy prices rise, the optimum moves toward lower energy use. The shift of optimum from O_D to O_E is exaggerated to show the effect of the environmental cost of energy on optimal use. Even if energy use were economically efficient (based on direct costs), inclusion of environmental costs, which tend to rise nonlinearly with increased energy use, would shift the optimal point of energy use toward slightly lower energy use. The arrows on the axes point to higher energy and dollar costs.

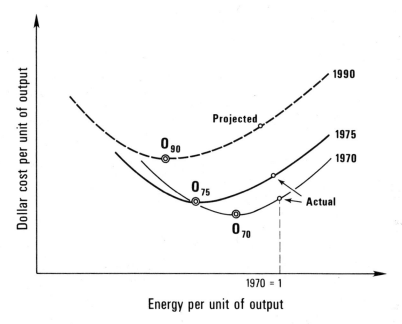

Figure 18 Optimizing energy use. Energy use was economically inefficient in 1970 and 1975, as symbolized by the differences between *actual* and O. If historical growth in use persists despite higher energy prices, use will be even more economically inefficient than use in 1975.

cooling will become economically efficient in more and more instances, in comparison with fueled systems[12] (see also 43, 49, 63).

Amassing a large number of curves such as in Figure 19 is a formidable task if one is to include all of the options available to the major energy-consuming systems in this country. However, such curves would then allow projection of energy needs both on the basis of desired goods and services and on the basis of the most economic, specific energy needs of these goods and services. The cost of conserving an additional Btu for each task, relative to today's consumption, could be compared with the cost of producing each additional Btu beyond some reference price and supply that is assured. Adding up these curves would indicate ranges and costs of total supply and demand. The area where these demand and supply curves cross would indicate the most economic amount of energy consumption in a future year, for a given mix of tasks. Other balances of supply and demand would probably cost more than the optimum. This is indicated in Figure 20. Non-market factors,

---

[12] See (82) for a comparison of costs of fueled and solar systems in various locations. It is generally anticipated that solar heating/cooling systems will fall in cost through economies of scale in manufacture as well as through technological advances.

such as environmental costs or social variables, tend to diffuse the optimal point somewhat, as would differences in judgement and uncertainties about future technologies. If policymakers or their constituents felt that a significant departure from this optimum were either desirable or dangerous, policies could be adopted that would aid or inhibit the economic and technical factors that shape energy use. In California, for example, the Energy Resources Conservation and Development Commission has been empowered to develop such policies.

At the same time, research and development of conservation applications to present and future patterns of energy use can change the shapes of Figures 19 and 20 by making energy use even more economically efficient. For example, improvements in the heat-transfer properties of materials could lower the electricity requirements of heat pumps, and development of new kinds of insulation could lower the cost of insulating structures. Such applications are complementary approaches to the same goal—aiding society in deriving more well-being from energy use.

Recent systematic studies of energy options and policies, such as the Ford Foundation–sponsored Energy Policy Project (13) or ERDA's "Creating Energy Choices for the Future" (16), reflect the kind of economic evaluations suggested here, as well as the prospects for increased energy savings made possible through

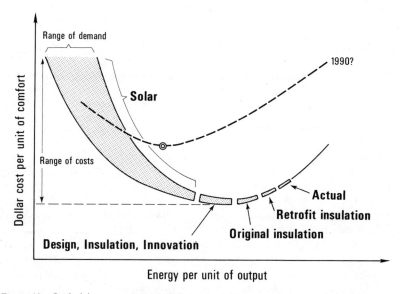

Figure 19 Optimizing energy use: buildings. Actual conservation strategies, such as those listed in Tables 9 and 10, can be displayed in curves similar to this one. Solar heating/cooling is not economic in many buildings at 1975 energy prices, but it would become more economic in the future (1990) as energy prices rise. The ranges for cost and energy savings of solar heating/cooling reflect uncertainties as well as the use of nonsolar backup systems. Comfort can be defined by using temperature, humidity, and other physical-physiological parameters.

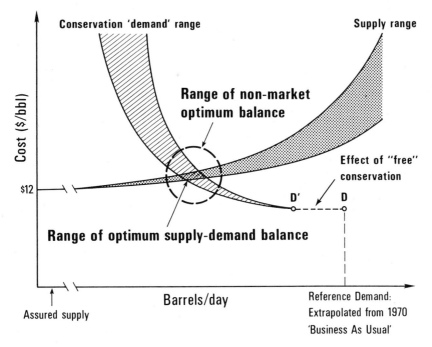

Figure 20 Generalized comparison of future costs of increasing supply and increasing efficiency. Compared with historical extrapolations, demand would be lower through conservation that takes place with immediate cost savings—"free" conservation. Beyond that point other economic factors substitute for energy (as explained in the text) until the cost of saving an additional barrel of oil, or barrel of capacity per day, exceeds the marginal cost of generating one. Because of qualitative and quantitative uncertainties, the optimal supply-demand balance is best represented by the darker shaded area, with the circle indicating the effect of non-market factors on this optimum.

R & D into energy use and conservation. Energy projections from these studies, which explicitly recognize the effects of more efficient energy use, are summarized in Figures 21 and 22. Although some observers feel that energy demand cannot be modified significantly by technical and economic changes (108, 109), it is clear that an enormous potential exists for raising the efficiency of energy use. It is important, however, to consider the long-term implications of realizing this potential.

LONG-TERM IMPLICATIONS OF EFFICIENT ENERGY USE

Total expenses for energy in the United States average around 10% of the GNP (13, 14), and this figure is expected to grow. Therefore, a substantial conservation program means that billions of dollars will be redirected from energy expenditures to non-energy expenditures. Policymakers considering efficiency standards, taxes, restrictions on energy use, or subsidies for energy efficiency will want to know what might happen to the economy when energy is used more efficiently. This discussion explores the effects of conservation on employment, pollution, climate for capital investment, and energy resources.

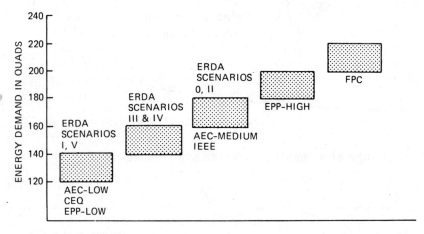

Figure 21 Comparison of forecasts for estimated US energy demand for the year 2000. ERDA scenarios are found in (16), EPP (Energy Policy Project) (13); others are discussed in (1, 4).

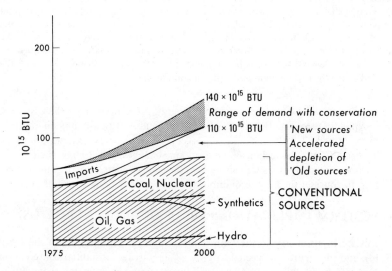

Figure 22 One possible approach to the "low" scenarios of Figure 21. Source: (14).

Employment

The use of energy is essential to nearly all employment in this country, but energy requirements of different goods and services vary greatly. Input-output techniques reveal the average amounts of energy and labor required to satisfy the demand for a product or service (50–52). These values include indirect energy and labor required by industries whose output was used by the producing industries, and the energy and labor required by suppliers of those industries, and so on.

Figure 23 displays the energy and labor requirements per dollar of demand for some goods and services, and Table 11 gives the intensities of some important personal consumption activities. Energy forms (and raw materials) have low labor intensity and high energy intensity. Manufacturing is intermediate, whereas services are labor-intensive but not energy-intensive. Closer inspection of the energy industries shows them to be very capital-intensive but far less labor-intensive than the economy as a whole [see Tables 3 and 4 in (14)].

Table 11 Energy and labor intensities of the top 20 (dollarwise) personal consumption activities in 1971[a]

Personal Consumption Expenditure— Sector Description	Energy Intensity (Btu/$)	Labor Intensity (Jobs/$1000)
Electricity	502,473	0.04363
Gasoline and oil	480,672	0.07296
Cleaning preparations	78,120	0.07332
Kitchen and household appliances	58,724	0.09551
New and used cars	55,603	0.07754
Other durable house furniture	45,493	0.08948
Food purchases	41,100	0.08528
Furniture	36,664	0.09176
Women and children's clothing	33,065	0.10008
Meals and beverages	32,398	0.08756
Men and boys' clothing	31,442	0.09845
Religious and welfare activity	27,791	0.08636
Privately controlled hospitals	26,121	0.17189
Automobile repair and maintenance	23,544	0.04839
Financial interests except insurance co.	21,520	0.07845
Tobacco products	19,818	0.05845
Telephone and telegraph	19,043	0.05493
Tenant occupancy, non-farm dwelling	18,324	0.03258
Physicians	10,271	0.03258
Owner occupancy, non-farm dwelling	8,250	0.01676
Average, including energy purchases	70,000	0.08000
Average, non-energy purchases only	52,000[b]	—

[a] Source: (100).
[b] 1967 figure. The corresponding 1967 figure for average including energy was 80,000 Btu/$. Source: R. Herendeen, private communications.

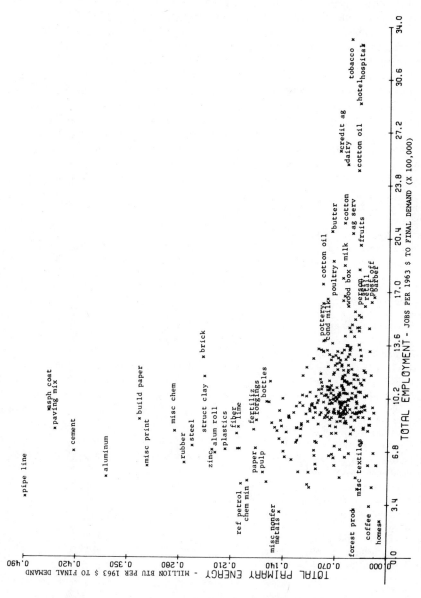

Figure 23 Total direct and indirect energy and labor intensities for 362 economic sectors in 1963. Source: (51). 1971 figures are similar.

The strategies of energy conservation considered here substitute capital, materials, labor, know-how, or management for energy. Compare, for example, two air conditioners of equal capacity, operating in similar homes under similar loads in the same climatic region, one requiring half the power of the other. If a consumer buys the more efficient unit, some of the money otherwise spent on energy is used for extra materials and labor, and this expenditure results in a more carefully constructed, more efficient air conditioner. Since manufacturing is generally more labor-intensive than electric utilities, the redirection of spending—from paying for electricity to investment in a more efficient unit—raises the total demand for labor per unit of air conditioners and still provides for the consumer's desire for comfort [see Table 11, or details in (50)]. When the consumer spends the money he saved by energy conservation, his new purchases will require increased labor, in contrast to buying electricity (50–52). The result is more goods or services and more employment, with less energy consumed. The notion that welfare or employment can only grow in step with energy use completely ignores the effects of strategies that increase efficiency.

In industries that conserve energy, employment will generally increase, since nearly every energy conservation strategy calls for energy specialists to monitor and adjust energy usage in the plant or building. Similarly, the implementation of long-range conservation plans (input juggling, thrifty technology) calls for equipment, consultants, architects and designers, and other specialists not otherwise required. The costs of changing to a more efficient use of energy are, of course, borne out of savings from energy bills, and the net dollar savings is either passed on to consumers, reinvested, or taken out in profits. In these and most other conservation applications, energy expenditures are replaced by non-energy expenditures, which generally increase employment.

One of the arguments for output juggling has been the observation that goods and services that are less energy-intensive tend to be labor-intensive (Table 11). If consumers preferred less energy-intensive items, whether in response to higher energy prices or from a desire to save energy for the public good [as suggested in (95) concerning gifts], the energy/GNP ratio could decrease even with no changes in the technologies of energy utilization in production of goods or in the ratio of services to goods. Some changes in consumer preferences, such as greater use of buses, could improve design efficiency by increasing load factors (100); changes in travel and tourism would affect transportation energy use; and changes in aesthetic and architectural preferences would alter energy demands in housing and buildings (110).

Hannon (50) concluded that output juggling cannot save a large percentage of US energy use. On the other hand, a study of Sweden (33) suggests that substitution of social services for large cars or other energy-intensive personal consumption does reduce energy consumption somewhat for a given level of GNP. But it may not be necessary to engage in output juggling to save energy, since the other techniques discussed here promise such large savings with little or no change either in life-style or in the mix of goods and services in final demand.

It is important to remember that many of the changing patterns of employment and intermediate demands effected by a policy of energy conservation would appear in any case as a result of natural market forces, although perhaps over a longer period of time. Certainly any changes in the structure of the economy, whether gradual or forced by an embargo, could create temporary unemployment and some social dislocation. The short-term effects of changing energy-use patterns will continue to raise issues, and one must insist that those affected not be asked to bear a burden that is too heavy. Society can and must cushion these effects during the transition to a period of more efficient energy utilization.

Conservation and Pollution

When energy-conserving practices are adopted, the net result is an increase of useful output per unit of energy input. This has important environmental consequences because energy use and harvesting are among the most polluting activities in our economy. If a homeowner in the Tennessee Valley Authority service area replaces electric resistance heating with a heat pump (saving approximately 50% of the electricity commonly used to heat homes there), less coal need be mined [about

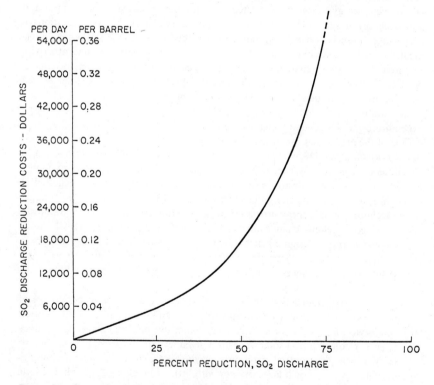

Figure 24 Costs of reducing discharge of SO_2 from a typical US petroleum refinery. Source: (21).

Figure 25 Symbolic representation of the double interaction of energy and well-being. The beneficial use of energy in the economy is partially offset by the adverse impact that energy use has on the environment, affecting well-being directly as well as through the economy itself. Drawing due to J. Holdren, private communication.

3–4 tons less per house per year (105)], fewer power plants need be built, less cooling water is used, and less air, land, and water pollution occurs for each night of comfort in the winter or day of cooling in the summer. Similar considerations apply to industries that conserve energy. Since the demands for materials required by energy-saving technologies are usually only slightly higher than correspondingly less efficient options, the reduction in all forms of pollution will be appreciable.

The effect of conservation on pollution is important, for as the number of various polluters grows and as the amount of pollutants grows in kind and total amount, more abatement technology will be required to hold constant the concentration of pollutants in the environment (107). But costs of pollution control tend to increase faster than the increments of control (Figure 24). The total bill for abatement of pollution could thus rise in the future faster than the GNP itself (106, 111). Pollution also means costs to health, welfare, and property (112–114). More efficient energy use reduces all of these costs and the pollution as well.

It is often said that "large amounts of energy will be needed to clean up the environment" (115). Actually the energy requirements for environmental control, while not insignificant, are not large. It has been estimated that energy requirements for pollution control raise total energy consumption by only 2–4% (116, 117). Reduced automobile size, increased use of mass transit, or longer-lived cars all reduce pollution per passenger-mile, as well as provide energy savings. In addition the percentage of fuel penalty imposed by pollution control on automobile emissions is generally smaller for lighter cars than for heavier ones (94, 116). Moreover, recycling, which reduces solid wastes, also saves energy (87–89, 117), and some solid waste can even be used as fuel.

Use of energy has two effects on well-being—it is positive through its application to satisfying human needs and negative through its adverse effects on the environment (see Figure 25). Increased economic and human well-being obtained per unit of energy, through conservation, reduces the environmental costs per unit of well-being, and this adds to our welfare.

Energy Conservation and Investment Requirements

New energy production facilities demand large, and ever increasing, amounts of capital per Btu produced or per unit of capacity. The energy sources most often cited as vital to our energy future—nuclear energy, shale oil, coal gasification, enhanced recovery of oil and natural gas—would, if the historical trend continued, require capital investments over the next 25 years totaling trillions of dollars (118–120), making the energy industry's share of all investment grow faster than the economy as a whole. This means that consumers, industry, and the government would have to forego both consumption and investment, through higher interest rates, higher prices, or higher taxes, in order to finance the expansion of the energy industry.

Higher energy productivity, on the other hand, slows the growth of investment in energy systems to a more manageable rate, easing pressure on interest rates and allowing more personal consumption of other investments, because conservation is cheaper. If the criterion of greatest marginal benefit is applied to investments, it can be shown that proper conservation techniques (see Table 7) save more energy than new energy sources can produce, per dollar invested (62). Some of the money that would have been invested in greater energy production should be invested instead in greater energy productivity—that is, in conservation practices. This point, that

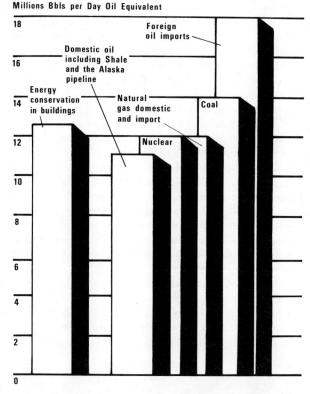

Figure 26 Comparison of the potential for energy conservation in buildings with some supply options, 1990. Source: (121), with supply options taken from (1).

conservation is a cheaper substitute for greater production (at least up to the point shown in Figure 20), is not sufficiently appreciated.

The cost of an investment in Btu, or capacity added, is especially important in the case of buildings, which tend to outlast most energy production facilities. The American Institute of Architects (121, 122) estimates that investments in extra efficiency in new and old buildings could economically replace the supply equivalent of 12 million barrels of oil per day by 1990 (Figure 26). Again, a Btu saved costs less than a Btu harvested.

Similar estimates have been made regarding industry. A recent study by Dow Chemical Company (47) estimated that if both electricity and heat were co-generated on site by large industrial power users, with power and steam being shared with existing public utilities, the required investment of $13 billion would replace an investment in utilities alone of $29 billion and would save the equivalent of 725,000 bbl/day of oil or equivalent. In the Technical Fix scenarios of the Ford Foundation's energy study, a reduction of energy demand by 33% in the year 2000, compared with historical growth, would mean a capital savings of $300 billion (13).

Growth in energy demand would have to be met by construction of energy systems not yet in existence. One can therefore estimate the energy-harvesting facilities that would not be needed if future energy-use technologies become more efficient. Electric resistance heating and large, inefficiently designed office buildings would have been major ingredients in the growth in electricity use. Poorly insulated homes, leaky industrial processes, and fuel-hungry autos make up much of the future demand for fuels based on historical growth [see (5) for a naive description of future demand]. An insulation program alone that would result in a savings of one third of the energy used to heat homes and buildings (more in new structures, and less in existing ones) would replace the equivalent energy output of oil refineries totaling 2 million bbl/day or 75 1000-MW nuclear power plants operating at 60% capacity factor.[13] Replacing resistance heating with heat pumps would reduce system capacity needs still further (64, 104, 105).

Therefore, what conservation means to investors is that relatively more energy-related investment should take place at the end point of energy use, displacing even more dollars on the energy production side. Since the ingredients in more efficient structures and industrial processes are usually highly dependent on careful engineering, quality work, and tender loving care, one expects that the employment requirements of conservation investments will be slightly higher than those for the additional investment in energy production. This point has already been made in the discussion of employment.

Conservation and Marginal Energy Sources

Conservation has another effect on the future of energy production, by slowing the rise in physical and dollar costs associated with marginal, less accessible, more

---

[13] J. Holdren, personal communication. These numbers are easy to calculate. Heat pumps are becoming increasingly common in areas where electric resistance heating is appreciable, like the TVA service territory (105).

energy-costly sources of energy. "Scarcity" of an energy resource really means "high entropy" (high degree of dispersal) of the energy fuel; increased amounts of high-quality, low-entropy energy fuels are needed to recover each Btu of marginal fuels.

The degree to which the real marginal cost of energy production from conventional sources is rising today is evident from Figure 27 (21). The rise in cost of drilling is nearly exponential with depth. Increased requirements for earth moving, drilling, and water and waste disposal forewarn of rising environmental spoilage per unit of net energy actually gained from all energy harvesting. Lignite, with only half the Btu content of bituminous coal, requires substantially larger environmental disruption per Btu recovered and produces more ash and sulfur per Btu of heat obtained in a boiler (58). A nuclear power program can consume a large fraction of its own output during excessively rapid expansion (59, 60), and tertiary oil recovery, shale oil, and coal gasification produce far less net energy than the actual Btu content of the fuel resource in situ (21). Vyas & Bodle (123) have estimated the net energy output from various synthetic fuel processes (Table 12).

Expensive or marginal energy resources, no matter how large they may be in Btu content, pose tremendous environmental problems if they are to be exploited on a

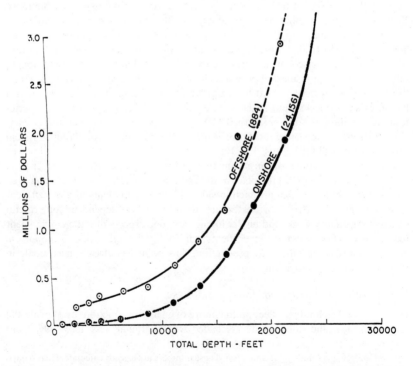

Figure 27 Cost of drilling and equipping oil and gas wells (including dry holes) in the United States, 1971. Source: (21).

Table 12 Net energy output from various synthetic fuel processes[a]

Process	Lurgi-Gas	Hygas	CSF-Coal Process	Coal-Methanol	Shale-Syncrude
Percentage of Btu's recovered in desired form	56.2	59.7	55.0	39.6	66.5
Other by-products by Btu content	15.3	8.2	12.2	1.5	7.4
Total	71	68	67	41	74

[a] Source: (123). The estimates do not include energy expenditures for capital equipment or harvesting (earth moving, crushing, water supply), nor are transportation energy requirements given. Figures are percentages of Btu-inputs.

large scale compared with 1975 energy demands in the United States (106).[14] Environmental questions are also important for the technology of nuclear power where the risks from mismanagement are great (124). Conservation, in promoting slower growth in energy harvesting, allows society to buy time for the testing and environmental engineering that are required to insure safe and clean recovery of net energy and other social benefits from these resources. "Income" sources such as solar or geothermal energy, which are large in supply but limited in utilization rate, are better suited to an energy budget reduced by conservation. Solar energy, in particular, offers environmental advantages because it makes use of already existing solar heat (125). A society's choices of which energy sources it will exploit depend greatly on what total rate of energy utilization will be required. Conservation makes that rate more manageable.

Social considerations related to future energy sources are also important. Certainly the cost to society of energy shortages is large. But there will be many social costs of the new energy sources, which are far more complex technologically and environmentally than energy conservation strategies. We must consider fully the social costs of building the Alaska pipeline, of offshore oil development with its hazard of oil spills, of creating and maintaining large-scale strip mining, shale oil, or coal gasification centers in the West. Distressing is the lack of understanding about the social implications of a large-scale commitment to nuclear power, including the social cost of the eternal vigilance that proponents and critics of nuclear power agree is required in order to manage the nuclear waste products, which long outlast the power plants themselves (124, 126, 127). These kinds of social issues must influence decision-making about energy use, especially since they favor more efficient energy use than would be dictated solely by microeconomic considerations.[15]

[14] See the discussions in (21) or (114). It is important to consider the various costs associated with different levels of "clean" in production from "new" sources. As Figure 24 suggests, these costs will also rise with the level of use of "new" sources.

[15] Page (23) discusses some of the difficulties of applying economics to environmental problems that stretch out over decades or centuries.

NONTECHNICAL BARRIERS TO EFFICIENT ENERGY UTILIZATION

Studies of insulation (63) and air conditioners (43) suggested that even before energy prices began to rise in the early 1970s energy use for space comfort was far from optimal. This section reviews some of the reasons why energy use in these and other applications today may not be economically efficient. These reasons suggest that important barriers to more efficient energy use in the future still exist.

Economists invoke the mechanism of the marketplace as the measure of efficient use of resources. Price determines the optimal use of resources, balancing the rate of supply of energy with the rate of demand. When the price of a commodity rises, the user may elect to use less of that commodity, substitute, or do without the

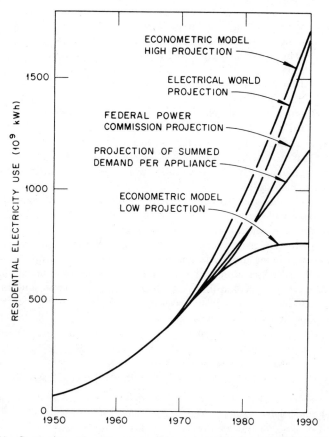

Figure 28 Comparison of some econometric projections of residential electricity use. Source: (41).

benefits of that commodity. If use is sensitive to price, the relationship between use and price is termed *elastic*. If, on the other hand, changes in prices do not induce changes in use, or induce relatively small changes in use compared with the change in price, the relationship is termed *inelastic* [for studies in elasticity of energy use, see (128–130)].

Elasticity is really a characteristic of human behavior, and economic models that predict this behavior are important. Some predictions are illustrated in Figure 28. But responses to higher prices are slow to come about: obsolete systems must be replaced, new technologies must be developed, or the fact must be "discovered" that energy is being wasted. That short-run elasticities to energy prices may be low should not discourage society from expecting that long-run changes in the patterns of energy use will in fact come about as a result of higher prices.

Unfortunately, few studies of elasticity evaluate the efficiency of energy use or model the technical options available to the consumer of energy. Physical analysis of changing economic conditions as well as the technologies involved reveals surprisingly effective options for energy conservation that are not predicted by conventional econometric studies. The *summed demand* in Figure 28, or the Brookhaven Model [results of which are shown in (16)], explicitly takes technologies into account. The projections of Makhijani and Lichtenberg (72), illustrated in Figure 29, show the often dramatic difference between efficient scenarios and pure exponential growth.

Energy Prices

As was suggested above, the price of various fuels plays a decisive role in determining which of the energy-use options is the most economically efficient. For many years, however, the real prices of energy fuels and electricity fell (13, 41), stimulating growth in energy demand while inhibiting concern for efficiency. Since 1970, however, energy prices have begun to rise dramatically, so it can be expected that the rate of growth of energy use, as well as energy-use patterns, will begin to change in response to changing energy prices. The response can lead to increased efficiency of energy use through application of the first five conservation strategies listed in Table 6 or through output juggling, as energy users turn to products, services, or materials less affected by rising energy costs. Additionally some output juggling may be expected in the form of reduced automobile travel, lower thermostat settings in winter, reduction in hot water use and its temperature, and so on.

There are, however, also distortions and imperfections in energy prices themselves. While the fuel and capital costs of electric utilities are rising, electric power is still sold on a declining block basis in most parts of the country: the more one uses, the less one pays per kilowatt-hour (131–133). As long as large amounts of electric power were cheap, the aluminum smelters ignored efficiency. Similarly, the attractive prospects for large savings in fuel from on-site cogeneration of process steam and electricity are severely inhibited by the present rate structure for electricity (47). Extra charges for peak-period usage, when production is most expensive, do not yet exist in the United States, although the maintenance of a peak reserve is expensive in terms of capital, and the peaking equipment is usually less efficient

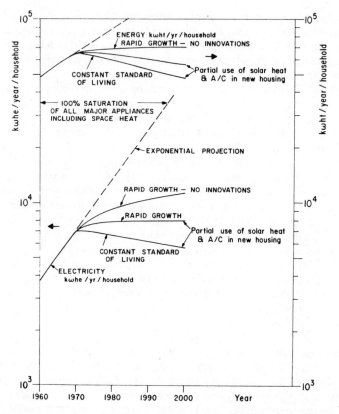

Figure 29 Some projections of residential electricity and energy demand. Higher efficiency and innovations make continued exponential growth (predicted by some models) superfluous. Source: (72).

than the baseload equipment (62). It now appears that there will be revisions in electricity pricing schedules in the near future (see the chapter by Sander in this volume).

Other price distortions are equally harmful. Price controls keep energy prices below market prices and sometimes below actual marginal costs. The depletion allowance on fuels allows some of the "risk" cost of exploration to be paid by taxpayers instead of the fuel users, by allowing producers a substantial tax benefit; other pricing practices for petroleum do the same (134). Elsewhere governmental policy subsidizes housing, highways, or air travel, regulates natural gas in interstate commerce but not in the state produced, and so on. The result is, of course, a price system wildly distorted from real costs. Economists have long warned that this leads to misallocation of scarce resources.

Further distortions in the price system are caused by the exclusion of environmental costs from market prices. Political battles take place in the US Congress over such environmental issues as a sulfur tax on coal, a reclamation tax on strip mining, smog devices on cars, and the cost of using low-sulfur oil or lead-free gasoline.[16] The environmental risks of nuclear power have not been quantified sufficiently to indicate whether nuclear power is underpriced, but the cost of safety is of concern to the nuclear industry, to regulators, and to opponents of nuclear power (124, 126, 127).

As public discussions of the risks and benefits of the various options indicate, environmental, safety, and social costs have come to dominate much of the concern (and research) regarding these technologies. Since these problems are largely excluded from the market system of prices, that system may be increasingly inappropriate to deal with energy problems. Internalizing the external costs of energy use will be a difficult political process because politicians and consumers alike will be faced with charging themselves for that which they apparently now get without costs: pollution.

It is difficult to imagine that appreciable conservation efforts will take place if energy prices continue to be controlled or rolled back, since few users will bother with the investment or the thought that is necessary to effect conservation. Voluntarism works well only for "the other guy." At the same time higher prices of energy, whether caused by real scarcity, monopoly power, fuel taxes, or environmental costs, are by no means an automatic cure by themselves for energy waste, because of the barriers to conservation discussed here.

Barriers to Efficient Energy Use

As costs of energy and systems change, different systems will be more economic (cheaper) than others. In a free market, each consumer will in theory adjust her energy use to the economic optimum, responding to price changes. Unfortunately economic systems, enterprises, entrepreneurs, and private consumers cannot readily respond to changes in energy prices. To be able to do so, energy users must have complete and accurate information about the energy and life-cycle costs of systems and about the cost of energy embodied in various products. This information has been difficult to find. Recently labeling of appliances and automobiles has begun to raise both the consciousness and the level of available information about energy (135). The government has also aided homeowners and building owners by publishing several books of suggestions and conservation procedures as well as displaying the results from efforts in its own buildings (67, 136); many utilities and fuel companies have done the same. Some industries using large amounts of energy, such as those in travel, plastics, and aluminum, take pains to measure and plan their energy use, but homeowners, small businessmen, and renters of office space

[16] Reader should consult the transcripts of hearings held by Congress. The Committees on Commerce or Interior and Insular Affairs are the most active in the Senate, whereas the Committees on Science and Astronautics, Ways and Means, Interior and Insular Affairs, and the Subcommittee on Energy and Power of the House hold most of the energy hearings where policy is debated. In addition to these committees, the Senate Committee on Public Works has debated most of the important environmental policy issues.

can rarely afford this practice individually, and few understand the theory and practice of life-cycle costing, upon which economic efficiency depends heavily.

Most important, energy use is rarely a goal in itself but occurs in conjunction with other processes toward personal or economic ends. The price of energy is

Table 13 The energy content of selected goods and services in 1971: partial list[a]

Product	Energy Content (Btu/$)	Gasoline Equivalent (gal)	Energy Value Content (¢/$)
Plastics	218,097	1.74	13.2
Man-made fibers	202,641	1.62	7.4
Paper mills	177,567	1.42	7.9
Air transport	152,363	1.22	12.0
Metal cans	136,961	1.10	7.3
Water, sanitary services	116,644	0.93	11.6
Metal doors	109,875	0.88	6.7
Cooking oils	94,195	0.75	7.1
Fabricated metal products	91,977	0.74	5.8
Metal household furniture	91,314	0.73	5.9
Knit fabric mills	88,991	0.72	6.5
Toilet preparations	85,671	0.70	5.1
Blinds, shades	81,472	0.65	6.3
Floor coverings	79,323	0.63	5.8
House furnishings	75,853	0.61	5.3
Poultry, eggs	75,156	0.60	7.3
Electric housewares	74,042	0.59	5.6
Canned fruit, vegetables	72,240	0.58	5.2
Motor vehicles and parts	70,003	0.56	5.9
Photographic equipment	64,718	0.52	3.8
Mattresses	63,446	0.51	4.5
New residential construction	60,218	0.48	4.5
Boat building	60,076	0.48	4.9
Food preparation	58,690	0.47	4.8
Soft drinks	55,142	0.44	4.5
Upholstered household furniture	51,331	0.41	4.1
Cutlery	50,021	0.40	4.0
Apparel, purchased materials	45,905	0.37	4.0
Alcoholic beverages	43,084	0.34	3.0
Hotels	40,326	0.32	5.4
Hospitals	38,364	0.30	5.4
Retail trade	32,710	0.26	4.4
Insurance carriers	31,423	0.25	4.4
Miscellaneous professional services	26,548	0.21	4.3
Banking	19,202	0.15	2.5
Doctors, dentists	15,477	0.12	1.9

[a] Source: (136a). These values are for producer's prices and do not take into account markup to retail price, about 66%.

Table 14 The cost of electricity as a percentage of production costs: US manufacturing industries 1939–1967[a]

Industry	Cost of Purchased Power (% of Product Value)	
	1939	1967
Primary metal industries	1.8	1.8
Fabricated metal products		0.6
Chemicals and allied products	2.0	1.7
Paper and allied products	3.9	1.9
Food and kindred products	1.1	0.4
Transportation equipment	0.7	0.4
Petroleum and coal products	1.2	0.8
Stone, clay, and glass products	3.2	1.5
Textile mill products	2.3	0.9
Electrical machinery	1.1	0.5
Machinery, except electrical	0.9	0.5
Rubber products	1.6	1.0
Lumber and wood products	1.8	0.9
Printing and publishing	0.8	0.4
Apparel and related products	0.4	0.3
Instruments and related products	—	0.4
Furniture and fixtures	1.0	0.5
Leather and leather products	0.6	0.4
Tobacco manufactures	0.2	0.2
Miscellaneous manufactures[b]	0.9	0.4
All manufacturing	1.41	0.79

[a] Source: Edison Electric Institute, 1973.
[b] Includes ordnance and accessories.

usually a small fraction of the cost of consumer goods (Table 13) or the cost of production (Table 14). The relatively low cost of energy compared with other expenses may mean that even where conservation is economic, the potential cost savings may be ignored.

The problems of sharply rising energy costs are particularly acute for energy-intensive industries addicted to unusually cheap or subsidized energy supplies (137). For airlines, the cost of fuel is now about 20% of the cost of doing business.[17] In these and other energy-intensive activities[18] the responses to higher energy prices—input juggling and thrifty technology—will be limited only by the time it takes to raise money and replace presently inefficient equipment. For example, the aluminum industry has a new process for producing aluminum from ore, which

[17] The major airlines sent telegrams to President Ford after his announced intention (January 1975) to "free" oil prices. The airlines asked for a percentage quota of their 1972 fuel use—at controlled prices—instead of the higher prices that the President's action would have allowed.

[18] The average dollar of expenditure for personal consumption requires about 80,000 Btu.

reduces energy requirements by one third (138). With the disappearance of low-cost electricity (the main ingredient in aluminum production besides the ore), smelters will turn more quickly to the new process even with its higher initial cost, because its life-cycle costs are lower.

The fact that energy consumption is so dependent on existing capital equipment is a general barrier common to all conservation strategies. If relative prices of certain foods change, consumers can alter their demands on a day-to-day basis, independent of past expenditures or present possessions. On the other hand, energy use is largely pre-determined by the existing stock of devices and structures, so that a time lag can be expected between the rise in the price of energy and the response to conserve it. For a homeowner, the physical condition and design of the house and heating system, and the weather, determine how much heating fuel is needed to maintain a given indoor temperature. The homeowner can belt-tighten and plug leaks through improved maintenance, retrofitting of insulation devices, and use of warm clothes and lower temperatures, but the savings are usually smaller than those available if the house and furnace had been optimized (input juggled) in the first place. The same applies to large buildings. It is a fact that the energy savings possible from building energy-efficient structures in the future are as much as the total energy consumption of today's automobiles (15, 62) (Table 7). But while automobiles and industrial equipment tend to be replaced within ten years, buildings remain in use for decades or centuries. Thus if wastage of energy is built into structures, only feeble responses can be expected to the rising price of energy.

Even with complete information, it may not be possible for energy users to juggle inputs or plug leaks. Homeowners may not be able to choose homes (new or used) on the basis of energy fitness alone: families who move often are not in one home long enough for conservation to pay off to themselves; renters cannot easily add insulation to property they do not own, nor can they force landlords to insulate buildings if they, not the landlords, pay the utility bills. The same dis-

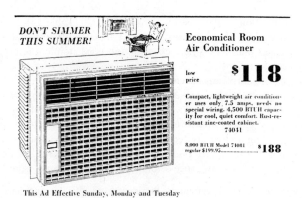

Figure 30 Advertisement depicting air conditioner with high operating cost and low first cost. A bargain?

advantage hits small businessmen in large structures, where the utility bills are hidden in the rent. Industries affected by rising energy costs can pass those costs on to the consumer, especially if competition is limited or if the energy costs are small compared with the total value added, as is often the case.

Incentive barriers also exist for the production of energy-efficient equipment. Manufacturers have no incentive to produce energy-efficient autos or appliances if advertising, social pressure, consumer habits, or marketing procedures (such as rebates) give apparent advantages to less-efficient equipment. This holds especially if the buyer sees only the first cost, not the operation cost or life-cycle cost (see Figure 30). For refrigerators the energy costs are much larger than the purchase price (69, 71), but models packed with every feature (except insulation) tend to be advertised and sold on the basis of first cost alone. It was reported (139) that as of early 1975 buyers were largely still ignoring efficiency in selecting appliances, probably for the reasons cited here. Performance standards, which set minimum design efficiencies for energy-using equipment and structures, seem necessary to assure that systems surviving several owners or several generations are built efficiently in the first place. The advantage of performance standards is that the systems in question operate efficiently, while the marketplace allocates the cost and savings of higher efficiency.

The problem of misplaced incentives hinders conservation in new buildings as well. Banks now have little incentive to lend extra money to make structures energy-tight, although utilities could refuse to provide services to inefficient structures. Developers will not risk the extra cost of insulation and other energy-conservation measures if competitors can omit the same, charge less, and obscure the differences with advertising (140, 141). Requiring efficiency standards on appliances and statements of heat loss through walls or limiting energy consumed per square foot in large buildings would assure that all developers and builders, as well as the banks that finance them, work under the same cost-effective constraint of energy effectiveness. The higher first costs, passed on to buyers and renters, would be more than repaid during the life of the structure or appliances.

The impact of higher prices for fuels on low-income groups cannot be ignored, and rebates have been suggested to aid these groups. But people earning more than about $10,000 per family require more energy indirectly for the goods and services they purchase than they do directly as heat, electricity, and gasoline. This means that the impact of energy costs is only mildly regressive with income (142) (see Figure 31). Some economists suggest that it is better to aid directly those who cannot afford expensive energy rather than make energy artificially cheap to all, which encourages everyone to waste energy.

President Ford has suggested an energy tax coupled with rebates that rise with income; this idea was lost in the debate over the antirecession program (143). Another alternative is a system of flat rebates, which distribute a nearly constant amount of money to all families, under the assumption that well-to-do groups have more opportunities to improve efficiency, allocate resources more carefully, or simply do without certain luxuries such as airplane travel, second homes, recreational vehicles, or swimming pool heaters. This assumption appears to be borne out by

Herendeen's work (Figure 31). Input-output data also suggest that price increases for energy will affect basic food and clothing and housing construction far less than luxury items such as large cars and travel.

Whether the economic barriers to efficient use discussed here can be overcome by legislative, institutional, or personal actions is of course an open question. Although economic models of energy use rarely consider these non-price variables in attempting to predict energy use, higher energy prices should stimulate energy users towards more optimal energy utilization. At the same time the effects of higher prices can be monitored and anticipated, as the input-output work cited here suggests. But society may want to conserve energy for non-market problems such as pollution, the effects of reliance on imported fuels, or the impact on the poor of expensive energy used inefficiently. Additionally, the nation may wish to implement conservation strategies at a faster pace than would occur in response to market forces alone. In these cases additional incentives for efficient use may be called for; some of these are discussed briefly in the next section.

Additional Incentives to Conserve

In addition to higher prices, incentives in the form of low-interest loans, tax benefits, rebates, or penalties, and other "carrots and sticks" should be considered carefully. The most common "stick" for encouraging more efficient energy use is higher prices, but bans on certain forms of consumption, rationing of fuels,

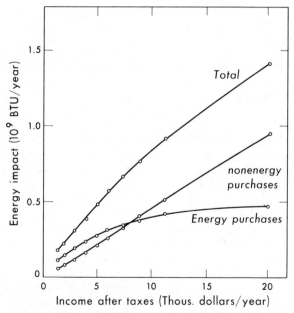

Figure 31 Direct and indirect use of energy as a function of household income. Direct means gasoline, heating, electricity; indirect is energy used to provide goods and services. Source: (142).

voluntary quotas, and so forth are often discussed (143). At the same time "carrots" are receiving increasing attention; some are shown in Figure 32. Additional incentives to invest in energy conservation can come from the tax system.

Figure 32 Incentives ("carrots") for energy conservation. **A**: low-interest loans for household improvements. **B**: rebates for more efficient air conditioners. **C**: a mass transit card from the greater Stockholm "Stockholms Lokaltrafik" mass transit district. (For $17.50/mo. the author was permitted to ride all buses, trains, subways, and many boats in an area the size of the San Francisco Bay Area.) **D**: direct appeal for higher efficiency—with offer of direct aid from the local utility.

Governments can allow tax benefits for investments in energy-efficient equipment, provided that the investments are clearly aimed at efficiency. These incentives would include tax forgiveness in the increased valuation of property upgraded through conservation, tax credits or accelerated write-offs, and direct grants to those whose incomes do not allow setting aside capital for energy conservation.

Part of the cost of investing in conservation is interest, which is tax-deductible. For homeowners this interest is paid out of saved fuel and electricity costs, which are not tax-deductible. If conservation investments were added onto mortgages, the payments for which are interest-intensive during the first few years, the tax benefits would accrue early in the life of the investment. On the other hand, owners of apartments or factories deduct energy costs or pass them on in rents; for them direct tax credits for installing more efficient equipment may be necessary. Tax benefits should rise with increased efficiency, so that investment in efficiency well beyond the minimum required by standards would be encouraged.

Other Barriers to Efficient Energy Use

Other economic difficulties inhibit energy conservation. The investment patterns required for the conservation measures discussed above mean that millions of small investments by consumers must take the place of relatively few large investments in energy facilities. The energy industry is experienced at accumulating capital, but the consumer and small businessman often struggle to make ends meet. This problem is especially acute for those on fixed or very small incomes.

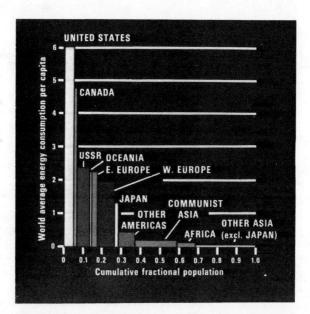

Figure 33 Per capita energy consumption as a ratio to the world average population, as a fraction of the world total. Source: *International Petroleum Yearbook,* 1972 ed.

Governments should engage in conservation campaigns in which low-interest loans or grants are made available for clearly defined conservation investments. Sweden has already embarked on such a campaign to help finance the insulation of buildings (144).

A further barrier to conservation, perhaps one of the most difficult, is that consumers do not directly perceive some of the "hidden benefits" of conservation outlined here. Instead, these benefits accrue to society as a whole, through the millions of individuals who find more employment, cleaner air, lower energy prices, less congestion, better mass transit, and diminished threats to national security. Perhaps these common benefits need to be reinforced by taxes or subsidies, so that individuals can also perceive direct economic benefits to themselves from energy savings. Other benefits from more efficient energy utilization include lower world energy prices, and this can only be of help to the two thirds of the world that has not yet begun to realize the great social benefits that careful energy utilization brings (Figure 33). Last but not least, fossil fuels that are saved now will be available to future generations for use as chemical feedstocks.

The most often discussed reasons for conserving energy are the "necessity" to reduce oil imports and the desire to minimize threats to national security presented by interruption of fuel supplies (143).[19] Whatever the merits or problems of these two reasons, the derived benefits are difficult to quantify economically and do not accrue directly to individuals. Policymakers include energy conservation as part of the overall national plan to reduce oil imports, yet this plan may not explicitly recognize that even without the problem of imports present energy use is far from economic.

Because economics favors more efficient energy use, it would seem that conservation has few opponents. On the other hand, the energy industry is threatened with slower growth than at any other time since the Depression (145). As noted above, some industry spokesmen or organizations have continually endorsed the need to continue historical growth in energy use (24, 25).[20] This is particularly true for some electric utilities. Under the name "People at America's Investor-Owned Electric Companies," one group has stated that growth in electric power consumption at historical rates is beneficial and inevitable, as Figure 34 illustrates. The claims of advertisements such as this have been challenged elsewhere (147–149), and it has been noted that they have seldom carried any suggestions about energy conservation. However, other utilities have now begun earnestly to tell subscribers how to conserve electricity and gas, as Figure 32D shows.

It has been suggested that if utilities were to engage in the selling of comfort systems, rather than only of energy, they would find a greater incentive to participate in energy conservation programs. In this idea, the utility would own and maintain the entire energy system in a structure and lease it to the firm or individual. One utility (150) now installs insulation in residential property and allows owners to pay off the investment on the same bill as for natural gas. Another arranges for

---

[19] See footnote 16.

[20] For a view supporting the aggressive promotion of load growth in electricity use, see (146).

outside financing and installation (W. Zitlau, San Diego Gas and Electric Company, 1974, private communication). The degree to which the energy industry participates in energy conservation R & D and implementation will appreciably influence the future demand for energy.

The very notion that energy is being used inefficiently, that Americans must husband resources and eke out the next unit of welfare from higher efficiency rather than higher gross inputs, suggests a confrontation between aspects of the traditional American way of life and the true finiteness of the world vis-à-vis the rates of use of resources. As is often noted, energy prices fell for many years while the side effects of cheap energy use—pollution, urban sprawl, decay of the mass transportation system, the endless substitution of energy for other production factors—became a part of that way of life. Now American society, and indeed the world, is faced with the prospects of intervening in these trends in order to use energy more efficiently, not because society has run out of energy, but because society is having difficulty even running at today's rate of use. The distribution of energy usage in the world (Figure 33) shows clearly that most of the world has not begun to realize the social

Figure 34 Advertisement from a group representing many utilities. The advertisement, which appeared in *Newsweek* in late 1974, expresses the opinion that electricity needs will grow.

benefits of energy, while part of the world struggles with the side effects and economic dangers of the wasteful use of energy (151).

To some, the challenge for changes in energy utilization might be interpreted as some kind of threat to the American economic system. To me, however, the need for energy conservation must be interpreted as a fortunate signal. Indeed while man's physical activities and uses of resources are rate-limited,[21] both technical and social changes in structure and operation of systems are possible which will allow us to win more social benefit from increasingly scarce resources. The hope for "cheap nuclear fission," "cheap nuclear fusion," or "cheap solar energy," none of which is in fact cheap, constantly distracts individuals and institutions from making economic adjustments now to more efficient energy use and obscures the possibility that nature has, in reality, imposed a kind of speed limit on our activity.

SUMMARY

The role of energy in an economy can only be understood through consideration of both an economic description of energy inputs and outputs and a physical analysis of the activities in the economy that use energy. Such analysis leads to several definitions of energy efficiency, waste, and conservation, definitions that are sometimes, but not always, close to those of traditional economics. Conserving natural resources for future generations, or for those who cannot afford them today, and preserving environmental quality are important reasons for conserving energy. But economic analysis of the physical options for energy conservation shows that saving 30–40% of the expected future total energy demand in the United States would be far less expensive than supplying the increased amounts of fuels and electricity dictated by naive extrapolation of historical trends.

Conservation strategies also tend to increase employment and decrease pollution, while saving energy and money. By easing demands on dollar and energy capital required to build and run energy-producing facilities, conservation slows the real rise in the cost of energy. However, conservation faces a full range of important nontechnical problems, which are rooted in the history of energy utilization at low energy prices, as well as barriers connected with defects in the pricing of energy, the control of the end use of energy, and the time necessary for society to adjust to sharply rising energy costs.

A variety of social, political, economic, and technical changes are often suggested as remedies for today's energy problems (96, 142, 143, 151–156).[22] These include decontrol of fuel prices, energy taxes, rationing or allocation, subsidies or low-interest loans for efficient use, bans on certain end uses or social activities, and educational programs designed to change people's attitudes. Energy policy designed to encourage efficient use of energy will probably have to incorporate many of these measures, using both traditional and novel market and non-market tools. Even before considering the question of what sources of energy to develop tomorrow, one must

---

[21] See (151).
[22] See also footnote 16.

confront energy conservation today: inefficient energy use means inefficient and costly malfunctions in the US economy. Perhaps recognition of the influential role of energy waste in exacerbating our economic and environmental problems will aid progress toward more efficient energy utilization.

ACKNOWLEDGMENT

I wish to thank J. M. Hollander and M. K. Simmons, of the Lawrence Berkeley Laboratory, Energy and Environment Division, for their help in preparing this manuscript. This research was supported in part by the Lawrence Berkeley Laboratory, under contract with the US Energy Research and Development Administration. Opinions are solely those of the author.

Literature Cited

1. US Congress, Joint Committee on Atomic Energy. 1973. *Understanding the Nation's Energy Dilemma.* Washington DC: Joint Comm. At. Energy
2. Battelle Northwest Laboratories. 1969. *Review and Comparison of Selected Energy Forecasts.* Washington DC: Office Sci. Technol.
3. Erickson, L. 1974. A review of forecasts for U.S. energy consumption. In *Energy and Human Welfare,* ed. B. Commoner et al. New York: Macmillan
4. Portola Institute. 1974. *Energy Primer.* Menlo Park, Calif: Portola Inst.
5. Chase Manhattan Bank. 1972. *Energy Outlook in the United States to 1985.* New York: Chase Manhattan Bank
6. Office of Emergency Preparedness. 1972. *Potential for Energy Conservation.* Washington DC: GPO
7. Shell Oil Company. 1973. *Energy Conservation Potentials.* Houston: Shell Oil
8. Kovach, E., ed. 1974. *Technology of Efficient Energy Utilization.* Scientific Affairs Division, NATO, Brussels, Belgium
9. Large, D. 1973. *Hidden Waste.* Washington DC: Conservation Foundation
10. US House of Representatives, Subcommittee on Science and Astronautics. May 1974. *Conservation and Efficient Use of Energy.* Pts. 1–4. Washington DC: GPO
11. US Senate, Committee on Interior and Insular Affairs. Aug. 1973. *Energy Conservation and S 2176.* Washington DC: GPO
12. US Senate, Committee on Commerce. May, June 1974. *Energy Waste and Energy Efficiency in Industrial and Commercial Activities.* Also US Senate, Committee on Commerce. Oct. 1974. *Industry Efforts in Energy Conservation.* Washington DC: GPO
13. Ford Foundation, Energy Policy Project. 1974. *A Time to Choose: America's Energy Future.* Cambridge, Mass: Ballinger
14. Schipper, L. 1975. *Energy Conservation: Its Nature, Hidden Benefits, Hidden Barriers.* Rep. No. LBL 3295. Lawrence Berkeley Lab., Berkeley, Calif. Also 1976. *Energy Communications.* In press
15. Schipper, L. 1975. *Efficient Energy Use.* Rep. No. ERG-75-08. Energy and Resour. Group, Univ. Calif., Berkeley
16. US Energy Research and Development Administration. June 1975. *A National Plan for Energy Research, Development, and Demonstration.* Publ. No. ERDA 48, Vols. 1, 2. Washington DC: ERDA
17. Keenan, J. H. 1941. *Thermodynamics.* New York: Wiley
18. Keenan, J. H. et al. 1974. The fuel shortage and thermodynamics. In *Energy: Demand, Conservation, and Institutional Problems—Proc. MIT Energy Conf., Feb. 1973.* Cambridge, Mass: MIT Press
19. Priest, J. 1973. *Problems of Our Physical Environment: Energy, Transportation, Pollution.* Reading, Mass: Addison-Wesley
20. Cottrell, F. 1955. *Energy and Society.* New York: McGraw-Hill
21. Cook, E. 1975. *Study of Energy Futures.* Chapel Hill, NC: Environ. Design Res. Assoc. Also Cook, E. 1971. The flow of energy in an industrial society. *Sci. Am.* 224:134
22. Barclay, J., Seckler, D. 1972. *Economic Growth and Environmental Decay.* New York: Harcourt-Brace-Jovanovich
23. Page, N. T. 1976. *Economics of a*

Throwaway Society. Resources for the Future, Inc. In preparation
24. National Petroleum Council. 1972. *U.S. Energy Outlook, A Summary Report.* Washington DC: Nat. Pet. Counc.
25. Cook, D., Vassell, G. 1974. Energy conservation within the framework of vital societal objectives. *Public Util. Fortnightly* 65(10): 1
26. Darmstadter, J. et al. 1971. *Energy and the World Economy.* Baltimore, Md: Johns Hopkins Univ. Press
27. Rubin, B. et al. 1972. *Energy: Sources, Uses, Issues.* Rep. No. UCRL 51221, Lawrence Livermore Lab., Livermore, Calif. Springfield, Va: Nat. Tech. Inf. Serv.
28. See Ref. 13, comments by D. Burnham
29. Meyers, J. Feb. 1975. *Energy Conservation and Economic Growth: Are They Incompatible?* In The Conference Board *Record,* p. 27. Reprinted in ERDA Authorization Hearings, Pt. 1, Feb. 18, 1975, Subcomm. on Science and Astronautics, US House of Reps. Washington DC: GPO
30. National Economic Development Office. 1974. *Energy Conservation in the United Kingdom.* London: HMSO
31. Elbek, B. 1975. *Energi—Energi—Energi Krise.* Copenhagen: Munksgaard
32. Stanford Research Institute. 1975. *Comparison of Energy Consumption Between West Germany and the United States.* Prepared for Fed. Energy Admin. Menlo Park, Ca: Stanford Res. Inst.
33. Schipper, L., Lichtenberg, A. 1975. *Efficient Energy Use: The Swedish Example.* Rep. No. ERG-75-09. Energy and Resour. Group, Univ. Calif., Berkeley
34. Makhijani, A., Lichtenberg, A. 1971. *An Assessment of Energy and Materials Use in U.S.A.* Mem. No. M-310, Energy Res. Lab., Univ. Calif., Berkeley. Also 1972. *Environment* 14(5): 10
35. Energi Prognos Committeen. 1974. *Energi 1985–2000.* Rep. No. SOU-64, Industridept., Stockholm, Sweden: Almanna Förlaget. 2 vols.
36. Swedish Federation of Industries. 1974. *Energy Conservation in Swedish Industry.* Stockholm: Industriförbundets Förlag
37. Mooz, W. E. June 1973. *Transportation and Energy.* Santa Monica, Ca: Rand Corp.
38. Hirst, E. 1973. *Energy Intensiveness of Passenger and Freight Transportation Modes.* Rep. No. ORNL-NSF-EP 44. Oak Ridge, Tenn: Oak Ridge Nat. Lab.
39. Federal Energy Administration. 1974. *Project Independence and Energy Conservation: Transportation Sector,* Vol. 2. Washington DC: GPO
40. Foster Associates. 1975. *Energy Prices 1960–73.* Energy Policy Project, Ford Foundation. Cambridge, Mass: Ballinger
41. Tansil, J. M. 1973. *Residential Consumption of Electricity.* Rep. No. ORNL-NSF-EP-51. Oak Ridge, Tenn: Oak Ridge Nat. Lab.
42. American Gas Association. Oct. 1974. *A Pilot Project in Homeowner Energy Conservation.* Washington DC: Fed. Energy Admin.
43. Moyers, J. 1973. *Room Air Conditioners as Energy Consumers.* Rep. No. ORNL-NSF-EP-59. Oak Ridge, Tenn: Oak Ridge Nat. Lab.
44. Allen, J. Oct. 1974. The craft of electric motors. *Environment* 16(8): 36
45. American Physical Society. 1975. *Efficient Use of Energy: A Physics Perspective.* 1974 Summer Study. New York: Am. Inst. Physics
46. Lawrence Livermore Laboratory. 1973. *U.S. Energy Flow Charts.* US Atomic Energy Commission. Rep. No. UCRL 51487. Springfield, Va: Nat. Tech. Inf. Serv.
47. Dow Chemical Company et al. June 1975. *Energy Industrial Center Study.* For Nat. Sci. Found., Grant No. OEP 74-20242. Washington DC: NSF
48. Hise, E. C. 1975. *Seasonal Fuel Utilization Efficiency of Residential Heating Systems.* Rep. No. ORNL-NSF-EP-82. Oak Ridge, Tenn: Oak Ridge Nat. Lab.
49. Petersen, S. R. 1974. *Retrofitting Existing Housing for Energy Conservation: An Economic Analysis.* Nat. Bur. Standards. Washington DC: GPO
50. Hannon, B. July 1975. Energy conservation and the consumer. *Science* 189:95
51. Hannon, B. Feb. 1974. Options for energy conservation. *Technol. Rev.* 76(4): 24
52. Bullard, C. W., Herendeen, R. A. Jan. 1975. *Energy Costs of Goods and Services.* Center for Adv. Computation, Univ. Ill., Urbana
53. Chapman, P. March 1975. The energy cost of materials. *Energy Policy* 3(2): 47
54. Berry, R. S., Fels, M. Sept. 1973. Production and consumption of autos. *Bull. At. Sci.* 27: 11
55. International Federation of Institutes for Advanced Study. 1974. *Energy Analysis Workshop on Methodologies and Conventions.* Nobel House, Storegatan 14, Box 5344 s-10246, Stockholm, Sweden

56. Pilati, D., Richards, R. Aug. 1975. *Total Energy Requirements for Nine Electricity Generating Systems.* Doc. No. CAC 165. Center for Adv. Computation, Univ. Ill., Urbana
57. Chapman, P. F., Leach, G., Slesser, M. Sept. 1974. The energy cost of fuels. *Energy Policy* 2(3):231
58. Rieber, M. 1974. *Low Sulfur Coal: A Revision of Resource and Supply Estimates.* Doc. No. CAC 87. Center for Adv. Computation, Univ. Ill., Urbana
59. Rieber, M. et al. Nov. 1974. *Nuclear Power to 1985.* Doc. No. CAC 137P. Center for Adv. Computation, Univ. Ill., Urbana
60. Price, J. Dec. 1974. *Dynamic Analysis of Nuclear Power.* Available from Friends of the Earth Ltd., London
60a. Chapman, P. Dec. 1974. The ins and outs of nuclear power. *New Sci.*, Vol. 64
60b. Leach, G. Dec. 1974. *Nuclear Energy Balances in a World with Ceilings.* London: Int. Inst. for Environ. and Dev.
61. Putnam, D. E. Aug. 1975. *Energy Benefits and Costs: Housing Insulation and the Use of Small Cars.* Doc. No. CAC 173. Center for Adv. Computation, Univ. Ill., Urbana
62. Ross, M., Williams, R. 1975. *Assessing the Potential for Fuel Conservation.* Rep. No. 75-02. Inst. for Public Pol. Altern., SUNY, Albany, NY
63. Moyers, J. C. 1972. *The Value of Thermal Insulation in Residential Construction: Economics, and the Conservation of Energy.* Rep. No. ORNL-NSF-EP-9. Oak Ridge, Tenn: Oak Ridge Nat. Lab.
64. Delene, J. Nov. 1974. *A Regional Comparison of Energy Resource Use and Cost to the Consumer of Alternate Residential Heating Systems.* Rep. No. ORNL-NSF-TM-4689. Oak Ridge, Tenn: Oak Ridge Nat. Lab.
65. Achenbach, P. 1973. *Effective Energy Utilization in Buildings.* Washington DC: Nat. Bur. Standards
66. *Technical Options for Environmental Conservation in Buildings.* 1973. NBS Tech. Note No. 789. Washington DC: Nat. Bur. Standards
67. Federal Energy Administration. Nov. 1974. *Energy Conservation: The Residential Sector*, Vol. 1. Proj. Independence. Washington DC: GPO
68. Berg, C. Feb. 1974. Technical basis for energy conservation. *Technol. Rev.* 76(4):14. Also Berg, C. 1973. *Energy Conservation through Effective Utilization.* Washington DC: Fed. Power Comm.
69. Goldstein, D., Rosenfeld, A. 1975. *Projecting an Energy Efficient California.* Rep. No. LBL 3274. Lawrence Berkeley Lab., Berkeley, Calif.
70. Mutch, J. May 1974. *Residential Water Heating: Fuel Conservation, Economics, and Public Policy.* Santa Monica, Calif: Rand Corp.
71. Center for Policy Alternatives. 1974. *The Productivity of Servicing Consumers Durable Products.* Rep. No. CPA-74-14. Cambridge, Mass: MIT Center for Pol. Altern.
72. Makhijani, A., Lichtenberg, A. 1973. *An Assessment of Residential Energy Utilization in the U.S.A.* Eng. Mem. No. ERL-M370. Univ. Calif., Berkeley
73. Appel, J., MacKenzie, J. 1974. How much light do we really need? *Bull. At. Sci.* 30(10):18
74. Dubin-Mindell-Bloom Associates. Feb. 1975. *Report to U.S. Home Corp. on Resources Saving House.* Project No. DMBA-PC-43-73. 42 W 39th St., New York, NY
75. Hammond, J. et al. 1974. *A Strategy For Energy Conservation.* Winters, Calif: Living Systems Inc.
76. Dubin-Mindell-Bloom Associates. 1974. *Energy Conservation for Existing Buildings.* Rep. No. ECM 1, 2. Prepared for Fed. Energy Admin. Washington DC: FEA
77. Dubin, F. Aug. 1973. G.S.A.'s energy conservation test building—a report. *Actual Specif. Eng.*
78. Stein, R. Oct. 1972. It's a matter of design. *Environment* 14(8):17
79. Berman, S., Silverstein, S. 1975. *Energy Conservation and Window Systems.* New York: Am. Inst. Physics
80. Dubin, F. Feb. 1973. Total energy for mass housing. *Actual Specif. Eng.*
81. Morrow, W. Dec. 1973. Solar energy, its time is near. *Technol. Rev.* 76(2):30
82. Löf, G., Tybout, R. 1973. Cost of house heating with solar energy. *Sol. Energy* 14:253–78
83. E. I. du Pont de Nemours and Co. 1973. *Du Pont Energy Management Services.* Educ. and Appl. Technol. Div., Wilmington, Del. 19898
84. Federal Energy Administration. 1974. *Energy Conservation in the Manufacturing Sector, 1954–1990.* Proj. Independence. Washington DC: GPO
85. Gatts, R. 1974. *Industrial Energy Conservation.* Pap. No. 74-WA/Energ-8. New York: Am. Soc. Mech. Eng.
86. Gyftopolous, E. et al. 1974. *Potential Fuel Effectiveness in Industry.* Cambridge, Mass: Ballinger
87. Bravaard, J. et al. 1972. *Energy Costs*

Associated with the Production and Recycling of Metals. Rep. No. ORNL-NSF-EP-24. Oak Ridge, Tenn: Oak Ridge Nat. Lab.
88. Lowe, R. A. 1974. *Energy Conservation Through Improved Solid Waste Management*. Rep. No. SW-125. Washington DC: US EPA
89. Hannon, B. 1972. *System Energy and Recycling: A Study of the Beverage Industry*. Doc. No. CAC 23. Center for Adv. Computation, Univ. Ill., Urbana
90. Berry, R. S., Makino, H. 1974. Energy thrift in packaging. *Technol. Rev.* 76(4): 32
91. Rubin, D. et al. 1973. *Transportation Energy Conservation Options*. Rep. No. DP-SP-11 (draft). Washington DC: US Dep. Trans.
92. Rice, R. 1974. Towards more transportation with less energy. *Technol. Rev.* 76(4):44
93. Marks, C. Dec. 1973. *Which Way to Achieve Better Fuel Economy?* Presented at Seminar, Energy Consumption of Private Vehicles, Present and Possible, Calif. Inst. Technol., Pasadena. Available from General Motors Corp., Detroit, Mich.
94. US Environmental Protection Agency. Oct. 1973. *A Report on Automobile Fuel Economy*. Washington DC: US EPA
95. Schipper, L. 1974. *Holidays, Gifts, and the Energy Crisis*. Rep. No. UCID 3707, Lawrence Berkeley Lab., Univ. Calif., Berkeley. Washington DC: Nat. Tech. Inf. Serv.
96. Fels, M., Munson, M. 1975. Energy thrift in urban transportation: options for the future. In *Energy Conservation Papers*, ed. R. Williams. Cambridge, Mass: Ballinger
97. Pilati, D. 1975. *Energy Conservation Potential of Winter Thermostat Reduction and Night Setback*. Rep. No. ORNL-NSF-EP-80. Oak Ridge, Tenn: Oak Ridge Nat. Lab.
98. Federal Energy Administration. 1974. *Lighting and Thermal Operations:* Case Studies. Washington DC: GPO
99. Fox, J. et al. Dec. 1973. *Energy Conservation in Housing—First Year Progress Report*. Rep. No. 16 (revised). Princeton Center for Environ. Stud., Princeton, NJ
100. Hannon, B., Puleo, F. 1974. *Transferring from Urban Cars to Buses*. Center for Adv. Computation, Univ. Ill., Urbana. Reprinted 1975 in *Energy Conservation Papers*, ed. R. Williams. Cambridge, Mass: Ballinger
101. Real Estate Research Corp. 1974. *Cost of Urban Sprawl*. Prepared for Counc. Environ. Qual. Washington DC: GPO
102. MacLean, R., Portland Cement Association. May, June 1974. In *Energy Waste and Efficiency in Commercial and Industrial Activities*. Committee on Commerce, US Senate, Washington, DC
103. Meyers, J. et al. 1975. *Energy Consumption in Manufacturing*. Cambridge, Mass: Ballinger
104. Dunning, R. 1973. *Comparison of Total Heating Costs with Heat Pumps vs. Alternative Heating Systems*. Rep. No. PSP-4-2-73. Westinghouse Electric Corp., Pittsburgh, Pa.
105. Tennessee Valley Authority. 1975. *A Heat Pump Program for the TVA Area*. Chatanooga, Tenn: TVA
106. Singer, S. Nov. 1974. Future environmental needs and costs. *Eos*
107. Holdren, J., Ehrlich, P. 1974. Human population and the global environment. *Am. Sci.* 62(3):282
108. Linden, H. R. Dec. 1974. Testimony before US Dep. of State, Seminar on US Energy Policy. Available from Inst. Gas Technol., Chicago, Ill.
109. Mobil Oil Company. 1974. *Energy Manifesto*. In *New York Times*. Available from Mobil Oil Co., New York
110. Stein, R. 1973. In *Energy Conservation—Implications for Building Design and Operation*, ed. D. Abrahamson. Univ. Minn. Counc. on Environ. Qual., Minneapolis
111. Ehrlich, P., Holdren, J. 1972. Impact of population growth. In *Population, Resources, and the Environment*, ed. R. Ridker. Washington DC: GPO
112. Chapman, D. et al. 1973. *Power Generation, Health, and Fuel Supply*. Submitted to Nat. Power Survey, Fed. Power Comm., Washington DC
113. Smith, K., Weyant, J., Holdren, J. 1975. *Evaluation of Conventional Power Systems*. Rep. No. ERG-75-5. Energy and Resour. Group, Univ. Calif., Berkeley
114. Epstein, S., Hattis, D. 1975. Pollution and human health. In *Environment*, ed. W. Murdoch. Sinauer Assoc., Sunderland, Mass.
115. See Ref. 5, page 26
116. Ross, M. et al. 1975. Energy needs for pollution control. In *The Energy Conservation Papers*. Cambridge, Mass: Ballinger
117. Hirst, E. 1973. *Energy Implications of Several Environmental Quality Strategies*. Rep. No. ORNL-NSF-EP-53. Oak Ridge, Tenn: Oak Ridge Nat. Lab.

118. Anderson, C. et al. 1974. *An Assessment of U.S. Energy Options for Project Independence.* Rep. No. UCRL-51638, Lawrence Livermore Lab. Springfield, Va: Nat. Tech. Inf. Serv. See also *Project Independence Reports.* Washington DC: Fed. Energy Admin.
119. Hass, J., Mitchell, E., Stone, M. 1975. *Financing the Energy Industry.* Cambridge, Mass: Ballinger
120. Bullard, C., Pilati, D. Sept. 1975. *Direct and Indirect Requirements for Project Independence Scenarios.* Doc. No. CAC 178. Center for Adv. Computation, Univ. Ill., Urbana
121. American Institute of Architects. 1974. *Energy and the Built Environment: A Gap in Current Strategies; 1974. A Nation of Energy Efficient Buildings by 1990.* Washington DC: AIA
122. American Institute of Architects. 1974. *Energy Conservation in Building Design.* Washington DC: AIA
123. Vyas, K., Bodle, W. 1975. Coal and oil shale conversion looks better. *Oil Gas J.* 73(12):45
124. Holdren, J. 1974. Hazards of the nuclear fuel cycle. *Bull At. Sci.* 30(8):14
125. Schneider, S., Dennett, R. 1975. Climatic barriers to long-term energy growth. *Ambio* 4:65
126. Weinberg, A. M. 1972. Social institutions and nuclear power. *Science* 177:27
127. Foreman, H., ed. 1972. *Nuclear Power and the Public.* Garden City, NJ: Anchor
128. Anderson, K. P. Oct. 1973. *Residential Energy Use: An Econometric Analysis.* Santa Monica, Calif: Rand Corp.
129. Chapman, D. et al. 1972. Predicting the Past and Future in Electricity Demand. Agric. Econ. Pap. No. 72-9. Cornell Univ., Ithaca, NY
130. Mount, T. D. et al. 1973. *Electricity Demand in the U.S.: An Econometric Analysis.* Rep. No. ORNL-NSF-EP-49. Oak Ridge, Tenn: Oak Ridge Nat. Lab.
131. Cicchetti, C. 1974. Electricity price regulation. *Public Util. Fortnightly* 94(5):13
132. Berlin, E., Cicchetti, C. J., Gillen, W. J. 1975. *Perspective on Power: A Study of the Regulation and Pricing of Electric Power.* Cambridge, Mass: Ballinger
133. Bierman, H., Hass, J. Jan. 1974. *Public Utility Investment and Regulatory Practices.* Cornell Energy Project, Pap. No. 74-1. Cornell Univ., Ithaca, NY
134. Lichtenberg, A., Norgaard, R. 1975. Tax Treatment of Oil and Gas Income and Energy Policy. *Nat. Resour. J.,* Vol. 14, No. 4
135. US Senate, Committee on Commerce. 1973. *Truth in Energy and Car Pooling.* Washington DC: GPO
136. Federal Power Commission. 1975. *Measures for Reducing Energy Consumption for Homeowners and Renters.* Washington DC: FPC
136a. Bullard, C. 1973. *Energy Conservation Through Taxation.* Doc. No. CAC 95. Center for Adv. Computation, Univ. Ill., Urbana
137. *The Wall Street Journal.* March 31, 1975. p. 3
138. Rubin, M. Dec. 1974. Plugging the energy sieve. *Bull. At. Sci.,* Vol. 30. See also Ref. 11, Vol. 2, p. 659
139. *The Wall Street Journal.* Jan. 28, 1975. Back page
140. Rauenhorst, R. 1973. See Ref. 110
141. Fraker, N., Shorske, E. Dec. 1973. *Energy Husbandry in Housing: An Analysis of the Developmental Process.* Rep. No. 5. Princeton Center for Environ. Stud., Princeton, NJ
142. Herendeen, R. Oct. 1974. Energy and affluence. *Mech. Eng.* Also Herendeen, R., Tanaka, J. 1975. *Energy Cost of Living.* Doc. No. CAC 171. Center for Adv. Computation Univ. Ill., Urbana
143. Ford, G. R. Jan. 1975. *The President's 1975 State of the Union Message, including Economy and Energy.* Washington DC: The White House
144. Department of Industry and Riksdagen, Stockholm, Sweden. Feb. 1975. *Energi-Hushållning m.m.*
145. *Towards Responsible Energy Policies.* 1973. Policy Statement of Nat. Coal Assoc., Am. Pet. Inst., Am. Gas Assoc., Atomic Indus. Forum, Edison Elec. Inst. Available from Edison Elec. Inst., New York
146. Rydbeck, V. Aug. 1975. Rx for utility financial vitality. *Public Util. Fortnightly,* Vol. 96, No. 4
147. Hirst, E. 1972. *Electric Utility Advertising and the Environment.* Rep. No. ORNL-NSF-EP-10. Oak Ridge, Tenn: Oak Ridge Nat. Lab.
148. Council on Economic Priorities. 1972. *The Price of Power.* Cambridge, Mass: MIT Press
149. Schipper, L. 1975. *The Efficient Energy Future and ERDA-48.* Testimony before Counc. Environ. Qual. on US ERDA, Los Angeles, Calif., Sept. 8–9, 1975. Washington DC: Counc. Environ. Qual.
150. Ralls, W. 1974. See Ref. 10, Vol. 2, p. 467

151. Lovins, A. 1975. *World Energy Strategies.* Cambridge, Mass: Ballinger
152. Seidel, M. et al. 1973. *Energy Conservation Strategies.* Rep. No. R5-73-021. Washington DC: US EPA
153. Acton, J. et al. 1974. *Electricity Conservation: The Los Angeles Experience.* Santa Monica, Calif: Rand Corp.
154. Wildhorn, S. et al. Oct. 1974. *How to Save Gasoline: Public Policy Alternatives for the Automobile.* Santa Monica, Calif: Rand Corp.
155. National Science Foundation. 1975. *Energy Conservation Research: Proc. NSF/RANN Conf., Feb. 18–20, 1974.* Washington DC: NSF
156. Darmstadter, J. July 1974. *Limiting the Demand for Energy: Possible? Probable?* Washington DC: Resources for the Future Inc.

Copyright 1976. All rights reserved

POTENTIAL FOR ENERGY CONSERVATION IN INDUSTRY

✤11018

Charles A. Berg

RFD 1, Box 165, Buckfield, Maine 04220

INTRODUCTION

In the winter of 1973–1974 energy conservation became a popular subject of discussion, so much so that the subject was often, and not entirely inaccurately, referred to as energy conversation. During this period those energy conservation measures which received serious attention in government were limited to such things as higher energy taxes, reduced speed limits, and reduced thermostat settings during the heating season.[1] Some have characterized these measures as fashionable suffering.

Gradually, as a stream of studies and reports on the subject poured into Washington, the federal government came to recognize that it was not absolutely necessary to confine one's attention to such measures to conserve fuel and that, in fact, significant opportunities existed to conserve fuel by improving the efficiency of fuel use. It was also brought home to the government—particularly to the Congress—that these improvements in efficiency were of dual significance. First, they could reduce substantially the fuel required by various sectors of the economy without requiring undue sacrifice of production employment or of delivery services. Thus, improved efficiency could be an important instrument of energy policy to save fuel in the intermediate term and even in the short term. But, in addition to this, it was also recognized that long-range energy prices have no way to go but up and that they are climbing at an unprecedented rate. An economy designed originally to function on high-quality fuels, the price of which has been held down by regulation, cannot long continue to function effectively with the coming high cost

---

[1] In one instance, an "energy conservation officer" responded to government appeals to conserve energy by reducing the thermostat setting in his office building . . . during the air-conditioning season.

of fuels.[2] The economy, in fact, must be redesigned to enable it to cope with the future era of high energy costs.[3]

It is clear that there are opportunities to improve the efficiency of fuel use that go considerably beyond merely tuning up present-day industrial operations. There are, for example, opportunities to integrate the use of high-temperature heat in industry with the generation of electricity. This is one example of thermal integration of industrial processes.[4] Considerable research will be required to screen the sound and useful ideas on improved fuel use from those that may finally be found to be impractical. In addition, to implement any of the useful ideas for improving the efficiency of fuel use in industry will require a substantial research and development effort to provide the necessary materials, electronic controls, suitable methodologies of controls, etc.[5]

The need to devise new systems to make more efficient use of fuel in industry will probably come upon our economy more rapidly than industry can reorganize fully its research, development, and construction efforts to meet this need. In this regard the need for sound, far-reaching scientific and technological research on better and more efficient ways to use fuels is fully as important as research and development for creating new forms of fuels and exploiting new forms of energy.

However, since the end of the Arab oil embargo there has been a general relaxation in the intensity of thought given to energy conservation research. There has been a clear tendency for the government, in particular, to revert to the patterns of the past. The energy research effort of the US government places extraordinarily strong emphasis on the development of nuclear power, the development of coal as a source of energy, the redevelopment of domestic oil fields through secondary recovery, and other such matters. However, when it comes to research and development to find new and more efficient means for making use of energy, government tends to take the view that research and development on energy use should be the private responsibility of the user. If this were a valid argument, it would be difficult to understand why research and development on coal gasification, for example, should not be the private responsibility of the coal industry or the oil industry. In response to the suggestion that such is the case, government officials correctly point out that gasified coal is needed even now and that radical technical changes may be required in industrial operations to bring gasified coal

---

[2] The policy of the US government to pass the subsidized cost advantage of hydroelectric stations on to the regional consumers of this power is an example of a policy of depressing the price of high-quality energy. In the estimation of a number of people (including myself), regulations have held the price of high-quality energy too low for too long.

[3] For some examples of newly reapplied technology see (1).

[4] Thermal integration of industrial processes is exemplified by the combined steam raising and electric power generation that is commonly practiced in the paper industry. However, the opportunities to use thermal integration go considerably beyond this measure and similar other measures in practice today.

[5] For example, it would be useful to have a well-posed methodology for control of thermal cycling of industrial furnaces.

into use at any significant level. The need is too great and the time is too short—and the risk of failure too great—for a responsible government to stand by and hope that private forces in the marketplace will solve all of the scientific problems of bringing gasified coal into general use quickly enough in the United States. The same is true, I believe, of energy conservation research. Better systems to make more effective use of energy are needed immediately. The failure of the government's energy research program to reflect comprehension of this point is, in my estimation, one of the more serious problems of energy research policy.

There is another related problem of energy conservation research that bears mention here. Examples of the technical measures that have been applied to conserve fuels are, in many instances, anything but stimulating to the imaginative scientist. These measures can be quite exciting to those with commercial interests, but, for example, the application of greater insulation on water-cooled furnace skid rails to save fuel is unlikely to stimulate greatly the curiosity of the young student physicist or engineer, or his professor. Much of the literature on energy conservation, having been focused upon the application of known and in some cases homespun technology, has denied to energy conservation the scientific challenge often required to attract imaginative scientific talent.

Although I can offer no guaranteed solution for the two problems mentioned above, perhaps it would be helpful to provide a simple concrete example of how a given industrial process might be redesigned to make more efficient use of fuel. The example offered below shows that significant changes in the design of a process could be made and indicates where advanced design and research are needed.

ENERGY CONSERVATION IN SPECIALIZED FERROUS HEAT-TREATING

The use of energy in specialized ferrous heat-treating is discussed. For the benefit of the unacquainted reader, this section describes certain of the significant aspects of heat-treating.

Heat-treating is a complex thermochemical process by which the intrinsic structure of metals is altered to yield desired properties. As a rule, heat-treating must be conducted in controlled atmospheres to prevent undesirable reactions. Ferrous heat-treating, in particular, requires exacting control of carbon in the heat-treating atmosphere.

Heat-treating facilities range in size from small plants with four or five furnaces to rather large facilities dedicated to the processing needs of a large production line. The operation of a given facility tends to be somewhat specialized and reflects the experience and ingenuity of the plant manager. Competition in this field depends strongly upon know-how, and, particularly among small installations, the operating procedures are somewhat idiosyncratic. Small plants offering specialized services can become key to the operation of large manufacturing facilities. For example, there is a plant that heat-treats approximately 6800 tons of steel per year and is virtually the sole source of heat-treating for a critical component in the steering system of one of the major automobile manufacturers.

The firms that supply equipment for heat-treating are nearly as strongly differentiated as the heat-treating firms themselves. There are suppliers of furnaces, burners, gas generators, gas distillation controls, instruments, etc. Each of these tends to concentrate upon perfecting the line of equipment in which it specializes. In certain cases, equipment-manufacturing firms were set up by individuals who saw an opportunity to refine or improve an individual item of equipment, such as a specific type of furnace or a specific type of burner. Some of these firms have become very successful by concentrating upon the refinement and perfection of specialized equipment. But, it is relatively seldom that an equipment manufacturer undertakes the design and construction of a heat-treating system. The potential for suboptimization in design of heat-treating plants is nearly unavoidable. Also, the present differentiation of equipment suppliers stands as a major constraint upon the rate at which new heat-treating processes and new equipment systems might be introduced to facilitate adjustment of the industry to the rapidly changing situation of energy supply and price.

Energy Use

In a typical small ferrous heat-treating plant the consumption of fuel and electric power will be as shown in Table 1. These data represent all energy used in the plant, including space heat, hot water, and other such services. An exact breakdown of the use of fuel for space heating is not available. In any event, space-heating requirements can vary widely from one installation to another and depend strongly upon the size of the installation. Large installations can use as much as 30% of the total purchased fuel for space heating, whereas in small installations lost heat from furnaces may carry nearly all of the space-heating load. In the example under consideration we will assume that virtually all of the purchased gas (or other similar fuel) is used for the heat-treating process itself.[6]

If we consider a heat-treating process in which a one-pound ferrous part is to be held, for example, at 1750°F for 12 hr, approximately 183 Btu of fuel are required to raise the part to the required temperature.[7] However, bearing in mind that one standard cubic foot (scf) of natural gas contains approximately 1000 Btu, one sees from the data above that approximately 5000 Btu of gas are used to process each pound of product. On this basis, one can say that heat-treating processes use about 27 times as much natural gas (or equivalent fuel) as is required to heat the product to the temperature required; this implies an efficiency of roughly 4%.

However, this notion of efficiency does not take into account the fact that the parts are eventually cooled back to room temperature. In principle, the net heating requirements in heat-treating are virtually zero.[8] Heat could be recovered from the work (i.e. the parts) after it leaves the furnace and reinjected into the furnace.

[6] We assume this because the data here reflect energy use in a small plant where this is substantially true.

[7] The temperature cited here, 1750°F, is at the upper range of ferrous heat-treating conditions.

[8] There are some small energy requirements for rearrangement of internal metallic structure, but these are minute.

Table 1 Use of energy and chemicals in a typical small ferrous heat-treating plant (per pound of product)

Natural Gas (or Propane or Equivalent Fuel)	Electricity	NH_3	Quench Oil
5.0 scf	0.08 kW-hr	0.01 lb	0.002 gal

This is not presently done. It is most difficult to ascribe an efficiency of any sort to a process for which the net energy requirement is zero.

Of course, the main reason that so much more fuel is required to heat-treat a part than would be required merely to raise the temperature of the part to the desired level is that the part must be held at temperature for some period of time. The higher the temperature of processing and the longer the time for holding the part at temperature, the greater will be the heat losses from the heat-treating furnace. Present ferrous heat-treating temperatures range from 1450°F to 1750°F, and the duration of treating ranges from 15 min to 12 hr. In addition, tempering temperatures range from 300°F to 1100°F.

The potential for reducing fuel requirements in heat-treating through improved furnace insulation, and through similar measures, is significant in any installation, but it depends strongly upon the type of heat-treating process actually being executed. In addition, diligent attention to the adjustment of combustion air in the burners can also contribute significantly to control of fuel consumption. However, fuel consumption in heat-treating can be affected by many more technical modifications in heat-treating systems than merely these well-known and properly emphasized steps, as suggested below.

Thus far I have compared the use of natural gas (or equivalent fuels) only with the fuel requirements of heating. However, natural gas is not used solely for heating. In many heat-treating operations it is necessary to provide a controlled atmosphere in the furnace to regulate carbon reactions in the stock being processed. These are called endothermic atmospheres. They are produced on site in most heat-treating shops, by reforming of natural gas in small catalytic reactors. The methane is reformed to yield a gas having approximately the composition shown in Table 2.

Substantial quantities of this gas are used as the atmosphere in carbonizing and

Table 2 Approximate composition of endothermic atmospheres

Constituent	Concentration
H_2	40%
N_2	40%
CO	20%
CO_2	trace ($\sim 0.2\%$)
H_2O	very small (dew point, 32°F)

nitriding furnaces. By adding small quantities of either air or natural gas to the endothermic atmosphere one can enhance or decrease the carbonizing potential of the furnace atmosphere. In general, additional air will reduce the carbonizing potential. Additional methane added to the endothermic atmosphere will be reformed in the furnace and will increase the carbonizing potential. Endothermic atmosphere gas is a very high quality fuel gas (it contains 40% H_2); the heating value of this gas is approximately 700 Btu/scf, nearly as great as the methane gas from which it is made.

According to estimates made within the industry, approximately 20% of all the natural gas used in ferrous heat-treating is used as feedstock for endothermic atmospheres. These atmospheres—or some equivalent method of chemical control— must be used in ferrous heat-treating. In a carbonitriding furnace (which is typical of equipment in which endothermic atmospheres are used) a part may be held at 1550°F for 30 min. In this case the endothermic atmosphere is flushed through the furnace and flared at the entry and exit of the furnace to prevent air from entering the chamber. Meanwhile, the interior of the furnace will be heated by burning natural gas, or some equivalent fuel, in radiant tubes. Thus, the combustion of the heating fuel is isolated from the chamber of the furnace. The total fuel value of natural gas used to heat the furnace, in an instance such as the one described here, is approximately equal to the fuel value of the endothermic atmosphere that is flared at the ports of the furnace. Again, while it is possible to substitute other fuels for natural gas in heating (propane or #2 distillate oil, for example), it is not

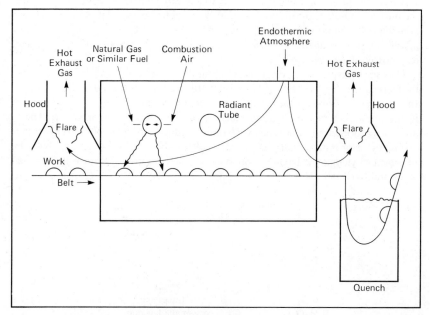

Figure 1 Schematic diagram of a heat-treating furnace with an endothermic atmosphere.

possible to execute a process such as carbonitriding without a controlled atmosphere; at present, natural gas in the only source of this controlled atmosphere.

A schematic diagram of a furnace of the type described above is shown in Figure 1.

The diagram of Figure 2 shows a radiant tube in somewhat more detail. The hot gases of combustion from the radiant tube are vented to the atmosphere. No attempt is now made to use the heat from the flared endothermic atmosphere in the process.

When one first sees an industrial heat-treating furnace glowing at 1750°F, for example, it is easy thereafter to focus one's attention entirely upon the use of fuel on site in heat-treating. However, heat-treating plants also use a great number of pumps, fans, and motors. Consider again the data in Table 1. Note that approximately 0.08 kW-hr of electrical energy is used for every pound of product. This electrical energy is, of course, purchased and is derived from the fuel consumed at central generating stations. Assuming that a central station can generate electric power at about 34% thermal efficiency (10,000 Btu/kW-hr), one sees that the total fuel required to generate the electricity used in a typical small heat-treating plant is approximately 16% of all the fuel consumed on site for heating and other purposes. It is sometimes surprising to learn that a heat-treating plant could require significant fuel consumption for electricity. But the use of circulating pumps, air fans, and conveyor belts in a typical small plant requires significant amounts of power.

Incidentally the cost of electric power looms large in the economics of heat-treating. Electricity rates and fuel prices vary widely across the country, so that no single statistic can be given to reflect the relative importance of fuel and power costs for the country as a whole. But for a typical midwestern plant, the costs of electric power can represent as much as 25% of the total energy costs of plant operation.

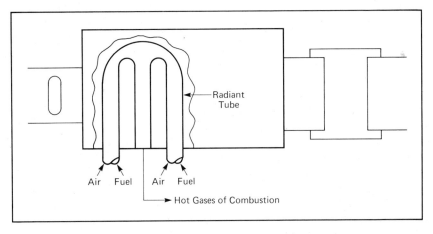

Figure 2 Top view of heat-treating furnace, with view of radiant tube.

Possible Improvements

Having discussed in limited detail the use of energy in ferrous heat-treating, I consider next the way in which energy requirements might be reduced, without interference in production.

In the very short range, the principal steps that could be taken to save fuel are chiefly of the housekeeping variety. Exacting adjustment of combustion air, for example, can reduce fuel requirements below present levels in many plants; the mechanism for doing so is straightforward. Most of the heat transfer in heat-treating occurs by radiation, and the radiant flux of energy from the flame is proportional to the fourth power of its absolute temperature. Excess combustion air quenches the flame and reduces the capacity of each unit of fuel to transfer heat to the work during the time the fuel passes through the combustion chamber. Industry experts have shown that proper attention to adjustment of combustion air, to hold the ratio of air to fuel close to stoichiometric conditions, can yield fuel savings as great as 30% in some plants.[9] Another short-term measure of importance is control of the logistics of material in the plant. For example, if a part is to be first heat-treated, then quenched and tempered, or is otherwise to be subjected to several thermal histories, there are often opportunities to save fuel simply by moving the part quickly from one processing step to another, so that the part does not cool off excessively in between. These and other similar measures come under the general heading of process control.

Another important element of process control that should be considered is optimal thermal programming of furnace equipment. For example, when the work schedule of a plant requires that a given furnace stand idle for some time, what should be the optimal fueling program for turning the furnace down, holding it at idle, and finally bringing it back on line? The exacting control of these matters alone has been shown to be capable of reducing the fuel requirements of a large steel-reheating plant by 25% (2). In the instance cited here, an on-line computer system was used to regulate stock motion, fueling rates in furnaces, and similar items. In a small heat-treating plant it would be difficult to justify the expense of a full-scale computer control system. But, if one could afford such a control system, the fuel savings it could provide would be roughly similar to those attained in large installations. There would appear to be an intriguing opportunity for controls engineers to devise methodologies for optimal control of small heat-treating plants and the miniature control systems that these plants would require. One simple example of a place where one might start on such systems would be automatic oxygen sensors with feedback control systems. With the materials available today for oxygen measurement via observation of changes in electronic properties (e.g. TiO_2 or ZrO_2) it is possible to build accurate, durable, high-temperature oxygen sensors capable of monitoring oxygen in the 2% concentration range. In fact, at least one such sensor is available commercially. However, because the cost of these units is still relatively

---

[9] In most high-temperature installations, combustion air should be regulated to provide approximately 2% excess oxygen.

high, it is difficult to justify installing them on small furnaces. We may assume that a combination of rising fuel prices and greater efforts through manufacturing technology and other measures to reduce the cost of these units will bring the cost of the units down to a level where their application can be justified. Then what will be needed is a small, inexpensive, automatic feedback control system that will accept the command of a sub-miniature computer in which one can insert programmed commands to govern furnace fueling rates, combustion air feed, and stock motion for optimal operation of the furnace during the various phases of its use.[10] To create such a system is not actually a demanding research task; it is more of a development task that could lead to application in the field in a very short time.

Longer-Range Measures

In considering the longer range I believe that several constraints should be recognized. First, ferrous heat-treating is essential to our industrial economy. Thus, some way must be found to continue operations in this field. On the other hand, this field depends now upon natural gas, which is the most critically scarce fuel in the US energy system. Certain substitutions can be made for natural gas as a heat-treating fuel. For example, propane or #2 distillate oil can be substituted. But all of these fuel substitutes derive from oil, and to use them in place of natural gas merely exacerbates national problems of oil imports.[11] One could substitute electric heating for combustion in heat-treating. But, this step would throw significant additional demands upon local utility systems, and it is not at all clear that the utility systems could fully absorb this demand in time to provide significant relief in the national requirements for critically scarce high-quality fuels. Wherever a utility system can absorb the demand of the heat-treating plants in its region, there is the potential to substitute coal or nuclear fuel as the primary energy resource for heat-treating.[12] But even where the electricity is available, there is not at present a large supply of electric heat-treating furnaces. Moreover, conversion of heat-treating to electric energy would require demolishing and rebuilding most of the heat-treating plants in the country, unless existing heat-treating furnaces are modified to use electric energy or combustion interchangeably. This opportunity for engineering design research may offer a rewarding possibility.

Even if the use of natural gas as fuel were eliminated, heat-treating plants would still need controlled atmospheres and electric power for motor drive. In addition, another important aspect of the use of energy and materials in heat-

[10] It is reemphasized that the systems mentioned here are for small plants, of which there are a large number. The needs of large plants are much more easily met because it is easier in a large plant to justify substantial investments in controls.

[11] This is one of the reasons why I feel that the Federal Power Commission's policy on curtailment of natural gas deliveries to industry is misguided.

[12] If one assumes that an electric heat-treating furnace might operate with 50% heat transfer efficiency, then the average fuel demand for heat-treating parts at, for example, 1750°F might be on the order of 1000 Btu per pound of product. This would be a substantial gain over the present average of approximately 5000 Btu per pound.

treating is that heat-treating apparatus require high-quality materials, which in themselves require substantial expenditures of fuel. For example, the Ni-Cr steel radiant tubes in a typical ferrous heat-treating furnace last only about one year. Other high-temperature alloy parts such as conveyor belts have similarly limited lifetimes. These parts must be remanufactured yearly. The remanufacturing requires not only additional expenditures of critically scarce materials such as chrome but also substantial additional expenditure of fuel.

With respect to the preceding discussion, we may now ask: What are the opportunities for longer-range research or development in energy conservation in relation to ferrous heat-treating? First, I hope to have clarified earlier the fact that there is no certain understanding of how ferrous heat-treating—or in fact any other disaggregated process industry—fits into our national energy system now, or how it might fit into a future modification of that system. There have been numerous attempts to devise energy accounting systems, ranging from input-output models derived from more recent economic theory to some more straightforward tracking of the costs of energy purchases. However, I am not aware of any system by which one could determine with any acceptable accuracy the answer to such a question as: If ferrous heat-treating should be converted to electric heating, what would be the net effect on our energy system and our economy? This, in my estimation, represents a fundamental deficiency to which government research efforts should be addressed. Our problem is that we are, as a nation, attempting to modify our use of fuels and energy to alter the extent to which our energy system depends upon imported oil and nonrenewable domestic fuels. However, we do not yet understand how the national energy system functions. There is a compelling need for government-sponsored research to characterize the national energy system of the United States—to provide the information required as a basis for sensible decisions on fuel conservation, energy use, directions for technical research in energy use, and other related matters.

I would like to offer a specific suggestion regarding the nature of energy systems studies. Too much effort has been expended in attempting to conduct energy analyses with conventional economic methodology. Energy use is fundamentally a thermodynamic subject. One of the basic dictums of thermodynamics is that whereas energy (properly defined) is a property of a given system, it is the interactions between systems by which one is able to make use of changes in the energy of various systems. The two thermodynamic interactions that occur between systems are heat and work. As has been pointed out earlier, the most valuable interaction is work, and the proper measure of the value of the energy of any system is the extent to which the energy may be converted to a work interaction.[13] Against this background of knowledge, one sees the development of energy accounting systems, which concentrate upon energy flows and the energy investment in various materials and commodities. These accounting methods attempt to determine the energy expenditure of various industrial processes by summing the fuel and power used by the industry together with the "energy invested" in the commodities pur-

---

[13] See, for example (3), for a partial summary of the literature on this subject.

chased by the industry. There are even some nascent attempts to allocate energy costs among joint products.[14] The basic thermodynamic fallacy in this approach is to ascribe an energy content to substances, as if energy were itself a substance that flows from place to place, either in a primal form (supposedly as fuel) or in some indirectly observable form that has pervaded other materials; in essence this is a modern application of phlogiston theory. In practice, this approach obscures the opportunities to control energy use in interactions, while concentrating on "flows" that occur between sites of interaction. In my opinion, this quasi-economic approach to the study of energy use is sterile. A system of energy analysis based upon post-phlogiston thermodynamics is urgently needed.

In terms of the heat-treating plant itself, what might be done in future heat-treating systems, and what opportunities for technological research suggest themselves? Several possibilities should be considered.

First, it would be very useful to find some substitute for natural gas as a basis for controlled atmospheres in industrial heat-treating. There are ancient arts by which ferrous carbonizing was effected through the use of charcoal. These might be reexamined to see if they could be reintroduced in a suitable form for modern day industrial use. In addition, there are opportunities to use certain inert gases as a basis for controlled atmospheres. The present structure of the heat-treating industry does not readily admit the comprehensive basic research or the necessary co-ordinated demonstrations to prove such possibilities. But, irrespective of industrial structure, these possibilities are important and merit support in energy conservation research. In addition, there is, in nearly every city, a daily source of feedstock from which a gas quite similar to "endothermic atmospheres" could be manufactured: garbage. The technology for manufacturing a fuel gas consisting, roughly, of equal parts of CO and H_2 from urban waste has been available for several decades. While several cities are considering the generation of fuel gas—and related by-products—from waste, the possibility of using this gas directly to replace natural gas feedstocks for furnace atmospheres seems not yet to have been considered. I believe it should be. It may develop that the best place to build a future heat-treating plant will be adjacent to the city waste disposal site.

Suppose that in the future, furnace atmospheres are to be combustible, rather than inert, gases. It would seem reasonable to try to use the heat of combustion of these gases to assist in furnace heating. This would require redesign of the furnace. This could very well be a major redesign in which a number of other aspects of energy use could be altered as well. One of the first steps in redesigning a combustion-heated furnace is to provide for recuperation of the heat of flue gases to assist in heating the furnace. One of the most direct ways to accomplish this is to use flue gases to preheat combustion air. To do this one needs two things that are not now readily available: (*a*) high-temperature heat exchangers that are

---

[14] Recently I discussed the notion of discounting future fuel consumption with an economist. While this appeared to be a meaningful question to the economist, I have been unable to understand where one could deposit a present quantity of oil so that at some future time it could be withdrawn with interest . . . in oil.

durable, effective, and not inordinately expensive, and (b) burners that can receive heated combustion air and also are durable and not inordinately expensive.

Most high-temperature heat transfer equipment built for use in furnaces today is made from nickel-chromium steels. As noted above, these materials are costly and not especially durable in furnaces. In addition, the service temperature for these materials is generally limited to 1800°F; temperature excursions above 1800°F can severely damage parts made of these materials. There are more recently developed nonmetallic materials that have shown great promise for use in high-temperature heat transfer service and for similar applications. One of these is silicon nitride, which appears to be the most desirable, durable, and versatile of the high-temperature materials in existence. The only problem with it is its cost, and its present high cost can be traced directly to the high cost of producing metallic silicon. There is a great need here—and a great opportunity—for research. Any researcher who can find a practical way to reduce the cost of metallic silicon will have done a great service to energy research in general and conservation research in particular.

Other materials are available to use while one awaits the advent of cheap silicon nitride. One of these is silicon carbide. As a rule, sintered silicon carbide does not withstand "thermal shock" well enough to endure in an industrial furnace. But there are some other castable varieties of SiC that appear to be capable of enduring furnace conditions at least as well as the Ni-Cr steels in present use. Silicon carbides can serve at temperatures above 2000°F, which, from the standpoint of thermodynamics and heat transfer, offers obvious advantages. In addition to the SiC family, there are glass ceramic materials that can serve at very high temperatures (> 1800°F) and may find application in high-temperature furnaces. In general the field of high-temperature materials—especially nonmetallic materials that do not require scarce Ni or Cr—offers many opportunities for research and development toward fuel conservation.

Suppose that all the materials required to execute high-temperature heat recuperation and high-temperature combustion are available. How might one begin to apply these to a heat-treating furnace, such as a carbonitriding furnace, to attain greater efficiency of fuel use?

In the first place the heating value of endothermic atmospheres should not be wasted. These atmospheres, if they are to be burned, should be used either for furnace fuel or to preheat the fuel and combustion air for the furnace.[15] Heat from flue gases should also be recuperated and used to preheat fuel and air for combustion. In addition, since the parts exiting from these furnaces must, as a rule, be quenched, it may also be useful to subject them to an air quench by blowing the combustion air over them, just prior to the oil immersion quench that is conventional today. The benefits of such a measure would be of questionable value on a furnace in use today because very little of the heat put into the furnace

---

[15] If it is possible to develop an inert atmosphere for such a furnace, then of course the measures cited here would not fully apply.

actually leaves in the treated stock. But in a future, more efficient, furnace, using the heat in the stock for the first-stage preheating of combustion air may make a good deal of sense.

Finally, the heat in flue gases exiting from a high-temperature industrial furnace contains a very high potential for generation of power. The flue gas temperature at the exit of a single-pass radiant tube can be above 2000°F. The measure of the potential of these gases to provide electric power—or some equivalent work interaction—is the thermodynamic availability A of the gases; this is related to the energy E, volume V, entropy S, and molal concentrations N_j, the constituents of the gas, as well as to the temperature T_0, pressure p_0 of the surrounding atmosphere, and the chemical potentials μ_j^0 of the flue gas constituents in the surrounding atmosphere by

$$A = E - T_0 S + p_0 V - \sum \mu_j^0 N_j \quad [\text{see}(4)].$$

The decrease in A of any system represents a capacity to provide useful work by interaction with the environment.[16] The terms $\sum \mu_j^0 N_j$ represent the work that could be obtained from diffusional processes. For example, when a fuel is burned to produce H_2O and CO_2, the CO_2 may be present in the flue at a partial pressure of 0.5 atmosphere. Its partial pressure (which in this case is the same as its chemical potential) in the surrounding atmosphere will be of the order of 0.03 atmospheres. If one were able to pass the CO_2 from the flue through a system of semipermeable membranes and a turbine to reduce it reversibly to its equilibrium partial pressure in the atmosphere, a significant amount of work could be extracted. The same is true of the H_2O vapor in the flue gas. This diffusional work is always available, in principle, in any chemical reaction. But the only practical device able to recapture the diffusional work of any reaction is the fuel cell, and fuel cells work almost solely for oxidation of hydrogen. Thus, although the final (diffusional) terms in the expression for A represent a significant fraction of the total availability (approximately 20–25% for oxidation of most hydrocarbons) and although they must be included in any complete expression of A to avoid basic thermodynamic confusion, these terms will not be considered here as offering any potential for power generation from furnace flue gases. Only the "thermal" components of A will be considered.

To calculate the "thermal" components of A for a typical flue gas mixture of CO_2, H_2O, and N_2 is a straightforward exercise, which will not be repeated here. We will point out only that the result of this exercise is that the thermodynamic availability of flue gases at atmospheric pressure and 2000°F is approximately 80% of the energy in the gases (5). In most small industrial furnaces, 60% or more of the heat supplied to the furnace exits up the flue. Thus the total theoretical potential for thermal power generation in ferrous heat-treating as it is presently conducted (see Table 1) would be on the order of 0.5–0.6 kW-hr per pound of

---

[16] The notion of interaction with the environment is crucial here. This is why the properties T_0, p_0, and μ_0 appear in A.

product through the furnace.[17] Modern-day power generation equipment can convert 50–60% of this potential to power if units of sufficient size can be used. Thus it may appear possible to produce somewhat more than 0.25 kW-hr per pound of product through the furnace. Since the heat-treating plant itself uses only 0.08 kW-hr per pound of product, there is the potential to produce somewhat more than 0.13 kW-hr surplus electrical energy per pound of product through the plant. This raises an intriguing possibility for the future direction of ferrous heat-treating plants and other similar installations. Perhaps rather than being electrified themselves, the heat-treating furnaces should be used to generate electricity.

Let us consider a small plant, for example, that processes 13–14 million pounds of product per year; the total electric generating potential that could be tapped, to provide surplus electric power, would be only about 200 kW.[18] This, clearly, is too small to justify the integration of furnace heating and electric generation in small plants. However, it might prove attractive in some locations to consolidate the operations of one hundred or so small plants in order to tap this potential for power generation. In such a case the electric power produced might support most of the costs of the plant, and the heat-treating services could be viewed as a by-product. In the case of the small plant, with a production of 13–14 million pounds per year and an electrical demand as indicated in Table 1, the total demand for electricity is approximately 130 kW. Even though the thermodynamic potential to generate electric power, which is lost in the flue gas, is more than sufficient to generate this amount of power, it is somewhat difficult to find an electric generator of this size, irrespective of the prime mover one might employ to convert flue gas heat to shaft power.

In the case of large plants (those having surplus generating capacity of 10–20 MW) it may be possible to use an externally fired gas turbine to generate power from flue gas heat.[19] To produce an external combustion gas turbine has long been a goal of engineers. The principal obstacle to achieving this goal has been the high-temperature heat exchanger required to heat the air in passage from the compressor to the turbine, with heat from an external source. With the advent of nonmetallic high-temperature heat exchangers for (for example) combustion air preheating, it appears that it may be possible to solve at one time the engineering problems of both high-temperature heat recuperation and external combustion gas turbines. Of course, I reiterate that the application of such devices to generate power from furnace flue gases will most probably be limited to large furnaces.[20]

[17] Here we consider the data of Table 1 that about 5 scf of gas are used per pound of product; of this 80% is for fuel and of this amount, 60% (approximately) exits up the flue.

[18] The plants considered here operate 24 hr per day, 7 days a week.

[19] One hundred heat-treating plants of the size considered here could provide about 20 MW surplus power for distribution. This would be inappropriate for commercial power generation.

[20] The smallest industrial gas turbine suitable for use with an external heat source seems to be about 500 kW. This is twice as large as what the typical small heat-treating shop needs or could drive.

And as one improves the efficiency of the furnace itself, one reduces the amount of heat available to generate power.

Summary

In summary, the following opportunities for longer-range technical research to conserve fuels in industrial ferrous heat-treating have been cited.

1. Natural gas should be eliminated as a basis for endothermic furnace atmospheres. Combustible mixtures of the same composition might be generated from urban waste, or furnace atmospheres based on inert gases might be developed.
2. Where combustible furnace atmospheres are used, they should also be used as fuel.
3. High-temperature heat recuperation can be used to save fuel. This measure, in particular, requires a substantial effort for development of high-temperature materials, particularly nonmetallics.
4. Heat-treating plants might be concentrated into larger facilities to permit the use of high-temperature flue gas heat for power generation. This possibility might entail an unacceptably drastic alteration in the ownership and structure of the heat-treating industry. I believe that on-site power production in small plants is unlikely to become attractive.
5. Finally, there is the possibility to switch heat-treating to electric furnaces. If this could be combined with use of inert gas furnace atmospheres, it would be possible to conduct heat-treating operations using coal or nuclear power as the primary energy resource. To exploit this opportunity will require significant rapid growth of utility capacity in some industrial cities. In addition, it will require some imaginative advanced technological design to permit conversion of present combustion-driven furnaces to efficient electrical furnaces; otherwise, the conversion from gaseous fuels to electricity will require the industry to scrap its present capital equipment. This would probably be unacceptably costly and would drastically inhibit the rate of conversion.

CONCLUDING REMARKS

In the above I have sought to illustrate, at least in part, the range of technical contributions that would be required in a research effort for conservation of fuel in the single field of specialized ferrous heat-treating. This field, which tends to be the province of the small independent technical expert/businessman, offers a wide range of intriguing technical research opportunities extending from advanced materials development through design of modifications of existing equipment— despite the highly specialized and strongly disaggregated character and the fairly small size of the installations involved. Certain of the research fields cited are also clearly of value in energy conservation efforts in a broader context. Because there is a unity to research in energy conservation, measures that are scientifically sound and useful in conserving fuel in one field are quite likely to be useful elsewhere as well.

In this chapter, I have sought to suggest to the reader, rather than to persuade, that certain fields of research need be pursued. Thus, I have not attempted to "prove" that certain measurable benefits in fuel savings must follow from the development of one measure or another. In my opinion it is too early in energy conservation research to make such claims; it is also unreasonable to require a researcher to make and substantiate such claims before permitting him to proceed on a promising idea. Nor do I accept the currently fashionable notion in government that research can or should be justified via a cost-benefit calculation.[21] Perhaps some of the suggestions offered above will stimulate some interest and ideas for research on fuel conservation among those who read this, and perhaps it will be possible for some of these readers to pursue their ideas, with whatever support may be required from either government or industry.

Literature Cited

1. Berg, C. A. 1973. Energy conservation through effective utilization. *Science* 181:128–38
2. Hollander, F., Huisman, R. H. 1971. *Computer Controlled Reheating Furnaces Directed to an Optimum Hot Strip Mill Process.* Presented at Ann. Conv. Assoc. Iron and Steel Eng., Chicago, Ill.
3. Berg, C. A. 1974. A technical basis for energy conservation. *Mech. Eng.* 96:30–42
4. Keenan, J. H. 1941. *Thermodynamics.* New York: Wiley
5. Hershey, R. L., Eberhardt, J. E., Hottle, H. C. 1936. Thermodynamic properties of the working fluid in internal combustion engines. *SAE J.* 39:409–24

---

[21] It has never been clear to me that any work that merits being called research could be susceptible to any accurate determination of ultimate benefits.

SOCIAL AND INSTITUTIONAL FACTORS IN ENERGY CONSERVATION

✕11019

Paul P. Craig
Energy and Resources Council, Systemwide Administration,
University of California, Berkeley, California 94720

Joel Darmstadter
Resources for the Future, Inc., 1755 Massachusetts Avenue NW, Washington DC 20036

Stephen Rattien
Directorate for Scientific, Technological, and International Affairs,
National Science Foundation, Washington DC 20550

This paper addresses the status of energy conservation activities—the forces that shape the demand for energy; the historic relationship between energy use and GNP and the questionable inexorability of this relationship; the framework in which to view energy conservation; recent trends in energy conservation; various strategies for encouraging conservation; and what the future might hold. We have sought to identify and discuss these issues in a nontechnical and not necessarily comprehensive framework. Rather, we have tried to impart an understanding of the opportunities and limitations that provide the context for energy conservation activities. For those readers who are seeking a more intensive investigation of various aspects of energy conservation, we have appended a brief bibliography that is representative of the rapidly expanding technical literature in this field.

SOME BASIC PROPOSITIONS

Energy conservation would seem to be both easy to define and attractive as a social goal. In the sense in which it currently figures as a public issue of growing prominence, energy conservation signifies the reduction or elimination of energy waste—and who can oppose the desirability of that objective? It would stretch out finite natural resources; it would minimize environmental damage; it would aid the balance of payments and blunt the insecurity of foreign dependence; it might help redistribute limited energy supplies toward regions or application of greater need; it might even provide moral uplift. With so many benevolent consequences, energy conservation would appear to be as noncontroversial as motherhood and apple pie.

In fact, energy conservation is not that easy to define—either in concept or, in all cases, as a self-evident policy imperative. Consider the following examples.

1. A household trades up to a self-cleaning oven or frost-free refrigerator. For the same amount of heating and refrigeration the family will require more electricity and thus will face not only a higher utility bill but the increased expense of a more costly appliance. Is this household's "anticonservationist" behavior unsound?

2. A manufacturing process permits substitution of manpower and/or other inputs for fuels and power to produce an equivalent output but at an increased cost. Is the switch desirable?

3. In spite of the greater expense, a commuter prefers riding alone in his car to using a more energy-conserving bus. Another commuter's company subsidizes his parking, enabling him to save money by driving his car. Assuming the adequacy of public transport (admittedly a heroic assumption in most places), what is the appropriate approach to a more energy-sparing commuting practice?

4. An apartment dweller shopping for a room air conditioner encounters competing—but identically priced—5000 Btu/hr models. One model uses 40% more wattage than the other. What is the rational energy-conserving behavior?

In the foregoing range of illustrations, the last is clearly the most unambiguous case, for it points unmistakably to both energy and monetary savings unencumbered by undesirable trade-offs and disutilities as encountered, for example, with the self-cleaning oven or frost-free refrigerator. (Even the air conditioner raises some questions. What should the consumer do if the more efficient air conditioner is more expensive? What should he do if he, as an apartment dweller, does not pay his own utility bills?) With the oven or refrigerator, a consumer willingly accepts the higher cost of the more energy-intensive appliance as the price of reduced drudgery. The manufacturing example also seems pretty clear-cut insofar as producers, guided by a dollar rather than Btu standard of business behavior, would reject the costlier, though energy-sparing, production process.

This is not to accord the monetary criterion a sacrosanct standing. The manufacturer's energy use might so pollute the environment that were he charged (as he should be) with these damaging emissions, his inducement to shift away from energy-intensive processes would rise. This defect in market pricing, which—at least historically—has failed to exact payment for such social harm, is more vividly underscored in the automotive case. A fair levy for auto-induced congestion, air pollution, noise, and urban disruption could well sway commuters to alternative, energy-efficient transport modes. (The automobile example betrays another fault in our pricing system: the boss' willingness to underwrite parking costs spares the commuter from confronting directly the costs of services furnished him and thus motivates him less actively toward energy-conserving habits.)

What these thoughts suggest is that a minimum condition prompting a voluntary shift to an energy-conserving practice is lower (or at least unchanged) monetary costs resulting from such a shift. (From a societal point of view, these monetary costs should be fully reflective of the environmental and other externalities that the energy user imposes upon such common-property resources as land, air, and water. Incidentally, if one is willing to accept this condition as a legitimate criterion for energy conservation, one should, conversely, permit intensification of energy use where such a change is also economically dictated.) But even where diminution of energy use is warranted by monetary savings, that fact alone may not be—and indeed should not be—sufficient to spur the energy user in that direction. Recall the self-cleaning oven. Consumers willingly pay for self-indulgence. One person's frivolity is another's necessity. Tastes, values, and affluence all join to put a subjective stamp on consumer preference scales. National priorities must sometimes outweigh individual preferences, and decisions taken for the common good may be perceived by an individual as working either for or against his own self-interest. In the final analysis, it may well be that there are rather few totally costless conservation measures. Even conservation involving only more efficient means of achieving the same level of end-use consumption is apt to carry with it an investment either in new capital or in R&D that diverts productive effort and personal income from the satisfaction of other current consumption.

ENERGY ACCOUNTING

The foregoing discussion evaluates energy conservation decisions within a framework in which economic considerations and social values are judged largely adequate for impelling whatever conservation actions are worthwhile. For a number of reasons to be explored later in this review (e.g. market pricing failure, deficient information) one may question whether rational energy conservation measures can be pursued without much more explicit and quantitative detailing of the energy implications of alternative actions and processes. In its extreme, rather than typical, expression, such an effort would accord an "energy standard of value" parallel status to the conventional monetary basis of decision-making.

Energy accounting, as this attempt at systematization is called, starts with the recognition that most energy use analysis to date has concentrated upon the direct energy requirements of processes—heating and cooling of buildings; provision of fuel for automobiles, trains, ships, and buses; and provision of energy to generate electricity. Comprehensive analysis of energy conservation options must take into account not only these direct uses of energy, but also the indirect energy inputs required to fabricate, maintain, and recycle particular elements of the energy system. Such comprehensive analyses have not generally been used for analysis of energy strategies. This situation results from several factors, among which are the relatively low price of energy until recently (which rendered careful analysis uninteresting) and the belief among many economists that energy accounting was unnecessary, since economics would generally automatically minimize the total cost of a particular societal system. If energy accounting has utility—and there is controversy on this score—it will be through its ability to counter market imperfections (such as those arising from price regulation), to account for externalities, and to illuminate the implications of certain types of conservation policy decisions, such as might be illustrated by policies designed to minimize reliance upon imported oil.

In a review of the fundamentals of energy accounting,[1] two general approaches

[1] International Federation of Institutes for Advanced Study, 1974. See suggested reading list.

can be identified. In the first approach (process analysis) all of the specific processes giving rise to an end product are analyzed in terms of their energy inputs and summed to give the total energy input.

The system boundary used is critical and must be specified carefully. Those system boundaries that permit comparison of alternative energy inputs for accomplishing particular societal objectives are the most useful.

A second approach is energy input/output methods. Energy requirements for particular functions are determined in terms of an input/output analysis based upon the energy use of the major sectors of the economy. For example, an input/output analysis for the automobile system includes many elements in addition to fuel: road construction, insurance, automobile construction, maintenance, and disposal. Input/output analysis has also been applied to evaluate the energy cost of producing energy in the United States and the United Kingdom (Table 1).

Process analysis and input/output analysis are complementary tools, which should yield similar results when applied to similar structures. Process analysis is better suited to exploring improvements possible with new or highly specialized technologies, for which input/output sectors and data are too highly aggregated (or often out of date) to encompass. Input/output methods are also difficult to apply to dynamic or nonlinear systems and to systems with changing technologies.

A number of ratios have been proposed to indicate relative efficiency. We believe that ratios should not be used but that attention should be concentrated on careful definition of system boundaries (so that the energy to manufacture a heat pump or insulation is included, along with the energy to manufacture a power plant, when an electric building heating system is completely analyzed).

ENERGY AND GNP: THE "DE-COUPLING" ISSUE

Energy conservation can be effected in any of three ways.

1. Given output patterns can be satisfied by a lessened input of energy resources. This result can, in turn, be achieved by technological advances, as in improved central-station-generating efficiency in electric power production; or by applying known technology more efficiently in particular end-use activities, as in enhanced thermal efficiency through more insulation.

Table 1 Energy requirement for energy (energy units per unit of energy made available)

	USA[a]	UK[b]
Coal	1.014	1.047
Refined Petroleum	1.19	1.23
Electricity	2.89	3.54
Natural Gas	1.16	—

[a] Source: R. Herendeen and C. Bullard, 1974, *Energy Costs of Goods and Services*, Center for Advanced Computation, University of Illinois.
[b] Source: P. Chapman, G. Leach, and M. Slesser, Sept. 1974, *Energy Policy*.

2. A shift in output patterns may be possible, such that goods and services embodying, or associated with, relatively heavy energy requirements are supplanted by the production of fewer energy-intensive things. A shift from passenger cars to public transport, or from aluminum to steel, illustrates the point, albeit in a highly simplistic fashion.

3. Finally, one could contemplate national economic-growth policy as a governmental lever to dampen energy consumption. We assume in this article that for the time span for which it makes sense to look ahead, any suppression of energy-consumption growth will not be brought about by a deliberate slowdown of economic growth in general, even though that may be an independent goal for other reasons and may ultimately be forced on us by resource and environmental constraints.

This is not to say that the level and growth of a country's energy consumption is predetermined by its level and growth of GNP. Even though per capita energy consumption and per capita GNP are highly correlated over time for a given country, and cross-sectionally among different countries, the link is by no means ironclad. Americans, for example, consume considerably more energy per capita than do citizens of countries—e.g. Germany and Sweden—with roughly comparable levels of per capita GNP. A climate of rising resource costs and promotion of resource-saving technology could markedly "de-couple" the historic energy/GNP relationship.

Indeed, US history bears witness to the extent to which the character of industrial development and technology affects the energy/GNP linkage.[2] During the period of intensive US industrialization between 1880 and 1920, the annual growth of energy consumption averaged 5.6%, substantially exceeding annual GNP growth of 3.4%. By contrast, over the ensuing four decades, US energy-consumption growth fell steadily below GNP growth, the annual figures for the period 1920–1960 averaging out to 2.1% and 3.2%, respectively. In both periods, of course, energy and GNP each grew more rapidly than population, yielding per capita growth for both indicators.

Sorting out the factors at work in the contrasting trends for these two periods of American history is a complicated business and cannot be done here, but some summary observations are worth making. The steeper rise of energy consumption than of GNP in the closing decades of the nineteenth century and in the early part of the twentieth century almost certainly reflects the disproportionately fast growth of manufacturing in the economy—a sector requiring far higher energy input per unit of activity than the agricultural component, which had dominated the economy in the past. This phenomenon is at work at present in developing parts of the world where consumption of energy rises at rates considerably in excess of output in general, unlike the situation in most of the advanced countries.

[2] Various ways and various terms are used to characterize the relationship between energy consumption and GNP. Energy-GNP *elasticities* or *coefficients* describe percentage changes in energy consumption divided by percentage changes in GNP. Energy-GNP *ratios* or energy *intensities* describe—at a point in time—the level of energy consumption divided by the level of GNP.

The forces making for relatively slower growth in energy consumption than in GNP in the United States after World War I (or, to put it differently, for progressively less energy used per constant dollar of GNP) are more difficult to unscramble, but at least several distinct elements appear to have played a role. An important contributing factor was the rapid rise of electrification, which greatly enhanced the efficiency of factory operations formerly dependent on, and constrained by, the limitations on plant layout imposed by the coal-burning factory steam engine. This occurred in spite of the fact that the generation of a kilowatt-hour of electricity required many times the caloric output of the electricity produced. This disadvantage was moderated in a very important respect, however: the thermal efficiency of electricity generation improved substantially over the years. In 1925 it took over 2 pounds of coal to produce 1 kilowatt-hour of electricity; by the 1960s it took less than 1 pound. Another important instance of improved energy conversion later in this period was the replacement of steam locomotives by more efficient diesel engines.

In the 1960s, US energy consumption once again tended to outpace GNP growth; a halt to efficiency improvements in the electricity-generating sector seems to have been an important factor. More recently, a parallel trend between energy and GNP has reasserted itself. It would be rash to preclude either ups or downs in the future energy/GNP ratio. One source making for a possible upward trend is the aggressive development of US coal resources to limit oil import expansion, which might mean more recourse to electricity conversion than otherwise; because of low thermodynamic conversion efficiency, a greater share of primary energy going into the utility sector might also signify greater overall energy-consumption growth than would otherwise be the case. On the other hand, market conditions and emphasis on conservation practices and policies could constrain energy-consumption growth to a rate well below GNP growth. Of these two sets of possibilities, the second—less energy per constant unit of GNP—is the likelier outcome. In any case, GNP as an important, but not unique, determinant of energy consumption is likely to endure. The "de-coupling" can only be partial, but it is an important consideration for the formulation of national energy policy.

In this respect, comparisons citing substantially lower energy/GNP ratios for other industrially advanced countries are deserving of study but are not, on the face of it, totally persuasive. Over time, other countries have tended—fortunately or not—to emulate the United States and its energy-intensive life-styles; the economic structure of some other countries evolved toward less energy-intensive patterns of activity because of high fuel and power costs; and then there are differences in climate, population density, and other factors that need to be evaluated for their impact on international differences in energy use relative to GNP. One of the most significant differences among nations is per capita energy use for passenger travel. Energy use for transportation has been remarkably high in the United States— largely as a result of cheap gasoline relative to other nations. This situation has been responsible, to some degree at least, for the present structure of our cities and suburbs. The extent to which this structure will change in the future as relative energy prices increase is at present a matter of speculation.

MOTIVATION FOR ENERGY CONSERVATION

We have already alluded to the fact that extravagance of energy use in the United States stems in large part from a history of cheap energy whose legacy will continue for some time. But high per capita energy use reflects on more than cost alone. Thus, lack of information to consumers about the energy consumption of the products they buy or the marketing problems of increasing the first cost of durable goods to produce more energy-efficient products has led to, and will continue to cause, energy use above an economically optimum level.

In the consumer appliance field, for example, until recently there was essentially no information on the energy use characteristics of various products. Even today, when air conditioners carry energy efficiency ratings, the consumer finds it difficult to determine his total costs. (And if the consumer is a renter who does not pay for his own utilities, then least first cost automatically becomes a prime incentive.) The problems consumers face because of lack of information from manufacturers are compounded because most retailers do not now have the information, expertise, or incentive to translate the available information into a form that is useful to most buyers.

There are many reasons for encouraging energy conservation. These include the adverse environmental impacts of energy production and use, the depletion of irreplaceable fossil and nuclear fuels, and the development of life-styles and land-use patterns that are both energy-intensive and consumptive of other resources. Further, the fact that so much of the energy we use comes from abroad causes significant problems in the areas of national security, balance of payments and world political stability.

To elaborate on the reasons for pursuing a conscious energy conservation policy, we begin by identifying some of the failures of market mechanisms to assure that economically rational decisions are made from the consumer or societal viewpoint. In the housing industry a prime objective of the builder is to hold down first costs. Much of the home-buying public is constrained by the amount of money a bank will lend. Therefore, no matter what the merits of increased-first-cost equipment— insulation, construction quality, efficiency of appliances, etc.—the buyer must often opt for the least first cost, and this is often dictated by the building code. In the absence of proof of the savings afforded by a better built home, perhaps it is reasonable for the consumer to choose the least-cost home. (These considerations suggest a need for substantial improvements and extensions in regulation- and standard-setting. Some states—e.g. California—are setting up new state agencies or commissions to accomplish this.)

The market mechanism has been further weakened of late by shortages in price-controlled natural gas and sometimes oil. This has meant that far more costly, and less energy-efficient, electricity has been substituted as a space-heating fuel. The problem of fossil fuel shortages is a primary example of market mechanism failures in every energy-consuming area. Industry has made major switches to expensive and inefficient electric power to assure uninterrupted energy supplies. Even the electric power generation industry, faced with difficulty in borrowing capital or obtaining rate increases, has taken advantage of fuel adjustment clauses and has invested heavily in inefficient peaking turbines burning expensive distillate fuel, usually from imported sources.

Reducing the adverse environmental impacts of energy production and use is another reason for encouraging conservation. Although more stringent environmental protection laws have been passed, there remain major environmental impacts of energy use. Land and water damages due to strip mining are proportional to coal use. More efficient automobiles would not only save gasoline but also reduce environmental impacts throughout the entire fuel cycle from the oil well to the refinery to the gas station. Exponential growth of energy use, if it continues at the present rate, will exacerbate all energy-related environmental problems, despite our best efforts at implementing control technologies.

The awareness that our fossil energy resources are finite is further reason to encourage conservation. While individuals and firms must apply some finite discount rate to all future streams of benefits and costs, it is apparent that as a nation we should be concerned about the transition from a fossil fuel economy and about the energy needs of future generations. Careful husbandry of inherently scarce energy resources will prolong their availability, making the transition less sudden and enabling us to explore a broader range of options.

These benefits of conservation occur no matter what the price of energy or the source of fuel. But the oil embargo and the dramatic increase in world oil prices suggest a number of other important reasons to promote conservation. The dramatic rise in world oil prices has had an adverse effect on our balance of payments situation. The importation of 5–6 million bbl of oil/day costs the United States about $60 million daily, or over $20 billion annually, which has to be matched with increased exports or acceptance of a high level of foreign investment in our economy, and an imported barrel of oil costs more than an average domestic barrel. This means that every barrel of oil saved through conservation is a barrel less of high-cost imports.

Along with the problem of dollar outflow is the problem of national security. Reliance on potentially unstable sources for foreign oil creates a national security problem. This can be reduced through a number of mechanisms—increased defense spending, a domestic oil storage program, increased domestic energy production—but conservation of oil works directly to cut dependence on foreign suppliers. Generalized to the world as a whole, the control of vital energy resources in the hands of a very few nations has real economic and political dangers. To the extent that conservation can cut the dependence and can shift oil from a seller's to a buyer's market, this is a significant motivation for encouraging conservation.

RECENT EXPERIENCE

The impetus for fundamental change toward less energy-intensive life-styles has appeared only within the last several years. The turnabout from falling real energy prices to drastically rising ones; politically actuated supply manipulation overseas;

exhortation, information, and some beginning policy moves in the direction of mandated conservation actions—all of these developments are too recent to be a valid guide to longer-term and enduring change in the behavior of energy users. It is nonetheless instructive to survey some of the observed short-term responses, as providing both possible clues to longer-range patterns and insight into deficiencies in policy-making. We examine, in turn, energy consumption trends surrounding the 1973–1974 embargo; the success of federal conservation efforts; a brief case study of what happened in a specific metropolitan area; and the result of an attitudinal survey into the public's thinking about energy and conservation matters, at least as viewed within the perspective of current events.

Impact of the Embargo

The steady energy growth rate that the United States has had over the past several decades appears to have subsided since the embargo, but the trend line from recent experience is as yet unclear.

In 1972 and 1973, prior to the OPEC embargo, total energy use in the United States was 72.1 and 75.5 quadrillion Btu, respectively, reflecting annual growth rates approaching 5%. In 1974, total energy input was 73.5 quadrillion Btu, a drop of 2.7%. Price changes undoubtedly played a role, but the fear of shortages early in the year was an exhortation to conserve, and the onset of economic recession also contributed.

Heating oil, which had cost 18–20¢/gal only a few years ago, was selling for 40¢ in the Washington DC area. Gasoline for automobiles, which had been selling in January 1973 for about 25¢/gal (excluding tax), was selling in December 1974 for 39.8¢/gal (excluding tax).

Since—as noted— it is not possible to separate unambiguously the various short-term developments interacting with energy use, judgments about the probability of an enduring break in the historic energy growth trend must be withheld.

Federal Energy Conservation Efforts

Since the embargo the federal government has undertaken a number of programs that, if implemented, would encourage energy conservation. For example, in November 1974 HUD published a policy that would set minimum standards for insulation in federally insured new housing covering one- and two-family units. HUD estimated the energy saving at 60,000 bbl of oil/day by 1985.

The Federal Energy Administration (FEA) is seeking voluntary agreement of manufacturers on minimum performance standards for their products. It is anticipated that by June 1976 there will be agreements reached with manufacturers of water heaters, refrigerators, freezers, and room air conditioners. FEA has estimated an energy saving of 118,000 bbl/day by 1980. There is a similar program to label appliance efficiencies. Further, FEA has undertaken an extensive advertising campaign under the "Don't Be Fuelish" banner. In addition to suggestions on thermostat settings, the federal campaign includes many suggestions on energy conservation through public transportation, car pooling, and reduction of lighting loads.

Aside from these limited federal efforts, the federal government has taken a strong lead by in-house action through the Federal Energy Management Program established by presidential order on June 29, 1973. The federal government was directed to reduce its anticipated use of energy by 15% during fiscal year 1974. In fact, during the succeeding 12 months, consumption was cut by 24%. The program included such elements as lighting standards, which were specified as reductions to 50 foot candles at work stations, 30 foot candles in nonwork areas, and 10 (but not less than 1) foot candles in nonwork areas. Heating and cooling specifications were also established, with office space to be set between 65° and 68°F during working hours and not more than 55°F during nonwork hours during the heating season. Constraints were placed upon the size of vehicles that could be purchased by federal agencies. Provision for bicycle parking was required in federally owned or operated buildings. The program is continuing in fiscal year 1975 with apparent success. Although much of the federal savings resulted from changes in heating, cooling, lighting, and travel, 50% was attributed to reduction of jet fuel consumption, mostly in the Department of Defense.

Another federally promoted program is the national 55 mph speed limit adopted in the winter of 1974, which has been estimated to save approximately 100,000 bbl of oil/day. The 55 mph speed limit has been enforced sporadically but will probably be a long-term element of national strategy.

Automobile efficiency has also been a subject of much discussion both by the Administration and the Congress. A voluntary program for fuel efficiency labeling was called for in the President's energy message of April 1973 and has been in effect for some time. But in the October 1974 presidential energy message, a 40% improvement in new car fuel economy for 1989 model cars (compared to 1974 cars) was called for. However, the 1974 model cars were markedly inefficient in energy use (about 13 mpg vs about 15 mpg in 1950). The Congress has supported similar fuel economy objectives.

The Los Angeles Experience

Much of what we know about our national ability to save energy is speculative. Fortunately, a few examples where energy conservation was accomplished do exist. As a result of an accident of contracting, virtually all of the oil to produce electricity for the Los Angeles Department of Water and Power was affected by the Arab oil embargo in 1974. Los Angeles responded by developing a precise and strong program for achieving energy conservation. Just before Christmas 1973 the City of Los Angeles told merchants and building managers that they would have to cut their electricity consumption by 20%. Within four days consumption dropped 11%, and by the end of the first week it was down 14%. During the first two months of the program, average electricity consumption was reduced 17% below the corresponding months of the year before; commercial sector usage in Los Angeles dropped almost 30%. These figures compare with a less than 5% reduction in total energy use for the rest of the United States.

The Los Angeles plan specified reductions in energy relative to a base period. Quotas were assigned to individual customers. Sanctions for failure to comply with the quotas were stiff. The penalty for using more electricity than permitted during the first billing period was a surcharge of 50% of the entire bill; the second violation called for a two-day power shutoff; a third violation called for a five-day shutoff. The Rand Corporation surveyed Los Angeles to determine the kinds of procedures that had been used to conserve electricity. A typical procedure in an office building that reduced energy electricity use in excess of 40% included the following: removing half of the fluorescent lighting tubes; extinguishing 75% of the lights in indoor parking facilities; turning off many exterior lighted signs; running the main plant air conditioning system less frequently; reducing the winter thermostat from 75° to 70°F; removing one of three elevators from service except during peak rush periods. Reduction of lighting load was a major factor throughout the city. This not only decreased the direct load for producing light, but also reduced the load upon the air conditioning system. A computer facility achieved a 23% cutback in its total energy use as a result of decreasing the number of lights in the rooms in which the computer facility was located. The major factor in the Los Angeles plan was that the importance of the problem and the severity of the penalties made almost everyone concerned with energy use aware of the problem. Consequently, there was improvement in maintenance and thermal load management in many cases.

Consumer Attitudes

We have discussed the changes in demand for energy that have occurred since the OPEC oil embargo. Unfortunately, it is extremely difficult to be precise about inducements for changes in energy use because these are very much dependent upon the attitudes of individuals. This section summarizes some of the attitudes as determined through several FEA public opinion polls. While these polls are far from a perfect guide, they do indicate factors that individuals regard as important.

Even at the height of the OPEC oil embargo, only two thirds of the population of the nation felt that energy shortage problems were serious. Approximately one third either felt that problems were not serious or else had no opinion. This relatively low percentage stems in part from the fact that in many parts of the United States the oil embargo had minimal effect. By the spring of 1975 the percentage of the population regarding energy problems as serious had risen to 79%, even though the embargo was no longer in effect. This reflects the lag times associated with altering public attitudes. Further, the percentage of the population that felt that the energy problem would become more severe within the coming months had increased from 11% in the spring of 1974 to almost 30% in the spring of 1975. The areas with which people were concerned changed in an expected direction. Among those who felt that there was an energy problem and who were affected by energy shortages, two thirds felt in the spring of 1974 that gasoline was the primary problem. This dropped to only 39% in the spring of 1975, with electricity, natural gas, and heating oil problems assuming increased importance.

There was a general reaction against price as a mechanism for allocation. In spring 1975, 51% of the population felt that rationing was the best approach, whereas 32% felt that price increases were the best approach (the remainder had different or no opinions). Automotive emission control equipment is generally perceived by the public as being correlated with decreases in fuel economy. Delays on the imposition of stricter automobile pollution control requirements were favored by 62% of the public.

The perceived impact of the 55 mph speed limit was relatively small. While 74% of car owners stated that they were driving less, only 10% stated that they were driving slower, only 9% had gotten tune-ups for energy conservation purposes, 8% had joined car pools, and 6% had bought a smaller car.

Even today there is relatively little awareness that the federal government has taken actions to structure itself internally to be responsive to energy policy needs. In spring 1975, 31% of the population felt that the federal government had established some type of agency with responsibility for energy policy; 28% said that the federal government had not established such an agency, and 41% didn't know. Only 3% were able to identify the FEA as the agency with principal responsibility for energy policy. On the other hand, 76% were familiar with the "Don't Be Fuelish" slogan, and 85% of those gave an accurate answer as to the meaning of the slogan.

The surveys point out, as one might expect, that energy conservation options that are convenient and have little cost appear to be practiced. In the spring of 1975, 75% of the population indicated that they turn off the lights more often, 63% turn down the thermostat, and 26% use appliances less often. On the other hand, only 2% have installed storm windows and only 2% have installed smaller light bulbs. There was, however, a positive response to the concept of rebates to provide incentives to install insulation and storm windows.

Actions designed to encourage recycling also received a positive response. A law requiring returnable bottles with deposits for soft drinks and beer was favored by 73% of the public. The favored price to be effective was 5 cents/container.

On balance, the FEA public opinion studies disclosed an American population that feels that the federal government should take action to promote energy conservation, but does not believe very much action has been taken at present. There is a general feeling that a policy encouraging energy independence is desirable and that energy conservation is an important and acceptable approach. Incentives to energy conservation are felt to be important. Energy conservation does, however, appear relatively low on the list of concerns compared to problems of unemployment and inflation, and our discussion of the uncertain relationship between energy and GNP suggests the reason for such a cautious approach. There does not appear to be a preponderance of opinion favoring energy conservation at the expense of economic and social well-being. If energy conservation can be obtained at a modest price, it appears to be worth it. If not, then energy conservation should be sacrificed, according to most Americans.

THE ROLE FOR GOVERNMENT POLICY

A host of energy-conserving practices would meet the criterion of economic rationality, and many of these would, if understood, probably not be perceived as intruding severely into life-style preference. Indeed, when such changes are induced by market forces, the question of encroachment into life-style preferences is rarely raised. These practices include, among others, building insulation, more efficient lighting, industrial waste heat recovery, and some of the less disruptive changes in the transport system. What are the principal measures that would lead to the adoption of these and other energy-conserving practices? We can distinguish among the following three instruments for influencing energy use decisions in the private sector: (*a*) the operations of impersonal market forces; (*b*) tools to facilitate more rational responses to given market forces; and (*c*) governmental policy tools to shape market forces to enhance their allocative performance.

Evolving market conditions—for example, the almost inexorable rise in real electricity prices that we are now witnessing—will inevitably induce a degree of retrenchment in energy use that would not have been indicated from past trends and that will occur independently of deliberate policy actions. Such sensitivity to market conditions is especially likely to take place in energy-intensive industrial activity. Simply because responses to market forces lag is no reason to despair of their efficiency in due time. Some inescapable lag exists because of the lifetime of the existing stock of energy-using capital assets, such as household appliances, and durable equipment, vehicles, and structures. Further, the development of consumer awareness of the problem and his choices has a certain lag time.

To be sure, the private market mechanism often may operate too slowly or require very extreme short-term price increases to meet desirable conservation targets—as, for example, a quick slowdown in fuel imports for national security reasons. Econometric analysis discloses relatively long response lag times. A typical lag time for electricity demand has been estimated at 8 years for 90% of the response to take place. The problem of response time is particularly serious in the present situation in which econometric estimation techniques are unreliable because we are in a new and unknown price domain.

Thus, the school of thought (especially among some economists) that price increases should be the primary, if not the only, mechanism by which energy conservation is accomplished seems to us somewhat misguided.

If the market mechanism needs to be reinforced by purposeful governmental policy measures, two forms of intervention can be envisaged. In the first and more passive case, governmental policies need to be fashioned that would guide consumption practices along a more informed path. This will help energy users respond knowledgeably to market conditions. Examples include mandatory information on energy efficiency and costs—given the degree of insulation—in the heating and cooling of newly constructed buildings, the operation of automobiles, or the use of room air conditioners. One might also cite the potentially effective role of governmental purchasing policy to spur well-conceived forms of energy conservation. Through demonstration projects in the construction of public buildings and acquisition of public transportation fleets, for example, such a role might prove beneficial.

Information collection has received considerable emphasis in the federal energy conservation program, but in many cases information on particular devices is simply not available or is extremely difficult to obtain. In certain cases, when the saving associated with buying an energy-efficient device is relatively low, the cost to an individual to gather information may be much more than that information is worth.

But for those energy-intensive consumer products where information is lacking, the federal government has a major potential role in gathering such information or assuring that reliable information is made available. Similarly, information programs —including research, development, and demonstration of energy-conserving opportunities—should be conducted in the industrial, commercial, and transportation sectors.

In our last category, governmental policies designed to affect the allocative outcome of market forces are brought into play. For example, the automobile exacts extensive burdens on society, which in many cases are inadequately (if at all) reflected in charges to users. A federal fuel economy tax would help sway owners toward smaller, more energy-efficient cars. The expansion of public transport, and mechanisms designed to encourage public acceptance, deserve earnest policy consideration. In housing, compulsory insulation standards and some changes in home financing arrangements favoring energy conservation practices suggest themselves. Some experimentation with federal regulations of this type has occurred. An example already noted is the FHA requirement on insulation in homes. To qualify for a loan through the FHA, the homeowner must comply with specified standards developed by the Bureau of Standards.

There is one additional respect in which the influence of public policy upon the functioning of energy markets has received, and deserves, prominent attention. That is the case in which, either directly through regulatory powers or indirectly through tax laws, pricing for fuels and power—hence, demand-supply responses— departs to a greater or lesser extent from "solutions" governed by the efficiency normally characterizing reasonably competitive markets. On a national basis, for example, many (though by no means all) competent observers believe that the Federal Power Commission's approach to regulation of the price paid by interstate pipelines to natural gas producers has led to a distortion of the public-interest objectives that the policy was designed to serve. From regulation of natural gas prices at a point judged by many to be far below the "market-clearing" level, an artificial stimulus to demand (i.e. a disincentive to conserve) has developed. This, coupled with a failure to encourage development of new supply, has at the very least contributed to the scarcity of natural gas prevailing now in many parts of the nation. What this suggests is that, whatever its intended or realized other objectives, a deliberate low-price policy for natural gas distorts market signals and hence energy use allocations.

Further, the extent to which certain tax benefits accorded the crude-oil-producing industry (principally percentage depletion and intangible drilling expenses) involved a burden shifted from the price mechanism and the individual consumer onto the general taxpayer, a similar encroachment on the efficacy of market processes resulted. In other words, oil prices tended to be lower than supply-demand forces would suggest. If one assumes that these low prices were intended to spur domestic production, it should be recognized that there are options for assuring that "the cost of customer security . . . be paid by customers"[3]—rather than through indirect

[3] G. M. Brannon, 1974, *Energy Taxes and Subsidies*, p. 74. For the Energy Policy Project of the Ford Foundation. Cambridge: Ballinger.

instruments, such as the depletion allowance. Again, as with natural gas (though for different reasons), the price should not provide misleading stimulus to demand above levels implied by a "true" market price.

In the context of energy conservation, the area in which the appropriateness of publicly regulated energy prices has evoked perhaps the most debate is the electric utility rate structures decreed by state regulatory bodies. At issue is the question of electric utility pricing and promotional practices—particularly the characteristic of the rate structure that provides for a progressively lower unit price (cents/kW-hr) as the volume of total energy consumed during the month rises. Critics of these so-called declining block rates argue that, whatever their past justification, prevailing rates in the large-volume blocks are not justified by economies of scale and are below marginal costs, as a result of which consumers are given incorrect price signals and encouraged in "uneconomic" consumption. A few state regulatory commissions, while not prepared to embrace unqualifiedly the inverted rate structure proposed by some critics of declining block rates, are nevertheless moving in the direction of a more "flattened" or neutral electricity tariff.

Quite apart from proposed fundamental reforms in the basic rate structure, there is one important and particular respect in which cost-of-service precepts are frequently ignored in utility pricing, and that is in the case of power use when the utility system operates at its peak load. There is evidence that peak power is much more costly to produce than off-peak output (by as much as a factor of 4 or 5), yet it is usually sold at no greater price, particularly when it involves residential sales. In New York City, for example, the peak summer demands caused largely by the concurrent use of air conditioning, rush hour subway transport, and early evening household use has meant the costly and energy-inefficient deployment of gas turbine generators for meeting the extra load. A shift toward a dual-peak and off-peak pricing system would therefore seem to be an economically justified approach that would cut the overall costs of service, reduce dependence on distillate and natural gas, and cut total fuel use. Peak and off-peak pricing, and even more sophisticated approaches to rate design, exist in other countries, particularly France. Initially, the cost of dual-meter installation would have to be faced (and economically justified), but this cost would be a legitimate charge in the interest of a presumably more rational and beneficial utility pricing system.

A less drastic approach than dual metering to penalize use at the time of the daily system peak involves seasonal peak surcharge; a number of utility systems now impose a flat kW-hr charge during the summer air conditioning months. Even this more modest proposal has not yet been embraced in principle, let alone practice, by all regulatory commissions.

Apart from reform of electric utility rates, a number of new financial mechanisms might be fashioned within the regulated utility area in the interest of energy conservation. In Michigan, the Public Service Commission has approved the servicing of loans on home insulation as part of the natural gas bill. The effect of this procedure is that the amount paid monthly by the consumer does not increase and may even decrease, but the bill is now composed in part of payments for gas and in part, repayment of a loan to provide improved building insulation.

A similar procedure is being explored by the Southern California Gas Company to provide hot water with solar heater assist. The concept is that the gas company would provide the service, namely hot water, partly through solar heaters and partly through gas. The solar units would be maintained by the existing maintenance crew of the gas company. For this highly promising experiment to be undertaken, approval by the public utility commission will be required (and is expected).

Probably as a result of uncertainty within the government as to what role it should play, decisions on federal activities in energy conservation have been fragmented and generally not well conceived. The first Office explicitly concerned with energy conservation was in the Department of Interior. With the establishment of the FEA, this Office was moved there, and programs in both research and development and public information were developed. When the Energy Research and Development Administration (ERDA) was formed, the Congress mandated that it undertake conservation research. Although conservation was to be a major element in ERDA, it has evolved very slowly, and now it appears that the primary emphasis in ERDA will be on technological opportunities. FEA shows signs of placing increasing emphasis upon public information and policy analysis. Demonstration experiments (e.g. of techniques for retrofitting houses with energy-conserving features such as insulation, storm windows, and time-switched thermostats) have been proposed by both FEA and ERDA, and it is unclear which agency, if either, will take the lead. The relation between demonstration experiments and changes in public response remains conjectural and will require clarification if such programs are to receive emphasis in federal programs. It seems likely that unless the national energy system receives some additional major shocks, federal energy conservation programs will not be accorded emphasis akin to that received by energy supply programs—despite the fact that in many areas the payoff (from a national point of view) is greater in terms of barrels of oil equivalent to investment in demand reduction than from investment in increased energy supply.

EPILOGUE

This review has identified the factors that have historically discouraged energy conservation, the changes that have occurred in the recent past, and the opportunities for encouraging energy conservation in the years ahead. We suggest that economic forces are likely to play a major role in spurring conservation but that there is need for governmental action to overcome a number of marketplace problems: lack of information; inadequate research and demonstration; failure to account for problems of resource depletion; adverse environmental effects of energy supply and use; limited capital availability; and difficulties in balance of payments and national security. Thus, energy conservation comes about through changes in public attitudes, market forces, and governmental actions. While energy prices are rising and may be expected to continue upward, the other forces depend on public and governmental perceptions, which have shifted partly toward a conservation ethic but have by no means made a major commitment. As of this writing (May 1975), it appears that within the United States the perception of an energy shortage is at a low level and the cost of energy, while rising, is perhaps judged to be not all that different from other costs during this inflationary period. We can anticipate that relatively limited action will be taken by either the federal government or the private citizen until energy shortages become far closer to reality than they appear at present or energy prices are driven markedly higher. A disquieting thought is that we rise—convulsively—to challenges in the midst of crises, only to relapse into a business-as-usual attitude when the immediate danger seems to have passed—at least that is what one senses from the post-embargo experience. Although the issue of conservation is indeed a subject of serious consideration and some actions are being taken, clear-cut resolve for, and implementation of, a longer-run conservation strategy still eludes us.

Suggested Reading

Berry, R. S., Fels, M. F. July 1972. *The Production and Consumption of Automobiles.* Rep. to Illinois Inst. Environ. Qual., Chicago

Berry, R. S., Makino, H. Feb. 1974. Energy thrift in packaging and marketing. *Technol. Rev.*

Citizens Advisory Committee on Environmental Quality. 1973. *Citizen Action Guide to Energy Conservation.* 1700 Pennsylvania Ave., Washington DC 20006

Federal Energy Administration. August 1974. *Electricity Rates and the Energy Crisis, Report of a Conference,* June 19

Federal Energy Administration. 1974. *Second Electricity Conference,* Sept. 19–20

Federal Energy Administration, Office of Conservation and Environment. Oct. 31, 1974. *A Pilot Project in Homeowner Energy Conservation*

Federal Energy Administration, Office of Conservation and Environment. Nov. 1974. *Federal Energy Management Program, First Annual Report, Fiscal Year 1974*

Federal Energy Administration. Dec. 1974. *Energy Conservation Study, Report to Congress*

Federal Energy Administration, Office of Conservation and Environment. 1974. *Lighting and Thermal Operations— Energy Management Action Program for Commercial, Public, and Industrial Buildings*

Federal Energy Administration, Office of Conservation and Environment. Jan. 1975. *How Business in Los Angeles Cut Energy Use by 20 Percent.* Rep. No. WN-8866-2-FEA

Federal Energy Administration. *Monthly Energy Review.* Springfield, Va: Nat. Tech. Info. Serv.

Hannon, B. Feb. 1974. Options for energy conservation. *Technol. Rev.*

Herendeen, R. Oct. 1973. *The Energy Costs of Goods and Services.* Oak Ridge Nat. Lab. Rep. No. ORNL-NSF-EP-SS

International Federation of Institutes for Advanced Study. Aug. 1974. *Energy Analysis Workshop on Methodology and Convention.* Nobel House, Storegatan 14, Box 5344 f-10246, Stockholm, Sweden

Macrakis, M. S., ed. 1973. *Energy: Demand, Conservation, and Institutional Problems, Report of a Conference.* Cambridge: MIT Press

Office of Emergency Preparedness, Executive Office of the President. Oct. 1972. *The Potential for Energy Conservation Patterns of Energy Consumption in the United States, A Report to the Office of Science and Technology.* Jan. 1972. Rep. No. 4106-0034. Washington DC: GPO

Peterson, S. R. Dec. 1974. *Retrofitting Existing Housing for Energy Conservation; An Economic Analysis.* Rep. No. C13.29:2/64. Washington DC: GPO

Seidel, M., Plotkin, S. E., Rect, R. O. May 1973. *Energy Conservation Strategies.* Environmental Protection Agency Rep. No. EPA-R5-73-021

Copyright 1976. All rights reserved

SOCIAL AND ENVIRONMENTAL COSTS OF ENERGY SYSTEMS

✳11020

Robert J. Budnitz[1]
Lawrence Berkeley Laboratory, University of California, Berkeley, California 94720

John P. Holdren
Energy and Resources Group, University of California, Berkeley, California 94720

INTRODUCTION

An issue of growing importance in contemporary society is the assessment of the environmental and social costs of various technological policies and activities. It is by now widely recognized, first, that these "secondary" costs have often not been negligible compared with the economic costs of the enterprises in question; second, that the growing power of technology and the growing number of people on whose behalf it is exercised are steadily enlarging the potential, frequency, and magnitude of actions that may later be deemed environmental or social mistakes; and, finally, that responsible decision-making requires accordingly the best possible prior analysis of environmental and social costs of proposed activities. The embodiment of this perception in laws requiring formal environmental impact statements on major technological enterprises is surely a step forward, even if the literary ingenuity applied to early examples of such statements has occasionally exceeded the technical competence.

Environmental and social costs cover a wide spectrum of concerns (for example, occupational safety, public health, economic productivity, environmental diversity, social stability), and each policy or action produces a different mix of impacts and costs. (We distinguish at the outset between impacts, meaning disruptive influences exerted on the physical and social environment, and costs, meaning measures of the response of the environment to those influences.) This wide spectrum has led to an equally wide variety of methods for analyzing and weighing the impacts and costs. In this chapter we review the available methods for such evaluations as they

[1] This work was performed under the auspices of the United States Energy Research and Development Administration.

relate to energy technologies. Of course, in many cases the methods have been developed specifically for assessment of particular impacts and costs of particular technologies, so it is only sensible to discuss methods in the context of these substantive examples.

TYPES OF IMPACTS AND COSTS

Some coherence and manageability can be provided to this immense subject by agreeing on some sort of logical structure with which to subdivide it. It seems useful for this purpose to distinguish among

1. the *origins* of impacts on the physical and social environment, meaning the fuel cycles used to supply energy (coal, petroleum, fission, etc) and, within each cycle, the various stages or operations (exploration, harvesting, transportation, etc);
2. the *character* of the impacts themselves, meaning what is added to or done to the environment (accidents; solid, liquid, and gaseous effluents; heat; noise; other environmental transformations);
3. the *costs* of the impacts, meaning the nature of the damage produced by what is done to or added to the environment (illness, loss of life, loss of economic goods and services, etc);
4. the types of *indices and criteria* by which the costs can be measured quantitatively or otherwise evaluated (days of life lost, dollars of economic damage, etc); and
5. the *methodologies* that can be used to arrive at the values of these indices.

A classification of impacts and costs based on the first three categories is given in

Table 1 Classification of impacts and costs

Fuel Cycles	Phases	Impacts	Costs
Coal	Exploration	Accidents	Death and disease
Oil	Harvesting	Gaseous effluents	Genetic effects
Gas	Concentration	Liquid effluents	Loss of economic goods
Oil shale	Refining	Solid effluents	and services
Tar sands	Transportation[a]	Heat	Loss of environmental
Fission	Conversion[a]	Resource consumption	("free") goods and
Fusion	Storage[a]	Environmental	services
Solar	End-use	transformation	Aesthetic loss
Geothermal	Management of final	Noise	Undesirable social and
	wastes	Altered opportunities	political change

[a] May occur more than once.

Table 1. In the remainder of this section, we elaborate on the character of these impacts and costs. Indices, criteria, and methodologies are taken up later.

Death and Disease

Of all environmental impacts and social costs of energy production and use, it might be expected that none would be easier to quantify than human mortality. Certainly, one would think that statistics should be readily available and easily compiled and that such a dramatic effect as death would command significant attention.

This expectation is only partly correct. Death from occasional catastrophic accidents or from more frequent small accidents can be quantified rather well and, where there is adequate operating experience, predicted. However, when the total deaths associated with or attributed to some technology must be quantified, the task is not nearly as easy because it is necessary to understand such effects as long-delayed deaths from earlier exposure to toxic substances and life-shortening by aggravation of existing morbidity.

This discussion points up one of the most important problems in analysis of environmental impacts: if mortality is difficult to quantify, how can any other phenomena, which are less dramatic and less easily measured, be understood quantitatively? For example, human illness (morbidity) is unarguably more difficult to quantify than mortality, even when the effect is direct causation of illness; when the effect is aggravation of existing morbidity, the task is even more difficult.

There are several different effects whose quantification will be discussed. Both mortality and morbidity effects can be divided into occupational and non-occupational categories (the latter often termed *environmental*). Within these categories, one distinguishes further among

1. acute mortality or morbidity from an accident;
2. latent (i.e. delayed) mortality or morbidity from an accident;
3. effects of chronic exposure to some pollutant;
4. effects in future generations from genetic damage.

The methods employed in quantifying the various effects are numerous; they can be conveniently divided as follows:

1. direct observations of specific individuals (i.e. after accidents);
2. epidemiological studies of suspect population samples;
3. controlled dose/effect studies with animals, or occasionally with humans;
4. biological-biomedical studies of physiological indices or system functions.

It is also important to differentiate between the quantification of effects that have already occurred and the attempt to predict possible effects that might occur from possible future environmental insults. Obviously, past experience must be the basis for prediction, but prediction also makes use of other information such as accident-probability analyses and pollutant-emission data.

In what follows, we discuss the quantification of death and disease by means of several examples, which serve to illustrate both the usefulness and the limitations of the available techniques.

DEATHS AND INJURIES FROM MASSIVE INDUSTRIAL FIRES This is the easiest class of effect to quantify. Usually, the incidents are well studied by investigatory teams after the fire has been put out, and both deaths and serious injuries are accounted for with good accuracy. Because these incidents are relatively rare, it should be possible in principle to gather data on the frequency with which they occur at various types of energy-related facilities (refineries, etc), and hence to predict the likelihood of occurrence at some individual facility—taking into account, of course, that fire-prevention and fire-fighting techniques are improving steadily from one year to the next.

DEATHS AND ILLNESS FROM ACUTE DOSES OF RADIATION This type of effect presents more difficulties than the case of industrial fires, for a number of reasons. Again, the incidents are quite rare, and a historical tabulation of their number, character, and severity is in principle as achievable as for the fires. However, much more is involved, if one is interested not only in the historical record but also in prediction of possible future acute radiation effects. Of course, if only the frequency with which one or two occupational workers receive radiation injuries is of interest, it is probably safe to extrapolate recent historical data, again taking into account improvements in industrial safety. However, the prediction of acute effects from an accident such as a major release of radioactivity from a nuclear-reactor accident is much more difficult. Even if the amount of radioactivity to which a postulated individual is exposed is known (or predicted), the dose actually received by the critical organs and the biological effects are not very well known. There are data on the amount of whole-body external radiation required to produce an acute fatality (death within 30 days of exposure), but even these data may not be usable directly; an exposed individual may be subject to several different types of exposure simultaneously (external whole-body, inhalation of several radioisotopes, ingestion, etc), and in many cases the interaction among these different doses is not understood. Neither human data (rare indeed) nor even animal data (which exist for a very few situations) are at present adequate to resolve the matter. Thus an investigator attempting to analyze impacts on health of possible massive acute doses of radiation must rely on incomplete information. This limitation is borne out by the calculations in the draft of the recent AEC *Reactor Safety Study* (1) in which acute fatalities were calculated for a variety of release scenarios from accidents in light-water nuclear power reactors. The authors themselves indicate that the range of uncertainty shows values three times smaller to three times larger than the figures they present for acute fatalities (assuming no uncertainty in the doses received), but other investigators have stated that the uncertainties could be much greater (2, 3).

DEATHS FROM ACUTE AIR POLLUTION EPISODES This type of effect is extremely difficult to quantify, despite a few famous examples in the lore of the air pollution field (Donora, Pennsylvania in 1948; London in 1952; the Meuse Valley, Belgium in 1930). Analysis of these incidents (4) has shown that the increased mortality, while statistically significant, is not of the type in which numbers of otherwise healthy individuals suddenly die of acute symptoms. Rather, much of the change in death rates reported in these episodes can be attributed to aged or infirm patients whose life expectancy would normally be considered very short. Also, in none of these few classic episodes has the air quality been well characterized, in terms of measurements of all of the many possible trace pollutants now known or suspected to be important (SO_2, NO, NO_2, CO, oxidants, particulate sulfate and nitrate, particulate metals and organics, etc). Morbidity in these acute episodes is easily documented (apparently, nearly everybody in Donora suffered respiratory symptoms in the 1948 episode, for example), but again quantitative data are lacking or poor. One complication is that in situations such as these, people tend to report symptoms when they "know" the air quality is particularly poor, causing a systematic bias in epidemiological studies.

BLACK LUNG DISEASE (PNEUMOCONIOSIS) IN COAL MINERS Black lung disease, prevalent in underground miners throughout the world, occurs with a reasonably long latent period and seems to be associated with long-term, chronic exposures to the poorly ventilated air in the mines. While much improvement has occurred in the mines almost everywhere in the world in recent years, significant numbers of miners will probably die in the coming decades from exposures already incurred, and many others will be exposed in the future, even with the improvements. This type of effect has been well studied, at least insofar as the effects are concerned; as in many other such situations, however, the exposures (expressed as concentrations of the various air pollutants) are poorly known at best.

LUNG CANCER IN URANIUM MINERS The toxic agents causing lung cancer in uranium miners are known quite well: they are the short-lived, chemically active, radioactive daughters of the inert radioactive gas radon-222, which emanates from uranium-bearing ores within the mines. The daughters become attached to dust particles, are inhaled, and deposit in the lungs, where their radioactive decay provides doses to the lung tissues. The doses from specific concentrations of these daughters are reasonably well understood (5), and epidemiological studies have established the association between radiation dose and the lung cancer effect (6). The association is shown in Figure 1. Within the last few years, actions have been taken which should reduce the doses by factors of at least ten and in some cases of a hundred or so (7). Such improvements will produce a corresponding reduction in illness and death—depending in detail on the (unknown) dose-response relationship at the much lower doses now involved.

CHRONIC EFFECTS OF GASEOUS AND PARTICULATE SULFUR COMPOUNDS We have discussed briefly the few, rare acute episodes of air pollution in which large numbers of deaths and reports of illness occurred. The much more common situation is human exposure, in urban environments, to nonfatal but significant concentrations of various air pollutants. According to the media today, the villain is often considered to be sulfur compounds, usually from combustion of fossil fuels such as coal, petroleum products, and natural gas. While a detailed discussion cannot be given

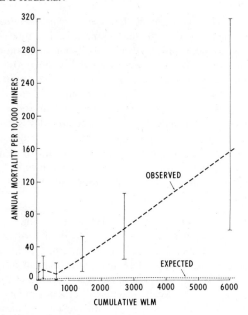

Figure 1 Observed and expected annual lung cancer mortality per 10,000 uranium miners, and 95% confidence limits, in relation to exposure in cumulative working-level-months (WLM). 1 WLM = exposure for 1 month (170 working hours) to a concentration of any combination of radon daughters in one liter of air that will result in the ultimate emission of 1.3×10^5 MeV of potential alpha energy. For details see (6).

here, it is certainly recognized by all workers in the field that the response of humans to air pollutants is a very complicated, possibly synergistic response whose etiology is only beginning to be understood. The Environmental Protection Agency's standard for gaseous SO_2 is based upon an analysis (8) that recognizes the possible role of other sulfur compounds, especially sulfuric acid mist and sulfate on particulate matter. However, there has never been definitive work to demonstrate how the various compounds, separately or together, produce the effects observed. Animal experiments are possible, and a series of important ones have been underway for many years (9), but definitive dose-response relationships still elude the investigators, partly because of the difficulties in generating realistic polluted air in a controlled laboratory. Epidemiological studies have produced associations with various respiratory diseases and impairment of pulmonary function, but no completely satisfactory studies have been performed; there are always intervening variables in air pollution parameters, socioeconomic effects, and/or other disease symptomatology. Recently, the entire situation has been cast into a new light by experimental data revealing that the chemical and physical properties and transformations of the sulfur compounds are much more complicated than previously supposed (10).

TRACE METALS IN ENVIRONMENTAL AIR AND WATER   It is by now well documented that many components of the energy-delivery system (e.g. petroleum refinery complexes, coal-fired electrical plants, coal mines, petroleum extraction) result in the presence of trace metals in environmental media such as air and water. However, rather little is known about the ultimate effects on humans in terms of disease and possible death from these pollutants. Only recently have there been studies attempting to associate burdens in selected body tissues (blood, urine, hair, etc) with metals in particulate aerosols, and little is known about the relationship between levels in the air and such body burdens. Even less is known about the effect of metals, brought into the body by inhalation or ingestion, on essential biochemical systems such as various enzymatic systems, cellular membranes, or genetic mechanisms (11). In this area research on a basic cellular and physiological level is greatly needed, along with epidemiological studies where appropriate.

Impact on Goods and Services

The costs in terms of human well-being that result from the adverse impacts of energy technology take many forms in addition to death and disease as direct effects of effluents and accidents. One such class of problems is interference with the production or enjoyment of economic goods and services. A parallel set of difficulties arises from disruption of environmental processes which, while "free" in the economic sense, perform a variety of functions supportive of human well-being. Interference with economic goods and services and with environmental processes may result from the same kinds of effluents and accidents that produce direct damages to human health, or from other forms of environmental transformation. In either case, the final consequences for human beings can range from nuisances and aesthetic impacts, to substantial destruction of property, to tolls of death and disease that in some instances may exceed those produced by the more direct pathways.

ECONOMIC GOODS AND SERVICES The most dramatic and visible losses of economic goods and services through damage to property are those potentially associated with major accidents at energy facilities. Even excluding damage to the energy facility itself, the property losses associated with a major dam failure, the explosion of a liquefied natural gas (LNG) tanker in port, or a catastrophic accident at a nuclear power reactor could in the worst cases reach hundreds of millions and even billions of dollars. Probably somewhat smaller in terms of economic impact outside the facility itself are fires at oil refineries, accidents at nuclear facilities other than reactors, and major oil spills. The damages in the smaller but much more frequent

Table 2 Property at risk in energy-related accidents

Accident	Property at Risk
LNG tanker explosion	docks, warehouses, commercial buildings
Hydro dam failure	farmland, towns
Nuclear reactor accident	farmland, residences (contaminated)
Refinery fire	adjacent chemical plants, rocks
Oil spill	beaches, pleasure boats
Radioactive waste leakage	farmland, ground water
Electrical fires, gas explosions	buildings

accidents that occur at the applicational end of energy flows—e.g. electrical fires and gas explosions in individual buildings—should also be recorded as a debit on energy's economic balance sheet. The kinds of property principally at risk in different types of accidents are summarized in Table 2.

Less dramatic but often more significant in integrated economic impact are the damages to property that arise from the routine effluents of energy technology. Dominant here, as in direct impacts on health, are the combustion products arising from use of fossil fuels, both in stationary sources and in transportation. Oxides of sulfur and nitrogen attack nylon, rubber, metal, and stone, shortening the lifetime of clothing, tires, structures, and works of art. Most seriously, plants are damaged by oxides of sulfur and nitrogen, by ozone, and by various hydrocarbons, at concentrations regularly recorded in and around urban regions in the United States (12).

Economic damage through environmental transformations other than effluents and accidents has until recently been less commonly discussed, but is often serious. A nearly ubiquitous example is ground subsidence resulting from underground coal mining and from the extraction of petroleum, natural gas, or geothermal steam. Principally at risk are residences and other structures, since agriculture can generally still be carried out on subsided land. Surface disruption by strip mining and open-pit mining of coal, uranium, tar sands, and (potentially) oil shales, including damage done by the motion of spoil banks, is another expensive type of environmental transformation, with agricultural and recreational values chiefly at risk. Hydroelectric dams, through the increased evaporative losses associated with their reservoirs, may decrease stream flow enough to aggravate salinity problems downstream, with expensive effects on agriculture. Unsightly facilities, such as oil derricks and offshore production platforms, refineries and port facilities, and electric transmission towers and wires, may sufficiently change the character of coastal and inland regions to impair recreational and property values. The effects of air pollution on visibility and the odors from refineries are other aesthetic impacts with potential economic consequences.

ENVIRONMENTAL GOODS AND SERVICES The class of impacts on goods and services that traditionally has received the least attention is perhaps the most important one— interference with environmental processes that provide essential services in support of human well-being. To evaluate the seriousness of this set of problems, one must know three things: (*a*) the nature of the environmental services and their links to well-being; (*b*) the mechanisms and extent of human disruption of these services; and (*c*) the possibility and costs of replacing disrupted natural services with technological substitutes. A good deal of qualitative understanding and a growing body of quantitative information exist concerning the first two subjects (13–15). The available evidence concerning the third has not been organized into a coherent picture, but even casual reflection suggests enormous economic and logistic barriers against substituting technology for basic natural processes on a global scale.

The greatest apparent potential for harm in the near future in the category of disruption of natural processes involves impact on agricultural productivity. Agriculture depends on natural systems for control of most potential crop pests (through natural enemies and environmental conditions), for maintenance of soil fertility (through natural nutrient cycles and regulation of the pH of surface water), and for maintenance of regional climatic conditions favorable to the crops now growing there. Production of protein in the sea, of great importance because of the shortage of protein in the global diet, depends on the integrity of estuarine habitats and on maintenance of appropriate chemical and structural characteristics of nearshore waters. Perversely, the productivity of the oceans is concentrated precisely where the potential impact of civilization is greatest—close to the continents.

Beyond loss of food production, the principal threats to human well-being through disruption of environmental services consist of accumulation of toxic substances (including carcinogens, mutagens, and teratogens) in the environment—owing to circumvention of, or intervention in, natural chemical cycles—and alteration of environmental conditions governing agents of epidemic disease and their vectors (16).

Energy technology in particular has the potential to disrupt essential environmental services in many ways. Global climate can in principle be influenced by the buildup of carbon dioxide and particulate matter in the atmosphere from the combustion of fossil fuels. Local and regional climates can be affected by increased humidity from hydroelectric reservoirs and cooling towers for electric power plants, by waste heat discharged to the environment by power plants, and by the ubiquitous end-use degradation to low-grade heat of essentially all the energy used by civilization.

Chemical cycles and especially the chemical balance of surface waters can be influenced over large regions by the oxides of sulfur and nitrogen produced in fossil fuel combustion, and over somewhat smaller areas by the salt-laden and/or acid runoff from surface mining operations and spoil banks. Chemical problems can also arise from the disposal of brines from oil-drilling operations, from exploitation of wet-steam and hot-water geothermal resources, and from the storage of solids or slurries produced by scrubbing sulfur from power plant stack gases. Hydrocarbons added to the oceans by drilling operations, tanker operations and accidents, refinery discharges, atmospheric fallout (originating largely as automobile emissions), and river discharges (crankcase oil and industrial effluents) can be directly toxic to marine organisms or disruptive of marine ecology in other ways (e.g. interference with chemical messages), and under some circumstances could influence climate.

The uses to which energy is put also have profound effects on ecological systems and processes. (That the environmental transformations brought about by the application of energy are intentional does not mean they are always beneficial in an ecological sense.) Abundant, cheap energy in the United States has had a major role in creating an agricultural system often characterized by overloaded or broken nutrient cycles (overfertilization and attendant eutrophication, feedlots), the inefficient pattern of settlement described by the term suburbanization, a transportation system whose backbone is land-gobbling highways rather than railroads, and an economy that has chosen to turn resources into pollutants after only a single use. The consequences have been a reduction in the areas of unexploited or lightly exploited ecosystems and an increase in the stresses on ecosystems of all kinds, reducing overall the capacity of these systems to perform their various services in support of human well-being. These generalized consequences of the pattern of end uses of energy are an important part of energy's environmental impact and deserve far more attention than they have yet received.

Consumptive Use of Resources

Part of the economic cost of energy technology is the value of the physical resources, other than fuels themselves, that are used in the construction and operation of energy facilities and in energy delivery and end-use systems. Land, water, and nonfuel mineral resources are three examples. It is not obvious that the apparent economic value of these resources—the price that energy enterprises must pay to use them—is always an adequate measure of the real cost to society of making the resources unavailable for other uses now or in the future. In other words, as is well known, the market as a determinant of costs is imperfect. If the prices of electric power and wheat are such that electric utilities can outbid wheat farmers for water in regions where water is scarce, heavy social costs—both to the farmers and to the nation that suddenly faces wheat shortages—may accrue. Energy operations that require land and can pay for it may pay far too low a price if the competing land use exerts no influence in the marketplace—as is the case with lightly exploited or unexploited land performing ecological services on which there is no price tag. Certain chemical elements with unusual properties (cesium, beryllium, helium) could in their scarcity eventually constrain specific advanced energy technologies and competing applications all out of proportion to the present price of these materials.

One cannot assume, therefore, that the economic cost of any energy technology subsumes all of the important resource questions. The demands of energy technology on resources subject to competing demands must, for completeness, be reckoned not only in the currency of dollars but also in the physical currencies of acres, gallons of water, tons of steel, and so on. These currencies, like dollars, often lend themselves to direct and instructive comparisons among the technological alternatives for supporting a given level of energy use.

Among the most interesting resource demands associated with energy technology is the demand such technology makes on energy itself. That is, the construction and operation of energy facilities—mines, refineries, pipelines, uranium enrichment plants, and so on—naturally require some energy. It might seem at first glance that variations in the energy inputs needed to obtain a unit of energy output in different forms would be reflected in a straightforward way in the price of the output—in other words, that economics makes superfluous a separate discussion of the energy costs of energy. That economics is in fact not sufficient is due in part to widely varying subsidies and other irregularities associated with how energy is used to get energy, and in part to differences in usefulness and thermodynamic quality among different kinds of energy. (That is, one Btu is not the same as another, either economically or thermodynamically, an idea made persuasive by the fact that a million Btu's of coal is worth about $0.80, a million Btu's of electricity $5–$10, and a million Btu's of hamburger about $700.)

It is characteristic of rich energy resources, such as thick coal beds near the earth's surface, that only a small amount of energy must be invested in exploration and harvesting in order to reap a large energy reward. If the resource is deeper or leaner, or if it must be processed extensively before use, the necessary energy investment increases. Naturally, society has tended to exploit first those resources that could be harvested with the smallest investment of energy, and the visible trend today is in the general direction of heavier energy investments. It is possible to envisage an energy resource so lean or so difficult to produce that more energy must be spent to obtain and process the fuel than is contained in the fuel itself. If the form of energy has especially desirable properties, this may still be an economically viable enterprise. This is the case with food—its production and processing, in the United States, consumes six to ten times as much energy (as fossil fuels) as the food contains (17). It is also the case with pumped-hydroelectric storage schemes, in which there is an energy debt paid for the benefit of availability during peak periods of demand.

Such examples aside, it is clearly desirable when comparing alternative technologies of energy supply to include the associated energy requirements as a criterion distinct from other economic parameters. The study of energy investments needed to obtain energy is termed "net energy analysis." One should distinguish in such analysis among three general kinds of investments and/or losses: (a) the part of the resource that is dispersed or left in the ground in nonretrievable form during extraction and processing operations; (b) the part of the resource that is directly used as energy to support extraction and processing operations; and (c) the inputs of other energy forms (fuel, electricity) needed to support these operations. Care must also be taken to account for the thermodynamic quality and the spatial availability of the energy involved at different stages.

Accurate figures for energy investment in the construction of facilities are difficult to obtain. These inputs are especially important when an energy system is growing rapidly. In such circumstances, the ratio of facilities under construction to facilities in operation is high, and, accordingly, a substantial part of the energy flows associated with the system is being invested in the facilities under construction, which will not yield an energy output until they are finished. Some analysts argue that light-water-cooled nuclear reactors as a system actually become net consumers of energy during periods when the system is growing very rapidly, even though each such reactor is a substantial net producer of energy over its operating lifetime (18). Other analysts have disputed this result in detail (although at some growth rate it would certainly be true). The resolution of the uncertainty is central to the issue of how rapidly reliance on fossil fuels can be reduced by means of the growth of nuclear power. The same question must be asked of other new technologies of energy supply (solar, geothermal, fusion) and indeed of the technology of energy conservation.

Political and Sociological Effects

It is generally agreed that energy production, conversion, and use have impacts in the political and social arenas. We discuss a few of these here. Two themes are apparent throughout the discussion: first, the range of potential impacts is extremely broad; second, in many cases the causal links between energy technologies and the impacts are not conclusively established at this time, or not well apportioned among energy technology and other putative causes.

In the broadest sense, energy's productive role in economic systems is not only an economic function but also a social impact—generally taken to be a positive one. Clearly, availability of energy is an essential element in the high productivity of the economies of industrial nations, in manufacturing, agriculture, transportation, and the provision of services. Whether continued growth of economic prosperity is contingent in a one-to-one way on growth of energy availability is not so clear; indeed there is growing evidence that the link between energy and prosperity is flexible, not rigid (19). Nor have the degree and kind of industrialization made possible by cheap and abundant energy always been, on balance, a benefit in social terms— perhaps there would be more numerous and more interesting job opportunities in a somewhat less mechanized society, for example. We dwell no further here on the complex question of the social impact of how energy is used, but confine ourselves instead to the social and political ramifications of how and where it is obtained.

INTERNATIONAL EFFECTS In a number of important cases in recent history, the political implications of energy availability have been far-reaching. The 30-year conflict in the Middle East is an example, wherein the presence of large petroleum reserves has been the main bargaining point of the otherwise weak Arab states in the conflict over Israel. The recent 1973–1974 oil embargo and price increase has important implications in world politics, as all recognize. Other examples include the German interest in Rumanian petroleum in World War II, which had a major effect on the course of that war, the French-German dispute over the Saar, and Japanese interest in oil-rich Indonesia.

The international spread of nuclear fission reactors for the generation of electricity is accompanied inevitably by the spread of the capability to manufacture nuclear weapons. Although it is argued by some that "the genie is out of the bottle" in any case in terms of the spread of nuclear weapons, there is no doubt that the proliferation of reactors is accelerating the spread beyond what would otherwise be possible or likely. To the extent that the likelihood of a nuclear conflict increases with the number of nations in possession of nuclear bombs, this acceleration threatens to deprive statesmen and political scientists of time to fashion an international political system that can permanently prevent a nuclear war. This is an awesome social cost indeed.

A rarely discussed but important social-political impact of the choices of energy technology made by industrial nations is the influence of technology on the prospects for narrowing the rich-poor gap between nations. To the extent that rich countries focus their own research and development on technologies that are heavily capital intensive, technically sophisticated, unforgiving of errors in operation and maintenance, and attractive economically only in large units, there will be minimal useful technology transfer to benefit developing countries. Were the rich countries, on the other hand, to devote some effort to development of durable, forgiving, perhaps labor-intensive energy technologies that are viable in small scale (naturally at some cost in efficiency and potential economies of scale), one result could be the great social benefit of a tangible contribution toward narrowing the demoralizing and destabilizing rich-poor gap that divides the world today.

DOMESTIC EFFECTS   Even within the boundaries of the United States, the character and distribution of energy sources have important political and social implications. For example, California, with 10% of the US population, is approximately self-sufficient in energy from its own oil, gas, hydro, and geothermal resources, and thus has vastly more economic and political power than would be the case if it had no native energy resources. On the other hand, energy-poor New England suffers politically because of its lack of native resources; much of its national political muscle has historically been devoted to maneuvering to avoid national energy policies (such as oil import quotas) detrimental to its energy interests.

The political impacts of energy reach into other spheres as well. For example, the regulation of interstate commerce in natural gas has been a vehicle of social control on a broad front, because the availability and price of natural gas have had major impacts on the distribution and scale of economic growth in the country. Another example is the possibility that plutonium might be diverted from nuclear fuel cycles for illicit use by terrorists, which may lead to the use of military, police, or quasi-police guards on an extensive scale. This potential extension of security concerns into the energy arena has been viewed with alarm by some civil libertarians, but it has been urged by analysts who believe that reducing the threat of diversion seems worth the cost.

Another social-political aspect of various energy technologies is their relative vulnerability to disruption by natural, purposeful, or accidental events. Separate from (but associated with) the vulnerability of a particular energy technology is the vulnerability of *society* to the consequences of such a disruption. Examples here are numerous. On one end of the range are routine (or statistically anticipated) disruptions, such as the abnormally low rainfall in the Pacific Northwest in late 1973, which led to power shortages on the hydroelectric network in early 1974. On the other end of the spectrum is the vulnerability of energy-poor regions (New England, Hawaii) to sabotage, labor strikes, or other purposeful interruptions in petroleum or electrical energy distribution. Associated with this vulnerability is the political pressure for control to prevent it, and sometimes even political blackmail. The consequences of disruptions, of course, may include not only loss of energy but also direct damages from the event—e.g. sabotage at a nuclear reactor or hydroelectric dam.

The development of abundant energy in a region, especially if done in too short a period or with inadequate planning, can have sociological impacts because of built-in social, demographic, or economic inequities. As an example, the American Southwest is one of the fastest-growing regions of the country, in both population and economic activity. One of the spin-offs of continued construction of coal-fired electrical capacity in the region is the fact that, with much of the coal on Indian lands, development of these resources might bring both employment and more widespread prosperity without forcing these citizens to leave their native lands. Unfortunately, unless planning is done with great care, both the social and the environmental impacts of this development may be borne disproportionately by these same Indians; existing mechanisms may be inadequate to enable them to protect their environmental and social integrity.

ANALYSIS OF IMPACTS AND COSTS: METHODS, CRITERIA, AND EXAMPLES

The variety of methods for analyzing environmental impacts and social costs is as diverse as the impacts and costs themselves. Indeed, many analysis methods are specifically tailored to individual problems. However, a few methodologies have such wide applicability, and are already in such common use, that they can be considered in a more general manner. We discuss here a number of these methodologies, usually in the context of substantive examples.

Critical Pathways Method for Trace Substances

This approach has as its philosophical basis the idea that one can isolate from among all possible effects a few that can be represented as the most important pathways of a trace pollutant substance in the environment. This method is now more or less institutionalized in the analysis of low-level radioactive emissions from nuclear electric power plants, having been required by the US Atomic Energy Commission in all environmental reports since 1972 (20).

Each critical pathway is a specific, identifiable, interconnected system in the environment that provides an important route for transport and transformation of some potential hazard from its source to man. There may or may not be successively larger concentrations of the hazard in the chain. Perhaps the best studied of the radionuclide pathways around nuclear power stations is the transport of radioactive iodine-131 in the chain: stack → air → grass → cow → milk → human consumption → thyroid. Extensive studies of this chain have shown that if a human drinks milk (1 liter per day) from a cow constantly grazing on grass exposed to iodine-131, his thyroid receives a dose about 1000 times greater than the thyroid dose received by merely breathing the same air. For this reason, measurements of the milk are now considered essential to assure that the iodine-131 emissions that occur are below permissible levels. The various elements in the chain are as follows:

1. THYROID DOSE FROM INHALATION   The International Commission on Radiological Protection (5) has set the occupational exposure limit for iodine-131 in air at 3000 picocuries/cubic meter (pCi/m^3), for continuous (around-the-clock) exposure; such an exposure will produce a thyroid dose of 15 rem per year in an adult. For children, the breathing rate, thyroid mass, and uptake fractions differ, but there is uncertainty about the exact values of the parameters; the best estimates (21) are that for ages 1–9, the dose is about the same for children (with a factor of perhaps 2 uncertainty either way).

2. CONCENTRATIONS IN AIR AND IN MILK   Depending on whether deposition is "dry" or "wet," it is reported (22) that if a cow eats grass grown under an air concentration of 1 pCi/m^3, the cow's milk will contain 700–1200 pCi/liter. Another worker (23) finds the value 560 pCi/liter.
3. DOSE FROM INGESTION OF MILK   Morley & Bryant (24) find that an infant consuming 1 liter of milk daily containing 400 pCi/liter of iodine-131 will receive a thyroid dose of 2100 mrem/year.

When it is all put together, an air concentration of 3000 pCi/m^3 produces an adult thyroid dose of 15 rem/year from direct inhalation (5). Drinking cow's milk, if the cow eats grass growing below that same air mass, will result in a dose of 10,000–20,000 rem/year (child's thyroid), and an adult dose of presumably about the same. Thus it can be seen that the dose is bigger by a factor of (10,000–20,000)/15 or a factor of about 1000, when the milk chain is considered.

Methods for Resources Subject to Competing Demands

The analysis of the true social costs of using resources subject to competing demands is complicated by the need to consider both explicit dollar costs and externalities (costs that are not now paid by the resource user and that sometimes cannot even be tabulated in dollars). There are two main types of resources to be considered: those for which replenishment is possible (whether or not practiced) and those for which it is essentially impossible. In the first category are such resources as wood, water, wildlife caught for food (e.g. fisheries), and some natural chemicals (e.g. some fertilizers and fibers); in the second are most mineral resources, the fossil fuels (replenishable only on a geological time scale), trace gases (e.g. helium), and endangered species of permanently damaged ecosystems (e.g. whales, tropical rain forests, dammed rivers).

For many replenishable resources, continued replenishment is regularly practiced: the water resources of the United States, crucial for all energy technologies, provide an example. In such cases, the true dollar cost of the resource can usually be determined, since the assurance of continuing steady-state availability has a dollar cost and value. However, it is much more difficult to do correct cost accounting when replenishment, though possible, is not practiced: the construction and operation of hydroelectric dams is an example, since in nearly all cases the silting up of the artificial lake limits the lifetime of the resource, transfers a large environmental impact to future generations, and could be prevented. In this case, not only is the direct balance sheet skewed, but the true lost future economic value cannot even be determined with much accuracy. Also, whether replenishment is practiced or not, there are social costs associated with such effects as the future loss of recreational opportunities in the affected regions; the possible harm to watersheds or wild fauna; or the long-range effects due to interruption of the natural growth/death/detritus cycles by the presence of dams and the transportation of water over great distances for other uses.

The case of depletable resources is quite different in kind, as well as in degree. Here an analysis might take into account the various competing uses for a particular resource. Thus it is sometimes claimed that the burning of fossil fuels is wrong because it is destroying a resource that should be saved instead for use as a petrochemical feedstock. The analysis of whether this is true is incomplete without answers to the questions of how much resource remains (or is likely to exist); at what rate it is being consumed (1% per year? 0.01% per year?); at what prices the remaining resources are extractable; and what competing uses exist or can be foreseen. The pitfalls here are obvious: depletion only happens once and lasts forever, while foresight into the future seldom extends very far. Thus, who in 1900 (when known petroleum reserves were miniscule by present standards, but usage was also low) could have foreseen the tremendous petrochemical industry of today? Similarly, who would have realized, in the 1930s, that the element niobium, now used for superconducting wire, might someday play a role in future electrical transmission or even electricity generation? (Even today, this possible future role is unclear, but at least its potential is recognized.)

Given these difficulties, how can one arrive at any net assessment of whether to deplete or to save a nonrenewable resource?

In harsh economic terms, an answer can of course be given: if the value of the resource as used now, including its value as capital to be invested for the future, exceeds its (presently apparent) economic worth over the long haul if saved for future use, then economics commands society to deplete all of it now. While quantifiable environmental impacts and their costs (the costs of providing by other means the environmentally disrupted services) must be properly accounted for in the economic balance sheet, there is, sadly, no way to put in the cost of an unforeseen technological opportunity.

Methodologies from Economics

There are a number of impacts and costs in the general area of economics, the discussion of which is beyond the scope of this review. Among them are

1. the economic costs of environmental pollutants;[2]
2. the impact of energy technologies on the national and world capital marketplace;[3]
3. the impact of energy technologies on labor markets, both directly through labor involvement in the fuel cycles and indirectly through secondary effects such as impacts due to availability of energy for industrial use; and
4. the impacts of energy use on the general levels or patterns (geographical or demographic) of economic activity.

Methods Viewing Environmental Systems as Systems

There is an important class of environmental impacts that is not amenable to analysis by the critical pathways approach, because the impacts affect an entire

[2] See "Economic Costs of Energy-Related Environmental Pollution" by Lave & Silverman in this volume.

[3] See "The Energy Industry and the Capital Market" by Pelley, Constable & Krupp in this volume.

system. The systems under consideration are of two broad kinds—ecological systems and social systems—the former including such physical subsystems as geological, hydrological, and atmospheric systems, as well as their interrelationships.

The methods of analysis of impacts on such systems must necessarily address the properties of the systems involved, conceived as complicated entireties. Such systems analysis methods, brought to maturity two decades ago for management of defense procurement and operational programs (e.g. the Polaris submarine program), have only recently begun to be applied to environmental systems. The methodology, now applied to many other problems in the field of operations research, involves development of a mathematical model (typically suited for execution on a high-speed computer), which treats each important but discrete segment of the larger system separately, characterizing it by selected variables and linkages significant for the issue at hand. The many important interrelationships and interactions between segments of the model are also represented mathematically. The analytical procedure consists first of describing the equilibrium or steady-state properties of the system through the identified variables, and then of describing the various modes of departure from equilibrium.

It seems clear that much more use of the systems analysis methodology in environmental impact analysis will occur in the future. Up to the present time, only a few examples exist.

Perhaps the most important recent manifestation of the maturity of systems analysis is the establishment in 1972 of IIASA, the International Institute for Applied Systems Analysis. IIASA is sponsored by an international consortium of governmental and quasi-governmental bodies, such as the National Academy of Sciences in the United States; it has gathered at its headquarters near Vienna a group of analysts who are now applying the methodology of systems analysis to a wide range of problems, including many related to energy and its environmental impacts. Examples of recent IIASA work in these areas are that of Holling (25) on environmental impact analysis and that of Avenhaus & Häfele (26) on environmental accountability. The latter describes the benefits of a materials-accounting approach for understanding of large, complicated environmental systems such as the global carbon dioxide cycle.

Criteria, Indices, and Examples

Criteria with which to characterize the impacts of energy technology fall into three broad categories: (*a*) those that are quantifiable (at least in principle) and amenable to comparisons among different technologies; (*b*) those that are quantifiable but difficult or impossible to compare from one technology to another; and (*c*) those that are difficult or impossible even to quantify for a single technology. Table 3 provides a listing of some of the most important criteria, arranged according to this scheme. For those in the two quantifiable categories, the indices that provide a quantitative measure of harm are also listed.

The basic raw material for a systematic environmental assessment of a given energy technology is a tabulation of the values of indices associated with the quantitative criteria, for all the phases in the fuel cycle (Column 2 of Table 1).

Table 3 Criteria for evaluating severity (indices follow colons)

Quantifiable, Readily Comparable

  Deaths: number, days of life lost
  Accidents, illnesses: number, days of productive activity lost
  Damages to economic goods and services: dollars
  Use of land, water, energy: square meters, liters, joules
  Material effluents: kilograms (of the same substance)
  Nonmaterial effluents: joules, decibels
  Dollar costs of reducing quantifiable impacts by specified degrees

Quantifiable, Difficult to Compare

  Use of nonfuel minerals: kilograms (of different substances)
  Material effluents: kilograms (of different substances), curies
  Magnitude of perturbation in a natural process: dimensionless fraction
  Spatial and temporal distribution of harm: area, time

Difficult or Impossible to Quantify

  Degree of irreversibility of harm
  Degree of voluntarism in risk
  Degree of coincidence of risks and benefits
  Quality of evidence of harm
  Political implications

Several such tabulations have been published in the past few years, covering most of the major fuel cycles existing or envisioned for the generation of electricity (27–32). These studies cover, in substantial measure, the impacts of many nonelectric energy flows as well, inasmuch as such general processes as coal mining, coal gasification, oil transportation, and oil refining are all treated. A useful format for the presentation of the most readily quantified information is the fuel-cycle flow diagram, a simplified example of which, for residual fuel oil (32), is shown in Figure 2. For a better example of the enormous amount of information that can be packed into this format, the reader should consult the article by Pigford in the 1974 *Annual Review of Nuclear Science* (33).

Perusal of the published data reveals that (*a*) wide discrepancies exist from one work to the next, often owing to different "accounting" procedures, and (*b*) the heaviest impacts in different fuel cycles occur at quite different stages (routine discharges to air in the coal–electric fuel cycle are most serious at the power plant itself, while in the fission fuel cycle they are most serious at the fuel-reprocessing plant). These points underline the importance of making comparisons on the basis of the entire relevant fuel cycles, and the desirability of establishing agreed-upon, consistent accounting procedures for the most frequently occurring indices.

In the remainder of this section, we discuss some of the intricacies and difficulties involved in using the criteria and indices summarized in Table 3, in the context of some data from the published literature. The ranges of data given in Tables 4–7 were compiled, except where otherwise noted, from References 27–32, and rounded to one or two significant figures. The figures are normalized to correspond to power plants of 1000 MWe capacity, operating at a load factor of 75% (i.e. delivering 6.57×10^9 kW-hr(e) per plant per year). Thermal efficiencies (electrical output ÷ thermal input) assumed at the power plants themselves are: light-water reactor, 32%; residual fuel oil, 37%; coal with lime scrubbing, 37%; combined cycle burning low-Btu gas from coal, 47%; and solar-thermal, 10%.

QUANTIFIABLE, COMPARABLE CRITERIA Even among criteria that are readily quantified and lend themselves to comparisons, there arise enough ambiguities and methodological problems to make using and comparing the literature of environmental impact a frustrating experience. Counting accidental deaths is straightforward enough, for example, but deaths from energy-related diseases may be concentrated in certain age groups, making it important to count lost days of life as well as numbers of deaths. Similarly, numbers of accidents or illnesses are not in themselves very instructive without the additional measure of severity provided by the number of days of work lost per event. Data for occupational accidental deaths and injuries in the electric-power fuel cycles for coal, residual fuel oil (RFO), and uranium [light-water reactors (LWRs)] are summarized in Table 4.

Some economic damages, such as damages to crops, may be relatively easy to quantify; others, such as loss of recreation, stimulate controversy about the proper methods of accounting. There is also the uncertainty about the appropriate discount rate for determining the present value of future damages.

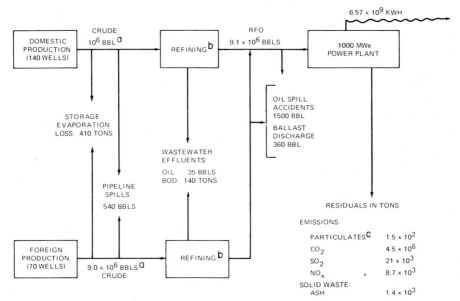

Figure 2 Residual fuel oil cycle for electric power generation. The basis is annual operation of one 1000 MWe electric power plant at 75% capacity factor. **a**: portion of refinery input attributable to residual fuel oil production, calculated on a Btu equivalent basis; **b**: residual fuel oil yield of US refineries 10%, foreign refineries 40–50%; **c**: assumes a 90% collection efficiency for electrostatic precipitators. Source: (32).

Table 4 Occupational accidental deaths and injuries in fuel cycles for electricity generation (one significant figure)

Fuel	Deaths per Plant-Yr	Injuries per Plant-Yr	10^3 Man-Days Lost per Plant-Yr[a]
Deep-mined coal	2–6	30–100	10–40
Surface-mined coal	1–4	10–60	7–30
Oil (RFO)	0.1–0.2	4–10	1–2
Uranium (LWR)[b]	0.1–0.3	5–10	1–2

[a] Evaluated at 6000 man-days/death and 50–100 man-days/injury, depending on fuel cycle and stage.
[b] Range encompasses surface and underground uranium mines.

Quantifying resource use also poses questions. Does one distinguish between water that is evaporated and water that is polluted but returned to the surface? In fuel cycles for electricity generation, evaporative cooling towers (if used) invariably dominate the water use, whether water polluted and returned is counted or not (Table 5). Concerning land use, accounting problems arise in discriminating between temporary and permanent commitments of land. It is probably useful to distinguish inventory commitments (km^2 per MWe installed, committed for the duration of the facility's operation, e.g. the land on which the plant sits), temporary commitments (km^2-years per MWe-yr of delivered electricity, e.g. km^2 strip-mined per MWe-yr, multiplied by the mean number of years required to restore the land to other uses), and permanent commitments (km^2 per MWe-yr, e.g. repositories for radioactive wastes). Very few land-use data are available disaggregated in this detail. Some figures are collected in Table 6. Another question related to resource use is how far one traces these impacts. In net-energy accounting, for example, one would usually ascribe to the energy costs of coal-fired electricity generation the fuel burned

Table 5 Use of water in fuel cycles for electricity generation (10^6 m^3/plant-yr)

Fuel	Evaporated in Wet Towers at Power Plant	Blowdown Water in Plant Cooling Towers[a]	Fuel-Processing Water Use[b]	Waste Management Water Use[b]
Standard coal[c]	11.	6.6	0.3	1.7
Coal gasification/ combined cycle[d]	6.6	4.0	0.5	—
Oil (RFO)	10.	6.0	1.5	—
Uranium (LWR)	17.	10.	0.5	0.01

[a] Returned to surface polluted.
[b] Some evaporated, some returned.
[c] Wet-lime scrubbing for SO_2 removal.
[d] Combined-cycle power plant efficiency = 47%; fuel-cycle thermal efficiency ≈ 37%.

Table 6 Land use in fuel cycles for electricity generation

Fuel	Inventory (km² per plant)[a]	Temporary Commitment (km²-yr per plant-yr)[b]	Permanent Commitment (km² per plant-yr)
Deep-mined coal	12–15	10–29	—
Surface-mined coal	12–15	20–240	—
Oil (RFO)	3–14	—	—
Surface-mined uranium for LWR	1	1–2	0.001
Solar-thermal[c]	56	—	—

[a] Includes facilities for processing and transportation, but not transmission.
[b] A 10-yr mean time for restoration to other use.
[c] Plant capable of delivering 1000 MWe-yr per yr at 100% load factor (18 MWe average per km²).

by trains hauling the coal. Should one also count the energy used to manufacture the trains? Or the gasoline used by workers commuting to work to manufacture the trains?

QUANTIFIABLE, HARD-TO-COMPARE CRITERIA Difficulties in comparability between different technologies arise even with easily quantified impacts such as use of non-fuel minerals. If construction of a solar power plant were to require 100 kg of aluminum per electrical kilowatt, for example, and a nuclear power plant required 10 kg of stainless steel per electrical kilowatt, how would one decide which is the more serious impact (aside from price, which as noted above may not reflect the full social costs)? Measurement of the material demands against known reserves, annual consumption for other purposes, and estimates of eventually recoverable resources provides indices that are a step toward comparability, but are still imperfect. (Resource estimates are flawed, and consumption for other purposes may change.)

The same problem arises with respect to material effluents. A kilogram of carbon monoxide is not equivalent in social costs to a kilogram of sulfur dioxide. A curie of tritium is not equivalent to a curie of plutonium. An increasingly popular index that supplies some measure of comparability in these instances is the number (in units of volume) obtained by dividing the quantity of the material by the maximum concentration permitted by applicable regulations. In this way, the impact of discharges is represented in terms of the volume of air or water needed to dilute the effluent to the permissible concentration. Thus a kilogram of SO_2, divided by the primary federal (US) standard of 80 μg per cubic meter of air, corresponds to a "dilution index" of 12.5 million cubic meters of air. A curie of tritium (about a tenth of a milligram), for which the Recommended Concentration Guideline, or RCG (formerly Maximum Permissible Concentration, or MPC) for public exposure is 0.2 μCi per cubic meter of air, has a "dilution index" of 5 million cubic meters of air. Dilution volumes for several fuel cycles are shown in Table 7. The main shortcomings of this approach to comparability are the following:

1. The standards for different substances are often neither equally well founded in terms of evidence of harm nor set with equal presumed margins of safety.
2. The very different physical and chemical properties of different effluents influence how rapidly and under what circumstances the indicated dilutions are actually achieved.
3. The *persistence* of the need for dilution varies greatly among different pollutants (e.g. some are transformed into innocuous substances, some leave the medium and enter another).
4. Even for similar effluents in two different fuel cycles, the availability of air or water to serve as the receiving body varies widely and may depend on technology-specific factors (e.g. acid-leaching from coal mines is spatially limited, but different from acid waste disposal from uranium mills).

When one is concerned with ecological disruptions, it is generally useful to compare the scale of the technological disturbance against the yardstick of the relevant natural process. For example one can compare additions of CO_2 to the atmosphere with the "natural" concentration of CO_2 or with natural flows into and out of the atmosphere, one can compare technological energy flows in a specified area with the natural energy flows that govern climate, and so on. Some comparisons of this kind are presented in Table 8. It is sometimes hard to know, however, which of several candidate natural yardsticks is most meaningful, and the comparison between completely different kinds of impacts is not straightforward in any case. For example, human inputs of sulfur into the atmosphere amount to about half of natural inputs (34); human inputs of CO_2 have increased the natural background concentration by about 10% (14); and human inputs of tritium to the atmosphere, mostly from nuclear weapons tests, have produced inventories almost a million times those naturally occurring (35). Which is the most serious

Table 7 Dilution volumes in air for routine effluents of fuel cycles for electricity generation (10^3 km^3/plant-yr)

Fuel cycle	Effluents	Dilution Volume (power plant only)[a]	Dilution Volume (all other steps)
Coal with lime scrubbing	NO_2, SO_2, HC, particles, heavy metals	200–550 23–48	7–8 29–370[b]
Coal-gas/ combined cycle	NO_2, SO_2, HC, particles, heavy metals	8–77 5–48	7–8 29–370[b]
Oil (RFO)	NO_2, SO_2, HC, particles, heavy metals	66–450 12–120	21–58 1–4
Uranium (LWR)	^{3}H, ^{85}Kr, ^{222}Rn, trans-U	0.0003–0.027 —	0.013–1.9 0.5–1.6

[a] Standards used, per m^3: NO_2 = 100 μg, SO_2 = 80 μg, HC = 160 μg, particles = 75 μg, heavy metal = 1.5 μg, ^{3}H = 0.2 μCi, ^{85}Kr = 0.3 μCi, ^{222}Rn = 0.003 μCi, transuranium nuclides = 5×10^{-8} μCi.
[b] High figure includes coal losses in transport, probably not comparable to other particulate emissions.

Table 8 Environmental inputs from energy cycles as fractions[a] of natural yardsticks

Energy-Related Input	Natural Yardstick	Input/Yardstick	References
Petroleum in oceans	Natural seepage	6–20	13
CO_2 in atmosphere	atmospheric CO_2 reservoir	0.1[b]	14
Particles in atmosphere	volcanoes, sea salt, dust	0.05–0.5	14
Sulfur in atmosphere	sea salt, biological processes	0.5	34
Nitrogen fixation ($N \rightarrow NO_x$)	biological processes, lightning	0.7	36
Heat dissipation at surface	sunlight absorbed at surface	<0.0001 (global) <0.01 (large urban regions)	14

[a] Ratio of annual flows on a global basis, unless otherwise noted.
[b] Cumulation perturbation in inventory.

ecological problem? Not tritium, as it turns out, but there is no way to be sure which of the other two is more serious.

It is important and often possible to specify the way in which social and environmental costs are distributed in space and in time. One distinguishes among local, regional, and global effects, and among effects that are borne essentially at the time of the causative event (e.g. accidental deaths), later in the life of the exposed person (e.g. cancer), or in future generations (e.g. genetic disease). In practice, people seem more impressed by costs that are concentrated in space and time (and society is usually willing to pay more to avoid them). As an ethical problem, however, perhaps more attention should be given to those cases in which the bearers of the costs are far removed in space and time from those who reap the benefits of the activity in question. How assessments that compare different technologies should weigh differences in the spatial and temporal distribution of impact is not at all clear.

CRITERIA DIFFICULT OR IMPOSSIBLE TO QUANTIFY The boundary between quantifiable and nonquantifiable criteria is a fuzzy one, as the foregoing paragraph illustrates. Areas and times affected can be specified quantitatively, at least in principle, but the associated issues of the degree of voluntarism in imposed risks and the degree of coincidence of risks and benefits lend themselves to no tidy index.[4]

Two other criteria that are clearly important but, at the same time, quite resistant to quantification are degree of irreversibility of harm and quality of evidence of potential harm. These aspects are not unrelated. The greater the degree of irreversibility potentially associated with a particular course of action, the heavier should be the burden of proof upon those advocating this action to show that the

[4] See "Philosophical Basis for Risk Analysis" by Starr, Rudman & Whipple in this volume for additional discussion of these points.

irreversible harm will not in fact materialize—or, in other words, the less conclusive the evidence against proceeding should have to be in order to stop the action. Some semblance of a quantitative index for irreversibility can in principle be supplied in the form of the time required to repair the damage, but this is enormously uncertain in many cases of greatest interest (e.g. nuclear war or climatic change). Quality of evidence can be characterized (e.g. as speculation, hypothesis based on limited data, theory with extensive empirical support, etc), but hardly quantified.

Comparing Apples and Oranges

The foregoing brief survey of criteria and indices underlines some of the difficulties of weighing the social and environmental costs of alternative energy technologies. Those aspects that lend themselves most readily to quantification may not be the most important ones at all, and often the problem seems to boil down to the proverbial one of comparing entities that are fundamentally incommensurable— apples and oranges. How does one weigh a small chance of a big disaster against a persistent routine impact that is significant but not overwhelming? How much social disruption should be tolerated to implement an energy technology that diminishes air pollution?

Often, attempts are made to compare apples and oranges by the expedient of forcing them into common units, such as dollars. Usually, such attempts have proven to be unsatisfactory, as the examples cited below reveal. The central thrust of this discussion, therefore, is that such comparisons require the utmost caution, with particular attention paid to the motivation for the comparison and the use to which it is put.

Consider one of the simplest of issues, the dollar value of a lost day of labor due to personal injury. When a large employer purchases insurance to cover the cost of this class of loss, he is paying the actuarially determined cost. The dollar value is well defined and accepted by society at large: it covers not only direct costs such as medical expenses, but also costs of occasional litigation, awards for liability, and other fees. This is considered satisfactory for the limited purposes of insurance, but it contains only a rudimentary way of accounting for items such as human suffering, dislocation, or the disruption of the labor that the individual would have performed had the injury not occurred. This rudimentary accounting comes about due to the dollar value of occasional tort awards for the suffering and dislocation, which is actuarially factored into the insurance premiums.

There are several other types of impacts to which, for better or worse, a monetary value can be assigned. Loss of life is one: recent large awards of substantial damages in tort cases reveal a remarkable upward trend, with the death of, for example, a nonworking housewife/mother sometimes resulting in $\sim\$500,000$ awards rather than the $\sim\$100,000$ awards common only a few years ago. While the value of human life may be indeterminate in philosophical terms, the dollar value measured in this way has some accepted range, depending of course upon age and social status.

In a similar way, the dollar value of an acre of despoiled land has been determined, for example, by the willingness of society to pay for its reclamation to some productive (or nonproductive) use. Thus, reclamation of strip-mined land may cost from a few hundred to a few thousand dollars per acre, varying with the type of land and the nature of desired use. Whether these costs are paid (and sometimes they have been) depends on the perceived value of the land in its spoiled vs restored state.

This technique of attempting to determine dollar values for various environmental effects has both uses and abuses. The uses arise, for example, when a decision must be made on expenditures to improve or prevent some particular environmental degradation. The abuses arise, as often as not, from just the same source: such decisions often consider only the direct dollar expenditures, indeed perhaps only the costs whose consideration is forced by external pressures or regulations. A good example is the land reclamation problem, in which a fairly low level of reclamation may be considered acceptable in some situations, thereby imposing hidden (if not indeterminate) ecological or other environmental costs.

There are a number of attempts in the literature to put a variety of environmental impacts on a common dollar basis. An example is the work of Morgan et al (37) on the various costs of producing electricity from coal in the United States. These authors cite a direct cost of 7–8 mills/kW-hr as the price paid in 1970 and calculate a social cost of 11.5 ± 2 mills/kW-hr (1 mill = $0.001). This social cost includes dollar values for SO_2 pollution, particulate air pollution, thermal pollution, land reconstruction, health hazards in coal mining and from air pollution, and several other effects. The direct costs are now paid by purchasers of electricity, while the social costs are borne by society at large.

This approach would be useful if it could guide some action or policy. Morgan et al address this point by determining an "optimal control strategy"—that which minimizes the sum of direct costs and social costs. Their analysis indicates that the optimal strategy costs perhaps 3 ± 1 mills/kW-hr, but reduces the social costs by perhaps $\sim 7 \pm 2$ mills/kW-hr. The control strategy is dominated by abatement of SO_2 and thermal pollution and by land reclamation, all of which are found to be cost-effective. The strength of this analytical approach is its ability to highlight areas whose social costs are greatest and to determine costs for various levels of abatement. The pitfalls arise if the numerical analysis is taken too seriously, since the whole range of nonmonetary effects and costs certainly cannot be represented by dollar equivalents.

Another example of an attempt to put an environmental impact on a dollar basis is that of several investigators who have studied the dollar cost of imposing on an individual a dose-equivalent of one rem of ionizing radiation. The typical calculation uses data such as those in the recent report of the National Academy of Sciences (38), in which estimates are derived of the probability of contracting latent cancers or genetic defects many years after delivery of rather high doses ($\sim$ hundreds of rems). Using the linear hypothesis that these probabilities per rem are equally valid at low doses, and using one or another "dollar value for human life" as discussed above, various investigators have quoted the "cost" of one rem dose-equivalent to one adult in the range from a few tens to a few hundreds of dollars (39). Recently, the US Nuclear Regulatory Commission has suggested as an interim measure "the conservative value of $1000 per total body man-rem for . . . cost-benefit evaluations" (40).

Besides dollars, there are a variety of other units in which comparisons of environmental impacts are feasible. One important class is pollution indices, in which each of several pollutants is assigned some weighting factor in the determination of an overall air pollution or water pollution index (41, 42). Since the uses to which these indices are put are usually rather limited, they are not often as susceptible to misuse as are the dollar-cost comparisons. However, some important distortions can still occur. Consider an air pollution index keyed to effects on human health. Consider also the widespread public image of air pollution in terms of visibility degradation. It is at least conceivable that public outcry would largely disappear if the particulate pollutants responsible for much of the poor long-range visibility were abated, even if no improvement were made in other health-degrading air pollutants. In that case, an air pollution index based on the effects on health would not reflect an important public concern, yet it would still have value if used properly.

Particularly knotty problems arise when comparisons in common units are either impossible or very controversial. This is the true "apples and oranges" problem. Examples are abundant in all areas where decisions must be made, and it is unlikely that much can be said to cast light on the problems involved.

Handwringing aside, there is one philosophical point of enormous importance: in such cases the most important role of technical analysis is the clarification of technical issues. This clarification takes the form not only of quantifying those items to which a number can be assigned, but also of determining ranges of uncertainties or of likely errors in the numbers. This is sometimes given short shrift when the analyst knows or feels that decisions will be made on other grounds, but it is no less important than in any other situation.

CONCLUSIONS

Perhaps the most salient feature of the discussion presented in this article is the apparent inadequacies in most of the methodologies now available for detailed analysis of environmental impacts of energy technologies. The inadequacies range over the entire spectrum of analytical tools: the criteria and indices by which impacts are judged and compared are under dispute; the methods for quantifying impacts and costs are in many instances poorly developed or seriously flawed; and the inability to compare apples and oranges makes the final goal unattainable in many (perhaps all) important situations.

Is the situation really all that bleak? In fact, it is not: the apparent inadequacies in methodology are counterbalanced in part by vigorous (and often fervent) activity in environmental analysis itself. While analyses using weak methods are often of dubious worth, the mere level of activity is providing an ever-larger data base, as well as continually refined understandings of which criteria and intercomparisons are most valid. These understandings are then being used iteratively to point toward inadequacies in the data bases. This stimulation goes full circle, and our understanding is indeed growing rapidly, perhaps at this time exponentially.

Despite the difficulties with detailed methodologies, there is enough information available to support at least four important points:

First, the available data suggest the possibility of significant interference in critical environmental processes, as well as direct effects on human health. Such interference is plausible in some cases at the present time, in many others in the immediate future (over the next few decades).

Second, there is no such thing in the energy business as a free lunch. No existing or proposed energy technology is so free of environmental liabilities as to resolve satisfactorily the central dilemma between energy's role in creating and enhancing prosperity and its role is undermining it through environmental and social impacts.

Third, where high degrees of irreversibility are possible, the burden of proof must be shifted from the opponents of further growth to the proponents. Although it will never be possible to eliminate environmental mistakes, we must strive to reduce the chances of irreversible ones.

Finally, the situation that civilization has reached the predicament where large-scale environmental disruptions are not only possible but perhaps likely, without having developed the knowledge to understand the possibilities in detail or to cope with them, gives reason to slow greatly the growth in energy consumption. Only such a slowdown can buy the time needed to obtain more knowledge of the threats and to develop and deploy more benign technologies.

ACKNOWLEDGMENT

We are grateful to Kirk Smith and John Weyant for their comprehensive review and distillation of the recent literature of the environmental impact of full cycles for electricity generation.

Literature Cited

1. US Nuclear Regulatory Commission. 1975. *Reactor Safety Study: An Assessment of Accident Risks in U.S. Commercial Nuclear Power Plants.* Rep. No. WASH 1400. Washington DC: US NRC
2. Kendall, H. W., Moglewer, S. 1974. *Preliminary Review of the AEC Reactor Safety Study.* Cambridge, Mass: Union of Concerned Scientists
3. US Environmental Protection Agency. 1974. *Comments on Reactor Safety Study.* Washington DC: US EPA, Office Radiat. Programs
4. See Ref. 8, Chap. 9
5. International Commission on Radiological Protection. 1964. *ICRP Publ. 6.* Oxford: Pergamon
6. Lundin, F. E., Wagoner, J. K., Archer, V. E. 1971. *Radon Daughter Exposure and Respiratory Cancer, Quantitative and Temporal Aspects.* Joint Monogr. No. 1. Washington DC: Nat. Inst. Occup. Safety and Health
7. US Secretary of Labor. 1968. *Fed. Regist.* 33: No. 252
8. US Department of Health, Education, and Welfare. 1969. *Air Quality Criteria for Sulfur Oxides.* Rep. No. AP-50. Washington DC: US EPA
9. See Ref 8, Chap. 6
10. Novakov, T., Chang, S. G., Harker, A. B. 1974. Sulfates as pollution particulates; catalytic formation on carbon soot particles. *Science* 186:259
11. Sax, N. I. 1968. *Dangerous Properties of Industrial Metals.* New York: Van Nostrand
12. Hindawi, I. J. 1970. *Air Pollution Injury to Vegetation.* US Dept. Health, Education, Welfare. Washington DC: GPO. 44 pp.
13. Report of the Study of Critical Environmental Problems (SCEP). 1970. *Man's Impact on the Global Environment.* Cambridge, Mass: MIT Press
14. Report of the Study of Man's Impact on Climate (SMIC). 1971. *Inadvertent Climate Modification.* Cambridge, Mass: MIT Press
15. Holdren, J. P., Ehrlich, P. R. 1974. Human population and the global environment. *Am. Sci.* 62:282
16. May, J. M. 1972. Influence of environ-

mental transformation in changing the map of disease. In *The Careless Technology: Ecology and International Development*, ed. M. T. Farvar, J. P. Milton. Garden City, NY: Natural History Press
17. Steinhart, J. S., Steinhart, C. E. 1974. Energy use in the U.S. food system. *Science* 184:307
18. Rieber, M., Halcrow, R. 1974. *Nuclear Power to 1985: Possible Versus Optimistic Estimates*. Rep. No. 137P. Univ. Ill. Center for Advanced Computation, Urbana, Ill.
19. Schipper, L. 1975. *Energy Conservation: Its Nature, Hidden Benefits, and Hidden Barriers*. Univ. Calif. Rep. No. UCID 3725/ERG 2. Energy and Resources Group, Univ. Calif., Berkeley
20. US Atomic Energy Commission. 1973. *Preparation of Environmental Reports for Nuclear Power Plants*. Regulatory Guide 4.2. Washington DC: US AEC
21. Lewis, H. W. et al. 1975. Report to the American Physical Society by the Study Group on Light-Water Reactor Safety. *Rev. Mod. Phys.* 47: Suppl. 1
22. VanAs, D., Vleggaar, C. M. 1971. Determination of an acceptable iodine-131 concentration in air when the critical intake is through milk. *Health Phys.* 21:114
23. Burnett, T. J. 1970. A derivation of the "factor of 700" for iodine-131. *Health Phys.* 18:73
24. Morley, F., Bryant, P. M. 1969. Basic and derived radiobiological protection standards for the evaluation of environmental contamination. In *Environmental Contamination by Radioactive Materials*. Vienna: IAEA
25. Holling, C. S., ed. 1975. How can a model help? In *Environmental Impact Assessment: Principles and Procedures*, Chap. 5, SCOPE Workshop on Impact Studies in the Environment. Toronto: UNESCO/SCOPE
26. Avenhaus, R., Häfele, W. 1974. Systems aspects of environmental accountability. In *Systems Analysis and Modelling Approaches in Environment Systems, Proceedings of IFAC/UNESCO Workshop, Zakopane, Poland*. Warsaw: Polish Acad. Sci.
27. Council on Environmental Quality. 1973. *Energy and the Environment: Electric Power*. Washington DC: GPO
28. Argonne National Laboratory. 1973. *Social Costs for Alternate Means of Electrical Power Generation for 1980 and 1990*. Rep. No. ANL-8093, Vols. 1–3. Springfield, Va: Nat. Tech. Inf. Serv.
29. Pigford, T. H., Keaton, M. J., Mann, B. J., Cukor, P., Sessler, G. 1974. *Fuel Cycles for Electric Power Generation*, Pts. I and II. Rep. No. EEED-103/106 for US EPA. Teknekron, Inc., Berkeley, Calif.
30. Hittman Associates, Inc. 1974, 1975. *Environmental Impacts, Efficiency and Cost of Energy Supply and End-Use*. Rep. No. HIT 593 for US Counc. Environ. Qual., NSF, and US EPA. Vol. 1 (Nov. 1974), Vol. 2 (Jan. 1975)
31. Battelle Memorial Institute. 1973. *Environmental Considerations in Future Energy Growth*. Rep. for US EPA. Columbus, Ohio: Battelle Mem. Inst.
32. US Atomic Energy Commission. 1974. *Comparative Risk-Cost-Benefit Study of Alternative Sources of Electrical Energy*. Rep. No. WASH 1224. Washington DC: US AEC
33. Pigford, T. H. 1974. Environmental aspects of nuclear energy production. *Ann. Rev. Nucl. Sci.* 24:515
34. Kellogg, W. W., Cadle, R. D., Allen, E. R., Lazarus, A. L., Martell, E. A. 1972. The sulfur cycle. *Science* 175:587
35. Jacobs, D. G. 1968. *Sources of Tritium and Its Behavior upon Release to the Environment*. Rep. No. TID-24635. Washington DC: US AEC
36. Institute of Ecology. 1972. *Man in the Living Environment*. Madison, Wisc: Univ. Wisc. Press
37. Morgan, M. G., Barkovich, B. R., Meier, A. K. 1973. The social costs of producing electric power from coal: a first order calculation. *Proc. IEEE* 61:1431
38. National Academy of Sciences. 1972. *The Effects on Populations of Exposures to Low Levels of Ionizing Radiation*. Washington DC: Nat. Acad. Sci.
39. Cohen, J. J. 1973. On determining the cost of radiation exposure to populations for purposes of cost-benefit analyses. *Health Phys.* 25:527
40. US Nuclear Regulatory Commission. 1975. Radioactive material in light-water-cooled nuclear power reactor effluents. *Fed. Regist.* 40:19439
41. Babcock, L. R. 1970. A combined pollution index for measurement of total air pollution. *J. Air Pollut. Control Assoc.* 20:653
42. Schneider, A. M. 1973. An effluent fee schedule for air pollutants based on Pindex. *J. Air Pollut. Control Assoc.* 23:486

Copyright 1976. All rights reserved

HEALTH EFFECTS OF ENERGY PRODUCTION AND CONVERSION

✕11021

C. L. Comar
Electric Power Research Institute, Palo Alto, California 94303

L. A. Sagan
Palo Alto Medical Clinic, Palo Alto, California 94301

INTRODUCTION

Historically, the development of technology has not been guided by particular concern for associated health effects. It is now becoming clear that future technological assessment will almost certainly require some prior attention to health impacts. This reflects the perception that the health effects of technology have not been internalized, i.e. they are either inadequately compensated for in the working population or are borne by segments of society not directly benefitting from the responsible industry. Black lung disease of coal miners is an example of the former; disease produced or aggravated by industrial smogs is an example of the latter.

There are a number of compelling reasons for attempting some assessment of health effects related to energy even though such estimates may, because of inadequate data, lack precision. The development of national energy policies and strategies, including difficult choices among fuel systems, must be guided not only by market economics but by careful consideration of health and environmental effects not necessarily reflected in pricing. Another reason for health-effects assessment is to permit rational and cost-effective safety and control procedures. Miscalculation of the risks involved can create severe dislocations, diseconomies, or hazards: for example, excessive expenditures on construction or equipment out of proportion to risk or, in the other extreme, inadequate expenditures to protect life and health. Finally, the attempted assessment of health effects can provide guidance as to research needs and priorities.

It is important that the scientific community not only provide data and interpretations for policymaking bodies but also assure that the broad implications are understandable by the general public. There have been major shortcomings in the general understanding of the impacts of health effects. One of these has been the emphasis on individual energy systems treated in isolation rather than comparisons of the various systems available, each of which has its own attendant risks. Knowledge of the health effects associated with a given type of power plant is by itself of little value. For example, if a given nuclear plant is not built, then for the near and midterm future a coal plant must be built in its place or no plant at all. (It is assumed that oil and natural gas will be in short supply and that other possibilities such as geothermal, fusion, solar, wind, tide, etc either have such limited application or are so uncertain that it would be highly imprudent to rely on them for the needs of the next 10–30 years. Each of these potential sources should, however, be researched as intensively as feasible and brought into commercial production whenever possible.) It is beyond the scope of this paper to consider the biological risks associated with inadequate supplies of electricity, but it is by no means certain that a coal plant or no plant at all would have less health impact than would a nuclear plant.

A second shortcoming is that health effects have usually been expressed in absolute terms as numbers of either premature deaths or serious disabilities. Assessments presented in this way are most difficult to relate to the real significance that they have for individuals. We propose that it would be much more helpful to convert this measure to the degree of enhanced risk to which individuals and the population are exposed. Still another technique for presentation of these risks is through conversion to a dollar cost. This concept, although particularly useful in evaluating safety expenditures, is beyond our scope and is not developed further here. Nor is there consideration of abatement and control procedures or their cost-effectiveness.

The data are limited primarily to the production of electricity because of the increasing importance of this energy form in our society and because various fuel options are available. Following discussions of some basic principles in consideration of health effects, the available health-effects data are presented for fuel extraction, transport, processing, and electricity generation from the major fuel systems with emphasis on the major uncertainties and controversies.

HEALTH EFFECTS: THEIR NATURE AND MEASUREMENT

General

The National Academy of Sciences has recently published a report (1) that contains numerous references to support the general statements of this section and is a perceptive treatment of what man needs to know to deal effectively with chemicals in the environment. References 2–6 are comprehensive reports that cover health effects of air pollution with emphasis on the role of fossil fuel combustion products. References 7–11 provide the basic documentation on the biological effects of radiation, and references 12 and 13 deal with concepts of benefit and risk.

Health consequences of technology cover a wide spectrum ranging from mere nuisance to violent death. They may be transient or permanent. They may be localized or may affect worldwide populations, as from nuclear fallout. The noxious agents may be chemical, physical, or biological; they may be synthetic, and therefore their presence identified easily, or they may be naturally occurring and appear only as an increased concentration of a natural element, e.g. mercury, lead, arsenic. The pollutant source may be obvious, as in the case of a noisy airport, or it may be obscure, as in the case of commingled combustion products in city air. The resultant health effects may be obviously and immediately related to the cause, as with asbestos-induced lung cancer; or the relationship may be obscure, as with the London smog of 1952, which was unrecognized until identified retrospectively by careful epidemiologic analysis of mortality statistics. Acute effects (those following soon after relatively large exposure) are generally more easily related to their causative factors than are chronic effects. Chronic exposures to routine releases of pollutants generally cause effects that are far more difficult to recognize; they often result in an increased frequency of a disease that occurs "naturally" in the exposed population.

There can be considerable individual variation in response to pollutants. For example, groups that appear to be more sensitive to air pollution include young children, possibly because of higher inspiratory rates, more mouth breathing, and frequent respiratory tract infection; aged persons, possibly because of decreased cardiopulmonary adaptive capacity; asthmatics; persons with preexisting chronic bronchitis and emphysema; certain occupational workers; and smokers. Generally, the fetus and the infant are more susceptible to toxic exposures than are adults. Effects from fetal exposure may also appear as congenital malformations, as from the use of thalidomide or from maternal rubella; these agents are known as teratogens.

Probably no health effect of pollutant exposure is of greater public concern than cancer. That cancer can follow environmental exposure to chemical carcinogens has been known with some certainty ever since the eighteenth century when Sir Percival Potts recognized scrotal cancer in the chimney sweeps of London, undoubtedly induced by the hydrocarbons of soot. Although it is not known to what extent cancers of all organ systems are the result of exposure to chemical carcinogens, experts ascribe the majority to this source. Probably the best recognized relationship observed on a large scale in the general public is between lung cancer and cigarette smoking, but a number of other agents are known or suspected as carcinogens. One of the difficulties hindering investigators is the long latent period between exposure and development of the cancer, which may appear decades later when occupation or exposure may have changed. Furthermore, better understanding is complicated by the possibility of synergistic or additive effects among various agents that individually may be weak carcinogens, difficult to identify. Although cancer is generally a "somatic" effect, i.e. occurring in the person exposed, it may be mediated by a genetic mechanism operating on the informational macromolecules at the cellular level. The implications, which are discussed later, are that since only one or a few target molecules may need to be modified to produce cancer there is some chance of harm at any exposure level above zero, although repair mechanisms may operate at low dose rates.

The classes of carcinogenic agents identified as present in products of fossil fuel combustion and nuclear effluents are: (*a*) polycyclic and other aromatic hydrocarbons, (b) trace metals, and (c) radionuclides. They are often found associated with particulate matter. It must be noted that all of these agents are present in routine effluents in amounts that are too small to produce detectable effects, so that there is considerable uncertainty as to their actual impact on man. Lung cancer has been more studied by epidemiologists than any other cancers that might be associated with air pollution. The evidence presently available does not support—nor does it deny—the conclusion that air pollution per se is a causative factor (6).

Genetic effects, i.e. those occurring in the progeny of the exposed persons, are caused by a class of agents called mutagens. Although mutations confer on species the ability to evolve and react to new environmental conditions, most often such new characteristics are detrimental to the individual. Mutational changes may vary in severity from lethal effects to subtle effects that are not associated with mortality. Uncertainty as to how genetic effects of environmental agents will express themselves makes their recognition and quantification difficult if not impossible. In fact, no genetic effects from any agent have yet been found in human populations that can be traced with any certainty to parental exposures. Because of the great amount of research done on the genetic effects of ionizing radiation it is possible to predict the types and numbers of effects that would be produced in human populations from low levels (about natural background) even though none have been detected.

Exposure-Response Relationships

Attention is directed towards a dilemma that arises from dissemination of pollutants where there are large populations exposed to detectable levels of a contaminant known to be harmful at some high level. The outstanding example is ionizing radiation produced from nuclear weapons testing. As a result of weapons testing, it has been demonstrated that practically every living organism on the earth contains detectable amounts of man-produced radioactivity, which is known definitely to be harmful at high levels. The potential for harm has caused great amounts of money and manpower to be spent for research since the middle 1940s, and it is fair to say that we know much more about ionizing radiation and its effects than about any other agent, chemical, or drug to which man exposes himself. The problem became visible because of the extremely sensitive means of detection. Since that time, there have been developed sensitive measures for various chemical pollutants in the biosphere that could be potentially harmful, so that we are faced with concerns and public reaction to such contaminants as mercury in fish, vinyl chloride from plastics, asbestos fibers, chlorinated hydrocarbons in purified water, pesticides, and sulfur compounds, nitrogen oxides, polycyclic organic materials, and trace elements from fossil fuel combustion. This public sensitivity understandably brings about great pressure to reduce the possible harm to zero or near-zero. To present the problem of both protecting the public welfare and persuading the public to realize that it must accept certain risks, we give a gross classification of the relationships between the level of a harmful agent and its effects. The solid lines represent levels of exposure ranging from high to zero; on the right are described effects caused by the corresponding level.

Level of Exposure		Effect
High		Death
	Range A	Clinical symptoms
	Range B	Undetectable effects at high probability
	Range C	Undetectable effects at low probability
Zero		No effects

For any pollutant, there will be some high level that will cause lethal effects and a large proportion of deaths in the population exposed. Below that will be a range of levels (Range A) in which only clinical signs in the population are observed and the proportion of deaths, if any, is much lower. These clinical symptoms, as in the case of radiation effects or a pollutant-produced persistent cough, may not themselves be long-term contributors to ill health, but may denote an associated high probability of serious late effects. At still lower levels (Range B) the effects may be undetectable either directly or statistically, but yet they occur at high enough probabilities that society usually decides not to tolerate them. These can be effects that are undetectable at high or low frequency because they are minor in character, or they can be serious health effects that are undetectable because they are of low frequency and occur either immediately or (with radiation, for example) at some later time or in a later generation; in this paper we deal with the latter. Ionizing radiation is an agent about which enough is known so that levels in this range can be specified. At the present guidelines for radiation protection, no effects can be clinically or epidemiologically observed in exposed populations, although on the basis of radiobiological theory it is postulated that effects do or could occur. Referring again to any pollutant, at even lower levels (Range C) the probabilities of effects may be so low that society is justified in accepting them if they are more than compensated by the associated benefits of the process that produces the harmful agent.

To further understand the need for risk acceptance, we must distinguish within Range C between two types of agents: the "threshold" and the "nonthreshold." For so-called threshold agents there is some level of pollutant below which no effects occur. For nonthreshold agents there are effects at any level of pollutant above zero. In terms of regulatory procedures, standards can be set for threshold agents that guarantee absolute safety. However, for nonthreshold agents there must be acceptance of some level of risk unless the agent can be completely eliminated. To protect the public and at the same time to provide the benefit of the process that produces this agent, it is desirable to establish regulations so that the population exposure falls somewhere within Range C. To do this it is usually necessary to ascertain the slope and nature of the exposure/effects relationship in the regions of interest so that there can be recommended public acceptance of certain levels of risk commensurate with associated benefits.

Risks of Death

As already mentioned, health effects range from the trivial to those causing severe disability and premature death. Attention in this paper is focused on the latter, mainly because they are the most important and we have no meaningful way of evaluating many of the less serious effects. It should also be noted that even premature deaths are not equivalent in their impact on individuals and society. The ethical ideal that every being has a right to be born without man-induced effects and to die from old age (the natural wearing out of the body) is fully recognized. However, it is obvious that all early deaths do not have the same societal and personal impact. For example, the death of an early embryo before anyone even knows about it or the death of an elderly person a few months before he or she would otherwise die of old age would penalize neither the individual, the persons left behind, nor society, as would the death of a young adult, with family responsibilities, on the brink of a productive career. In addition, genetic effects can produce over a lifetime a variety of suffering and health effects, which, while they do not necessarily cause premature deaths, do constitute a social cost. At present it is not possible in the formulation of risk values to estimate the number of person-years lost because of exposure to the agent and to weigh them so that societal and personal impacts are taken into account. In this discussion the numbers are rounded and concepts are presented in a simple form that hopefully will lead to correct interpretation as well as facilitate understanding. The attempt is primarily to assess the health consequences associated with energy in terms of effects on individuals in the general population.

Table 1 Normal risk of death per year in the United States as related to age (1969 data) and comparison of high and low additional risks

Age	Normal Risk of Death per Year (14)
5	1 in 1587
10	1 in 3846
25	1 in 690
35	1 in 465
45	1 in 205
55	1 in 88
65	1 in 39
All ages	1 in 100
	Added risk of death per year
High risk	1 in 100 or less to 1 in 10,000
Low risk	1 in 10,000 to 1 in 1,000,000 or more

Table 2 Effects of accepting additional risks

	10-Year-Old	50-Year-Old
Risk of death/year	1 in 3800	1 in 100[a]
Normal risk/year plus additional risk/year of 1 in 10,000	1.38 in 3800	1.01 in 100
Normal risk/year plus additional risk/year of 1 in 1,000,000	1.0038 in 3800	1.0001 in 100

[a] This corresponds to the risk faced by the US population at large: 2×10^6 deaths per year/200×10^6 persons.

In the United States, with a population of about 200 million, about 2 million people die each year from all causes, such as old age, sickness, accidents, homicide, and suicide. This means that for the population at large in the United States there is a probability of dying in a given year of 0.01 $[=(2 \times 10^6)/(200 \times 10^6)]$, or expressed another way, a risk of death in any year of 1 in 100. This risk is age-dependent, as can be seen in Table 1. It is usually considered that an additional man-produced risk of death per year, consciously accepted, is high if it falls in the range of 1 in 100 or less to 1 in 10,000 and low if it falls in the range of 1 in 10,000 to 1 in 1,000,000 or more.

To convey more easily the implications of accepting additional risks, we propose that data be calculated in terms of what the enhanced risk per year from a given activity will be. This is illustrated in Table 2. The 10-year-old who accepts an additional risk of 1 in 10,000 will change his overall risk from 1 in 3800 to 1.38 in 3800, a considerable proportional increase (38%). An older person (about 50 years of age) who accepts the same additional risk will change his overall risk, for example, from 1 in 100 to 1.01 in 100, a change of much less significance (1%). As indicated in Table 2, the acceptance of an additional risk of 1 in 1,000,000 produces changes of little significance regardless of age. These additional risk values can also be used to estimate the total number of individuals in an exposed population who may be affected. For example, an industrial process that imposes an additional risk per year of 1 in 10,000 on the population of the United States could cause 20,000 casualties per year $[(200 \times 10^6)/10,000]$.

The following comparisons may be cited not to justify any additional risks, but to relate these probabilities to everyday risks. A person while riding in a car in the United States has an additional probability of death per year of about 1 in 100[1]; this is a risk we accept, but for the 10-year-old, as an example, this represents about a fortyfold increase of normal risk. The risk of death from being struck by lightning or from a snakebite is about 1 in 1,000,000 per year.

[1] The physical meaning of this risk estimate is confusing because of the time units; the absolute risk would be 1 in 100 if a person spent full time during the year riding in a car. The risk of death per hour while riding is 1 in 876,000, which is the same as the normal risk of death per hour of the 50-year-old person. Essentially, this means that while riding in a car, the risk of death of this person is about double.

Table 3 Premature deaths per year associated with operation of a 1000-MWe power plant (values are lowest and highest estimates from cited references)[a]

	Coal	Oil	Natural Gas	Nuclear
Occupational				
Extraction				
Accident	0.45–0.99 (15, 17, 18, 21, 22)	0.06–0.21 (15–18, 22)	0.021–0.21 (15–18, 22)	0.05–0.2 (15, 17, 18, 20, 22)
Disease	0–3.5 (17)	—	—	0.002–0.1 (17, 19, 20, 22)
Transport				
Accident	0.055–0.4 (15, 17, 18, 22)	0.03–0.1 (15–17, 22)	0.02–0.024 (15, 17, 18, 22)	0.002 (15, 18, 22)
Processing				
Accident	0.02–0.04 (17, 18)	0.04–1 (15–18, 22)	0.006–0.01 (15, 17, 18, 22)	0.003–0.2 (15, 17, 18, 20, 22)
Disease	—	—	—	0.013–0.33 (17, 19, 20, 22)
Conversion				
Accident	0.01–0.03 (15–18, 22)	0.01–0.037 (15–18, 22)	0.01–0.037 (15–18, 22)	0.01 (15, 17, 18, 22)
Disease	—	—	—	0.024 (20)
Subtotals				
Accident	0.54–1.5	0.14–1.3	0.057–0.28	0.065–0.41
Disease	0–3.5	—	—	0.039–0.45
Total	0.54–5.0	0.14–1.3	0.057–0.28	0.10–0.86
General Public				
Transport	0.55–1.3 (15, 17, 21, 22)	—	—	—
Processing	1–10 (17)	—	—	—
Conversion	0.067–100 (17, 21)	1–100 (17)	—	0.01–0.16[b] (15, 17, 19, 20, 22)
Total	1.6–111	1–100	—	0.01–0.16
Total Occupational and Public	2–116	1.1–101	0.057–0.28	0.11–1.0

[a] Note: Dashes indicate no data found; effects, if any, are presumably too low to be observed; and no theoretical basis for prediction.
[b] For processing and conversion.

Table 4 Occupational injuries per year associated with operation of 1000-MWe power plant (values are lowest and highest estimates from cited references)

Occupational Injuries	Coal	Oil	Natural Gas	Nuclear
Extraction				
Accident	22–49 (15, 17, 18, 22)	7.5–21 (15–18, 22)	2.5–21 (15–18, 22)	1.8–10.0 (15, 17, 18, 22)
Disease	0.6–48 (17, 21)	—	—	—
Transport				
Accident	0.33–23 (15, 17, 18, 22)	1.1–9 (15–17, 22)	1.2–1.3 (15–17, 22)	0.045–0.14 (15, 18, 22)
Processing				
Accident	2.6–3 (17, 18)	3–62 (15–18, 22)	0.05–0.56 (15–17, 22)	0.6–1.5 (15, 17, 18, 22)
Conversion				
Accident	0.9–1.5 (15–18, 22)	0.6–1.5 (15–18, 22)	0.6–1.5 (15–18, 22)	1.3 (15, 17, 18, 22)
Totals				
Accident	26–77	12–94	4–24	4–13
Disease	0.6–48	—	—	—

HEALTH EFFECTS FROM ELECTRICITY GENERATION

Quantitative Data

Several reports have been published that contain estimates of the health effects associated with electricity production (15–22). We have constructed Tables 3 and 4 to summarize the available estimates for each phase of the fuel cycle for each of the four fuels: coal, oil, natural gas, and nuclear. By and large, the estimates relate to contemporary technology and existing circumstances. In each case the data have been adjusted to represent the number of premature deaths or occupational impairments produced per year by processes associated with a 1000-MWe power plant, which is roughly that required for a population of 1,000,000 people. The values given represent the lowest and highest from the cited references. The references should be consulted for an understanding of the methodology and detailed assumptions; limitations are discussed in later sections.

Consider first from Table 3 the effects on workers. For coal-fired plants the values range from 0.5 to 5 premature deaths per year and for the other fuel sources they range somewhat lower, from 0.06 to 1.3. Most of these effects are due to accidents in coal mines, to conditions that cause black lung disease, or to activities in oil refineries, uranium mining, and nuclear fuel reprocessing.

Consider now from Table 3 the effects on the general population. It has been estimated that the transport of coal required for a year's operation of a 1000-MWe plant is responsible for 0.6 to 1.3 premature deaths by accidents at railroad crossings; no estimates are available for truck or barge transport. The comparative values for the other fuel systems are insignificant.

The data so far discussed have a reasonable statistical base of past operation and are to that extent reliable. The number of premature deaths among the public from power plant operation (conversion or generation of electricity) results primarily from dissemination of air pollutants and, as discussed later, these effects are a matter of great uncertainty. The upper-limit estimates for coal and oil are about 100 premature deaths per year compared with 1 or less for natural gas and nuclear.

Table 4 presents data on the number of nonfatal occupational injuries per year associated with the operation of a 1000-MWe power plant. These have been defined as injuries serious enough to cause loss of working time for several days or more. These effects are roughly the same for coal and oil, ranging from about 12 to 100 cases per year, and somewhat lower for natural gas and nuclear. Most of these effects are associated with mining, well digging, coal transport, oil refining, and nuclear reprocessing.

Fossil Fuels

The data presented on the health effects associated with fossil fuels suffer from certain limitations and uncertainties. First, genetic effects are not included because our present state of knowledge does not allow even an approximate estimate for such effects. Second, the data do not adequately discriminate between premature deaths that may occur early in life, such as from accidents, and those that may shorten life only slightly, as seen, for example, in increased mortality among persons hospitalized for chronic disease, who already have high mortality rates. Perhaps of greatest importance is the uncertainty about the validity of the upper estimates for the effects on the general public from burning coal and oil. Not only is there the problem of the magnitude of the effect, but lack of knowledge about the causative agents makes it difficult to institute effective control procedures.

The primary data come from epidemiological studies. Major episodes (Meuse, Donora, London, New York City, etc) clearly showed that air pollution, sufficiently severe, could cause illness and premature death. During the 1950s and 1960s the major issue was whether air pollution in concentrations usually existing over industrial cities would cause adverse health effects. The emphasis shifted next to quantifying pollution relative to effects produced and more recently to the effects of low levels of pollution and the effects of interactions.

From a methodological standpoint epidemiological, animal, and experimental human studies are needed. Epidemiological studies are important in uncovering possible associations that can then be tested under controlled conditions; they are also needed for evaluation of human risks suggested by laboratory experiments. Animal studies are used to determine efficiently the sites of effects, mechanisms, and dose-response relationships, and they are more easily adapted for chronic studies than are human investigations. Because of species differences, controlled studies on humans are needed to establish specific responses and to determine the influence of disease or of various physiological states on the effects of pollution.

In this review, it is not possible to do more than present some general conclusions. Attention is called to a recent and comprehensive document (3, Chaps. 1–4) that references about 380 articles in this subject area.

In 1970, air quality standards for selected pollutants were mandated by the United States Clean Air Amendments. Emphasis was placed on sulfur dioxide because of the evidence that ambient levels were associated with health effects of air pollution disasters. Subsequent studies indicated that sulfur dioxide by itself could not be the primary causative agent and it was postulated that a combination of sulfur dioxide and particulates was responsible (23–25). More recent evidence suggests that oxidation products of sulfur dioxide (i.e. sulfuric acid and particulate sulfates)—possibly acting synergistically with sulfur dioxide and other pollutants such as nitrates, particles, and ozone—are primarily the causative agents (26–28). It must be emphasized that although suspended sulfates are now being used as an indicator of health effects and there appear to be correlations between them and such effects, there is no firm evidence as to which substance or substances in polluted air are the causative agents. Without such knowledge, air pollution control strategy based on reduction of sulfur alone does not have a valid scientific basis.

The major categories of health effects associated with air pollution are (a) chronic respiratory disease; (b) symptoms of aggravated heart-lung disease; (c) asthma attacks; (d) children's respiratory disease; and (e) premature death. It would be most useful to understand the quantitative relationships between exposures to specific agents and these health effects in order to know how much investment is justified for control measures, to know which chemical effluents to control, and to make comparisons with biological costs of nuclear power.

In a recent report of the National Academy of Sciences–National Academy of

Engineering–National Research Council (3, Chap. 13), illustrative calculations were made of the health effects associated with sulfur oxide emissions for representative power plants in the Northeast. The results are presented in Table 5. They were derived from models that related ambient levels to emissions including factors for conversion of SO_2 to sulfates; health effects from ambient levels were calculated by using dose-response curves from epidemiological data from studies of the Environmental Protection Agency (EPA). It must be emphasized that the numerical estimates of Table 5 are controversial, relying on limited information and numerous arbitrary assumptions, and cannot be regarded as proven results. A critique in the same document from which Table 5 was derived (3, Chap. 4) suggests that the estimates could be low by a factor of two or high by a factor of ten. What can be concluded from Table 5 with reasonable assurance is that the effects listed are produced at detectable levels by factors associated with air pollution, with power plants most likely making a significant contribution. It should also be noted that a cost-benefit assessment of the data in Table 5 indicates that the economic impact of the nonlethal effects is much greater than that of the premature deaths.

Nuclear Power

Health risks to both the general public and occupational personnel from the nuclear fuel cycle are considerably better estimated than those from fossil fuel combustion. This is because (*a*) there is a single causative agent released from the nuclear plant—ionizing radiation—whereas there are literally hundreds of individual species released from fossil fuel combustion; (*b*) since the first nuclear weapons tests in the 1940s, about a billion dollars have been spent on research on the effects of ionizing radiation; (*c*) radiation exposures are easily and precisely measured; and (*d*) there is a great body of knowledge from natural background exposure and from accidental, industrial, and military exposures of populations.

The uncertainties and limitations in regard to nuclear fuel are somewhat different from those for fossil fuel combustion. The primary concerns cannot be completely answered by scientific evidence or technical solutions. They include such factors as

Table 5 Health effects associated with sulfur oxide emissions[a]

	Remote Location	Urban Location
Cases of chronic respiratory disease	25,600	75,000
Person-days of aggravated heart-lung disease symptoms	265,000	755,000
Asthma attacks	53,000	156,000
Cases of children's respiratory disease	6,200	18,400
Premature deaths	14	42

[a] Source: (3, Chap. 13). Illustrative calculations based on distributive models, postulated conversions of SO_2 to SO_4, and EPA epidemiological data for representative power plants in the Northeast emitting 96.5×10^6 pounds of sulfur per year—equivalent to a 620-MWe plant.

(*a*) validity of predictions of effects and frequency of catastrophic accidents; (*b*) the need for long-term and adequate management of radioactive wastes; and (*c*) problems of malevolence. The upper-limit values quoted in Table 3 are most likely overestimates for reasons discussed later in connection with the report by the Advisory Committee on the Biological Effects of Ionizing Radiation (BEIR) (7); although genetic effects are not included, there are enough data to indicate that the values given for nonaccidental premature deaths would not be increased by more than about 50% in the first generation or by more than severalfold after hundreds of years if genetic effects had been taken into account.

Generally speaking, the biological effects of ionizing radiation can be correlated with the dosage received. The effects on human individuals and populations can be considered most meaningfully in terms of two levels of exposure:

High-Level: Acute, whole-body exposure greater than tens of rems,[2] which could result from nuclear warfare or from catastrophic nuclear or other radiation accidents.

Low-Level: Partial or whole-body exposure, usually chronic and of the order of natural background (1 rem/year) or less, which could result from normal occupational exposure, fallout to date, a normally operating nuclear power industry, and miscellaneous sources such as television, jet travel, and luminous watch dials.

High-level exposures can and have produced directly observable manifestations. Acute exposures of about 1 rem can be harmful to the fetus. Low-level exposures, as defined above, have not produced detectable deleterious effects in human individuals or populations or in other living organisms. This does not mean that such effects do not occur; it means that if they do occur, they do so at such low frequencies as to be undetectable.

HIGH-LEVEL EFFECTS High-level doses of ionizing radiation may produce both acute and delayed effects in humans or other organisms. The gross, acute effects of whole-body exposure to high-level doses are summarized in Table 6. Delayed effects may be noted after partial-body exposure to high-level doses; usually they are a result of hundreds to thousands of rems administered therapeutically to localized areas of the body. Such delayed effects on the reproductive organs, for example, range from decreased fertility or temporary sterility at about 25 rems to the gonads to permanent sterility at 600 rems or more.

Exposure to high levels of radiation also may produce late effects, those that may not develop until some years after the exposure. These effects can occur either in the exposed individual alone (somatic effects) or in the offspring of the exposed individual (genetic effects). (The probabilities that late somatic or genetic effects will occur can be estimated from the relationships discussed in the next section.) The

[2] Rem: a special unit of dose equivalent. The dose equivalent in rems is numerically equal to the absorbed dose in rads multiplied by the quality factor, the distribution factor, and any other necessary modifying factors. The rem represents that quantity of radiation that is equivalent—in biological damage of a specified sort—to 1 rad of 250 kVp X rays.

Table 6 Representative dose-effect relationships in human beings for whole-body, acute irradiation[a]

Nature of Effect	Rads of X or γ Radiation
Minimal dose detectable by chromosome analysis or other specialized analyses, but not by blood changes	5–25
Minimal dose readily detectable in a specific individual (e.g. one who presents himself as a possible exposure case)	50–75
Minimal dose likely to produce vomiting in about 10% of people so exposed	75–125
Dose likely to produce transient disability and clear blood changes in a majority of people so exposed	150–200
Dose likely to cause death in about 50% of people exposed	300

[a] Source: (8).

principal somatic effects are the induction of leukemia and other forms of cancer. Cataracts also may develop if the lens of the eye receives a heavy dose of X rays, gamma rays, beta particles, or neutrons; neutrons are believed to be particularly damaging. In addition, research on experimental animals indicates that high-level, whole-body irradiation shortens life spans, even when other effects do not appear. Such exposure may somehow accelerate an aging mechanism or it may weaken the body's defense mechanisms, increasing susceptibility to the usual causes of death.

LOW-LEVEL EFFECTS As stated earlier, low-level effects have not been observed. Nevertheless, current understanding of the mechanisms of interaction of radiation with biological systems requires that we assume that any level of radiation may be harmful to some degree. That is, we assume that there is no dosage threshold below which no damage occurs. Many have taken for granted that exposure to radiation in addition to the natural background and of the same or lower magnitude represents a risk so small compared with other hazards of life that any associated, nontrivial benefit gained by such exposure would far offset whatever harm resulted from it. But there has been public pressure to estimate the probabilities or frequencies of the effects of such exposure. The argument is that, since any level of radiation may cause some harm and since entire populations of nations or of the world could be exposed to additional low-level radiation, if care is not taken, the extent of the absolute harm could increase even though the damage might not be detectable.

The National Council on Radiation Protection and Measurements (NCRP) (8) has taken the view that estimates of quantitative risk are not useful and has expressed a philosophy as follows:

> In particular, it is believed that while exposures of workers and the general population should be kept to the lowest practicable level at all times, the presently permitted exposures represent a level of risk so small compared with other hazards of life, that such approbation will be achieved when the informed public review process is completed.

The BEIR committee of the National Academy of Sciences–National Research Council (7) decided, however, that there was an advantage in considering quantitative risk estimates, despite the recognized uncertainties in the data and calculations. The overall numerical values from the BEIR report can be summarized as follows:

1. It is estimated that exposure of the parents to 170 mrems per year (or 5 rems over the 30 years of the usual reproduction period) would cause in the first generation between about 150 and 3600 serious genetic disabilities per year in the US population, based on 3.6 million births per year.

2. It is estimated that the same exposure of the US population as above *could* cause from roughly 3000 to roughly 15,000 deaths from cancer annually, with 6000 being the most likely number. (*Could* is used in the preceding sentence because many scientists feel that as a result of the efficiency of the body's repair mechanisms at the very low dose rates involved, the true effects might approach zero production of cancer.)

The above numerical values are in essential agreement with those reported by the International Commission on Radiological Protection (10) and the United Nations Scientific Committee on the Effects of Atomic Radiation (11). The latter report stresses that the risk estimates are valid only for the doses at which they have been estimated (high levels), whereas the BEIR report suggests that the values are useful as upper-limit estimates in assessment of effects at low levels. A recent NCRP report (9) discusses this matter critically and concludes that the BEIR values have such a high probability of overestimating the actual risk that they are of only marginal value, if any, for purposes of realistic risk-benefit evaluation. At this time, we judge the consensus to be that the BEIR values are most likely overestimates by a considerable margin, but if used with that understanding, then there are important comparisons that can be made.

OTHER ISSUES   There are several issues of public concern, some of which involve technical solutions and others that are more related to institutional and sociological factors. Many of these are discussed elsewhere in this volume. It is our intention only to comment briefly on accidents because the risks can usefully be compared to those already discussed and on plutonium because the biomedical aspects have not been generally appreciated.

Accidents   The basic document in regard to reactor accidents is the "Rasmussen Report" (29), the final version of which has not yet been published. It attempts to predict the probabilities and consequences of a total spectrum of conceivable reactor accidents. Critical reviews of this report have been made by the American Physical Society (30) and the Union of Concerned Scientists (31). The essence of these analyses is that the Rasmussen estimates would have to be low by 3–5 orders of magnitude in order for the risks from catastrophic accidents to be comparable to those from normal operations of the coal, oil, or nuclear fuel cycles (see Table 7). It is a matter of conjecture whether the public would accept the probability, although very small, of a single nuclear event causing an immediate loss of hundreds of lives as preferable to or in place of the loss of a larger number of lives from fossil fuel combustion occurring in driblets and therefore unnoticed.

Table 7 Summary of implications of quantitative assessments of health effects in the general population associated with electricity production (all values rounded)

		Coal and Oil	Natural Gas	Nuclear
Premature deaths/year/1000-MWe plant		2–100	0	0.01–0.2
Added risk/year[a]		1 in 10,000	0	1 in 5,000,000
Age	Normal risk of death/yr	\multicolumn{3}{Enhanced risk of death per year because of electricity production[a]}		
10	1 in 3800	1.38 in 3800[b]	1 in 3800	1.0008 in 3800
25	1 in 700	1.07 in 700[b]	1 in 700	1.0001 in 700
45	1 in 200	1.02 in 200[b]	1 in 200	1.00004 in 200
65	1 in 40	1.004 in 40	1 in 40	1.000008 in 40
All ages	1 in 100	1.01 in 100	1 in 100	1.00002 in 100
Number of premature deaths in 30 years associated with routine operation of 300 plants[c]		20,000 to 1,000,000	0	100 to 2,000
Number of deaths statistically predicted from catastrophic accidents in 30 years from 300 plants [Rasmussen estimate (29)][d]		—	—	10

[a] Upper estimates.

[b] These estimates are undoubtedly quite high because premature deaths from fossil fuel combustion products fall almost exclusively in the older age groups.

[c] This represents the total operation for a generation of power plants that would supply about 300 million people.

[d] Based on 1 chance in 10^6 of an accident per reactor-year causing 1000 immediate and delayed casualties.

Plutonium Of all the radionuclides involved in the nuclear fuel cycle, plutonium has aroused the greatest public concern in regard to potential hazard. A great deal of experimental work has been done over the years on the biological effects of plutonium (32, 33); but of course as with other toxic substances it is not possible to predict precisely the effects of low levels in the range of exposure that would produce undetectable effects.

Following is a discussion of those factors that tend to cause plutonium-239 to be hazardous, and then of those that tend to reduce its hazard. Plutonium, as any alpha-emitting radionuclide, is very biologically effective in producing cancer when it is located within the body in direct contact with living tissue. When it is inhaled it comes into direct contact with living tissue, and when it enters the blood it is deposited in such tissues as bone, liver, and lymph node; once deposited it remains for a long time during which it irradiates the tissue. Because of its long physical half-life (24,300 years) it must be regarded essentially as a permanent contaminant just as are many other stable industrial chemicals that pollute the biosphere.

Because alpha radiation will not even penetrate the dead layer of skin, plutonium is not a hazard when it exists outside of the body. Contrary to popular conception, plutonium when swallowed remains essentially outside the body because it is extremely poorly absorbed, does not enter the bloodstream in significant proportions, and being mixed with intestinal contents does not irradiate the surface of the intestines as it passes through the gastrointestinal tract. Plutonium does not become concentrated in the food chain. These characteristics result in large part from the low solubility of plutonium in water and biological fluids and its tendency to remain fixed in soil.

It appears that inhalation of plutonium is the most hazardous route of exposure. Because plutonium deposited in the lung may be present as small particles, a question has been raised as to whether a given amount of such radioactivity deposited in the lung would be more hazardous if present as small particles rather than being uniformly deposited. This is presently a matter of controversy. One group of workers (34) claims on the basis of theoretical considerations that small particles would be more hazardous (hot particle theory) and therefore that existing standards, which are based on uniform distribution, should be made more stringent. Other workers and several official groups claim that experimental data support existing concepts and that there is no reason for any drastic change of standards (35-38).

The problem of malevolent use of plutonium cannot be logically assessed; this matter has been discussed by Cohen (39). It appears that except for an unreasoning widespread public fear, terrorist purposes could be much more readily achieved by using other more easily available chemical or biological agents.

In general it can be stated that plutonium when inhaled is a toxic carcinogen and great care should be taken to prevent its access to the biosphere. Essentially none would be released from normal operation of the nuclear fuel cycle. Estimates of risks from it as a component of nuclear fuel cycles and the experience of the past 30 years indicate that they are lower than from other parts of the cycle and from other fuel systems.

DISCUSSION AND CONCLUSIONS

Certain qualitative conclusions seem justified. Occupational deaths from the use of coal are considerably greater than for the three other classes of fuel. The explanation for this is undoubtedly the very large volume of mass that must be physically extracted from the earth and the attendant risks associated with underground work. Although the extraction of ore is also required for nuclear fuel, the volume required, and therefore the labor required, is approximately 25 times smaller for nuclear fuel than for coal.

Premature deaths among the general public are very much more likely to result from the use of coal and oil than from natural gas or nuclear fuel. The reason for this is that the use of coal and oil causes very large releases of combustion gases and particulates that cannot practically be contained, whereas releases from gas combustion are small and radioactive releases from nuclear plants are relatively insignificant.

A summary of the implications of the quantitative assessments is presented in Table 7 using upper estimates and rounding all values.[3] It is noted that for nuclear fuel the added risk of death per year among the general population from electricity production is less than 0.1% (e.g. 1.0008 in 3800 compared to 1 in 3800 for 10-year-olds). For coal and oil the added risk using upper estimates appears greater (e.g. 2% at age 45), but the values for the younger age groups are undoubtedly lower than those given because no account was taken of the fact that premature death from fossil fuel combustion products fall heavily in the older age groups. In addition to risks, values illustrative of absolute numbers of premature deaths are predicted for the operation of 300 plants for their typical lifetime of 30 years. For nuclear effects to be comparable to those of coal and oil the upper estimates would have to be in error by a factor of about 500, with coal and oil effects having been overestimated or correctable by improved technology and/or nuclear effects having been underestimated. The issues have recently been raised of population exposures from ^{14}C (40) produced in nuclear reactions and from radon-222 (41) from uranium mill tailings. The magnitude of these exposures and the fact that ample time is available to mitigate them if they do turn out to be important problems indicate that they would have no significant impact on the general relationships as presented.

The Rasmussen estimates of premature deaths from catastrophic nuclear accidents would have to have been underestimated by factors of 200 or 100,000 to be comparable to the upper estimates of effects of routine operations of nuclear or coal and oil systems, respectively. It should be noted that of the ten statistical deaths from catastrophic accidents, one represents immediate fatalities and nine represent delayed cancer and genetic effects.

Because of the uncertainties in the upper estimates for coal and oil it is concluded that such analyses are of little help in decision-making in a cost-effective manner for control processes or fuel selection in coal-burning plants. It is clear that large-scale expensive research to provide such evidence is justified. In our judgment, however, a satisfactory estimation of risks to health from coal and oil combustion is not likely to be available for many years because of the enormous complexity of dealing with the effects and interactions of the many chemical agents released.

Despite the uncertainties in the numerical values, they are of interest as a starting point in thinking about what risks society is willing or not willing to accept in order to avoid the acknowledged technical difficulties of handling nuclear power and the chance of catastrophe, or to avoid the biological risks of inadequate electricity.

---

[3] The estimated values of enhanced risk are calculated for illustrative purposes of comparison. They do not take into account age distribution of deaths associated with the processes dealt with and other factors and therefore should be used with care in any further calculations.

NOTE ADDED IN PROOF

The final version of the Rasmussen report (29) is now available and is cited as follows: US Nuclear Regulatory Commission, 1975, *Reactor Safety Study: An Assessment of Accident Risks in U.S. Commercial Nuclear Power Plants,* Executive Summary. Rep. No. WASH-1400 (NUREG-75/014), Washington DC: US NRC.

The numerical values used in the text of this article differ from those quoted in the final version but not enough to change the sense of the comparisons made.

ACKNOWLEDGMENT

Appreciation is expressed to Dr. Ronald Wyzga for valuable assistance in the collection and interpretation of the quantitative data.

Literature Cited

1. National Academy of Sciences/National Academy of Engineering, Environmental Studies Board; Committee for the Working Conference on Principles of Protocols for Evaluating Chemicals in the Environment; National Research Council, Committee on Toxicology. 1975. *Principles for Evaluating Chemicals in the Environment.* Washington DC: Nat. Acad. Sci.
2. National Academy of Sciences/National Research Council, Assembly of Life Sciences. 1973. *Proceedings of the Conference on Health Effects of Air Pollution.* Prepared for Committee on Public Works, US Senate, Serial No. 93-15
3. National Academy of Sciences/National Academy of Engineering/National Research Council, Commission on Natural Resources. 1975. *Air Quality and Stationary Source Emission Control.* Prepared for Committee on Public Works, US Senate, Serial No. 94-4
4. National Academy of Sciences/National Academy of Engineering, Coordinating Committee on Air Quality Studies. 1974. *Air Quality and Automobile Emission Control.* Prepared for Committee on Public Works, US Senate, Serial No. 93-24
5. Electric Power Research Institute. 1975. *Conference Proceedings: Workshop on Health Effects of Fossil Fuel Combustion Products.* Rep. No. EPRI SR-11. Palo Alto, Calif: EPRI
6. Goldsmith, J. R., Friberg, L. 1976. Impact of air pollution on human health. In *On Air Pollution: A Comprehensive Treatise,* ed. A. C. Stern, Chap. 16. New York: Academic. 3rd., in preparation
7. National Academy of Sciences/National Research Council, Division of Medical Sciences, Advisory Committee on the Biological Effects of Ionizing Radiations (BEIR). 1972. *The Effects on Populations of Exposure to Low Levels of Ionizing Radiation.* Washington DC: Nat. Acad. Sci./Nat. Res. Counc.
8. National Council on Radiation Protection and Measurements. 1971. *Basic Radiation Protection Criteria.* NCRP Rep. No. 39. Washington DC: Nat. Counc. Radiat. Prot. Meas.
9. National Council on Radiation Protection and Measurements. 1975. *Review of the Current State of Radiation Protection Philosophy.* NCRP. Rep. No. 43. Washington DC: Nat. Counc. Radiat. Prot. Meas.
10. International Commission on Radiological Protection. 1969. *Radiosensitivity and Spatial Distribution of Dose.* ICRP Publ. No. 14. Oxford: Pergamon
11. United Nations Scientific Committee on the Effects of Atomic Radiation. 1972. *Ionizing Radiation: Levels and Effects.* New York: United Nations
12. National Academy of Engineering, Committee on Public Engineering Policy. 1972. *Perspectives on Benefit-Risk Decision Making.* Report of a Colloquium, April 26–27, 1971. Washington DC: Nat. Acad. Eng.
13. Starr, C. 1969. Social benefits versus technological risk. *Science* 165:1232–38
14. US Department of Commerce, Bureau of the Census. 1972. *Statistical Abstract of the United States.* 93rd ed. Washington DC: GPO
15. Argonne National Laboratory. 1973. *A Study of Social Costs for Alternative Means of Electrical Power Generation for 1980 and 1990.* Argonne, Ill: Argonne Nat. Lab.

16. Battelle Memorial Institute. 1973. Environmental Considerations in Future Energy Growth. Columbus, Ohio: Battelle Mem. Inst.
17. Hamilton, L. D., ed. 1974. The Health and Environmental Effects of Electricity Generation—A Preliminary Report. Upton, NY: Brookhaven Nat. Lab.
18. Council on Environmental Quality. 1973. Energy and the Environment: Electric Power. Washington DC: GPO
19. Lave, L. B., Freeburg, L. C. 1973. Health effects of electricity generation from coal, oil, and nuclear fuel. Nucl. Saf. 14:409–28
20. Sagan, L. A. 1972. Human costs of nuclear power. Science 177:487–93
21. Sagan, L. A. 1974. Health costs associated with the mining, transport, and combustion of coal in the steam-electric industry. Nature 250:107–11
22. US Atomic Energy Commission. 1974. Comparative Risk-Cost-Benefit Study of Alternative Sources of Electrical Energy. Rep. No. WASH-1224. Washington DC: US AEC
23. Amdur, M. O. 1957. The influence of aerosols upon the respiratory response of guinea pigs to sulfur dioxide. Am. Ind. Hyg. Assoc. J. 18:149–55
24. Amdur, M. O. 1959. The physiological response of guinea pigs to atmospheric pollutants. Int. J. Air Pollut. 1:170–83
25. Amdur, M. O., Underhill, D. W. 1968. The effect of various aerosols on the response of guinea pigs to sulfur dioxide. Arch. Environ. Health 16:460–68
26. Amdur, M. O., Corn, M. 1963. The irritant potency of zinc ammonium sulfate of different particle sizes. Am. Ind. Hyg. Assoc. J. 24:326–33
27. Amdur, M. O. 1971. Aerosols formed by oxidation of sulfur dioxide: review of their toxicology. Arch. Environ. Health 23:459–68
28. US Environmental Protection Agency. 1974. A Report from CHESS 1970–1971. Rep. No. 650/1-74-004, ORD, NERC/RTP. Research Triangle Park, NC: US EPA
29. US Atomic Energy Commission. 1974. An Assessment of Accident Risks in U.S. Commercial Nuclear Power Plants. Rep. No. WASH-1400, draft. Washington DC: US AEC
30. American Physical Society. 1975. Report to the American Physical Society by the study group on light-water reactor safety. Rev. Mod. Phys. Suppl., Vol. 47
31. Sierra Club/Union of Concerned Scientists, Joint Review Committee. 1974. Preliminary Review of the AEC Reactor Safety Study. Cambridge, Mass: Union of Concerned Scientists
32. Stannard, J. N. 1973. Biomedical aspects of plutonium (discovery, development, projections). In Uranium, Plutonium, Transplutonic Elements. New York: Springer Verlag
33. US Atomic Energy Commission. 1974. Plutonium and Other Transuranium Elements: Sources, Environmental Distribution and Biomedical Effects. Rep. No. WASH-1359. Washington DC: US AEC
34. Tamplin, A. R., Cochran, T. B. 1974. Radiation Standards for Hot Particles: A Report on the Inadequacy of Existing Radiation Protection Standards Related to Internal Exposure of Man to Insoluble Particles of Plutonium and Other Alpha-Emitting Hot Particles. Washington, DC: Nat. Resour. Def. Counc.
35. Medical Research Council. 1975. The Toxicity of Plutonium. London: HMSO
36. Bair, W. J., Richmond, C. R., Wachholz, B. W. 1974. A Radiobiological Assessment of the Spatial Distribution of Radiation Dose from Inhaled Plutonium. Rep. No. WASH-1320. Washington DC: US AEC
37. National Council on Radiation Protection and Measurements, Ad Hoc Committee on Hot Particles. 1975. Alpha Emitting Particles in the Lung. NCRP Rep. No. 46. Washington DC: Nat. Counc. Radiat. Prot. Meas.
38. Sanders, C. L., Dagle, G. E. 1974. Studies of pulmonary carcinogenesis in rodents following inhalation of transuranic compounds. Symposium on Experimental Lung Cancer, Carcinogenesis and Bioassays, June 23–26, 1974, Battelle Seattle Research Center, Seattle, Wash.
39. Cohen, B. L. 1975. The Hazards in Plutonium Dispersal. Oak Ridge, Tenn: Inst. for Energy Analysis
40. Magno, P. J., Nelson, C. B., Ellet, W. H. 1975. A consideration of the significance of carbon-14 discharges from the nuclear power industry. Proc. 13th AEC Air Cleaning Conf. August 12–15, 1974, Vol. I, CONF 740807, UC-70
41. US Environmental Protection Agency. 1973. Environmental Analysis of the Uranium Fuel Cycle. Pt. I (003-B), Fuel Supply; Pt. II (003-C), Nuclear Power Reactors; Pt. III (003-D), Nuclear Fuel Processing. EPA Rep. No. 520/9-73-003 A-D. Washington DC: US EPA

Copyright 1976. All rights reserved

ECONOMIC COSTS OF ENERGY-RELATED ENVIRONMENTAL POLLUTION

✕11022

Lester B. Lave
Department of Economics, Carnegie-Mellon University, Pittsburgh, Pennsylvania 15213

Lester P. Silverman
Assembly of Behavioral and Social Sciences, National Academy of Sciences, Washington DC 20418

INTRODUCTION

The numerous vital services provided by the environmental resources of air, water, and land include a sink for the deposit of the residuals of production and consumption activities. Polluting, or venting residuals into the environment in a manner that affects the other services provided by these resources, is cheap for the producer, but not for society. Businessmen and consumers historically ignored pollution, noting that smoke smelled like money and full employment.

The environmental implications of alternative energy sources and uses are clearly of concern to policymakers and the public. Cheap, abundant energy used to be considered the basis for economic prosperity and individual well-being; now some contend that energy has been too cheap, in part because environmental damage has been under-assessed, and that even the sharply increased prices of recent times are too low.

The mechanics of supply and demand are recognized to work quite well in the field of energy. Economic analysis provides a powerful tool for understanding the implications of the Arab oil boycott, the effects of price controls on natural gas, and industry enthusiasm for the Alaskan pipeline. Much is different in the "market" for environmental resources; most obviously, the resources are rarely priced. However, we will find the tools of economic analysis useful in understanding the relationship between environmental and energy resources and in discussing the implications of this relationship for public policy.

In this chapter we review a large literature on the economics of environmental pollution, with special reference to that resulting from energy. In the first section, we outline an economic theory of the market, incorporating environmental considerations and the problems of exhausting natural resources. We then consider the theory and methods of benefit-cost analysis, applying these concepts to determine the cost of venting residuals into the environment. The specific effects of venting residuals into the air and water and on land are considered under "The Social Costs of Effluents." Public policy goals and strategies with respect to environmental management are then addressed. This is followed by an illustrative analysis of the fuel cycle costs of electricity generation with alternative fuels. We conclude with a summary that includes suggestions for future research.

MARKETS, COMPETITIVE EQUILIBRIA, AND EXTERNALITIES

Neoclassical Economic Theory

In examining the economic costs of environmental pollution, economists start with the neoclassical paradigm of the production and allocation of goods and services [see any intermediate microeconomic text, such as (1)]. The standard theory has tremendous power to explicate the implications for social welfare of the exceedingly complicated process by which an economy employs almost 100 million workers in millions of firms to produce tens of millions of products and services.

THE EQUILIBRIUM OF A MARKET ECONOMY The fundamental insight of Adam Smith in 1776 was that each individual—as consumer, worker, or employer—could pursue his or her individual goals of self-interested profit and simultaneously increase social welfare in the most efficient way. Altruism or cooperation, traits presumably in short supply, are not necessary in order to have an economy work well.

The modern mathematical expression of this insight consists of two theorems (2). The first is a statement about efficiency; the equilibrium of a competitive market is efficient (Pareto optimal) in the sense that no individual can be made better off without making some other individual worse off. While a competitive equilibrium is efficient (under the appropriate assumptions), there are generally a great many efficient points; indeed, there should be a different equilibrium for each distribution of income. The second theorem is that any efficient equilibrium (Pareto optimum) *can* come from a competitive market, with a suitable redistribution of income. Thus, whatever notion of equity one has in mind, a competitive equilibrium can achieve it, and achieve it efficiently.

These theorems are powerful and mathematically elegant. They provide insights as to the organization and management of as complicated a machine as the US economy; they also provide insights for correcting pathologies (3). However, all of the underlying assumptions are not satisfied by a real economy, and in particular, not by markets for the allocation of environmental resources.

THE ECONOMIC THEORY OF THE ENVIRONMENT The environmental resources of air, water, and land are vital inputs to each stage in the process of making energy available and disposing of its by-products. Air, water, and land receive these wastes and may become polluted; if receiving waste were their only service to man, the pollution problem would be of little importance (4). The problem arises because these resources provide multiple services, including the sustenance of life, provision of recreation and aesthetic values, and enhancement of commercial and industrial interests. All these services are "economic" in the sense that people or firms are willing to pay to receive more of them or to avoid a reduction in the quantity or quality of the services they provide. Ayres & Kneese (5) argue that waste disposal is central to human activity rather than an occasional problem. Freeman, Haveman & Kneese (4) note that much of the environmental problem arises because we are not sure, and cannot ever expect to be sure, that subtle, long-range, and indirect damages (deleterious effects on the capacity of a resource to provide non-waste services) are occurring in situations about which we are currently ignorant. [For introductory material on the economics of pollution, see (4, 6–8); for more advanced material, see (9, 10).]

Were the assumptions of the neoclassical paradigm to hold, policy with regard to pollution would not be an issue. Clean air, clean water, and unpolluted land would be bought and sold in markets established by the interaction of those wishing to buy and those willing to sell the services of these resources. In such markets the marginal cost (or "price"), which would have to be borne by the polluter of an environmental resource, would be greater than or equal to the marginal benefit that a slightly cleaner environment would have provided for recreation, aesthetics, etc.

Carrying this hypothetical discussion somewhat further, prices associated with environmental resources would vary across relatively short periods of time (depending on climatic and hydrological conditions, for example), across relatively long periods of time (as preferences for alternative services and goods shifted), and across geographic areas (for example, where alternative sources of the same services were available). Clean water might be valued more highly, for example, in the congested Northeast where there is a large population demanding recreation than in sparsely populated parts of the West; tastes for an unpolluted environment may differ as well. In addition, the potential interactions of other pollutants must be considered. The effects of any given pollutant, and hence its price, may be dependent on other conditions of the air or water. Further, the optimal price associated with discharging a particular effluent into a stream may be dependent on the nature of the air pollution in a particular place. The implications of such varying prices for public policy in this area are discussed in the section on public policy alternatives.

Markets do not exist for all environmental resources because fundamentally most resources are not privately owned; mineral rights are owned while disposal rights are not. The lack of enforceable property rights for some environmental resources results in a price perceived as zero by those firms, municipalities and, to a lesser degree, consumers who use them, particularly for waste receptor services. This private benefit of polluting is generally much smaller than the social loss in terms of the foregone possibilities of alternative services derived from the resource. Externalities, or divergences between private and social values, arise whenever property rights are not clearly defined or enforceable. The waste receptor service of oceans, rivers, and air around factories appears free to the polluter. Since the price of the service is zero to them, they will expand their use of it to the point where the marginal value of the service is zero to them, even though the marginal social cost may greatly exceed zero. [For a review of the literature on externalities see (11, 12); note that the existence of a general equilibrium solution to the allocation problem in the presence of externalities is a generally unresolved issue, with only a small, highly technical literature (2).]

Extensions of the Neoclassical Model

Of the extensions of the neoclassical model, two (one theoretical and one primarily empirical) are particularly relevant to a discussion of energy and the environment.

INTERTEMPORAL ALLOCATION OF RESOURCES Intuition has little to offer in examining questions about the long-term implications of saving or about attempts to bequeath a high standard of living to future generations (13, 14). A central problem has to do with renewable versus exhaustible resources. Baumol and Oates (9) distinguish four archetypes: (a) resources that are producible at constant or diminishing costs (e.g. liquid oxygen), (b) resources that are producible at rising costs (e.g. going to less rich ores), (c) resources that autonomously regenerate (e.g. the stock of new fish this year depends on the stock of fish last year), and (d) resources whose stocks are fixed so that using some today means that less will be available tomorrow [see Solow (15)]. While upper bounds can be placed on the total quantity of any element in the earth's crust, the more relevant limit is provided by extraction costs—cases a, b, or c.

The theory sketched above shows that the market is capable of dealing with all four cases. However, to the taste of some, the market does not take a long enough view in allocation in case b, resources whose extraction costs increase sharply. There is disagreement as to whether current market prices take too short or too long a view, i.e. whether the prices of case b resources are too high or too low (9); the disagreement revolves around the amount of technological change that can be expected. Of course, central planning may also be myopic (15). Whether one relies on individual preferences expressed through the market or some mechanism of political choice, a series of prices stretching into the indefinite future can be derived that will have the property of allocating resources in the market so that we can bequeath whatever standards of living we desire to future generations (within limits and at the cost of reduced consumption today) [see Baumol and Oates (9)].

A large theoretical literature has resulted from the examination of the long-term implications of policies toward growth, such as savings behavior and technological changes (16). In recent years this literature has been extended to the relationship between economic growth and environmental quality (17–20) and the study of exhaustible resources (21, 22). The relatively few attempts to quantify the implications of long-term economic growth on energy resources and the environment [most notably, Meadows et al (23); for comments on Meadows, see (24, 25)] are extremely sensitive to assumptions about technological change, the formation of expectations, population growth, and resource availability (at a range of future prices).

SPECIALIZED EMPIRICAL MODELS The general equilibrium models of the economy extended to include environmental (exhaustible and inexhaustible) resources suggest the importance of empirical studies assessing the implications of changes in economic structure, such as increasing prices of raw materials or tightening emission standards.

The most widely used empirical model, input-output analysis, makes enormous simplifications about technology (fixed proportions) and the number of inputs and outputs (26). However, input-output analysis has been of immense help in tracing out the implications of various emission standards and resource constraints. Another model of economic activity focusing directly on the residuals of production and consumption is that of materials balance (5, 27). The assumption is that matter is neither created nor destroyed, only its form is changed. Thus, all human activity involves the production of residuals. The problem is to find chemical forms and physical locations for venting the residuals that do little damage to or actually help man. Materials balance models have been derived for the entire US economy as well as for a number of industries (27, 28).

BENEFIT-COST ANALYSIS AND DECISION-MAKING

These models, together with data on the working of the economy, serve to identify and quantify deviations from the assumptions behind the powerful theorems of neoclassical theory. In particular, the focus is on externalities, indicated by the divergence of private and social cost of some economic activity. Once detected, externalities can be ignored as unimportant or can lead to restructuring of the market in some way, ranging from taxes (or otherwise altering market prices) to central control of economic activity. To evaluate the importance of an externality, and alternative proposals for correcting it, benefit-cost analysis can be applied (29–31). All effects of a project are classified as either benefits or costs and are then translated into dollars; the project is judged by the relation of dollar benefits to costs. The tool is basically a sound one, although there are a number of potential dangers; in particular, the monetization of factors that are not bought and sold directly in a market is difficult, e.g. time, health, etc. The subjective nature of the valuation of these factors has made it easy to bias the analysis so that a bad project can be made to look good, or the reverse.

Since project effects are evaluated in dollar terms and the criteria for evaluation is a comparison of dollar benefits with dollar costs, the tool is focused on economic efficiency (32). A more balanced look at such other social goals as improving the distribution of income or goods over time and location could conceivably be encompassed within the framework, although this is not done in practice. Thus, benefit-cost analysis is an invaluable aid to decision-making, but not the sole criterion.

There are several important issues in the use of benefit-cost analysis for quantification of the social cost of pollution. The first is the implications of project interdependencies (10) and, in particular, the importance of the timing of projects with irreversible consequences (33). The second is the valuation of the intangible or indirect services of environmental resources. Although it can be argued that the prices of some goods and services do not measure their social cost because of monopolies or other deviations from competition, this is not the real problem. Rather, the issue is the valuation of goods or services not traded in the market at all. Bishop and Cicchetti (34) note that the weaker the analogy between a nonmarketed (or public) good or service and its best market counterpart, the more likely the analyst is to proclaim the output as an intangible and not treat it quantitatively in the analysis. For example, air pollution soils clothing and the resultant extra expenditures are valued, but improved visibility does not have a close market counterpart and so is treated as an intangible. By viewing the consumer's utility function as defined over a limited number of attributes or consumption activities rather than over an infinite list of unique products, Bishop and Cicchetti (34) draw out the commonalities between things that are sold in markets and the characteristics of public goods that are not.

Some of the benefits from environmental improvement most difficult to evaluate are:

Nonuser benefits A nonuser benefit is measured by the amount individuals would be willing to pay to preserve their option to enjoy a beautiful, unpolluted area at some time in the future [this is typically referred to as option value (35)] and is primarily associated with unique resources and irreversible processes (33). Additional types of nonuser benefits are based on existence value (benefits derived from the sheer knowledge that an area is kept unpolluted) and bequest motivation (benefits derived from insuring that an area will be unpolluted and available to one's grandchildren). [For a discussion of benefits associated with the preservation of botanical specimens, see (36).] Of these, only option value has been addressed by economists.

Postponement of death Although the implicit valuation of human life takes place constantly in the public decision-making process, its explicit valuation to improve decision-making evokes strong protest. The postponement of death has been valued by using the difference between projected earnings and consumption, by gross earnings, and by individuals' willingness to pay for reduced risk of death (37–40). Note that the key element of risk to be valued is that which is involuntary; smoking cigarettes carries a greater risk of death than does breathing air pollution, but the former is voluntary while the latter is largely involuntary. We know little about the psychic benefits of postponing death, both for the individual and those who care about him or her (37, 38).

Morbidity Valuing reductions in the risk of illness is even more difficult empirically than is valuing the risk of death. Definitional problems abound and welfare loss is extremely difficult to evaluate. Theoretically, we wish to obtain estimates of the demand for engaging in activities, depending on the state of health. One could then infer the effect of health on an individual's activities (39). A more direct approach is to determine how much individuals are willing to pay to reduce the probability of having a particular disease.

Aesthetics, life-style, and the general quality of life Some effects of pollution are unpleasant but not harmful to health or property. It is extremely difficult to quantify or monetize these effects mitigating the general quality of life. By purchasing housing and office space with a view, or by using air fresheners, people demonstrate that correcting these effects is important and worth a substantial amount. However, quantifying and then evaluating these effects is nearly impossible, since they are generally situation specific, and rarely are analogous services sold in the market.

Benefit-cost analysis has an important bias (41). A well-done analysis is inherently biased toward overestimating costs and underestimating benefits. For a carefully done analysis, costs are overstated because no account is taken of technological change and there is assumed to be little adjustment to lower the cost of the new constraints; benefits are understated since such categories as aesthetics are invariably not quantified. Thus, when a carefully done benefit-cost analysis of pollution abatement concludes the project should be done, it certainly should be done; when costs appear to be slightly higher than benefits, more careful consideration is warranted because of the inherent bias.

THE SOCIAL COSTS OF EFFLUENTS

Considerable attention has been given to measuring the effects of effluents on human health, animals, plants, materials life and maintenance, and general ecology [for air, see (42–44); water (45–47); land (48)]. Here we summarize this literature, focusing on the methods used to quantify the social cost of energy-related pollution.

The major insight of the materials balance approach (5, 27) must be emphasized again. We cannot stop emitting residuals. We can only find chemical forms and locations for emission that are harmless, do the least damage, or even are beneficial to man. For example, burning 1 million tons of 3% sulfur coal will liberate 30,000 tons of sulfur. Without any controls, most of the 30,000 tons will be emitted into the air as SO_2 and other gases; of the remainder, some will be emitted into the air as part of the ash, and some will remain in the boiler ash. The simplest control is to wash the coal prior to combustion and to use a wet scrubber to absorb and react SO_2. Although these techniques curtail emissions into the air, they produce a substantial quantity of sulfur compounds that must be discharged into a river (causing water pollution), dumped on land (causing land pollution), or further treated. In particular, lime-limestone scrubbing produces a gypsum sludge (44). A large power plant would produce millions of tons of sludge annually, resulting in a large waste disposal problem. Thus, as the materials balance approach emphasizes, all 30,000 tons of sulfur have to be dealt with, either as gaseous, liquid, or solid waste or as a useable chemical.

Similarly, power plants vent a great deal of heat (between 2.5 and 3 times the electricity generated). This waste heat is most efficiency carried off by cooling water, but cooling towers could be used to vent the heat into the atmosphere. Alternatively, one could design the facility to use more of the heat for power generation or space heating. The prices of fuel and equipment and discharge regulations will determine what is done.

Public policies to manage the environment will affect the form in which residuals are vented. Such policies must be based on assessments of the value of air, water, and land in terms of the non-waste receptor services they provide. We now discuss procedures for quantifying these values for each environmental resource.

Emitting Effluents into the Air

The principal effects of air pollution abatement, classified as benefits or costs, are listed in Table 1. As the discussion below will make clear, only two of the costs (new capital and operations and maintenance) are generally handled and only five categories of benefit are even discussed in the literature. Thus, many of the relevant costs and benefits remain unanalyzed.

The Clean Air Act (as amended) requires the Environmental Protection Agency (EPA) to submit an annual report on the cost of air pollution abatement. The 1972 report (49) is particularly good in setting out estimated costs and benefits in a comparable fashion through 1977. In this analysis, EPA assumed that both current stationary and mobile source emission standards would be fully in force in 1977; delay in implementing standards reduces both the costs and benefits. Shown in the last column of Table 2 are the (annualized) 1977 costs for mobile sources, solid waste, stationary fuel combustion, and industrial processes (in 1970 dollars). Mobile sources (essentially motor vehicles) would cost $8385 million; all other abatement costs would run $3913 million. The 1977 investment in new equipment would be greater than this amount, since many of the conversion costs would just be occurring. However, discounting the investment over its useful life leads to the annualized cost (after maintenance and operating costs are added). By 1977, the Clean Air Act is projected to result in an additional 39% decrease in emissions of suspended particulates, an 81% decrease in SO_x, a 57% decrease in CO, a 17% reduction in HC, and a 60% reduction in NO_x.

There are many reasons to believe that these costs are overestimates of the true cost of abatement. They are based on currently available technology, while research and development in this area is progressing rapidly.

We can be much less certain of the estimates of the benefits of abatement (by source and by three classes of effect) shown in Table 2. A basic problem is that

Table 1 Benefits and costs of air pollution abatement

Benefits		Costs	
Direct	Indirect	Direct	Indirect
Health improvements	Better quality of life	Cost of new capital equipment	Temporary unemployment
Plants and crops			Migration
Animals		Operating cost	Capital losses—
Cleaning		Impaired design	private and social
Materials life			
Odors			
Visibility			
General psychic conditions			

Table 2 Projected national annual benefits (damage cost reduction) by source class in fiscal 1977 (1970 dollars in millions)[a]

	Benefit Class				
Source Class	Health	Residential	Materials and Vegetation	Total Benefit	Control Cost
Mobile	$ —[b]	$ —[b]	$ 945	$ 945[b]	$ 8385
Solid waste	172	145	119	436	224
Stationary fuel combustion	3812[c]	3267	2366	9445	2476
Industrial processes studies	1413	1302	734	3350	1213
Total benefit[d]	$5397	$4615	$4164	$14,176	$12,298

[a] Source: (49, Tables 1–4).
[b] Value of benefits from reducing CO, NO_x, and HC emissions not available (lack of data).
[c] Health damage cost due to NO_x, from stationary fuel combustion, not included (lack of data).
[d] Benefit computation based on proportional reduction of damage costs excludes "miscellaneous" source damage costs since these are generally not controllable and therefore cannot become benefits.

natural pollution is difficult to discriminate from man-made pollution; winds cause suspended particulates, volcanos emit sulfur, and trees vent hydrocarbons. A second basic problem is estimating the physical damage resulting from pollution. The third basic problem is monetizing the physical damage. In addition to these problems, the EPA estimates are suspect on other grounds, such as the double-counting of residential property benefits. A comprehensive review of these issues is found in Lave and Seskin (50). Some of the major methodologies for assessing benefits are discussed below.

AIR POLLUTION AND HUMAN HEALTH   A large literature documents the association between air pollution and ill health (43, 44, 50–52). Based on regression analysis of mortality rates across US cities, Lave and Seskin (52) find that a 50% decrease in air pollution would lower the mortality rate by an estimated 4.5%. If we assume that the reduction in air pollution would have the same effect on morbidity as on mortality (which is certainly a very conservative estimate), we could reduce the economic cost of morbidity and mortality by just under $9 billion per year.[1] This number is not necessarily a good estimate of the value of the reduction in illness and death (37, 38); an affluent society seems to be willing to pay much more than the saving in health care costs and foregone earnings to lower morbidity and mortality. Thus, this figure represents a *lower* bound for the value of the reduction in illness and death.

[1] The estimate is 4.5% of the total expenditures on health care plus the loss in earnings due to premature death and illness (assuming a 6% discount rate) in 1963 [see Rice (40, p. 110)]. It is adjusted to 1970 dollars by applying an annual cost increase of 8% and to 1977 magnitudes by applying a 4% annual growth rate thereafter.

AIR POLLUTION AND THE VALUE OF PROPERTY   Recent summaries of the literature on the effect of air pollution on cleaning expenditures and on the life of materials are found in (44, 53–56). Michelson and Tourin (57) compare the expenditures on cleaning in two cities and conclude that $84 more per capita was spent in a highly polluted city than in a relatively clean one.

One might expect that real estate in unpolluted areas ought to carry a higher price than property in polluted areas, controlling for other factors that determine real estate value. Numerous studies have attempted to infer the effect of air pollution on property values (58–64). There are many difficulties with these studies (10, 41, 65). The results are highly sensitive to assumptions about the model and specifications of the equations. There is also the problem that estimates reflect a vast amount of double counting, since they must embody valuation due to health, cleaning, and materials damage, all estimated explicitly elsewhere. Finally, these estimates assume that individuals implicitly or explicitly perceive differences in pollution levels and act on these perceptions in determining their demand for real estate.

AIR POLLUTION DAMAGE TO PLANTS AND ANIMALS   There are a number of documented cases where plants or animals were damaged by air pollution. Good summaries of the literature can be found in (43, 44, 66, 67). Crocker and Rogers (68) give an extended description of the effect of phosphate reduction on nearby citrus farms and cattle ranches. For experimental results, see (69–71). Recent work by Benedict et al (66) implies that agricultural damage due to sulfur oxides is small, certainly less than $20 million per year; total damage to vegetation is around $100 million per year. However, attempts to assess air pollution damage on plants in the field are plagued by the fact that pest damage, drought, and disease can produce leaf damage quite similar to that caused by most pollutants. Thus, it is difficult to tell how much of the leaf damage was caused by air pollution and how much was caused by other forces (72).

Emitting Pollutants into the Water

Water resources are inextricably linked with the fuel cycle of virtually every source of energy. Water is used in pretreatment of fuel (e.g. washing coal) and in generating electricity (principally as a coolant), and hydroelectric power accounts for a small proportion of electricity. Various aspects of this relationship are discussed in (73). However, of interest here are the costs imposed on society by the undesirable consequences associated with the use of water. The conceptual benefit-cost framework for air shown in Table 1 also serves well as a framework for evaluating abatement of water pollution.

Water-related consequences of energy production, extraction, transportation, refining, and disposal constitute a small percentage of the water pollution problem (74–76). The primary fuel-cycle effects on water quality come from oil spills in the oceans, acid drainage from surface and deep mines in rivers and streams, thermal discharges, and emission of residuals that are slow to degrade from the cooling process in steam electric plants. Oil spills have harmful effects on the recreation value of beaches and commercial value of fishing water, as well as potential health hazards from the entry of hydrocarbons into the marine food chain (77). Acid mine drainage results in increases in acidity, hardness, and trace metals in the water. This can result in increased corrosion of vessels, bridges, and water-using equipment. The potential changes in water color and degraded surrounding land, rather than acidity of water itself, can be a significant factor affecting recreation uses (78). Residuals that are slow to degrade can result in toxicity, cloudiness, unpleasant taste, or hardness of the water, all of which have implications for alternative uses of water resources (79). [Petroleum refining releases organic wastes into waterways (80), but this is not a major problem.]

Estimates of the national costs for municipal and industrial water treatment at various levels of removal are presented in Table 3. EPA estimates that to achieve 85–90% removal of all residuals from point sources (not just energy-related by-products) will cost $4.1 billion in 1981. Moreover, treatment costs rise dramatically as additional increments of water quality are to be achieved. Further, the potentially enormous costs of controlling water pollution from non-point sources are excluded from the calculations of Table 3.

Attempting to estimate the percentage of these costs associated with energy-related pollution is extremely difficult. EPA has estimated (76, p. 516) that the capital costs for water pollution control associated with acid mine drainage and oil and hazardous spills might constitute 9–15% of the total capital costs for water pollution control. Additional costs would be incurred in abating thermal discharges and slow-to-degrade residuals; thus energy-related costs are a small but significant proportion of the costs of water pollution control. Power generation has been singled out with regard to thermal discharges—by 1983 no thermal discharges into waterways will be allowed. The cost of controlling thermal discharges is large compared to control costs of other water-related pollution (ignoring disposal of scrubbing sludge). For a discussion of the costs of alternative strategies for abatement of water pollution, see the papers in (45).

Table 3 National costs for municipal and industrial water treatment at various levels of removal[a,b]

Percentage of Wastes Treated	Level of Removal (%)	Annualized Costs in 1981 (Billions of dollars)
100	100	21.1
80	95–99	12.4
20	100	
100	95–99	8.4
100	85–90[c]	4.1

[a] Source: (79a, p. 156).

[b] Excludes potentially enormous costs to control pollution from storm and combined sewers, agricultural wastes, and other dispersed sources.

[c] This is roughly the pre-1972 "uniform secondary treatment" standard.

Water resources provide a range of important services including domestic and industrial use, commercial fishing, aesthetics, and recreation. The satisfaction derived in each use of water is likely to vary depending on a number of factors, one of which is water quality (which itself is a multidimensional concept). The benefits from cleaner water are easy to identify but quite difficult to measure and evaluate quantitatively. Some of the problems follow.

1. In order to associate social costs with a particular aspect of a fuel cycle, we must be able to identify its contribution to the overall quality-related characteristics of the water resource. It is difficult to distinguish man-made from natural water quality problems and even more difficult to distinguish the losses from man-made water pollution and to identify control alternatives (79, 81). For example, hard water may be traced to natural processes of erosion and groundwater conditions, as well as to various industrial and mining processes. Abatement of energy-related water pollution will have little effect on general water pollution levels; since dollar benefit is unlikely to be directly proportional to percentage of abatement, the benefit of abating energy-related water pollution will depend on abatement activities associated with other sources and definitive analysis is difficult.

2. The effects of particular aspects of water quality on the uses of water are difficult to establish. The effects may be conflicting; thermal discharges that raise the water temperature may extend the season for swimming while adversely affecting some species of fish; industrial discharges of acid may adversely affect recreation and fishing but may improve water for industrial use by retarding growth of algae (82). Kneese notes that we know very little about the quality characteristics of water that human beings find attractive for recreation (83). Clearly, a multi-disciplinary research effort is required here.

3. The net effect of the changes on human behavior and satisfaction has to be valued. The benefit-cost framework adopted by economists for addressing this problem is widely accepted but still subject to debate. (Some of the generic issues have been discussed above.)

As with air quality, water quality is often a public good in the sense that only one level of water quality can be provided at a time; thus individuals and small firms cannot improve the quality of the water they use. It is important to distinguish cases where the individual has no choice about water quality from other cases (e.g. recreation and fishing) where the individual can choose a clean lake rather than a polluted one. The first set of uses can be valued by examing "defensive" expenditures or other methods independent (for the most part) of a specific location. Valuation of recreation or fishing services of a particular water resource, on the other hand, is greatly influenced by alternative (substitutable) opportunities that are available.

A comprehensive review of empirical studies is found in Tihansky (81). We review briefly some of the prominent methods for evaluating benefits of water quality improvement in terms of human health, industrial and domestic uses, fishing, aesthetics, and recreation.

WATER POLLUTION AND HUMAN HEALTH   The effect of water quality on human health is enormously difficult to assess. Energy-related pollutants are suspect of harmful health effects in terms of hydrocarbons from oil spills entering the food chain (77), potential carcinogens present in emission from power-generating stations (84), and potential heart problems related to acidic water (85). As has been noted, economic valuation of a marginal increase in the probability of early death or increased morbidity is a difficult issue.

WATER POLLUTION AND INDUSTRIAL AND DOMESTIC USES The major energy-related water pollution problem for industrial users is the acidity and hardness associated with acid mine drainage. These factors accelerate corrosion and scale formation in water-using equipment. Another related problem is that dissolved concentrations of iron and manganese can affect certain types of industrial equipment. Although production processes were not altered as a result of the presence of acid mine drainage, various methods (e.g. substitution of corrosion-resistant materials) presumably involving higher costs were used to adjust to this problem (78). Tihansky (86) shows that the most economically damaging characteristics of water for household appliances and water distribution facilities are hardness and total dissolved solids. Because these constituents are partly natural in origin, however, their impacts are difficult to interpret in evaluating the abatement of energy-related pollution.

WATER POLLUTION AND COMMERCIAL FISHING Losses in revenue from fishing have been evaluated by assigning an arbitrary price per fish to the estimated loss in fish catch (81). Although this is an acceptable method of valuation (as long as the calculation of loss accounts for the availability of alternative fishing opportunities), the difficulty in estimating the effect of energy-related water pollutants on fish yields is substantial. The interaction of water quality characteristics appears to be important in affecting fish reproduction. Data on the effect of acidity on reproduction of fish indicate that the effect is directly related to the temperature of the water; high acid concentrations have substantially worse effects on fish in warm water than in cold (78). Indeed, heat speeds biological reactions to any type of pollution.

WATER POLLUTION AND AESTHETICS The aesthetic benefits of water quality improvement, other than those discussed below with regard to recreation, are difficult to quantify. The dimensions of aesthetic appreciation of a resource (by users, much less nonusers) are enormously complex and certainly related to numerous aspects of the physical, social, and cultural environment. Aesthetic benefits have been inferred from examinations of property values (87, 88) and tourist expenditures (89). These studies suffer from problems of measuring the aesthetic benefits for those who are not users of the resource, as well as consideration of the offsetting gains in utility from use of substitute resources.

WATER POLLUTION AND RECREATION There is substantial agreement among the available evidence that the benefits in the above categories are small in comparison to greater availability and higher quality of water-based recreational services provided by the resource (4, 81). A method for inferring the total net willingness to pay for a recreation resource has been developed by Clawson and Knetsch (90, 91) (based on an idea originally suggested by Hotelling). This procedure, which infers the value attached to a resource from the travel and time costs incurred in transportation to the site, implicitly recognizes that access to the resource and the availability of alternatives must be considered as well as inherent natural qualities of the site. Other methods of valuation are discussed in (92, 93).

The practical problems of implementing the Clawson and Knetsch methodology are numerous (94), but the major issue for implementation is finding a comparable site to allow identification of the explicit effect of water quality (82). There have not been attempts to infer the marginal recreational benefits of marginal improvements in water quality; it has been assumed that improvement in water quality allows a full range of water-based activities where none existed before (95). (The objective function implied by such a "threshold effect"—if it indeed exists—could be quite difficult to achieve in practice due to hydrological uncertainties.)

Land Pollution

The pollution of land areas is an ill-defined subject encompassing a broad range of undesirable land uses or disposal problems. These include the disposal of ash and radioactive waste. The ash problem is small and the radioactive waste problem is a major one. We do not discuss them here; see (96, 97) for further treatment of ash and radioactive wastes. We have already alluded to the problem of disposing of sludge from lime-limestone scrubbing. Two major effects on land areas associated with various fuel cycles are strip mining of coal and transportation of oil across land (in particular, the Trans-Alaska Pipeline). Since these cases provide some illuminating contrasting features, they are worthy of closer examination.

The ravages of strip mining are not irreversible; at some cost, land can be reclaimed. The Council on Environmental Quality has estimated (98) that the costs of reclamation in the Appalachian region range from 15 to 90 cents per ton of coal mined (depending on production of coal per acre and physical conditions at the site). Since the costs of producing coal are more than $4 or $5 per ton, reclamation costs are not overwhelming. A committee of the National Academy of Sciences estimates (99) that surface-mined areas of the western United States can be reclaimed for 3–10 cents per ton on top of production costs of $1.25–$2.50 per ton. The rate at which reclamation can take place is very much dependent on the level of rainfall. Thus, the cost of nearly complete restoration of strip-mined areas (which is the accepted practice in Germany) is substantial, but not intolerable if imposed on the industry as a whole (100).

The social costs not reflected in the above estimates are the foregone opportunities for recreation and insults to the aesthetic appreciation of the area. To the extent that substitute areas can provide the same recreation or aesthetic values, which is likely to be the case in the western and Appalachian coal-rich regions, these costs are not likely to be high. Further, the costs should only be counted for the years until reclamation is complete (this could be as few as 2–3 years in the East, although it would probably be substantially more for arid western lands).

Assessment of the environmental impact of the Trans-Alaska Pipeline is considerably more difficult. Hundreds of miles of the largest undeveloped area of the United States will be affected and exposed to the probability of leaks, which could have severe, long-term implications for the Alaskan tundra. The uniqueness of this land area and long-term (accounting is in terms of decades, rather than years) consequences of development and despoliation render economic assessment of benefits considerably more difficult than in the case of strip mining. For a discussion of the issues and analysis of economic options with regard to the pipeline, see (101).

PUBLIC POLICY ALTERNATIVES FOR MANAGING THE ENVIRONMENT

Kneese and Schultze (102) note that the problem with the use of environmental resources is not that the price system does not work but rather that it works with marvelous efficiency in the wrong direction. The signals that air and water are free goods encourage firms and consumers to overuse them. To achieve efficiency, the goal of public policy should be to equalize the private marginal costs of energy use with its true social costs (including valuation of the deleterious effects not valued by the market, as we have discussed above). The question remains, however, how to achieve this equalization. There is a large literature on public policy alternatives—on an introductory level see (4, 6–8, 102); for more advanced discussion, of interest primarily to economists, see (9, 10). Before addressing public policy alternatives, we review the criteria for public decision-making.

The Criteria for Public Policy in Environmental Protection

There are at least three criteria that must be used to judge whether collective (government) action is warranted and desirable: efficiency, equity, and administrative costs. Our focus thus far has been on efficiency. The criterion of administrative costs is at the heart of evaluating such abatement strategies as effluent charges and emission standards, as argued in the next section. We now turn to the criterion of equity. It is useful to distinguish at least three distinct notions of equity (the distribution of the proceeds of economic and social activity): (*a*) across socio-economic groups (and even the inequality of income per se), (*b*) across generations (providing for the welfare of unborn generations), and (*c*) across geographical regions and areas. Intergenerational and intergeographical trade-offs have been alluded to previously in this paper and are discussed again below. The socioeconomic dimension of equity with regard to environmental resources merits further discussion.

Policymakers are at least as much concerned with equity, or the distribution of the proceeds of economic activity, as with economic efficiency. Indeed, the notion of willingness to pay is often abhorrent to the non-economist because of its implied acceptance of the current distribution of income. The economist, invoking the theorems (discussed earlier) of the traditional paradigm, asserts that the questions are separable; given any distribution of resources and income, an efficient (Pareto optimal) allocation of goods and services can be achieved, and the issue of income distribution can be solved by other means. However, equity issues rarely are separated in the political process (32). Even when the effects of public decision-making are easily quantified and monetized, the equity (or fairness) question remains. When uncertainty in these effects (cost and benefits) is present, the issue of fairness becomes even more complicated (103).

Freeman (104) notes that several of the strategies for managing environmental resources involve the allocation of enforceable property rights; thus, payments for polluting or cleaning the environment must be made by one group to another. Therefore, the equity issue becomes fundamental, both across groups (including income and geographical) within society and across generations. The efficacy of the political process in reflecting public interest in equity is discussed by Maass (105) and Wildavsky (106).

The distributional effects of environmental pollution abatement programs are not easily analyzed (104, 107). Cleaning the polluted air of our central cities and the water of our dirtiest rivers would seem to provide relatively greater benefits for inner-city (predominantly poor and black) residents. However, the displacement (particularly of jobs) costs and continuing costs of environmental measures, as well as the subsequent increase in property values, are all likely to affect the inner-city residents to a relatively greater degree. One possibility is that the central city environment could be so improved that poor residents are forced out, at great cost to them. Baumol and Oates (9) review the theoretical and empirical literature on this subject and conclude that measures to improve environmental quality may have an uneven pattern of incidence, which on balance might be expected to be somewhat regressive. The major redistributive consequences are likely to be job displacement during periods of transition and the payment of the continuing cost of environmental measures (which, like general sales taxes, are likely to be regressive). Freeman (104) contends, and Baumol and Oates (9) agree, that environmental programs are not generally well suited to achievement of equity objectives and, therefore, the primary purpose of environmental programs must be to increase the efficiency in the use of environmental resources. Roberts (108) argues that since the federal tax system is more progressive than most state and local taxes, it should be used to finance aspects of environmental programs that are basically regressive.

Alternative Strategies for Abatement

THEORETICAL MODELS Public policy with regard to managing the environment involves choice among a range of alternatives for each type of resource. Water quality management can be achieved by alternatives such as shutdown of waste-generating activity, end-of-pipe treatment of waste, internal process change, recovery and recycling of residuals, centralized treatment, flow augmentation, etc. (46, 109). Note that there is often a choice as to the chemical form of the effluent, e.g. ozone instead of chlorine as oxidizing agent. In the management of air pollution, the alternatives include physical-chemical recovery in home or work place or abatement at the source (42, 49). The administrative costs of enforcing policy are an important component of total cost. An appropriate policy would be one choosing the combination of measures such that, at the margin, the additional cost of disposal (in air, water, and land) was just equal to the marginal benefits provided by each resource in its most valuable non-disposal-receptor use. Inter-regional issues, closely related to the equity notions discussed above, are also crucial. An interesting alternative, although one not permitted by current regulation, is to make some bodies of water quite unpolluted while allowing others to become sewers. The choice of which of two rivers should become polluted, or whether to have one very dirty river or several marginally dirty rivers, often must be decided strictly on equity grounds.

The most frequent response in our society to perceived environmental problems has been to legislate authority for standard setting and enforcement. Following Davies (110), it is useful to distinguish three kinds of standards: community goals (program objectives), ambient water or air quality standards (which translate goals into operational environmental quality objectives), and emission standards (which prescribe the quantity of each pollutant to be allowed from a given source). Davies (110) argues that the establishment of goals is basically a political question, with at most only an intuitive benefit-cost calculation. [The formal equivalence of an environmental quality standard to some benefit-cost ratio is shown by Thomas (111).] If there were complete knowledge of the effects of all pollutants, the transition from goals to quality standards would be straightforward. Uncertainty is compounded in the subsequent attempt to derive emission or effluent standards to achieve a given quality objective (due to complex biological and physical dispersion processes and environmental chemical changes).

Emission standards in existing legislation are tied both to ambient water quality (for existing polluters) and to technological feasibility (best-practical-technology and best-available-technology are commonly used criteria) for new sources. Such technology-based strategies can involve huge costs of negotiation (high administrative costs) for virtually every large individual source of waste discharge, and they also provide no incentive (and in some cases, a negative incentive) to develop innovations to reduce pollution (102); stating explicit goals for emissions should stimulate innovations.

Emission standards are administratively understandable and politically viable (in part because they avoid the politically difficult issue of defining property rights). However, it should be recognized that many difficulties remain. Just as the discussion earlier indicated that the marginal social cost of pollution is a function of location (relative to people, other polluters, commercial interests, etc.), time (including weather and hydrological conditions as well as longer-term population shifts), and presence of other pollutants, so emission standards theoretically should vary over these conditions.

Another strategy for managing the environment focuses directly on the cause of the problem, the divergence between the private and social cost of the fuel cycle uses of environmental resources that are not priced in the market. It has been suggested that dischargers of residuals be charged for their use of the waste assimilative capacity of the environment. [See Fisher and Petersen (10) for an historical review of the literature. For a discussion of the advantages to such proposals, see (4, 102, 112–114). For a critique, see (115).] Such charges, equal to the marginal damage for each unit of residuals, would provide an incentive for the discharger to economize on the use of the environment. Dischargers would decrease their flows of residuals as long as the marginal cost of doing so was less than the charge for discharging; equilibrium would occur at a point where marginal treatment costs were equal to the charge (presumably set at the marginal social damage). Such a system of residuals charges implies that the property rights in the environment have been vested in the users of the non-waste-disposal services of the environment, with the state acting as agent for the sale of rights to the waste assimilative capacity of the environment. Variations on this strategy include the auctioning of a fixed number of pollution rights (116) and bidding operations for property rights between dischargers and a centralized public authority (117). We have already alluded to the necessity for charges (in order to result in Pareto optimal levels of pollutions) to vary across geographical areas (118, 119) and across time (120). [See Lerner (121) for a discussion of the treatment of old vs new plants.]

It should be noted that some economists, notably Coase (122), do not believe that quality standards are appropriately defined by the government. With reliance on models of externalities of one producer affecting another producer (e.g. upstream factory polluting waters used by downstream factory), the possibility of achieving optimal levels of pollution by bargaining ("bribes" from one producer to the other, with no government interference) has been advanced. Numerous real-world objections to this solution, primarily the administrative complexity of bargaining when externalities are generated on a large number of diverse consumers, argue strongly against the bribes solution. Bribes, in the form of subsidies to undertake certain pollution control steps, are a commonly used mechanism for public management of the environment (102). An important distinction between bribes and the effluent or emission charge strategy is that activities not profitable under the latter could become profitable under the former (123). Thus, as shown by Baumol and Oates (9), the effect of a system of bribes might be to reduce emissions per firm, while increasing emissions at the industry level. (They note that there is no contention that this mechanism is reasonable for other than the case where there is a very small number of emitters or receptors.)

A final strategy for management includes regulation of specific processes, e.g. limits on sulfur content of fuel oil or a requirement for secondary treatment of water waste discharges. These are reviewed and criticized as "piecemeal," inefficient, and not likely to lead to efficient abatement or to the desired level of environmental quality (4, 102).

The issue of "standards" versus "charges" as a strategy for management of environmental resources merits further discussion. Weitzman (124) argues that it is neither easier nor harder to estimate the right prices (residuals or effluent charges) than the right quantities (standards on emissions or effluents) because in principle exactly the same information is needed to specify either correctly. Based on a model including uncertainty in the cost and benefit function, Weitzman (124) develops conditions under which price or quantity signals are preferred.

The question appears to be whether, in the presence of uncertainty, a 20% error in charges lends to a better equilibrium than a 20% error in standards. Standards that are "too strict" impose higher costs on older plants, possibly forcing them out of business "too soon." A strategy of charges, however, may well lead to "too clean" or "too dirty" an environment. Another issue is the difficulty of estimating the correct charge or standard. If the benefit function is linear, the charge would be constant, and there is likely to be neither uncertainty nor error. [Lave and Seskin (50) find the effect of air pollution on health is linear.] A highly nonlinear benefit function implies that the proper charge depends on the range of emissions; in addition, there is the possibility of a nonconvex demand for environmental dumping, in which case charges would not lead to the desired level of environmental quality, even in theory. See (9, 125) for a discussion of this issue.

Difficulty in specifying diffusion models makes the step from emissions to ambient quality an uncertain one; this is a problem for ascertaining either correct charges or standards. Systems of charges on polluters thus add a step to what is already an uncertain process (the resulting emissions can only be estimated in advance). The traditional economic paradigm is based on the assumption of prices fluctuating until equilibrium is reached, a system that could impose enormous costs on industrial and municipal dischargers. In many circumstances even a single adjustment would be prohibitively expensive and so adjustment is impossible. For a discussion and demonstration of the problems in implementation of a pricing mechanism, see Dorcey (126), who asserts that this system results in bargaining and negotiating behavior, the end of which is often cited as an advantage by advocates of pricing.

EMPIRICAL STUDIES OF ALTERNATIVE STRATEGIES   As discussed above, the cost of pollution control will be dependent on the administrative mechanisms established to achieve it. In particular, costs will be a function of the stringency of the goal, the speed with which the goal is to be met, and the efficiency of alternative control programs (102). There is little empirical work to guide the assessment of alternative strategies to achieve desired pollution goals. Several case studies are discussed by Kneese and Bower (46), one showing that to achieve any given level of water quality, a strategy based on effluent charges would cost 40–50% less than uniform treatment requirements (standards). Gutmanis reports similar efficiency in achieving reduction of biodegradable discharge among industries in the Great Lakes region (127). [The effect of municipal sewer charges as a method of pollution control, an issue not directly relevant to the topic at hand, is discussed in (46, 128, 129). Methods for the computation of the optimal taxation for abatement of water pollution are presented in (130–132).]

Crocker (133) presents an interesting case of producer-producer externalities in which the bribes or trade solution achieves an economically desired solution, while demonstrating the importance of transaction costs (including costs of information, contracting, and policing) and the effect of the assignment of property rights on those transaction costs.

Various proposals have been made for establishing an emission tax on sulfur discharges. In a thorough analysis of the potential efficiencies of such an emission tax, Chapman (134) concludes that a sulfur emission tax would cause significant reductions in emissions and damage, have little effect on electricity demand growth and nuclear power generation growth, and result in greater social benefits than costs. Kneese and Schultze (102) quote an EPA memo assessing the cost of reducing sulfur emissions by emission charges to be 35–40% less expensive than uniform regulation. On this topic, see also (135).

COMPARISON OF FUELS FOR GENERATING ELECTRICITY: AN EXAMPLE

The preceding discussion can be illustrated by comparing the health and environmental effects of fuels used to generate electricity. As emphasized by Lave and Freeburg (136) and by Sagan (137, 138), the comparison must be on the basis of the entire fuel cycle, including exploration, extraction, transport, processing, generation, and waste disposal or reprocessing. The focus has been on the health effects of each fuel (136–138), although some attention has been given to environmental effects (139). We now review the health and environmental effects of each stage of the fuel cycle for alternative sources of electricity. These are summarized in Table 4.

Exploration is unlikely to have substantial health or environmental effects because of its small scale. The extraction stage, however, involves major effects. The environmental effects of mining coal, either by deep mines or strip mines, have received much attention (47, 139); the former results in acid mine drainage and subsidence while the latter scars the land. Pumping oil can lead to subsidence and, for undersea wells, raises the risk of spills (such as that at Santa Barbara, California). Extraction of natural gas involves few environmental problems, save for the possibility of explosion. Uranium mining has the same environmental problems that characterize any mining; in addition, tailings are highly radioactive and present a disposal problem.

The health effects of extraction are important, especially for fuels requiring mining. In addition to hazards from accidents, coal miners are subject to pneumoconiosis (140) while uranium miners are threatened with silicosis and lung cancer due to radon gas. Occupational diseases of oil and natural gas workers appear to be less important. Lave and Freeburg (136) calculate accident and occupational hazards for the period 1965–1969 and state them in terms of disability days/10^6 MW-hr of electricity: 1545 disability days/10^6 MW-hr for coal mining; 157 for uranium mining and milling; and 135 for oil drilling, production, and refining. It is estimated (136) that coal miners have 18 times greater excess mortality from chronic disease than uranium miners, per kilowatt-hour of electricity. (These figures reflect the experience of a period when mining regulation was considerably less strict and are not necessarily representative of what will prevail in future periods.)

Transport of fuels involves risks of accidents. Whether transport takes place by ship (oil, liquified natural gas), pipeline (oil, gas), or rail or truck (coal, uranium), the number of accidents is likely to be proportional to the number of ton-miles of transport. Since coal involves many more ton-miles per kilowatt-hour than the other fuels, it has the greatest health risk. Sagan (137) estimates that a major number of fatalities and injuries can be ascribed to the transport of coal by rail. The transport of oil poses much larger environmental dangers than the transport of uranium, coal, or natural gas. Oil spills in the ocean, lakes, or rivers are particularly damaging to aquatic life, and the potential effects of the Trans-Alaska Pipeline have been discussed above.

The environmental effects of processing are likely to be large only for oil. With the exception of washing coal to get rid of sulfur, there is no large-scale processing at present. In the future, there may be large-scale processing of oil shale or of coal in the form of gasification or liquefaction. Indeed, this pregeneration processing of coal should allow cleaning of flue gas to be dropped. However, there are significant effluent problems with these coal-processing plants (140). Some recent research indicates that pregeneration treatment of coal can be made efficient by controlling the partitioning of the output between gas, solid, and liquid phases,

Table 4 Significant health and environmental effects of electricity generation

Stage of Fuel Cycle	Coal	Oil	Gas	Uranium
Exploration				
Extraction	mine drainage, subsidence, scarred land, pneumoconiosis of miners, accident hazards	spills, subsidence	risk of explosion	subsidence, scarred land, radioactive tailings, silicosis of miners
Transport	accident risk	spills	risk of explosion	accident risk
Processing	pneumoconiosis of workers	residuals in air (SO_2, NO_x, hydrocarbons), residuals in water (degradable and non-degradable)		silicosis of workers
Generation	sulfur, nitrogen	air pollution (SO_x, NO_x, hydrocarbons, trace elements), water pollution (heat, non-degradable residuals)		accident risk, waste heat, radioactive waste, radionuclides in air
Waste Disposal	sludge			reprocessing or disposal of fuel
Ranking[a]	4	3	1	2

[a] Based on health and environmental effects per kilowatt-hour.

with relatively easy control of the effluent (141). However, little can be said at this time about the health or environmental effects of these processes. Refining oil vents residuals into both the air (SO_2, NO_x, and hydrocarbons) and water (biodegradable and non-biodegradable); indeed, refineries are alleged by EPA to be a major source of air pollution.

The generation phase is the best studied. Environmental effects for all fuels (to varying degrees) consist of waste heat (vented into water or air) as well as chemical pollutants released into the air and radionuclides into the air and the water. Another environmental problem with coal concerns large storage piles; there is an acid runoff from rain and the possibility of spontaneous combustion with associated smoke, sulfur, and nitrogen.

Effluents from the combustion of coal, gas, and oil (particulates, SO_x, NO_x, some hydrocarbons and radionuclides and trace elements in waste gas and fly ash) and from light water reactors (small quantities of radionuclides released into the air and cooling water) have important public health implications. Using dose-response relationships estimated for chemical pollutants and radionuclides, Lave and Freeburg (136) conclude that for power plants typical of the late 1960s, a coal-burning plant has a health cost 18,000 times greater than that of a pressurized water reactor. A smaller factor was estimated for the comparison of coal and a boiling water reactor. Low-sulfur oil appears to have approximately the same health cost as the nuclear reactors. Possible generation accidents, either for fossil fuel or nuclear plants, were ignored in the analysis. However, the differences in the health costs of coal versus nuclear fuel are so great that it is extremely unlikely that accident risks would vitiate the conclusion.

Waste disposal or reprocessing after the generation phase is relevant for either the nuclear reactor or a fossil fuel plant where limestone is used to curtail sulfur emissions. In the latter case, the sludge presents an environmental problem. The reprocessing of spent nuclear fuel poses health problems, but it is difficult to make a detailed estimate of the extent of the problem, since no commercial plant is currently in operation.

The above illustrative analysis makes it apparent that in deciding on how to generate electricity, there are options for what type of effluent to produce, whether to vent a particular residual in the air or water, and whether to have an environmental or health problem. We have not attempted to estimate the dollar value of the health and environmental effects summarized in Table 4; as the earlier discussion of this paper indicated, the techniques of benefit assessment are still unrefined. Based on our assessment of this analysis, it appears evident that, per kilowatt-hour of electricity generated, nuclear fuel in a light water reactor has a much lower health and environmental cost than does coal. Since natural gas and oil have fewer health and environmental effects than coal, comparing them to a light water reactor does not give such clear-cut results. The results suggest that natural gas is preferred to uranium, which is preferred to oil. However, the limited supplies of oil and gas suggest that the relevant alternatives are coal and uranium.

In appraising the sources of energy for any use, it is essential to examine the entire fuel cycle, including the amount of equipment required. Thus solar or wind power would appear to be ideal, looking only at the generation phase. But if an inordinate amount of expensive equipment is needed, the environmental and health cost of equipment production may make other fuels preferable (142).

SUMMARY AND CONCLUSION

We began with the neoclassical economic paradigm for the allocation of goods and services in a competitive economy. Externalities, the economic interaction of producers and consumers outside the market, are a major qualification to the efficiency of a competitive market. The theory was extended to cover intertemporal trade-offs, including the theory of exhaustible natural resources and the long-term implications of saving. The application of input-output analysis and the materials balance model to environmental pollution was described. The importance of externalities was next assessed, and programs designed to alleviate them were evaluated. Benefit-cost analysis was described and some of its problems discussed.

The effect of pollutants in the air, water, and land was described along with the basic insight from the materials balance approach that we cannot destroy an element such as sulfur, but we can choose to émit the effluent in one medium rather than another or change its chemical form. The principal methods for assessing the benefits of abatement of air, water, and land pollution were reviewed.

Environmental management was also discussed. The most common approach is the use of standards (e.g. ambient air or emission standards), but effluent fees present an alternative advocated by many economists. A host of difficulties plague environmental management, including the possibility of irreversible damage and complicated trade-offs between environmental quality for air, water, and land, among various locations and over time. The setting of standards or effluent fees or even the determination of detailed goals is an exceedingly complicated process. Last, examining the entire fuel cycle was stressed in the context of a comparison of the pollution problems of fuels used to generate electricity.

As this review makes evident, some areas, such as the effect of air pollution on health, are important and have received a great deal of attention. Others, such as the effect of water pollution on health, are of less importance and have received less attention. There has been reluctance to deal with many of the problems of environmental degradation simply because they are difficult; however, the importance of these issues dictates that better approaches must be found.

We have suggested throughout this paper areas of relative uncertainty concerning the environmental costs of energy-related pollution. Some of the major areas in which further research is desirable are as follows:

1. The design and evaluation of pollution control programs will inevitably proceed in the presence of uncertainty. The appropriate criteria for public decision-making in the presence of individual and collective attitudes toward risk need to be better understood [see Slovic et al (143)].
2. Greater emphasis on strategies for management of non-point sources of pollution, particularly with regard to water, is necessary.

3. The rapidly growing theoretical literature on exhaustible resources emphasizes the need for empirical investigations of resources that are presumably exhaustible (e.g. iron ore).
4. The relative degree of "myopia" of market and central planning systems in accounting for the concerns of future generations is an empirical question for which there is still very little evidence.
5. The materials balance framework provides an analytic tool for assessing the implications of alternative regulatory strategies and should be extended in this manner.
6. While the importance of nonuser benefits of environmental resources has been noted, we still have precious little evidence or methods with which to quantify these benefits.
7. Better evidence is needed on the damage function, for both air and water pollution, as well as valuation techniques for assessing the benefits of abatement.
8. Better use should be made of "natural" experiments resulting from recent dislocations in energy-related markets, as well as the aftermath of recent environmental legislation.
9. Effective management of environmental resources will depend on a better understanding of the incidence of the benefits and costs of environmental protection.
10. More empirical evidence is necessary on the efficiency and equity effects of alternative regulatory strategies.
11. Fuel cycle analyses, such as that sketched briefly for electricity generation in the preceding section, would be extremely useful if carried out for other energy uses.

ACKNOWLEDGMENTS

We thank Stanley W. Angrist, Robert W. Dunlap, Raphael Kasper, and Francis C. McMichael for comments and Wendy H. Siniard and Sarah M. Streuli for typing numerous drafts. The work of Lester B. Lave in preparation of this review was supported by grant 5R01HS01529 from the National Center for Health Services Research. The authors are responsible for opinions and errors.

Literature Cited (by major section)

Markets

1. Henderson, J. M., Quandt, R. E. 1958. *Microeconomic Theory: A Mathematical Approach.* New York: McGraw-Hill
2. Arrow, K. J., Hahn, F. H. 1971. *General Competitive Analysis.* San Francisco: Holden Day
3. Bator, F. 1958. The anatomy of market failure. *Q. J. Econ.* 72:351–79
4. Freeman, A. M., Haveman, R. H., Kneese, A. V. 1973. *The Economics of Environmental Policy.* New York: Wiley
5. Ayres, R. U., Kneese, A. V. 1969. Production, consumption, and externalities. *Am. Econ. Rev.* 59:282–97
6. Hite, J. C., Macaulay, H. H., Stepp, J. M., Yandle, B. Jr. 1972. *The Economics of Environmental Quality.* Washington DC: American Enterprise Institute
7. Enthoven, A. C., Freeman, A. M. 1973. *Pollution, Resources, and the Environment.* New York: Norton
8. Seneca, J. J., Taussig, M. K. 1974. *Environmental Economics.* Englewood Cliffs, NJ: Prentice-Hall
9. Baumol, W. J., Oates, W. E. 1975. *The Theory of Environmental Policy: Externalities, Public Outlays, and the Quality of Life.* Englewood Cliffs, NJ: Prentice-Hall
10. Fisher, A. C., Petersen, F. M. 1975.

Natural resources and the environment in economics. Working paper. Univ. Maryland
11. Mishan, E. J. 1971. The postwar literature on externalities, an interpretive essay. *J. Econ. Lit.* 9:(1):1–28
12. Meade, J. E. 1973. *The Theory of Economic Externalities*. Geneva: Institut Universitaire de Hautes Études
13. Arrow, K. J. 1974. Limited knowledge and economic analysis. *Am. Econ. Rev.* 64(1):1–10
14. Koopmans, T. C. 1957. *Three Essays on the State of Economic Science*. New York: McGraw-Hill
15. Solow, R. M. 1974. The economics of resources or the resources of economics. *Am. Econ. Rev.* 64(2):1–14
16. Hahn, F. H., Matthews, R. C. O. 1964. The theory of economic growth: a survey. *Econ. J.* 74:779–902
17. Uzawa, H. 1974. The optimum management of social overhead capital. In *The Management of Water Quality and the Environment*, ed. J. Rothenberg, I. Heggie. New York: Macmillan
18. D'Arge, R. C. 1972. Economic growth and the natural environment, pp. 11–34. In *Environmental Quality Analysis: Theory & Method in the Social Sciences*, ed. A. V. Kneese, B. T. Bower. Baltimore: Johns Hopkins Univ. Press
19. D'Arge, R. C., Kogiku, K. C. 1973. Economic growth and the environment. *Rev. Econ. Stud.* 40(1):61–77
20. Keeler, E., Spence, M., Zeckhauser, R. 1972. The optimal control of pollution. *J. Econ. Theory* 4(1):19–34
21. Arrow, K. J., Fisher, A. C. 1974. Environmental preservation, uncertainty, and irreversibility. *Q. J. Econ.* 88(2):312–19
22. Heal, G. M., ed. 1974. Symposium on the Economics of Exhaustible Resources. *Rev. Econ. Stud.*, pp. 1–152
23. Meadows, D. L., Randers, J., Behrens, W. W. 1972. *The Limits to Growth*. New York: Universe
24. Passell, P., Roberts, M. J., Ross, L. 1973. The limits to growth: a review. See Ref. 7, pp. 230–34
25. Nordhaus, W. D. 1973. World dynamics—measurement without data. *Econ. J.* 82:1156–83
26. Leontief, W. 1970. Environmental repercussions and the economic structure. *Rev. Econ. Stat.* 52(3):262–71
27. Kneese, A. V., Ayres, R. U., D'Arge, R. C. 1970. *Economics and the Environment: A Materials Balance Approach*. Baltimore: Johns Hopkins Univ. Press

28. Noll, R. G., Trijonis, J. 1971. Mass balance, general equilibrium, and environmental externalities. *Am. Econ. Rev.* 61(4):730–35

Benefit-Cost (10)*

29. Dasgupta, A., Pearce, D. 1972. *Cost-Benefit Analysis*. New York: Macmillan
30. Prest, A., Turvey, R. 1965. Cost-benefit analysis: a survey. *Econ. J.* 75:683–735
31. Haveman, R., Harberger, A., Lynn, L., Niskanen, W., Turvey, R., Zeckhauser, R., Wisecarver, D. 1973. *Benefit-Cost and Policy Analysis*. Chicago: Aldine
32. Wildavsky, A. 1967. The political economy of efficiency. *Public Interest* 8:30–48
33. Krutilla, J. V. 1967. Conservation reconsidered. *Am. Econ. Rev.* 57(4):777–86
34. Bishop, J., Cicchetti, C. 1975. Some institutional and conceptual thoughts on the measurement of indirect and intangible benefits and costs. See Ref. 45, pp. 105–26
35. Weisbrod, B. A. 1964. Collective consumption services of individual consumption goods. *Q. J. Econ.* 77:71–77
36. Kreig, M. B. 1964. *Green Medicine: The Search for Plants that Heal*. Chicago: Rand McNally
37. Schelling, T. 1968. The life you save may be your own. In *Problems in Public Expenditure Analysis*, ed. S. Chase. Baltimore: Johns Hopkins Univ. Press
38. Acton, J. P. 1973. *Evaluating Public Programs to Save Lives: the Cost of Heart Attacks*. Rep. No. R-950-RC. Santa Monica, Calif.: Rand Corp.
39. Acton, J. P. 1973. *Review and Critique of Evaluation Methodology, Mortality, and Morbidity Reduction*. Rep. No. WN-8429-1-NHLI. Santa Monica, Calif.: Rand Corp.
40. Rice, D. 1966. *Estimating the Cost of Illness*. PHS Publ. No. 947-6. Washington DC: US GPO
41. Lave, L. B. 1972. Air pollution damage: some difficulties in estimating the value of abatement. In *Environmental Quality Analysis*. See Ref. 18, pp. 213–42

Social Cost (4, 5, 10, 27, 36–38, 41)

42. Stern, A., ed. 1968. *Air Pollution*. New York: Academic Press
43. National Academy of Sciences. 1974. *Air Quality and Automobile Emission Control, Vol. 1, Summary Report*. Washington DC: US GPO
44. National Academy of Sciences. 1975.

* References from previous sections.

Air Quality and Stationary Source Emission Control. Washington DC: US GPO
45. Peskin, H. M., Seskin, E. P., eds. 1975. *Cost-Benefit Analysis and Water Pollution Policy.* Washington DC: The Urban Institute
46. Kneese, A. V., Bower, B. T. 1968. *Managing Water Quality: Economics, Technology, Institutions.* Baltimore: Johns Hopkins Univ. Press
47. Kneese, A. V., Smith, S. C., eds. 1966. *Water Research.* Baltimore: Johns Hopkins Univ. Press
48. National Academy of Sciences. 1975. *Mineral Resources and the Environment.* Washington DC: Nat. Acad. Sci. Printing & Publishing Office
49. US Environmental Protection Agency. 1972. *The Economics of Clean Air.* Ann. Rep. of Admin. of EPA to US Congress. Washington DC: US GPO
50. Lave, L. B., Seskin, E. P. 1975. *Air Pollution and Human Health.* Baltimore: Johns Hopkins Univ. Press
51. Anderson, D. 1967. The effects of air contamination on health. *Can. Med. Assoc. J.* 97:528–36, 585–93, 802–806 (in three parts)
52. Lave, L., Seskin, E. 1970. Air pollution and human health. *Science* 169:723–32
53. Waddell, T. E. 1974. *The Economic Dangers of Air Pollution.* Washington DC: EPA
54. Ridker, R. 1967. *Economic Costs of Air Pollution.* New York: Praeger
55. Haynie, F. H. 1974. The economics of clean air in perspective. *Mater. Perform.* 13(4):33–38
56. Yocom, J. E., McCaldin, R. O. 1968. Effects of air pollution on materials and the economy. See Ref. 42, pp. 617–51
57. Michelson, I., Tourin, B. 1966. Comparative methods for studying the costs of controlling air pollution. *Public Health Rep.* 81:505–11
58. Anderson, R., Jr., Crocker, T. 1969. *Air Pollution and Residential Property Values.* Presented at the Econometric Soc. Meet., New York
59. Harris, R., Tolley, G., Harrell, C. 1968. The residence site choice. *Rev. Econ. Stat.* 50:241–47
60. Nourse, H. 1967. The effect of air pollution on house values. *Land Econ.* 43:181
61. Ridker, R., Henning, J. 1967. The determinants of residential property values with special reference to air pollution. *Rev. Econ. Stat.* 49:246
62. Freeman, A. M. 1971. Air pollution and property values: a methodological comment. *Rev. Econ. Stat.* 53(4):415–16
63. Freeman, A. M. 1974. On estimating air pollution control benefits from land value studies. *J. Environ. Econ. Manage.* 1:74–83
64. Small, K. A. 1975. Air pollution and property values: further comment. *Rev. Econ. Stat.* 57(1):105–107
65. Polinsky, A. M., Shavell, S. 1975. The air pollution and property value debate. *Rev. Econ. Stat.* 57(1):100–104
66. Benedict, H. M., Miller, C. J., Smith, J. S. 1973. *Assessment of Economic Impact of Air Pollutants on Vegetation in the United States: 1969–1971.* Menlo Park, Calif.: SRI
67. National Air Pollution Control Association. 1970. *Air Quality Criteria for Sulfur Oxides.* Washington DC: US GPO
68. Crocker, T., Rogers, A. 1971. *Environmental Economics.* Hinsdale, Illinois: Dryden
69. O'Gara, P. 1922. Sulfur dioxide and fume problems. *Ind. Eng. Chem.,* Vol. 14
70. Guderian, R., Van Haut, H., Strettman, H. 1960. The estimation and evaluation of the effects of atmospheric gas pollutants on vegetations. *Z. Pflanzenkr. Pflanzenschutz* 67(5):257–64
71. Heck, W. W., Brandt, C. S. 1974. Impact of air pollutants on vegetation: crops, forests, native. In *Air Pollution,* Vol. 1, eds. A. C. Stern, B. J. Steigerwald. New York: Academic Press
72. Jacobson, J. S., Hill, A. C., eds. 1970. *Recognition of Air Pollution Injury to Vegetation.* Pittsburgh, Penn.: Air Pollution Control Assoc.
73. Stork, K. E. 1973. The role of water in the energy crisis. *Proc. Conf. Nebraska Water Resourc. Res. Inst.,* Lincoln, Nebr.
74. Wollman, N., Bonem, G. W. 1971. *The Outlook for Water: Quality, Quantity, and National Growth.* Baltimore: Johns Hopkins Univ. Press
75. Freeman, S. D. 1974. *Energy: The New Era.* New York: Vintage
76. National Water Commission. 1973. *Water Policies for the Future.* Washington DC: Nat. Water Comm.
77. Blumer, M., Sanders, H. L., Grassle, J. F., Hampson, G. R. 1971. A small oil spill. *Environment* 13(2):2–12
78. National Academy of Sciences. 1969. *Acid Mine Drainage in Appalachia.* Washington DC: Nat. Acad. Sci.
79. Kneese, A. V., Bower, B. T. 1968. *Managing Water Quality: Economics, Technology, Institutions.* Baltimore: Johns Hopkins Univ. Press
79a. US Environmental Protection Agency.

1972. *The Economics of Clean Water,* Vol. 1. Washington DC: EPA
80. Russell, C. S. 1973. *Residuals Management in Industry: A Case Study of Petroleum Refining.* Baltimore: Johns Hopkins Univ. Press
81. Tihansky, D. 1975. A survey of empirical benefit studies. See Ref. 45, pp. 127–72
82. Freeman, A. M. 1975. A survey of the techniques for measuring the benefits of water quality improvement. See Ref. 45, pp. 67–104
83. Kneese, A. V. 1968. Economics and the quality of the environment: some empirical experiences. In *Social Sciences and the Environment,* ed. M. E. Garnsey, J. R. Hibbs. Denver: Univ. Colorado
84. Marx, J. L. 1974. Drinking water: another source of carcinogens? *Science* 186:809–11
85. Schroeder, H., Kraemer, L. A. 1974. Cardiovascular mortality, municipal water and corrosion. *Arch. Environ. Health* 28:303–11
86. Tihansky, D. P. 1973. *Economic Damages to Household Items from Water Supply Use.* Washington DC: EPA
87. Beyer, J. 1969. *Water Quality and Value of Homesites on the Rockaway River.* Water Resourc. Res. Inst., Rutgers Univ., New Brunswick, NJ
88. David, E. L. 1968. Lake shore property values: a guide to public investment in recreation. *Water Resourc. Res.* 4:697–707
89. Krutilla, J. V. 1971. *Evaluation of an Aspect of Environmental Quality: Hells Canyon Revisited.* Resources for the Future, Inc., Washington DC
90. Clawson, M., Knetsch, J. L. 1966. *Economics of Outdoor Recreation.* Baltimore: Johns Hopkins Univ. Press
91. Knetsch, J. L. 1974. *Outdoor Recreation and Water Resources Planning.* Washington DC: American Geophysical Union
92. Knetsch, J., Davis, R. 1966. Comparisons of methods for recreation evaluation. In *Water Research,* eds. A. V. Kneese, S. Smith. Baltimore: Johns Hopkins Univ. Press
93. National Academy of Sciences. 1975. *Assessing Demand for Outdoor Recreation.* Washington DC: Nat. Acad. Sci.
94. Smith, V. K. 1975. *The Estimation and Use of Models of the Demand for Outdoor Recreation.* Prepared for the Committee on Assessment of Demand for Outdoor Recreation Resources, Nat. Acad. Sci., Washington DC
95. Davidson, P., Adams, F. G., Seneca, J. 1966. The social value of water recreational facilities resulting from an improvement in water quality: the Delaware estuary. In *Water Research,* ed. A. V. Kneese, S. C. Smith. Baltimore: Johns Hopkins Univ. Press
96. Ford Foundation, Energy Policy Project. 1974. *A Time to Choose: America's Energy Future.* Cambridge, Mass.: Ballinger
97. Rose, D. J. 1974. Energy policy in the U.S. *Sci. Am.* 230(1):20–29
98. Council on Environmental Quality. 1973. *Report to the Senate Committee on Interior and Insular Affairs.* Washington DC: US GPO
99. National Academy of Sciences/ National Academy of Engineering. 1974. *Rehabilitation Potential of Western Coal Lands.* Report to the Ford Foundation, Energy Policy Project. Cambridge, Mass.: Ballinger
100. Garvey, G. 1972. *Energy, Ecology, Economy.* New York: Norton
101. Cicchetti, C. J. 1972. *Alaskan Oil: Alternative Routes and Markets.* Baltimore: Johns Hopkins Univ. Press

Public Policy (4, 9, 10, 32, 42, 46, 49, 50)

102. Kneese, A. V., Schultze, C. L. 1975. *Pollution, Prices, and Public Policy.* Washington DC: Brookings Inst.
103. Turvey, R. 1966. Side effects of resource use. In *Environmental Quality: In a Growing Economy,* ed. H. Jarrett. Baltimore: Johns Hopkins Univ. Press
104. Freeman, A. M. 1972. The distribution of environmental quality. In *Environmental Quality Analysis.* See Ref. 18, pp. 243–78
105. Maass, A. 1970. Public investment planning in the United States: analysis and critique. *Public Policy* 18(2):211–43
106. Wildavsky, A. 1964. *The Politics of the Budgetary Process.* Boston: Little, Brown
107. Haveman, R. H. 1973. Efficiency and equity in natural resource and environmental policy. *Am. J. Agric. Econ.* 55(5):868–78
108. Roberts, M. *Who Will Pay for Cleaner Power.* Mimeograph. See Ref. 9, p. 209
109. Davis, R. K. 1968. *The Range of Choice in Water Management.* Baltimore: Johns Hopkins Univ. Press
110. Davies, J. C. 1970. *The Politics of Pollution.* New York: Pegasus
111. Thomas, H. 1963. The animal farm: a mathematical model for the discussion of social standards for control of the environment. *Q. J. Econ.* 77(1):143–48
112. Freeman, A. M., Haveman, R. H. 1972.

113. Freeman, A. M., Haveman, R. H. 1972. Residuals charges for pollution control: a policy evaluation. *Science* 177: 322–29
114. Ruff, L. E. 1970. The economic common sense of pollution. *Public Interest* 19: 69–85
115. Rose-Ackerman, S. 1973. Effluent charges: a critique. *Can. J. Econ.* 6(4): 512–28
116. Dales, J. H. 1968. *Pollution, Property and Prices.* Toronto: Univ. Toronto
117. Ferrar, T. A., Whinston, A. 1972. Taxation and water pollution control. *Nat. Resourc. J.* 12(3): 307–17
118. Tietenberg, T. H. 1973. Specific taxes and pollution control. *Q. J. Econ.* 87(4): 503–22
119. Tietenberg, T. H. 1974. On taxation and the control of externalities: comment. *Am. Econ. Rev.* 64(3): 462–66
120. Baumol, W. J., Oates, W. E. 1975 (forthcoming). The instruments for environmental policy. In *Economic Analysis of Environmental Problems,* ed. E. S. Mills.
121. Lerner, A. P. 1972. Pollution abatement subsidies. *Am. Econ. Rev.* 62(5): 1009–10
122. Coase, R. H. 1960. The problem of social cost. *J. Law Econ.* 3: 1–44
123. Bramhall, D. E., Mills, E. S. 1966. A note on the asymmetry between fees and payments. *Water Resourc. Res.* 2(3): 615–16
124. Weitzman, M. L. 1976. Prices vs. quantities. *Rev. Econ. Stud.* In press
125. Starrett, D., Zeckhauser, R. 1974. Treating external diseconomies—markets or taxes? In *Statistical and Mathematical Aspects of Pollution Problems,* ed. J. W. Pratt. New York: Marcel Dekker
126. Dorcey, A. H. J. 1973. Effluent charges, information generation and bargaining behavior. *Nat. Resourc. J.* 13(1): 118–33
127. Gutmanis, I. 1972. *The Generation and Cost of Controlling Air, Water, and Solid Waste Pollution: 1970–2000.* Washington DC: Brookings Inst.
128. Elliott, R. D. 1973. Economic study of the effect of municipal sewer surcharges on industrial wastes and water usage. *Water Resourc. Res.* 9(5): 1121–31
129. Ethridge, D. 1972. User charges as a means for pollution control: the case of sewer surcharges. *Bell J. Econ. Manage. Sci.* 3(1): 346–54
130. Haimes, Y. Y., Kaplan, M. A., Huser, M. A. 1972. A multilevel approach to determining optimal taxation for the abatement of water pollution. *Water Resourc. Res.* 8(4): 851–60
131. Hass, J. E. 1970. Optimal taxing for the abatement of water pollution. *Water Resourc. Res.* 6(2): 353–65
132. Upton, C. 1971. Application of user charges to water quality management. *Water Resourc. Res.* 7(2): 264–72
133. Crocker, T. D. 1971. Externalities, property rights, and transactions costs: an empirical study. *J. Law Econ.* 14(2): 451–64
134. Chapman, D. 1974. A sulfur emission tax and the electric utility industry. *Energy Syst. Policy* 1(1): 1–30
135. Griffen, J. M. 1974. An econometric evaluation of sulfur taxes. *J. Polit. Econ.* 82(4): 669–88

Comparison (47)

136. Lave, L. B., Freeburg, L. C. 1973. Health effects of electricity generation from coal, oil, and nuclear fuel. *Nucl. Saf.* 14(5): 409–28
137. Sagan, L. A. 1974. Health costs associated with the mining, transport, and combustion of coal in the steam-electric industry. *Nature* 250: 107–11
138. Sagan, L. A. 1972. Human costs of nuclear power. *Science* 177: 487–93
139. Dials, G. E., Moore, E. C. 1974. The cost of coal. *Environment* 16(7): 14–24
140. Rubin, E. S., McMichael, F. C. 1975. Impact of regulations on coal conversion plants. *Environ. Sci. Technol.* 9(2): 112–17
141. Nakles, D. V., Massy, M. I., Forney, A. V., Haynes, W. P. 1975. *Influence of Synthane Gasification Conditions on Effluent and Product Gas Production.* Pittsburgh: Energy Research Center, ERDA
142. Sorensen, B. 1975. Energy and resources. *Science* 189: 255–60

Summary

143. Slovic, P., Fischhott, B., Lichtenstein, S. 1976. Cognitive processes and societal risk taking. In *Cognition and Social Behavior,* ed. V. S. Carroll, V. W. Payne. Potomac, Md: Lawrence Erlbaum Assoc.

Copyright 1976. All rights reserved

PHILOSOPHICAL BASIS FOR RISK ANALYSIS

×11023

Chauncey Starr, Richard Rudman, and Chris Whipple
Electric Power Research Institute, Palo Alto, California 94303

INTRODUCTION

All societies and individuals have recognized exposure to personal risk as a normal aspect of our life. Presumably such risk exposures have been accepted as necessary to attain a compensating benefit. This relationship has been explored previously, and several hypotheses have been suggested to explain the historically accepted balances of risk and benefit (1, 2).

When individuals have "voluntarily" taken risks for personal pleasure or profit, they appear to be willing to accept relatively high risk levels in return for rather modest quantifiable benefits. For example, sportsmen frequently explore the outer physical limits of their chosen activity—with many resultant statistical accident records of their risk-taking propensities. Assuming that intangible benefits do exist, the controlling parameter appears to be the individual's perception of his own ability to manage the risk-creating situation. If he believes he can do so, he is likely to take the chance.

The situation changes markedly when the individual no longer believes he can control his risk exposure. In such "involuntary" situations, the risk management is in the hands of a societal group usually remote operationally from the individual and his risk exposure. Major societal technical systems all create such involuntary risk exposures—for example, transportation systems, energy supply systems, public utilities, and food supply systems. Whereas in the voluntary case the feedback loop of "control of risk–benefit balance" is very tightly coupled by the individual, in the involuntary case this loop is usually very weakly coupled, dimly perceived, and with each of its elements usually dispersed in time, geographically, politically, and managerially. Under these circumstances, the individual exposed to an involuntary risk is fearful of the consequences, makes risk aversion his goal, and therefore demands a level for such involuntary risk exposure as much as one thousand times less than would be acceptable on a voluntary basis (1, 2).

Inherent in all operational major technical systems is an implicit choice of such an acceptable level of risk for the involuntary exposure of the public. The study reported in (1, 2) has suggested the hypothesized relation shown in Figure 1 between risk and benefit as a typical basis for such decisions. The figure shows the approximate relation between the per capita benefits to real income of a system and the acceptable risk as expressed in deaths per unit time (i.e. time of exposure in units of a year). The highest level of acceptable risks that may be regarded as a reference level is determined by the normal US death rate from disease (about 1 death/year per 100 people). The lowest level for reference is set by the risk of death from natural events—lightning, flood, earthquakes, insect and snake bites, etc (about 1 death/year per 1,000,000 people).

Between these two bounds, the public is apparently willing to accept involuntary exposures—i.e. risks imposed by societal systems and not easily modified by the individual—in relation to the benefits derived from the operations of such societal systems.

In common familiar terms, an involuntary risk may be considered *excessive* if it exceeds the incidence rate of disease, *high* if it approaches the rate, *moderate* if the risk is about 10–100 times less, *low* if it approaches the level of natural hazards, and *negligible* if it is below this. Events in these last two levels of risk have historically been treated as acts of God by the public generally—in recognition of their relatively minor impact on our societal welfare compared with the effort required to avoid the risk.

Thus any risk created by a new sociotechnical system is acceptably "safe enough" if the resultant risk level is below the curve of the figure. If, as is usually the case, a new technical system has a range of uncertainty in its risks, a design target may

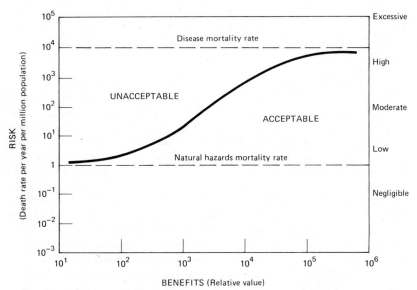

Figure 1 Benefit-risk pattern of involuntary exposure.

be set below the curve by an equivalent amount—possibly as much as a factor of 10 or 100.

Although the relationship hypothesized in Figure 1 appears reasonable, its quantitative aspects should be considered as very tentative and primarily useful for illustrative purposes. The evaluation of the comparative benefits derived from the availability of technical systems is complex, difficult, and presently more heuristic than analytic. Because most new technical systems initially appear to be replacements for existing systems (for example, nuclear for coal power), public policy is generally concerned with comparative risk levels rather than comparative benefits. Perhaps it is not usually foreseen that new technology is likely to affect societal systems profoundly—the automobile became more than a substitute for the horse, although it was originally perceived only as a horseless carriage.

Thus, because comparative risk analysis is the issue of primary national concern, this chapter focuses its discussion on the risk portion of the risk-benefit balance.

EVALUATION OF RISK

The study of risk analysis has as its objective the development of a methodology for predictive evaluation of future risks. Unfortunately, the literature on "futures" is apt to be obfuscated by a mix of personal value-system assessments of present trends and imaginative scenarios of alternative futures, usually written to dramatize the author's bias. For this reason it is useful to recognize the existence of four different evaluations of future risk, as follows:

1. *real risk,* as will be determined eventually by future circumstances when they develop fully;
2. *statistical risk,* as determined by currently available data, typically as measured actuarially for insurance premium purposes;
3. *predicted risk,* as predicted analytically from system models structured from historical studies; and
4. *perceived risk,* as seen intuitively by individuals.

Air transportation illustrates the above types of risks. To the flight insurance company, flying constitutes a statistically known risk; to the passenger purchasing the insurance at the airport, a perceived risk. In this case the perceived risk usually exceeds the statistical risk.[1] To the Federal Aviation Administration, anticipated changes in air traffic patterns and equipment involves predicted risk as an approximation to future real risk.

Perception

In some cases, the risk that an assessor predicts by careful analysis of experimental and historical data, the use of scientific laws, transfer of experience from related

---

[1] The rate for life insurance at airports is $0.25 per $7500 insurance per flight. By 1971 data, this rate is approximately 30 times the actuarially fair value. Regular life insurance is usually available at somewhat less than twice the actuarially fair rate.

data, or a combination of these methods does not correlate well with either an individual or a societal perception of risk. Perhaps the single most important factor in risk perception is risk manageability or controllability, where an individual feels safer if given some control over the degree of risk from an activity. Examples of this may be found in transportation where individuals perceive more risk from flying than driving, or in skiing, when an individual expresses more fear riding a chair lift up a mountain than skiing down. In both cases, the perception is not borne out by accident statistics. This is possibly the significant issue that results in voluntary activities having acceptable levels of risk several orders of magnitude higher than the levels associated with involuntary activities (1). It is usually difficult to differentiate between a faulty perception of risk and its distortion by a personal value system that results in a preference for a statistically higher risk, even when the risks are known, as in sports.

An important factor in the perception of risk is the probable severity of injury if an accident were to occur. A complete assessment of risk requires that the potential effects of an accident be combined with a probability of occurrence. Because the accident probabilities are usually not given significant weight in an individual perception, concentration on accident consequences can lead to distortions of perception. Again the perceived safety of commercial air transportation suffers relative to the automobile (although the real risk is less) because the results of an average aviation accident are far more severe than the effects of the average automobile accident. Similarly, motorcycling is perceived as far more dangerous than automobile travel because of this perceptual factor. In this case, the perception is accurate, as accident statistics reveal that motorcycle riders are exposed to a greater risk than automobile passengers.

Another very significant factor is the episodic nature of some risks. On an individual level, a risk is usually perceived as a probability of death, with less severe effects such as disability, either temporary or permanent, considered insignificant in comparison to the risk of fatalities. On a societal level, the episodic preoccupation is evident in the common concern with disasters claiming many lives, while in fact single-fatality accidents are responsible for the majority of all accidental deaths. It thus seems that the size of the potential accident is given more weight than the probability. This factor dominates the popular perception of risk from nuclear power plants, and can be recognized when the criticisms of nuclear power are examined, as the potential results of a major accident are frequently cited with no mention of the associated probabilities. This is probably representative of societal values to a great degree, and activities capable of producing catastrophic accidents therefore must meet more stringent societal standards than higher-frequency, individual risks. The failure to consider probability is, however, a perceptual factor.

One time-dependent effect that occurs in risk perception is the discounting of future risk. The risks of smoking would seem far less acceptable if the risk were experienced immediately, rather than developed over a number of years. This perception does not seem to apply when nuclear power is discussed, probably because of the dramatic scenarios that have been developed for hypothetical nuclear accidents. Because the principal health effect of low-level radiation exposure is increased likelihood of cancer in the future, occurring perhaps 15 to 45 years after exposure, this risk should also be discounted. In most adversarial discussions of nuclear power, these are treated as immediate fatalities.

Perception of Benefit

Just as risk may be perceived to be different from the assessed value, so too may societal perceptions of benefit vary. In a previous paper (1), an attempt was made to relate risk data to a crude measure of public awareness of the associated social benefits (see Figure 2). The "benefit awareness" was arbitrarily defined as the product of the relative level of advertising, the square of the percentage of population involved in the activity, and the relative usefulness (or importance) of the activity to the individual. These assumptions may be too crude, but Figure 2 does support the reasonable position that advertising the benefits of an activity increases the public acceptance of a greater level of risk. Smoking has been cited as an example of a perceived benefit due in large part to advertising (3). The decision to ban cigarette advertising on television indicates a tacit understanding of this phenomena.

Perception versus Reality

When a gap appears between the perceived and assessed risks and benefits, the societal decision-makers are called upon to resolve the conflict. A recent example of such a resolution was the requirement that automobiles be equipped with a seat belt interlock system. In this case, the individual perception underestimated the risk, as the majority of people were not using seat belts. The regulatory decision was made that the reduction in risk was substantial enough to justify the

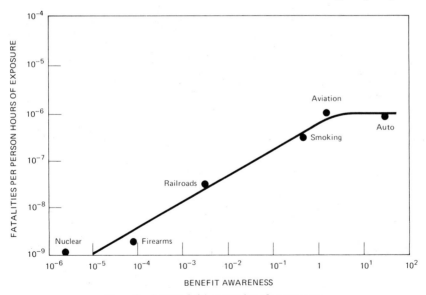

Figure 2 Accepted risk versus benefit awareness.

cost of the safety system. This regulation was subsequently repealed because the total social cost of the system was underestimated, that is, the cost of inconvenience was apparently significantly large.

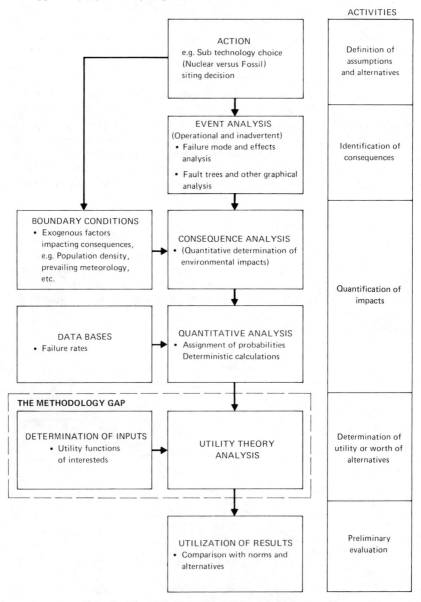

Figure 3   Illustrative flow chart: risk analysis. Source: (4).

No general method of dealing with differences between perception and technical assessment exists in the societal decision-making process. This problem is not directly within the realm of the technologist, as the technologist is not generally the decision-maker. The role of the technologist is to provide an assessment of risk and benefits. The societal perceptions and values are made visible through normal political inputs. The technologist should, however, indicate to the decision-maker that reliance on societal perceptions will usually result in a misallocation of resources. The difficult task of separating faulty perceptions from societal values has not been resolved.

Situations frequently arise where the individual member of society simply does not have the information to relate perception to reality. As a case in point, we may consider the perception of the solar system. To those who are not astronomers, the visible evidence is that the sun travels around the earth. The only reason that the population as a whole believes otherwise is that the word of the scientists has become part of the common wisdom. What seems significant about this situation is the duration of the perceptual lag, that is, the time between scientific discovery and cultural acceptance.

Three levels of perceptual problems can be identified. First, there are the real-time problems of perception. These include the factors previously identified as controllability, episodic preoccupation, and other similar effects. The second problem is perceptual lag. In general, perception eventually catches up with reality, but the time required may result in a massive misallocation of resources in the interim. The third level of perceptual problem is assessment of hypothetical future risks. If decisions were based entirely upon public perception, many technologies would not exist now because an exaggerated fear of new risks would have prohibited their introduction.

Societal Value System for Risk

The incorporation of risk estimates into national decision-making is in its early stages of methodological development. The use of a societal value system for risk acceptance must rank as the major unresolved issue in this part of the decision process. While methods have been proposed (4), none has yet been put into practice. Figure 3, taken from (4), is representative of a proposed method of dealing with this issue.

In the past, and to a lesser extent at present, the scientific community has tended to view risk-taking decisions based on factors other than the expected value of the risk as irrational. Because society responds to more factors than just expected value, a conflict has arisen. An outgrowth of this conflict has been that groups concerned or threatened by risk have rejected the estimates developed by the risk analysts. In seeking an input to the decision-making process, these groups have challenged the estimated risk, because on intuitive grounds the risk seemed less acceptable than the expected value would indicate. The need to consider these public perceptions is clear.

Although the source of conflict may be recognized, the resolution of the societal decision-making process is by no means simple. The societal value system fluctuates with time, and the technological capability to follow fast-changing societal goals does not exist. As an example, the emission requirements for automobiles were established at a time when air quality was considered far more important than fuel economy or cost. The societal goals in this area have seemingly changed to some degree, as these two latter considerations are now given more weight.

This fluctuation of social values makes the decision process more difficult and serves as perhaps the strongest argument for analytic decisions. Resolving the gap between societal perception and analytically expected value is of prime importance and requires a balanced consideration of both concepts. The aim of such a process is to maintain consistency within the framework of a value system.

The particular method referred to in Figure 3 is called *utility theory*; briefly stated, its estimates measure the social values associated with various objectives on a common scale. In theory, the diverse social costs associated with energy production, and with a lack of energy, may be estimated for comparison. The key question in such a method is that of estimating a societal utility function. This process occurs in a nonempirical manner at present. The social value system is made known through various channels of public expression such as elections, letters to congressmen, and the many polls. Such a process is incapable of separating values from perceptions.

Use of Risk Data

A decision-maker is frequently called upon to compare different types of risk (and more generally, different types of social cost) resulting from any of the various options under consideration.

At one end of the spectrum is the social cost function that follows expected value rather closely. This approach is taken by the US Public Health Service in decisions regarding their budget allocations. In this case the goal is to minimize disability days because the social cost is assumed directly proportional to the total number of days of disability.

A second type of social cost function results if an individual's value system is used. Perhaps this explains the charitable organizations that raise money for research to develop cures for multiple sclerosis or muscular dystrophy. In terms of disability days to society, these diseases are probably far less significant than the common cold. The catastrophic nature of the effects upon the individual results in a personal evaluation of a very high social cost.

One way to represent the difference between the two systems is to assume that personal value systems charge an increasing social cost to each successive disability day. As an example of this proposed representation of the personal value system, the social cost of one individual being disabled for a year is greater than the cost of 365 separate individuals each sick for one day. An example of such a representation would be

$$\text{total social cost} = NC_1(1+i)^t, \qquad\qquad 1.$$

where N is the number of individuals involved, i is a daily interest rate, t is the time in consecutive days of disability, and C_1 is the personal cost of one disability day. If such a representation is used, the curve in Figure 4 is obtained.

The mix of personal and societal cost scales selected is dependent upon the decision-making agency. Autocratic countries and military organizations are more likely to use a societal scale because for these agencies the general welfare of society is placed above that of an individual. Democratic governments are more likely to use a mix of both scales; the congressional decision to spend extensively for cancer research is probably a result of a personal value system.

The choice of a cost system will, in some cases, determine whether an option with a high probability of a few fatalities is preferable to an option with a lower probability of a greater number of fatalities. While there are compelling arguments for both systems, the choice can only be made by a reading of the public's preference.

The term *value of life* appears frequently in the risk-benefit literature, and the meaning is derived in a variety of ways (5). The original approach was to discount future earnings at an appropriate rate, to arrive at some value. Such a method has obvious drawbacks, for the life value of retired persons goes quickly to zero as does the value for those who work without direct compensation (e.g. housewives). Court awards for accidental deaths are not representative, for the compensation to an individual's survivors is probably representative of lost earnings. The approach taken by Sagan (6) is to charge a flat rate ($50) per day of disability and to use the National Safety Council equivalent of 6000 disability days for a death, which gives a life value of $300,000. This is a valid method if the societal value system is used, but it would not work for the personal system. A method

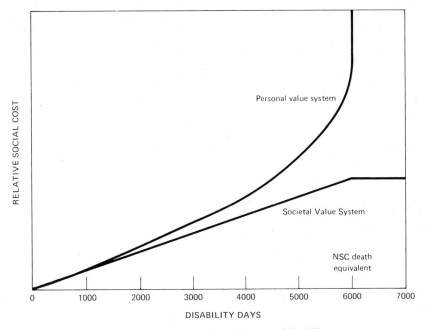

Figure 4   Value systems for social costs of disability.

frequently employed is called *self-evaluation*. In this case, the premiums paid to workers in hazardous occupations are compared with the increased risk, and the marginal risk-benefit trade-off for an individual is scaled upward to a life value. Implicit in such a calculation is the assumption that the individual knows the level of risk and that employment in a similar but safer occupation is available at a slightly lower wage. In addition, the increase in risk must be small, or nonlinear effects (risk aversion) will occur. In general, the value of life remains a loosely defined term; the social values for risk should be more clearly defined before such a definition is made.

Comparison of Risk and Benefit

It is rare to find a technology that provides a new benefit. The high-technology energy and transportation systems frequently used as examples in risk-benefit studies are always competitive with other energy and transportation technologies. When a new technology arises—for example, a medical treatment for a previously incurable condition—the risk-benefit analysis weighs the social cost against the benefits to gain insight into the criteria for acceptability.

A more common procedure is to compare several technologies through a risk-benefit viewpoint. In an analysis of the risks associated with nuclear power, it is more common to weigh these risks with the social costs of using coal or oil than to balance risks and benefits. The reason for this is simple. Coal and oil have been acceptable fuels for power plants for years. If the social costs of nuclear power compare favorably with those of coal and oil, then nuclear power automatically satisfies the risk-benefit criteria. Under these circumstances, the concept of benefit remains simple: the benefit is expressed in terms of electrical energy. Such a situation proves enormously helpful in an assessment of an energy technology because no widely accepted method of assessing benefits exists.

The typical problems of assessing benefits of energy, and electrical energy in particular, stem from the fact that prices are set according to costs, and a free market condition does not apply (7). Attempts to improve the benefit-assessment methodology frequently include additional benefits such as tax revenues and local employment, but these techniques usually represent double accounting. An alternative method from classical welfare economics is to use the concept of the consumer surplus, but such a method produces severe measurement difficulties. Fortunately, an assessment of benefits is not necessary for the general analysis of energy technologies. The historical record of energy usage provides proof of the acceptability of the technologies associated with the use of fossil fuels to generate power, with the possible exception of strip mining. If the total social costs of new energy technologies are lower than the social costs associated with the traditional production methods, then societal acceptability has been demonstrated.

A method commonly employed in comparison of energy technologies is predicting the health costs from various types of power plants with the assumption that each plant is of a given power rating and the plants are sited at equivalent locations. In this case the benefits are truly identical: the generation of power to serve the needs of a given population. The quantification of social costs may be simplified under these circumstances, since all social costs need not be translated into a dollar amount. Instead, each technology is responsible for various types of health effects, which may be expressed in more convenient units such as injuries, fatalities, disability days, or years of life shortening. Due to the societal distinction between voluntary and involuntary risk, the analyses are frequently divided into occupational and public health effects. A risk-benefit analysis includes all the social costs associated with a given technology. In the electrical power field, this generally includes all health costs associated with the mining, transport, and preparation of fuel, combustion of fuel, and waste disposal. The combustion of fuel includes effects from both normal operations and potential accidents. In addition, an analysis of nuclear power requires the consideration of fuel reprocessing. The one health cost generally not considered in the analysis is the effect of the use of electricity, namely electrocution and electrically caused fires, because this is independent of the method of generation. Additional social costs include burdens on the environment and most frequently are expressed in terms of land-use requirements and quantities of pollutants emitted.

The methodology of assessing social costs due to environmental burdens is less developed than the comparative health cost methods. The reason for this is that the environmental burdens are of an aesthetic nature, whereas health costs may usually be translated into medical costs, days of disability, and fatalities. For this reason, risk-benefit studies have focused upon the health effects associated with the production of power, with less detailed comments offered regarding the environmental burden.

An even more difficult externality to quantify in the energy field is the potential social cost of undependable supplies of a particular form of energy. The current emphasis applied to reducing oil imports and establishing a goal of energy self-sufficiency indicates that this externality is now recognized. On a smaller scale, some industries are now switching from natural gas to electricity, due to the threat of nonavailability. An example of this is the glass industry, which is now building electric furnaces for glass manufacturing. Reliability of fuel supply is a significant factor in electric utility planning, although cost is also a major factor.

Interpreters of Risk-Benefit Analysis

The roles of the various participants in a risk-benefit decision process depend largely upon the societal structure. It seems that the scientist's role includes those activities up to and including the quantification of impacts, regardless of the decision-making process (see Figure 3).

In an autocratic society, the social costs of accidents are assumed directly proportional to the estimated expected values of direct accident costs. In this case, a social value system relating to risk characteristics (shown as *worth of alternatives* on Figure 3) is not applied, and the decision process is rather automatic. The risk assessor may in fact be the decision-maker.

In more egalitarian societies, the scientist's role ends with the quantification of impacts. The worth of alternatives has generally been added to the decision process by special-interest groups, letters to decision-makers, editorials by opinion-makers, and other inputs of political measure to the decision-maker. The degree to which the decision-maker (in this case, an elected official) accurately interprets and responds to the political input determines, to some degree, the success of the decision-maker. Two ways exist in which such a process can result in decisions inconsistent with societal values and scientific assessments. First, the input to the decision-maker may not accurately reflect the overall social values of the public. Second, the input is usually expressed as a recommendation, either pro or con, for a particular alternative of the decision process. In this case the input may reflect a faulty perception of the situation, rather than an accurate expression of values.

The issues are clouded because some degree of scientific disagreement exists about virtually every controversial topic. To provide a measure of consistency, the risk-assessor or members of the public can do several things. First, the assessor can indicate that certain courses of action are more consistent with prior societal behavior and, second, the character of the risk may be explored to indicate which risks are likely to be misperceived. Through the use of this information, the decision-maker can more accurately interpret the social values implicit in the political input. The goal of such a method is a separation of the debates over societal goals and scientific estimates.

Use of Risk-Benefit Analysis

Risk-benefit analysis can be applied in three principal areas. The first area relates to the allocation of resources for safety expenditures. In general, a law of diminishing returns applies to all types of controllable risk: risk reduction becomes progressively more expensive. If an accepted methodology for determining the social cost of risk is developed, safety allocations would be determined as demonstrated in Figure 5.

Ideally, safety expenditures should be made until the marginal exchange between social cost and control cost is equal; that is, the expenditure of one dollar for safety is expected to reduce the social cost by one dollar. This results in the most efficient allocation of resources available to reduce all risks.

The second area of application for risk-benefit analysis is in the setting of standards, whether for occupational safety or for public health risks. This will provide an analytic basis for trade-offs such as that between automotive emissions and mileage, and electric power plant emissions and electricity cost.

The final and most important application is in societal risk-taking decisions. Risk-benefit analysis provides a background of consistency through which decisions can be carefully examined. The consistency produced will ensure that the path of minimum social cost will be selected.

NATURE OF RISK

Before a methodology for dealing with risk on a societal level can be developed, it is first necessary to examine the nature of risk and the history of societal risk-taking. Each major element in such a process is described below.

Risk is defined here as the probability per unit time of the occurrence of a unit cost burden. The cost burden may be measured in terms of injuries (fatalities or days of disability) or other damage penalties (expenses incurred) or total social costs (including environmental intangibles). Risk thus involves the integrated combination of (*a*) the probability of occurrences, (*b*) the spectrum of event magnitudes, and (*c*) the spectrum of resultant personal injuries and related costs.

As thus defined, risk is directly related to the familiar insurance premium determination for any class of activity. Thus, automobile insurance premiums are based on the frequency of collisions, the distribution of their physical magnitude, and the distribution of the resultant damage. Better roads are addressed to the first point, elastic bumpers to the second, and seat belts to the third. This separation of parameters is often glossed over in many discussions of the subject, but it is important for analytic studies.

Characterization of Risk—Exposure Probabilities

The probability of exposure to risk (the first item above) has three major elements: space, population, and time dependency.

The spatial character of exposure probabilities can range from quite localized risks to global hazards. For example, the production and utilization of energy

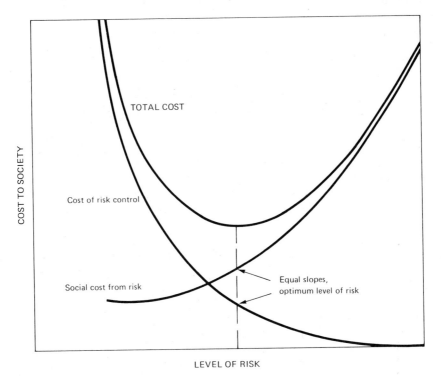

Figure 5 Social costs and control costs versus level of risk.

causes effects over a wide range of space; hydroelectric dams and some pollutant emissions represent localized risks, whereas the production of carbon monoxide, the increase in global turbidity, and nuclear safeguards are of global concern.

A second aspect of exposure probability is determined by population dependency. For many types of risk, specific groups may be identified as the population fraction that bears a specific risk. While many such groups exist, certain group characteristics may be identified as the determinants of the risk-bearing population. The following list includes some types of group-dependent risks:

1. heredity—genetic diseases;
2. age—childhood or geriatric diseases, certain accident probabilities that are higher for certain age groups, such as poisons (children) and falls (older persons);
3. occupation—pneumoconiosis (black lung disease) among coal miners, burns to fire fighters, and many other occupational hazards;
4. sex—incidence of heart attacks in males, risk to women in childbirth;
5. activity groups—smokers, skiers, and gun owners.

The time-dependent exposure probability may be classed into three major groups:

1. continuous—describes most diseases and accidents that can occur any time;
2. periodic—certain risks that are periodic in nature such as hurricanes, tornadoes, influenza, and some earthquakes;
3. cumulative—the exposure probability of risk that is in some cases related to previous exposure such as smoking, pneumoconiosis, asbestosis, and radiation effects.

Magnitude and Impact of Events

The magnitude of a risk event is a physical measure and does not consider the consequences of an event to individuals. In this sense, an event's magnitude is not directly a measure of risk, but rather one of event characteristics. Examples of such measures are magnitude of an earthquake, height of a flood, or energy release from an explosion.

Impacts measure the effects of risk events upon exposed populations. As such, the impact is one key part of risk (the other is probability). This distinction between magnitude and impact is made because estimates of risk (probability of a specified impact) are frequently derived from a combination of two estimates: probability of an event of a given magnitude and the impact of an event versus magnitude. The relationship between magnitude and impact is explored in the next section under rare events and risk.

HYPOTHESIZED LAWS OF RISK

Risk has certain usually repeatable characteristics, which may be termed laws of risk. An exploration of these characteristics is useful for the purpose of comparing risks from different sources.

Logarithmic Relationship between Probability and Magnitude

Risk is properly defined as the probability per unit time of a unit cost burden (injury) occurring. It consists of both the probability that some event will occur and the magnitude of damage that results if the event occurs. From data for various types of risk (see Figures 6 and 7) it is clear that the probability and magnitude are not independent but that for many risks the high-impact events are much less common than the lower-impact events. This result is in itself not surprising where societal choice existed, for it seems clear that in response to both natural and man-made hazards, successful societies have selectively evolved in a manner designed to avoid catastrophic events. What is surprising is that even for natural phenomena (earthquakes) the probability, or frequency, of the event usually falls off at least as fast as an exponential function of the magnitude for many types of risks (see Figures 8 and 9). Mathematically, this is

$$\log f = a - bm, \qquad 2.$$

where f is the frequency, m the magnitude, and a and b are constants.

If estimates of the probability of high-magnitude events are desired, the data base is usually sparse because the events are rare. The probability distribution for the general type of risk can, however, be estimated from the available data for less extreme cases. A class of statistical distributions, known as asymptotic distributions, have proven useful in estimating extreme probabilities for natural hazards such as floods and earthquakes. The asymptotic distributions come from the theory of extreme value distributions (9) and are useful because less restrictive conditions are required for the use of the asymptotic distributions than for the exact distributions of the extremes of functions (10). Extreme value distributions deal with the distribution of the maximum or minimum values in large samples of independent values drawn from an initial distribution. This approach requires knowledge of the initial distribution and the sample size.

The asymptotic-type distributions do not require as detailed information; the sample need only to be large and the initial distribution must satisfy more general requirements. $F(x)$ is the probability that some random variable is less than x (the cumulative distribution). If $F(x)$ tends to unity at least as fast as an exponential as $x \to \infty$, then the maximum values of large samples of independent events follow the Type I asymptotic distribution of maximum values. This condition is satisfied if the initial distribution is gamma, exponential, chi-square, normal, or lognormal, among others. The Type I asymptotic distribution for maximum values is

$$F(x) = \exp[-e^{-a(x-b)}]. \qquad 3.$$

From this it can easily be shown that

$$f(x) = a\exp[-a(x-b) - e^{-a(x-b)}], \qquad 4.$$

which reduces asymptotically to

$$f(x) = ae^{-a(x-b)}, \qquad 5.$$

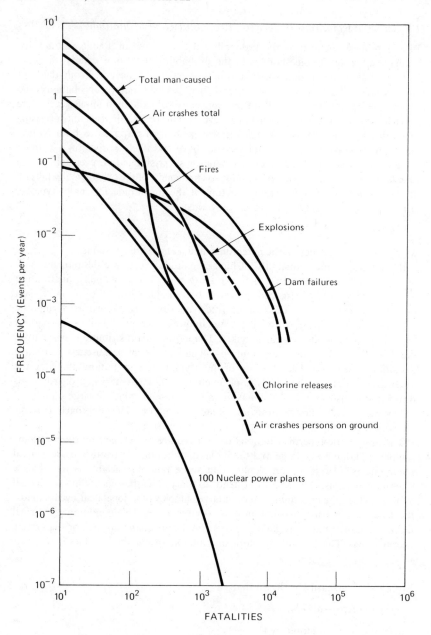

Figure 6 Frequency of fatalities due to man-caused events. Source: (8).

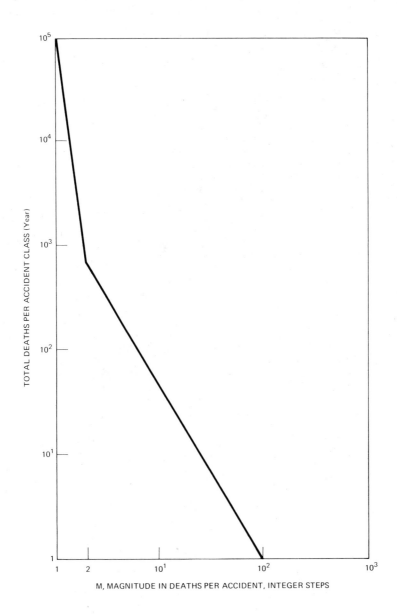

Figure 7 Total deaths arising from accidents that kill M people versus magnitude.

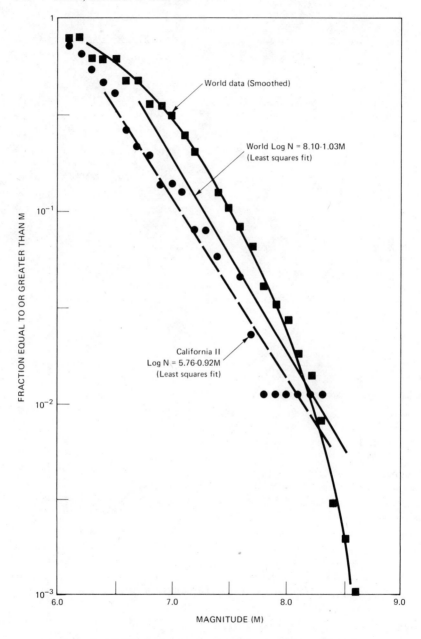

Figure 8 Magnitude distribution of shallow earthquakes 1904–1952.

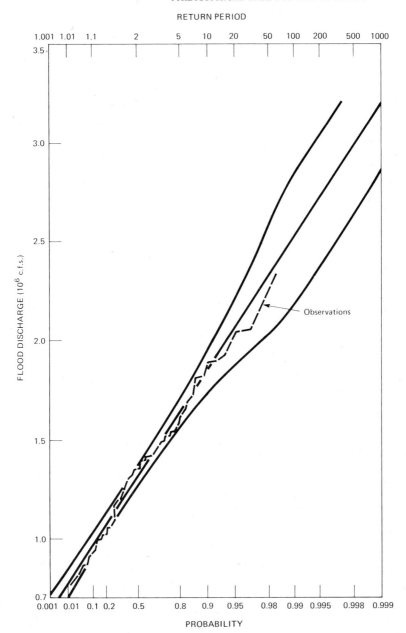

Figure 9 Mississippi River floods, Vicksburg, Mississippi, 1898–1949. Source: (9).

similar to equation 2. The return period for rare events is defined as

$$T(x) = \frac{1}{1-F(x)}, \qquad 6.$$

and asymptotically it converges to

$$T(x) = e^{a(x-b)}. \qquad 7.$$

If, as an example, large earthquakes of magnitude m are considered, the common notation is (10)

$$F(m) = \exp[-\alpha e^{-bm}], \qquad 8.$$

which can be written in Type I asymptotic form as

$$F(m) = \exp\left\{-\exp\left[-b\left(m - \frac{\ln \alpha}{b}\right)\right]\right\}, \qquad 9.$$

so the return period is

$$T(m) = \frac{1}{\alpha} e^{bm} \text{ (years)}, \qquad 10.$$

which means that it takes an average of $T(m)$ years to observe an earthquake of magnitude m.

This type of distribution is frequently applied to floods (9, 10), and the nature of the result may be seen from Figure 9.

A general list of natural hazard distributions has been compiled (11) (see Table 1)

Table 1 Examples of natural phenomena described by probability distributions[a]

Poisson Distribution	Rayleigh and Weibull Distribution
1. Meteorite strikes on (potential) human targets (12)	1. Wind speed (17)
	2. Wave heights: trough-to-crest (14)
Negative Binomial	Lognormal Distribution
1. Frequency of tornadoes (13)	1. Tsunamis (18)
	2. Hydrologic series (various examples) (19)
Gamma Distribution	3. Tornadoes: dimension of damage swath (13)
1. Sea waves: height (14)	4. Flood damage magnitude: USA (20)
2. River levels: recurrence of levels being exceeded (15)	5. Earthquakes: magnitude and frequency (21)
3. Precipitation: drought occurrence (16)	
Exponential Distribution	
1. River levels (15)	

[a] Source: (10).

and demonstrates that most natural hazards do satisfy the conditions necessary to be of the Type I asymptotic form.

Rare Events and Risk

An estimation of risk from rare events entails three steps. First, the probability of an event of given physical magnitude must be estimated, as described above. Second, the manner in which such an occurrence can produce injuries or fatalities must be developed, and finally the population distribution in the vicinity of an occurrence must be known. In general, the population distribution is easily measured or estimated; estimates of injuries from an event are more difficult to perform.

All rare events of this type have one common feature, the release of energy. To translate this into an estimate of risk, two mechanisms must be quantified: the nature of energy transport through the population and the relation between energy exposure and injury. These mechanisms may be either probabilistic or deterministic, and it does not seem that either mechanism is capable of being described in a general sense for a large class of risks. As examples of energy transport mechanisms, several hazards may be considered. Earthquake and explosion energies are best described deterministically; earthquake propagations can be estimated from measurable soil characteristics, and explosions transmit energy that varies with the inverse of the square of the distance, to a first-order approximation. A man-made hazard that also produces an energy release that propagates in a deterministic method is a flood due to a dam failure. In this case, the flood plain may be predicted.

Probabilistic transport mechanisms apply for estimates of hazard from events such as forest fires and nuclear power plant accidents, because in both cases the energy transport is strongly dependent upon the wind direction and speed, where the wind can only be characterized in a probabalistic sense. In any case, a general rule is that individual energy exposures are proportional to the total energy release.

The relationship between individual energy exposure and the degree of injury is frequently more difficult to quantify but does have some general characteristics. For many types of hazards, two thresholds may be defined. The first threshold places a lower bound on certain energy exposures; this level cannot produce injuries. Mechanistically, this lower bound corresponds to exposures to which the body responds elastically. The upper threshold defines a level of individual energy exposure that will cause death, if physical injuries are considered, or complete destruction in the case of property damage.

To consider the meaning of these thresholds, an example of an airplane accident may be used. In a crash, the individual energy exposures in general are in excess of the upper threshold, and thus death is the most likely result. A bump caused by slight turbulence or a rough landing rarely exceeds the lower threshold, and no injuries result.

For those exposures in which the two thresholds are quite close in value (such as the effects of electric shock or drowning) the body is said to be quite resilient. This implies that recovery from a nonlethal level of energy is usually rapid. The response to other damage mechanisms, such as mechanical injury, is one of lesser resilience.

Between these two thresholds, the damage is a monotonic function of energy, although frequently in a probabilistic sense.

An example of a single mechanistic model of the damage due to a localized

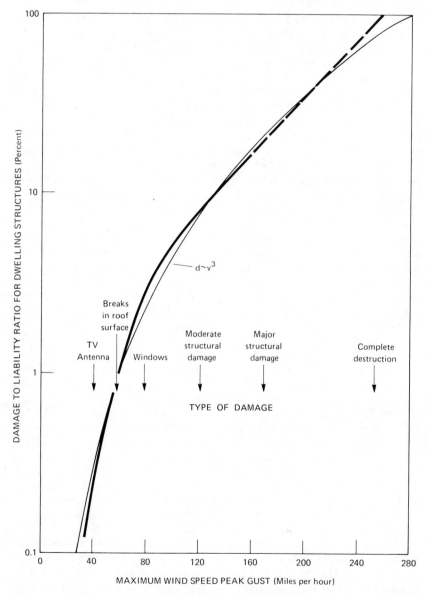

Figure 10 Dwelling damage versus wind speed relationships. Source: (23) except for $d \approx v^3$ curve.

energy exposure can be postulated for the property damage due to wind gusts. In this case, the damage is assumed to vary as the cube of the velocity, because that is the rate at which wind transmits energy to structures. This model is in excellent agreement with experimental evidence (22, 23) (see Figure 10). In the case of wind, the upper damage threshold occurs at approximately 250–280 mph.

While a general response to hazards cannot be found, the empirical evidence is that the damage resulting from a range of event magnitudes also varies as the log of the frequency (see Figures 6, 7, and 11). From this empirical evidence, it is possible to construct a single, three-dimensional surface for the number of accidents, A, as a function of d disability days lost per person and n people injured per accident (see Figure 12). Because the data are difficult to obtain, Figure 12 is only an approximate representation of the actual surface. The surface is defined by

$$A(n, d) = \frac{a}{n^b d^c},$$ 11.

where a, b, and c are constants.

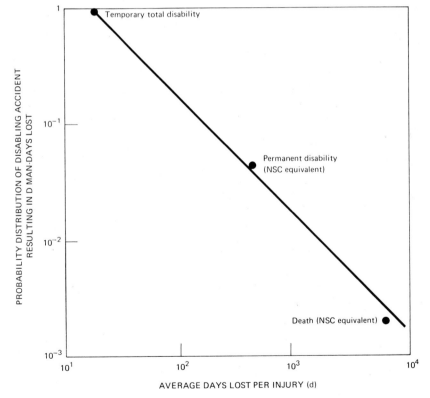

Figure 11 Probability distribution of disabling and fatal accidents versus average days lost (US manufacturing industries, 1969).

In Figure 12, $A(n, d)$ represents all US accidents, a is 8×10^6, b is between 1.5 and 3 (depending upon whether $n = 1$ is included), and c is between $\frac{1}{2}$ and 1. Such a surface may be defined for any risk, and in many cases the coefficients b and c

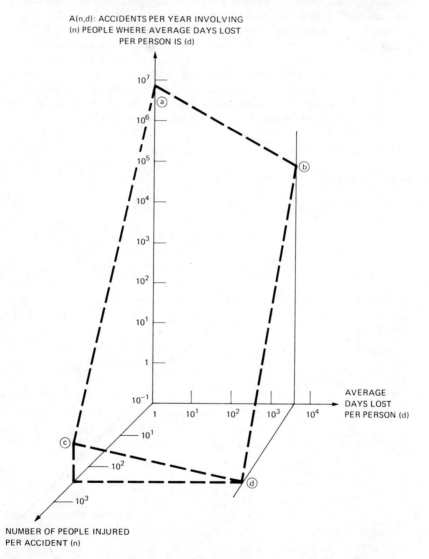

Figure 12 Distribution of fatal and disabling injuries by magnitude and severity. ⓐ Single injury per accident (1 day disability); ⓑ single fatality per accident (6000 day equivalent); ⓒ maximum group per accident with 1 day average disability; ⓓ maximum fatalities per accident.

appear to be similar. Some rather special cases may be considered that fit the general form of the relation, although the constants will be quite different from those for average risks. Two examples mentioned previously are drowning and electrocution, which would be expected to have a very high value for c. This is due to the nature of these risks in which the damage thresholds are quite close. To this end, c is a measure of the individual resiliency to a risk. The constant b (related to the number of people involved per accident) will vary with the distribution characteristics of the population. In the case of aircraft accidents, the population density is quite high at the point of energy release, which would suggest a higher value for b. The usefulness of such a surface is indicated for those risks resulting in a distribution similar to that for all accidents: a great deal of information can be inferred from a slight bit of statistical information. From a societal point of view, the total number of disability days T (and hence the social cost) resulting from a particular risk is simply the volume integral of the surface:

$$T = \int_1^\infty \left[\int_{d_1}^{d_2} A(n,d)\,dd \right] dn, \qquad 12.$$

where d_1 and d_2 are the damage thresholds previously defined.

In the case of US accidents, d_1, the lower threshold for damage, would be one disability day (accidents resulting in no disability days do not contribute to the total social cost). The National Safety Council equivalent of 6000 disability days per death can be used for d_2. For $b > 1$, and $0 > c > 1$, the total number of disability days is

$$T = \frac{a(d_2^{1-c} - D_1^{1-c})}{(b-1)(1-c)}. \qquad 13.$$

From this result, and the estimates for a, b, and c given previously, it is quite evident that the most significant contributions to total disability came from accidents involving only one individual. It should be noted that T is a measure of social cost in disability days and is not representative of personal value systems.

Uncertainty Principle

One major factor in the acceptance and perception of risk is the degree of uncertainty associated with the probability estimates, particularly of future hypothesized situations, discussed earlier in the section on perception of risk. It is valuable to examine the sources and nature of the uncertainty in a quantitative manner to put the discussion of perception in perspective.

Accident probability estimates for complex systems are usually made by estimating the failure probabilities of individual components of the total system, then combining the probabilities of the various failure modes by use of fault trees or event trees into an overall probability estimate. Uncertainty arises from two sources in such an estimate. The first source is that due to statistical variations in the failure probabilities of the individual components, and it results simply because the assessor is limited to a finite number of experiments (or the operating histories of the components is finite). Such uncertainty is reducible by increased experimentation, although the cost associated with producing more precise estimates is frequently considered excessive. This type of uncertainty is referred to as *experimental uncertainty*.

The second source of uncertainty seems more significant in its influence in risk perception. This uncertainty is due to modes of failure not previously considered. These failures have been termed *hypothetical* (24), for they are events that have not been observed. The hypothetical uncertainty is due to the possibility of unanticipated occurrences. While the uncertainty resulting from the possibility of unanticipated events is not removable, some justification can be provided for the engineering judgment that the unanticipated event risk is not expected to be high.

In both cases, a simple relationship may be developed to reveal the characteristics of the uncertainty as they relate the probability estimate to operating experience (which in the first case is equivalent to the number of experiments and in the second case is related to the cumulative operating time). If U_a represents experimental uncertainty, U_n hypothetical uncertainty, P the assessed probability, N_1 the number of experiments that have been run (assumed proportional to the investment I), and N_2 a relative measure of societal experience with a particular system, then the relationship is

$$U_a \approx \left(\frac{1}{PN_1}\right)^{1/2} \approx \left[\frac{1}{(PI)}\right]^{1/2}, \qquad 14.$$

$$U_n \approx \frac{1}{PN_2}, \qquad 15.$$

and thus the total,

$$U = \frac{a_1}{(PN_1)^{1/2}} + \frac{a_2}{PN_2} \qquad 16.$$

The full derivation of these results is in the Appendix. This indicates that two types of experiments give information to reduce uncertainty. The first type, N_1, carries a cost that increases with decreasing probability of an accident. The second type of experiment, N_2, is a free by-product of normal operating experience.

The second relation, that for unanticipated uncertainty, perhaps lends some credence to the intuitive statement of an experienced engineer who claims to be "confident that everything of significance has been considered." While this is not always true (the designer of the Boeing 727 certainly never considered the possibility of a hijacker parachuting from the rear door), an upper bound on the probability of these new events does exist.

While it would be foolish for anyone to claim omniscience in any new technical field, nevertheless there is a professional basis for believing that the estimates of end-event probability are reasonably correct, i.e. within a few orders of magnitude either way. This is because the event-tree analyses indicate that while there may be many initiating circumstances, there are a limited number of key effects, that is, few outposts result from the event sequence with many inputs. As an extreme example, the most serious public hazard from a nuclear power plant would be an extensive sequence resulting in large amounts of radioactive emissions to the atmosphere. For this to occur, a substantial fraction of the radioactivity contained in the nuclear core would have to be released. We do know that the only way this can happen physically is by a melting of the core fuel. Thus, the core melting becomes a key point that all initiating circumstances would have to reach before significant public hazard could occur. Thus, even if there are some unknown failure modes that might eventually lead to a core meltdown, that particular phenomenon acts as a common output for many such event sequences. Thus, if the nuclear power plant is designed to handle such an end-point event as a core meltdown, the existence of a few unknown sequences presumably will not alter significantly the public hazard probabilities.

The validity of this assumption depends on the ratio of unknown to known failure modes of "like kind." For example, if the unknown are roughly equal in number and of roughly the same importance as the known, then the actual probability of an end effect would be twice that calculated from the known event tree. Because of the uncertainties in the probabilities of failure of an individual component or subsystem, the final end-effect probability in a nuclear system calculated from known event trees is likely to be uncertain to at least one order of magnitude. Thus, an estimated end-release probability of 10^{-7} (one in ten million per year) may actually be 10 times lower or higher. Therefore, a missed failure mode is not important unless it is differs radically from any considered in the design and also effectively bypasses many of the plant protective systems. Clearly, some experience with, and insight into, plant operating experience and failures is necessary to provide the designer with a reasonable base for event-tree probability analysis.

Resilience

The concept of resilience to a particular event's impact was introduced briefly in earlier sections, in particular in the discussion of the elasticity of individuals to the hazards of drowning or electrocution. This concept can be expanded to the societal level to consider those events capable of causing a lasting social impact.

Two principle types of societal resilience should be examined, relative to two different types of damage. First is the accident such as an earthquake or flood that causes a great deal of immediate damage. The threat associated with this type of event is that the loss of substantial portions of a population and resources will require a substantial effort and time for recovery. The second type of risk to a society is that which threatens the societal structure and culture, rather than large losses of life. This second type of risk and its relations to societal resilience is of great importance, for large-scale physical disasters have been routinely tolerated by societies for years, and populations continue to live in areas of high risk from flooding or earthquakes.

The importance of the resilience factor is the apparent nonlinearity of societal response to accidents as the impacts increase. The following quotation (25) perhaps dramatizes the issue:

> Small accidents throughout the world kill about 2 million people each year, or 4 billion people in 2000 years. This is "acceptable" in the sense that society will

continue to exist, since births continually replace the deaths. But if a single accident were to kill 4 billion people, that is, the population of the whole world, society could not recover. This would be unacceptable even if it only happened once in 2000 years.

The concept of resilience deals primarily with the survivors' point of view, unlike individual risk acceptance.

An example of the risk to societal stability is war, in which the loser suffers a dramatic change in structure and culture, as in the South following the US Civil War. A more localized threat would be the case of increasing water pollution in an area where the predominant industry is fishing, as has happened with mercury in some areas of Japan. The key factor in both of these cases is the long time required for cultural recovery. In the production of electricity, this is probably a key factor in the perception of risk from nuclear power plants. While a number of studies have indicated that the risk of nuclear power is lower than that of fossil-fueled electric power of hydroelectric power, the time frames associated with waste management and the fear of long-term contamination represent a type of risk different from a dam collapse or oil tanker explosion. While these risks have been exaggerated in many instances, the nature of this type of risk is subject to a set of social values different from the more direct threats. This is perhaps the reason for increasing concern over acid rainfall and trace metal contamination resulting from the combustion of coal. As a further example, a prolonged energy shortage could produce a severe economic impact, resulting in substantial social changes. It is for this reason that the factors of societal resilience need to be considered in national planning.

This study of societal resilience is an approach to the evaluation of the elasticity of a society to high-impact events and to the identification of those characteristics of societies that determine such resilience. One such factor is the level of technological development of the society. The more technically advanced a society is, the lower its resilience. This is due to a complex societal infrastructure that cannot operate with a sizable portion of its subsystems missing. In less developed areas, with less division of labor, the culture is more immune to accidents. An example of the lesser resilience of technological cultures may be seen from the bombing of ball bearing factories in Germany in World War II. While the ball bearing industry was small in comparison to the total industrial capability, the infrastructure was such that the loss of ball bearings would have been a substantial blow to the society at that time. Conversely, the severe flooding that caused over a hundred thousand fatalities in Bangladesh was not a major test of societal resilience because the social structure is dependent only upon the quantity of food produced, not upon vital industries or key population segments. In such a case, the population recovers rapidly to a level constrained only by the food supply. A striking example of the sensitivity of the US economy occurred with the oil embargo of 1973–1974, and the effects were far reaching even though the imported oil represented only a small portion of the total economic expenditure of the United States.

The key factors that seem to determine the societal resilience to a given accident or event may be arranged in the following dimensional form; if the resilience is denoted as R, it follows that

$$R = \frac{PL}{B(T)S^a t_r^b t_d^e M^f t_c^g} \qquad 17.$$

seems reasonable. In this relation, P is the societal population, L its average lifetime, B a monotonically increasing function of the level of technology T, S the size of population affected by the event, t_r the average recovery time for an injured population group, t_d the time to discover the source of damage, t_c the time to correct the situation, and M the average disability per person affected. The factors a, b, e, f, and g are constants. Although there are unquestionably other significant factors, this relationship may provide a starting point for work on the subject of societal resilience. Ecologists have studied a similar problem, that of species resilience (26). The topic of resilience as it relates to a societal value system for risk is explored in the discussion of that topic. It should be recognized that this relationship is quite crude and serves more as an example than as a proposed societal model.

The impact of considerations of resilience upon the societal value system for risk has not been explored in any detail, although a suggestion has been made by Wilson (27) to estimate the social costs of accidents of n fatalities as proportional to n^2 rather than n. An additional factor for involuntary risk is referred to as the "cost of public confidence."

Some inference can be drawn from empirical risk data for the social values appropriate to high-impact events. If f_n represents the frequency of accidents in which n lives are lost, it follows that

$$f_n = \frac{a}{n^b}, \qquad 18.$$

as demonstrated in a previous section. Typical values of b lie between 1.5 and 3.0. If one were to make the common overt assumption that all lives lost in accidents are of equal value, then the social cost due to the risk of an accident with magnitude n is

$$f_n nl = \frac{al}{n^{b-1}} = \frac{\text{const}}{n^{b-1}}, \qquad 19.$$

where l is the societal value of each lost life. The conclusion that must be drawn from the assumption that all fatalities are of equal cost is that the high-frequency, low-magnitude accidents are of greatest social cost because the observed values of b are greater than unity.

An analytic interpretation of the empirical data supports the assumption that for fatalities resulting from any technical system, the relationship between accident frequencies and number killed has been adjusted by investment in safety so that each frequency point has social value equal to any other, i.e.

$$f_n n l_n = \text{const}, \qquad 20.$$

in which l_n represents the societal value of each life lost in an accident of n

fatalities. From equation 20 is derived

$$f_n n l_n = \frac{a l_n}{n^{b-1}} = \text{const} \qquad 21.$$

and

$$\frac{l_n}{l_1} = n^{b-1}. \qquad 22.$$

The conclusion resulting from such an assumption is that each additional life lost in an accident is increasingly more costly to society. The cost of an accident of n fatalities is then

$$f_n n l_n = f_n n^b l_1. \qquad 23.$$

For the case $b = 2$, this corresponds directly to Wilson's result (27).

Objectives of Risk Analysis—Identification

Three methods are employed to identify risk: empirical evidence, statistical inference, and postulation based upon a transfer of experience or the laws of nature. Risk identification through empirical or statistical means is straightforward, but the third method is of more interest in the energy field because the types of risk associated with new or developing technologies are frequently postulated. This course is usually preferable because action may be taken before a statistically significant portion of the population is injured. In the energy area, the most widely studied risks are those from nuclear power. While no members of the public have been killed or injured by commercial nuclear plants, a number of possible risks have been identified and studied, by using a transfer of experience from radiation effects upon animals and from people exposed to radiation from sources other than power plants.

Risk Estimation

The estimation of risk contains both estimates of exposure probabilities and event impacts. These estimates generally require different skills. As an example in the energy field, the risk from accidents in light-water reactors is estimated by computing the probability of the release of a specified amount of radioactive material, estimating the subsequent exposure of a population to this material, and then estimating the health effects due to that level of exposure.

In this case, the first two estimates (release probability and subsequent exposure) are within the realm of the engineer or physical scientist. The estimate of health effects due to the postulated exposure is made by medical specialists, usually health physicists. For this case, the greatest uncertainties seem to lie within the estimates of the probabilities of release; the estimates of exposure due to a release and predicted effect of exposure are more generally accepted. The situation is quite opposite for the estimates of health effects due to the routine emissions of fossil-fueled power plants and automobiles. The exposure levels are directly measurable, but the health effects are not well known. The two roles—probability estimation and impact estimation—are not completely separable, for the probability varies with the postulated amount of impact. This relationship was previously explored under the first hypothesized law of risk, the logarithmic relationship between probability and impact.

Estimates can be developed in four ways. First, the analyst can apply the laws of nature directly, even though these laws may be empirically based. An example of this would be an estimate of the probability of failure of a component due to metal fatigue. Second, the estimates can be developed from historical or experimental data. This is generally the source of estimates for failure or accident rates of widely used devices such as pumps and relays, or for activities such as automotive travel, skiing, or coal mining. The third method is that of transfer of experience, where probabilities of failure are assumed similar to those of similar devices. The final method, systems modeling, is actually a combination of the previous methods and usually includes the estimation of accident trees. This last method provides estimates for complex technologies without established accident rates. Estimates of risk from nuclear reactors are made in this manner.

CONCLUSION

Risk-benefit analysis is an increasingly important part of technology assessment. As the capability to estimate the external effects of technical systems improves, and as the ability to modify and reduce risk from these systems develops, a methodology to deal with questions such as "How safe is safe enough?" is needed to efficiently allocate societal resources.

Three principal areas of application of risk-benefit analysis have been identified. The most fundamental application concerns the societal choice of expanding or curtailing the development of our existing technological capabilities. For example, the relative emphasis on mass and individual transportation systems is such an issue. A second area for risk-benefit analysis is in setting performance targets for existing or new technologies where trade-offs between safety, environmental effects, and cost can produce major social and economic impacts—for example, air quality criteria. The third application involves decisions between competitive technical systems that produce a similar beneficial function but with differing societal costs. For example, the selection of either coal or uranium as a fuel system for a new electrical power plant must be based on a comparison of their social costs, since the same output of electricity would be generated by both.

A step-by-step procedure to deal adequately with such issues is not yet available. Rather than try to describe the various suggested methodologies in detail, we have reviewed the general philosophic features common to most risk-benefit decisions. In many cases we have deliberately simplified analytical descriptions of the important dimensions of risk, with the hope that this might seed future work in these areas.

Appendix: Uncertainty Principle

If experiments and analyses are applied to the assessment of a probability, uncertainty will accompany the assessment for two reasons. First, the usual sort of uncertainty due to a finite (rather than infinite) number of experiments will be present, and second, a possibility exists that events will occur through mechanisms not considered in the analysis. What is usually of primary interest is an upper bound on the estimate of uncertainty, and attempts to estimate the difference between the assessed value and this upper bound refer to the bias in a system. The word *bias* unfortunately carries different connotations to the nonstatistician readers from the meaning of the word to statisticians. Because it is hoped that the reference to this topic will interest both groups, the word *uncertainty* is used and defined in a rather arbitrary manner. In addition, the derivation is not strictly rigorous but serves to demonstrate the functional form of the relationship.

If the assessed failure probability is termed P_a, the uncertainty U, and the true failure rate that would in theory be observed with unlimited operating experience P_t, it follows that

$$P_a(1+U) = P_t. \qquad \text{A.1}$$

This definition was chosen because the uncertainty associated with low-probability events is frequently expressed as a product, e.g. the probability is P_a to within a factor of 20. The quantity $(1+U)$ was chosen so zero uncertainty would refer to $P_a = P_t$.

As mentioned above, the uncertainty arises from two sources. The first source, experimental uncertainty, is usually expressed as confidence limits for a number of standard deviations, σ, around the assessed probability. The second source is the probability of new events, termed P_n. For the upper bound, we find that

$$P_t = P_a + \alpha\sigma + P_n, \qquad \text{A.2}$$

where α is a constant. From equation A.1 is derived

$$P_a(1+U) = P_a + \alpha\sigma + P_n \qquad \text{A.3}$$

and

$$U = \frac{\alpha\sigma}{P_a} + \frac{P_n}{P_a} = U_a + U_n. \qquad \text{A.4}$$

Uncertainty Principle for Experimental Testing

The value of U_a may be computed from the common method of defining confidence limits. In this case P^* is the upper bound threshold probability and $\Phi(y)$ is the cumulative probability of a normal distribution.

To compute the one-sided confidence limit, the cumulative distribution is

$$\Phi(y) = \frac{1}{(2\pi)^{1/2}} \int_{-\infty}^{y} \exp(-\tfrac{1}{2}x^2)\, dx. \qquad \text{A.5}$$

The confidence is defined by

$$\frac{P^* - P}{\alpha} = y, \qquad \text{A.6}$$

where the law of large numbers justifies the choice of a normal cumulative function. For a binomial process (experiments that result in either success or failure) the standard deviation for the estimated mean is

$$\sigma(P) = \left[\frac{P(1-P)}{n}\right]^{1/2} \qquad \text{A.7}$$

where n is the number of tests.
For

$$P^* = (1 + U_a)P, \qquad \text{A.8}$$

we find that

$$\frac{(1+U_a)P - P}{\sigma} = y \qquad \text{A.9}$$

or

$$U_a = \frac{y\sigma}{P}. \qquad \text{A.10}$$

This is the same as the previous definition from equation A.4.
For this case we find that

$$U_a = \frac{y\left[\frac{P(1-P)}{n}\right]^{1/2}}{P}, \qquad \text{A.11}$$

and for $P \ll 1$, this reduces to

$$U_a \approx \frac{1}{(Pn)^{1/2}}. \qquad \text{A.12}$$

New-Event Uncertainty Principle

It is assumed that some probability of system failure can be due to the occurrence of a "new" event, that is, one for which statistics are not available and estimates of failure due to this event were not included in the assessment.

This situation is roughly similar to a series of experiments (in this case operating experience and transfer of experience) during which no failures were observed as a result of this particular cause. If this is considered in the framework of Baysian analysis, with a diffuse prior probability, then the probability distribution for such an event is

$$f(P_n) = (N+1)(1-P_n)^N, \qquad \text{A.13}$$

where N represents the operating experience. If one considers events such as "giant meteor destroys earth," N is 4×10^9 years (the age of the earth). The expected value of the probability in this case is

$$E(P_n) = \frac{1}{N+2} \approx \frac{1}{N}. \qquad \text{A.14}$$

As a result, the probability that a giant meteor will destroy the earth is expected to be 2.5×10^{-10}/year, if only the experimental evidence with the absence of such meteors is considered. This simplistic approach assumes that the probability of meteors has not changed over the last 4×10^9 years.

From the definition of new-event uncertainty (equation A.4) is derived

$$Un = \frac{P_n}{P_a} \approx \frac{1}{NP_a}. \qquad \text{A.15}$$

From the brief example it seems clear that risks or system failure probabilities of less than 2.5×10^{-10} are not really worth considering, as an unremovable risk threshold of 2.5×10^{-10} has been defined.

ACKNOWLEDGMENT

We would like to acknowledge assistance from Dr. Boyer Chu.

Literature Cited

1. Starr, C. 1969. Social benefit versus technological risk. *Science* 165:1232
2. Starr, C. April 1971. Benefit-cost studies in sociotechnical systems. In *Perspectives on Benefit-Risk Decision Making*, p. 17. Washington DC: Nat. Acad. Eng.
3. Tamerin, J. S., Resnik, H. L. P. April 1971. Risk taking by individual option—Case study: cigarette smoking. See Ref. 2. p. 73
4. McGrath, P. E., Papp, R., Maxim, L. D., Cook, F. X. Jr. 1974. A new concept in risk analysis of nuclear facilities. *Nucl. News* 17(14):62
5. Zeckhauser, R. June 1975. *Procedures for Valuing Lives*. No. 29D, 2nd version. Discuss. Pap. Ser. Public Policy Program, Kennedy Sch. Gov. Harvard Univ., Cambridge, Mass.
6. Sagan, L. A. 1972. Human costs of nuclear power. *Science* 177:487
7. Marcus, M. 1974. The economic benefits of nuclear power plants. *Public Utilities Fortnightly* 93(13):26
8. US Atomic Energy Commission. 1974. *Reactor Safety Study: An Assessment of Accident Risks in U.S. Commercial Nuclear Power Plants*. Rep. No. WASH-1400 (draft). Washington DC: US AEC
9. Gumbel, E. J. 1958. *Statistics of Extremes*. New York: Columbia Univ. Press
10. Apostolakis, G. E. Sept. 1974. *Mathematical Methods of Probabilistic Safety Analysis*. Rep. No. UCLA-ENG-7464. Los Angeles: UCLA
11. Hewitt, K. 1969. *Probabilistic Approaches to Discrete Natural Events: A Review and Theoretical Discussion*. Presented at Invitational Conf. Comm. on Quant. Methods, Ann Arbor, Mich. [taken from (9)]
12. La Paz, L. 1958. *Adv. Geophys.*, Vol. 4
13. Thom, H. C. S. 1963. *Mon. Weather Rev.*
14. Longuet-Higgins, J. S. 1952. *J. Mar. Res.*, Vol. 2
15. McGilchrist, C. E. et al. 1968, 1969. *Water Resour. Res.*, Vols. 4, 5
16. Barger, G. L., Thom, H. C. S. 1949. *Agron. J.*, Vol. 41
17. Davenport, A. G. 1968. *Wind Effects on Buildings and Structures*. Toronto: Univ. Toronto Press
18. Van Dorn, W. G. 1963. *Adv. Hydrosci.*, Vol. 2
19. Chow, V. T. 1954. *Proc. Am. Soc. Eng.*, Vol. 80
20. American Insurance Association. 1952–1955
21. Asada, T. 1957. *J. Phys. Earth*, Vol. 5
22. Brinkmann, W. A. R. 1975. *Local Windstorm Hazard in the United States: A Research Assessment*. Rep. No. NSF-RA-E-75-019. Inst. Behav. Sci., Univ. Colo., Boulder
23. Friedman, D. G. 1971. *The Weather Hazard in South Dakota*. Hartford, Conn: Travelers Insurance [Taken from (11)]
24. Häfele, W. July 1974. Hypotheticality and the new challenges: the pathfinder role of nuclear energy. (Int. Inst. Appl. Syst. Anal., Schloss Laxenburg, Austria, Res. Rep. No. RR-73-14, Dec. 1973.) *Minerva*, p. 303
25. Wilson, R., Jones, W. J. 1974. *Energy, Ecology, and the Environment*. New York: Academic
26. Holling, C. S. 1973. Resilience and stability of ecological systems. *Ann. Rev. Ecol. Syst.* 4:1–23
27. Wilson, R. Oct. 1975. Examples in risk-benefit analysis. *Conf. Adv. Energy Syst. June 1974, Denver, Colo. Chemtek* 5:604

Copyright 1976. All rights reserved

SAFETY OF NUCLEAR POWER

✕11024

J. M. Hendrie

Department of Applied Science, Brookhaven National Laboratory, Upton, New York 11973

INTRODUCTION

The safety of nuclear power is currently a subject of some public interest. The aim of this chapter is to provide a brief overview of nuclear safety matters from the standpoint of accidents or events that might lead to release of a significant amount of radioactive material. The review deals with nuclear power plants, where the radioactive products of the fission process are generated, and with facilities where fission product wastes from spent fuel are processed and stored. The general safety basis for nuclear facilities is the starting point. Next, specific safety aspects of the various types of facilities are reviewed, and then accident probabilities and consequences are considered with emphasis on the limiting cases. Finally, some related special topics are noted.

A limited number of references are cited to provide a starting point for those interested in the extensive and recondite literature on nuclear safety. As a general source of useful data, the 1973 Atomic Energy Commission (AEC) report on nuclear facility safety (1) is still the best single source book. The extensive hearings in 1973–1974 of the Joint Congressional Committee on Atomic Energy (2) include statements from many different individuals and groups and indicate the wide diversity of opinion on the subject. A recent issue of the *Bulletin of the Atomic Scientists* (3) includes a series of articles of similarly diverse viewpoints. Reports are available from groups critical of the nuclear enterprise (4, 5) as well as from those favoring nuclear power (6, 7): the references cited should provide entry to other literature from these sources. The recent report of the American Physical Society study group on light-water reactor safety (8) is of interest because it is the work of scientists outside the fields of nuclear engineering and reactor safety.

GENERAL SAFETY BASIS FOR NUCLEAR FACILITIES

The basic safety philosophy of nuclear plant design and operation is that the design should be conservative, there should be ample safety margins in operation, and redundant components and systems should be provided to cover the residual uncertainties.

Defense-in-Depth Design Concept

The safety philosophy is carried out in the design by using a "defense-in-depth" basis, which involves three levels of safety-oriented effort. The first is based on the objective that nuclear facilities should function with a high degree of reliability. This objective involves designing the plant to conservative standards and engineering practices so there will be a large tolerance for system transients, operator errors, and off-normal operation. Plant systems are designed to allow monitoring of components for signs of wear or incipient failure.

The second level is based on the fact that failures and malfunctions must be anticipated during the service life of the plant. Measures to forestall or cope with such events are provided. All important safety systems are required to be redundant to assure that no single failure will cause the loss of a needed safety function.

The third level involves additional margins in the plant design to protect the public against highly unlikely accidents. To establish these additional margins, major failures of plant components are postulated and the accident sequences that would follow these events are analyzed. Safety systems and other plant design features are then provided to mitigate the consequences of these major failures. In addition to plant component failures, environmental phenomena such as tornadoes, floods, and earthquakes are hypothesized and plant safety elements are designed to withstand these events. In most cases a limited number of these postulated major failures or natural events turn out to be limiting from the standpoint of the plant safety design; these are called the design basis accidents and natural events.

Rules, Guides, Codes, and Standards

The matters important to safety in nuclear facilities are covered by governmental rules and guides and by a large body of industrial codes and standards. The basic requirements are set forth in the rules of the Nuclear Regulatory Commission (NRC), the federal agency that assumed the regulatory function of the AEC in 1975. The NRC rules have the force of law, and compliance with them is required. Regulatory guides are published by the NRC staff to implement the rules and provide detailed guidance. The guides identify design approaches and solutions to safety issues that are acceptable to the staff, but also allow other approaches that give an equivalent level of safety.

In addition to NRC rules and guides, nuclear facility design, construction, and operation are governed by codes and standards published by the American National Standards Institute, the American Society of Mechanical Engineers (ASME), the Institute of Electrical and Electronics Engineers, and a number of other professional engineering societies and standards organizations.

Review and Inspection

Adherence to the various rules, guides, codes, and standards is monitored by vigorous review and inspection efforts by the NRC staff. A company wishing to construct a nuclear facility must first obtain a construction permit from the NRC. A design and safety analysis for the proposed facility is assembled in an application document that may run to a dozen thick volumes for first submission. The application is reviewed in considerable detail by engineers and technical specialists on the NRC staff and their conclusions are incorporated in a public Safety Evaluation Report. The application is also reviewed by the independent Advisory Committee on Reactor Safeguards. The application is examined in formal proceedings before an Atomic Safety and Licensing Board appointed by the NRC. Testimony on the application is given under oath at these hearings and parties to the hearing have the right to cross-examine witnesses.

If a construction permit is recommended by the Licensing Board, and after review by an Appeals Board, the permit may be issued and the applicant can construct the facility. Before operation can start, however, a new application detailing the final design and safety analysis must be filed. Reviews for the operating license by the NRC staff and the Advisory Committee on Reactor Safeguards are carried out in even more detail than before. The hearings before a Licensing Board are repeated, this time with regard to the issuance of an operating license.

Throughout the process, starting even before the construction permit application is filed, the NRC staff inspects the applicant's quality assurance provisions, the design procedures, all phases of facility construction, and the start-up experiments preparatory to operation. After issuance of an operating license, the plant operation is monitored and inspected periodically by the staff.

NUCLEAR POWER PLANTS

Nuclear power plants in the United States are planned or currently exist in three forms, depending on the type of reactor system used. Most of the plants now operating are of the light-water-cooled reactor (LWR) type. The other plant types are those using the high-temperature gas-cooled reactor (HTGR) or the liquid-metal-cooled fast breeder reactor (LMFBR).

Water-Cooled Reactor Plants

The LWR generating units are divided between pressurized water reactor (PWR) and boiling-water reactor (BWR) plants. In both PWRs and BWRs the fuel is made of ceramic pellets of slightly enriched uranium oxide, sealed in zirconium alloy cladding tubes to form fuel rods. These are grouped into fuel assemblies to form the reactor core, which is mounted in a large pressure vessel. Control rod drives attached to the vessel move neutron-absorbing rods out of the core to allow the chain reaction to start or into the core to shut it down.

In LWRs, purified ordinary water is the coolant and neutron moderator. Water is pumped through the core to remove the fission energy and transfer it as high-pressure steam to the plant turbine generator. In BWRs, the primary coolant boils in the core at a pressure of about 1000 psi and the steam is piped directly to the turbine. In PWRs, the primary system pressure is about 2200 psi, to suppress boiling, and hot water from the core is passed through steam generators where steam to drive the turbine is generated in a secondary loop.

The reactor systems are housed in a containment structure, typically built of steel and reinforced concrete. The containment is a large pressure vessel, designed for pressures of 15–60 psi depending upon the internal volume and particular type of containment system.

SAFETY-RELATED PLANT CHARACTERISTICS   The most significant safety characteristic of LWR plants is that the reactor systems are maintained at high internal pressures and at temperatures (550–600°F) at which the coolant water would flash to steam if the pressure were reduced substantially. In the event of an accidental opening in the primary pressure boundary, coolant would be driven out of the reactor system at high velocity and most of the water that might otherwise remain in the system after the pressure was relieved would boil away.

The oxide fuel pellets operate at very high center temperatures and are much hotter than the fuel rod cladding. This large stored energy in the pellets, together with the afterheat resulting from fission product decay, means that even after reactor shutdown it is quite important that the fuel rods be in contact with the coolant. The afterheat corresponds to a few percentage points of full power at one minute after shutdown and is still about one-half percent of full power one day after shutdown.

OPERATING EXPERIENCE   The operating experience to date for LWR plants has been reasonably satisfactory. The plant availability and capacity factors have been less on the average than were projected by the designers, although they are almost exactly the same as the factors for oil-fired and coal-fired generating plants of comparable size. The record includes some items of the sort to be expected in the initial operating phases of a new technology, as well as the assorted component malfunctions and failures that routinely occur in large and complex industrial installations.

The operation of each plant licensed for commercial service is governed by a thick book of Technical Specifications, which sets forth requirements for reporting all manner of plant conditions and malfunctions. As a consequence, a large number of reports are filed each year with the NRC. The great majority of these concern events that have little or no significance with regard to the safety of the plant, but that meet the reporting requirements of the Technical Specifications.

Typical events are valve malfunctions, instrument failures or off-normal settings, minor leakage of radioactive waste, valve packing and pump mechanical seal failures, and electrical switchgear malfunctions. There have been interruptions in the off-site electrical power supply at several plants, and during the periodic tests that are required, there have been a number of failures of one of the redundant emergency power diesel generators to start and assume load.

Among the more significant events that have occurred are stress-corrosion cracking in stainless steel piping in a number of BWRs, steam generator tube corrosion and leakage in a number of PWRs, and vibration and fretting wear of core structures in both types. An unexpectedly large densification of fuel pellets early in operation with fresh fuel was discovered in 1972. The effect required recalculation of the heat transfer between the pellets and the fuel rod cladding and resulted in temporary power reductions at a number of plants. Fires in electrical cabling have occurred at several plants, the most recent and serious being the fire in 1975 just outside the cable spreading room in the Brown's Ferry units.

ACCIDENTS AND EVENTS CONSIDERED IN PLANT SAFETY ANALYSES   The plant conditions considered in the safety analyses for LWRs range from relatively mild events to design basis accidents and natural events that have potential for serious consequences. This range of accidents and events can be divided into general groups as follows:

Anticipated operating transients   These plant events and system transients may reasonably be expected to occur sometime in the service life of a plant. These events lead to no significant releases of radioactivity. A partial list includes turbine trip, loss of electrical power from off-site sources, partial loss of feedwater flow, and control rod withdrawal errors.

Minor accidents   The events in this group may result in small releases of radioactivity. They include such things as significant leakage from the radioactive waste system, malfunctions leading to leakage of gaseous fission products from fuel rods into the reactor primary coolant, steam generator tube failures, instrument tubing failures, and temporary loss of all electrical power.

Design basis accidents and natural events   The events in this group are postulated in order to establish performance requirements for the safety systems of the plant. They include refueling accidents, loss-of-coolant accidents, steam-line-break accidents, and rod drop or ejection accidents. In addition, very severe postulated natural events form part of the design basis conditions. These include a tornado substantially more severe than any yet observed; flooding to a level corresponding to the Probable Maximum Flood as established by the Corps of Engineers; the Probable Maximum Hurricane as defined by the National Weather Service, if the plant is located along the Atlantic or Gulf coasts; and for all sites a very large earthquake.

The potential radiological consequences of all of the events and accidents considered in the safety analysis are expected to be well within the NRC siting guidelines. The reason for this low level of consequences from the design basis accidents and natural events is that engineered safety features are provided to control them. Engineered safety features include such systems as emergency core-cooling systems, the various designs of reactor containments, hydrogen control systems, containment atmosphere sprays, and special filtering systems. The net effect of the engineered safety features in a plant is to provide a series of overlapping safety systems and physical barriers to prevent significant releases of radioactivity.

LOSS-OF-COOLANT ACCIDENTS AND EMERGENCY CORE-COOLING SYSTEMS   The postulated loss-of-coolant accident (LOCA) begins with the assumption of a break in a primary coolant system pipe. The coolant discharges from the break and flashes to steam in the containment building surrounding the reactor. At the end of the blowdown, which would take only 20 or 30 seconds for the largest pipe breaks, the reactor system is filled mostly with low-pressure saturated steam. The loss of water from the reactor core stops the nuclear chain reaction, and insertion of the control rods and neutron-absorbing solutions keep the reactor shut down after the core is reflooded.

The emergency core-cooling system (ECCS) supplies water to the core after the blowdown to maintain acceptable temperature conditions in the fuel rods. The ECCS must continue to function in a recirculation mode, with heat rejection to sinks outside the containment, for a long period after the accident.

The typical ECCS for PWRs is composed of three independent subsystems, each with equipment and flow path redundancy. This means that the ECCS will function in a fully effective way even with a major component failure in any subsystem. In fact, a number of component failures can occur in the various ECCS subsystems and not result in overall failure of the cooling function, provided at least one of the redundant trains in each subsystem remains operative. The first subsystem is an accumulator injection system that operates by forcing borated water stored under gas pressure in the accumulator tanks into the primary system. The second subsystem is a high-volume, low-pressure pumping system that comes into operation following blowdown and injection of water from the accumulator tanks. The third system is a high-pressure pumping system provided to supply coolant if a small break occurs and the primary system pressure does not decrease enough for the accumulator system or low-pressure pumping systems to operate effectively.

The typical ECCS for BWRs is also composed of three subsystems with equipment and flow path redundancy. The first subsystem is a high-pressure spray system that sprays water on the reactor core. For small pipe breaks, the high-pressure spray can maintain the water level in the reactor vessel so that the core remains covered. For larger breaks, this system cools the core by spraying water on the fuel rods. The second subsystem is a low-pressure spray system that delivers a large-volume water spray to the top of the core. The third subsystem is a low-pressure core injection system that provides a large flow of water to the reactor vessel to reflood and cover the core.

Considerable emphasis has been placed on critical reviews and analyses of ECCS designs to determine how effectively they would work. The approach that has been used is to measure the parameters of the various phases of a LOCA in a series of separate experiments. The results of the experiments then provide input parameter values for analytical models that tie the individual LOCA phases together and track the complete accident.

These analytical models, in the form of large and complex computer programs, are used to evaluate the effectiveness of ECCS designs. There has been argument over whether the computer programs are sufficiently conservative for that purpose. The great majority of nuclear engineers and NRC staff experts believe that they are, particularly with the more conservative ECCS requirements established after public hearings in 1972–1973. The independent review of the American Physical Society study group (8) led the group to agree that ECCS would probably work as intended, but they felt that a really complete and quantitative basis for ECCS performance and evaluation is not yet in hand:

> We have no reason to doubt that the ECCS will function as designed under most circumstances requiring its use. However, no comprehensive, thoroughly *quantitative*

basis now exists for evaluating ECCS performance, because of inadequacies in the present data base and calculational codes. In addition, it is not clear that the present approximate calculations, even though based on generally conservative detailed assumptions, will in all cases yield conservative assessments of ECCS performance.

INTEGRITY OF REACTOR PRESSURE VESSELS Reactor pressure vessels are built to the rigorous standards of the nuclear plant components section of the ASME Boiler and Pressure Vessel Code. There have been no failures of any kind in nuclear pressure vessels in service. This experience, amounting to something less than 2000 nuclear vessel–years, is not great enough to be statistically significant in showing the very small probability of failure that is required. There is, however, considerable operating experience with the large number of high-pressure vessels built to the non-nuclear component sections of the ASME code. The data base for non-nuclear vessels covers some 68,000 high-pressure vessels manufactured to the Code since 1943. No significant failures have occurred in these vessels. Detailed examinations of that record by the AEC regulatory staff (9) and the Advisory Committee on Reactor Safeguards (10), together with a comparison of the code requirements for nuclear vessels and non-nuclear vessels, result in an upper-limit probability of a disruptive failure event in the range of one in one million to one in ten million per nuclear vessel–year. Thus, the likelihood of a significant failure of a reactor pressure vessel is diminishingly small.

The American Physical Society study group noted in its summary of conclusions (8):

> Although we have not been able to analyze all of the many possible failure sequences for light water reactors, one which we have studied in detail is the possible failure of the integrity of the primary reactor pressure vessel. We find that reactor vessels are constructed of materials chosen with care and are designed with substantial safety factors. The reactor vessel is subject to careful scrutiny and testing. Based on our study, we believe that catastrophic failure of the primary pressure vessel is not likely to be an important contributor to accident initiation; however, this is dependent upon maintaining a strong quality assurance program.

THE RASMUSSEN REPORT The Rasmussen report (11), prepared by a group headed by N. C. Rasmussen of MIT, is the only extant complete and authoritative work on LWR plant safety that includes a full range of possible accidents, including those beyond the design basis. The study deals with both the probabilities and the consequences of accidents and natural events.

The detailed designs of one PWR plant and one BWR plant were examined in the study. These two were the 24th and 34th large nuclear plants to come into operation in the United States. The results of the study are considered applicable to about the first 100 LWR plants only, since later plants will have profited from a safety standpoint by substantial evolution of the designs.

The probability calculations are based on extensions by the Rasmussen group of event-tree and fault-tree methods originally developed by reliability engineers in the aerospace industry. The starting point is to identify the various ways in which radioactive materials in the reactor core could be released to the environment. Event trees are used to define these accident sequences, of which there are a very large number. An event tree starts with an initiating event and proceeds through many branching points to a large number of possible consequences. The branching points correspond to points at which various functions or systems may operate successfully or fail. The accident sequences are carried through various possible modes of containment failure to determine the amount of radioactivity released to the atmosphere for each sequence. To arrive at a probability of occurrence for a sequence, the probability of success or failure at each branching point must be determined.

A separate set of analyses, based mostly on fault-tree methods, are used to determine probabilities at individual branching points. The fault trees have to be carried down in level of detail to elements for which failure-rate data are available. The failure rates are based on experience data for component failures, human error, and testing and maintenance contributions. Fortunately, most of the components of interest in reactor systems are the same sorts of valves, pumps, piping, and controls used in many other types of industrial plants. Thus, meaningful data on component failure rates are available for these components, in contrast to the situation in aerospace systems where many of the elements are first-of-a-kind configurations.

The problems of selecting initial events and having confidence that all significant accident sequences have been included are made tractable by the fact that about 98% of the radioactive material is locked in the oxide fuel pellets and stays there unless the core melts. Thus, events and accident sequences that do not lead to general melting of the core cannot lead to significant consequences. In turn, melting of the core requires either an extended loss of cooling or an extended over-power condition.

The various accident sequences that include core melting, each having a specified release of radioactivity to the atmosphere and probability of occurrence, are divided into a number of "release categories." The release categories are characterized by the nature and amount of radioactivity released to the atmosphere. The overall probability of a given release is an appropriately formed sum of the probabilities of the individual accident sequences that fall within that category.

Seven release categories involving core melting were used for the PWR and four for the BWR. The overall results of the probability calculations are tables or histograms in which the probability of a given release category is plotted against the severity of the release. The probability histograms show the familiar characteristic of such frequency distributions; the probability decreases as the magnitude of the release becomes larger. For both PWRs and BWRs, the most severe release category has a probability of about one chance in one million per reactor-year. This category includes such extreme events as a sequence in which core melting is followed by failure of the reactor vessel, rupture of the containment dome by heavy missiles, and ejection of a quarter of the core radioactivity into the atmosphere.

There are two very significant results of using event trees and grouping the individual accident sequences into a small number of release categories. First, the event-tree format makes it relatively easy to sort out those accident sequences that do not lead to core melting or are illogical in view of the design of the reactor. Elimination of such sequences reduces the number of mathematically possible sequences from a hundred thousand or so in some of the event trees to a few hundred. Of these, only 10–20% will have probabilities that make them possibly significant contributors to one of the release categories. These potentially significant sequences are then few enough in number so that it is feasible to examine each in considerable detail for interdependencies of the probabilities assigned to the branching points. This is important, since it would be nearly impossible to examine all of the mathematically possible sequences in full detail for probability dependencies. Dependent probabilities, e.g. from common mode failures, must be identified and the appropriate values assigned if the total probability for an accident sequence is to be meaningful. It has been possible to make this analysis in the study by virtue of the limited number of individual sequences that must be examined.

The second result follows from the fact that in each release category there are a few accident sequences that dominate the overall probability by virtue of their high individual probabilities. This means that accident sequences that have been overlooked cannot affect the results unless they have probabilities that are as high as those of the few dominant sequences in that category. Although there is no absolute guarantee that there are no such sequences, it is highly unlikely that there are any in view of the many years and great effort that have gone into identifying reactor safety problems. There are bound to be a number of low-probability sequences that have not been included in the study tabulation, but their omission cannot change the overall probabilities for the release categories. A case in point is the electrical cable fire that disabled several core-cooling systems for a few hours at the Brown's Ferry units. Fire sequences were omitted from the comment draft of the report on the basis of an estimate that fires plus the LOCA or other failures that would be required to lead to core melting would form sequences of lower probability than the dominant sequences. The final report includes the results of more detailed calculations on fires. These confirm the original judgment and show that inclusion of fire sequences does not change the overall accident probability results.

There would be large differences in the amounts of radioactivity released to the atmosphere in the various accident sequences. The action of containment sprays and the mode of containment failure have major influences on the amount of the release. In some accident sequences, containment failure would result from the molten core mass working its way through the concrete bottom mat of the containment. The release path would then be through the soil beneath and around the containment, and the final release of radioactivity to the atmosphere would be relatively small. The consequences beyond the plant boundary depend on the amount and time sequence of radioactivity release, and, as might be expected, there would be a very large range in consequences between the smallest releases to the atmosphere and the largest.

The effects of the releases were calculated by using a consequence model that involves six sets of weather conditions and population distributions characteristic of those at actual plant sites. Four different types of consequences that might be associated with a single accident were calculated. These are early fatalities, early illnesses, long-term health effects, and property damage. Each type of consequence was calculated for each combination of release category, weather condition, and population distribution. The resulting large numbers of outputs of the consequence model were then combined to form cumulative probability density functions, which are given as plots of the probability of occurrence per reactor-year versus the magnitude of a particular type of consequence, e.g. early fatalities.

The final plot of probability versus consequence for early fatalities shows that a core melting accident has a probability of about 1 in 20,000 per reactor-year, and that 99 out of 100 core melting accidents would cause no early fatalities. About 1 in 170 core melting accidents are predicted to cause more than 10 early fatalities, and only one core melting accident out of 500 is predicted to cause more than 100 early fatalities.

The limiting-case accident, combining the severest release category, the most unfavorable weather conditions, and the largest population density, has a probability of about one in a billion per reactor-year and results in about 3300 early fatalities. Thus, the maximum possible accident for LWRs is a very serious accident, but it has a vanishingly small chance of happening at all, and has no worse consequences than very large earthquakes, storms, floods, or industrial accidents.

High-Temperature Gas-Cooled Reactor Plants

Current HTGRs operate on the uranium-thorium fuel cycle, with highly enriched uranium. The fuel is a mixture of uranium carbide and thorium oxide particles, coated with multiple layers of pyrolytic carbon and silicon carbide, and embedded in graphite blocks. The entire core structure in the HTGR is made of graphite. The core is supported on graphite blocks in a central cavity in a large prestressed concrete reactor vessel. Steam generators and coolant circulators are located in other cavities arranged around the core cavity and within the same concrete vessel.

The coolant in the HTGR is helium gas, pressurized to about 700 psi. The helium is heated in the core to an outlet temperature of about 1400°F and is forced through the steam generators and back to the core by axial-flow compressors. Auxiliary cooling loops provide the capability for safe shutdown and removal of afterheat in the event that all of the main coolant loops are inoperable.

SAFETY-RELATED CHARACTERISTICS The HTGR has several notable features from a safety standpoint. First, the entire primary coolant system of the reactor is enclosed within a single large vessel. The vessel itself is a massive structure, heavily reinforced, and tied together by a great number of steel prestressing cables. Prestressed and reinforced concrete structures of this type have great tolerance for fabrication flaws because the number of strength-providing elements is very large and many of these could fail before the structural integrity of the vessel would be threatened.

The second important safety characteristic has to do with the very large mass and high-temperature capability of the reactor core. This great graphite block has an enormous heat capacity, so that in any accident resulting in a loss of cooling there is a long period of time before any emergency heat removal systems have to operate. The core block simply absorbs the afterheat and its temperature rises slowly for many hours. This situation is in notable contrast to that in LWRs, where ECCS action after a LOCA must occur and be effective within a few minutes or fuel melting may result.

DESIGN BASIS ACCIDENTS The largest accidents in the HGTR design basis are the depressurization accident and the water entry accident. In the depressurization accident the seal structure around a major penetration in the vessel is assumed to fail. The resulting hole is limited in size by the way the penetrations are constructed. The helium gas expands into the containment until the pressures inside and outside the vessel are equalized at about 30 psi. The reactor is shut down automatically, the main circulators trip, and heat removal is accomplished by the auxiliary cooling system's circulators and heat exchangers. Some air from the containment atmosphere can be expected to enter the vessel and may react with the hot graphite of the core structure to form a limited amount of carbon monoxide. There is not enough air in the containment building, however, to make much impression on the very large amount of graphite in the core. The off-site radiological consequences of the accident are negligible.

The water entry accident results from a leak or tube failure in the steam generators. A leak is significant because the water may react with the hotter portions of the graphite core structure to form carbon monoxide and hydrogen. The safety systems to deal with this accident involve sensors that detect water or steam leakage, fast-acting valves to shut off the water supply, and arrangements to dump the water content of the steam generators so that it is not available for leakage. The reactor is shut down automatically upon detection of a water or steam leak. A ton of water might be released into the vessel from the most serious of such failures, but only a small portion of this is calculated to react due to the relatively low graphite temperatures following reactor trip. The off-site radiological consequences of this accident sequence are negligible.

LIMITING CASES In looking beyond the design basis accidents to limiting cases, it appears that more extreme versions of the water entry and depressurization sequences are the most likely candidates.

A water entry accident more severe than that in the design basis can occur if the safety systems that cut off the water supply do not function. A large amount of water could be leaked into the reactor's internal cavities in this case. However, the water-graphite reaction is strongly endothermic and has a substantial cooling effect on the graphite. Thus, although much more water might be available to react, the reaction itself would quite rapidly decrease the temperature of the graphite core structure. The reaction rate of water with graphite is strongly temperature-dependent and is quite slow below 1300°F. It seems likely that a large water entry accident would cause some core damage, but it would be self-extinguishing and would not lead to a large release of fission products even if the evolved gases caused a penetration failure and a containment leak.

The severest accident sequence that might occur appears to be a large depressurization accident coupled with complete failure of all of the main and auxiliary cooling circuits. The core mass would eventually rise in temperature to the point where first the gaseous fission products and then the more volatile solid fission products would be driven out of the fuel. Containment failure due to internal overpressure would be projected in this sequence as the temperature of the atmosphere in the containment is increased by the afterheat of the fission products. In view of the very large heat capacity of the core graphite mass and the high-temperature capability of the core materials, this process would take several days. Short-lived fission products would have decayed substantially in this time, and solid fission products vaporized by the high core temperatures would condense on the containment walls and structures.

The eventual release of fission products to the atmosphere should be much smaller than in the severe release categories considered in the Rasmussen study for water reactors. Further, although the efficacy of evacuation measures might be questioned in some of the short time sequences for water reactor accidents, such measures should surely be more effective with several days to carry them out. It seems fair to judge on these grounds that upper-limit accident consequences for the HTGR will be substantially smaller than those calculated for water reactors. The probability that might be assigned to the limiting-case accident is a more speculative matter. My opinion is that it will be comparable to or slightly higher than that for water reactors.

Liquid-Metal-Cooled Fast Breeder Reactor Plants

All current designs of LMFBRs use a fuel composed of mixed oxides of plutonium and uranium, clad with stainless steel. The coolant is sodium and the reactor operates at temperatures well above the sodium melting point. Reactor heat carried by the primary sodium stream is transferred to an intermediate sodium loop. This arrangement is used to confine the radioactive primary sodium to a limited volume and to avoid transfer of radioactivity into the steam circuit should a leak occur in the steam generators.

SAFETY-RELATED CHARACTERISTICS   From a safety standpoint, the use of sodium as the primary coolant has several advantages. The sodium has a low vapor pressure so that the primary system pressure in an LMFBR is quite low compared with LWR and HTGR systems. Pressure stresses on vessels and piping are therefore relatively small, and the pressure forces expelling coolant in the event of a pipe break are also relatively small. In the event of a leak or pipe break, the sodium does not flash to vapor. Sodium is compatible with the fuel oxides and is an excellent heat conductor, so that fuel fragments can be easily cooled if contact can be maintained with the sodium.

A disadvantage of sodium is that the material is chemically reactive with water or with the oxygen in air. Inert gases must be used to maintain separation of the sodium from air, and if sodium leakage occurs into air or water-containing spaces, a sodium fire can result.

The prompt power coefficient of LMFBRs is negative, which means that the core reactivity is reduced as the power level rises and the fuel increases in temperature.

However, the reactivity associated with sodium heating or boiling is positive in the central regions of most LMFBR cores. Any substantial overheating in the core central regions has the potential for boiling the sodium and voiding the central coolant channels, thus producing reactivity and power pulses.

LMFBR cores are comparatively compact and operate at high power density. These factors tend to increase the likelihood of sodium boiling in the hot regions of the core from power transients or sodium-flow perturbations. The cores also contain large loadings of the fissionable material and are sensitive to relocation of the fuel material into a more reactive configuration.

DESIGN BASIS ACCIDENTS The design basis accidents currently considered include local fault conditions in the core, LOCA, and core disruptive accidents. Local fault conditions generally postulated are random failure of fuel pins, debris in the primary system that may lead to blockage of flow in an assembly, and overpower in a few fuel pins. The interest in these events is whether they can lead to a propagation of failure outside the fuel assembly in which they initially occur. The assembly structural shell in most designs is a fairly heavy and rigid metal structure, and evidence to date indicates that propagation of local failures is not likely to occur.

LOCA due to pipe breaks may have severe consequences, depending on the break location and the associated assumptions about reactor shutdown. Pipe breaks on the outlet side of the reactor vessel do not have severe results if the shutdown systems function. Breaks on the inlet side of the vessel result in a rapid reduction of flow leading to boiling, voiding, and some clad melting even with reactor shutdown. If the shutdown systems do not function, both inlet and outlet pipe breaks lead to core damage and are rather similar in consequences to the core disruptive accidents.

The core disruptive accidents, at the extreme end of the range of design basis accidents, are a loss-of-flow accident and a transient overpower condition, both without action of the shutdown system. In these accidents sodium boils and voiding occurs in the central regions of the core, giving a large reactivity pulse that results in a large power pulse. Extensive disruption of the core is calculated to occur in this circumstance.

This projected course of events is based on the current calculational models for these accidents, which are extremely complex and do not model all phases of the phenomena in a realistic way. Over the years, as these calculational models have improved and the results of experiments on various aspects of the accident phenomena have been incorporated, the severity of the projected consequences has tended to diminish. There is a reasonable prospect that detailed modeling of individual phenomena of fuel pin failure, together with an accurate treatment of the incoherence of pin failures, may show that present estimates of the degree of core disruption from these accidents are grossly exaggerated.

Another aspect of these extreme design basis accidents that needs careful consideration is the assumption that shutdown systems will not function at all. Most LMFBR designs now include two separate shutdown systems, each with its own independent set of sensing elements, rod drives, and control rods. The consequences of the accidents are very much reduced by action of the rods in either shutdown system, and it is a hotly debated point whether the likelihood of total failure of both systems is great enough to make this assumption a reasonable design basis. The more conservative approach is probably appropriate for the first few large reactors of this new type to be built, but it should be reassessed rather carefully before being accepted as a general long-term design basis for LMFBR power plants.

LIMITING CASES   With regard to possible limiting cases, the current design basis already includes about the worst possible set of circumstances related to the initiation and disruptive phases of the accident. Since the plant would be designed to contain and deal with these events, the likelihood of further failures and releases to the atmosphere should be very much reduced. All that can happen beyond the accidents currently hypothesized is for the afterheat removal systems to fail, with the subsequent eventual failure of the outer containment either by overpressurization or by melt-through of the core debris into the ground beneath the reactor. The release of radioactivity to the atmosphere should be comparable to that calculated for LWRs for similar release modes. The probability of such a major release to the atmosphere for LMFBRs should be substantially smaller than that calculated for LWRs because the LMFBR design basis extends to much more severe circumstances.

OTHER FUEL CYCLE FACILITIES

The other fuel cycle steps that have substantial inventories of fission products include fuel reprocessing plants, final waste storage facilities, and the transportation steps between these facilities (12, 13). The range of possible accident consequences in this part of the fuel cycle is much reduced from that in the power plants. The main reason is that spent fuel is not shipped from the power plants for at least six months, so that the short-lived, high specific-activity fission products have largely decayed. The fission-product afterheat decreases by a factor of about 100 in the first six months, from about 48 W/MW-day of equilibrium fission products at reactor shutdown to about 0.5 W. The need for prompt-acting cooling systems, such as the ECCS in water reactors, is thus not present in this part of the fuel cycle. Also, the high temperature and pressure conditions characteristic of the power plants are not present to drive the fission products out into the environment in the event of an accident.

Fuel Reprocessing Plants

Designs for commercial fuel reprocessing plants are adaptations of the chemical processing stages used in government-owned plants that have operated for more than 20 years. The reprocessing has two main purposes: to recover the uranium and plutonium in the spent fuel and to reduce the fission-product wastes to a form suitable for long-term storage. Plutonium is stored for possible future use as fuel, and the uranium is sent either to an enrichment plant for recycling or to a fuel fabrication plant for direct reuse.

The waste solutions resulting from the chemical processing contain the fission products, except those that are gases at ambient conditions. The separation of fission products from the uranium and plutonium is highly efficient, but it results in a loss to the waste stream of about 1% of the uranium and plutonium. The waste solutions also contain the other actinides formed by neutron capture in the reactor.

The high-level liquid waste, containing the fission products and actinides, is kept in stainless steel storage tanks in underground concrete vaults to await processing to solid form. Redundant water-cooling circuits are provided for these tanks, which are the only components in the reprocessing plant that require continued and reliable cooling. Other liquid wastes containing lower levels of radioactivity are collected from various points along the process line. These streams are treated to reduce the liquid volume to be handled and are stored in tanks until they are ready to be converted to solid form.

The amount of radioactivity in any single stage of the chemical process line is limited, and the only point of collection of substantial amounts of fission products is in the final high-level liquid waste storage tanks. The various process streams in the plant are contained within shielded cells with multiple filtering on the ventilation air exhausts. The plant structure itself is typically of heavily reinforced concrete, built to withstand the extreme tornado, hurricane, flood, and earthquake conditions used as design bases for nuclear facilities.

The design basis events and accidents considered for reprocessing plants, in addition to the usual set of environmental phenomena, include an assortment of failures and accidents in the chemical process steps and containment features of the plant. The more significant events include fuel assembly ruptures, criticality accidents, solvent fires at various stages in the process, plutonium and waste concentrator explosions, and leaks or pipe breaks at the various points along the process line. All of these events have expected consequences outside the plant that are well within the siting guidelines of the NRC regulations.

Even if multiple failures beyond those considered in the design basis for these sorts of accidents were to occur, the limited amounts of fission products and actinides at any stage along the process line would keep off-site consequences to levels very much below those possible from the power plants.

Beyond the range of design basis accidents, the only remaining points of interest are the high-level liquid waste storage tanks. Here the longer-lived fission products and unrecovered actinides are accumulated from many reactors. The heat generation rate in these tanks, although small by comparison with reactor afterheat levels, is still enough to cause a tank's contents to boil dry and melt, if cooling is lost for an extended period. Maintenace of cooling on the storage tanks is not very difficult, and as noted, redundant cooling circuits are provided. If all of these circuits fail and the solution begins to boil, all that needs to be done to stabilize the situation is to add water directly to the tank at a rate equal to the boil-off rate. Since there is ample time to make emergency connections, if the normal ones are inoperative, the probability of an extended loss of all cooling must be exceedingly low.

If all cooling is lost and cannot be restored, the tank would boil for about a week until the solution began to dry out and oxidize. Eventually the dry mass would rise in temperature, melt through the tank and vault floor, and stabilize in the earth beneath the tank vault. Vaporized fission products would be trapped in the off-gas filter banks for a time. However, the fission-product loading on the filters would become large enough to break down the filter medium, and a release to the atmosphere would result. The off-site consequences would be fairly well localized since all of the fission products being released would condense as soon as they reached an ambient temperature region and only the small amount that might form aerosols would continue airborne for any distance.

Waste Storage Facilities

Governmental regulations require that reprocessing plant inventories of high-level liquid waste be limited to those produced in the preceding five years. The liquid wastes are then to be converted at the reprocessing plant to dry solid forms and placed in sealed containers. No later than ten years after separation of the fission products and fuel, the containers containing the dry waste are to be transferred to a federal repository for permanent storage.

Developmental work on various solidification methods for high-level waste has resulted in several methods that have been operated at pilot-plant scale to produce ceramic or glasslike solids. These final solid forms for the high-level waste are chemically inert and resistant to leaching in water and provide a substantial volume reduction from the liquid. Plans for federal waste storage facilities involve two stages: an interim storage facility and a permanent storage facility. Use of the interim facility allows the time needed for extended pilot-plant operation to prove out the permanent storage facility. At the interim facility, sealed waste cannisters would be stored in water-filled basins or in vaults cooled by natural circulation of air. The most likely choice for a permanent disposal facility is in cavities mined into a geologically stable, deep-bedded salt deposit. This method for permanent storage of wastes was recommended by a National Academy of Sciences committee some years ago.

High-level wastes from the current designs of reprocessing plants contain a number of long-lived radioactive species. Of the fission products, ^{90}Sr and ^{137}Cs, with half-lives of about 30 years, are the prominent contributors to the waste activity after 50–100 years. The small quantity of actinides left in the high-level waste by current processing methods includes several species that are very long-lived; ^{239}Pu, for instance, has a 24,000-year half-life. If the actinides and the strontium and cesium in the high-level liquid waste could be separated out with good efficiency, the remaining bulk of the waste would become relatively inert in about 100 years.

The range of possible accident consequences at either the interim storage facility or the permanent storage facility is limited. At this stage in the fuel cycle, the fission products and residual actinides are fixed in very stable and insoluble ceramic forms, the heat generation rate is modest, and the heat is easily removed by natural conduction and convection. Even if cannisters were to be broken open and the contents exposed to the atmosphere, very little of the radioactivity would be dispersed beyond the immediate site. Cannisters sealed in deep salt deposits would be largely removed from man's activities on the surface, protected from groundwater, and cushioned against geologic forces by the plasticity and self-healing properties of the salt deposit.

Transportation

The transportation steps of most interest involve the movement of materials containing substantial quantities of fission products. Such shipments are made at two stages in the fuel cycle: spent-fuel assemblies are moved from power plants to fuel reprocessing plants, and solidified high-level wastes are subsequently to be transported to waste storage facilities. The shipping containers for spent fuel assemblies are massive shielded casks with provisions for heat dissipation to the atmosphere by natural convection. Depending on the number of assemblies carried, these casks may weigh from 20 to 180 tons each. The packaging and transportation of radioactive materials are regulated by the NRC and the Department of Transportation. Spent-fuel shipping casks are certified by these agencies after passing rigorous tests involving high-speed impact tests, puncture tests, fire tests, and water immersion tests.

Statistics for transportation accidents involving both truck and rail shipments show that an accident will occur once every million vehicle-miles and that one or two accidents in every hundred will be very severe. Severe accidents involving shipments of spent fuel might then be expected to occur every five to ten years in a fully developed nuclear power industry. The cask would not be breached and the consequences would be negligible in most cases, since the cask design basis includes the impact, puncture, and fire loads associated with severe accidents.

The limiting case for a spent-fuel transportation accident would be one involving an impact great enough to break open the cask and damage the fuel assemblies, coupled with a large fire. Even this eventuality would have modest consequences. The fission products released to the atmosphere would consist of the small fraction of gaseous fission products contained in the fuel rod plenums and gaps and a small amount of material vaporized by the fire. The total release to the atmosphere in a spent-fuel transportation accident would be small compared with releases for power plants in the severe-release categories as calculated in the Rasmussen report. The principal hazard from the accident would be the direct radiation in the immediate vicinity of the unshielded fuel assemblies. The radiation field would extend about 100 yards from the accident site, and this localized area around the radioactive material would have to be evacuated.

The planned shipment of solidified high-level waste in sealed cannisters to a final waste storage facility would pose much less risk than the shipment of spent-fuel assemblies. This is because there are fewer shipments involved, the level of radioactivity and heat generation of the waste is much reduced by the additional five to ten years of decay time, and there would be no gaseous fission products in the solidified waste.

SPECIAL TOPICS

Sabotage

Nuclear facilities are relatively unfavorable targets for a saboteur intent on creating substantial public damage. Nuclear facilities are heavily built and shielded installations. They are more difficult to penetrate than normal industrial installations, and there is greater difficulty in obtaining access to critical areas and components. The power plants, which have the greatest potential for large consequences in the event of a successful sabotage attempt, are particularly well protected by the provisions for containment and shielding of the reactor system.

The probability of a successful sabotage attempt is a speculative matter. The Rasmussen study group, for example, examined the sabotage question and concluded that they could not estimate the probability and that a methodology for such an estimate would have to await future development. The maximum possible consequences of successful sabotage are known, even if the probabilities are not. The maximum consequences cannot be greater than those for the limiting-case accidents discussed in previous sections, and the much more likely results of a successful sabotage attempt would be consequences comparable to those for the design basis accidents.

Safeguards

The term *safeguards* refers to the provisions used to prevent illegal diversion of plutonium and highly enriched uranium (14). Weapons-grade plutonium and uranium have been produced in this country for national defense purposes for some 30 years. Raw materials and the weapons themselves have been transported between processing facilities, weapons fabrications centers, and military depots and units. The safeguards measures used in the defense program have been highly effective to date in preventing any illegal diversions.

In the nuclear power fuel cycle, highly enriched uranium is not used in LWRs, which are the most common type, but plutonium is produced during operation and is recovered from the spent fuel at reprocessing plants. Plans have been made to recycle the plutonium as fuel in power plants and these proposals are under consideration by the NRC. The points of interest with regard to safeguards are storage facilities for plutonium in reprocessing plants, transportation to the fuel fabrication plant, the fabrication plant itself, and the final transportation step to a power plant. The most vulnerable point is the transportation of plutonium oxide from the reprocessing plant, and it has been suggested that if recycle fuel fabrication facilities were located adjacent to reprocessing plants this step could be eliminated.

In considering the use of recycled plutonium as a fuel, the regulatory staff concluded in 1974 that safeguards in the commercial fuel cycle should be substantially upgraded to achieve an adequate level of protection for plutonium (15). The measures proposed for upgrading of the safeguards systems fall generally under the headings of physical security and materials accountability. The NRC subsequently delayed consideration of plutonium recycle until several special studies on safeguards matters requested by Congress could be completed.

It should be noted that the possibility of illegal uses of nuclear materials does not depend upon whether or not plutonium recycle fuel, or even nuclear power, is used in this country. Other nations are committed to nuclear power and there are already a large number of nuclear weapons in existence in the world. The United States should maintain a leadership role in the field and use its considerable influence in matters relating to nuclear energy to obtain strong standards and practices for international safeguards.

Plutonium

The potential hazards of plutonium were recognized soon after its discovery in 1941, and a portion of the first small quantity of plutonium produced in the Manhattan Project was used in experiments to determine the safety measures that would be needed in handling large amounts of the material. In the years since, the radiobiological effects of plutonium have been studied extensively, and knowledge of the effects of plutonium on plant and animal life is considerably greater than that for many toxic chemicals and materials in current use.

Plutonium is quite similar to radium, which occurs naturally in substantial quantities, but it is slightly less toxic than radium. The biological hazard of plutonium is associated with radiation-induced cancer from material that is inhaled, introduced into the bloodstream through wounds, or ingested with food. Plutonium outside the body does not constitute a significant radiation hazard.

Cohen has recently calculated the most probable consequences of plutonium dispersal (16). Using data of the National Academy of Sciences Commission on Biological Effects of Ionizing Radiation and the International Committee on Radiological Protection, he estimates that the effect of dispersal of reactor-grade plutonium in a city would be to cause one fatal cancer, occurring 15–45 years later, for each 15 g of material released if there is no warning and no protective actions are taken. The calculation is crude, because of the broad assumptions used, but it is based on well-known meteorological and particle-deposition processes and on the available radiobiological data, and it is certainly more realistic than some widely circulated statements implying many millions of fatalities per gram of plutonium released.

There were some accidental exposures to plutonium in the early years of the Manhattan Project that are of interest with regard to the hazards of the material. The medical case histories of 25 men heavily exposed to plutonium at that time have been followed very carefully. The body burdens of plutonium in these cases ranged up to ten times the current maximum permissible body burden for occupational workers. After 27 years the men were judged to be in good health and there have been no cancers from the plutonium exposures (17).

None of this is to suggest that plutonium is not a toxic and potentially hazardous material. Plutonium is a material that has to be handled with great care, as are many other chemicals in current use. There are methods for handling plutonium safely just as there are methods for handling other toxic materials safely. That these methods work and are effective is shown by the fact that plutonium has been used in national defense programs for 30 years in amounts of thousands of pounds. It has been produced, chemically processed, converted to oxide and metal, transported, fabricated into weapons and reactor fuel assemblies, and used in all manner of experiments. In all of this long experience of handling and using large quantities of plutonium, there is no case where the toxic nature of the material has resulted in a single fatality. This excellent experience should be ample evidence that suitable precautions can be taken to avoid adverse health effects.

SUMMARY

Nuclear facilities are designed, built, and operated to stringent government standards. Component failures and malfunctions are guarded against by redundant and overlapping systems. Additional safety systems and special design features are provided to deal with a wide range of hypothesized design basis accidents and extreme natural phenomena such as great earthquakes. The public consequences of all of the design basis accidents and natural events are small, with no fatalities or acute illnesses from radiation exposure.

A recent detailed study of the safety of water-cooled reactor power plants typical of those now in operation has considered all accident possibilities, including those beyond the design basis. The probabilities and consequences were calculated for accidents leading to core meltdown, which is a necessary condition for any significant release of radioactivity. It was found that the chance of core meltdown was about 1 in 20,000 reactor-years of operation and that 99% of core meltdown accidents would cause no direct fatalities.

Comparable studies of probabilities and consequences for accidents in other nuclear facilities are not available at this time. As a matter of judgment, it is considered likely that high-temperature gas-cooled reactor plants would have accident probabilities comparable to those of water reactor plants but would have accidents of lesser consequence, whereas liquid-metal-cooled fast breeder reactors would have lower probabilities of large accidents, but consequences comparable to those of water reactor plants for the largest accidents possible.

The public consequences of accidents in other phases of the nuclear fuel cycle, including fuel reprocessing plants, final waste storage facilities, and transportation, would be much smaller than those possible from reactor plant accidents because the shorter-lived fission products have decayed away at these stages and because the high temperatures and pressures characteristic of the reactor systems are not present as driving forces for release of the remaining radioactive material.

The probability of a successful sabotage attempt at a nuclear facility, or of a successful theft of weapons-grade nuclear material for illegal purposes, is judged to be small but has not been quantified. The consequences of a successful sabotage attempt cannot be any greater than those of the largest accidents already considered for nuclear facilities and are very likely to be much smaller. The safeguards systems for protecting weapons-grade uranium and plutonium in the national defense program have been highly effective, but the government's nuclear regulatory staff has concluded that safeguards measures used in the nuclear power industry should be substantially upgraded if there is to be extensive use of plutonium as a reactor fuel.

Finally, plutonium is characterized as a very toxic material because of the hazard of radiation-induced cancer if ingested, but it is also one for which the range of hazards is comparable to that of other toxic materials used in industry or occurring in nature. The long and successful handling of large amounts of plutonium in the national defense program is cited.

Literature Cited

1. US Atomic Energy Commission. 1973. *Safety of Nuclear Power Reactors and Related Facilities.* Rep. No. Wash-1250. Washington DC: US AEC
2. Joint Congressional Committee on Atomic Energy. 1974. *Hearings on Reactor Safety.* Washington DC: GPO
3. Primack, J. et al. 1975. Nuclear reactor safety. *Bull. At. Sci.* 31:15–51
4. Hocevar, C. J. 1975. *Nuclear Reactor Licensing—A Critique of the Computer Safety Prediction Methods.* Cambridge, Mass: Union of Concerned Scientists. 94 pp.
5. Ford, D. F. 1975. *Nuclear Fuel Cycle: A Survey of the Public Health, Environmental and National Security Effects of Nuclear Power.* Cambridge, Mass: MIT Press
6. Lapp, R. E. 1974. *Nuclear Energy: The Nuclear Controversy.* Greenwich, Conn: Fact Systems, Inc.
7. Westinghouse Electric Corporation, Advanced Reactor Division. 1975. *Comments on the "Statement of Ralph Nader on the Plutonium Breeder Reactor Program before the Joint Economic Committee, United States Congress, Washington, D.C., May 8, 1975."* Madison, Pa: Westinghouse Electric Corp. 64 pp.
8. Lewis, H. W. et al. 1975. *Rev. Mod. Phys.* 47: Suppl. No. 1, pp. S1–S124
9. US Atomic Energy Commission, Regulatory Staff. 1974. *Analysis of Pressure Vessel Statistics from Fossil-Fueled Power Plant Service and Assessment of Reactor Reliability in Nuclear Power Plant Service.* Rep. No. WASH-1318. Washington DC: US AEC. 84 pp.
10. US Atomic Energy Commission, Advisory Committee on Reactor Safeguards. 1974. *Integrity of Reactor Vessels for Light-Water Power Reactors.* Rep. No. WASH-1285. Washington DC: GPO. 83 pp.
11. US Nuclear Regulatory Commission. 1975. *Reactor Safety Study—An Assessment of Accident Risks in U.S. Commercial Nuclear Power Plants.* Rep. No. WASH-1400 (NUREG-75/014). Washington DC: GPO
12. American Nuclear Society. 1974. Waste management symposium. *Nucl. Technol.* 24:263–456
13. US Atomic Energy Commission. 1974. *Environmental Survey of the Uranium Fuel Cycle.* Rep. No. WASH-1248. Washington DC: US AEC
14. Willrich, M., Taylor, T. B. 1974. *Nuclear Theft: Risks and Safeguards.* Cambridge, Mass: Ballinger. 252 pp.
15. US Atomic Energy Commission. 1974. *Generic Environmental Statement on the Use of Recycle Plutonium in Mixed Oxide Fuel in LWR's.* Rep. No. WASH-1327 (draft). Washington DC: US AEC
16. Cohen, B. L. 1975. *The Hazards in Plutonium Dispersal.* Dep. Physics, Univ. Pittsburgh, Pittsburgh, Pa. 43 pp.
17. Hempelmann, L. H., Langham, W. H. Richmond, C. R., Voelz, G. L. 1973. Manhattan Project plutonium workers: a 27-year follow-up study of selected cases. *Health Phys.* 25:461–79

ENERGY SELF-SUFFICIENCY ✵11025

Peter L. Auer
Sibley School of Mechanical-Aerospace Engineering,
Cornell University, Ithaca, New York 14850

INTRODUCTION

Historical Background

The outbreak of war in the Middle East during October 1973 and the subsequent oil embargo imposed on the United States brought a new slogan to the forefront—"energy self-sufficiency." As a description of US energy policy, the concept of energy self-sufficiency has undergone a variety of transformations in the ensuing 18 months, subject to a variety of interpretations and even changes in name. By 1974 we began to hear instead about energy independence, and by 1975 the slogan-coining voices were becoming more and more muted. Energy policy is far too complex an issue to be described in terms of a single notion, be it self-sufficiency, independence, or other high-sounding calls to action. Indeed, it is my task here to chronicle some of the various attempts that have taken place during the past two-year period (starting with the first quarter of 1973 and ending with the first quarter of 1975) in trying to formulate a consistent national policy on energy.

A brief historical perspective may be in order before beginning the task proper. It would be somewhat difficult to argue that the United States has ever enjoyed an all-embracing energy policy at the national level; it might be argued just as well that there was little need for one in the past. That is not to say that the federal government has had little influence on the conduct of energy business in this country. Much of this is manifest through the regulatory responsibilities assigned to the federal government by statute. Thus, regulation by the Federal Power Commission (FPC) of prices for natural gas in interstate markets·derives from the government's constitutional authority over interstate commerce. The fact that this authority in the case of gas has been extended all the way to the producing wells in the field is somewhat of an historical accident. As a result, there exists a degree of anomaly in that the intrastate sale of natural gas follows "free market" trends while interstate sales must conform with the mandatory rules of the FPC.

The government's attitude toward oil, at least until recent times, has been rather different. Domestic oil production has always been encouraged and allowed to operate under terms that usually favored the producer. When domestic production appeared to exceed demand, threatening to drive down prices, the producing states were allowed to introduce prorationing rules to limit production and encourage reservoir conservation. On oil issues the federal government saw no need to interfere on behalf of the consumer, while at the same time it was regulating natural gas to protect the consumer's interests. In nature, gas is often found along with oil. As a result, a producer would in many instances find himself in a position where, on a heating value basis, he would obtain three to four times as much money for his oil as for his gas.

Natural gas is relatively expensive to transport over long distances. Thus there is sound reason why markets in the Northwest and Midwest purchase their gas from the closer Canadian fields rather than from the more distant Southwest. On the other hand, the US net import level of oil and refined products was imperceptible until the early 1950s, and we remained a net exporter of oil and petroleum products up to 1950. As the balance of oil trade began to threaten to move in the direction of net US imports and there was a danger of US markets succumbing to cheap foreign oil, the Eisenhower administration moved quickly to impose import quotas in order to protect the domestic petroleum industry. US oil prices were allowed to exceed world oil prices by substantial factors. Domestic production continued to increase but was unable to stay even with growing demand. The original quota system had to be revised periodically; the heavy demand in the Northeast for fuel oil and residual oil used to fire steam boilers could only be met by imports, and these items were eventually removed from quota restrictions.

Public attitudes are liable to change, however, and public opinion these days is against the large oil companies as a reaction against the seemingly large profits accumulated during the hectic rise in world oil prices following the embargo. When in early 1973 the Nixon administration removed all controls over prices and wages, the price of domestic oil remained under control. Subsequently a two-tier price structure was installed. New oil, brought into production after 1972, has not been controlled, whereas old oil, from wells in production during 1972, must be sold under controlled conditions. There are some exceptions to these rules, meant to encourage small producers; but in effect, federal laws exist at the time of this writing whose purpose is to control the price of oil and interfere with the so-called free market forces.

Not only does the federal government influence the market in oil and gas by direct price regulation, but it can achieve similar effects through its taxation policies. The recent removal of the depletion allowance from large oil producers is an example and one that is in keeping with the punitive attitude the Congress has recently adopted with regard to the major US oil producers. Both price regulation and taxation policies along with import restrictions can act as subsidies benefiting the producer and the consumer, depending on how such policies are formulated. Traditionally, governmental policy has tended to favor the oil producer, maintaining domestic prices artificially high with respect to world prices. The situation is currently inverted and average domestic oil prices are below world prices. The American motorist receives what amounts to a federal subsidy, enabling him to pay one half to one third less for his gasoline purchases, per unit amount, than his European counterpart. In terms of real disposable income, the disparity is even greater. The removal of the depletion allowance, however, will tend to moderate this subsidy. One may conclude on this basis that present US policy is to stimulate the use of petroleum products by rules that maintain them at price levels less than the free market would establish. The same remarks apply with equal validity to natural gas in interstate commerce. Interestingly enough, the stated policy of the executive branch is exactly opposite to the conclusions drawn above. This example is introduced here merely to illustrate the difficulties of trying not only to establish national policy but even to determine definitively the nature of the policy.

In addition to direct and indirect regulatory measures, as illustrated above, the federal government has other means for exerting influence on energy markets. One of its most important instruments is the offering of mineral rights for lease on federally held public lands. The various land acts of the last century left to the federal government vast tracts, mostly in the West, that are rich in coal, lignite, oil shale, and other bitumens. Some are known to contain oil and currently form the so-called Naval Petroleum Reserves; some are known to contain gas, although most of this is in a form not economically recoverable by present techniques; some contain uranium and thorium; and still others may contain geothermal resources. The mineral-bearing land owned by the federal government that is by far the most significant for present purposes lies underwater on the continental shelf, beyond the three mile limit to individual state ownership. How the government chooses to dispose of its mineral rights, what leasing policies it adopts, and on what time scales it makes tracts available for bids can determine in large measure the pace at which domestic energy production will proceed.

I have enumerated two primary roles that the federal government plays in the energy sector—its regulatory role, including its powers to set taxes and govern imports, and its role as the proprietor of mineral-bearing public lands. However, the government's regulatory role has increased considerably in recent years for reasons not yet mentioned. Over the past decade the public has become increasingly concerned with environmental quality. A number of laws have been placed on the statute books which require the federal government to set environmental standards and see to it that these are enforced either at the federal, state, or local level. Aside from the automobile industry, the coal industry and the electric utilities are most significantly affected by new environmental regulations.

Coal has been relatively unaffected in the past by government rules except that, as mentioned above, much of western coal lies on federal land and must be obtained through lease. However, the safety of coal mining does come under federal scrutiny, and the rules aimed at reducing injury and health hazards to coal miners have been tightened very considerably over the past five years. Coal mine operators claim that there is a direct correlation between the onset of these rules and the relatively sharp decline in miner output or productivity during the same period. The coal mining industry is about to be faced with still another regulation that is an outgrowth of the environmental movement, namely, restrictions on strip mining or surface mining. By now, slightly more than one half of the total annual coal production is surface mined. Many examples abound of past irresponsibility where land stripped of coal has been left in destitute condition. The Congress is determined that such practices be discontinued and that federal standards on land reclamation be observed. Although executive and legislative branches have had difficulty in agreeing on the proper phrasing of strip mining legislation, there is little doubt that the coal industry will be forced to abide by the new rules now in the making.

Perhaps an even more significant consequence of the new environmental rules has been their effect on markets for coal. Most coal is used and mined in the East. It is used primarily to raise steam in boilers (the form often referred to as steam coal). Unfortunately the average sulfur content of eastern coal is high. When burned directly, the combustion products of high-sulfur coal release sulfur oxides into the atmosphere in concentrations that exceed the air quality standards mandated by federal regulations. As a result the electric utilities, the principal users of coal, have found it increasingly difficult to guarantee coal a firm future market. Certain eastern utilities, mostly in the large metropolitan areas, have phased out coal completely in favor of oil. In order to continue relying on coal, other utilities are being forced to introduce new and not yet thoroughly tested scrubbing equipment designed to prevent the escape of sulfur oxides into the atmosphere. Until the utilities can be convinced that a reliable technology for sulfur removal will allow them to use high-sulfur coal without violating environmental quality standards, the future market for coal must remain somewhat uncertain.

The above condition illustrates another complicating factor in the search for consistent energy policy at the national level. Coal represents our most plentiful domestic resource of conventional fuels. Nothing would please many of our policy formulators more than if coal could be used in ever larger proportions as a substitute for our less abundant oil and gas. Yet, as of now, coal's traditional market in the electric sector is inhibited. This is but another example of internal conflicts that seem to beset the search for national policies. In fact, some critics have challenged the basic notion that energy policy should be set at the national level. In a variety of ways, often quite uncoordinated, governmental action can have profound influence on energy markets. But energy is a highly diversified industry, and some would like to see it continue operating with a minimum of interference, each element interacting with its federal counterpart through the patchwork of arrangements already in existence. This has been the traditional way of doing business, and the rules of the game in the past have rarely been elevated to what could be appropriately called a national energy policy.

There is one rather important exception to the above observation, and it has to do with the government's involvement in promoting the rise of commercial nuclear power. Here we have a clear example of the federal government adopting, of its own choice, a consistent policy aimed at making a place for nuclear energy in the commercial sector and stimulating its growth. Nuclear energy was a wartime development and at its outset was a government monopoly. The notion that there were potentially significant peaceful civilian applications to what had been exclusively a military program occurred to many, even during the war years. It was the Eisenhower administration, however, that made it a matter of urgent national policy to insure the emergence of a civilian nuclear industry. The promotion of nuclear energy has been, until recently, the major preoccupation of the federal government's energy research and development program.

Whether by intent or not, the federal government has also shown a tendency in favor of electrification. Rural electrification policy was made by design. But, government's role in promoting nuclear power is of direct benefit to the electric utilities as well. In fact, if one examined the structure of federal support in fields related to energy research and development up through fiscal year (FY) 1974, one would find better than three fourths of total funds spent on projects related to electricity. In addition, the several federally mounted regional development programs that started out as flood control projects and became hydropower construction projects have invariably led to increases in electrification. As a matter of fact the Tennessee Valley Authority (TVA) is now the single largest electric utility in the country. It is hardly a hydropower complex any longer, since it is also the single largest consumer of coal in the country. According to recent reports, TVA would have liked to become the largest coal producer in the country also via the purchase of the Peabody Coal Company.

The federal government is involved in the nation's energy business in a number of ways, some, but certainly not all, of which have been listed here. However, in many respects federal authority stops at state boundaries and the individual states also have considerable influence over the workings of the energy market. The individual states primarily regulate the local monopolies operating within their area, for example, the electric and gas utilities. They may, in some instances, also have siting authority, requiring utility power plants, petroleum refineries, gas processing plants, and others to obtain licensing from state boards in addition to any federal licensing boards (e.g. for nuclear plants, hydropower units, and interstate transmission lines). Coastal states have leasing rights up to three miles offshore, and several western states have substantial coal lands for lease. More recently there has emerged a regional concern on the parts of states that are destined to become energy producers through the exploitation of virgin lands. Thus land planning, a matter traditionally left to state and local authorities, is becoming intertwined with energy planning.

Concern over energy planning in a coordinated manner at the national level is a matter of relatively recent origin. As long as energy was abundant and cheap there seemed little need for serious national interest. Isolated issues, such as oil import restrictions, could be dealt with on a case-by-case basis, and no attempt was made to coordinate these resulting decisions within the framework of a coherent policy. Energy oversight, to the extent it exists, is exercised at various levels of government and by a multiplicity of agencies. Most important perhaps is the fact, not stressed until now, that the production of energy and its delivery to the ultimate customer is largely in the hands of the private sector. This sector is organized into a variety of constituencies, each of which has found through experience how best to achieve a working relationship with its governmental counterparts, whether they are regulating agencies at the federal or state level or appropriate committees of the Congress. Orchestrating this diffuse and fragmented collection of institutional arrangements to a common purpose under the banner of "national energy policy" is difficult at best.

Origins of the "Energy Crisis"

The current popularly held view that there is a need for a single, all-embracing, national energy policy, centered about a somewhat ambiguously defined concept of self-sufficiency, was slow in surfacing. Eventually it did arrive with something of a thunderclap. Historically, energy has always been cheap and plentiful, if not by design, then by good fortune. Trends of interfuel substitutions and the rapid growth of electricity could be readily accommodated at the market's dictate. As the demand for energy increased and indigenous supplies began to diminish, the slack was taken up by imports. During this period the price of energy relative to the cost of other goods tended to decline steadily. This was the pattern observed in the United States for most of the postwar years, and it has held true for most of the other consuming nations. Under these circumstances it is not very surprising that relatively little coordinated attention was paid to energy policy. The market appeared to be more than adequate for taking care of all the pertinent issues. By 1969, however, significant changes affecting both energy policy and its research and development began to appear, particularly in the United States.

These forces of change first appeared in terms of environmental concern. Although it has been recognized for some time that pollution was degrading the quality of life in the industrialized nations, it took traumatic events to move governments to take radical action in favor of environmental protection. In England it was the London fog episode of 1952 that finally triggered new measures for air pollution control. In the United States it was the Santa Barbara blowout that catalyzed the government to adopt the sweeping measures of the National Environmental Policy Act (NEPA).

The production, transport, and utilization of energy soon came into serious conflict with what had been largely unanticipated environmental issues. The situation in the United States was considerably more critical and more vocalized than elsewhere. Witness, for example, the long delay in the Trans-Alaska Pipeline, which was originally conceived in an atmosphere where environmental concern was not paramount. As a result, its proponents incurred considerable criticism and had to make major design changes and undergo extensive litigations in the courts before a final construction permit was issued. The landmark Calvert Cliffs court decision, interpreting the National Environmental Policy Act of 1969, is another example wherein environmental considerations have resulted in extending nuclear reactor licensing procedures in the United States by one to one and a half years. Unfortunately there has been, particularly in the United States, a certain amount of overreaction by both those who would protect the environment at all costs and those who would delay environmental safeguards as long as possible; a strong polarization of views has resulted.

By June 1971, events had progressed to the point where President Nixon felt it appropriate to issue the first message ever given by a US president that was devoted exclusively to the subject of energy (1). Although the message expressed concern about potential shortages in energy supply, the principal emphasis was on the need for clean forms of energy, that is, the need to resolve conflicts between environmental care and energy supply. Accordingly, the message stressed the federal government's support of three research and development programs: sulfur oxide control technology, which would permit stationary power plants to burn high-sulfur coal and oil; the liquid-metal fast breeder reactor (LMFBR) as the best hope for economical clean energy in the future; and coal gasification as a means for supplementing dwindling supplies of natural gas with abundantly available coal. It should be noted, however, that the lion's share of expenditures went to the LMFBR and that each of these programs called for nominal (to a varying extent) cost sharing by industry. The need for an integrated national energy policy was recognized in this message, and a new cabinet-level Department of Natural Resources was proposed as its focal point.

Within the next year and a half the specter of energy shortages became more stark, and in April 1973, President Nixon delivered his second energy message (2). The main thrust of this one was to announce the demise of the Mandatory Oil Import Program and prepare the United States for greater reliance on foreign oil. No new research and development initiatives were announced, but the message did stress that a 50% expansion had taken place in federal efforts over the previous two years and an additional 20% increase in funds was programmed for the coming year. The proposed Department of Natural Resources was now renamed Department of Energy and Natural Resources in order to emphasize its proposed central role in energy policy. In addition, a new Office of Energy Conservation was created within the Department of Interior to reflect the emerging importance of decelerating energy demand growth, and the State Department was called upon to begin examining ways to increase cooperation with other nations on energy research and development (R&D).

The April message fell on some skeptical ears and was soon followed by a third message in June 1973 (3). From the point of view of R&D, the June message was of historical importance, for in this instance President Nixon singled out R&D as a major instrument of energy policy, proposing that 10 billion dollars be spent by the government over a five-year period beginning with FY 1975 to assure the nation's long-term needs through the development of new forms of energy. In order to lend impetus to the accelerated pace of research, an additional 100 million dollars (eventually to be 115 million) was proposed for the energy R&D budget in FY 1974. In effect the President was proposing that an effort receiving approximately 20% increases per annum for the past several years now receive a quantum jump of nearly a 100% increase in one year to reach a new level of federal expenditures.

Why this increase in support for energy R&D by the federal government and how were the numbers determined? Fundamentally, the figure of 10 billion dollars was not meant to be a fixed commitment but rather an indication of the strength of the commitment. In 1960, President Kennedy committed the nation to go to the moon by the end of the decade and indicated that the price would be 10 billion dollars. The first US astronauts did indeed reach the moon within the allotted time, but the Apollo Program cost closer to 30 billion dollars. The Nixon commitment to energy R&D can be interpreted in the same vein. But why in the summer of 1973, when the United States was facing prospective gasoline shortages and possible fuel oil shortages the following winter and perhaps even brownouts of electricity, was R&D singled out for special attention? The US public had become accustomed to marvels of technical accomplishments. Nuclear weapons had brought security to the nation, or so it seemed. The success of the space program had brought prestige and more power. Now a dramatic pledge had been made that America's technological prowess would be brought to bear on solving the nation's energy problems.

During the first half of 1973 the United States perceived that it had a serious energy problem. It was obviously heading for greater and greater dependence on foreign oil. Essentially all increases in oil imports would have to come from the Middle East, which until then had not been a significant supplier of American markets. It was predicted that by 1980 about one half of the US oil requirements would have to be met by imports. Meanwhile there was a shortage of domestic refinery capacity, problems in the distribution system, and uncertainties about prices.

Events soon came to a head with the October conflict in the Mideast and the imposition of oil embargoes, and the full power of the oil cartel unfurled. The United States found its energy policies overtaken by international considerations; its position in world affairs as a major power would be seriously compromised if its economic well-being became the hostage of some foreign oil-producing nations. It was inevitable that the Nixon administration announce a program promising energy self-sufficiency in the near future, and so Project Independence came into being. When this seemed to cast the United States into an isolationist position, the modifier *Interdependence* was appended to the project and active cooperation on the part of the major consuming nations was solicited. This was dramatized when Secretary of State Kissinger called a conference of foreign ministers from the major consuming nations in February 1974. The conference led to the creation of an Energy Coordinating Group and subsequently to the recently formed International Energy Agency (IEA).

The situation now confronting the industrialized nations is divided roughly into three categories. For the United States the goal of energy self-sufficiency (this term will have to be defined more precisely) appears to be a matter of protecting its major power position internationally. For the members of the Common Market, or European Economic Community, and Japan, the matter of greatest concern is the sharp rise in oil prices and consequent balance of trade deficits. On the other hand, the USSR and its close trading partners remain relatively unaffected. The call for sharply increased cooperation and coordination of energy policies along with R&D activities is, therefore, of varying interest to these groups of nations.

The conventional wisdom that seems to be developing among the consuming nations is that for the immediate future the oil storm caused by gyrating prices and uncertain availability must be ridden out. A principal weapon is energy conservation. If the market can be given free reign, it should establish an equilibrium without creating international financial havoc. Meanwhile all appropriate steps must be taken to increase the role of coal as an energy source and to bring nuclear power on line. Longer-range research efforts should examine the potential of alternative energy sources. Translating conventional wisdom into policy and political implementation often follows a tortuous path, however, particularly in a country such as ours with its diversity in institutional structures and pockets of vested interests.

Policy Studies and Results at the Federal Level

The cry of "wolf" is no stranger to the US energy scene. It is conceivable, though not easy to document, that over a hundred years ago some students of energy viewed with alarm the potential future inability of the whaling industry to stay abreast with the growing demand for whale oil. (Today this is no longer a laughing matter since the United States treats the sperm whale as an endangered species, yet its oil is essential in many applications where high-pressure lubricants are required.) More to the point, many experts were convinced in the early 1930s that the United States was well on the way to depleting domestic crude oil supplies. Then came the discovery of the fabulous East Texas fields and the market soon had an oil glut.

Concern over the adequacy of resources, including energy resources, arose periodically at the presidential level. President Truman appointed a commission to examine this matter (4); their findings, commonly referred to as the Paley report, warned the country as early as 1952 that serious shortages in energy supplies could develop in coming years. The warning went largely unheeded, for it came at a time when domestic fuel production showed a continuous rise and supply and demand appeared to be in good balance, at least over the short term. It is in general difficult for the body politic to respond to long-range forecasts when there is no clear and present danger.

Studies on energy policy matters at the federal government level continued under succeeding administrations. Some were institutionalized, for example, the periodic FPC reports concerned with electric power or natural gas. Many other studies on energy supply and demand were published under both private and public sponsorship. Space does not permit a review of these early studies here; this chapter is more concerned with how government policy evolved. The actions of the Eisenhower administration have already been mentioned. As they relate to energy policy, these actions resulted in the United States adopting a protectionist attitude with respect to oil and a commitment to the growth of commercial nuclear power. The instrument for the latter was the US Atomic Energy Commission (AEC) under the vigorous leadership of its chairman, Admiral Lewis Strauss. In retrospect there is now clear evidence from the nuclear experience that the federal government could exert considerable influence over the conduct of the private energy market through federal intervention at the R&D level.

Responsibility for the formulation of energy policy and management of energy R&D at the federal level has been dispersed throughout a number of agencies within the US government. Within the past decade, however, a number of occasions have arisen when an integrated assessment has been made of the entire field encompassed by energy R&D. The first comprehensive report of this nature was the interdepartmental study (5) of 1963–1964. This study was prepared in response to a directive by President Kennedy, issued February 15, 1963. At that time the President had in hand a white paper prepared by the AEC in 1962 setting forth its proposed program for civilian nuclear energy. He also had surveys and projections on future national electrical requirements, which were periodically prepared by the FPC. The President felt that it would be appropriate to view these various recommendations in the framework of a coherent energy R&D assessment, noting that "the amount and allocation of federal research and development in the energy field will affect the efficiency of various components of our energy system, and, consequently, the rate and pattern of our national economic growth." He went on to say, "I direct that a comprehensive study be undertaken of the development and utilization of our total energy resources to aid in determining the most effective allocation of our research and development resources" (5).

Implicit in these statements was the understanding that the federally supported effort was not balanced by design, yet could exert large leverage on how the energy system evolved. The study concluded that although the United States did not appear to be resource limited, several deficiencies existed.

1. There was no long-range, integrated plan for civilian energy R&D.
2. Initiation of essential research with long-range benefits was being delayed.
3. Insufficient support was being given to devices and systems whose promise had not been demonstrated.
4. Certain existing resources were being neglected.

No apparent changes resulted in overall federal policies as a result of this study. At an earlier date the Kennedy administration had already established a new Office of Coal Research within the Department of Interior, and one could say this was a first, albeit very small, step in trying to diversify the federal role from its previous heavy concentration in the nuclear field. As a result, by 1971 President Nixon could announce that the Office of Coal Research's joint program with industry on converting coal to synthetic pipeline gas ranked along with the LMFBR as one of the nation's high priority efforts (1).

Although a number of studies were undertaken at both private and public expense subsequent to 1964 which explored the spectrum of energy R&D activities and opportunities, the next serious examination at the federal level was undertaken soon after the energy message of June 1971. The study was conducted under the auspices of the Office of Science and Technology. Unlike the earlier 1964 study, this one made specific budget recommendations in all possible areas not already highlighted in the President's earlier message. Thus, for example, the LMFBR program was not reviewed. The final summary report of this study (6) was issued in November 1973 and is commonly referred to as the FCST (Federal Council for Science and Technology) study.

The FCST study was carried out in response to President Nixon's request of June 4, 1971 that his Science Advisor "make a detailed assessment of all of the technological opportunities... and... recommend additional projects which should receive priority attention." By the time the study was completed, events had overtaken it, for the United States was rapidly becoming enmeshed in its "energy crisis." As mentioned earlier, President Nixon had already decided by June 1973 to call for rapid acceleration in energy R&D, supplemental appropriations for FY 1974, and stepped-up funding for FY 1975. He went on to say,

> I am directing the Chairman of the Atomic Energy Commission to undertake an immediate review of Federal and private energy research and development activities, under the general direction of the Energy Policy Office, and to recommend an integrated energy research and development program for the Nation. By December 1 of this year, I am asking for her recommendations for energy research and development programs which should be included in my fiscal year 1975 budget.

The resulting report (7) has set the pattern for existing R&D policies in the federal government and represents the first serious effort in developing an integrated federal support budget.

In parallel with the above events, the Nixon administration took steps to put its energy policy-making institutions in order. By early 1973 a need for more central coordination over energy matters became apparent. A troika consisting of the Secretaries of State and Treasury plus the President's economic adviser were placed in charge. This ad hoc arrangement lasted until June 1973 when the President announced the creation of a new Energy Policy Office within the White House and placed John Love, the former governor of Colorado, in charge. The Energy Policy Office was replaced by the Federal Energy Office (FEO) in the fall of that year and Governor Love was replaced by Deputy Secretary of the Treasury, William Simon. The press quickly dubbed Mr. Simon the nation's "energy czar." All of these rapid changes were made by executive action and reminded one of Lincoln searching for a general.

By early 1974, the Congress had acted on the administration's request to create a new Federal Energy Administration (FEA). John Sawhill, who had been Simon's deputy earlier, became the first head of FEA, Mr. Simon having been named Secretary of the Treasury sometime earlier. Congress felt the FEA was needed primarily to allocate oil and its products and to mitigate the adverse effects caused by the oil embargo. However, the FEA, like its predecessor, the FEO, saw itself as not only a regulatory agency but also the focal point for energy policy formulation. It therefore launched an ambitious study program initially termed the Project Independence Blueprint; the implication was that this study would chart the course toward self-sufficiency and independence as proclaimed in various presidential announcements and messages.

Meanwhile the administration was moving to centralize its energy R&D functions. By the end of 1974, legislation had been enacted to separate the regulatory role of the AEC and transfer it to the newly created Nuclear Regulatory Commission (NRC). The remaining AEC functions were absorbed by the newly established Energy Research and Development Administration (ERDA) and the AEC was disbanded. ERDA also absorbed the Office of Coal Research and the energy centers of the Bureau of Mines, both previously in the Department of Interior, along with certain energy programs previously with the National Science Foundation and the Environmental Protection Agency. Thus, ERDA became the focal point for federal support of energy R&D and inherited some 85% of the government's outlays in this sector.

By the time the FEA study was completed in November 1974 it was no longer a blueprint; instead it was issued as the Project Independence Report (8). Also, the administration had changed hands with President Ford's assumption of office in August. The new administration implemented Congress' desire that a high-level coordinating group for energy policy be established by creating the Energy Resources Council (ERC) with the then Secretary of Interior, Rogers Morton, as its head and Frank Zarb, who subsequently replaced Sawhill at FEA, as its executive director. Robert Seamans, Jr., the newly appointed head of ERDA, also was named a member of ERC. The federal government had finally achieved its goal of trying to bring some organizational order to the process of energy policy formulation with the FEA and ERDA as new foci and the ERC as the principal coordinator. However, the dream of a cabinet-level Department of Energy and Natural Resources, which would combine under one roof both FEA and ERDA and most of the functions of the Department of Interior, remains just that, a dream. This is the situation in early 1975; I examine next the country's progress toward energy independence.

ELEMENTS OF ENERGY POLICY

Units

The terms *energy policy* and *fuels policy* are often used interchangeably and this can lead to some confusion. Inanimate energy, as required by a highly industrialized society such as ours, has been traditionally provided by primary fuels (e.g. wood, coal, oil, natural gas, uranium). Consequently, one normally counts the amount of gross energy input in terms of quantities of fuels used (e.g. tons of coal, barrels of oil, etc). What is sometimes overlooked is that some of these minerals are used for non-energy purposes. For example, at current levels of consumption, approximately 10% of the coal mined or oil produced annually goes to non-energy use (e.g. metallurgical processes, lubricants, petrochemicals, etc); a comparable amount of natural gas is consumed as feedstock for the manufacture of fertilizers, drugs, etc. This follows the more or less conventional accounting method of treating gross energy input in terms of fuel equivalent, without disaggregating into non-energy and energy quantities. Another element of confusion is how to count electricity produced by hydropower inasmuch as this is useful energy derived from gravity and not primary fuels. The practice adopted here, dubious as it may be, is to treat hydroelectricity in terms of fuel equivalent—that is, to convert to heat units using the heat rate of steam plants (i.e. relative efficiency) rather than the physical ratio of electrical energy units to heat. This method of accounting will have to be scrutinized more carefully in future years when geothermal, solar, and other nonprimary fuel sources become more significant contributors to the energy sector.

The unit of accounting is based on British thermal units (Btu), where the unit of preference is $Q = 10^{18}$ Btu or 1.05×10^{21} joules. Rates of consumption are given in $mQ = 10^{-3}Q$. For approximate comparisons, one can readily convert from heat units to fuel units by noting: $1\ mQ = 1\ Tcf(10^{12}$ cubic feet) of natural gas, or 0.5 Mb/d (10^6 barrels per day) of oil times 365 days or 50 million tons of an appropriate mixture of eastern and western coal. Similarly, $1\ mQ = 10^{11}$ kW-hr of electrical energy, assuming a heat rate of 10^4 Btu/kW-hr. This amount of electrical energy could be produced by approximately 20 GW (10^9 watts) of installed steam-generated electric capacity operating for 365 days at a capacity factor of about 65%, that is, 40 GWe is roughly equivalent to 1 Mb/d of oil. These approximate conversions are valid within a 10% margin of error, given current efficiencies of fuel utilization and plant reliabilities.

For purposes of comparison with the above figures, the gross energy consumption in the United States was 72 mQ in 1972 and that of the entire world was about

225 mQ. A midrange estimate of the total recoverable crude oil of the world is 12 Q, and of this 1.5 Q has already been consumed. The amount of recoverable gas in the world is a comparable figure. The reason for particular concern with oil and gas is that these liquid and gaseous fuels have been supplying roughly three fourths of the energy requirements in the United States and throughout the world. Coal is often mentioned as being far more plentiful than oil or gas. While in principle the statement is correct, if we consider only those deposits economically recoverable by present techniques, the worldwide figure is about 25 Q. This can be doubled, in effect, to 50 Q if we include the so-called para-marginal deposits. Cumulative production of coal in the United States alone has been about 1 Q to date. The United States is believed to have somewhat less than 10% of the world's oil and gas resources and perhaps one third of its coal resources. One more factor should be taken into account in this brief primer; if the growth rate of energy demand remained constant and equal to the values attained in the early 1970s, then the world's demand for energy would double about every 15 years. As shown below, the US demand growth rate has been less rapid than the worldwide average; furthermore, it is quite unrealistic to extrapolate exponential growth rates ad infinitum.

Demand

In the above some of the major ingredients to any energy policy have been mentioned. At its apex, policy must concern itself with the central issue of balancing supply and demand. In a free economy, as opposed to a centrally planned economy, one would normally assume that the market dictates how demand and supply come to equilibrium. In fact, however, governmental actions may interfere with the workings of the marketplace, as shown in examples cited earlier. A fundamental fact to bear in mind, therefore, is that for energy the concept of a free market is a theoretical abstraction and not a realistic fact.

While the identification of what factors lead to final demand is still a poorly developed science and worthy of serious scholarly study, it is possible to list a few major categories, each of which might contain a long list of subfactors.

Energy Demand:
  population
  economic activity and composition
  societal life-style
  efficiency of conversion and utilization
  government policy

For the United States the growth of energy demand with that of population does not appear to be a crucial issue since our population growth seems well stabilized at this juncture. On a worldwide basis, however, the facts are quite different. There is ample evidence that economic activity and energy demand are strongly coupled to each other; however, the composition of the economy will in turn determine its degree of energy intensiveness. For example, a country like Luxembourg whose economy is dominated by heavy industry would be expected to consume more energy per unit of economic productivity than a country like Switzerland whose economy is dominated by light industry, commerce, and tourism. This may be seen from the data in Figure 1. The relationship of economic prosperity to energy consumption is an essential element to energy policy, for it couples the latter to economic policy and the general national welfare. The degree of dependence, economy vs energy, can be influenced by a variety of factors.

Gradual and steady improvements in the efficiencies of energy conversion and utilization have shown up as historical trends indicating decreasing ratios between per capita energy consumption and per capita real GNP, as indicated by Figure 2. Changes in the nation's fuel mix can reflect similar trends if in fact it takes less energy to produce and deliver gas and oil than coal or wood, for example. Some evidence for this is shown in Figure 3. The life-style preferred by a given society will also enter into this equation. The level of comfort in terms of temperatures maintained during summer and winter plus the levels of illumination in residential, commercial, and industrial structures is but one example. The preference for suburban living, mobility, and decreased reliance on mass transport furnishes another, and both examples, in their own way, can lead to increases in per capita energy consumption without a compensating increase in economic productivity. One analyst (9) has proposed that there is a direct correlation between increased per capita energy consumption and the increase of women in the labor force. There are far too many other potential factors that could be listed here—the growing energy intensiveness of agriculture and the food industry, the displacement of natural products by synthetics, etc are but some examples.

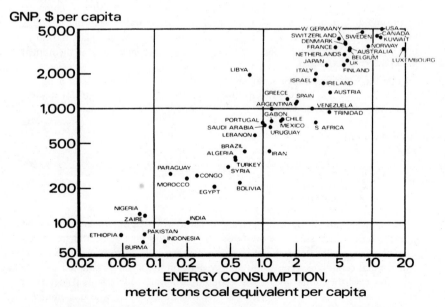

Figure 1 Relationship between GNP and energy consumption per capita for 52 countries—1971 data. Source: Institute of Gas Technology, Chicago.

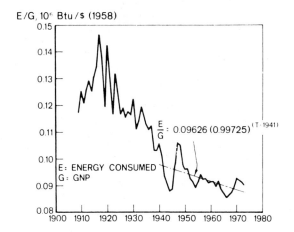

Figure 2 Trend of energy consumption per unit of GNP in the United States. Source: Institute of Gas Technology, Chicago.

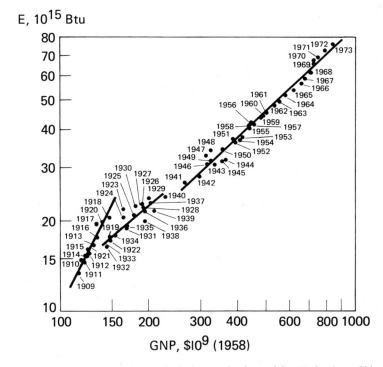

Figure 3 US energy consumption vs GNP. Source: Institute of Gas Technology, Chicago.

Finally, as observed earlier, governmental policy can exert significant influence on demand for energy. Depletion allowances and price controls can lead to lower costs for the ultimate customer and can stimulate demand. Others will argue that regulations in favor of the customer are necessary to protect him from the absence of sufficient competition in the market (virtual monopolies). There are always two sides to such arguments and their logical resolution is difficult if not impossible. Recently more attention has been given to identifying regressive regulatory practices that promote the wasteful expenditure of energy—for example, Civil Aeronautics Board regulations that have the effect of depressing the load factors of air lines in the attempt to maintain effective competition over given routes. Other examples could be furnished involving the interstate transport of goods, etc. An attempt to reform the multitude of governmental regulations in order to eliminate any unnecessary demand for energy has been started by the present administration.

Supply

The other side of the equation, supply, must now be examined briefly. Here, again, some broad categories encompass the large number of factors affecting supply.

Conventional Energy Supplies:
    resource limitations
    production limitations
    capital availability
    environmental and safeguard constraints
    market forces and interfuel competition
    technological and economic limitations
    government policy
Alternative Energy Supplies:
    potential impact
    technological and economic obstacles
    government policy

Among conventional energy supplies are listed the fossil fuels, hydropower, such geothermal sources as dry or wet steam, and, of more recent vintage, nuclear fuels. The fossil fuel and nuclear resources, along with mineral resources in general, are finite and exhaustible. Hydropower and geothermal sources are replenishable but are rate limited (i.e. by water flow, heat flow, etc) and opportunity limited (i.e. there are just so many watersheds or thermal anomalies that can be usefully exploited). Before any of these sources can contribute meaningfully to the energy supply they must be discovered and developed into producible reserves. This takes time and money, the amount varying to a considerable degree depending on the source. As resources become depleted, the discovery rates will diminish and the attendant costs will increase (i.e. amount discovered per number of feet drilled, etc will decrease with time). The time lag between discovery and production is also a widely varying function of source; that is, it may take three years to open a surface mine for coal and five years for an underground mine, five years to develop an oil field onshore and eight years for one offshore, eight to ten years to develop a uranium mine and milling complex, and so forth. Consequently, bringing fresh energy supplies to market requires a long planning horizon and may, at any given time, be production limited rather than resource limited if earlier planning was faulty.

None of the above can occur, of course, unless sufficient capital is attracted to the specific project in mind. Capital requirements also tend to vary greatly. It may take ten times the money to develop an offshore oil field in the coastal waters of the United States than one onshore in the Persian Gulf area, for a fixed unit of production capacity. At any given time capital is either limited and allocated by an unhindered market or influenced by government intervention. All other things being equal, capital tends to flow to ventures where return on investment appears relatively high and risk appears relatively low.

Productivity, which takes many of the above factors into account, can be seriously affected by environmental and safeguard considerations. More stringent standards for air quality, thermal discharge, and radiation release can add significant costs to the construction of stationary power plants. Environmental concerns can add significant delays to construction schedules for energy conversion plants, also resulting in added costs, or to the exploitation of virgin lands for the production of mineral fuels both onshore and offshore. Concern over nuclear safeguards is delaying the commercial utilization of plutonium as a nuclear fuel. It is neither possible nor desirable in today's world to plan for future supplies of energy without due regard for their potential environmental impact or hazard to the public.

In principle, the market in some perfect way should be the final arbiter of energy supplies. In fact, however, the market can never be completely free of government influence. As a result of a combination of forces, we have witnessed the gradual displacement of coal from its markets by the cleaner, more easily handled liquid and gaseous hydrocarbons, petroleum and natural gas; we have seen that the influx of foreign oil could be controlled until finally domestic production could no longer satisfy demand; we have observed electricity generation grow more rapidly than other forms of energy and capture a relatively greater share of the market; and, finally, we have experienced the advent of commercial nuclear power.

Civilian nuclear power is a perfect example of how government policy in the form of public expenditures on research, development, and demonstration along with guaranteed fuel supply (i.e. enrichment and stockpiled uranium) plus other incentives can stimulate the commercialization of a brand new technology. There is a long list of other alternative energy sources that in the minds of many could follow a similar road, given adequate government support. The purpose here is not to examine the relative merits of the various proposals since this is a subject covered more thoroughly elsewhere. It should be noted, however, that the commitment of the government to nuclear energy was made in the early 1950s and the impact of this commitment was hardly measurable in the commercial sector for 20 years thereafter. It will be at least another 20 years before nuclear power can grow to a commanding position in the energy market, if then. Thus, any of the alternative energy sources now under consideration and still in early stages of development can hardly be expected to make significant contributions to energy self-sufficiency or independence before the end of the present century.

If we are to limit our time horizon to no more than 10–15 years in search of an energy policy consistent with national goals, then on the supply side, energy policy will still amount to a fuels policy or some mixture of policies governing fuels. The fuels of interest will be oil, gas, coal, uranium, and perhaps oil shale. On the demand side the basic problem is how to dampen growth rates while avoiding measures that could lead to economic disruptions. Having defined some goal for energy independence, the policy formulator must devise means for expanding domestic energy supplies on the one hand while curbing demand growth on the other and must find tactics for implementation that are politically acceptable and economically sound.

FUTURE PROSPECTS

Historical Perspective

The art of being able to predict the outcome of future events as a consequence of policy actions adopted today lies at the heart of policy formulation. Unfortunately, forecasting is hardly a well-developed science, for either the energy sector or the general economy. Most policy analyses, nevertheless, express their arguments largely in terms of certain forecasts. It is instructive to view these forecasts in terms of historical trends.

The so-called historical trend in the growth of energy demand is usually associated with the 25-year period from 1947 to 1972, following economic adjustments after World War II. During this period the average annual compound growth rate of energy consumption in the United States was 3.2%. This relatively low figure, however, does not reflect the more rapid rise in demand experienced over the last decade of that period. Table 1 lists the relevant data for demand in terms of gross and per capita consumption. It can be seen from this information that most of the total demand growth is due to increased per capita consumption rather than an increase in the population. These rates continued through 1973, until finally the effects of the oil embargo and the deep downturn in the general economy became apparent in 1974.

Table 1 Historical growth of US energy consumption[a]

Year	Gross Energy Inputs in mQ (Average Annual Percentage Change)		Gross Energy Per Capita in Millions of Btu (Average Annual Percentage Change)	
1950	34.0	(—)	223.2	(—)
1955	39.7	(3.1)	239.3	(1.4)
1960	44.6	(2.3)	246.8	(0.6)
1965	53.3	(3.6)	274.4	(2.1)
1970	67.4	(4.8)	329.1	(3.7)
1971	68.7	(1.9)	331.9	(0.9)
1972	72.1	(4.9)	345.3	(4.0)
1973	74.7	(3.6)	355.2	(2.9)
1974	73.1	(−2.2)	346	(−2.8)

[a] Source: Dupree and West (10).

Table 2 Historical growth of US energy imports in mQ/year[a]

Year	Crude Oil	Petroleum Products	Natural Gas	Sum
1950	0.8	0.4	—	1.2
1955	1.6	0.4	—	2.0
1960	2.1	1.4	0.1	3.6
1965	2.5	2.4	0.4	5.3
1970	2.7	4.2	0.8	7.7
1971	3.4	4.4	0.9	8.7
1972	4.5	5.1	1.0	10.6
1973	6.6	6.6	1.0	14.2
1974	7.1	5.9	0.9	13.9

[a] Source: Dupree and West (10).

In terms of energy supplies, fossil fuels have accounted for the bulk of our needs as well as the energy needs of most other industrialized nations. During the period 1947–1972, one finds that about 95% of gross energy inputs in the United States was accounted for by fossil fuels. The mix of fuels, however, has not remained constant with time. In 1947, for example, coal represented 48% of the US fuel mix, and by 1972 its share had decreased to 17%. Natural gas experienced the most rapid rise, from 13% of the fuel mix in 1947 to 33% in 1972. The remainder of the fossil fuel mix has been made up of crude oil, its derivatives, and natural gas condensates, the composite usually being referred to as liquid hydrocarbons. Demand for liquid hydrocarbons has generally kept pace with the growth rate in total energy demand up until recently. Natural gas reserves in the United States peaked around 1969 and its supply has been in a shortfall position since. For this reason, plus others arising from environmental considerations (i.e. inability to burn high-sulfur coal, nuclear slippage, inefficiencies introduced by automobile emission controls), the demand for oil increased to the point where its annual growth rates reached 6–7% in the early 1970s.

The ever-growing demand for energy—the preference for oil and gas coupled with decreasing domestic production for these fuels—resulted in a rapid rise of imports. This is reflected in Table 2, which does not take into account the small but steady export market for coal (mostly but not entirely metallurgical grade) and the relatively small amount of trade in electricity between Canada and the United States. The effect of the oil import quota system, which was finally abandoned in early 1973, may be inferred from the data in Table 2. Import restrictions on distillates were relaxed well before then; a significant fraction of the petroleum products imported was residual oil for electric power generation. After 1972, the import of crude oil could grow without inhibition of quotas. The decreases between 1973 and 1974, again, can be traced to the effects of the temporary embargo and economic recession.

Virtually all of the imported natural gas indicated in Table 2 comes from Canada. Considerable attention is being given, however, to importing liquified natural gas in the future from North Africa, Siberia, and other distant places. Until relatively recently, imported oil and its derivatives came to the United States primarily from the Western Hemisphere. Table 3 shows in detail the sources of US oil imports in 1973. However, one must try to interpret these figures to the extent that origins of petroleum products simply designate the country where the oil refinery is located. The flow of oil that results in distillate imports from Western Europe and Japan, for example, comes from Africa, the Middle East, or Indonesia, while most but not all Caribbean distillates originate from Venezuelan crude oil.

It would now appear that an increasing fraction of any future US oil imports will have to come from the Eastern Hemisphere, specifically from the Middle East. Canada has already served notice that it intends to phase out oil exports to the United States by 1982, while Venezuela has announced a policy of oil conservation and does not at this time appear interested in expanded exploration and production. Thus, a very significant portion of the imports previously obtained from the Western Hemisphere seems destined to be lost in future years.

Forecasts

As mentioned earlier, the Project Independence study undertaken by FEA and its contractors resulted in an elaborate econometric analysis of a number of policy options. The results are summarized in the Project Independence Report (PIR) previously cited (8). Fundamentally, four broad strategies were examined:

1. *business as usual,* in which the government takes a minimum of new initiatives;
2. *accelerated supply,* in which the government adopts a number of measures intended to stimulate domestic energy production, including accelerated leasing of federal lands onshore and offshore, diversion of the naval petroleum reserves to commercial production, and introduction of regulatory changes where current practice impedes production (as in nuclear licensing delays);
3. *conservation,* in which the government adopts mandatory regulations or offers incentives that will lead to a reduction in energy demand growth, a concept also referred to as demand management;
4. *emergency preparedness,* in which the government takes steps to insure against severe economic disruption in case of interruptions in imported energy supplies, basically through a stockpiling program plus provisions for standby authority to curtail demand and allocate supply.

In the various cases studied, the forecast of future demand and supply of energy was expressed in terms of world oil prices (at constant mid-1973 dollar value)—for example, $4, $7, or $11 per barrel delivered at point of entry, corresponding roughly to $3, $6, and $9 at a point of origin in the Persian Gulf. This approach implied, but did not necessarily assume, that the dominant influence on future energy demand and supply was price elasticity (the slope of a log quantity vs log price curve). While PIR does not make any firm policy recommendations, it does imply that on the time scale of the next ten years the United States can readily achieve increased domestic production and curb demand growth so that the goals of energy independence can be met, providing only that energy prices as determined by the world price of oil are kept sufficiently high.

A relatively new issue, still not fully resolved, arose while PIR was in preparation.

Table 3 US imports of crude oil and products, 1973[a]

Origin	Crude Oil			Refined Products			Total Imports		
	Per Day	Total for Year	Percentage of Total	Per Day	Total for Year	Percentage of Total	Per Day	Total for Year	Percentage of Total
Canada	1,001.0	365,370	30.9	311.9	113,854	10.6	1,312.9	479,224	21.2
Caribbean (includes Mexico)	1.3	489[b]	0.1[c]	1,196.2	436,601	40.4	1,197.5	437,090	19.3
Other Latin America	455.8	166,379	14.0	994.1	362,838	33.6	1,449.9	529,217	23.4
Western Europe				254.9	93,046	8.6	254.9	93,046	4.1
North Africa	285.1	104,041	8.8	45.9	16,755	1.6	331.0	120,796	5.3
Sub-Saharan Africa	497.1	181,440	15.3	13.4	4,900	0.5	510.5	186,340	8.2
Middle East	802.7	292,988	24.7	53.5	19,539	1.8	856.2	312,527	13.8
Japan				8.7	3,171	0.3	8.7	3,171	0.2
Far East	200.8	73,289	6.2	45.6	16,635	1.5	246.4	89,924	4.0
USSR & Eastern Europe				33.4	12,188	1.1	33.4	12,188	0.5
Total	3,243.8	1,183,996	100.0	2,957.6	1,079,527	100.0	6,201.4	2,263,523	100.0

[a] Given in thousands of barrels. Source: (11, p. 34, Table 30).
[b] Entirely from Mexico.
[c] Actually 0.04% of total.

Under the leadership of the United States, the major oil-consuming (also oil-importing) countries joined together to form the IEA. France, the only major exception, chose not to join. Under the accords, members of the IEA will share shortfalls in energy supplies, establish stockpiles in proportion to their import levels, and adopt appropriate energy conservation measures. It is unclear at the time of this writing to what extent the United States has entered into binding arrangements that could affect our degree of self-sufficiency in the future. The IEA was formed to provide a counterbalance against the OPEC oil cartel. It remains to be seen whether this goal is realized, since the binding interests of the OPEC nations are far more solid than those of the member states of the IEA.

The Ford Foundation, not the federal government, was the first to launch a massive energy policy study. The Foundation commissioned a sweeping examination of issues related to energy in December 1971, some 28 months before the FEA formally launched its study. The Ford Foundation project has resulted in a number of interesting reports on specialized topics within the energy area, many of which are appearing in published book form (11a–11s). The Energy Policy Project (EPP), which acted as

Table 4 1985 forecasts of demand[a]

Forecast	1985 Energy Demand in mQ in the United States		1975–1985 Annual Average Compound Growth Rate (%)	
PIR	(BAU)	(AS & C)	(BAU)	(AS & C)
$4/bbl	118.3		3.8	
$7/bbl	109.1	99.7	3.2	2.5
$11/bbl	102.9	96.3	2.7	2.3
EPP				
HG	116.1		3.6	
TF	91.3		1.9	
ZG	88.1		N.A., model approaches 100 mQ/year demand assymptotically	
Dupree & West[b]	116.8		3.7	
NPC[c]				
Low	112.5		3.4	
Med.	124.9		4.1	
High	130.0		4.4	
NAE[d]	108		3.1	

[a] Abbreviations used: PIR = Project Independence Report; BAU = business as usual; AS = accelerated supply; C = conservation; EPP = Energy Policy Project; HG = historical growth; TF = technical fix; ZG = zero growth; NPC = National Petroleum Council; NAE = National Academy of Engineering.
[b] See (10).
[c] See (13).
[d] See (14).

the coordinating office for the Foundation, has published its own final report (12), which appeared about the same time that PIR was completed. The report is organized around three basic scenarios—historical growth, technical fix, and zero growth—whose names are more or less descriptive of the differences in the basic assumptions made.

At first the similarities between the two analyses, PIR and EPP, seem somewhat surprising. On reflection, however, they may simply be due to the time frame in which these reports were written (soon after the onset of the Arab oil embargo) and to the fact that they both rely heavily on rather similar econometric methodology. Both reports conclude, in effect, that the historic relationship between rate of economic growth and energy demand growth does not follow from fundamental considerations and that the gross economy can be effectively decoupled from energy. This is a central and most significant thesis. In details the two reports differ to a considerable degree, and in general EPP is more inclined to embrace radical departures from conventional wisdom than PIR. More important, perhaps, is that PIR essentially stops short at 1985 while EPP extrapolates to the year 2000.

The conclusions of these two reports regarding possible growth of energy demand (demand forecasts) with more conventional pre- and post-1973 forecasts are compared in Table 4. Table 5 compares forecasts for various energy supplies, with domestic production sources listed first. It is fairly evident that the difference between pre- and post-1973 forecasts lies largely with the expectations that increased energy prices and the creation of a favorable climate will lead to significantly lower demand growth for the remainder of this decade and thereafter, coupled with increased production of oil and/or natural gas from domestic sources.

Critique of Project Independence

As the results of PIR were being disseminated, and for some time subsequently, considerable criticism was leveled at it from a variety of sources. Unfortunately there is little agreement among the critics. One school of thought, represented by certain energy analysts at Massachusetts Institute of Technology (15), states that PIR is unduly pessimistic about the price response of oil and gas. On the basis of econometric models they argue that indigenous supplies of gaseous and liquid hydrocarbons can be increased quite rapidly providing price controls and regulations are relaxed and federal land leasing is liberalized.

The exact opposite conclusion is reached by the Department of Commerce's Technical Advisory Board (CTAB). The CTAB Panel on Project Independence Blueprint feels that future production figures assumed for the United States for oil and particularly gas are unattainable and that PIR should be faulted for not having given sufficient attention to coal and nuclear energy (16). The CTAB report includes a number of specific policy recommendations, rather than a list of policy options, which would require the federal government to take firm actions toward achieving self-sufficiency, under conditions predicted to be most difficult.

The response of industry has been generally skeptical; a recent projection released by Exxon (17) is fairly representative. The Exxon forecast predicts demand for energy growing at an average annual rate of 2.1% for the remainder of this decade

Table 5 1985 forecasts of energy supplies in mQ[a]

	PIR				EPP		Dupree & West	NPC Med.	NAE	1971 Actual Supplies (For Comparison)
	$7/bbl		$11/bbl							
	BAU	AS	BAU	AS	HG[b]	TF				
Oil	23.1	30.5	31.3	38.0	32–27	30	23.8	28.5	23.5	22.6
Gas	23.9	24.7	24.8	25.5	29–26	27	22.5	27.3	32.8	21.8
Coal	19.9	17.7	22.9	20.7	28–21	16	21.5	21.4[c]	21.2	12.6
Nuclear	12.5	14.7	12.5	14.7	10–12	8	7.9	25.2[c]	17.4	0.3
Other	4.9	4.8	5.1	5.3	6–4	4	5.7	7	4.6	2.9
Oil & gas imports	24.8	17.1	6.5	0	11–26	7	33	24.5	8.5	8.8
Total	109.1	109.6	102.9	104.2	116	92	116.8	124.9	108	69.0

[a] Abbreviations used: see footnote *a*, Table 4.
[b] Range of variations depends on degree to which imports are de-emphasized and compensated for by domestic supplies.
[c] 9 mQ of coal and nuclear availability not utilized.

and at 3.4% from 1980 to 1990. Over the next 15 years it sees our fuel mix changing gradually so that coal and nuclear power will supply 34% by 1990, in about equal proportions, while gas and oil combined will account for 63%. This is to be compared with the anticipated mix for 1975, consisting of 2% nuclear, 18% coal, and 76% oil and gas, combined. In each case, hydro- and geothermal power makes up the remainder.

Exxon expects domestic natural gas production to continue decreasing until the early 1980s and then to begin to increase, but to values well short of the historic heights of past years. Oil production remains fairly level until the mid-1980s and then grows to 11.8 Mb/d by 1990 compared with 10.6 Mb/d in 1975 (these figures include both crude oil and gas liquids, presumably). Quantities of imported oil and gas increase throughout this time period, with oil going from 6.3 Mb/d in 1975 to 12 Mb/d in 1990, while gas goes from 1 Tcf/yr in 1975 to 3.5 Tcf/yr in 1990. Suffice it to say, this is not quite what the authors of Project Independence had in mind.

Public reaction to the administration's call for energy self-sufficiency has been apathetic as judged by the performance of their elected representatives. During the early part of 1975 President Ford submitted an omnibus energy legislative package to the Congress which embraced many of the concepts in PIR and the CTAB recommendations. Congress has yet to take any action on the administration's proposals. Both branches of government became preoccupied with the demand side of the energy question—that is, finding acceptable mandatory measures that could lead to cutbacks in oil imports. But the notion of imposing higher gasoline taxes and import tariffs or decontrolling oil and deregulating gas is not a popular one in the political arena when unemployment figures continue to rise and the economy is in a recession. The warnings of PIR—that if oil imports reach 12.4 Mb/d by 1985, the economic consequences of a one-year interruption would be $205 billion, or the annual outflow of funds from the United States could reach $31.7 billion (both numbers are in terms of mid-1973 dollars) just to pay for oil—fall on somewhat deaf ears while all of our economic indicators are moving downward.

In fact, there is a school of thought urging that the government de-emphasize its

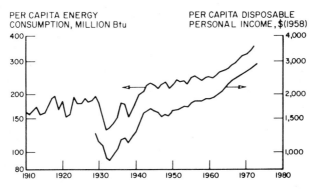

Figure 4 Comparison of trends of per capita energy consumption and per capita disposable personal income. Source: Institute of Gas Technology, Chicago.

apparent preoccupation with energy conservation lest it lead to continuing economic weakness. An outstanding spokesman for this point of view has been Henry Linden from the Institute of Gas Technology, who bases his arguments on certain macroeconometric analyses of J. D. Parent (18). Figure 4 illustrates the historical similarity between per capita energy consumption and per capita disposable income. An explanation for this phenomenon has been offered recently (18), in which stepwise regression analysis indicates a strong correlation between the energy consumed and the size of the labor force [reminiscent of the earlier observation made by Fisher (9)]. The formula is given as

$$E = 0.01619 F^{-0.1997} M^{1.891} P^{0.1791},$$

where E is annual energy consumption, in mQ; F is fossil fuel and power price index (1958 = 100) adjusted by the implicit price deflator based on GNP (1958 = 100); M is employment in the civilian labor force, in millions; and P is populations, in millions. A similar correlation is found between E/G, the ratio of energy to real GNP in units of 10^6 Btu/1958 dollars, and the size of the labor force,

$$E/G = 2.1714 F^{-0.2071} (M/P)^{1.458} (G/P)^{-0.9289} (1.01428)^{T(1947)},$$

where T is calendar year and G is GNP in billions of 1958 dollars.

The foregoing analysis tends to imply that, based on historical behavior, the E/G ratio cannot change significantly from its current value near 0.09 if unemployment is to be reduced from present levels of greater than 8% to future levels in the range of 4–5%. Should this prove to be correct, it would refute the elaborate econometric arguments of both PIR and EPP, which indicate that energy consumption and economic welfare can be decoupled without penalty.

SUMMARY AND ASSESSMENT

It is virtually impossible at this time to predict what set of policies the United States will be willing to adopt in the near term and what their future consequences might be. There is a great aura of uncertainty surrounding energy at the time of this writing. President Ford seems determined to achieve some measure of reduction in oil imports. He has already imposed a $2 tariff per barrel of imported oil and has promised to raise the tariff to $3 per barrel unless Congress acts on his submitted proposals. Meanwhile FEA is exploring ways to convert some oil-fired steam plants back to coal and also the means to decontrol the price of domestic oil by executive action. The Congress is in a position to block many of these measures and there is an air of confrontation between the two branches of government. One cannot guess with any confidence what the outcome will be.

The administration is engaged in a complicated set of negotiations and wishes to keep as much bargaining power as it can. On the one hand it is trying to be an "honest broker" in the Mideast disputes between Israel and its Arab neighbors. On the other hand, it is trying to lead the consortium of oil-consuming states (IEA) in their confrontation with the producing states (OPEC). The general worldwide economic decline has already caused OPEC to cut back severely in production from the

30 Mb/d levels of 1973–1974 to 25 Mb/d in 1974–1975, while world prices of oil remain firm. The IEA nations can try to exert more pressure on OPEC by further decreasing import requirements through continued conservation measures. It is somewhat unlikely that the consumers can break OPEC's cartel power by these actions unless economic recovery becomes very slow or major oil discoveries are developed outside of OPEC. It is more likely that some sort of accommodation will be reached between the consumers and producers, in which event world trade in oil will adjust to new equilibrium conditions. There is probably more than sufficient oil in the world to satisfy its needs for most of the remainder of this century; only its geographical distribution is highly uneven.

Under ideal conditions and an equitable distribution of the world's resources, one would expect to see a gradual transformation of the world's energy economy from its present state of heavy dependence on oil and gas to one dependent on more plentiful energy supplies. Such a transformation might be accomplished over several decades. The steady depletion of oil and gas would, in time, raise the inherent production costs of these mineral fuels to a level at which substitutes derived from coal, oil shale, or tar sands would be fully competitive with the natural product. Such heavily capital-intensive energy sources as the synthetic fuels, nuclear power, or solar-generated electricity would eventually be on a par with the capital needed to develop the dwindling resources of oil and gas. And so the readjustment of the world's fuel mix would be determined largely by competitive forces, cartel or no cartel. Similarly, the price of energy would continue to increase relative to the price of other goods; thus, user patterns would be expected to adjust naturally, with energy-saving methods evolving in response to market forces.

The policy of the United States, in response to what is seen as the political threat of OPEC and its controlling Arab forces, is to foreshorten significantly the time span of this transformation. To do so, the government must impose a number of artificial constraints on its energy system, the sum total of whose effect must be to raise the cost of energy production and use. This extra cost is balanced against the measure of security we gain. In a sense then, the energy policy we are trying to devise is no different from our defense policy. Both impose extra burdens on our society in the name of national security. For the time being, however, the body politic has been unable to rationalize or accept the essence of energy policy implications in these terms.

If there is a consensus within the government regarding energy matters, it is that more federal expenditures should be spent on energy R&D, although there may be disagreement on priorities: some favor more non-nuclear at the expense of nuclear or accelerating solar, geothermal, or wind energy, and many policy formulators seem to have their own pet project. The natural extension of this consensus is that the government should do all that is possible in each area of energy production or utilization.

Obviously some order and relative priorities must be established in any sensible national energy R&D program, and considerable attention is being given to this matter. The Congress has mandated ERDA to submit a number of plans setting forth the national energy R&D objectives and their implementation schemes. Less clear at this point is how these goals fit in with an overall national policy. In a sense the fundamental purpose of any R&D program is to develop and demonstrate certain technical options. The construction of the Shippingport reactor in the 1950s may be cited as an example of federal involvement in developing and demonstrating the commercial feasibility of light water nuclear reactor technology. Plans to build the Clinch River Breeder Reactor is meant to accomplish a similar purpose for the LMFBR technology. But Shippingport in itself did not result in the commercialization of nuclear reactor technology; that was done only as a result of private industry's willingness to commit venture capital and skilled manpower to the effort.

There is a certain degree of willingness on the part of the government to support development and demonstration in a number of technical areas. There is also a determination to do this in some not yet clearly defined mode of partnership with private industry. What is even less clear is the point at which government involvement will finally cease and what role, if any, it will play in the commercialization of new or innovative technology. In accelerating the pace of technological evolution beyond its natural pace, there is always the specter of white elephants to be faced. What should be done with a synthetic fuel technology, for example, requiring plant investments in the billion dollar range and producing products at the equivalent cost of $20/barrel of oil in the face of world oil prices ranging around $10/barrel? The conventional response is that a certain level of demonstration is necessary to evaluate commercial feasibility realistically and that no one really knows what world oil prices will be 10–15 years hence. To an extent, one is buying an insurance policy.

Perhaps such arguments can stand up to careful scrutiny, but it would seem more reassuring if the R&D program were integrated with a well-defined set of policies regarding fuels. Our European and Japanese trading partners are in the process of trying to do just that. Hopefully, the United States will be able to arrive at a more encompassing and comprehensive plan as well—a plan that spells out in some detail our anticipated degree of reliance on coal, nuclear fuel, oil, and gas, and alternative energy supplies as a function of time, and that indicates how the various obstacles in the path toward final commercialization will be tackled and eliminated and how the various potential environmental, economic, and social constraints will be accommodated.

In summary, the goals of Project Independence and energy self-sufficiency have been defined, subject to some latitude of interpretation. It is difficult to judge whether we have yet taken the first firm step toward these goals, save for the commitment to expand our R&D effort. Based on performance to date there is ample reason to be skeptical about the outcome, but for observers of energy policy there promise to be more interesting times ahead.

Literature Cited

1. *Clean Energy.* June 1971. Message from President of the United States to House of Representatives, 92nd Congress, 1st Session. Doc. No. 92–118. Washington DC: GPO
2. *Concerning Energy Resources.* April 1973. Message from President of the United States to House of Representatives, 93rd Congress, 1st Session. Doc. No. 93–85. Washington DC: GPO
3. Nixon, R. M. 1973. *Presidential Documents* 9 (6): 867–75
4. President's Materials Policy Commission. 1952. *Resources for Freedom,*

A Report to the President. Washington DC: GPO
5. Cambel, A. B. June 1964. *Energy R&D and National Progress.* Washington DC: GPO
6. Balzhiser, R. E. 1973. *An Assessment of New Options in Energy Research and Development.* Rep. No. AET-9. Brookhaven Nat. Lab., Upton, NY
7. Ray, D. L. 1973. *The Nation's Energy Future.* Rep. No. WASH-1281. Washington DC: GPO
8. Federal Energy Administration. 1974. *Project Independence Report.* Washington DC: GPO
9. Fisher, J. C. 1974. *Energy Crisis in Perspective.* New York: Wiley
10. Dupree, W. G. Jr., West, J. A. December 1972. *U.S. Energy through the Year 2000.* Announcements Bur. Mines, Dep. Interior
11. Bureau of Mines, US Department of Interior. January 1974. Crude petroleum, petroleum products, and natural-gas liquids. In *Mineral Industry Surveys*, p. 34. Washington DC: Dep. Interior
11a.* Berlin, E., Cicchetti, C. J., Gillen, W. J. 1975. *Perspective on Power: A Study of the Regulation and Pricing of Electric Power.* 204 pp.
11b. Boesch, D. F., Hershner, C. H., Milgram, J. H. 1975. *Oil Spills and the Marine Environment.* 144 pp.
11c. Brannon, G. M. 1975. *Energy Taxes and Subsidies.* 204 pp.
11d. Brannon, G. M., ed. 1975. *Studies in Energy Tax Policy.* 400 pp.
11e. Cicchetti, C. J., Jurewitz, J. L., eds. 1975. *Studies in Electric Utility Regulation.* 224 pp.
11f. Duchesneau, T. 1975. *Competition in the U.S. Energy Industry.* 448 pp.
11g. Gray, J. E. 1975. *Energy Policy—Industry Perspectives.* 144 pp.
11h. Foster Associates. 1975. *Energy Prices 1960–73.* 292 pp.
11i. Gyftopoulos, E. P., Lazaridis, L. J., Widmer, T. 1975. *Potential Fuel Effectiveness in Industry.* 112 pp.
11j. Hass, J. E., Mitchell, E. J., Stone, B. K. 1975. *Financing the Energy Industry.* 160 pp.
11k. Hollomon, J. H., Grenon, M. 1975. *U.S. Energy Research and Development Policy.* 192 pp.
11l. Myers, J. G., et al. 1975. *Energy Consumption in Manufacturing.* 656 pp.
11m. Makhijani, A., Poole, A. 1975. *Energy and Agriculture in the Third World.* 168 pp.
11n. National Academy of Sciences/National Academy of Engineering. 1975. *Rehabilitation Potential of Western Coal Lands.* 228 pp.
11o. Newman, D. K., Wachtel, D. D. 1975. *The American Energy Consumer.* 384 pp.
11p. Schoen, R., Hirshberg, A. S., Weingart, J. M. 1975. *New Energy Technologies for Buildings.* 224 pp.
11q. Williams, R., ed. 1975. *The Energy Conservation Papers.* 416 pp.
11r. Willrich, M., Taylor, T. B. 1975. *Nuclear Theft: Risks and Safeguards.* 272 pp.
11s. Yager, J. A., Steinberg, E. B. 1975. *Energy and U.S. Foreign Policy.* 515 pp.
12. Ford Foundation Energy Policy Project. 1974. *A Time to Choose: America's Energy Future.* Cambridge, Mass: Ballinger. 511 pp.
13. National Petroleum Council, Committee on US Energy Outlook. 1972. *U.S. Energy Outlook.* Washington DC: Nat. Petrol. Counc.
14. National Academy of Engineering. 1974. *U.S. Energy Prospects—An Engineering Viewpoint.* Washington DC: Nat. Acad. Eng.
15. MIT Energy Laboratory, Policy Study Group. May 1974. Energy self-sufficiency: an economic evaluation. *Technol. Rev.*
16. US Department of Commerce. March 1975. *CTAB Recommendations for a National Energy Program*
17. Exxon Company. *Energy Outlook 1975–1990.* Public Affairs Dep., P. O. Box 2180, Houston, Texas 77001
18. American Gas Association. April 1975. *Gas Supply Rev.*, Vol. 3, No. 4

---

* Refs. 11a–11s published by Ballinger, Cambridge, Mass.

Copyright 1976. All rights reserved

ENERGY REGULATION: ✕11026
A QUAGMIRE FOR ENERGY POLICY

William O. Doub
LeBoeuf, Lamb, Leiby & MacRae, 1757 N Street NW, Washington, DC 20036

INTRODUCTION

The ultimate effectiveness of any policy is largely dependent on the individual efficacy and coordination of the agents or agencies that implement it. There are ample illustrations of the truth of this premise in the recent attempts by the Administration and Congress to formulate and implement a national energy policy; as a result, that policy, irrespective of any intrinsic soundness, could inevitably become trapped in a quagmire of regulatory policies and practices.

The difficulties that energy policymakers in the United States have experienced in 1974 and 1975 are in many respects symptomatic of the very problem that they have intended to resolve—the lack of a comprehensive and coordinated national energy policy. Decisions concerning energy supply and general policy that have been made over the years have contributed to the creation of areas of special concern and interest, institutionalized them, and nourished them through dedicated sponsorship by either the Congress, the Executive Branch, the independent federal agencies, or industry. The difficulties that stymied congressional consideration and executive implementation of an effective energy policy in 1974 and the first half of 1975 mirror this state of affairs.

In 1974, energy policymaking literally resembled a Rube Goldberg production with an abundance of action and few real results. More than 2000 energy-related bills and resolutions were introduced in the 93rd Congress. More than a thousand days of exhaustive hearings were held; and finally, about forty laws were enacted. Among these was the extremely important Energy Reorganization Act that established the Energy Resources Council and separated the functions of the Atomic Energy Commission into the Nuclear Regulatory Commission and the Energy Research and Development Administration. This measure consolidated most of the government's energy R & D efforts and removed the conflict-of-interest stigma that had stymied the AEC's efforts to both develop and regulate nuclear energy. In the final analysis, however, the legislative record of the 93rd Congress with respect to energy matters reflected neither a guiding philosophy nor a strategy for coordinated and comprehensive action.

The 94th Congress, on the other hand, has in some ways begun to approach the energy situation as a composite whole, rather than as a collection of distinct and separate issues. And, in 1975, the Administration seemingly settled on an energy strategy and, through the Energy Resources Council, has provided some continuity of purpose and support to its energy proposals.

These developments, while often bearing the seeds of productive action, have sometimes resulted in political confrontation and conflict. Where once there was no proposal detailing a comprehensive national energy policy, at least three competitive plans emerged—the President's (S 594), the House Ways and Means Committee's (HR 5005) and the Democratic Majority's. Congress became embroiled in a surplus of energy proposals and internecine competition for oversight to such a debilitating extent that Senate Majority Leader Mansfield suggested the creation of a new congressional committee to which all energy-related bills could be referred. Minority Leader Hugh Scott called the existing situation a "legislative nightmare." An example of this congressional congestion is the fact that issues that were once the almost exclusive province of the Joint Committee on Atomic Energy are now subject to consideration before several congressional committees. In late April, for example, Congressman Morris Udall's Subcommittee on Energy and the Environment of the House Committee on Insular Affairs began hearings to review the entire range of issues associated with nuclear energy. In the Committee's view, it has been given, through the jurisdictional rules of the House of Representatives, special oversight functions to review and study activities in this area. Other committees too are competing for this role at a point in time when prompt congressional action on nuclear issues is needed if utilities are to proceed with the kind of advance planning necessary to assure a reliable electrical supply and if industry is to attract and commit substantial financial resources to further development of the nuclear fuel cycle.

The irony in the present situation is that there essentially *is* agreement on the goals of a national energy policy. The three major energy plans before Congress all center on four distinct goals:

1. Reduction of US dependency on energy imports
2. Protection against disruption of foreign supplies
3. Increased conservation and efficiency
4. Increased domestic energy supplies

The differences among the various proposals are over means rather than ends.

The opportunity to formulate national energy policy with enough staying power to outlast the next energy crisis does exist. However, there are some obvious impediments to the achievement of such a policy. For almost every piece of positive legislation that has been put forward, there seems to be other legislation with either a countervailing purpose or effect. What sense does it make to talk about legislative action to increase domestic fuel supplies when other proposed legislation, if enacted, would have the effect of prohibiting such an increase? Such competition between congressional goals can in no way be construed as supportive of the goal of a comprehensive national energy policy.

The same need for cohesive and coordinated action is apparent when one looks at the broad field of energy regulation. Quite obviously, the energy industry in this country is significantly affected by the actions of regulatory agencies. Thus, how effectively energy is regulated becomes an inseparable factor in determining how close we come to achieving energy self-sufficiency. That is why in early 1973 the President called for a "comprehensive study" of all energy regulatory activities. Submitted to the President in April 1974, the Federal Energy Regulation Study concluded that the present energy regulatory framework lacks overall policy direction and coordination of effort. In the face of today's energy challenges, energy regulatory mechanisms cannot keep pace; concerned regulatory agencies fail to act in concert or cooperation; and some, in effect, have the powers of veto over others. These fundamental deficiencies in the regulatory framework represent the Achilles' heel of the nation's hopes for energy self-sufficiency.

REGULATORY PHILOSOPHY

For too long energy regulation has been an institutionalized concept in which principles, procedures, and organization are for the most part reviewed in the context of a narrow case or a discrete regulatory issue. As a result, the demands of a coordinated policy have been irresponsibly avoided. Regulation has not taken place with sufficient emphasis on its real goal, which is the protection of the total public interest while assuring that essential services are provided as needed.

Walter Lippman, in his book *The Public Philosophy,* defines the public interest as "what men would choose if they saw clearly, thought rationally, and acted disinterestedly and benevolently." From the perspective of regulation, it is an awesome challenge to meet these qualifications. And yet, notwithstanding the difficulty of perceiving the public interest, this is exactly what the regulator must do. Thus, it is not surprising to find that regulation, particularly at the federal level, is and has always been a controversial subject. Indeed, it has often been perceived as a restraint on our free enterprise system.

The path of the regulator since the passage of the Interstate Commerce Act in 1887 has transversed a thicket of criticism. The regulatory system by its very nature has traditionally been an easy target. Those bent on change argue that the regulatory system perpetuates the status quo, that regulatory agencies are the compliant captives of the industries they are supposed to regulate, and that money and access to high places coupled with the personal career ambitions of the regulator usually guarantee to the industrial interest favored treatment. Industry, on the other hand, today poses the same argument it has always made—that regulation, or overregulation as industry sees it, is counterproductive. It stridently claims to be tangled in a worsening maze of rules and nonsensical procedures that bureaucratic types have developed and nurtured through the years; to industry, the regulator is essentially a timid soul who fears to do the "right" thing lest he be denounced by public interest representatives. The consensus of both groups is that the energy regulatory system is confused, indecisive, expensive, duplicative, excessively pliable to pressure, and partial to narrow interests.

It is perhaps unnecessary to observe that many of these criticisms contain some truth. Yet, the fact is that the regulatory function is and was intended to be one of the major "pressure points" in government. Regulatory processes are rightfully an arena for contests involving high stakes. But they were never intended to be the spit on which the public interest is roasted.

REGULATION AND PRACTICALITIES

Today there exists a compelling need for regulators to reexamine the question of what a philosophy of serving the public interest really means. This need became undeniable in 1974 when the bonds ratings of many utilities declined to less favorable ratings, forcing cancellation or deferral of major construction projects. Unfortunately, the propensity of many regulators to remain wedded to patterns and approaches that were devised and perfected during a period of declining costs for electricity has become one of the factors that is perpetuating, and even exacerbating, the current phase of difficulties for utilities. For example, there is the often-cited tradition of regulatory lag—a prolonged time lapse between rate requests and rate actions. Prior to 1968, when consumer rates continued to decline, utilities found economic value in regulatory lag. But today, with rates on an upswing, lag is a grave threat to the economic health of those same utilities.

In a sense, regulation has become part of a problem, rather than part of a solution. If utilities are to garner and maintain the amounts of capital required for the construction of generating capacity, bold steps to eliminate regulatory lag are essential.

Although the current climate of consumer activism makes even the most stoic regulators uneasy at the thought of higher utility bills, the alternatives to rate increases are even more undesirable. The costs of postponing capital plant expansion are twofold. Decreased reliability of services becomes a distinct possibility, and the costs of installing additional capacity at a later date will escalate. The history of the regulation of the nation's railroads provides some idea of where this course might eventually lead. From a public interest perspective, the level of profitability of the utility industry is designed primarily to serve only one function: to assure that a utility's financial position is sufficiently vigorous to permit optimization of services to the consumer. At present, there exists increasing evidence that traditional rate treatment is failing to meet this test.

The increase in investment tax credit for utilities for three years, proposed by the Administration early in 1975, represents a means to reduce capital expansion costs, while the provisions to allow deduction of preferred stock dividends could stimulate equity rather than debt financing. But the most significant initiatives have been those recommending reforms in regulations. Without a uniform, timely, and realistic basis for consideration of rate applications by the diverse state and federal regulatory authorities, there is little hope of restoring financial stability to the utility industry.

The Administration's Omnibus Energy Bill is designed to meet this need for reform by aiming at the heart of what some have come to view as a sacrosanct area—the activities of state regulatory commissions. In essence, that proposed legislation would require state public utility commissions to: (*a*) act on requested rate increases within five months after a utility has made an application; (*b*) conform with provisions for pass-through of fuel adjustment clauses; (*c*) permit higher pricing for electricity produced for peak-load consumption; and (*d*) permit utilities to include capital costs of environmental controls and current construction in the rate structure.

The reactions of state regulatory bodies to these problems have been predictable and perhaps understandable in light of the fact that many of those bodies interpret any attempt at such reforms as an infringement upon fundamental state prerogatives. Some observers have gone so far as to characterize the problem posed by this attitude as purely political. Thus, they argue that there is little chance that the situation will improve unless legislation such as the Administration's proposal is enacted. Lacking such legislation, there is little the federal government can do directly to effect any reform in the regulation of electric utility rates since federal control of rates is limited to wholesale energy supplies that move across state lines—less than 15% of the power generated in this country. However, given the lack of support for the Administration's proposal from any segment of the electrical energy community, its chances for survival in Congress seem slim. Of course, the very fact that such suggestions were made at all may stimulate the states themselves to take the kind of action needed to remedy, if only partially, the present situation.

While many favor federal intervention in state controls of electric utility rates, others like L. Manning Muntzing, former Director of Regulation of the US Atomic Energy Commission, argue for a middle ground—a position more likely to succeed in the present political climate. He has recommended a federal program designed to *guide* the states toward providing successful capital attraction incentives for the utility industry. Any implementation, as related to individual utilities, should be the continuing responsibility of state public service commissions. Mr. Muntzing has also suggested that "an agency such as the Bureau of Labor Statistics, which has credibility in the public and private sectors, should provide a judgment—broadly developed—of the cost of capital needed by utilities to meet the nation's energy goals." This guideline approach appears more amenable to all the parties involved than the prescriptive form of reform contained in the Omnibus Bill now before Congress.

FEDERAL ENERGY REGULATION: AN OVERVIEW

Energy-related regulatory problems do not stop at state lines. Regulation at the federal level teems with even more controversies. In the present economic circumstances, federal regulators have become the scapegoats for almost everyone—consumers, Congress, industry, the Administration, and even fellow regulators. These attacks upon federal regulation have produced much that is merely polemical. However, in at least one sense they have also been productive since they have resulted in a consensus of the need for what is generically termed "regulatory reform." In the case of energy regulation this consensus has resulted, as mentioned above, in specific legislative proposals to reform state regulatory practices, legislation to improve licensing procedures for energy facilities and, more broadly, proposals to create a National Commission on Regulatory Reform that would investigate all aspects of regulation in the US economy. Virtually all of these initiatives, discussed more fully below, reflect the recommendations and positions set forth more than a year ago in the Federal Energy Regulatory Study (FERS).

FERS began with the recognition that federal energy policy must result from concerted efforts in all areas dealing with energy, not the least of which was the manner in which energy is regulated by the federal government. Energy self-sufficiency is improbable, if not impossible, without sensible regulatory processes, and effective regulation is necessary for public confidence. Thus, the President directed that "a comprehensive study be undertaken, in full consultation with Congress, to determine the best way to organize all energy-related regulatory activities of the government." An interagency task force was formed to study this question.

With 19 different federal departments and agencies contributing, the task force spent seven months deciphering the present organizational makeup of the federal energy regulatory system, studying the need for organizational improvement, and evaluating alternatives.

More than 40 agencies were found to be involved with making regulatory decisions on energy. Although only a few deal exclusively with energy, most of the 40 could significantly affect the availability and/or cost of energy. For example, in the field of gas transmission, there are five federal agencies that must act on siting and land-use issues, seven on emission and effluent issues, five on public safety issues, and one on worker health and safety issues—all before an onshore gas pipeline can be built. The complexity of energy regulation is also illustrated by the case of Standard Oil Company (Indiana), which reportedly must file about 1000 reports a year with 35 different federal agencies. Unfortunately, this example is the rule rather than the exception.

Despite the involvement of a multitude of agencies, there is no central organizational mechanism to coordinate these scattered operations. Until a few years ago, this was not an unworkable situation because, traditionally, regulatory agencies were primarily concerned with economic questions, like rates and competition. Today, however, a different situation prevails. At almost every stage of the energy cycle new technologies and aroused public concern bring forth complex issues—land use, air quality, water pollution, recreational and aesthetic needs, and public health and safety, together with equally valid requirements for economic growth, new sources of energy that may cost more and confront us with higher levels of environmental risk, and more efficient methods of obtaining and using vital fuel resources.

The primary problem of today's compartmentalized approach to energy regulation is not one of duplicated legal authority and excessive overlapping of jurisdiction among the agencies involved. Applicable laws and directives have been drawn very carefully, and responsibilities among agencies have largely been clearly defined in an historical sense. Ironically, the problem is that agencies are, for the most part, very diligently pursuing narrowly conceived goals that are legitimate and appropriate to their mission. In this pursuit, the major agencies often exercise what almost amounts to an effective veto power over the regulatory actions of each other and ultimately over a particular project or at least its schedule.

One of the key recommendations of FERS was the formation of a permanent National Energy Council to formulate national energy objectives and provide policy guidance to energy regulatory agencies. This concept was largely reflected in the Energy Resources Council established by the Energy Reorganization Act of 1974.

Experiences with regulating natural gas prices and emissions from fossil fuel plants have illustrated the overriding need for such an approach. In the early 1960s the Federal Power Commission, spurred by a series of court decisions, instituted a program of area rate-pricing for field sales of natural gas. This program is believed to have dampened the incentive to explore for new natural gas supplies, and the resulting reduction in the rate of new discoveries prompted large consumers of natural gas to look to alternate fuels, like coal and oil. But when the environmental movement gained momentum in the early 1970s and the Environmental Protection Agency began a program directed at limited emissions of sulfur dioxide, these same large consumers were further limited in fuel options and were forced to purchase low-sulfur oil. The concentrated switch to low-sulfur oil, in combination with other factors, necessitated an increase in imports and growing dependence on foreign sources. This, in turn, exacerbated balance-of-payments problems and raised questions about the national security implications of the projected large-scale dependence on imported oil. Thus, in 1975, the Federal Energy Administration instituted an effort to have utilities switch from oil to coal in any boilers that could make such a conversion. Today, all three regulatory objectives are being pursued simultaneously.

Admittedly, there are logical justifications and considerations that have resulted in the independent stature of a number of regulatory agencies. To ignore the validity of these considerations would be foolhardy; rather, it seems sensible to provide the regulatory agencies with a backdrop of broad national policies and goals, against which they can make their decisions with full appreciation of the collateral implications of what at first glance might appear to be narrow decisional issues. The Energy Resources Council is a step in this direction.

The FERS project also investigated the conflicts that exist between federal, state, and local regulatory activities. Local and state governments are deeply involved in the evaluation of proposed energy sites and facilities. Their regulatory functions embrace rate-making, zoning approvals, construction permits, safety inspections, and pollution controls. And while a few of these regulatory matters are the exclusive prerogative of state and local officials, many involve areas in which the federal government shares an interest or responsibility. But despite the interdependence between these levels of government, there is no effective, sustained mechanism to coordinate action and reconcile objectives. This has contributed to the tendency of energy regulatory agencies to adopt procedures and actions that sometimes embrace purely local and parochial concerns. To help remedy this situation FERS recommended a three-pronged program to coordinate energy regulation at all governmental levels and to assure that energy projects are treated from an integrated regulatory perspective, rather than as a series of loosely connected actions. Essentially those recommendations were:

1. Establishment of a federal office to improve the operational relationships between federal and state energy regulatory agencies—to facilitate the meeting of regional and national energy needs, largely by providing the guidance and assistance required to achieve timetables that will satisfy regional and national needs for energy facilities.
2. Establishment of a permanent organization of state representatives to work intensively and continually with this federal office, perhaps built upon already existing Regional Reliability Councils. Such an organization would require a strong working relationship with, and the support of, such groups as the National Governors Conference, the National Association of Regulatory Utility Commissioners, and the Council of State Governments.
3. Development of an extensive energy-related information exchange system between the federal and state levels. This recommendation has in some ways been reflected in the development of a data base by the Interagency Task Force on Energy Information and the Federal Energy Administration.

REGULATORY REFORM INITIATIVES

During testimony before the Senate Government Operations Committee in November 1974, at hearings on legislation to establish a National Commission on Regulatory Reform (S 4145), I stated that it was imperative, especially in the area of energy matters, that such a Commission incorporate a review of both federal and state regulatory entities within its scope. I remain convinced that the disparity among local, regional, and national interests in energy facility siting and licensing is among the most serious issues facing energy regulation today. The question of federal-state-local relationships in the siting and licensing of energy facilities is as compelling as the need to have agreement between state practices and national policies in the matters that impinge on the reliability of the nation's electric utility system—especially when consideration must eventually be given to the ramifications of locating large energy centers in the future. The difficulties are legal, political, and philosophical. It is nevertheless clear that a concentrated and coordinated effort is needed at both the federal and state levels if current and future conflicts are to be resolved.

While the Energy Reorganization Act of 1974 mandates that the President submit to Congress no later than mid-1975 any possible executive recommendations related to a "consolidation in whole or in part of regulatory functions concerning energy," it is unlikely that the report will go as far as suggesting a federal energy regulatory super-agency. What appears more likely are proposals for selective regulatory reform measures to improve mechanisms for federal-state interaction, especially on questions of facility siting and licensing.

While recent actions in this area do not go as far as the recommendations in the FERS report, they do show promise of minimizing many of the delays associated with the siting and construction of power plants. For example, the Administration's proposed "Energy Facility Planning and Development" bill, would (a) require the Federal Energy Administration to set forth the number, type, and approximate location of needed energy projects; (b) require each state to submit a State Energy Facility Management Program to meet its energy needs; and (c) establish a five-year matching grant program to aid states in preparing their energy plans.

In addition to this comprehensive siting bill there are several bills pending in this Congress concerning siting of nuclear power plants. These proposals, which are in many ways similar to legislation introduced on behalf of the Atomic Energy Commission and the Joint Committee on Atomic Energy in the 93rd Congress, offer an opportunity to perform virtually all of the needed safety, environmental, antitrust, and safeguards reviews for nuclear power projects early, before any decision to add power is made. The proposed bills deal essentially with procedural matters and would not modify substantive standards for protecting the public health and safety or the requirements of the National Environmental Policy Act.

By utilizing the concepts of predesignated sites and standardized plants, they provide an effective way to reduce the completion times for a nuclear project from the current 7–8 years to about 5–6 years, without decreasing the quality of the regulatory review. In fact, the proposed bills entail every precaution included in the present approach to licensing—but they also provide the structure to complete regulatory considerations earlier and with enhanced opportunity for the kind of meaningful public participation needed to help achieve public acceptance. Other features of one or more of the pending bills that commend them as examples of positive regulatory reform include hearings only when requested, interim operating licenses for power reactors prior to completion of hearings, and strengthened federal-state cooperation.

There are also promising nonlegislative initiatives in the nuclear energy area. For example, the Nuclear Regulatory Commission, which is responsible for licensing an increasingly significant share of new power facilities, is pursuing efforts to foster improved federal-state cooperation in licensing matters and has been joined by other involved federal agencies, such as the Environmental Protection Agency and the Federal Energy Administration. These initiatives encourage the varied groups involved to work jointly through the decision-making process from the start, thus reducing the likelihood for a project veto by some involved regulatory entity when an ultimate decision is reached.

Given the apparent lack of support for the proposal for a National Commission on Regulatory Reform, it appears that actions such as those described above, other ad hoc manipulations of specific regulatory mechanisms, or self-reform initiatives by regulatory agencies themselves or through the efforts of such institutions as the National Association of Regulatory Commissioners, represent the only tangible forms of regulatory reform that can be reasonably expected for the immediate future.

CONCLUSIONS

The President's Council of Economic Advisors estimated in its January 1975 report that the cost of compliance with federal regulations represents about $14 billion annually. Energy regulation inevitably contributes to these costs, but the idea of the complete deregulation of energy has become moot. Rapidly changing pressures and world political as well as economic developments so affect the domestic energy situation that they have compelled innovations in regulatory philosophy and practice. In many respects, regulation has become a response as well as a control mechanism. Many seemingly settled concepts of regulation have been challenged and some already discarded. Issues considered in the past to be irrelevant to regulation have often been made paramount by courts and legislatures. A decade ago, for example, environmental impact was seldom at issue. Now its evaluation is institutionalized in regulatory processes by laws such as the National Environmental Policy Act of 1969. Legal suits on environmental questions have become the rule rather than the exception. Likewise, economic conditions prior to 1968 supported efforts to maximize energy production and consumption at the lowest price possible. Today, a different situation prevails; shortages in fuel supply threaten curtailments and regulatory responsibilities have assumed the dimension of determining the equitable distribution of these curtailments. And, while a decade ago traditional cost-of-service rate-making was a workable formula, this no longer is the case, and other cost allocation methods have been fashioned upon assumptions once considered unacceptable.

The need to reevaluate regulatory goals and approaches has become clear. Unchecked, the trends toward further complications in the existing system of state and federal regulation will in all likelihood continue. And the challenges to this system will only intensify throughout 1975 and the immediate years ahead.

Fifty-five years have passed since the federal government first entered into energy regulation as licensors of hydroelectric facilities in 1920. Subsequent years brought increasing federal regulation that included control of monopolies, maintenance of interstate transmission efficiencies, protection of public health and safety, controls of environmental impact, and management of fuel supply. Each such federal intrusion into the energy market was carefully calculated to remedy a specifically perceived difficulty (or crisis); however, as the regulatory system grew, it became less and less responsive to changing situations—it became like a tradition that people faithfully practice without knowing why. In essence, this *is* the case against energy regulation. Corrective action probably lies somewhere between continued tolerance, or creating new regulations to undo the old, and complete rejection of the system. The Federal Energy Regulatory Study recommended certain measures to enhance the responsiveness of energy regulation; the Energy Reorganization Act contains the ingredients for a central policy guiding force; legislative initiatives represent ad hoc remedies; and a National Commission on Regulatory Reform, while not a total solution, is surely a constructive suggestion. Beyond the immediate situation, however, there is a legislative, as well as an administrative, lesson to be learned—governmental regulation should not be permitted to become so institutionalized that it is considered permanent. In many respects, the priority goal of regulatory agencies, except those involved with ongoing health and safety concerns, should be to put themselves out of business. There is no service to the public interest when regulation occurs for the sole benefit of either the regulated or the regulator. Each fiscal year should bring close scrutiny to the questions: Is this regulatory function still required? Does it serve the public interest? The legislation to establish the Federal Energy Administration intentionally made this kind of approach mandatory and it seems to have sharpened congressional oversight. Perhaps similar legislatively built-in appraisal requirements for other areas of federal regulation would prove as effective.

A status quo, business-as-usual approach to energy regulation in the United States could lead to failure on one level to assure the continued service provided by electric utilities and, on another level, to a failure to foster the nation's goal of energy self-sufficiency. Neither result could by any conceivable stretch of the imagination serve the public interest.

Copyright 1976. All rights reserved

FEDERAL LAND AND RESOURCE UTILIZATION POLICY

✳11027

John A. Carver, Jr.

University of Denver College of Law, Denver, Colorado 80204

> The existence of the public domain inescapably cast on the law a function of economic planning.... The massive fact of the public lands created a challenge to social order most striking and unusual in a society little given to philosophy, but—so far as it had a program—deeply committed to a broad dispersion of power with emphasis on a free flow of decision from countless centers of private decisionmaking.
>
> J. Willard Hurst, *Law and Economic Growth*

The federal government in its sovereign capacity has a policy role with respect to land and resource utilization. It shares this role with the states and by no means occupies the entire field. For example, zoning laws are traditionally a part of the state police power.

The federal government has a distinct policy role with respect to its own lands and resources and their utilization—a role that is, by and large, not shared with states. The Public Land Law Review Commission (PLLRC) studied the dimension of this separate policy role and identified its magnitude by entitling its final report *One Third of the Nation's Land*,[1] to reference the fraction of the nation's total land area in this category.

In general, the federal government's status as a sovereign landowner applies to the national forests, the national parks, wildlife refuges, and all of the public domain lands managed by the Interior Department's Bureau of Land Management.

The federal government's entry into the areas of land and resource policy formulation for nonfederal lands tends to displace or preempt state police power functions. Where sustained by constitutional warrant, such as the commerce clause, the property clause, or the public welfare clause, its laws are supreme compared with inconsistent state laws.[2]

---

[1] *One Third of the Nation's Land,* June 1970, A Report to the Congress and to the President by the Public Land Law Review Commission, Washington DC: GPO (hereinafter cited as *One Third*).

[2] Cases under the supremacy clause, article VI, clause 2, of the Constitution are numerous. For recent cases applicable to point cited, see *Askew* v. *American Waterways Operators, Inc.,* 411 U.S. 325 (1973), and *Northern States Power Co.* v. *Minnesota,* 447 F.2d 1143 (CA8 Minn. 1971).

It is important to differentiate the two bases for formulation of policy. The general regulation provisions and the public land provisions do not work the same way. One example may be given for introductory purposes: the veto by President Gerald R. Ford of a comprehensive Surface Mining Control and Reclamation Act (the strip-mining bill)[3] was greeted with a threat from its congressional sponsors to enact a moratorium bill to forbid the issuance of leases on coal owned by the federal government until strip-mining legislation was passed and signed.[4]

The vetoed bill was general regulatory legislation, which would have been applicable to the mining of coal on all private and public lands. The threatened alternative and retaliatory legislation would have been applicable only to the sovereign proprietor's program for leasing coal either on lands that it owned and controlled under article IV, clause 3, section 2 of the US Constitution or on lands patented under federal law subject to a reservation of the subsurface minerals.[5]

Antedating the Constitution, the proprietor's role is the older of the two bases for policy enunciation. The Land Ordinance of 1785 and the Northwest Ordinance of 1787, both enacted by the Continental Congress under the Articles of Confederation, were land-use planning acts for the infant nation; it was specified therein that the United States would continue to hold lands for disposal under national rules even after the Northwest Territory and all of the new lands to the West achieved statehood. An important corollary idea was that states should be created from these territories as promptly as possible.[6]

The Land Ordinance of 1785 contributed the rectangular survey system, which has survived as a lasting tool of land planning in the United States.[7] Some ideas in these two ordinances did not survive, such as the provisions that alternate townships should be sold intact and that a "third part of all gold, silver, lead and copper mines to be sold or otherwise disposed of as Congress shall hereafter direct" should be retained by the United States.[8] But other ideas survived and grew: for example, the idea of reserving one section per township "for the maintenance of public schools."[9] Article IV, section 3 of the US Constitution reflects the main ideas of these precursory land ordinances; clause 1 provides for the admission of new states; clause 2 provides that the Congress shall have power to

---

[3] A conference report on H.R. 25, the House bill, was agreed to by the Senate on May 5, 1975 (Cong. Rec., May 5, 1975, p. S7455) and by the House of Representatives on May 7, 1975 (Cong. Rec., May 7, 1975, p. H3774). It was vetoed by the President on May 20, 1975, and the veto was sustained by the House of Representatives on June 10, 1975 (Cong. Rec., June 10, 1975, p. H5205).

[4] Bills to carry out this threat were introduced by Senator Metcalf, S. 1703, and by Congressman Melcher, H.R. 6880, 94th Cong., 1st sess, 1974.

[5] Much of the coal underlies lands patented under the provisions of the Stockraising Homestead Act of December 29, 1916, 39 Stat. 862, as amended and codified, 43 U.S.C. secs. 291–301.

[6] See, for example, Gates and Swenson, 1968, *History of Public Land Law Development*, PLLRC Study Report, pp. 59–74, Washington DC: GPO.

[7] Ibid, p. 65.

[8] Ibid.

[9] Ibid.

dispose of and make rules respecting the territory or other property belonging to the United States.

The outlines of the philosophical and political cleavages that mark much of the current debate about policy were present in our country's earliest debates. Should land policy contribute to the agrarian democracy favored by Madison and Jefferson, or should it maximize the revenues for the country's scant treasury, as Hamilton and Gallatin urged? When the country was bent on raising revenue, lead mines were leased and not sold or granted, but the leasing act was repealed in favor of a measure authorizing sale of such mines at auction, and a general policy on leasing did not return until 1920. A general mining law was passed in 1872,[10] which reflected a line of thought expressed by President Fillmore in 1851:

> I am inclined to advise that they (the gold mines) be permitted to remain as at present, a common field, open to the enterprise and industry of all our citizens, until further experience shall have developed the best policy to be ultimately adopted.[11]

"Further experience" has given us mineral leasing acts,[12] but mining law for the metalliferous minerals on public lands has been essentially unchanged during the intervening century; valuable mineral deposits in lands belonging to the United States can be acquired by locating a claim, marking and posting it, and if a valuable mineral deposit is discovered and if compliance with formalities is evidenced, a valid possessory interest can be converted into full legal title.

The process is by no means as simple and easy as the foregoing may indicate, because the accompanying administrative system has become increasingly restrictive. In this fact, we see another key element of public land policy, the growth of the process of administration and management. Nevertheless, the location system enacted in 1872 is still the only way in which access may be gained, under existing laws, to many minerals required by our industrial economy. We may or may not at last be about to apply our "further experience" to develop "the best policy to be ultimately adopted" in the words of President Fillmore. Congress has apparently yielded to the executive branch its leadership in driving to replace the General Mining Law of 1872 with a comprehensive mineral leasing system. A brief flurry of hearings on various proposals along these lines occurred in the 92nd Congress, but there was almost no action in the 93rd Congress or in the early days of the 94th Congress.

In the western states particularly, federal policies with respect to the public lands exert leverage on policies of land and resource utilization in the private sector. In some areas, municipal expansion depends on the availability of public land to the city.[13] The creation of a national park or other major development such as a flood control or power-generating facility significantly affects the development of

---

[10] Act of May 10, 1872, 17 Stat. 91, as amended and generally codified at 30 U.S.C. secs. 22–24, 26–30, 33–35, 37, 39–42, 47.

[11] See footnote 6, p. 713.

[12] The General Leasing Act was the Act of February 25, 1920, 41 Stat. 437. See, generally, 30 U.S.C. secs. 181–184, 185–193, 201–214, 223–229, 241, 251, 261–263.

[13] See, for example, *One Third* (footnote 1), pp. 226–227.

surrounding private land, as does the leasing of a large coal deposit or oil shale deposit or the establishment of a uranium-processing facility.

Both by statute and by contract, the federal government can regulate the use of private land when that use is intertwined with a federal project or requires access over public lands. A purchaser of federal timber may be required to grant reciprocal rights-of-way over his private lands, even those remote from the contract area.[14] A licensee of the Federal Power Commission wishing to develop the hydroelectric potential of a dam on a navigable stream or on public lands is subject to extensive federal regulation of rates, accounting procedures, and provision of recreational and other public benefits.[15]

The Supreme Court has determined that Congress can grant, and has granted, broad and virtually uncontrolled authority to the Secretary of the Interior to allocate shortages in the flow of the Colorado River among competing claims of upper basin and lower basin states exceeding the flow of the stream.[16] Such authority can potentially displace established private water rights, without compensation.

The entry of the federal government into regulation of land use and resource utilization apart from activities related to the public lands or to purely governmental functions such as making war or providing for defense is historically related to the growth of the conservation movement. At first, there was a strong utilitarian cast to the conservation movement, but later, its preservationist aspect became dominant. In 1903, Theodore Roosevelt's veto message of the Second Muscle Shoals Bill made clear that the issue of a federal role in the development of the Tennessee River was to be fought on the basis of what would be most efficient economically and in the interest of the whole country, not whether the states or the federal government should control:

> The recent development of the application of water power to the production of electricity available for use at considerable distances has revealed an element of substantial value in streams which the Government is or is not liable to be called upon to improve for purposes of navigation, and this value, in my judgment, should be properly utilized to defray the cost of the improvement.[17]

In the 1920s, the US Forest Service gained authority to assist private forestry;[18] in the 1930s the Congress authorized a soil conservation program that looked toward cooperation between private landowners and the federal government;[19] in the 1940s the Corps of Engineers and the Bureau of Reclamation sensed and responded to the public demand for recreational opportunities to be added to their water projects;[20] and in the 1950s, as a logical extension of the idea of building

---

[14] 36 C.F.R. sec. 212.9(b) (1974).
[15] Part I of the Federal Power Act, particularly section 10(a), 16 U.S.C. sec. 803(a).
[16] *Arizona v. California,* 373 U.S. 546 (1963).
[17] Cong. Rec., 1903, vol. 36, p. 3071.
[18] Clarke-McNary Act of June 7, 1924, 43 Stat. 653, 16 U.S.C. sec. 564 et seq.
[19] Act of April 27, 1935, 49 Stat. 164, 16 U.S.C. sec. 590 (e).
[20] Act of December 22, 1944, 58 Stat. 889, 16 U.S.C. sec. 460 (d), for the Corps of

federal projects to include a recreation component came the work of the Outdoor Recreation Resources Review Commission[21] for a federal program of furnishing recreational opportunities even where no federal project existed. This led to the creation of a Bureau of Outdoor Recreation,[22] the establishment of a Land and Water Conservation Fund,[23] and other statutory and executive adjustments to the then unquestioned demand for more and more recreational opportunities as a legitimate function of the federal government. It also led to the enactment of the Wilderness Bill[24] and the Wild and Scenic Rivers Bill.[25]

Federal involvement in land and resource utilization policies beyond its own lands under the umbrella of the interstate commerce clause took the form of economic regulation beginning comparatively early. Certificates of public convenience and necessity from the Interstate Commerce Commission, for example, displaced state control over locations of rail and pipeline facilities just as, later, similar authority was granted to the Federal Power Commission with respect to interstate natural gas pipelines. The Corps of Engineers, under the same constitutional clause, has long had control over activities in or affecting navigable waters, territorial seas, and the outer continental shelf.

The specialized war-related history of the Atomic Energy Commission (AEC) accounts for an extremely broad federal role in connection with the uranium resource and all phases of mining, production, fuel fabrication, fuel reprocessing, transportation, and power generation in connection with nuclear energy.

Many of the federal government's regulatory activities are responsive to perceived failures of the system of dispersed decision-making with which the country first approached its land-use problems. Others reflect the growth of a conservation ethic, in its turn reactive to abuses of resource utilization—the denuding of forests, overgrazing, underutilization of water resources, and scandals in resource management such as the Teapot Dome affair.

The resulting uneven pattern seemingly identified yet another evil—an absence of planning. In the 1960s and 1970s, therefore, the federal government sought to achieve improvements in land and resource utilization policy by facilitating the process of planning *qua* planning. This idea was not exactly new. In the 1930s, the federal government experimented with planning when it enacted the Submarginal Land Retirement Program of the New Deal, which contemplated the purchase and retirement of unprofitable, eroded, thin-soiled and exhausted farmland and the transfer

---

Engineers; various project authorizations for the Bureau of Reclamation. Cf Federal Water Project Recreation Act, P.L. 89–72, 79 Stat. 213.

[21] Authorized by the Act of June 28, 1958, P.L. 85–470, 72 Stat. 238. The ORRRC Report, *Outdoor Recreation for America,* was issued in January 1962 and is reprinted as a document of the Senate Committee on Interior and Insular Affairs, *A Nationwide Outdoor Recreation Plan,* 93d Cong., 2d sess., June 1974, pp. 7–268 (hereinafter cited as ORRRC Report).

[22] P.L. 88-29, 77 Stat. 49.

[23] P.L. 88-578, September 3, 1964, 16 U.S.C. sec. 460 L5.

[24] Act of September 3, 1964, P.L. 88-577, 78 Stat. 890, 16 U.S.C. secs. 1131–1136.

[25] Act of October 2, 1968, P.L. 90-542, 82 Stat. 906, 16 U.S.C. secs. 1271–1287.

of the farmers to more promising areas where they could be rehabilitated and taken off the relief rolls. The National Resources Board planned to retire 454,200 farms comprising over 75 million acres, but appropriations were not forthcoming.[26]

The Department of Housing and Urban Development was authorized early in its existence to administer a comprehensive planning program to assist in land-use planning at the community-metropolitan level and to help the states plan their own land use and development.[27]

The act establishing the Bureau of Outdoor Recreation required that federal departments and agencies carry out their recreation responsibilities in conformance with a nationwide plan. Fourteen years after that act was passed, a volume was issued by the Senate Committee on Interior and Insular Affairs bearing the title *A Nationwide Outdoor Recreation Plan*,[28] which contains a document promising the statutorily specified plan some time in the future. A continuing concern was expressed that the government

> provide Federal leadership in recreation and land-use planning and classification; make available Federal assistance to all levels of government in planning land-use for orderly growth, including provision of adequate open space and recreation areas; provide for Federal encouragement in coordination and joint planning among and between public and private agencies.[29]

Before the United States was deeply involved in the Vietnam War, President Lyndon B. Johnson used typically expansive language to state an attitude about the nature of federal responsibility:

> The tradition of our past is equal to today's threat to that beauty. Our land will be attractive tomorrow only if we organize for action and rebuild and reclaim the beauty we inherited. Our stewardship will be judged by the foresight with which we carry out these programs. We will rescue our cities and countryside from blight with the same purpose and vigor with which, in other areas, we moved to save the forests and the soil.[30]

Legislative and executive activity of the 1960s focused on water. The President had said in his 1965 State of the Union Message that "we will seek legal power to prevent pollution of our air and water before it happens."[31]

---

[26] The National Resources Board, an interagency and interdisciplinary group attached to the Office of the President in the administration of President Franklin D. Roosevelt, claimed "planning as a distinctly American idea," citing the Constitutional Convention, Hamilton's "Plan of Manufacturers," Jefferson's and Gallatin's "Internal Improvements," Clay's "American System," the American Homestead Policy, the Conservation Movement, and the "economic mobilization" of the World War (I). Cf Report of National Resources Planning Board, *Plan for Planning*, quoted in Hagman, 1973, *Urban and Land Development*, p. 44, St. Paul, Minn: West Publishing Co.

[27] Housing Act of 1954, section 701, P.L. 83-560, 68 Stat. 590, 640, 40 U.S.C. sec. 461.

[28] ORRRC Report (footnote 21), p. 6.

[29] Ibid, p. 1180.

[30] Message to Congress, February 8, 1965. The message served as the "conference call" of a White House conference on natural beauty: *Proceedings, Beauty in America*, May 24–25, 1965, p. 15, Washington DC: GPO.

[31] Delivered to joint session of the Congress on January 4, 1965.

A Federal Water Quality Act was passed in 1966, and an Air Quality Act was passed in 1967. Later, both were substantially strengthened,[32] but the most significant development came in the current decade with the National Environmental Policy Act (NEPA) of 1969, which was signed into law on January 1, 1970.[33] That act cut the ties with the past rationale of federal incursions into land-use control policy and adopted an entirely new and different rationale. In NEPA there was no longer the dichotomy between the activities of the federal government as steward of public resources and its activities as regulator of private activities or as assistant in financing of state and local activities. Instead, the act was framed broadly to require a procedure for decision-making, including legislative decision-making, on any subject that could affect the environment. Also, a Council on Environmental Quality (CEQ), patterned after the Council of Economic Advisers, and an Environmental Protection Agency (EPA) were created to have primary responsibility for protecting the public health and welfare from the effects of pollution. Substantive laws, including the Clean Air Act as amended in 1970 and the Federal Water Pollution Control Act as amended in 1972, gave the new EPA extremely broad authority and responsibilities. Under the Clean Air Act, EPA can establish national ambient air quality standards for air pollutants. The mechanism for achieving these standards is the state implementation plan, under which the states are required to set and enforce limits on emissions from pollution sources, using standards at least as stringent as those established by EPA. The plan is under the potential sanction of the federal agency itself, which sets the required emission limits for the states. EPA also establishes uniform national air pollution control standards for new plants and factories, emission limitations for hazardous air pollutants, and motor vehicle emission standards.

With respect to water, EPA establishes effluent limitations applicable to polluting entities and issues guidelines and criteria to help the states set water quality standards for specific use objectives. As in the case of air quality, EPA is required to set water quality standards if a state does not do so, or if a state's proposed standards do not meet federal requirements.

During the five years since its enactment, NEPA has centralized operational responsibility for air and water quality programs in a new government agency, has required the staffing of environmental offices in virtually every federal office and agency, and has altered traditional relationships among the branches of the federal government. The distinction between federal action on programs such as the building of a dam and the federal function of adjudicating competing applications of private parties for licenses to build a dam has been blurred. The courts have an expanded role in measuring conformance of an agency's action with the substantive as well as procedural requirements of NEPA.

It is true that Congress relieved the judicial branch of any further responsibility to evaluate the adequacy of the Interior Department's environmental impact state-

---

[32] The Clean Air Amendments of 1970, P.L. 91-604, 84 Stat. 1676, 42 U.S.C. secs. 1857–1858 (a), and the Federal Water Pollution Control Act Amendments of 1972, P.L. 92-500, 86 Stat. 816, 33 U.S.C. sec. 1311 et seq.

[33] 42 U.S.C. sec. 4321 et seq. (83 Stat. 852).

ment for the Trans-Alaska Pipeline,[34] but the exceptional circumstances of its doing so reinforce the conclusion that both Congress and the executive branch have accepted the idea that the language of NEPA concerning environmental quality has substantive as well as procedural effect.

The federal land-managing agencies, which are accustomed to operating within the broad discretionary range of their statutory charters and are generally exempt from the requirements of the Administrative Procedure Act, have been subjected to judicial scrutiny of their policies and programs earlier and more often than in the past. Conformance with the requirements of NEPA has been tested often in energy projects ranging from lease sales to hydroelectric projects.

In transmitting to Congress the *Fifth Annual Report* of CEQ, the policy agency created by NEPA to oversee the entire range of governmental activity concerning the environment, President Ford said that it has become "clear that we cannot achieve all our environmental and all our energy and economic goals at the same time."[35] CEQ responded to the exigencies of the Arab oil embargo of 1973 and to the national goal of energy self-sufficiency (called Project Independence) of 1974 by cooperating with EPA, AEC, and the National Science Foundation, to develop a data base for quantifying the environmental effects of each link or component of a large number of energy systems.

Under the acronym MERES (Matrix of Environmental Residuals for Energy Systems), CEQ hopes to help energy decision-makers understand systemwide environmental effects, including those to be expected in emerging technologies such as coal gasification, oil shale projects, and others.[36] It may be anticipated that as the national concern mounts about the extent of national reliance upon foreign supplies of energy, the environmental establishment within the government will react with additional projects of this kind designed to contribute to, rather than resist, national programs for energy development.

The broadly centripetal nature of governmental activity in the environmental area, which followed the establishment of CEQ and EPA, has been matched by the government's response to the Arab oil embargo and other events of the energy crisis.

There is an Energy Council at the cabinet level chaired by Rogers C. B. Morton, who first assumed the role when he was Secretary of the Interior and kept it when he became Secretary of Commerce. The Federal Energy Administration (FEA) was authorized by Congress[37] after its predecessor, the Federal Energy Office, which was created by executive order, had effectively centralized many diverse energy-related activities. Frank Zarb, the administrator selected by the President to head the new agency, quickly assumed the role of policy spokesman as well as chief operating officer in spite of the continued existence of the Energy Council.

The mandate to the FEA is broadly stated—the administrator is to make such

[34] By amendment of the Mineral Leasing Act of 1920, in P.L. 93-153, July 1, 1973 (87 Stat. 576) 30 U.S.C. sec. 185.
[35] Council on Environmental Quality, 1974, *Fifth Annual Report*, p. v.
[36] Ibid, pp. 298-301.
[37] P.L. 93-275, May 7, 1974, 88 Stat. 96, 15 U.S.C. sec. 761 et seq.

plans and direct such programs related to production, conservation, use, control, distribution, rationing, and allocation of all forms of energy as are appropriate in connection with the functions transferred to his agency or vested in it by the President under other authorities, or by congressional authorization. The administrator is required to prepare and submit a "comprehensive energy plan designed to alleviate the energy shortage," and to coordinate with state governments and provide technical assistance to them.[38]

The energy crisis not only centralized governmental regulatory activities in the FEA but also led to a review of policies concerning public lands and their resources. A notable example was the proposal that the Naval Petroleum Reserves be leased to alleviate shortage.[39] Another, somewhat more complex, proposal was to create a federal oil and gas corporation to compete with private enterprise in producing oil and gas from public lands in order to furnish a yardstick price—a form of regulation to control the anticompetitive effects of ownership concentrations and vertical integration of the petroleum industry.[40]

When the Nuclear Regulatory Commission was spun off from AEC, it was commanded by Congress to make a national survey to locate and identify possible nuclear energy center sites. The survey was to include not only an environmental impact evaluation but also the possible use of federally owned property and other property designated for public use. Significantly, the agency was ordered not to consider "national parks, national forests, national wilderness areas, and national historic monuments."[41]

The activity of the federal government that presented the greatest potential for impact on US land and resource utilization policies grew out of the decision of the US Supreme Court in 1963 in *Arizona v. California*,[42] which enunciated the reserved rights or implied reservation doctrine with respect to water. Under such doctrine, reservations of public domain land for particular purposes such as wildlife refuges, national forests, national recreation areas, and petroleum reserves carried an implied reservation of sufficient water unappropriated at the date of the reservation to satisfy the reasonable requirements of those reservations without regard to the provisions of either substantive or procedural state laws. Since 1963, the US Forest Service and the military departments have ceased complying with state laws in acquiring rights to use waters, and in a suit filed by the federal government in Colorado in 1972, which is now pending in the US Supreme Court,[43] the procedural implications of this doctrine are being tested specifically to determine whether or

---

[38] Ibid, secs. 779, 781.

[39] The Senate passed S. J. Res. 176, on Dec. 19, 1973. If enacted this would have authorized production from Naval Petroleum Reserve (NPR) No. 1 (Elk Hills, California) for one year at 160,000 barrels per day; established a fund to finance exploration of NPR No. 1 and NPR No. 4 (North Slope, Alaska); authorized an appropriation; and assured competition in bidding to participate in development.

[40] S. 2506, 93d Cong., 2d sess. (Stevenson).

[41] Energy Reorganization Act of 1974, P.L. 93-438, 88 Stat. 1233, sec. 207.

[42] 373 U.S. 546, 597–601 (1963).

[43] *Akin v. United States*, 504 F.2d 115 (CA10 1974) *cert. granted*, 95 S. Ct. 1674.

not the United States must assert its claim in state courts, and, if it is not so required, how the federal rights are to be integrated with adjudicated rights under state water laws. So far, legislative proposals to resolve these problems have enlisted scant enthusiasm.

The development of energy from federally owned coal and oil shale deposits in the West will require enormous amounts of water. If the United States elects to insist on contentions that the water necessary for these developments may be called from the owners of water rights adjudicated under state water laws, additional friction will develop between the states of the energy-rich West and the federal government, the proprietor of many of these resources.

The attitude of the western states about water rights is as well-known as it is strong. In a political sense, an equivalent attitude is developing with respect to energy resources, particularly coal and oil shale, of the federal lands in the western states. The governors of the Rocky Mountain states (and a few neighbors) have formed an energy agency, candidly aimed at influencing federally directed resource activities by their combined efforts.[44]

In the past, states have sought to impose regulatory controls, which federal licensees have deemed incompatible with the terms of their federal licences.[45] Land-use regulations have proliferated as the states have enacted land-use planning legislation that is at least nominally applicable to federal lands. Some state laws are explicitly authorized by applicable federal statutes in the areas of air and water quality.[46]

After the blackout of 1965 in the Northeast, Congress had before it a number of measures that would have preempted state regulation of the siting of electric facilities, particularly transmission lines and generating stations. No measure of this kind was ever adopted, but during the subsequent ten years, a metamorphosis occurred in which the electric utility industry began to demand rather than resist federal legislation that would permit faster resolution of siting problems. Congress succeeded in passing legislation for federal certification of port facilities sufficient to handle supertankers.[47]

Shortly after the Public Land Law Review Commission had issued its 1970 final report calling for a "continuing, dynamic program of land use planning," aimed principally at the public lands, Congress began perennial consideration of bills to

---

[44] On April 28, 1975, governors of ten intermountain and plains states became the incorporators of a nonprofit corporation authorized under the laws of the State of Colorado, known as Western Governors' Regional Energy Policy Office. Initially, the corporation is funded by grants from the Old West Regional Commission and the Four Corners Regional Commission, federal organizations under the Commerce Department, of which all the states except Nevada are members. The executive director of the Office is William L. Guy, former governor of North Dakota, and the chairman is Governor Apodaca, of New Mexico.

[45] *First Iowa Hydro-Electric Cooperative* v. *Federal Power Commission,* 328 U.S. 152 (1946).

[46] Cf *Northern States Power Co.* v. *Minnesota,* 447 F.2d 1143 (CA8, 1971), affirmed without opinion, 405 U.S. 1035 (1972).

[47] P.L. 93-627, 88 Stat. 2126; 33 U.S.C. 1501 et seq.

establish a national land-use policy. In the US Senate, bills for this purpose passed twice by large majorities, but the House of Representatives did not take action after the Interior Committee reported it in February 1974.[48]

Part of the problem with respect to such legislation was related to the question of conformance of the bill to the recommendations of the PLLRC. When the PLLRC was authorized in 1964, Congress passed stopgap legislation on sales of public lands and classification of public lands for disposal and other uses; both of these measures expired when the commission's life expired.[49] As a result, Congress has also been considering the so-called Bureau of Land Management Organic Act to give to the bureau the authority to manage, protect, and develop public lands.[50] This act, too, failed to receive final action in the House of Representatives.

However, as mentioned earlier, it was the House of Representatives that took the leadership for strip-mining control legislation. The President pocket vetoed the bill passed by the 93rd Congress on December 30, 1974, and he vetoed the bill passed in the 94th Congress. Neither veto was overridden. One feature of the bill is worthy of special note on the subject of this paper. When the federal government owns mineral rights but a private individual owns the surface rights, this provision would have required the private landowner's permission before mining could proceed (even when under a federal lease). The Senate's version of the bill would have prohibited any development of coal in such an instance.

The legislative proposals for land-use planning were displaced as a focal point for executive and legislative tension by the strip-mining proposal. It is possible that the so-called comprehensive land-use planning legislation will be a casualty of the energy crisis; nevertheless, it is worthwhile to review the general outline of the land-use planning bill[51] and also the strip-mining bill.[52] In the former, findings, policy pronouncements, and a statement of congressional purpose assert that it is in the national interest that there be "competent land-use planning" that heretofore has been generally absent. Also "while the primary responsibility and constitutional authority for land-use planning . . . rests with state and local government," failure to plan or bad planning at such levels "poses serious problems." The bill invokes the welfare clause (not the property or commerce clauses of the constitution) in which the congressional purpose is "to promote the general welfare and to provide full and wise application of the resources of the federal government in strengthening the environmental, recreation, economic, and social well-being of the people of the United States."

Title II of the land-use planning bill sets up a grant-in-aid program that con-

---

[48] This was H.R. 10294, 93d Congress. Reported to the House with Amendments, House Report 93-798.

[49] P.L. 88-607, September 19, 1964, 78 Stat. 986, the Classification and Multiple Use Act; and P.L. 88-608, September 19, 1964, 78 Stat. 988, the Public Sale Act.

[50] A bill that would have been called National Resource Lands Management Act (S. 424) was approved by the Senate on July 8, 1974.

[51] S. Report 92-869, 92d Cong., 2d sess., accompanying S. 632, p. 93 (minority views of Senators Fannin and Hansen), June 19, 1972.

[52] See footnote 3.

templates that within three years, grantee states will have developed an adequate land-use planning process consistent with fourteen separate substantive requirement paragraphs, six separate procedural paragraphs, and special provisions for land developers. It is anticipated that after five years, the states will have developed an adequate state land-use program that will conform with detailed requirements.

Title III creates a new office in the Interior Department to work with an interagency advisory board and to administer policy. Title IV mandates coordination between federal agencies and the states on programs affecting federal lands. Title V sets up a separate program for Indian lands.[53]

The principal criticism of the measure was that under the guise of assistance, the federal government would take from the states "one of the last vestiges of state police power."[54]

The Surface Mining Control and Reclamation Act of 1975—the strip-mining bill—represented a somewhat different pattern of asserting a federal policy in the area of land utilization. In this very long bill, Congress would find a burden on commerce and the public welfare from surface mining operations, which destroy or diminish the utility of land in a variety of ways by creating hazards dangerous to life and property and by degrading the quality of life in local communities. Congress would also find that the primary responsibility for devising and enforcing regulations should rest with the states, and it would declare one of its purposes to be exercising whenever necessary the full reach of federal constitutional powers to insure the protection of the public interest through effective control of surface coal mining operations.

Administrative responsibilities would be assigned to the Interior Department in a new office of surface mining reclamation and enforcement. Coal mining states would be granted money for mining and mineral resources research institutes. Procedures for interagency coordination would be specified. The bill would create a trust fund financed by a reclamation fee on each ton of coal mined (ten cents for coal mined by underground methods and thirty-five cents for aboveground methods), principally to reclaim lands not subject to continuing reclamation responsibility under state or federal law. A coal area analogue of soil conservation districts would be established, with the Secretary of Agriculture authorized to enter into agreements with the landowners for necessary conservation treatment of rural lands affected by coal mining operations and to provide loan and grant assistance as well as technical assistance.

In certain instances, the Secretary of the Interior would be authorized to acquire abandoned and unreclaimed mined lands by eminent domain, to provide for necessary reclamation work. A regulatory structure and procedures for submission and approval of state programs would be required. Written concurrence would be required from EPA with respect to air and water quality standards. Any state's program would have to be approved as showing the state capable of carrying

---

[53] From S. 268, as reported to the Senate, amended on June 7, 1973, Senate Report 93-197, 93d Cong., 1st sess. H.R. 3510 (94th Cong., 1st sess.) was introduced by Congressman Udall on February 20, 1975.

[54] See footnote 51.

out its responsibilities through a state law providing for regulation of surface mining in accordance with the requirements and regulations of the federal act, with appropriate sanctions, an adequate administrative staff, a process for designating unsuitable areas and a mechanism for coordinating with other federal or state regulation of surface mining.

Failure of a state to meet these requirements would enable the Secretary of the Interior to promulgate and implement a federal program for that state after 30 months. State laws inconsistent with the federal law would be superseded.

The federal legislation would specify in detail the contents of an approved reclamation plan, both operational and environmental. Elaborate procedural requirements would assure public participation, appellate review at several levels, and authority for citizen suits including suits against the United States for persons having an interest adversely affected.

Federal lands are to be regulated separately but at least as stringently. There would be no governmental immunity from the provisions of the act because federal, state, and local governments, as well as publicly owned utilities or publicly owned corporations, would be specifically subject to the act. As noted above, surface owners would have to consent to the removal of federally owned coal beneath their lands, with provisions for determination of any compensation in the event agreement is not reached.[55]

It may be seen that since the passage of NEPA, Congress has devoted time and attention both to general land-use planning legislation and to an extreme example of comprehensive regulation of one particular phase of energy production, surface mining of coal. The fact that no legislation has been finally enacted cannot detract from the drawing of strong inferences as to the changed perceptions of the Congress and how it should react to land and resource utilization issues that are recognized as national problems.

Before attempting to synthesize some of these inferences into some sort of guide for the future, I should return to the work of the PLLRC. The Commission made 137 specific recommendations in its final report in 1970. Land and resource utilization policy with regard to the energy system was subsumed in many of these recommendations, almost entirely in connection with the public domain lands. The retention of these lands in public, federal ownership was a key theme, although not clearly stated in those terms. Another key theme was that the planning process should be formalized with congressionally specified standards, public participation, and effective state and local participation. "As a general rule," according to recommendation 13, "no use of public land should be permitted which is prohibited by state or local zoning." Regional commissions were favored as the administrative mechanism for most of the recommended planning.

In terms of the interaction of public lands with private lands, the report recommended that public land policy be designed to enhance a high quality environment "both on and off the public lands,"[56] and that rights and privileges to use public

---

[55] H.R. 25, as agreed to in conference and as passed (Cong. Rec., May 5, 1975, p. S7423).
[56] *One Third* (footnote 1), recommendation 16, p. 68.

lands be conditioned upon compliance with environmental control measures governing operations off public lands closely related to the right or privilege granted.[57]

Mineral resources, including energy mineral resources of the public lands, would continue to be excluded from development in parks, wilderness, and similar areas, but, by a divided vote, the Commission rejected repeal of the location mining system in favor of a leasing system.[58] In one of the few recommendations identifiably adopted, it was recommended that oil shale leases be issued on an experimental basis.[59]

The Department of the Interior "should continue to have sole responsibility for administering mineral activities on all public lands, subject to consultation with the department having management functions for other uses,"[60] according to another recommendation, but it has since turned out to be inconsistent with the Department of Agriculture's drive to establish its own rules and regulations controlling mining and mineral activities on US Forest Service lands.[61]

The PLLRC recognized the problem of the implied reservation doctrine and recommended that the federal government quantify its water claims for future use and compensate any pre-1963 claims valid under state law that might be taken under such doctrine for public use.[62]

The recommendations with respect to the outer continental shelf did not convey the sense of urgency that has since developed, but they did emphasize that the environmental aspects of development are central to expansion of these activities.[63] A significant recommendation was that the federal government undertake its own expanded program of collecting and disseminating basic geological data.[64]

Land management agencies should have the authority to require a reciprocal right-of-way as a condition for grants of rights-of-way across public lands, under recommendation 96, but the federal government should not utilize without specific statutory authority its position as landowner to accomplish indirectly any policy objective unrelated to protection or development of public lands.

The PLLRC recommended that the federal government refrain from disposition until adequate state or local zoning laws were in place or, alternatively, that the deed contain covenants to protect public values.[65]

From these diverse strands, the policy of the federal government on land and resource utilization in the context of energy seems amorphous and inconsistent; it is, however, quite possible to make an assessment of the direction in which policy is moving. Almost all of the policy controversies relate to quite basic tensions in our pluralistic system—tensions between the executive and the legislative branches

[57] Ibid, recommendation 23, p. 81.
[58] Ibid, p. 130.
[59] Ibid, recommendation 52, p. 135.
[60] Ibid, recommendation 54, p. 136.
[61] 36 C.F.R. secs. 251.4, 251.10, 251.11, 251.12.
[62] *One Third,* recommendation 56, p. 146.
[63] Ibid, recommendations 72 and 73, pp. 188–190.
[64] Ibid, recommendation 77, p. 193.
[65] Ibid, recommendation 117, p. 266.

of the federal government, those among component agencies of the executive branch for dominance over some or all of the phases of energy or environmental programs or over the economy generally, and those between the states and the federal government.

The Doub study[66] on federal energy regulation identified these three public policy objectives of federal energy regulation: "first, to ensure adequate and economical energy supplies; second, to employ sound land management in assuring wise and efficient use of public lands and the energy resources underlying them; and third, to protect public health and safety and protect and enhance environmental quality."

Thus stated, almost no controversy should be expected. How to ensure adequate and economical energy supplies is a different matter, as are the questions of what constitutes sound land management and how to protect and enhance environmental quality, particularly when the steps to be taken seem at odds with the first objective.

The Doub study did not address itself to the tension between the executive and the legislative branches, whereas the PLLRC was preoccupied with that aspect of policy and policy mechanisms.

PLLRC clearly believed that the responsibility for devising federal land and resource utilization policy ought to be recaptured by Congress. A fair reading of its report indicates that the Commission believed that Congress had abdicated its policy role and that there was sound reason to prefer a legislative to an executive dominance in making choices for the allocation of the dwindling resources of the public lands.

Whatever gains Congress has made in asserting its role have not come in the legislation concerning the public lands but rather in the broader regulatory legislation such as NEPA, the Air and Water Quality Acts, and similar laws. The measures that stated congressional policy concerning classification of public lands and sale of public lands were permitted to expire. An organic act for the Bureau of Land Management in the Interior Department has not been adopted. Revisions of the mineral leasing laws have been spotty, principally the enactment of the Geothermal Steam Act and a congressional solution to the impasse created by a judicial interpretation of the right-of-way provisions that blocked the Trans-Alaska Pipeline. During this period, the executive branch was beset with other problems, from winding down the Vietnam War to Watergate, and if the Congress could do no better than it did, it is unlikely that it will do better in the future.

From 1965 to 1975, the focus of land and resource utilization policy shifted from recreation and natural beauty to the environment and then to energy. For the next decade it seems certain that the demands of our industralized system for energy to fuel a suddenly stagnant economy will strengthen the role of the executive branch of the US government.

---

[66] W. O. Doub, April 1974, *Federal Energy Regulation, An Organizational Study.* Washington DC: US AEC.

Copyright 1976. All rights reserved

INTERLATIONAL ENERGY ISSUES AND OPTIONS

✻11028

Mason Willrich
School of Law, University of Virginia, Charlottesville, Virginia 22901

INTRODUCTION

Energy has been and will always be a vital factor in human affairs. Never before, however, has it played such a decisive role in international relations (1, 2).

This review is intended to provide a framework for analysis of the world energy situation from a political viewpoint (3). Perhaps the most fundamental energy issues are political in nature and international in scope. Moreover, the political process is the means for making key trade-offs, nationally and internationally, between, for instance, cheap energy and secure supplies or between self-sufficiency and environmental quality.

A political perspective is at once immediate and long term, changeable and enduring. A nation's policy can be reversed or its government changed overnight. Yet a nation's foreign policy may be determined by perceptions of basic interests and by cultural aspirations that endure for centuries, despite wars and domestic revolutions.

First, the international political structure is described as it relates to the evolving world energy situation. Thereafter, the major international energy issues are discussed under three broad problem headings: national security, the world economy, and the global environment.

ENERGY AND THE INTERNATIONAL POLITICAL STRUCTURE

The world political system is decentralized, potentially unstable, and open to attack from many directions. The primary political units are nation-states, and the main actors are national governments. Governments are rarely of one mind, however, and their policies result from the interplay of contending political factions. Moreover, every government is more or less vulnerable to revolutionary political forces. International organizations play more or less important roles in the world political system, depending on the national policies and capabilities of their member states. Similarly, private multinational corporations enjoy more or less autonomy, depending on the capabilities and policies of the national governments of the territories in which they operate.

The vast bulk of the earth's land is now parceled out among well over one hundred nations that are defined geographically in terms of land and people. Moreover, coastal states appear determined to expand their control over resources in and under expansive ocean areas bordering their shores. Largely as a result of political boundaries, all the basic ingredients of political power—territory, population, resources, technology, wealth—are distributed unequally.

Disparities among countries are glaring and important in the case of energy (4, 5). The United States has less than 6% of the world's population and uses more than 30% of its energy, while India contains 15% of the people and consumes less than 2% of the energy. Japan is an industrial country with a population of 107 million, ranks third in gross national product, and yet imports 85% of its energy requirements. Abu Dhabi is a tribal society of 120,000, and yet the per capita income from government oil revenues was $17,500 in 1974.

The opaque front that a national government attempts to present to the rest of the world must be penetrated if we are to take proper account of the dynamic balance of forces operating within each nation and the transnational links among them. Saudi Arabia's oil prices and production rates may be influenced by the monarchy's hostility toward communism and fear of domestic insurgency, as well as by its assessment of the international economic outlook. Domestic political factors also affect energy consumption policies. Fewer and fewer countries in the world have workable democratic political institutions, and most democracies are large energy importers. It remains to be seen whether democratic procedures are luxuries to be enjoyed only in times of economic growth and abundant energy supplies, or whether democracy can also deal effectively with chronic scarcities of resources.

Until the 1970s, private multinational oil companies operated with considerable freedom of action in the world arena.[1] The formation in 1960 of the Organization of Petroleum Exporting Countries (OPEC) as an intergovernmental oil producers' cartel, and the subsequent widespread nationalization of oil reserves and production capacity, has prompted consumer-country governments to respond in a variety of ways. Well before OPEC's success as a cartel, a few had launched their own petroleum corporations in order to bypass the private companies and bargain directly with producer governments for a share of their oil output and the right to develop additional reserves. In the early 1970s, the number of consumer countries with national petroleum corporations increased and the activities of such corporations are expanding. However, the shrinkage so far in autonomy of the multinational oil companies still leaves these corporate giants with substantial

---

[1] The eight multinational oil companies that are commonly known as the "majors" include five that are based in the United States and privately owned—Exxon, Gulf, Mobil, Standard Oil of California, and Texaco. The three others are Royal Dutch Shell, which is privately owned by British and Dutch interests; British Petroleum, which is partly owned by the British government together with private British interests; and Compagnie Francaise des Petroles (CFP), which is partly owned by the French government.

economic and political power to influence on a worldwide scale who gets how much oil at what price (6, 7).

The role of the private sector and its relationship to government in the nuclear field is different from oil in two major respects. First, government control is pervasive in view of the health and safety hazards and the security implications of nuclear power. Second, research and development in nuclear power requires financing and direction on a scale that only government can provide. Diffusion of nuclear power technology is now occurring among the industrial countries mainly through licensing and other arrangements to transfer know-how, so that several countries are acquiring their own nuclear industrial capabilities. The industrial countries are, moreover, competing with each other for sales of power reactors to less-developed countries lacking their own nuclear industrial capabilities. The resulting decentralization and competition considerably complicates the task of governments seeking to control exports of nuclear materials and equipment as part of broader policies intended to inhibit the spread of nuclear weapons (8, 9).

Compared with the political power of nation-states or the power of private multinational energy companies, international organizations are weak indeed. OPEC, which controls 85% of oil moving in international commerce, may appear to be a major exception.[2] However, OPEC's strength lies mainly in the weaknesses of the oil-importing countries. The Organization for European Cooperation and Development (OECD), composed of the industrial countries of Western Europe, North America, and Japan,[3] and the recently expanded European Common Market[4] have so far proved to be considerably weaker than OPEC.

In the nuclear energy field, the International Atomic Energy Agency (IAEA) was established in 1957 as an outgrowth of the US Atoms for Peace proposals in 1953. The IAEA is an organization of over one hundred members intended to promote international cooperation in nuclear development. The Agency provides technical assistance in civilian uses of nuclear energy to less-developed countries, and it administers safeguards on civilian nuclear activities in a growing number of countries to ensure against diversion from civilian to military use (10). Several regions also have special international organizations to promote nuclear power development, and some of these also administer safeguards to ensure that nuclear fuel is not diverted. The European Atomic Energy Community (Euratom), established in 1957 in the wake of the Suez crisis, is the most important example of regional nuclear cooperation. Whether globally or regionally organized,

---

[2] The original members of OPEC were Iran, Iraq, Kuwait, Saudi Arabia, and Venezuela. It has expanded in recent years to include Algeria, Indonesia, Abu Dhabi, Libya, Qatar, Nigeria, Ecuador, and Trinidad.

[3] OECD member countries are Australia, Austria, Belgium, Canada, Denmark, Federal Republic of Germany, Finland, France, Greece, Iceland, Ireland, Italy, Japan, Luxembourg, Netherlands, New Zealand, Norway, Portugal, Spain, Sweden, Switzerland, Turkey, United Kingdom, and United States.

[4] The European Common Market now includes the original six—Belgium, Federal Republic of Germany, France, Italy, Luxembourg, and Netherlands—and an additional three—Ireland, Denmark, and United Kingdom.

the authority of these organizations to act independently or over the objection of any of its member states is severely circumscribed.

Finally, a new International Energy Agency (IEA) has been launched, largely as a result of a US initiative in February 1974 in response to the Arab oil embargo (11).[5] At this writing (May 1975), it is unclear whether the Agreement on an International Energy Program, which would establish the IEA, will become effective. The IEA membership would include most of the OECD countries, except France (which may become informally associated). The IEA's initial purpose would be to establish and implement, if necessary, an international program for emergency sharing of oil supplies and demand restraint. However, the Agency is also intended to serve as a mechanism for long-term energy cooperation, including conservation, development of alternative sources, research and development, and uranium enrichment. Decisions by the Agency would be based on a voting system weighted to reflect oil consumption. It remains to be seen whether the IEA will function effectively either in an oil embargo or in overall energy cooperation.

Thus, we return to nation-states as the basic political constituents of the world community and to national governments as the main actors.

ENERGY AND NATIONAL SECURITY (12)

Security is a paramount concern in international energy relations. This is due to the vital role energy plays in both industrial and developing societies; the large inequalities among nations in the distribution of energy resources, technology, and capital; and the world political structure, which lacks a central authority capable of allocating resources. In a world community without supranational government, national security requires freedom of action and bargaining power.

Many countries import one fuel and export another, and some import a particular fuel into one region and export the same fuel from another. Moreover, the ideal posture toward security in fossil energy may be self-sufficiency, but one country's self-sufficient nuclear power program may be another's security threat. With these qualifications in mind, it is still useful to classify countries as energy importers or exporters.

Importing-Country Security

An importing country is primarily concerned with the security of its energy supplies. Each importing country may view foreign energy supplies as more or less vulnerable to interruption.

The most obvious security risk is an embargo instituted for political reasons by a foreign supplier or suppliers' group. The Arab oil embargo of "unfriendly" countries during the winter of 1973–1974 is an example. Countries that rely on the

---

[5] The signatories of the agreement include Austria, Belgium, Canada, Denmark, Federal Republic of Germany, Ireland, Italy, Japan, Luxembourg, Netherlands, Spain, Sweden, Switzerland, Turkey, United Kingdom, and United States.

United States for uranium enrichment may also fear politically motivated interruptions of their nuclear fuel supplies.

In addition to an exporter-imposed embargo, interruptions of fuel supplies may result from actions by a third party that may be hostile to either the exporter or the importer. A hostile government or terrorist organization may seize or destroy a key production facility or block a transport link. It is difficult to imagine a more strategic energy link than the Straits of Hormuz joining the Persian Gulf and the Indian Ocean. Twenty million barrels of oil per day, or more than 40% of current oil consumption in the noncommunist importing countries, pass through this narrow waterway in tankers destined for ports all over the world.

With secure supplies in mind, the problem each importer faces revolves around three issues: What is *energy security*? What basic strategies will enhance security? What specific measures will implement various strategies?

CONCEPTS From an importing country's viewpoint, at least three different concepts of *energy security* are possible. First, energy security may be viewed narrowly as the guarantee of sufficient energy supplies to permit a country to function during war. Second, energy security may be viewed very broadly as the assurance of adequate energy supplies to maintain the national economy at a normal level. If economic growth is considered normal, security may require some expansion in guaranteed energy supplies over time. A third, intermediate concept is the assurance of sufficient energy supplies to permit the national economy to function in a politically acceptable manner. Such a concept makes all-important the question of what is politically acceptable. Although this may appear to be begging the question, it does point in a useful direction.

As to political acceptability, international and domestic perspectives may be distinguished. From the viewpoint of other countries and international order in general, there would seem to be a common interest in assuring sufficient energy to an importing country to prevent its sinking into political chaos. From the viewpoint of the particular importing country, where it will draw the line between energy security and insecurity depends on its economic and political organization and on its stage of industrial development. In any case, energy security is closely linked to general economic security.

OPTIONS (13) In seeking to enhance its energy security, an importing country may consider a variety of specific options. First are measures to decrease damage from possible supply interruptions. These include standby rationing plans and stockpiling. Second are measures to enhance energy security through strengthened guarantees of foreign supply. Such measures include not only diversification of supply sources, but also increased interdependence through exporter-country investments in the importer and assistance in industrial development by the importer to the exporter. Third are measures to enhance energy security through increasing self-sufficiency. As a practical matter, these measures are limited to industrial countries that also possess relatively large undeveloped domestic energy resources.

Government rationing plans may enhance a country's energy security in several ways. In normal times, such plans would not be operative in countries with market economies. However, their existence—and political acceptance in advance—may provide a useful signal. An exporter contemplating supply interruptions would know that the importer is reasonably prepared and, therefore, determined to resist such coercion. In the event of a serious supply interruption, a well-designed rationing scheme would minimize damage to the economy. The major issues in developing a mandatory rationing plan are: the trigger events; the basic approach— a system of priorities or percentage of cutbacks in use; and responsibility for administration.

In any importing country, there are practical limits on the government's ability to cut back energy consumption if the result would be large reductions in employment and economic activity. Especially in a market economy, there is always danger of a downward spiral destroying the economy or bringing about the downfall of the government, or both.

A fuel stockpile for use in an emergency may reduce an importing country's vulnerability to interruption of supplies by providing a cushion against its effect. The existence of an adequate stockpile may also serve as a deterrent to an embargo and strengthen the bargaining position of the importing country in diplomatic efforts to lift it.

The circumstances in which an emergency stockpile may be desirable depend on the significance of the particular fuel within the nation's energy economy; the degree of reliance on imports; the number of sources of foreign supply and the political reliability of each; the world market outlook and the likelihood of cartelization; and the substitutability of another form of fuel. The practicality of a stockpile will vary substantially depending on the fuel, the form in which it is set aside, and the geography of the importing country.

Among the fossil fuels, oil stockpiling has received the most attention because of oil's importance in total energy consumption and the heavy reliance in many industrial countries on imports from the Middle East. Stockpiles of oil may be established by setting aside installed domestic production capacity, or by storing crude oil in surface tanks or underground in geological formations, or by stockpiling refined products. Storage is likely to be cheaper than setting aside production capacity.

Stockpiling of nuclear fuel seems quite practical. Whereas development and maintenance of only a 90-day emergency oil reserve would be a major task, a 5-year strategic stockpile of fuel for a large nuclear power program would be relatively easy. The striking difference is, of course, due to the compactness of refined uranium. Nuclear fuel may be stockpiled either as natural uranium oxide—"yellow cake"— or as enriched product. Plutonium for recycling in light water reactors or use in breeder reactors may also be stockpiled. If breeder reactors can be successfully developed for commercial use, they will eventually reduce to a minimal level the amount of uranium input required to sustain a growing nuclear power capacity.

In assessing costs and benefits of stockpiling, the first issue is whether the same or greater gains in energy security may be achieved for less cost by investing an equivalent amount of money in some other measure, such as developing additional domestic energy resources or establishing new foreign energy relationships. A second issue is whether the security gains outweigh the welfare gains of investing an equivalent sum in any area not related directly to energy security.

Stockpiling of nuclear fuel has a unique aspect that merits attention. A small fraction of a fuel stockpile for a nuclear power program would be enough material to produce a substantial stockpile of nuclear weapons. What is a security gain for the importing country may thus be a substantial security loss for other countries.

An importing country may enhance its energy security by diversifying its foreign sources of supply. If there is idle world-production capacity, an importer may quickly achieve considerable diversification. An importing country may also achieve rapid diversification if it has sufficient market power to preempt supplies originally committed to other countries. However, the more interesting options for diversification are long term. Two bottlenecks presently motivating importers to diversify their foreign sources of energy supplies are the OPEC cartel and the predominance of the United States in uranium enrichment.

There is already substantial diversity of oil suppliers within OPEC. There is little likelihood that all members of the cartel will agree to embargo oil exports to any particular country. The main risk is an embargo by Arab members of OPEC, and this risk is linked firmly to the Arab-Israeli conflict. Therefore, an oil-importing country may attempt to assure itself of continued supplies from OPEC in one of two ways: it may make special arrangements with non-Arab members of OPEC, or it may support the Arabs in their conflict with Israel.

The operations of private multinational oil companies in the world oil market may offer an importing country two security advantages. The existence of a market with many buyers and sellers served by private companies acting as middlemen creates more possibilities for supply diversification than a series of country-to-country arrangements. Under normal conditions, moreover, countries selling in the market do not know where much of the oil they export is ultimately to be delivered. Despite these advantages, a privately controlled world oil market could well be a casualty of attempts by importing countries to gain more secure oil supplies through government-to-government deals.

In addition to diversification of oil supply sources within OPEC itself, there may be opportunities for the development of additional energy supplies outside OPEC. Such diversification may occur either by the addition of countries to the list of large oil exporters, or through the substitution of different fuels for oil by the importers. The time scale for either type of development is quite long.

An importing country may also wish to enhance its energy security by diversifying its sources of nuclear fuel. The type of commercial power reactor that is most widely used requires low-enriched uranium fuel. Until recently the United States enjoyed a monopolistic position in the supply of enrichment services to noncommunist countries, whereas the Soviet Union monopolized nuclear fuel supplies to Eastern Europe. Even though the Soviet Union is now competing quite successfully for enrichment contracts in noncommunist countries, many nations are likely to see risks in long-term reliance on such a duopoly. A country importing nuclear fuel and seeking further diversification may participate directly in the enrichment stage. The tripartite British-German-Dutch effort and the French-led Eurodif venture are two multinational enrichment ventures moving ahead in Western Europe. Alternatively, the importer may bypass the need for enrichment by adopting power reactors that use natural uranium fuel.

Countries that lack domestic uranium reserves may be concerned about the risk of a cartel emerging among the uranium-exporting countries (14). The low cost reserves presently available for the world market are highly concentrated in Canada, South Africa, Australia, Gabon, and Niger. Though uranium has been abundant and the price depressed until recently, demand is increasing rapidly in order to fuel nuclear power reactors that are already operable, under construction, and planned. The few countries in a position to export uranium in large quantities appear unlikely to embargo exports on political grounds. However, the possibilities seem strong that uranium exporters will, either in concert or alone, bargain for the transfer of enrichment technology and adopt strategies to make their operations in the world uranium market highly profitable after many lean years of oversupply.

From an importing country's viewpoint, energy insecurity is created not merely by the necessity of reliance on foreign supply sources, but also by the fact that the particular foreign sources relied on enjoy a substantial measure of autonomy. The oil importer's energy insecurity may thus be largely a function of OPEC's actual or perceived freedom of action. Similarly, countries depending on the United States for uranium enrichment services may feel insecure because the US government may unilaterally dictate the price and main contract conditions for nuclear fuel enrichment.

An importing country may seek to strengthen its security through interdependence with its present supply sources. This would involve a willingness on both sides to negotiate deeper forms of mutually dependent relations. Neither the importer nor the exporter would dominate the relationship. Interdependent relations of this sort may be very difficult to reverse, and hence they are not likely to be created quickly or easily. Whether an energy-importing country will view deeper interdependence with certain fuel suppliers as strengthening or threatening its position will depend largely on how ambiguous circumstances are perceived.

An importing country may seek to achieve greater interdependence in two different ways: long-term investments by the exporter in the importing country; and assistance in industrial development by the importer to the exporter. The fact that such economic measures have an important security aspect to them simply reflects the interweaving of energy security and economic issues.

An exporter's investments may be viewed as strengthening the importer's security in a number of ways. The investments may give the importer leverage on the exporter through the threat of nationalization or the ability to freeze the exporter's assets. The investments may also give the exporter a stake in the importer's economy so that an interruption of supplies would amount to a self-inflicted wound. The most effective form of investment from the standpoint of the importer's energy security may be downstream in the fuel cycles of the energy industry. Such an investment would give the exporter a direct economic stake in keeping the fuel supply line open.

An importing country may also enhance its energy security by creating interdependence between the flow of energy from the exporter to the importer and the flow of developmental assistance from the importer to the exporter. Because the importer has a vital need for energy, it would be important for the reciprocal flow of developmental assistance to be equally vital to the exporter and for substitute sources of supply not to be readily available.

It is discouraging that a large part of the flow of goods from the United States and West European countries into the Middle East in exchange for oil has been composed of armaments. Rather than improving energy security, the arms trade may create substantial military security risks for the United States and Western Europe.

Of the various ways for an importer to enhance its energy security, self-sufficiency or independence has received by far the most attention in the United States (15). Energy independence as an objective has several possible meanings. It may mean actual and exclusive reliance on domestic energy resources; the potential to rely indefinitely on domestic resources after some transitional period; or the capability to rely exclusively on domestic resources for a limited period of time only.

Complete self-sufficiency would amount to a self-imposed embargo on energy imports in order to assure that an embargo would not be externally imposed. The capability for this self-sufficiency, achieved after a period of time, may be useful in slowly deteriorating circumstances, but of little value in a world where political action can result in rapid changes in global energy flows. The capability to be self-sufficient for a period of time in the event of supply interruptions seems achievable through emergency stockpiles, as discussed above.

In any event, energy independence would involve substantial economic penalties for most countries capable of achieving that solution to their energy security dilemma. Expansion of domestic energy production for security reasons implies the development of a production capacity that may not be competitive economically. Self-sufficiency may also cause serious environmental consequences for the country concerned.

A general benefit of energy self-sufficiency is that one country will, by reducing its level of energy imports, reduce competition among other importers and increase it among exporters in the world market. Thus if the United States enhances its security by reducing its oil imports, the energy security of the rest of the OECD countries will also be improved.

Nuclear power is, however, a specific case in which the pursuit of national self-sufficiency may destabilize international security. For a country without nuclear weapons, self-sufficiency in nuclear power means that it would be impossible to deter or prevent diversion of nuclear fuels from civilian to military purposes by foreign interruption of nuclear fuel supplies or critical services. Consequently, a nation without nuclear weapons that, at great cost, develops a self-contained nuclear power program may appear thereby to threaten the security of neighboring countries (9).

Exporting-Country Security (16)

While an importing country is concerned primarily with access to resources and supply security, an energy exporter is preoccupied with access to markets and security of demand. The exporter's security problem also contains distinctive features such as the basic concern about national sovereignty over its natural resources. Moreover, an exporter's particular viewpoint, and hence its security requirements, will depend on the extent of its energy resource base and the stage of its industrial development.

CONCEPTS Sovereignty over natural resources is perhaps the minimum energy security concept for the exporter. The concept was only belatedly and grudgingly accepted by the oil-importing countries in the 1970s. Sovereignty in this respect has a double aspect: security against military intervention to deprive a nation of control over its energy resources, and freedom from external interference in national decision-making regarding exploitation of those resources. The concept does not preclude a variety of more or less important roles for foreign enterprise, either private or government owned. It does require, however, that the appropriate role be a matter of national decision, not foreign dictation.

Recognition and observance of national sovereignty over natural resources may not, however, be sufficient as a security concept for many energy exporters. Since World War II, the world oil market has suffered decades of glut and only a few years of artificial scarcity; and excess production capacity has been a feature of the world uranium market. Therefore, an exporter may adopt, as a corollary to sovereignty over its basic raw materials, a concept of energy security that includes guaranteed access to foreign markets. This raises possibilities for mutually beneficial negotations between exporters and importers, based on a shared interest in maintaining stability and equilibrium.

Beyond sovereignty and market access, an exporter may extend the concept of energy security to cover financial security for the investments made with its export earnings. Energy resources below ground are a precious national heritage. Once extracted, that heritage can easily be lost by improvident government investments or eroded by inflation. Most energy exporters will be in a posture in which they, like most importers, have no alternative to interdependence.

OPTIONS The post–World War II drive in less-developed nations to gain sovereign control over their natural resources has been largely successful. The process of supplanting the old oil concessions with new service contracts is nearing completion (17). In the future evolution of the world energy situation, this political aspect of national sovereignty is likely to be relatively unimportant as an international issue.

The possibility of Western military intervention in the Persian Gulf in the event of another Arab-Israeli war or Arab oil embargo has been mentioned often. If the United States intervened alone militarily, it might find itself in a confrontation with Soviet armed forces in a location where escalation would be a serious possibility and would work in favor of the Soviet Union. Moreover, the NATO allies and Japan might be suspicious of unilateral action by the United States, since US interests both in regard to Israel and to Middle East oil are not congruent with the interests of the other OECD countries. If certain West European countries intervened without the United States, again the lack of a congruent set of transatlantic interests in the Middle East might prompt the United States to object strongly. The reaction of the United States would be especially strong, if, as is likely, the Soviet Union threatened counteraction that would generally involve NATO security interests. Finally, the likelihood of the United States and a group of West European countries launching a coordinated military intervention in the Middle East seems remote.

In addition to the risks of intra-NATO and East-West conflicts, military intervention in the Persian Gulf would have repercussions throughout the Third World. The precedent that such action would establish might be perceived as a threat to all suppliers of raw materials. Though powerless to react militarily, the Third World countries would be likely to develop a very hostile diplomatic response.

Beyond these adverse consequences of intervention, it is difficult to see how Western seizure of certain Persian Gulf oil fields could accomplish anything positive. If a more secure flow of oil to the OECD countries were the objective, armed intervention would be a short-term expedient that would be self-defeating in the long run. Moreover, it is difficult to imagine a more pro-Western group of rulers than the present Shah of Iran and the Arab oil sheiks. Intervention might unleash the very revolutionary forces that the present rulers of most Middle Eastern countries have long feared and sought to avoid. It is also difficult to see how military intervention in the Persian Gulf could help Israel.

The Arab states on the Gulf could also take two steps that would greatly raise the cost of Western armed intervention. One is to ensure that oil production capacity would be rendered inoperable for a period of time following intervention and military takeover. Emergency stockpiles of oil in the OECD countries may or may not be sufficient to offset the stoppage that would occur following intervention. A second step is to build up armed forces capable of resisting. This is underway as a result of massive Western arms sales to the Gulf states in exchange for oil. If the Western countries ever seriously contemplated military intervention in the Gulf states, they have acted in a self-defeating manner. Taken together, these factors appear to rule out a Western military takeover of oil fields in the Persian Gulf region as a rational course of action.

An exporting country's interest in guaranteed access to foreign markets is the reciprocal of an importer's interest in access to energy resources. In a seller's market, of course, the buyers want long-term contracts at guaranteed prices, but in a buyer's market it is the sellers who want firm arrangements. An oil exporter with a large population is likely to plow back most of its oil revenues into domestic industrialization. As this process advances and gains momentum, the country may become more, rather than less, dependent on a stable flow of oil revenue. Here we have a security problem that grows out of economic success.

An exporter may seek to guarantee its access to foreign markets through any one or more of the interdependent arrangements we considered earlier from the importer's viewpoint. Long-term supply contracts may be negotiated, investments may be made downstream, and so forth. The important point is that once an energy exporter perceives that its particular security interests are served by pursuing interdependence, there will emerge a wide area for fruitful negotiation with importers who similarly perceive their security interests. Such a common perception seems most likely to emerge between Japan and the industrial countries of Western Europe, on the one hand, and the more populous and rapidly developing OPEC countries, on the other.

ENERGY AND THE WORLD ECONOMY (1, 2)

A world economy of sorts exists, although the destructive behavior of an insecure nation or group of nations may still pull it apart. From the viewpoint of an evolving world economy, there are four key energy issues: the economic efficiency of energy use; the role of energy in economic development; the role of private multinational corporations in the energy sector; and management of the financial aspect of the international oil trade. OPEC's quadrupling of world oil prices in 1973–1974 pushed these four issues to the forefront. We discuss them below in a nontechnical sense in order to perceive their political implications. A related issue, energy conservation, is considered in the following section.

Energy Efficiency (18, 19)

The earth is endowed with a rich variety of energy resources. What is an efficient way for the world community to develop and use these resources? In economic terms, efficiency is equated with costs, and the problem of doing anything efficiently becomes a problem of doing it in the least costly way possible.

The efficient development and use of the earth's energy resources involves a continuous effort aimed at the least cost alternatives. If oil is cheaper than coal for generation of electric power, oil will be used. But if electricity can be produced more cheaply with nuclear fuel than with oil, nuclear power will replace oil-fired generation. Costs tend to rise as nonrenewable natural resources are depleted. However, such increases may stimulate a search for additional resources of the same fuel or for a substitute fuel. Moreover, technological improvements may reduce costs. Over time, technological change may or may not fully offset cost increases resulting from depletion of resources and other factors. Energy costs must also be viewed in relation to other cost trends in the economy. If energy costs are increasing, but more slowly than other factors, then it would be efficient to use energy more intensively in production.

Cost as the criterion of economic efficiency is a simple concept that becomes both complicated and imprecise in its application. First, marginal or incremental costs, not original or historical costs, should be used to determine efficiencies. The practical application of marginal-cost principles involves making assumptions about an uncertain future, and the degree of uncertainty is reflected in cost as a risk premium.

Second, it is difficult to determine what items are to be included as costs. Of course, all operating costs and capital charges, including a reasonable return on capital investment, are included. But we should also take appropriate account of externalities or social costs, such as environmental pollution and health and safety hazards. The quantification of social costs is an imprecise art, and rarely is an environmental or social cost fully internalized.

A third problem of cost has to do with time of return on investments. Investments in energy resources are long term. Present values may be imputed to future benefits and costs through the use of discount rates. The choice of discount rate is often decisive in determining whether a particular energy investment is worthwhile and also in determining how much of an energy resource to produce currently. Yet the appropriate rate is a factor about which reasonable persons may disagree.

Finally, economic efficiency is indifferent to the problem of income distribution. However, distribution of wealth through energy prices that are either too high or too low in relation to costs results in an economically inefficient use of energy resources.

Difficulties with the cost criterion have led to suggestions of a variety of other approaches to judging energy efficiency internationally. The large-scale relationship between energy consumption and gross national product (GNP) is sometimes used in comparisons. According to this criterion, a national economy with increasing energy efficiency would be indicated by a decreasing ratio between the growth rate in energy use and the growth rate in GNP.

The ratio of energy to GNP may be useful in comparing mature industrial economies with relatively stable price relationships. However, it does not apply to a developing economy. In a changeover from a predominantly agricultural society to an industrial one, energy growth is likely to outpace growth in the GNP. Even in mature industrial economies, preferences for environmental quality or technological barriers can cause a declining energy/GNP ratio to reverse.

Per capita energy consumption is also sometimes used as the basis for comparing efficiency of energy use in various industrial countries. Per capita energy consumption in the United States is currently about twice that in several large West European countries. Differences in such indicators as per capita energy consumption may be traced to specific economic or structural factors and to income disparities. The relationship between levels of energy use and quality of life is far from clear. A very broad frame of reference is necessary to evaluate the role of energy in society.

Finally, use of energy resources may be suggested as a measure of the efficiency of specific processes or activities. Efficiency is a much more complex matter, however, than simple application of the criterion "less is better."

Options to Reduce World Oil Prices

In early 1975 the multinational oil companies were paying roughly $10 per barrel for oil from sources in the Persian Gulf. The cost of production is 10–20 cents per barrel, and the rest of the $10 price is government revenue. Therefore, current world oil prices bear no relation to costs and are plainly exorbitant.

The world's known reserves of low-cost oil appear adequate to meet projected worldwide demand for a few decades at least. In these circumstances, pressing for substantially lower world oil prices may be viewed as not just a tactic but a long-term economic strategy. How might such a strategy be carried out?

One way would be to discover additional oil provinces that are comparable in magnitude and production costs to those centered in the Persian Gulf. This seems unlikely, although not inconceivable.

A second way to lower oil prices would be to break up the OPEC cartel, restoring price competition in the world market for crude oil. One suggestion is that an absolute ceiling might be imposed on oil imports, and import tickets could then be auctioned off in secret bidding. This would provide a way for oil exporters to lower their prices without fear of disclosing their identities, creating pressure for lower oil prices. Another suggestion is to break up the integrated operations of multinational oil companies. Such action would be intended to achieve a more competitive industrial structure overall, which would make it more difficult for the governments of OPEC countries to use the oil companies as their agents for collecting rents or taxes from the consumers.

The US government has suggested that the OPEC price agreement might be breached by concerted action of the OECD countries to develop alternative supplies and to restrain demand.

On the supply side, at or near current world oil prices, development of a range of alternative fossil energy resources appears economical. These include oil and gas resources under the outer continental shelves and in arctic regions, natural gas liquefaction and long-distance transportation, coal gasification and liquefaction, and oil extraction from tar sands and shale. Moreover, high oil prices tend to accelerate the installation of nuclear power, although this effect in the short term has been more than offset by reduced demand for electricity due to the general economic recession and by financial difficulties of electric utility industries arising largely out of rapid inflation.

The outstanding fact about options for increasing energy supplies is that each one will require large capital investments and many years before it will make a substantial difference in the world energy picture. Moreover, many of them require the development of new technologies before commercial exploitation can proceed. Finally, the environmental costs of many of the options appear substantially greater than use of existing low-cost fossil fuel resources.

The cost of effective substitutes for Persian Gulf oil establishes a ceiling on the monopoly price that OPEC may extract. However, the OPEC price, or a price somewhat below it, may also be viewed as a floor on which private investments in alternative energy resources must rest. Governments will either have to underwrite private investments and guarantee them against a future reduction in oil price, or make direct investments themselves in the development of higher-cost energy resources. The costs of these actions will be high, and they will ultimately be borne by consumers in the prices or taxes they pay.

Economic Development

The oil price increases in 1973–1974 halted economic development worldwide, except in the OPEC countries where development was rapidly accelerated. In this section we consider options for economic development for less-developed countries (LDCs) that either import or export oil.

OIL-IMPORTING LDCs   In an oil-importing LDC, the high price of oil makes a direct hit on the economic vitals—agriculture and industry. The increases in oil prices have been translated into higher costs for the fuel and fertilizer required for domestic food production and into higher costs for boiler fuel needed for industrial production and generation of electric power. Moreover, the rises in oil prices have increased the prices of LDC imports from industrial countries and shrunk the market for LDC exports in industrial countries. Finally, the flow of development aid from industrial countries to the LDCs has been reduced.

Faced with a quadrupled price, an oil-importing LDC has three basic options. First, it may simply absorb curtailed imports through a reduction in its GNP. Second, it may attempt to offset the increase in its oil import bill with increases in the prices of its exports. And, third, it may seek increased development assistance from foreign sources. In the context of any of these options, an LDC may also redesign its development plan specifically to emphasize energy conservation.

Some reduction in GNP is almost certain in every oil-importing LDC. The extent will vary widely, depending on particular national economic circumstances. A country's development strategy may rest on assumptions that are wiped out by large increases in the price of energy. Moreover, the government may be too weak to take the decisive actions required to meet the challenge. However, the high price of oil might stimulate modernization in order to resume overall development as soon as possible. This could include adoption, at the outset, of energy-conserving forms of industrial technology, as well as development of any high-cost domestic energy resources that were available, and perhaps heavy emphasis on nuclear power.

In the past, the LDCs have vigorously pressed claims for higher prices for their commodity exports and for preferential access to markets in industrial countries. In this regard, oil has been one of a few success stories so far. It seems unlikely that LDCs would be generally successful in charging the industrial countries a great deal more for their commodity exports to offset increases in their oil import bills. Indeed, many commodity prices slumped in early 1975.

A particular LDC, or the LDCs as a group, may seek to offset increases in oil import bills by increases in the flow of development aid granted on concessionary terms. The LDCs may turn toward either the industrial countries or the OPEC oil exporters for assistance. And, of course, they may try both.

The industrial countries recognize that the new energy situation raises special problems for the LDCs. However, governments of industrial countries are largely preoccupied with their own economic difficulties. The most that some of them have been able to accomplish is to prevent substantial cutbacks in the monetary amount of development aid. Meanwhile, inflation is reducing the real amount being transferred.

Following their triumph in the oil price revolution of 1973–1974, many OPEC countries quickly recognized a special duty to aid the hard-hit oil-importing LDCs. Political self-interest was perhaps a stronger factor than humanitarian motives. Various OPEC members could see opportunities to use aid to enhance their power regionally as well as within the Third World generally.

Oil-importing LDCs may seek either oil at concessional prices or financial assistance that would enable them to pay their oil import bills. A number of OPEC members have reportedly lowered their prices for certain LDC importers. It is not likely, however, that OPEC will formally adopt a two-tier pricing system, i.e. one price for industrial countries and another for LDCs.

Despite the damage that the exorbitant oil prices of OPEC are causing to their economies, the oil-importing LDCs have sided with OPEC against the industrial countries. The LDC stance may be based on a political calculation along the following lines. For the first time in history, a nonindustrial group of countries has forced the noncommunist industrial world into a redistribution of wealth on a large scale. Politically, the fact that the redistribution is occurring is more important than the fact that a large number of LDCs are being hurt while only a few, the OPEC members, are being benefited enormously. The possibilities for obtaining compensatory or concessionary aid from OPEC are greater than the prospects for obtaining more aid from the industrial countries in a world economy where inflation was rampant well before the oil crisis.

OIL-EXPORTING LDCs   One useful way to categorize the oil-exporting LDCs is according to the capacities of their respective economies to absorb oil revenues at current price levels. In this respect, OPEC includes countries that can productively absorb all present and future oil revenues and countries that are piling up large surpluses. Only a few Arab members of OPEC are in a long-run surplus position. Countries with large absorptive capacities may be further divided into those that can absorb all their oil revenues and, other things being equal, still remain poor; those that can develop a viable and stable national economy through use of their oil revenues; and those that have the potential for affluence.

An oil-exporting LDC may seek to develop into a modern industrial state. This will require a long-term effort, however, because a viable industrial economy presupposes a relatively high level of literacy and education among the people. Such a level may take a generation or more to achieve, even with a massive effort.

A number of intriguing possibilities arise with respect to industrial diversification within an oil-exporting LDC. A diversification plan may be across-the-board, or it may build on energy-intensive industries and petrochemicals. Thus, a large share of the world's output of refined metals, such as steel or aluminum, and petrochemicals, such as fertilizer, may come in the future from the current oil-exporting countries.

In any event, an oil-exporting LDC will have a choice in emphasis in its economic planning between self-sufficiency and interdependence. Self-sufficiency would stress the development of a balanced and diversified economy, while interdependence would stress economic specialization, comparative advantage, and efficiency. Interdependence may make the world community as a whole better off. However, in the absence of a practical and effective international mechanism for redistribution of wealth, self-sufficiency may be worth the price. Moreover, a degree of self-sufficiency is necessary for bargaining power in an interdependent relationship.

Role of Private Multinational Oil Companies (7, 20, 21)

Until the early 1970s the major private multinational oil companies were the most powerful actors, politically and economically, in the world petroleum market. The

OPEC governments are now in the process of completing their takeovers of crude oil production. On the OPEC side, the private company will serve primarily as a contractor, supplying technical know-how and marketing skills. The remaining unresolved issues are largely on the buyers' or consumers' side of the world oil market.

Present relationships between an importing country and a multinational oil company are very complex. The mixture of rewards and punishments, subsidies and regulatory controls, for private activity varies from one country to another. The basic choice now before every importing country is, however, the same: whether to adjust its own government-industry relationship toward greater or less government intervention. In this respect, the world oil market and the various domestic energy markets are inseparable. There is a widespread tendency toward greater government intervention.

In the direction of increased government intervention, a variety of specific steps may be considered. On the demand side, these include price controls and various rationing schemes in the short run, and public-utility-type regulation of oil prices and operations in the longer term. On the supply side are subsidies for research and development, financial assistance for demonstration projects, tax incentives for energy investments, guaranteed prices for output, and protection against competition from cheaper foreign energy supplies. Beyond measures that government may apply to private industry are steps to give the government a proprietary role of its own. These range from the establishment of a government corporation that would compete with private industry to outright takeover of oil industry operations by the consumer-country government.

However, a government may consider less, rather than more, intervention into energy markets. Possible actions in this direction include decontrol or deregulation of prices, such as the wellhead price of natural gas, and repeal of tax advantages, such as depreciation allowances, deductions for intangible drilling expenses, or tax credits for royalty payments to foreign governments.

More broadly, the structure of the energy industry may be reviewed from the antitrust viewpoint. First, should the trend toward the formation of energy companies be stopped? Second, now that multinational oil companies no longer control the main sources of foreign supplies of crude oil, should the disintegration process be carried further, requiring, for example, refineries and pipelines to be divorced from marketing operations?

The industrial importing countries may have a mutual interest in developing a common understanding about the values presently contained in the private sector of the world oil industry that are important to preserve. The main values seem to be know-how and capital. If those values were lost or dissipated in the near future, they would be difficult, if not impossible, for the world community to replace.

Petromoney Flows (22, 23)

At the end of 1974, the revenues of OPEC governments were roughly $10 per barrel, and during that year almost $90 billion flowed from the rest of the world into OPEC treasuries in payment for the cartel's oil output. Of the $90 billion in revenues, the OPEC governments spent about $30 billion in 1974 on everything from guns to butter, leaving them with a surplus of about $60 billion. The massive transfer of wealth from the rest of the world to OPEC continues in 1975 largely unabated, and it will continue until something occurs to restrict or alter the flow. The international petromoney problem and options for managing it are considered below, first, from the viewpoint of the oil importer and, thereafter, from that of the exporter.

OIL IMPORT PAYMENTS   The money problem of the importing country is essentially how to pay for oil purchased from foreign sources for domestic consumption. The difficulty of solving the problem in a particular country will depend on the overall strength of that country's economy, as well as the size of its oil bill.

If the oil-importing country pays its bills currently, consumers in that country will have less money to spend on other goods and services. Consumer demand will drop, and, as this occurs, production will drop and unemployment is likely to increase. At the same time, prices will increase for all products in which oil is a factor. Consumers will be able to buy even less, and producers will, therefore, reduce output further and more unemployment will occur.

Faced with this depressing set of circumstances, what options does the importing country's government have? In the first place, it can do nothing and let the national economy adjust to the higher oil prices. This will be a viable option only for a country with an exceptionally strong, flexible economy and a highly mobile work force.

The government of an oil-importing country may, therefore, be tempted to offset the increase in oil payments by reducing imports of other goods, especially imports of goods for which there are substitutes produced domestically. But the government of a second oil-importing country would view the first country's import controls with grave concern. The effect would be to disrupt the second country's balance of trade. Moreover, the restriction on the second country's exports to the first would occur at the worst possible time; namely, when the government of the second oil importer would be attempting to solve its own problem of oil import payments.

If an oil importer cannot reduce its other imports without risk of retaliation and general reduction in world trade, it may seek to increase exports. This might be done by increasing government subsidies for exports, or by reducing the international prices of its exports by a currency devaluation. Export promotion, like import controls, will be viewed with alarm abroad. At the very time when consumers in other oil-importing countries are spending enormous amounts on oil, they are being induced to spend more on other imports that are promotionally priced.

Turning to options to deal with the problem at home, the government may finance its country's oil bill by deficit spending. The spending program may be designed to pump back into the domestic economy roughly the same amount as foreign oil payments are draining out. In order to run a deficit, however, the government of the importing country has to borrow. Borrowing from its own economy will drive interest rates up, making it more difficult and costly for other large borrowers like the energy industry to obtain the investment funds needed to operate or expand their activities. To avoid competing with its own private sector, the government may attempt to borrow money abroad. Up to some limit, the other oil importers with stronger economies may help to finance the first importer's deficit in support of the common interest in maintaining an interdependent world economy. But that limit may be short of what the first importer needs from external sources in order to finance its deficit.

Another option the government may consider is to stimulate domestic investment. Current consumption would be reduced by the increased investment, but economic growth would provide the eventual solution to the oil payments problem. However, it may be difficult to lower the interest rate to encourage investment, while at the same time the government is borrowing to finance a deficit. Another problem is the lag between investment and actual increases in production.

The government may increase the money supply. This may stimulate domestic demand in the short run or it may merely inflate the economy further. It would reduce the foreign exchange rate for the importer's currency, having an effect comparable to devaluation.

The government of a country with a seriously weakening economy may seek to postpone payments for its oil imports. This is the same as requesting a loan from its OPEC suppliers. The suppliers may or may not be willing to underwrite a bailout. Finally, there is the possibility of repudiating foreign debts and trying to start over.

PETROMONEY RECYCLING   Investment options for domestic development were considered previously. We are now concerned with investment of surplus funds, that is, oil revenues in excess of an exporter's absorptive capacity. For some sparsely populated oil-producing countries, the problem of investing surpluses may continue to grow indefinitely. If surplus oil revenues are invested wisely, they too will soon begin to yield substantial returns. Returns of investments, as well as revenues realized from current oil production, must then be spent, reinvested, or given away.

This leads to a central question: Why would a country produce more oil than necessary to generate the revenue it currently needs for its own economy? The economic incentive is that the producer expects to receive a higher return on petromoney investments than if the oil were produced and sold later rather than currently. The main economic incentive for some of OPEC's largest exporters to continue to produce oil is the expectation of ample attractive investment opportunities elsewhere in the world economy.

One option is for the oil exporter simply to let its foreign currency reserves pile up. The problem is that the larger the reserve, the more vulnerable it becomes to inflation or devaluation. In investing surplus funds from oil revenues, the government, like any investor, emphasizes returns or security in its investment strategy. Alternatively, the government may, for various reasons, wish to help other countries in need of assistance.

To obtain maximum security, the exporter would remain as liquid as possible, investing in government or private short-term debt. In the longer term, OPEC surplus oil revenues may be invested in countries with strong industrial economies or in projects in developing countries. Probably, some investments will be made in countries of each type. Reasons for investment in relatively strong industrial economies, such as West Germany, Japan, and the United States, may be the likelihood of a reasonable rate of return; the needs of these economies for financing their oil-induced deficits in payments; and the dependence of the world economy as a whole on the continued viability of these national economies. On the other hand, reasons prompting investment in developing economies in Latin America or Africa may be the potential for relatively high returns and the needs of the country concerned for assistance in development of capital.

The main point is that those OPEC governments that run large surpluses in the years ahead may have the leverage to reshape the world economy along lines that are more favorable for the less-developed countries generally. The industrial countries may themselves be better off in the long run if a generous proportion of the OPEC surpluses are directed toward the less-developed countries. In order to advance their own welfare, the industrial countries need expanding markets for their goods and services and a relatively open world economy. Evolution in this direction requires the growth of purchasing power in the less-developed countries, as well as the maintenance of consumer spending and current living standards in the industrial countries.

ENERGY AND THE GLOBAL ENVIRONMENT (24, 25)

As we have seen, the world political structure is fundamentally unstable and insecure, and the world economy is far from perfect and complete. Within existing national social structures, drives for either energy security or low-cost methods of energy use may cause widespread and perhaps irreversible environmental damage. Therefore, the impact of energy-related activities on the global environment requires special attention.

Approach to International Environmental Problems

In facing long-term environmental risks, like any other set of implications or projections about the future, we confront basic uncertainty. Some of our present uncertainty about the management of environmental problems is due to the fact that many risks have been recognized only recently. However, some important environmental consequences of human activity may prove to be inherently not much more predictable than political or social consequences.

A decisive policy issue regarding the environmental implications of human activity thus concerns presumptions. Confronted by uncertainty, is a particular human activity presumed safe or unsafe? Most activity is not presumed environmentally harmful. However, in a number of circumstances the presumption has been reversed. The trend toward greater caution has followed not only from an expanding public concern about environmental degradation, but also from the knowledge that pollution thresholds, where forces are set in motion in an irreversible direction, are difficult, if not impossible, to anticipate.

Environmental controls applicable to production and consumption of fossil energy have proceeded thus far primarily from an after-the-fact understanding of pollution effects. Activities related to nuclear energy, however, have proceeded on a more conservative premise, perhaps largely because of the violent birth of the nuclear age. From its inception, the nuclear power industry has borne the burden of showing that radioactive emissions from the nuclear fuel cycle do not constitute an undue environmental or health hazard.

What events resulting from energy-related activities are environmentally significant? What are the effects of these events? What are the causes? All three questions must be tackled in the management of environmental problems at any level of government—local, national, regional, or world. Moreover, action may be necessary on the basis of imperfect or partial answers.

Efforts to deal with the environmental problems may be divided into three generic areas: problem assessment, standard-setting, and administration. The main international issues with respect to the three areas are: What is the appropriate allocation of authority among national governments and international or supranational bodies? And, how are the international or global dimensions of energy-related environmental problems to be managed?

In dealing with particular environmental problems related to global production and consumption of energy, national governments are likely to play the primary roles in the foreseeable future. On the one hand, government regulation is the only way to ensure that energy industries will internalize environmental costs to an acceptable degree. On the other hand, national governments will be reluctant to delegate to international authorities the power to make binding decisions regarding environmental standards. They may be more willing to authorize international organizations, such as the Inter-Governmental Maritime Consultative Organization or the International Atomic Energy Agency, to make recommendations, which are optional or which may become binding only upon consent of the government concerned.

Two categories of environmentally significant events concern us: those that may occur accidentally in the energy industry and those that occur during the normal course of operations in the energy industry. The main political problem with respect to accidents is to determine an acceptable probability of occurrence, while the chief problem for normal industry operations is to establish a permissible level of emissions. We consider these questions briefly in the context of several major environmental areas below.

Catastrophic Accidents

The energy industry contains a variety of risks of accidents of potentially catastrophic magnitude. A very large crude-oil carrier may break up near a beautiful coastline or in arctic waters; natural gas may leak from a liquefaction plant into the atmosphere and explode, causing a firestorm; or a malfunction in a nuclear power plant may result in the contamination of a large area with radioactive debris. The risks of such accidents in the energy industry cannot be reduced to zero. Nor can they be reduced to very low levels without, in many cases, incurring substantial costs.

The problem of ensuring public and environmental safety against catastrophic accidents in the energy industry is complicated by a number of factors when it is dealt with in an international setting. Governments hold widely differing views about the probability of occurrence of particular catastrophic accidents and the amount of compensation required in the event of an accident. Thus, the energy industry faces less stringent safety requirements in some countries than in others. In the event of a catastrophic accident, those injured will receive more adequate compensation in some countries than others. Unfortunately, low safety standards tend to coincide with low levels of compensation, a fact that compounds the problem from the viewpoint of the individual citizen.

In an international market, stringent domestic safety requirements imposed on manufacturers of energy equipment may penalize them in competition for export sales in countries where standards are lower. However, the reverse effect may also occur. Safety measures incorporated in energy technology in order to meet high standards in one large national market are likely to be included in similar technology used in countries where standards may be lower. Thus, tendencies toward leveling up, as well as leveling down, may exist in an international setting.

Land Use

In the energy field, two areas of strategic land use decisions are extraction of resources and siting of facilities. These decisions are usually viewed as matters solely within national jurisdiction, but they have major environmental ramifications for the entire world. Moreover, the energy industry is moving more and more of its activities offshore from the continental land masses and into the oceans beyond the traditional limits of national jurisdiction.

The impact of decisions regarding the extraction of energy resources is truly momentous for the entire global environment. For example, a decision to prevent or delay development of US coal resources in order to preserve the environment would affect not only the world economy, because of the impact on world prices of fossil energy, but also international security, because of the impact on the world petroleum and nuclear markets. Despite their far-reaching international implications, decisions about resource extractions are the jealously guarded prerogative of the resource owner, whether Texas cattle rancher, Arab sheik, or socialist state.

However, international environmental regulation of offshore extraction of energy resources seems feasible. The oceans, over which no nation has unfettered sovereignty, will be directly affected by offshore production of energy resources. Moreover, the potential for use of the overlying waters for other purposes will be restricted. There is no environmental rationale for any particular boundary line on a continental shelf or ocean floor in order to divide national from international regulatory regimes. Coastal states are tending to exert control over a widening band of the most productive ocean areas. Concurrent national and international authority may be a sensible approach to environmental regulation of a substantial part of the extraction of offshore energy resources in areas where coastal states are granted primary control over exploitation.

Large energy-processing or conversion facilities have major impacts on the local environment. In addition, such facilities can be sources of air and water pollution that, in the aggregate, adversely affect the entire global environment. In many industrial countries, decisions about siting energy facilities have become very difficult to make. The particular facility may be viewed in the context of the entire fuel cycle of which it forms an integral part. An energy facility may be located near the place of resource extraction, near the place of consumption, or at some convenient intermediate location.

The international environmental ramifications of decisions on facility siting are complex. Decisions that may be treated as national affairs may contain strong international components. For example, whether by land or sea, the transportation of oil from Alaska to the lower 48 United States will inevitably affect Canada's environment. Though the United States consulted Canada, the US government never viewed its decision on the Trans-Alaska Pipeline as a joint one.

A region within a country, or an entire country, may refuse on environmental grounds to permit construction of an energy facility. This type of exclusion is most likely to occur in highly industrialized and affluent regions and countries, or in pristine areas that derive considerable economic benefit from recreational activities. If the facility is not built and energy use declines, the result may be beneficial. However, if the consequence of failure to build the facility is to shift demand for energy to a more polluting form of energy, then the result is likely to be detrimental. Often the result will be to shift the needed facility to some other location where it is politically acceptable.

The issue of pollution havens in less-developed countries arises here. Shifting pollution sources from industrial to less-developed countries implies reducing pollution concentrations in industrial countries and dispersing pollutants globally over a much wider area. It also implies industrialization and economic growth for less-developed countries. Industrial countries need not become more and more polluted while less-developed countries remain pristine and poor.

A final issue with international ramifications concerns the offshore location of energy facilities, such as superports and nuclear power plants. Like drilling platforms and offshore oil and gas development generally, the ocean location of such facilities will restrict the use of the oceans by other nations and may have widespread environmental impacts. The governments of coastal states may take adequate account of international concerns in their regulation of these facilities by agreeing to a regime of concurrent jurisdiction.

Air Pollution

Fossil fuel combustion causes an environmental problem of global dimensions, although the most acute effects are likely to be felt locally or regionally. Depending on the amount and nature of industrial activity in a particular region and the number of separate countries involved, air pollution may be more or less effectively dealt with at the national level.

Countries may be classified into those in which atmospheric emissions create substantially harmful effects that are confined largely within their own national boundaries and those from which emissions create substantially harmful effects in other countries. One difficulty with such a classification is that emissions from one country may worsen an existing air pollution problem in another country. A second difficulty arises with a country such as the United States, from which atmospheric pollution along the Atlantic coast is swept out to sea, causing harmful impacts on the marine environment.

Whether a country considers its air pollution problem as primarily national in scope, or as caused by external sources and thus international, there is much to be gained through international cooperation in problem assessment. Standard-setting is a function of the source country, however, unless harmful effects substantially and adversely affect another country. Rather wide variations in acceptable levels of air pollution may exist among industrial countries. In Europe, however, international cooperation in establishing standards for air quality may be necessary. Similar cooperation seems required in any heavily industrialized transnational region, such as the eastern Great Lakes region between Canada and the United States.

Water Pollution

The basic approach to global issues in water quality control has much in common with issues of air quality. As with air quality, it is useful to classify countries according to whether their water pollution problem is primarily national, or whether they are the source of a problem for one or more other countries. Here again, the need for action is likely to be acute in industrial countries, and the need for action on an international basis is likely to exist in Europe and in areas where large bodies of water are bordered by more than one country.

Two energy-related water quality problems that merit particular international attention are thermal pollution of rivers by electric power generation and oil pollution of oceans by tankers. Thermal pollution is a major international problem when a large river, such as the Rhine or Danube in Europe, flows through more than one country. Probably, thermal discharges will be only one of a variety of pollutants that must be dealt with in the environmental management of an international river. Environmental assessment, standard-setting, and administration are likely to be accomplished by the countries directly concerned, rather than by a broadly based international body.

Oil pollution of the oceans presents a different type of problem. Although the largest sources of marine oil pollution are land-based activities, considerable pollution results from oil transport. There are strong economic incentives to transport oil across the oceans as cheaply as possible. Increasing pollution of the oceans by oil tankers, coupled with the economic disincentives for effective environmental controls, has led to increasing interest in stronger regulation.

The basic issue is how regulatory authority should be allocated among flag states, where oil tankers are registered; coastal states, which are located along tanker routes; port states, where oil transported by tanker is unloaded; and an international body.

Given the array of conflicting interests, there seems to be no alternative to international problem assessment and standard-setting. Moreover, the administrative role of the international body would have to be stronger than in the case of many other environmental problems. Nevertheless, authority to take enforcement action against tanker polluters on the high seas is likely to remain primarily with the flag state.

Beyond particular problems, pollution of the ocean environment highlights fundamental issues that are also embedded in other environmental problems. One is the problem of management of a valuable resource that no one owns. All countries enjoy a right to free use of the high seas, and they may each exercise that right, subject only to noninterference with the exercise of comparable rights by others. From an environmental viewpoint, however, whenever a resource is shared by everyone, there are incentives to misuse it. Some countries overfish an ocean, while others turn it into a cesspool. Those exercising self-restraint in the common interest simply lose out to those that do not. The oceans thus exemplify what is aptly termed the "tragedy of the commons."

A second issue is how to manage a valuable resource for which there are multiple and often conflicting uses. The same body of water may serve as a commercial thoroughfare, a scientific laboratory, a food factory, a theater of war, a mining venture, a garbage dump, and a vacation paradise. How do we allocate among ourselves the right to use the oceans for particular purposes and how do we decide among uses when they are in conflict? A national government has difficulty in deciding how to use publicly owned natural resources within its jurisdiction. The oceans challenge us to solve this most difficult political problem at the international level, but without institutions capable of decision-making.

Radioactive Waste Management (26)

The high-level radioactive wastes from nuclear power programs will pose what would appear to be the ultimate in a long-term management problem. The international issues raised have two basic dimensions: space and time. Where may radioactive wastes be placed, and under what conditions, so that containment or isolation from the biosphere is achieved? And for how long must isolation be assured?

The radioactive wastes from nuclear power create three distinct problems. First is the problem of temporary storage of wastes that are separated from plutonium and uranium during the chemical reprocessing of irradiated nuclear fuels. After sufficient decay of materials with short radioactive half-lives has occurred, the problem of permanent storage of wastes arises. Then there is the problem of the radioactive component parts of nuclear power reactors and fuel cycle facilities after these plants are retired from use.

Temporary storage of high-level waste must occur at the chemical reprocessing plants where the waste is first separated because the materials involved are too "hot" to do anything else.

A local environmental risk arising during temporary storage of radioactive wastes is that materials may accidentally leak into the surrounding environment. An international risk is that the storage facilities may be blown up in a war or terrorist attack. The world community thus has an interest in the implementation of strict safeguards in the design and operation of temporary storage facilities for high-level radioactive wastes at chemical reprocessing plants.

A wide variety of techniques for permanent storage and disposal of radioactive wastes are being studied. These include solidification and storage in underground salt formations, solidification within glass and storage in drums spaced out on the surface in remote areas, and using rockets to shoot wastes into deep space or into the sun. One or more of the earthbound techniques seems likely to prove feasible.

Countries will be in very different positions as to long-term storage of radioactive wastes from reprocessed nuclear fuels. Many countries with nuclear power programs may ship their irradiated fuels to another country for reprocessing, thereby shifting the waste disposal problem to the recipient. However, some countries with chemical reprocessing plants are likely to have difficulty finding permanent storage sites for wastes on their territories.

Future generations in the world as a whole are likely to be safer if permanent storage sites are established in a few remote locations under optimum conditions. This assumes, of course, that safe methods of transportation between temporary and permanent storage facilities are implemented. International cooperation may thus be highly desirable, although difficult to bring about, in all phases of management of this environmental problem.

The time dimension of the problem of radioactive-waste management is at least as troublesome as the spatial dimension. Depending on the characteristics of the waste involved and the disposal technique adopted, radioactive material may be a lethal hazard for hundreds, or thousands, or hundreds of thousands of years. Lethality for only a few hundred years would exceed the expected lifetime of all governments and many nations. Lethality for thousands of years would transcend civilizations, and hundreds of thousands of years may reach into a different geological era. Governments and private industries using nuclear power thus assume what is virtually an eternal obligation to assure that radioactive wastes are effectively isolated from the biosphere and do not somehow find their way back to poison life.

Thermal Limit

There is a limit to the amount of heat that human energy consumption can add to the environment without serious disturbance to the earth's climate. Humans may try to avoid, but they cannot repeal, the second law of thermodynamics.

Two questions thus arise: What is the limit? And, how do we avoid transgressing it?

It is difficult to imagine an area of environmental-problem assessment in which the stakes are higher than in ascertaining the thermal limit, since so much of the future human condition hangs in the balance. Yet the world community may have to make the momentous political decisions necessary to avoid transgressing the thermal limit on rather vague grounds.

The problem may simply prove too difficult to solve with any precision, or it may be insoluble without actually observing the limit from the other side. The limit may look more like a wide fuzzy band than a thin, bright line. Finally, the inertia behind the growth rate in global energy use and its rate of closure on the thermal limit may mean that the most painful decisions will be necessary when we are still rather far beneath whatever limit exists and can perceive it only dimly. Ascertaining the thermal limit will thus be a mixed scientific-political matter. It is as important for the results of scientific research on this question to be politically accepted as it is for them to be scientifically accurate.

If and when a thermal limit is established, the world community may face social changes, in trying to avoid it, that would be almost as profound as the turmoil that would flow from a substantial alteration of the earth's climate. Procrastination would itself be a very effective form of decision. If we want the possibility of choice, we will have to develop in time the international political institutions that will give us the capacity for decision.

Any global allocation scheme covering the international distribution of thermal discharge rights would require for its implementation a supranational authority with enormous political capacity. The international political system and world economy would pivot around the decisions of such an authority. At least within its jurisdiction, the supranational body would need plenary and unchecked powers for its decisions to be effective.

An alternative to an unpalatably strong dose of world government, however, may be substantial self-restraint and self-discipline on the part of the largest energy users. This leads us to consider energy conservation as a strategy of avoidance.

Energy Conservation

The economist will argue that optimum energy use, including conservation, would result from the operation of a market with prices based on full costs. But an economically efficient pricing mechanism does not exist and, therefore, a specially developed energy conservation effort may be justified.

Energy conservation efforts may have multiple payoffs. First, conservation could ease national security problems by reducing energy import requirements. Second, the lifetime of energy reserves would be extended, and resources that are conserved could be consumed later in uses for which there are no substitutes. And, third, energy conservation would reduce the environmental effects not only of energy consumption, but of the totality of activity in the energy industry in support of that consumption. In short, energy conservation may be a way for industrial countries to eat their cake and still have it—to increase their security and preserve their environment.

But what will be the economic effects? To the extent that conservation efforts reduce actual waste, the efficiency of a national economy would be improved. However, to the extent that such efforts lead to overconsumption of non-energy resources or overinvestment in capital equipment, energy conservation would reduce economic welfare.

The degree of "waste" in the energy economy involved would vary from one country to another, as well as the political acceptability of the dislocations that might be necessary. In an industrial country where an energy-intensive economy had developed over the years in response to low prices, there would likely be some opportunities for significant and immediate reductions in energy demand, which would not entail any substantial economic disruption. The benefits of many important measures will take 5 to 10 years to show up, however, because of the large existing stock of energy-consuming technology.

A less-developed country may not confront directly, or to the same extent, the obstacles to energy conservation that are posed in an industrial country by an existing energy-intensive technological base. A less-developed country may have the option of redesigning its development plan to minimize energy inputs and maximize use of available energy-efficient technology from the start of industrialization.

In any case, what is involved in a major energy conservation effort is a substantial change in direction for an entire national economy. The new direction would be away from energy-intensive processes and products and toward more durable products and, perhaps, a service-oriented economy. Thus, the world energy situation may accelerate the emergence of a post-industrial society.

One final point about energy conservation is essential to an understanding of this problem in a worldwide context. Energy conservation by industrial countries does not necessarily benefit the less-developed countries or provide the world's poor people with more energy. The general problem of the world's poor is that they do not have enough money to buy what they need. If industrial countries conserve energy, the less-developed countries do not receive, as a consequence, more money with which to pay their oil import bills. In fact, if energy conservation in industrial countries leads to economic recession, the less-developed countries may suffer the most. If we are concerned about the plight of the poor in the new world energy situation, the way to help them is not through energy conservation, but through a redistribution of the wealth and income that command energy and other resources.

CONCLUSION

The tension between the particular interests of nation-states and the general interests of the international community, between the principles of interdependence and independence, is manifest in every aspect of the world energy situation—oil prices, production levels, and embargoes, for example. Consumer nations are inclined to characterize the producers' behavior as myopic, parochial, and nationalistic. The oil producers generally look upon their own actions as vindicating sovereign rights that were trampled by the multinational oil companies and the industrialized world for over a quarter century. The same tension lies at the heart of nuclear power development in which, due to the close connection between nuclear power and nuclear weapons, one nation's pursuit of self-sufficiency may threaten international stability and security.

In their efforts to deal with the energy crisis, many national leaders have preached interdependence and practiced independence; they have advocated multilateral cooperation and made bilateral deals. Such contradictory behavior simply reflects the unresolved tensions between the specific political interests of nations and the broader interests of the world community. Yet the world has no institutional framework within which the tensions can be resolved and the conflicting interests harmonized. In the final analysis, therefore, the energy crisis is an institutional challenge.

Since World War II all sorts of international organizations have been built to float above national interests. There are enough of these already. The world is crowded with international agencies, many of them eager to do what they can to help solve the world's energy difficulties. However, these organizations can in fact do very little because they lack the essential ingredient: political capacity.

Premature efforts at building new international institutions to deal with the energy problem could create dangerous illusions of progress. At best, the development of a new World Energy Organization would be a pleasant diversion from the painful necessity for nations to widen the area of common interest in the energy field. The scope of common interest must first be defined so that it may thereafter become possible to endow an international institution with the political capacity to harmonize particular national interests with certain overriding general interests. The process of developing well-defined, clearly perceived, common interests among the nations involved will require slow and painstaking exploration through the labyrinth of conventional, largely bureaucratic, diplomacy. International energy problems have too many important details and they engage too many weighty domestic interests to be solved by a few strokes of high-level statesmanship, no matter how brilliant.

Energy has become the cutting edge of a broad transition in international politics. The transition was brought on by the successful culmination of the main historical trends of the first quarter century after World War II: the growth of affluent consumer societies in the Western industrial countries; the growth in military power and industrialization of the communist countries; the rise of nationalism and recognition of the political value of natural resources in the less-developed countries; and the stalemate of the cold war between the United States and Soviet Union. In the energy field, the transition is exemplified by the growing politicization of energy, as consumers and resource owners alike realized the vital dependence of modern societies on abundant energy supplies; the end of rapid growth in energy demand, based on declining prices relative to other goods and services; the introduction into commercial use of nuclear power as the primary alternative to fossil energy; and the emergence of deep concern about the environmental impacts of energy production and use.

If we would apply one lesson from history to the current world energy situation, it would seem to be self-restraint. As there is a lack of political institutions to define the general interests of the world community as a whole, national self-restraint is mandatory. With restraint, there is opportunity: the world energy situation would be more manageable in the short run, and it might also be possible to lay the foundations for more effective institutional structures that will surely be needed in the long run.

Literature Cited

1. Organization for Economic Co-operation and Development. 1974. *Energy Prospects to 1985.* Paris: OECD. 2 vols. 224 pp., 211pp.
2. Yager, J. A., Steinberg, E. B. 1974. *Energy and U.S. Foreign Policy.* Cambridge, Mass: Ballinger. 473 pp.
3. Willrich, M. 1975. *Energy and World Politics.* New York: Free Press. 234 pp.
4. United Nations, Department of Economic and Social Affairs. *World Energy Supplies,* Series J. New York: United Nations. Issued annually
5. Darmstadter, J. 1971. *Energy in the World Economy.* Baltimore: Johns Hopkins Univ. Press. 876 pp.
6. US Congress, Senate, Committee on Foreign Relations. 1975. *Multinational Oil Corporations and U.S. Foreign Policy,* report together with individual views (Committee print). Washington DC: GPO
7. Adelman, M. A. 1972. *The World Petroleum Market.* Baltimore: Johns Hopkins Univ. Press
8. Nau, H. R. 1974. *National Politics and International Technology: Nuclear Reactor Development in Western Europe.* Baltimore: Johns Hopkins Univ. Press. 287 pp.
9. Willrich, M. 1971. *Global Politics of Nuclear Energy.* New York: Praeger. 204 pp.
10. Willrich, M., ed. 1973. *International Safeguards and Nuclear Industry.* Baltimore: Johns Hopkins Univ. Press. 307 pp.
11. US Congress, Senate, Committee on Interior and Insular Affairs. 1974. *Agreement on an International Energy Program* (Committee print). Washington DC: GPO
12. Bohi, D., Russell, M. 1975. *Policy Alternatives for Energy Security.* Baltimore: Johns Hopkins Univ. Press
13. National Petroleum Council. 1973. *Emergency Preparedness for Interruption of Petroleum Imports into the United States.* Washington DC: Nat. Petrol. Counc.
14. Willrich, M., Marston, P. M. 1975. Prospects for a uranium cartel. *Orbis* 19:166–84
15. Federal Energy Administration. 1974. *Project Independence Report.* Washington DC: GPO
16. Connelly, P., Perlman, R. 1975. *The Politics of Scarcity: Resource Conflicts in International Relations.* London: Oxford Univ. Press
17. Mikesell, R. F. 1971. *Foreign Investment in the Petroleum and Mineral Industries: Case Studies of Investor-Host Country Relations.* Baltimore: Johns Hopkins Univ. Press
18. Nordhaus, W. D. 1973. The allocation of energy resources. *Brookings Papers on Economic Activity* 3:529–76
19. Solow, R. M. 1974. The economics of resources or the resources of economics. *Am. Econ. Rev.* 64:1–14
20. Penrose, E. 1971. *The Growth of Firms, Middle East Oil, and Other Essays.* London: Cass
21. Jacoby, N. H. 1974. *Multinational Oil: A Study in Industrial Dynamics.* New York: Macmillan
22. Chenery, H. B. 1975. Restructuring the world economy. *Foreign Affairs* 53:242–63
23. Brookings Institution. 1974. *Cooperative Approaches to World Energy Problems,* tripartite report by 15 experts from the European Community, Japan, and North America. Washington DC: Brookings Inst.
24. Caldwell, L. K. 1972. *In Defense of Earth: International Protection of the Biosphere.* Bloomington, Ind: Indiana Univ. Press
25. Massachusetts Institute of Technology. 1970. *Man's Impact on the Global Environment, Report of the Study of Critical Environmental Problems.* Cambridge, Mass: MIT Press
26. US Atomic Energy Commission. 1974. *High-Level Radioactive Waste Management Alternatives.* Rep. No. 1297. Springfield, Va: Nat. Tech. Info. Serv.

AUTHOR INDEX*

A

Aamodt, R. L., 168
Achenbach, P., 472
Acton, J., 511
Acton, J. P., 606, 609
Acton, R. G., 318
Adams, A. A., 362
Adams, F. G., 436, 614
Adams, G. R., 433, 435
Adams, R. G., 350, 351
Adelman, M. A., 395, 399, 745, 758
Adlhart, O. J., 362
Ahern, W. R., 317
Ahner, D. J., 349
Alff, R. K., 324
Alich, J. A., 140, 148, 150
Allen, E. R., 574, 575
Allen, J., 463, 472
ALPERT, S. B., 87-99
Ambs, L., 359
Amdur, M. O., 591
Anderson, D., 409, 431, 437, 609
Anderson, J. H., 143, 169
Anderson, J. H. Jr., 143
Anderson, K. P., 434
Anderson, L. L., 262
Anderson, R. Jr., 610
Angelino, G., 360
Apostolakis, G. E., 643, 648
Appel, J., 472
Appell, H. R., 268
Archer, D. H., 349
Archer, V. E., 557, 558
Argall, B. M., 108
Armijo, J. S., 106
Armstead, H. C. H., 162, 165, 172
Armstrong, D., 168
Arrow, K. J., 602, 604
Asada, T., 648
AUER, P. L., 685-713
Austin, A. L., 171
Austin, C. F., 168
Avenhaus, R., 569
Ayers, D. L., 324
Ayres, R. U., 425, 603, 605, 607

B

Babcock, L. R., 578
Bair, W. J., 597
Baker, B. A., 362
Baker, N. R., 304
Balasko, Y., 409
Balcomb, J. D., 146

Balzhiser, R. E., 694
Bammert, K., 350-52, 357, 359
Barber, R. E., 350, 360
Barclay, J., 459, 509
Barger, G. L., 648
Barger, H. J., 362
Barkovich, B. R., 577
Barnea, J., 161
Barnert, H., 331
Barnett, H. J., 145, 440
Bates, J. L., 355
Bator, F., 602
Baughman, M. L., 430, 434, 435, 438, 439, 441
Bauman, E. R., 272
Baumol, W. J., 392, 603, 604, 615, 616, 618
Baxter, R. E., 434
Baylis, J. A., 333
Beccu, K. D., 329
Beckman, G., 326
Beckman, W. A., 135, 136
Behling, D. J., 446
Behrens, W. W., 604
Bell, F. R., 350, 351
Bendersky, D., 266
Benedict, H. M., 610
Benemann, J. R., 141
Benenson, P., 431
Berenson, J. A., 141
Berg, C., 472, 481, 483
BERG, C. A., 519-34; 520, 528
BERG, D., 345-68
Berkner, K. H., 239
Berkowitz, D. G., 316
Berkowitz, J., 231
Berlin, E., 411, 706
Berman, E. R., 28
Berman, S., 472
Berry, R. S., 467, 472, 551
Beyer, J., 613
Biasin, K., 319
Bienstock, D., 353, 355
Biggs, F., 324
Bilgen, E., 281
Billings, R. E., 305
Binder, H., 327
Birk, J. R., 327
Bishop, J., 606
Bloom, S., 138
Bloomster, C. H., 173
Blumer, M., 611, 613
Bodin, L., 443
BODLE, W. W., 65-85; 26, 33, 349
Boer, K. W., 138
Boesch, D. F., 706

Bohi, D. R., 746
Boiteaux, M., 409
Bonem, G. W., 610
Boom, R. W., 333
Borgstrom, G., 140
Bos, P. B., 137, 144, 145, 147, 150
Bowen, R. G., 173, 174
Bower, B. T., 607, 611, 612, 616, 619
Brady, N. C., 150
Bramhall, D. E., 618
Brandt, C. S., 610
Brannon, G. M., 145, 395, 548, 706
Bratton, R. A., 349
Bravvard, J., 469, 472, 491
Brewer, G. D., 306
Brewer, W. A., 360
Brinkmann, W. A. R., 651
Brite, D. W., 108
Brody, A., 429
Brown, D., 168
BROWN, H., 1-36; 5, 29
Brown, J. T., 316
Brumm, J., 380
Bryant, P. M., 567
Budd, C. F. Jr., 163, 165, 166
BUDNITZ, R. J., 553-80
Budwani, R. N., 113
Bullard, C., 144, 538
Bullard, C. W., 446, 467, 469, 470, 489, 491, 502
Bundy, F. P., 326
Burda, E. J., 134
Burger, J. M., 296, 332
Burnham, J. B., 168
Burnett, T. J., 567
Burright, B., 433, 435
Busch, H., 317
Busi, J. D., 317

C

Cadle, R. D., 574, 575
Caille, P., 409
Cairns, E. J., 329
Caldwell, L. K., 762
Callihan, C. B., 273
Calvin, M., 140, 141
Cambel, A. B., 693
Camp, R. N., 362
Carlson, P. S., 140
Carter, A. P., 429
CARVER, J. A. JR., 727-41
Casazza, J. A., 314, 332
Caywood, R. E., 402

773

AUTHOR INDEX

Cazalet, E. J., 442
Chan, Deh Bin, 260, 271
Chang, S. G., 558
Chapman, D., 491, 499
Chapman, L. D., 434
Chapman, P. F., 467, 470, 538
Charatis, G., 247
Charpentier, J. P., 431
Chenery, H. B., 759
Cherniavsky, E. A., 332, 443, 446
Cherry, B. H., 103
Chilenskas, A. A., 317, 330
Chow, V. T., 648
Christensen, C. L., 61
Christian, V. L. Jr., 402
Chubb, W., 108
Cicchetti, C. J., 411, 606, 615, 706
Ciliano, R., 431
Claude, G., 143
Clawson, M., 613
Coase, R. H., 618
Cochran, T. B., 597
Cochrane, T. S., 47
Coensgen, F. C., 244
Cohen, B. L., 122, 597, 681
Cohen, J. J., 577
COMAR, C. L., 581-600
Combs, J., 162
Connell, J. W., 359
Connelly, P., 751
CONSTABLE, R. W., 369-89
Cook, A. R., 327
Cook, D. C., 456
Cook, E., 458, 464, 467, 470, 483, 492, 496, 497
Cook, F. X. Jr., 634, 635
Corcoran, J., 50
Cormack, A. III, 324, 325
Corn, M., 591
Cortez, D. H., 169
Cotton, E. S., 136
Cottrell, F., 458, 467, 470, 473
Council, M. E., 324
Cox, K. E., 331
CRAIG, P. P., 535-51
Crocker, T., 610
Crocker, T. D., 619
Cukor, P., 570
Culbertson, W. C., 184, 186
Currin, C. G., 147

D

Dair, F. R., 272
Dales, J. H., 618
Dambolena, I. G., 321
Daniel, J. L., 108
Daniels, F., 136
Dann, R. T., 324
Dantzig, G. B., 428

D'Arge, R. C., 604, 605, 607
DARMSTADTER, J., 535-51; 459, 511, 744
Dasgupta, A., 605
Davenport, A. G., 648
David, E. L., 613
Davidson, P., 614
Davies, J. C., 617
Davies, J. H., 106
Davies, J. M., 136
Davis, N. C., 108
Davis, R., 614
Davis, R. K., 616
Davis, W. K., 113, 119
Day, W. H., 324, 349
Deam, R. J., 429, 432
Debanne, J. G., 442
De Beni, G., 141, 280, 331
DeChazeau, M. G., 396
Defeche, J., 265
Dehn, W. T., 264, 265, 269
Delene, J., 472, 481, 495
DeLisle, A., 321
Demeter, J. J., 353, 355
Deuster, G., 352
deWinter, F., 135
Dials, G. E., 620
Dicks, J. B., 353, 355
Dickson, E. M., 151
Donovan, P., 132, 133, 135
Dorcey, A. H. J., 619
DOUB, W. O., 715-25; 741
Douglas, D. L., 327
Driver, C., 307
Dubin, F., 472, 476, 479
Duchesneau, T. D., 401, 706
Dudley, J. C., 320
Dudley, J. D., 325
Duff, M., 178
Duffie, J. A., 135, 136
Dugger, G. L., 143, 149
Duncan, D. C., 184
Dundas, R. E., 349
Dunlap, C. E., 273
Dunning, R., 482, 495
DuPree, W. G. Jr., 440, 702, 703

E

East, J. H. Jr., 188
Easton, C. R., 137, 147
EATON, D., 183-212; 184, 186
Eaton, D. R., 140, 148
Eberhardt, J. E., 531
Edmundson, W. B., 143, 149
Ehrlich, P. R., 483, 491, 493, 560
Ehrlich, S., 349
Eiermann, A., 351
Eklund, L. G., 307, 360
El-Badry, Y. Z., 314, 316,
332
Elbek, B., 460
Elder, J. W., 159
Elkins, R. B., 103, 112
Ellet, W. H., 598
Elliott, R. D., 619
Ellis, N. M., 317
Emmett, J. L., 217
English, R. E., 352
Enns, J., 433, 435
Enos, G. R., 353, 355
Enthoven, A. C., 603, 615
Epstein, S., 491, 497
Erickson, L., 455, 456
Escher, W. J. D., 304, 308
Ethridge, D., 619

F

Farris, P. J., 331, 361
Fast, C. R., 168
Feduska, W., 328
Fels, M., 467, 472, 473, 511, 551
Fernandes, R. A., 314, 316, 321, 332, 360
Ferrar, T. A., 618
Ferreira, A., 322
Ferrell, D. T. Jr., 317
FICKETT, A. P., 345-68; 360
Finney, J. P., 166, 169, 175
Fischhott, B., 623
Fisher, A. C., 603-5, 610, 615, 617
Fisher, F. M., 434
Fisher, J. C., 9, 106, 648
Ford, D. F., 663
Ford, G. R., 505, 507, 509, 511
Foreman, H., 497, 501
Forney, A. V., 622
Forrester, J. W., 430
Fournier, R. O., 163
Fowler, T. K., 253
Fox, J., 474
Fox, J. A., 140, 148
Fraas, A. P., 358, 359
Fraker, N., 505
Freeburg, L. C., 588, 590, 619, 620, 622
Freedman, S. I., 320, 325
Freeman, A. M., 602, 603, 605, 607, 610, 612-18
Freeman, S. D., 610
Freshley, M. D., 108
Friberg, L., 582, 584
Friedman, D. G., 650, 651
Frink, C. R., 12
Fritz, K., 326
Fryling, G. R., 260, 267
Fu, Y. C., 268
Fujishima, A., 141
Funk, J. E., 288, 331

AUTHOR INDEX

G

Galanis, N., 321
Gannon, R. E., 353, 355
Gardner, E. D., 188
Garvey, G., 614
Garvey, T., 322
Gatts, R., 472, 477
Gay, F. W., 324
George, J. H. B., 318, 327
Ghosh, S., 271
Gibbs, M., 140, 141
Gilbert, J. S., 333
Gildersleeve, O. D., 314
Gillen, W. J., 411
Gilli, P. V., 326
Gillis, J. C., 306
Gilmore, J. S., 208
Giramonti, A. J., 324
Glasser, A. D., 361
Glendenning, I., 324
Goldsmith, J. R., 582, 584
Goldstein, D., 472, 481, 505
GOLUEKE, C. G., 257-77; 258-60, 271, 273, 274
Goodrich, W. F., 265
Gordon, R. L., 59
Gordon, S., 409
Gordon, T. J., 321
Gotaas, H. B., 273
Gouse, S. W., 60
Gowan, J. G., 359
Graham, H., 433, 435
Grant, F., 47
Grassle, J. F., 611, 613
Gray, H. F., 265
Gray, J. E., 706
GREGORY, D. P., 279-310; 142, 280, 281, 288, 289, 291, 296, 331
Grenon, M., 706
Grienpentrop, H., 351, 352, 357, 359
Griffin, J. M., 430, 434-36, 438, 619
Gruver, G. A., 362
Guderian, R., 610
Gumbel, E. J., 643, 647, 648
Gutmanis, I., 619
Gyftopolous, E. P., 472, 477, 479, 483, 706

H

Haal, S., 326
Hadlow, M. E. G., 333
Häfele, W., 30, 34, 332, 569, 654
Hahn, F. H., 602, 604
Haimes, Y. Y., 619
Halcrow, R., 563
Hals, F. A., 355
Hamilton, L. D., 588-90
Hamilton, R. M., 174
Hammond, J., 472, 474
Hampson, G. R., 611, 613
Hamstra, J., 122
Hanneman, R. E., 141, 288, 332
Hannon, B., 467, 470, 472, 473, 477, 489-91, 551
Harberger, A., 605
Harboe, H., 324
Harker, A. B., 558
Harrell, C., 610
Harris, R., 610
Hart, P. E., 108
Haselbacher, H., 351
Hass, G. M., 138
Hass, J. E., 382, 387, 414, 619, 706
Hassenzahl, W. V., 333
Hatch, A. M., 355
Hattis, D., 491, 497
Hausz, W., 332
Haveman, R. H., 602, 603, 605, 613, 615-18
Hayashida, T., 172
Haynes, W. P., 622
Haynie, F. H., 610
Hazen, T. E., 272
Heal, G. M., 604
Heath, C. E., 361
Heck, W. W., 610
Hedley, D. G. F., 47
Heichel, G. H., 12
Hein, R. A., 333
Hellman, R., 402
Hemingway, E. L., 376
Hempelmann, L. H., 681
Henderson, J. M., 602
HENDRIE, J. M., 663-83
Henning, J., 610
Henry, J. M., 353
Herbst, H-C., 324
Heredy, L. A., 330
Herendeen, R. A., 151, 429, 445, 451, 467, 469, 470, 489, 491, 505, 506, 511, 538
Hershey, R. L., 531
Hershner, C. H., 706
Hewitt, K., 648
Hickel, W. J., 160, 161, 174
Higgins, G. H., 171
Hildebrandt, A. F., 137
Hill, A. C., 610
HILL, G. R., 37-63
Hindawi, I. J., 560
Hirshberg, A. S., 706
Hirst, E., 12, 346, 461, 472, 477, 491, 509
Hise, E. C., 466
Hite, J. C., 603, 615
Hocevar, C. J., 663
HOFFMAN, K. C., 423-53; 297, 332, 441-43, 446
Holden, A. N., 349
HOLDREN, J. P., 553-80; 253, 483, 491, 493, 497, 501, 560
Hollaender, A., 140, 141
Hollander, F., 526
Holling, C. S., 569, 657
Hollomon, J. H., 706
Holt, B., 169
Honda, K., 141
Honea, F. I., 266
Hoover, D. Q., 324
Hopper, R. E., 258
Hottle, H. C., 531
Houthakker, H. S., 434
Howard, G. C., 168
Howard, J. H., 171
Hubbert, M. K., 18, 19, 22, 26, 28, 216
Hudson, E. A., 429, 446, 447
Huff, J. R., 362
Hughes, E. E., 151
Huisman, R. H., 526
Huse, R. A., 314, 332
Huser, M. A., 619
Hutchinson, A. J. L., 169

I

Illig, E. G., 268
Inman, R. E., 140, 148, 150
Ioffe, M. S., 231
Isler, R. J., 332
Isting, C., 294, 295
Ivins, R. O., 330

J

Jacobs, D. G., 574
Jacobson, J. S., 610
Jacoby, N. H., 396, 758
Jarvis, P. M., 324
Jensen, M. E., 266
Jernigan, T. C., 239
Jessup, E., 133
Johnson, J. E., 292
Johnson, L., 415
Johnston, J., 429
Jones, H. E., 264, 265, 269
Jones, P. H., 161
Jones, W. J., 655
Jordan, D. J., 352
Jordan, K. R., 107
Jordon, R. C., 136
Jorgenson, D. W., 429, 446, 447
Joskow, P. L., 430, 435, 438, 439
Jurewitz, J. L., 706
Just, E., 59
Just, J. E., 445

K

Kaganovich, S. A., 352
Kahn, A. E., 392-94, 396-98, 411, 412
KALHAMMER, F. R., 311-43; 315-17, 334, 337

AUTHOR INDEX

Kalra, P., 324
Kamen, M. D., 141
Kaminsky, F. C., 321
Kaplan, M. A., 619
Kaplan, N. O., 141
Karadi, G., 322
Katell, S., 376
Kaufman, A., 172
Kaufman, S., 168
Kaysen, C., 434
Keairns, D. L., 349
Kearley, J., 264
Keaton, M. J., 570
Keeler, E., 604
Keenan, J. H., 456, 457, 531
Kelleher, A. J., 314
Kellogg, W. W., 574, 575
Kendall, H. W., 556
Kennedy, M., 439
Kennedy, W. F., 402
Kern, E. A., 333
Kilgore, L. A., 324
Kilp, G. R., 108
Kincaide, W. C., 284
King, J., 364
Kirkham, F. S., 307
Kirkwood, T. F., 433, 435
Kivel, B., 355
Klaff, J. L., 346
Klass, D. L., 271
Klein, R., 350
Klein, S. A., 262, 270
Kneese, A. V., 403, 602, 603, 605, 607, 611-13, 615-20
Knetsch, J. L., 613, 614
Knoche, K., 331
Kobayashi, H., 353, 355
Koenig, J. B., 160, 161
Koerper, E. C., 258
Kogiku, K. C., 604
Kok, B., 140, 141
Kolba, V. M., 330
Konopka, A. K., 281, 291, 292
Koopmans, T. C., 145, 604
Korsmeyer, R. B., 324
Kovach, E., 456, 473
Kraemer, L. A., 613
Kramer, F. W., 103
Krampitz, L. O., 140, 141
Kreig, M. B., 606
KRUGER, P., 159-82; 159, 161, 164, 166-68, 175
KRUPP, H. W., 369-89
Krutilla, J. V., 605, 606, 613
Ku, W. S., 314
Kucera, G., 328
Kummer, J. T., 329
Kunz, H. R., 362
Kurtzrock, R. C., 353
Kydd, P. H., 349
Kyle, M. L., 329

L

Lacennec, Y., 330
Lacey, J. J., 353, 355
Lady, G., 433, 435
Lambert, R. W., 129
Landsberg, H. H., 9
Lange-Hüsken, M., 319
Langham, W. H., 681
La Paz, L., 648
Lapides, M. E., 109, 112, 121
Lapp, R. E., 663
Large, D., 456, 461, 473
LaRocca, S. J., 168
Larson, W. E., 324
LAVE, L. B., 601-28; 588, 590, 607, 609, 610, 618-20, 622
Lavi, A., 143
Law, S. H., 307, 360
Lazaridis, L. J., 706
Lazarus, A. L., 574, 575
Leach, G., 470, 538
LEE, B. S., 65-85
Lee, S. Y., 349
Lee, W. H. K., 159
Leonard, G. W., 168
Leonard, J. W., 57
Leontief, W., 605
Leontief, W. W., 429
Le Phat Vinh, A., 136
Lerner, A. P., 618
Lessard, R. D., 324, 359
LEVENSON, M., 101-30
Levy, S., 109
Levy, S. J., 268, 269
Lewis, A. E., 184-86, 191, 195
Lewis, H. W., 566, 663, 668, 669
Lewis, P. A., 314, 316
Ley, R. D., 314
Ley, W., 143
Lichten, N. M., 141
Lichtenberg, A., 460, 461, 464, 472, 473, 479, 480, 491, 499, 500
Lichtenstein, S., 623
Liebman, J. C., 258, 270, 271
Limaye, D. R., 431
LINDEN, H. R., 65-85; 487
Lindley, B. C., 333
Lissaman, P. B. S., 142
Little, C. W., 324
Loane, E. S., 314
Löf, G. O. G., 146, 472, 476, 485
Long, G. M., 142
Longuet-Higgins, J. S., 648
Lorsch, H. G., 325, 326
Lovins, A., 511
Lowe, R. A., 262, 266, 472, 491
Lueckel, W. J., 307, 331, 360, 361
LUNDBERG, R. M., 87-99
Lundin, F. E., 557, 558
Lyman, R. A., 434

Lynn, L., 605

M

Maass, A., 616
Macaulay, H. H., 603, 615
MacAvoy, P. W., 399, 426, 437
MacKenzie, J., 472
MacLean, R., 480
MacPherson, R. E., 359
Macrakis, M. S., 551
Magno, P. J., 598
Makhijani, A., 460, 472, 499, 500, 706
Makino, H., 472, 551
Malinvaud, E., 429
Mann, B. J., 570
Manne, A. S., 443, 444
Manning, G. B., 350
Marchetti, C., 141, 280, 288
Marcus, M., 638
Marcuse, W., 443, 446
Margen, P. H., 326
Marks, C., 476
Marshall, J., 359
Marshall, R. K., 108
Marston, P. M., 750
Martell, E. A., 362, 363
Martinez, J. D., 324
Marx, J. L., 613
Maslan, F., 321
Massy, M. I., 622
Matras, J., 1
Matsuo, K., 165
Matthews, R. C. O., 604
Matzick, A. R., 188
Maxim, L. D., 634, 635
May, J. M., 561
May, T., 349
Mayer, P. M., 352
McCaldin, R. O., 610
McCarty, P. L., 270
McCreath, D. R., 322
McCurdy, W. A., 349
McDonald, C. F., 350, 351
McFarland, J. M., 258, 260, 271
MCGAUHEY, P. H., 257-77; 258-60, 271
McGilchrist, C. E., 648
McGillivray, R. G., 433, 435
McGrath, P. E., 634, 635
McInteer, B., 168
McKay, R. A., 171
McLamore, R. T., 168
McMain, A. T., 331
McMichael, F. C., 620
Meade, J. E., 604
Meadows, D. L., 437, 449, 604
Meidav, T., 172
Meier, A. K., 577
Mencher, S. K., 317
Mesarovich, M., 449

AUTHOR INDEX 777

Metz, W. D., 184
Meyer, C. F., 326
Meyers, J., 460, 467, 479, 480
Michelson, I., 610
Mikesell, R. F., 752
Milgram, J. H., 706
Miller, C. J., 610
Miller, D. R., 359, 360
Miller, F. B., 273
Miller, J. S., 437
Mills, E. S., 618
Mills, R., 168
Mishan, E. J., 604
Mitchell, D. R., 57
Mitchell, E. J., 382, 387, 414, 706
Mitra, G., 273
Moglewer, S., 556
Moir, R. W., 251
Moor, B. L., 358
Moore, B. J., 144
Moore, E. C., 620
Moore, E. S., 49
Mooz, W. E., 461, 472
Morgan, M. G., 577
Morgenstern, O., 430
Morley, F., 567
Moroni, V., 360
Morrison, R. G. K., 58
Morrison, W. E., 440
Morrow, W., 472, 476
Morse, C., 145
Morse, D. C., 350, 351
MORSE, F. H., 131-58
Moskvicheva, V. N., 170
Mount, T. D., 434
Moyers, J., 463, 466, 472, 481, 482, 485, 498
Muffler, L. J. P., 161, 162, 167, 174
Mulholland, J. P., 399, 401
Mull, H. R., 359, 360
Mumford, S. E., 349
Munson, M., 473, 511
Murray, R. G., 191
Mutch, J., 472, 481
Myers, J. G., 706

N

Naill, R. F., 437
Nakles, D. V., 622
Nau, H. R., 745
Nelson, C. B., 598
Nelson, P. A., 317, 330
Netschert, B. C., 10
Newman, D. K., 706
Newton, C. L., 331
Ng, D. Y. C., 142
Niskanen, W., 605
Nixon, R. M., 691
Noll, R. G., 605
Nordhaus, W. D., 418, 443, 604, 754
Norris, T. G., 376
North, W. J., 140

Norwick, A. S., 364
Notti, J. E., 324, 325
Nourse, H., 610
Novakov, T., 558
Nuckolls, J. H., 217
Nuttall, L. J., 128

O

Oates, W. E., 603, 604, 615, 616, 618
O'Brien, T., 322, 323
Odell, P. R., 22
Oehms, K.-J., 317
O'Gara, P., 610
Olds, F. C., 147
Olien, N. A., 331
Olmsted, L. M., 346, 381
Olsson, E. K. A., 324
Oswald, W. J., 140, 273, 274
Othmar, D. F., 150
Otte, C., 159

P

Pace, J. D., 414
Page, N. T., 459, 492, 496, 497, 509
Paleocrassas, S. N., 333
Palmedo, P. F., 441
PANGBORN, J. B., 279-310; 288, 289, 303, 304, 306
Papp, R., 634, 635
Parker, A., 28
Parkins, W. E., 330
Passell, P., 604
Pasteris, R. F., 292
Pearce, D., 605
Pearson, E. A., 260, 271
PELLEY, W. E., 369-89
Penner, S. S., 186
Penrose, E., 758
Perlman, J., 751
Peskin, H. M., 607, 611
Pestel, E., 449
Petersen, F. M., 603, 605, 610, 615, 617
Petersen, S. R., 466, 480, 485
Peterson, H. A., 333
Peterson, S. R., 551
Petrick, M., 357
Petty, S. W., 353, 355
Pfeffer, J. T., 258, 270, 271
Phen, R. L., 357
Phillips, J. E., 188
Phillips, J. G., 59
Pietsch, A., 350
Pigford, T. H., 570
Pilati, D., 470, 474
Pimentel, D., 12
Pindyck, R. S., 426, 437
Pitman, J. K., 184, 186
Plotkin, S. E., 551
Plust, H. G., 317, 328

Podolny, W. H., 360, 361
Polacco, J. C., 140
Polinsky, A. M., 610
Poole, A., 706
Popov, A. E., 170
Porter, J. T., 331
POST, R. F., 213-55; 220, 251, 253, 324, 325
Potter, B., 168
Powell, J. C., 137, 147
Prest, A., 605
Price, J., 470, 496
Priest, J., 457
Prigmore, D. R., 360
Primack, J., 663
Proebstle, R. A., 106
Puleo, F., 477, 489, 491
Putnam, D. E., 470
Putnam, P. C., 142, 148

Q

Quade, R. N., 331
Quandt, R. E., 602

R

Rabenhorst, D. W., 324, 325
Raghavan, R., 166, 167
Rai, C., 268
Ralls, W., 509
Ramey, H. J., 166, 167
Ramsey, F. P., 145
Randers, J., 604
RATTIEN, S., 183-212; 535-51
Rauenhorst, R., 505
Ray, D. L., 135, 694, 695
Readling, C. L., 440
Reardon, W. A., 429
Rect, R. O., 551
Rees, R., 434
Reilly, J. J., 297, 331
Reistad, G. M., 171
Rensink, M. E., 244
Resnik, H. L. P., 633
Ribe, F. L., 251
Rice, D., 606, 609
Rice, R., 472
Richards, R., 470
Richmond, C. R., 597, 681
Ridker, R., 610
Rieber, M., 41, 46, 470, 496, 563
Rikkers, R. F., 321
Ringland, W. C., 324
Risser, H. E., 51
Roberts, M., 616
Roberts, M. J., 604
Robertson, J. A. L., 108
Robertson, N., 58
Robinson, E. R., 168
Robinson, N. II, 58
Roels, O. A., 150
Rogers, A., 610

AUTHOR INDEX

Rogers, F. C., 324
Rosa, R. J., 353, 355
Rose, D. J., 614
Rose-Ackerman, S., 617
Rosenberg, N., 409
Rosenbluth, R. F., 273
Rosenfeld, A., 472, 481, 505
Rosengarten, W. E., 314
Ross, L., 604
Ross, M., 137, 472, 473, 479, 481, 491, 494, 500, 504
Ross, W. E., 260, 262, 270
Rossbach, R. J., 359
Rowe, J. J., 163
Rowland, T. C., 106
Rowley, J., 168
Rubin, B., 459
Rubin, D., 472, 477
Rubin, E. S., 60, 620
Rubin, M., 504
Rudisel, D. A., 324
RUDMAN, R., 629-62
Rudolph, P. F. H., 349
Ruff, L. E., 617
Rurik, J., 351, 352, 357, 359
Russell, C. S., 611
Russell, J. L. Jr., 331
Russell, M., 746
Russell, P. L., 58
Rutkin, D. R., 106
Rydbeck, V., 509

S

SAGAN, L. A., 581-600; 588, 590, 619, 620, 637
Sakurai, T., 136
Salihi, J. T., 317
Salkind, A. J., 317
Salzano, F. J., 314, 332
SANDER, D. E., 391-421; 410, 412
Sanders, C. L., 597
Sanders, H. L., 611, 613
Sandquist, G. M., 176
Sandstede, G., 327
San Pietro, A., 378, 379
Sargent, S. L., 135
Savage, G., 258, 267
Savino, J. M., 142, 148, 320, 321
Sax, N. I., 559
Schelling, T., 606, 609
Schiffmacher, S. A., 331
SCHIPPER, L., 455-517; 456, 458, 460, 461, 464, 471, 473, 479, 480, 487-89, 491, 504, 509, 564
Schlimmelbusch, J. S., 322
SCHMIDT, R. A., 37-63; 151
Schmill, W. C., 324, 325
Schneider, A. M., 578
Schneider, C., 328

SCHNEIDER, T. R., 311-43; 314
Schnetzer, E., 358
Schoen, R., 706
Schrag, M. P., 266
Schroeder, H., 613
Schulten, R., 331
Schultze, C. L., 403, 615, 617-19
Schurr, S. H., 9, 10, 391
Schwartz, H. J., 327
Schwegler, R. E., 272
Schweitzer, S. A., 59
Schwieger, R. G., 265-68
Searl, M. F., 431
Sebald, A. V., 446
Sebramm, L. W., 188
Seckler, D., 459, 509
Seidel, M., 511, 551
Seneca, J., 614
Seneca, J. J., 603, 615
Seskin, E. P., 607, 609, 611, 618
Sessler, G., 570
Shannon, T. J., 266
Sharer, J. C., 288, 303, 304
Sharko, J. R., 431
Shavell, S., 610
Sheehan, D. P., 434
Sheldon, R. C., 349
Shepherd, B. P., 168
Shinnar, R., 141
Shorske, E., 505
Shurcliff, W. A., 134, 146
Silver, L. T., 29
SILVERMAN, L. P., 601-28
Silverstein, S., 472
SIMMONS, M. K., 131-58
Simpson, R. D. H., 22
Sindermann, F., 359
Singer, S., 483, 493, 497
Singh, Ram Bux, 270
Skinrood, A. C., 137
Slesser, M., 470, 538
Slovic, P., 623
Slusarek, Z. M., 359
Small, K. A., 610
Smith, B. E., 148
Smith, D. H., 331
Smith, J., 350
Smith, J. S., 610
Smith, K., 491
Smith, M., 168
Smith, S. C., 607, 620
Smith, V. K., 614
Solow, R. M., 604, 754
SOMERS, E. V., 345-68; 349
Sonju, O. K., 353
Sorensen, B., 623
Sparks, F. L., 205, 206
Spence, M., 604
Spencer, A. M., 168
Spencer, D. F., 320, 321
Sprankle, R. S., 171
Stannard, J. N., 596

Starkman, E. S., 319
STARR, C., 629-62; 582, 629, 632, 633
Starrett, D., 618
Steffgen, F. W., 268
Stein, R., 472, 476, 491
Steinberg, E. B., 706, 743, 754
Steinhart, C. E., 12, 563
Steinhart, J. S., 12, 563
Stelzer, I., 414
Stephenson, J. W., 266
Stepp, J. M., 603, 615
Stern, A., 607, 616
Sternlicht, B., 360
Steunenberg, R. K., 317
Stewart, D. H., 168
Stickler, D. B., 353, 355
Stoker, A., 175
Stone, B. K., 382, 387, 414, 706
Stone, L. K., 350
Stone, R. T., 176
Stonehart, P., 362
Stoner, C., 272
Stork, K. E., 610
Straton, J. W., 47
Stratton, L. J., 318
Strettman, H., 610
Strickland, G., 297
Strombotne, R. L., 317
Stuart, A. K., 281
Sulzberger, C. L., 314
Sulzberger, V., 314
Sulzberger, V. T., 332
Sussman, D. B., 263, 269
Sutterfield, G. W., 266
Sutton, O. G., 149
Sverdrup, E. F., 361
Swanson, V. E., 184
Sweeney, J., 433, 435
Swet, C. J., 331
Swift, M., 446
Symons, P. C., 329
Szego, G. C., 140, 148

T

Taiganides, E. P., 272
Takagaki, T., 327
Tamerin, J. S., 633
Tamplin, A. R., 597
Tansil, J. M., 461, 473, 498, 499
Taussig, M. K., 603, 615
Taylor, H., 260
Taylor, L. D., 431, 433-35
Taylor, R. J., 324, 325
Taylor, T. B., 125, 680, 706
Teno, J., 119
Tessmer, R. Jr., 446
Theil, H., 429
Thekaekara, M. P., 133
Thom, H. C. S., 648
Thomas, H., 617
Thomas, R., 142

AUTHOR INDEX

Thompson, Q. E., 125, 126
Threlkeld, J. L., 133
Tietenberg, T. H., 618
Tihansky, D. P., 612, 613
Tilton, J. E., 347
Titterington, W. A., 284
Todd, D. K., 326
Tolley, G., 610
Tolman, S. F., 260, 262, 270
Tongiorgi, E., 163
Tourin, B., 610
Towle, W. L., 330
Trezek, G. J., 258, 267
Trijonis, J., 605
Trombe, F., 136
Truesdell, A. H., 163
Tuller, H. L., 364
Turner, L. R., 317
Turvey, R., 392, 409, 605, 615
Tybout, R. A., 146, 472, 476, 485
Tyrrell, T. J., 434

U

Uhlmann, J. R., 355
Underhill, D. W., 591
Upton, C., 619
Uzawa, H., 604
Uziel, M., 274

V

Vakil, H., 332
VanAs, D., 567
Van Cleave, D., 319, 325
Van Dorn, W. G., 648
Van Haut, H., 610
Vant-Hull, L., 137, 147
Vanvucht, J. H., 331
Van Wijk, W. R., 149
Vasant, C. A., 317
Vassell, G. S., 456
Vaughan, C. M., 402
Verleger, P. K., 433-35
Veziroglu, T. N., 141, 142, 308
Vleggaar, C. M., 566
Voelz, G. L., 681
von Neumann, J., 430
VYAS, K. C., 65-85; 26, 33, 306, 349, 496, 497

W

Wachholz, B. W., 597
Wachtel, D. D., 706
Waddell, T. E., 610
Wagner, H. M., 428
Wagoner, J. K., 557, 558
Wald, H., 404
Walsh, J., 148
Warde, C. J., 361
Warman, H. R., 22
Warnoch, J. G., 322
Warshay, M., 329
Washburn, D. C. Jr., 324
Way, S., 353, 355
Weare, N. M., 141
Webbink, D. W., 399, 401
Weber, N., 329
Webster, D. S., 329
Weeks, R. A., 151
Weinberg, A. M., 497, 501
Weingart, J. M., 706
Weinstein, N. J., 268
Weisbrod, B. A., 606
Weitzman, M. L., 618
Wentorf, R. H. Jr., 141, 288, 332
Werth, J. J., 330
West, J. A., 440, 702, 703
Weyant, J., 491
Whan, G. A., 176
Whinston, A., 618
WHIPPLE, C., 629-62
White, D. E., 159-61, 164, 167, 176
Whitehouse, G. D., 324
Whiting, R. L., 166
Widmer, T., 706
Wildavsky, A., 605, 615, 616
Wildhorn, S., 511
Wilke, C. R., 273
Wilkinson, J. P. D., 109
Willett, D. C., 322
Williams, C. F., 284
Williams, D. L., 164
Williams, J. R., 356
Williams, R., 472, 473, 479, 481, 494, 500, 504, 706
Williamson, H. E., 103, 112
WILLRICH, M., 743-72; 51, 53, 58, 59, 102, 125, 680, 706, 743, 745, 750, 751

Wilson, D. G., 272, 273
Wilson, J. S., 168
Wilson, J. W., 434
Wilson, R., 655, 657, 658
Wilson, R. E., 142
Windhorn, S., 433, 435
Winston, R., 137
Wisecarver, D., 605
Wisely, F. E., 258
Wiswall, R. H., 297, 331
Witcofski, R., 306
Wolf, M., 147, 151
Wolf, M. J., 138
Wolfe, B., 129
Wollman, N., 610
Wood, B., 166, 348
WOOD, D. O., 423-53
Wood, L. L., 217
Wrights, L. O., 329
Wu, Y. C. L., 353, 355
Wurm, J., 291, 292

Y

Yager, J. A., 706, 743, 754
Yandle, B. Jr., 603, 615
Yang, B., 317
Yao, N. P., 327
Yeck, R. G., 263
Yellott, J. I., 133, 134
Yocom, J. E., 610
Young, W. E., 349, 353, 355
Yushmanov, E. E., 231

Z

Zankl, G., 353, 355
Zauderer, B., 356
ZEBROSKI, E., 101-30; 109, 112, 127
Zeckhauser, R., 604, 605, 618
Zelitch, I., 140
Zemkoski, J., 314, 316, 332
Zener, C., 143
Zerhoot, F. S., 434, 439
Zygielbaum, P., 316

---

* Names of organizations, governmental agencies, etc are listed in the Subject Index.

SUBJECT INDEX

A

Abu Dhabi
 per capita income of, 744
Accidents
 at energy facilities, 559-60, 595, 667-76, 763-64
 and risk analysis, 632, 639, 645, 651-53
Acid mine drainage, 610-11
Advisory Committee on Reactor Safeguards, 669
Advisory Committee on the Biological Effects of Ionizing Radiation, 593, 595
Aerospace Corporation, 139, 320
Agricultural productivity
 disruption of, 560-62, 610
 energy for, 12-13
Agricultural waste, 262-63, 272
Air conditioning systems
 conservation and, 472, 481, 498
 coolness storage for, 320, 325
 geothermal energy and, 171
 insulation and, 547
 solar energy and, 131, 134-35
 specifications for, 544
Aircraft
 fuel for
  cost of, 503
  hydrogen as, 306-7
Air pollution
 see Pollution
Air Quality Act, 733, 741
Alaska
 oil pipeline in, 22, 497, 614-15, 690, 733-34
 petroleum reservoirs in, 16
Alcohol
 as product of bioconversion, 132
Algae ponds
 for solar energy fuel, 140
Algeria
 and OPEC, 11, 375
Alkaline fuel cells, 362
Aluminum
 in fluidized-bed reactors, 267
American Chemical Society, 308
American Gas Association, 398, 462, 473, 710
 on coal reserves, 65

American Insurance Association, 648
American Nuclear Society, 676
American Physical Society, 463-65, 467, 472, 473, 476, 477, 479, 481, 595
American Society of Mechanical Engineers
 on incineration, 260
 on nuclear safeguards, 126
Ammonia
 as gas by-product, 77, 85
 from hydrogen, 280, 300
 power plants using, 359
Anaerobic digestion
 for solid waste conversion, 262, 264, 270-72
Animal manure
 anaerobic digestion of, 270, 272
Arab-Israeli conflicts
 and oil prices, 11, 749
Argentina
 nuclear reactors in, 103
Argon gas
 in MHD power plants, 356
Argonne National Laboratory, 357, 570, 588-90
Arizona v. California
 on resource utilization, 735
Asbestos
 for cell diaphragms, 282, 284
 lung cancer and, 583-84
Ash
 disposal of, 614
 as fuel, 267
 removal from coal, 72, 74-75, 78, 83, 92-95, 350, 607
Asia
 energy consumption in, 7
 nuclear capacity in, 102
ATGAS process
 of coal gasification, 84
Atlantic Richfield Oil Company
 oil shale research by, 191
Atlantic Seaboard formula, 399-400
Atmospheric burner, 301-2
Australia
 energy consumption in, 5
 grain production in, 13
 solar energy research in, 34, 153
Automobiles
 battery-powered, 317-19

efficiency of, 542, 544
emissions control and, 545-46, 636
fuel for
 costs of, 373
 hydrogen as, 304-6
 weight of, 473, 476-77, 493

B

Barge
 for coal shipment, 53-54
Barry, Theodore and Associates, 55
Battelle Memorial Institute, 121, 123, 455-56, 570, 588-90
Batteries
 for energy storage, 327-30, 335
 for vehicle power, 317-19
Bed reactors
 for coal gasification, 74-75
Belgium
 nuclear reactors in, 102
Benefit-risk pattern, 630-31, 637-40
Benzene
 as gas by-product, 85
Beryllium
 scarcity of, 562
BI-GAS process
 for coal gasification, 79-80, 82-83
Binary fluid power plant
 geothermal energy and, 161, 167, 169-71, 173
 potassium-steam and, 358-59
Bioconversion, 132
Biomass, 132
Biophotolysis, 140-41
Bipolar electrolyzer, 283-84
Bituminous coal
 air pollution and, 46
 for electric power plants, 352
 for gasification, 65, 72, 75, 78-79, 93
 reserves of, 41
Boiling-water reactors (BWR)
 fuel cycle of, 104, 108-11, 665-66, 668
 safety study of, 669-71
Boric acid
 from geothermal waters, 163

SUBJECT INDEX 781

Boron
  and geothermal energy production, 174
Bottoming plants, 348, 351, 353, 357
Brazil
  and oil shale deposits, 27, 188
Breeder reactors
  for clean energy, 690-91, 694
  core of, 674-75
  fusion reaction for, 220
  safety of, 674-75
  in United States, 103, 712
  uranium supply and, 29
Britain
  see United Kingdom
British Gas Corporation
  and coal gasification research, 98
Brookhaven Energy System Optimization Model, 443, 499
Brookhaven National Laboratory, 442
  research on hydrogen energy at, 332
Brookings Institution, 759
Btu's
  daily use of, 462
  energy accounting and, 696
  yearly requirement of, 369-75
Bureau of Land Management, 727
Bureau of Mines, 39, 45, 49, 53, 705
  coal research by, 40-41, 56-57, 93
  on geothermal resources, 171
  on oil shale mining, 188
Bureau of Natural Gas, 377
  see also Federal Power Commission
Bureau of Outdoor Recreation, 731-32
Bureau of Reclamation, 730
Bureau of the Census, 408, 586

C

California
  Geothermal Resources Board of, 172
  petroleum reservoirs in, 16
California Institute of Technology, Jet Propulsion Laboratory, 138-39, 152, 172, 320-21
Canada
  Alaska pipeline and, 765
  exports from, 703-4
  nuclear reactors in, 102
  refuse-energy facilities in, 265
Cancer
  from chemicals, 583-84
  from radiation, 593-97, 620, 681
Capital market
  and energy financing, 383-88, 493-95, 701-2
Captive hydrogen, 300-1
Carbon
  in ferrous heat-treating, 521
Carbonate, molten
  fuel-cell concept using, 362-64
Carbon dioxide
  excess in atmosphere, 27-28, 561, 574
  in gasified coal, 75
Carbon electrode structures
  in phosphoric acid research, 362
Carcinogens
  chemical, 583-84
  radioactive, 593-97, 681
Cartelization
  of oil industry, 395-97
  of OPEC, 744
Catalytic combustion
  of hydrogen, 302-4
C-b Shale Oil Project, 199
Cell diaphragms
  for electrolysis, 282-83
Cellulose hydrolysis systems
  for waste conversion, 272-73
Cement production
  energy consumed for, 479-80
Ceramic brick heaters, 319, 325-26
Cereals
  consumption of, 13-14
Cesium
  scarcity of, 562
Cesium-lithium working fluid
  for MHD power, 357
Char
  see Coal char
Charcoal beds
  for cooling water, 111
  ferrous carbonizing and, 529
Chase Manhattan Bank, 456, 495
Chemical energy
  cycles of, 561
  storage of, 327-33
China
  petroleum reserves in, 23
  plutonium production in, 119
Chrome
  for ferrous heat-treating, 528
Citizens Advisory Committee
  on Environmental Quality, 551
Civil Aeronautics Board, 700
Cladding
  see Zirconium cladding
Clean Air Act of 1970, 209, 591, 608, 733
Climate
  effects of carbon dioxide on, 561
Clinch River Breeder reactor, 103, 712
Closed-cycle MHD, 354-56
Closed-loop chemical systems
  for energy storage, 332-33
CO_2 Acceptor
  and coal gasification, 79-82
Coal
  costs of, 32, 376, 400-2, 577
  electric utility industry and, 37-39, 61-62, 87-88, 374
  in fuel cell technology, 364
  gasification of, 65-85, 95-99
  gas turbine power plants and, 348-52
  government control of, 687
  heating value of, 41, 46-47
  hydrogen production from, 280, 289-91
  industry
    deaths in, 588, 597-98
    models for, 437, 444
  liquefaction of, 88-95
  in MHD power plants, 353-57
  for non-energy use, 696
  nuclear power vs, 113-19
  oil shale vs, 184
  pollution by, 26-27, 46, 557, 607, 620
  properties of, 76
  refuse and, 266-67
  reserves of, 21, 25, 39-47, 56-57, 697
  supply for future, 376
  transportation of, 52-54, 376-77
  use in Japan, 11
  use in United States, 2, 9-10, 37-39
Coal char
  and coal gasification, 74-75, 81-84, 290
Coal gasification
  bed reactors for, 74-75
  by-products of, 84-85
  chemistry of, 72-74
  for clean fuel, 46, 215, 349, 690, 694
  electric utility industry and, 95-96
  for hydrogen production, 289-90

SUBJECT INDEX

processes of, 77-84, 97-99
Coal liquefaction
  for clean fuel, 46
  hydrogenation methods of, 89
  plants for, 94
  technology for, 90-95
Coal Mine Health and Safety Act of 1969, 401
Coal miners
  and black lung disease, 557, 581, 590, 620
Coal mining
  costs of, 59, 376
  future of, 50-52
  hazards of, 54-56, 557, 581, 590, 620, 687
  labor for, 58-59
  laws for, 401, 729
  methods of, 27, 47-49, 55, 58
  research and development of, 59-61
Cobalt-60, 128
COGAS process
  of coal gasification, 83
Colombia
  petroleum reserves in, 23
Colony Development Corporation
  and oil shale research, 188-90
Colorado State University, Solar Energy Applications Laboratory, 137
Combustion Power Company, 267
  and coal-fired gas turbines, 349-50
Committee on Mineral Resources and the Environment, 20
Compressed air storage, 323-24, 335
Conoco Coal Development Corporation
  and coal gasification, 79-82
Conservation
  see Energy conservation
Continental Oil Company
  and coal liquefaction, 93
Continuous mining method, 48-50
Conventional mining, 47-48, 50, 59
Cooling water
  for nuclear reactors, 665
  radioactivity in, 110-11, 667
Copper
  costs in processing, 31
Cornell University, 308
Council of Economic Advisors, 393
Council on Economic Priorities, 509

Council on Environmental Quality (CEQ), 570, 588-90
  creation of, 733-34
  on strip mining, 614
Crude oil
  see Oil
Cryogenic storage
  of hydrogen, 298
Czochralski process
  of silicon production, 138

D

Deep mining, 47-48, 55
Department of Energy and Natural Resources, 691
Department of Natural Resources, 691
Deuterium
  in nuclear fusion, 29-30, 215-16, 253
Diaphragms
  in electrolysis, 282-83
Dolomite
  in coal gasification, 75, 82
Dow Chemical Company, 464, 472, 495, 499
Drilling technology
  for geothermal research, 165
Dual metering
  for electric utilities, 549
Dubin-Mindell-Bloom Associates, 472, 474, 476, 482
du Pont de Nemours, E. I. and Company, 472, 477, 479

E

Earthquakes
  and risk analysis, 646
Econometric models
  for electricity, 433-35, 438, 547
  for energy systems, 706-7
  for refining industry, 436-37
  regression analysis and, 429-30
Economy
  and energy
    capital costs of, 375-81, 701-2
    capital market analysis and, 383-88
    conservation and, 487-97, 537-40, 769-70
    deregulation and, 403-9, 759
    econometric models for, 429-30, 433-36, 438, 445-51, 547, 710
    external financing for, 382-83

government regulation of, 393-409, 500-1, 548-49, 565, 685-87, 700, 759
    incremental pricing and, 409-14
    international issues and, 747, 754-62
    utility prices and, 416, 549
    and pollution, 501, 576-78, 601-24
Ecuador
  and OPEC, 11
Edison Electric Institute, 408
  electricity pricing studies by, 412
Electricity
  conservation of, 334, 544-45
  consumption of, 373-75, 500
  costs of, 353, 409-16, 499, 503
  distribution of, 402-3
  for ferrous heat-treating, 525, 527, 531-33
  fuel cycles for
    health effects of, 590-98, 620-23
  homes run on, 482-83
  as universal energy, 279-80
  see also Electric utility industry
Electric Power Research Institute, 364, 582
  breeder reactors and, 103
  coal liquefaction and, 93
  electricity pricing studies by, 412
  Office of Coal Research and, 353, 355-56
Electric utility industry
  capital market and, 382-88
  coal use in, 37-62, 88-99, 380-81
  energy conversion for, 345-65
  energy storage and, 296, 312, 322, 324, 332
  environmental standards for, 687-88
  geothermal energy use by, 159-73, 179-80
  government regulation of, 718-19, 723, 736
  hydrogen use in, 307-8
  hydropower and, 381, 689
  models for, 433-35, 437-39
  nuclear power and, 29, 33, 101, 113-14, 216, 218, 249, 279, 380-81
  pricing of, 409-16, 549
  risk analysis for, 638-39
  solar energy use in, 131-32,

137, 141-43, 146-47,
  320-21
thermochemical water splitting and, 289
wastes used by, 265, 267,
  270
weekly load curve of, 313
Electric vehicles, 317-19,
  328, 334, 338
Electrochemical devices, 327,
  360
Electrodes, 282-85
Electrolysis
  for hydrogen production,
    280-85
  vs thermochemical water
    splitting, 288-90, 331
  solar energy and, 141-42
Electrolyte, 283-85, 329
Electrolytic cell, 281-86,
  296
Electrolytic hydrogen plants,
  281
Electrolyzer, 281-86, 296
Electrolyzer/hydride storage/
  fuel cell system, 332
Electrothermal gasifier
  for coal, 81
Emission standards, 373, 545-46, 608-9, 615-19, 636,
  721
Employment
  and energy use, 489-92
Endothermic atmospheres,
  523-25, 530
Energy
  consumption of, 1-5, 301,
    369-75, 393, 696
  forecasts of, 7-8, 440, 488,
    698-99, 706
  "gap", 455-56
  international issues of, 743-71
  national policy for, 403, 427,
    440, 447, 685-712, 751
  net energy analysis and, 538,
    562-63
  thermodynamics and, 456-57
  "trap", 21-22, 32
Energy conservation
  attitudes toward, 545-46
  definition of, 456-57, 535-37
  economy and, 487-97, 537-40
  in ferrous heat-treating
    industry, 519-34
  government policy on, 543-44, 546-50, 704
  international implications
    of, 769-70
  methods of, 470-87, 538-40
  motivations for, 371, 403,
    458-59, 541-43
  in Sweden, 460-61

Energy/economic models,
  445-49
Energy management
  for conservation, 471
Energy planning
  see Energy system models
Energy plantations
  environmental impact of,
    150
  for solar energy fuel, 139-40
Energy Policy Office, 695
Energy Policy Project
  see Ford Foundation
Energy Reorganization Act of
  1974, 178, 715, 721-22,
  724
Energy Resources Council,
  715
Energy system models
  applications of, 425-27
  for coal industry, 437
  for conservation, 537-38
  for electricity use, 433-35, 437-39
  for natural gas industry,
    437-38
  for oil industry, 432-33,
    436-37, 439-40
  for total energy system,
    440-45, 528, 569
  for descriptive purposes,
    424-25
  econometric approach to,
    429-30, 445-49
  engineering/process
    approach to, 430
  for predictive purposes,
    424-25
Engineering-Science Incorporated, 200
Entrained-bed reactor
  for coal gasification, 74,
    79, 83, 97-99, 289
Environmental impacts
  of Alaska pipeline, 614-15,
    690, 733-34
  of coal, 46, 49, 54-56, 621-22
  of conversion systems,
    353
  criteria used to characterize, 569-76
  dollar value of, 576-78
  electricity prices and, 409-10
  of electric vehicles, 317
  energy models and, 427
  of energy production, 482-83, 486, 500-1, 541-42,
    687
  of fusion power, 253
  of geothermal energy, 173-77
  of hydroelectric dams, 567
  of marginal energy sources,
    496-97

of nuclear power plants,
  566-67, 621, 767-68
of oil, 610-11, 620, 766-67
of oil shale, 197-208
social costs of
  analysis of, 566-78
  types of, 554-66
of solar energy, 149-51
of solid waste, 260, 265-66,
  274
of strip mining, 27, 48-49,
  55-56, 542, 560
EPA
  see US Environmental Protection Agency
Equity Oil Company
  oil shale research by, 191
ERDA
  see US Energy Research and
    Development Administration
Ethanol
  in cellulose hydrolysis, 272-73
Euratom, 745
Europe
  closed-cycle gas turbine
    plants in, 351-52
  economic status of, 13
  energy in
    consumption of, 5-7
    self-sufficiency of, 692
  nuclear capacity in, 102
  oil imports and, 10-11,
    24
European Atomic Energy
  Community (Euratom),
  745
Exothermic chemical reactions,
  214
Exothermic nuclear reactions,
  214, 218-21
Exporting country
  energy security in, 751-54
  petromoney recycling and,
    761-62
  see also Organization of
    Petroleum Exporting
    Countries
Exxon Company
  and coal liquefaction research,
    93
  forecast for energy by, 707-9

F

FCG-1
  fuel-cell development and,
    360, 362
Federal Coal Mine Health and
  Safety Act of 1969, 47,
  55
Federal Council for Science
  and Technology, 694

784 SUBJECT INDEX

Federal Energy Administration (FEA), 164, 172, 185, 188-91, 195, 196, 199, 202, 204-6, 373, 377, 380, 391, 404, 416, 427, 434, 435, 437, 461, 472, 474-77, 501, 551, 704, 751
conservation and, 550
creation of, 695-96, 725, 734
electricity pricing studies by, 411
energy projects and, 723
minimum performance standards and, 543
Federal Energy Regulatory Study, 720-21, 724
conservation and, 543-50
energy crisis and, 689-92
energy policy of, 403, 427, 440, 685-712
energy regulation by cost of, 723-24
environmental quality standards and, 373, 545-46, 608-9, 617-18, 636, 721
Federal government
energy regulation by
price controls and, 393-409, 500-1, 548-49, 565, 685-87, 700, 759
mineral rights and, 687, 729, 737, 740
resource regulation by, 727-41
Federal Power Commission, 314, 400, 405-7, 416, 501
and natural gas, 65, 372, 398-99, 548, 685, 721, 730
Federal Trade Commission, 409
Federal Water Pollution Control Act Amendments of 1972, 202
Feedstock, 300-1, 696
Ferrous heat-treating
fuel for
conservation of, 526-34
consumption of, 522-25
temperatures necessary for, 523
Filter-press electrolyzer, 283-84
Finland
nuclear reactors in, 102
Fires
industrial, 556, 559, 666-67, 671
Fishing
and water pollution, 613
Fission
see Nuclear fission
Fixed-bed reactor
for coal gasification, 74, 77, 97-98, 290
Flash systems

and geothermal energy, 161, 167-69
Flat-plate collectors
for solar heating, 131, 135, 146
Floods
and risk analysis, 647-48
Fluidized-bed reactor
for burning organic wastes, 267
for coal gasification, 74-75, 79, 82-84, 96-97, 99, 290
gas turbine and, 349
for power plants, 357, 359
Flywheels, 324-25, 335, 338
FMC Corporation
and coal gasification research, 99
Ford Foundation, Energy Policy Project
energy policy study by, 10, 263, 391, 411, 456, 461, 473, 486-88, 495, 499, 614, 706-7
Fossil fuels
agricultural productivity and, 12-13
alternative energy from, 65-85, 87-99, 183-211, 756
consumption of, 7, 371-72, 703
costs of, 416
as depletable resource, 567-68
health effects of, 583-84, 590-92
nuclear power vs, 101, 115-19, 146, 159, 216
reserves of, 25
see also specific fuels by name
Foster Associates Inc., 393, 405, 461, 706
Fracturing
in oil shale extraction, 190-91
France
nuclear reactors in, 102, 119, 214, 235
sodium/sulphur battery research in, 329-30
solar energy research in, 153
utility costs in, 549
Freon
and power conversion, 170
Fuel cell
hydrogen energy and, 307, 329, 331-32, 531
for power plants
technology of, 360-64
total energy concept of, 361
Fuel rod
see Nuclear fuel rod
Fumaroles

and geothermal energy, 163
Furnaces
for ferrous heat-treating, 524-33
Fusion
see Nuclear fusion
Fusion-fission hybrid systems, 251, 253

G

Gabon
and OPEC, 11
Garbage, 259, 274
Garrett Research and Development Corporation
research on oil pyrolysis by, 268
Gasoline
energy systems models for, 433, 435
Gas turbine
in closed-cycle power plants, 350-52, 359-60
in compressed air storage, 323-24
in electric utility systems, 312-15
ferrous heat-treating plants and, 532
hydrogen energy and, 331-32
in open combined-cycle power plants, 348-50
General Electric Company, 127, 320
HYCO energy storage system of, 332
water-electrolysis system of, 284-85
General Electric Space Division, 146
Generic Environmental Statement on Mixed Oxide (GESMO), 119, 121
Geothermal energy
costs of, 172-73
for electricity, 159-73, 179-80
environmental impact of, 173-76
geofluids and, 162, 166, 169-71, 173
legal factors concerning, 176-77
power plants using, 169-71, 360
production of, 177-80
resources of, 28-29, 35, 159-68
steam turbine and, 163, 165, 169
Geothermal Energy Research, Development, and Demonstration Act of 1974, 178
Geothermal Resources Board

SUBJECT INDEX 785

California, 172
Geothermal Steam Act of 1970, 176, 179, 741
Geothermometry, 163
Germany
  energy conservation in, 461, 539
  heat storage in, 319, 324, 332
  hydrogen pipeline in, 294-95
  nuclear reactors in, 102-3, 126, 214
  solar energy research in, 153
GESMO, 119, 121
Geysers, The
  and geothermal energy, 163, 165-66, 174-75
Glucose
  from cellulose hydrolysis systems, 272-73
Gordian Associates, 319
Government
  see Federal government; State government
Grain reserves, 13
Graphite
  in high-temperature gas-cooled reactors, 672-73
Green Revolution, 15
Gross national product (GNP)
  energy consumption and, 370, 459-62, 487, 491, 538-40, 698-99, 710
  projected capital market and, 383-85
  in world economy, 755
Gulf General Atomic
  nuclear research at, 238, 351
Gulf Oil Company
  and coal liquefaction, 93

H

Health
  energy production and, 555-59, 581-98
  pollution and, 609, 612-13, 620-23
  risk analysis and, 629-62
Heating
  see Space heating
Heat-treating system, 521-34
Heavy hydrogen
  see Deuterium
Heavy-water reactors, 102-3
Helical rotary screw expander
  and geothermal energy, 171
Helium
  in high-temperature gas-cooled reactor, 351-52, 672
  in MHD power plants, 356
  in nuclear fusion, 215, 219

scarcity of, 562
High-temperature gas-cooled reactors (HTGR)
  in Britain, 103
  and closed-cycle gas-turbine power plant, 350-52
  and closed-cycle MHD power plant, 355-56
  in energy models, 444
  for hydrogen energy, 289, 321, 331
  and potassium cycle, 358
  safety of, 672-74
  temperatures of, 285, 287
Hittman Associates Inc., 191, 200, 202, 570
Hot springs
  and geothermal energy, 163
Hot-water heaters, 319, 325-26
Human wastes
  anaerobic digestion for, 270
HYCO concept
  for energy storage, 332
Hydrane process
  for coal gasification, 84
Hydriding
  and nuclear fuel failure, 103-7
Hydrocarbon Research Inc.
  and coal liquefaction, 93
Hydrocarbons
  alternative energy sources and, 26
  as declining resource, 20-23
  hydrogen and, 280, 299-300
  in ocean, 561
Hydroelectric pumped storage, 314, 322-23, 334, 563, 610
Hydrogasification
  of coal, 74, 75, 79-84
Hydrogen
  in coal liquefaction, 92-94
  energy from
    production of, 280-91
    storage of, 295-99, 321, 330-33
    transmission of, 291-95
    utilization of, 299-308, 335, 338
  heating value of, 291
  in nuclear fuel failure, 106-7
Hydrogen cyanide
  as gas by-product, 85
Hydrogen sulfide
  and geothermal energy production, 174-75
Hydropower
  electricity from, 28, 373, 381, 567, 689
  in Sweden, 460-61
Hydrothermal convective

systems
  for electric power production, 159-60
  liquid-dominated reservoirs, 160-61, 164, 166, 168
  vapor-dominated reservoirs, 160-61, 163, 166, 168
Hydroxide, molten
  and fuel-cell concept, 362-64
HYGAS process
  for coal gasification, 79-81

I

IIASA, 569
Importing country
  energy security in, 746-51
  less developed, 756-58
  payments by, 760-61
  private oil companies and, 759
  United States as, 10, 22, 88, 370, 375, 397, 509, 686, 709-10
Incremental cost pricing
  for electricity rates, 411-14
  for natural gas, 405
India
  energy consumed in, 744
  nuclear reactors in, 102
  petroleum prices in, 14-15
  solar water heaters in, 134
  waste disposed in, 270
Indian lands
  coal on, 565-66
Indonesia
  and OPEC, 11, 375
Industrial feedstock
  hydrogen as, 300-1
  natural gas and, 696
Industrial Revolution
  technological competitions in, 2-3
  emergence of, 5, 9
Industrial wastes
  energy from, 264, 266
Industry
  accidents in, 556, 559-60, 595, 651, 667-76, 763-64
  energy in
    conservation of, 472, 477, 495, 519-34
    consumption of, 372
  on government regulation, 717
  and reliance on fuels, 408
Industry market models, 436-40
Input juggling
  definition of, 471
  energy consumption and, 473-74, 476, 482, 503-4
Input-output models

for economy, 604-5
for energy systems, 429, 445-49, 528, 537-38
Institute for Research on Land and Water Resources, 262
Institute of Ecology, 575
Institute of Gas Technology
and coal gasification, 79-81, 99
on energy conservation, 710
Insulation
of buildings, 474, 481, 486, 498, 504
and FHA, 548
Internal combustion engine, 3, 9, 21
electric engine vs, 317
hydrogen-fueled, 304-6
International Atomic Energy Agency (IAEA), 745
International Commission on Radiological Protection, 557, 566-67, 582, 595
International Energy Agency (IEA), 692, 746
International Federation of Institutes for Advanced Study, 467, 537, 551
International relations
and energy issues
of exporting countries, 751-54
of importing countries, 746-51
and global environment, 762-70
and world economy, 762-70
International Research and Technology Corporation
on organic waste, 262
Interstate Commerce Act, 717, 731
and natural gas regulation, 406-7, 565, 685-87
InterTechnology Corporation, 29
research on energy plantations by, 148
Iodine isotopes
in nuclear coolant, 111
Iodine-131
thyroid dose of, 566-67
Iran
nuclear reactors in, 102
and OPEC, 11
Iraq
and OPEC, 11
Iron-air battery, 329
Iron ore
processing of, 31, 75, 301
Iron-titanium hydride system, 299, 305
Israel
conflicts with Arab states, 11, 749

solar water heaters in, 134
Italy
nuclear reactors in, 102, 126, 214

J

Japan
economic status of, 13
energy consumption in, 5-7, 744
energy self-sufficiency in, 692
nuclear reactors in, 102-3, 214-15
oil imports and, 10-11, 24
solar energy research in, 134, 153
Jet Propulsion Laboratory
see California Institute of Technology
Joint Congressional Committee on Atomic Energy, 663

K

Kellogg gasifier, 74, 79-80, 99
Kinetic energy
flywheels and, 324-25, 335
in nuclear fusion, 219, 221, 239
Kinetic temperature
for nuclear combustion, 217, 219, 221-22, 229, 233, 238
Koppers-Totzek process
for coal gasification, 97-98
Korea
nuclear reactors in, 102-3
Krypton
and nuclear fuel defects, 110-11
Kuwait
and OPEC, 11

L

Landfill
for solid waste disposal, 265
gas from, 271-72
Landgard system
of gas pyrolysis, 268-69
Land Ordinance of 1785, 728
Land use
in fuel cycles, 572-73
government regulation of, 727-41
planning for, 689
international implications of, 764-65
pollution and, 614-15

Laser beams, 233-34, 247, 249
Lawrence Berkeley Laboratory
see University of California
Lawrence Livermore Laboratory
see University of California
Lawson criterion
of nuclear fusion power, 221, 227, 232
Lead-acid battery, 327-28, 338
Leak plugging
for conservation, 471, 474-75, 477, 504
Less-developed countries
economic development of, 756-58
Libya
and OPEC, 11
Light-water reactors (LWR)
energy models and, 444
energy needed for, 563
fuel cycle of, 110, 119-21
safety of, 665-72
types of, 101-3
see also Nuclear energy; Nuclear fuel
Lignite
environmental impact of, 496
gasification of, 75, 79, 82
reserves of, 41
Linear programming
and energy system models, 428, 432, 436-37, 443, 449
Line pack storage, 297
Liquid hydrogen
as fuel, 305-7
storage of, 298
Liquid-metal-cooled fast breeder reactor (LMFBR)
see Breeder reactors
Liquid-metal cycle
for MHD power plants, 353, 357
Liquified natural gas, 20-23, 375, 377-78, 559, 703
Lithium
in nuclear fusion, 30, 215, 220, 253
Lithium/iron sulfide battery, 330
Longwall mining method, 48-50, 58-59
Los Alamos Scientific Laboratory
energy storage research at, 333
theta-pinch program at, 240-41
Los Angeles Department of Water and Power
conservation and, 544-45

SUBJECT INDEX 787

Loss-of-coolant accident (LOCA)
  in nuclear fuel cycle, 112-13, 667-69, 675
Louisiana
  petroleum reservoirs in, 16
Low-temperature aqueous systems
  for fuel cells, 361-62
Low-temperature closed-cycle power plants, 359-60
Lucas, Joseph (Industries) Ltd., 328
Lummus Company
  and coal liquefaction, 93
Lurgi process
  of coal gasification, 77-79, 97-98
Lysholm gas compressor, 171

M

Macroeconomic energy model, 445-47, 449, 710
Magnetic confinement
  definition of, 216-17
  experiments in, 234-46
  hydromatic instability and, 226-32, 242
  plasma pressure and, 221-26
Magnetic fields
  technology of, 248
Magnetic mirror
  see Mirror machine
Magnetic well, 231, 243
Magnetohydrodynamic (MHD) power plants, 347, 353-57
Magnets
  for MHD generator, 355
Marginal cost pricing
  of electricity, 411-14
  of natural gas, 404, 406
Massachusetts Institute of Technology, 560, 574-75, 762
  Energy Laboratory, Policy Study Group of, 447, 707
Mathematical programming
  of electric utilities, 437
  for energy systems models, 428
Matrix of Environment Residuals for Energy Systems (MERES), 734
Medical Research Council, 597
Mercury
  in fish, 584
Metal-gas battery, 329
Metal hydrides
  for hydrogen storage, 298-99, 305

Metals
  ferrous heat-treating of, 521-34
  production costs of, 469
Methane
  from coal gasification, 72-85, 96, 290
  found with petroleum, 20
  geothermal energy and, 168
  heat-treating and, 523-24
  from refuse, 262, 270-72, 274
Methanol
  from coal, 349
  from hydrogen, 280, 300
Mexico
  nuclear reactors in, 102
  petroleum reserves in, 23
MHD
  see Magnetohydrodynamic power plants
Middle East
  oil in, 10-11, 32, 564, 685, 692, 744, 752-53
Milk
  iodine-131 in, 566-67
Mineral rights, 687, 729, 737, 740
Mirror machine
  electrical energy and, 249-50
  magnetic confinement and, 228-32, 234, 241-46, 251
Mobil Oil Company, 487
Models
  see Energy system models
Mode mixing
  for energy conservation, 471
Molten bath reactor
  for coal gasification, 74
Molten electrolyte fuel cells, 362-64
Molten salt process
  of coal gasification, 84
Monsanto Enviro-Chem Systems
  research on gas pyrolysis by, 268-69
Moving-bed reactor
  for coal gasification, 74-75
Municipal wastes, 259, 263-65, 529
Mutagens, 584

N

Naphthas
  in hydrogen production, 280
  in oil refining, 432
National Academy of Engineering (NAE), 56, 58, 706
  Committee on Public Engineering Policy, 582

Task Force on Energy, 345, 376
National Academy of Sciences (NAS), 12, 20, 22, 26, 577, 607, 609-11, 613-14, 678
NAS/NAE, 614, 706
NAS/NAE, Coordinating Committee on Air Quality Studies, 582
NAS/NAE, Environmental Studies Board, 582
NAS/NAE, National Research Council (NRC), 582, 591, 592
NAS/NAE, Assembly of Life Sciences, 582
NAS/NAE, Division of Medical Sciences, 582, 593, 595
National Aeronautics and Space Administration, (NASA), 347, 365
  see also National Science Foundation
National Air Pollution Control Association, 610
National Association of Regulatory Utility Commissions, 415
  electricity pricing studies by, 412
National Center for Resource Recovery
  on waste processing, 262-63
National Coal Association, 52, 53, 185
National Commission on Regulatory Reform, 720, 722, 724
National Council on Radiation Protection and Measurements, 582, 594-95, 597
National Economic Development Office, 460
National Economic Research Associates
  and electricity pricing studies, 411-12
National Energy Council, 721
National Environmental Policy Act (NEPA), 690, 733-34, 739, 741
National Fuel Cell Program, 364
National Petroleum Council, 22, 237, 377, 378, 440, 456, 747
  Coal Task Group of, 56-58
  Committee on US Energy Outlook of, 184, 186, 191, 706
  Industry Advisory Council of, 391
  Oil Shale Task Group of, 205, 206

788 SUBJECT INDEX

National Resources Board, 732
National Safety Council, 637
National Science Foundation (NSF), 173, 176, 178, 511
  and Economic Research Unit, Queen Mary College, 431
  and ERDA, 135
  and NASA, Solar Energy Panel, 320
  and Office of Coal Research, 353
National Water Commission, 610, 611
Natural gas
  availability of, 9-10, 16, 20-23, 372, 697
  as bioconversion by-product, 132
  conservation of, 270, 406
  consumption of, 372
  cost of, 32, 146, 312, 378, 393, 398-99
    government regulation of, 394-95, 397-400, 403-7, 426, 541, 548, 565, 685-87, 721
  for electric power, 87, 373-74
  for feedstock, 696
  gasified coal and, 73
  geothermal reservoirs and, 161, 168
  heat-treating and, 522-27
  hydrogen and, 291-94, 299-300
  imports of, 375, 703
  for industrial fuel, 301
  industry for
    health problems in, 588-89, 621
    models for, 437-38
  in potassium power plants, 358
  shortage of, 65, 377, 399, 541, 548, 700, 721
  storage of, 296
Natural gas liquids, 20-23, 375, 377
Naval Petroleum Reserves, 687, 735
Net energy analysis, 538, 562-63
Neutral beam injection
  for plasma heating, 243-45
  technology of, 248-49
New Jersey Environmental Protection Agency, 258
New York State Public Service Commission, 410, 411
New Zealand
  energy consumption in, 5
Nickel/iron battery, 328
Nickel/zinc battery, 328-29
Nigeria
  and OPEC, 11
Niobium, 285, 568

Nitrogen oxides
  environmental impact of, 560-61
  from hydrogen combustion, 299-300, 302-5
Nonthreshold agents, 585
Northwest Ordinance of 1787, 728
Nuclear Assurance Corporation, 101
Nuclear energy
  closed-cycle gas turbine plants and, 351-52
  closed-loop chemical systems and, 332
  conservation of, 159
  conversion to hydrogen, 285
  electricity from, 29, 33, 88, 101, 113-14, 216, 218, 249, 279
  government involvement in, 688-89, 701, 716, 731, 745
  hazards of, 501, 592-97
  for MHD power plants, 353-57
  risk-benefit analysis of, 632-33, 638
  safety of, 663-82
  world capacity of, 102
Nuclear fission
  costs of, 113-17
  environmental requirements for, 119, 121
  fossil plants vs, 117-19
  fuel cycle of, 121
  fusion vs, 220-21, 253
  safeguards for, 124-29, 663-82
  waste disposal and, 121-24
  see also Nuclear fuel; Nuclear fusion
Nuclear fission reactors
  core of, 104, 110-12, 665, 669, 674-76
    emergency cooling system for, 667-69
    meltdown of, 113, 654-55, 670-72
  fuel cycle of, 113-24
  fusion reactors and, 251
  geothermal energy and, 164, 321
  safety of, 124-29, 592-97, 663-82
Nuclear fission reactors
  spread of, 564
  types of, 101-4
Nuclear fuel
  consumption of, 373
  denaturing of, 127-28
  reprocessing of, 119-24, 676-78
  stockpiling of, 748-49
  storage of spent, 121-22

  see also Uranium oxide pellets
Nuclear fuel cycle
  costs of, 113-19, 381
  failure of
    cladding creepdown and, 108-9, 666
    core meltdown, 113, 654-66, 670-72
    effects of, 110-12
    hydriding and, 103-7
    loss of coolant and, 112-13, 667-69
    pellet-clad interaction and, 109-10
  for fusion, 29-30, 215-16
  prepressurization of, 108
Nuclear fuel rod, 104, 106-7, 109-12, 665-69
Nuclear fusion
  cost of, 216
  deuterium and, 29-30, 215-16
  discovery of, 213-14, 239
  exothermic reactions of, 218-21, 239
  experiments in, 216-18, 234-47
  hydromagnetic instability and, 226-34
  plasma pressure in, 222-27
  technology of, 247-53
Nuclear fusion reactors, 219-52
Nuclear power plants
  construction of, 375, 445, 663-65
  health effects of, 588-89, 621-22
  iodine-131 and, 566
  licensing of, 690
  radiation from, 592-97, 654-55, 658-59
  sabotage of, 124-26, 679
  safety of, 126-29, 663-82
  see also Nuclear energy
Nuclear reactors
  see Nuclear fission reactors; Nuclear fusion reactors
Nuclear safety, 126-29, 588-89, 592-97, 621-22, 654-55, 658-59, 663-82
  risk-benefit analysis of, 632-33, 638
Nuclear weapons
  safeguarding of, 128

O

Oak Ridge National Laboratory
  nuclear fusion research at, 239
  study of potassium boiler at, 359
Occidental Petroleum Company

SUBJECT INDEX 789

and coal gasification
research, 99
oil shale research by, 190
Occupational safety, 553-59,
571-72, 576, 581-98
Ocean farms
for solar energy fuel, 140
Ocean thermal conversion
costs of, 148-49
for electricity, 132, 143
Oceans
resources in, 744, 756, 764-65
Office of Coal Research, 349, 350, 356, 694
and Electric Power Research Institute, 353, 355-56
and National Science Foundation, 353
Office of Emergency Preparedness, 456, 461, 472, 473, 477, 483
Office of Energy Conservation, 691
Office of Science and Technology, 694
Oil
conservation of, 312-13
geothermal energy in, 164
incinerated wastes in, 262
nuclear power in, 118, 216
solar energy in, 135
consumption of, 374
displacing coal, 9-10
embargo on, 11, 257, 372, 467, 543, 564, 685, 746
environmental impact of, 610-11, 620-21, 766-67
government control of, 394-97, 403, 685-86
imports of, 10-11, 22, 88, 370, 375, 397, 509, 686, 709-10
industry for
models for, 432-33, 436-37, 439-40
occupational injuries in, 588-89, 621
and international issues, 743-71
non-energy use of, 696
price of, 114, 194, 372, 378, 403-4
and OPEC, 10-11, 32-33, 195, 439, 710-11, 744-45, 754-62
production of, 87, 377
reserves of, 15-24, 697
stockpiling of, 748
storage of, 296
transportation of, 378-79, 426, 614
Oil shale
costs of, 191-97, 210
environmental impact of, 197-209

extraction of, 186-96
facilities for
model for, 445
resources of, 21, 26-27, 183-86
syncrude from, 380
use of, 32-33
waste disposal and, 189-90
Oil Shale Corporation
and retorting process, 188-89
OPEC
see Organization of Petroleum Exporting Countries
Open-cycle MHD, 353-55
Organic wastes
anaerobic digestion of, 271
Organization for Economic Cooperation and Development, 11, 395, 743, 754
Organization for European Cooperation and Development, 745
Organization for European Economic Cooperation, 395
Organization of Petroleum Exporting Countries (OPEC)
international relations with energy security and, 746-54
world economy and, 754-62
oil prices and, 10-11, 23, 32, 195, 439, 710-11, 744-45
Orogenesis
and geothermal energy, 160
Outdoor Recreation Resources Review Commission, 731
Output juggling
definition of, 471
energy use and, 477, 491, 499
Oxygen
and hydrogen for electricity, 285-89, 308
Oxygen gasifier
for coal, 81
Ozone
as pollutant, 560, 591

P

Pacific Gas and Electric
electricity pricing studies by, 412
Pakistan
nuclear reactors in, 103
Paley report
on energy crisis, 693
Paraho
and oil shale research, 188

Parsons, Ralph M. Company, 359
Pellet-clad interaction, 109-10
Pellet fusion
for nuclear power, 217, 221, 232-34, 247
Pennsylvania Department of Health, Bureau Housing and Environmental Control, 258
Pennsylvania Power and Light Company
research on low-energy house by, 474, 476
Persian Gulf, 752-53, 755
Petrochemicals, 33, 46, 696
Petroleum
see Oil
Petroleum Allocation Act, 403
Phenols
as gas by-product, 77, 85, 98
Philippines
nuclear reactors in, 102
Phosphoric acid fuel cells, 361-62, 364
Photolysis, 140-41
Photolytic reactions, 332-33
Photosynthesis
as energy source, 8, 132
solar energy fuels and, 139-41, 147-48
waste conversion and, 273-74
Photovoltaic conversion
of solar energy, 30, 132-33, 137-39, 147-52, 321
Pipelines
for hydrogen, 291-94
for liquified natural gas, 378, 398
see also Trans-Alaska Pipeline
Pittsburgh and Midway Coal Mining Company
and coal liquefaction, 93
Plasma
confinement of, 226-34
definition of, 217-19
in fusion experiments, 234-46
Lawson criterion and, 221
pressure of, 222-26
technology for, 247
Plutonium
in breeder reactors, 29, 674
health hazards of, 596-97, 681
in nuclear fuel cycle, 119-21, 126, 129, 676-77
stockpiling of, 748-49
terrorist use of, 124, 565, 680
Politics

SUBJECT INDEX

and energy availability, 564-66
Pollution
  agriculture and, 560-62
  of air, 46, 556-60, 574-75, 591-92, 607-10, 690, 733, 765-66
  conservation and, 492-93
  costs of, 501, 576-78, 601-24
  of ecological systems, 568-69
  electricity prices and, 409
  government regulation of, 373, 558, 592, 608-9, 617-18, 732-36
  health and, 581-98, 609
  international effects of, 762-69
  of water, 202-3, 558-59, 561, 578, 610-14, 765-67
  see also Environmental impacts
Portola Institute, 148, 152, 456, 488
Potassium
  for energy conversion, 357-59
Potassium hydroxide
  in electrolyzer, 283-84
Potassium-steam binary-cycle power plants, 358-59
Pressurized water reactors (PWR)
  fuel cycle of, 104, 107-9, 111
  malfunctions of, 666, 668, 670-71
  safety study of, 669
Price controls
  deregulation and, 403-7
  for energy industries, 393-403, 426, 500-1, 548-49, 565, 685-87
Process/econometric models, 436-37, 447
Process models, 431-32, 441, 537-38
Project Independence Blueprint
  on coal deposits, 40
  as energy model, 427, 447-49, 704
  for energy self-sufficiency, 692, 695, 707-10
Prorationing
  of natural gas, 397
  of oil, 396, 685
Public Land Law Review Commission, 736-37, 739-41
Public Service Commission of Wisconsin, 411
Public Service Electric and Gas Company
  study of energy storage by, 316-17

Pyrolysis
  of solid waste, 268-69
  used in MHD power plants, 353
Pyrolysis oil, 83, 268
Pyroteck process
  of pyrolysis, 268-69

Q

Qatar
  and OPEC, 11
Queen Mary College, Economic Research Unit
  and National Science Foundation, 431

R

Radiation
  dollar value of, 577
  from fusion, 253
  illness from, 556
  from nuclear power plants, 566, 592-97, 654, 666-67, 669-76
  from weapons testing, 584
Radioactive waste
  disposal of, 121-24, 593, 614
  storage tanks for, 676-78
  international implications of, 767-68
  liquid form of, 676-77
  solid form of, 678
  transportation of, 679
Radiolytic reactions, 332-33
Radium, 681
Railroads
  coal and, 9, 37, 52-53
Rand Corporation
  and energy conservation, 545
R&D Associates, 251
Rankins cycle
  potassium in, 358
Real Estate Research Corporation, 477
Recycling
  for conservation, 259, 262, 493, 546
  of metals, 469
  for pollution control, 32, 34, 616
Redox battery, 329-30
Reference energy system, 441-42
Refuse coal, 49-50
Reserve base
  for coal, 40-41
Residuals charges
  for pollution, 617-19
Retorting
  for oil shale recovery, 186-91, 195
Risk analysis, 629-62
Rubblization

in oil shale extraction, 190-91

S

Safety expenditures
  and risk analysis, 640
Saudi Arabia
  and OPEC, 11, 744
Scyllac experiment, 241-42
Sectoral models, 431-35
Seismic activity
  and geothermal energy production, 175
Shale oil
  see Oil shale
Shell Oil Company, 56, 456, 466, 473
  oil shale research by, 191
Shippingport reactor, 712
Sierra Club/Union of Concerned Scientists, Joint Review Committee, 595
Silicon nitride
  for heat-transfer equipment, 530
Silicon solar cell
  in photovoltaics, 138-39, 147
Sludge
  from lime-limestone scrubbing, 607, 614, 622
Sodium-cooled nuclear reactors, 327, 674-75
Sodium/sulphur battery, 329-30
Solar energy
  in closed-loop chemical systems, 332
  collection systems for, 131-32, 476
  compared to fossil fuels, 33
  compared to fusion, 216
  conversion systems for, 30, 136-39, 142-43, 146-47
  costs of, 143-49
  environmental impact of, 149-51, 153-54
  fuel production and, 139-42, 152
  for heating and cooling, 21, 134-35, 145-46, 215, 320-21, 482
  hydrogen production and, 297
  in low-temperature closed-cycle power plants, 360
  model for, 445
  thermal storage and, 325-27
  for waste conversion, 273-74
Solar Energy Industries Association, 152
Solar thermal conversion, 136-37, 146-47, 149,

SUBJECT INDEX 791

320-21
Solfus
  as ultimate energy source, 34-35
Solid oxide
  fuel-cell technology, 364
Solid polymer electrolyte
  fuel-cell technology, 284-85
Solid waste
  definition of, 258-61
  energy and, 260-63
  energy conversion systems for
    biological, 270-73
    solar, 273-74
    thermal, 264-69
  incineration of, 260, 262, 265-66
Southern California Gas Company
  and solar heating, 550
Southern Services Company
  and coal liquefaction research, 93
Soviet Union
  see USSR
Space heating
  conservation of, 472-73
  insulation for, 474, 481, 486, 504
  pumps for, 476, 486, 492
  by electricity, 541
  energy consumed for, 461-62
  by fuel cells, 361
  by geothermal energy, 171
  by solar energy, 131, 134-35, 215, 325, 484-85
  specifications for, 544, 547
Spain
  nuclear reactors in, 102
Stanford Research Institute, 171-72, 460-61
  and energy plantations, 148
  and energy system model, 442
State government
  energy regulation by, 719-21
  land-use regulation by, 736-739
Steam engine, 2-3
Steam-iron process
  for coal gasification, 81
Steel
  from hydrogen process, 301
Stellerator, 235, 241
Storage systems
  for energy, 311-33
Strip mining
  environmental impacts of, 27, 48-49, 55-56, 542, 560
  production of, 51-52
  reclamation of land after federal standards for, 401,
501, 576-77, 614, 687-88, 737-39
Subbituminous coal, 41, 75, 79
Submarginal Land Retirement Program, 731
Substitute natural gas plants, 290
Sulfonic acid fuel cells, 361-62
Sulfur
  in coal, 26, 41, 46, 381, 607, 620, 688
  pollution and, 557-58, 560-61, 574, 591
  in process wastes, 267
Sulfuric acid fuel cells, 361-62
Sulfur oxide
  air pollution and, 557-58, 560, 574, 591-92, 608, 610, 688, 690
  emissions tax on, 619
  in gasified coal, 65, 72-73, 77, 81, 85, 347, 349, 353
Superconducting magnetic energy storage, 333
Surface mining
  see Strip mining
Suspension-bed reactor
  for coal gasification, 74-75
Sweden
  energy consumed in, 460-61, 479, 491, 539
  nuclear reactors in, 102
Swedish Federation of Industries, 460
Switzerland
  nuclear reactors in, 102
Syncrude, 379-80
Syngas, 379-80
Synthane process
  for coal gasification, 79-80, 82
Synthetic fuels, 87, 375, 377, 379-80, 497

T

Taiwan
  nuclear reactors in, 102
Tank electrolyzer, 282-85
TARGET
  and fuel cell development, 360, 362
Tars
  in coal gasification, 77, 85, 98
Tar sand oil, 26
Tax laws
  for energy investments, 759
  for energy producers, 383, 394-402, 500-1, 507-8, 548, 686, 718
Tectonic plate boundaries
and geothermal energy, 160
Tennessee Valley Authority, 482, 483, 495
Texas
  petroleum reservoirs in, 16, 693
Thermal energy
  in ferrous heat-treating, 531
  storage of, 319-20, 325-27, 334, 338
  see also Geothermal energy; Solar energy
Thermal waters, 161
Thermochemical conversion
  for solar energy fuel, 141
Thermochemical hydrogen
  production, 281, 285-89, 321, 331
Thermocline power plants, 359
Thermodynamics, 456-57, 462-65, 528-29, 768-69
Theta-pinch
  electrical energy and, 249-50
  magnetic confinement and, 234, 241-43, 246, 251
Thorium, 29, 672
Threshold agents, 585
Titanium, 285
Tokamak
  magnetic confinement and, 231-32, 234-38, 243, 246, 251
Total flow system
  and geothermal energy, 167, 169, 171
Trace metals
  as carcinogens, 584
Trans-Alaska Pipeline
  environmental impact of, 497, 614-15, 690, 733-34
Transportation
  battery-powered vehicles as, 317
  conservation and, 472
  electric vehicles as, 317-19, 328
  fuel-cell power for, 361
  hydrogen as fuel for, 304-7
  petroleum consumption by, 317, 372-73
Tritium
  in atmosphere, 574-75
  in nuclear reactions, 30, 219-20, 253
TRW Systems Group, 146, 320

U

Uncertainty principle
  in risk analysis, 659-62

SUBJECT INDEX

Underground pumped storage, 297-98, 322-23, 335
Underground reservoir
 for energy storage, 297-98, 322-23
UNESCO, 163, 166
Union Carbide—Battelle process
 of coal gasification, 83-84
Union Oil Company
 and oil shale research, 188-89
United Arab Emirates
 and OPEC, 11
United Kingdom
 energy consumption in, 5
 incineration facilities in, 265
 nuclear research in, 103, 119, 214, 244
United Nations
 Department of Economic and Social Affairs, 744
 Scientific Committee on the Effects of Atomic Radiation, 582, 595
United States (US)
 energy consumption in, 10, 369-74, 459, 541, 696, 702-3, 744
 energy self-sufficiency in, 685-712, 751
 imports of oil to, 10-12, 22-24, 88, 370, 375, 397, 509, 686, 709-10
US Army Natick Laboratories
 cellulose hydrolysis research at, 273
US Atomic Energy Commission, 113, 116, 119, 121, 123, 126, 190, 566, 570, 571, 588-90, 595, 596, 644, 663, 669, 676, 680, 693, 695, 731, 767
US Congress
 House of Representatives
  Subcommittee on Science and Astronautics, 456, 473
 Joint Committee on Atomic Energy, 455, 488, 494
 Senate
  Committee on Commerce, 456, 473, 477, 501
  Committee on Foreign Relations, 745
  Committee on Interior and Insular Affairs, 393, 456, 473, 746
US Department of Commerce, 429
 Technical Advisory Board of on Project Independence, 707, 709
 see also Bureau of the Census
US Department of Health, Education, and Welfare, 558
US Department of Housing and Urban Development, 732
US Department of State
 and international energy research, 691
US Department of the Interior, 184, 205, 206
 on energy growth, 691
 on mineral rights, 740
 research on geothermal resources, 163, 176, 179
 on strip mining, 737-39
 see also Bureau of Land Management; Bureau of Mines; Office of Coal Research
US Department of Transportation
 Transportation Energy R & D Goals Panel, 317
US Energy Research and Development Administration (ERDA), 142, 152, 473, 486, 488, 499
 on coal gasification, 349
 conservation and, 550
 creation of, 695
 on energy needs, 456
 on energy storage systems, 314-15
 on national research and development, 711-12
 on nuclear energy, 103, 121, 123, 234
 on solar heating, 30, 135, 153
US Environmental Protection Agency (EPA), 60, 260-62, 381, 477, 479, 491, 556, 591, 598, 616
 emission standards and, 373, 558, 592, 608-9
 on pollution, 608, 611, 622, 733
 power plant licensing and, 723
 on shale development, 197
 on waste materials, 258
US Forest Service
 private forestry and, 730
 rights to water and, 735-36
US Geological Survey, 26, 31
 Resource Appraisal Group of, 185
 studies of natural resources by, 40, 87, 179
US Nuclear Regulatory Commission, 121, 556, 577, 599, 664-65, 669, 695, 715, 723, 735
US Secretary of Labor, 557
Universal Oil Products Company
 and coal liquefaction, 93
University of California
 Lawrence Berkeley Laboratory
  cellulose hydrolysis research at, 273
  nuclear fusion research at, 239
 Lawrence Livermore Laboratory, 463, 482
  magnetic mirror program at, 241, 243-46
  Solar Energy Group at, 137
 Sanitary Engineering Research Laboratory, 259-60
University of Miami, Clean Energy Research Institute, 308
University of Oklahoma, Science and Public Policy Program, 184
Uranium
 in breeder reactors, 29
 deuterium and, 216
 exports of, 750
 illegal diversion of, 124, 680
 International Energy Agency and, 746
 in nuclear fuel cycle, 119-22
 production of, 401
 in spent fuel, 676-77
 see also Uranium oxide pellets
Uranium miners
 and lung cancer, 557-58
Uranium oxide pellets
 fission products in, 110
 fuel fabrication and, 103, 665
 hydrogen in, 106
 microstructures of densification in, 108, 666
 porosity of, 107
 radioactivity of, 670
 stockpiling of, 748-49
Uranium-thorium fuel cycle
 for high-temperature gas-cooled reactors, 672
Urban wastes
 as fuel, 266, 529
USSR
 coal in, 27
 energy consumption in, 5-7
 energy self-sufficiency in, 692
 petroleum in, 23
 research in
  on geothermal energy, 170
  on MHD, 355
  on nuclear energy, 103, 126, 214, 231, 234-35

SUBJECT INDEX 793

on solar energy, 134, 153

V

Vacuum technology
  for fusion research, 248
Venezuela
  oil in, 10-11
Vermont Public Service Board
  on electricity pricing, 412
Vinyl chloride, 584
Volcanism
  and geothermal energy, 160-61, 163

W

Waste disposal
  geothermal energy and, 174
  radioactive, 121-24
  see also Solid waste
Water
  desalination of, 34, 159, 165
  energy from
    costs of, 31
    hydrogen in, 280-91, 299-300
  in fuel cycles, 572
  geothermal, 163, 169, 175
  heating of, 319
    by geothermal energy, 171
    by solar energy, 131, 134
  pollution of, 202-3, 558-59, 561, 578, 610-14, 765-67
  regulation of, 732-36
  supply of
    for oil shale development, 202-8
Water Pollution Control Act of 1972, 209
Water Quality Act, 733, 741
Water Resources Council, 204
Westinghouse Electric Corporation, 137, 320, 364
  Advanced Reactor Division at, 663
  coal gasification research and, 99
  Special Systems Division at, 146
Wind energy, 132, 142, 148, 321
Winkler process
  for coal gasification, 96
Winston collector
  for solar energy, 137
Wisconsin Commission
  electricity pricing studies by, 411
World Energy Organization, 771

X

Xenon
  and nuclear fuel defects, 110-11

Y

Yeast
  in cellulose hydrolysis systems, 272-73

Z

Zinc-air battery, 329
Zinc-chlorine battery, 329
Zirconium cladding
  and fuel failure
    by creepdown, 108-9
    by hydriding, 103-7
    by loss of coolant, 112-13, 667-69
    by pellet-clad interaction, 109-10